Nanotechnology Based Strategies for Combating Antimicrobial Resistance

Mohmmad Younus Wani •
Irshad Ahmad Wani • Akhilesh Rai
Editors

Nanotechnology Based Strategies for Combating Antimicrobial Resistance

Springer

Editors
Mohmmad Younus Wani
Department of Chemistry
College of Science, University of Jeddah
Jeddah, Saudi Arabia

Irshad Ahmad Wani
P.G. Department of Chemistry
Government Degree College Anantnag,
University of Kashmir
Srinagar, Jammu and Kashmir, India

Akhilesh Rai
Center of Neuroscience and Cell Biology
University of Coimbra
Coimbra, Portugal

ISBN 978-981-97-2022-4 ISBN 978-981-97-2023-1 (eBook)
https://doi.org/10.1007/978-981-97-2023-1

This Springer imprint is published by the registered company Springer Nature Singapore Pte Ltd.
The registered company address is: 152 Beach Road, #21-01/04 Gateway East, Singapore 189721, Singapore

In Loving Memory of Akhilesh Rai

This book stands as a testament to the brilliance and dedication of Akhilesh, our esteemed colleague, and friend, whose sudden departure has left an indelible void in our hearts and in the realm of nanotherapeutics.

As the head of the Nanotherapeutic subgroup within the advanced therapies group at CNC-University of Coimbra, Portugal, Akhilesh was a visionary leader, steering the team through the uncharted waters of nanotechnology with unparalleled expertise. His relentless pursuit of excellence and commitment to advancing the understanding of the interaction of nanoformulations with biological barriers and bacterial membranes set a benchmark for innovation in nanotechnology.

Akhilesh's remarkable contributions were not confined to the laboratory alone; his accomplishments extended far beyond, garnering recognition through the prestigious Bluepharma-University of Coimbra Innovation award in 2016. With over 40 research publication in international peer-reviewed journals and the inventorship of three patents, Akhilesh's impact on the scientific community was profound and enduring. Beyond the accolades and achievements, Akhilesh was a mentor and guide, shaping the academic journey of numerous Ph.D. and master's students who were fortunate to be under his tutelage. His generosity of spirit, intellectual curiosity, and unwavering support inspired those around him to reach new heights.

In dedicating this book to Akhilesh, we honor not only his professional legacy but also the warmth, kindness, and friendship that made him an integral part of our lives. Though he may no longer walk among us, his influence continues to resonate in the echoes of scientific inquiry and the hearts of all who had the privilege of knowing him.

May this dedication stand as a tribute to Akhilesh's enduring spirit, a beacon of inspiration for future generations in the ever-evolving landscape of nanotherapeutics and beyond.

Editors: Mohmmad Younus Wani and Irshad Ahmad Wani

Preface

WHO has declared that antimicrobial resistance (AMR) is one of the top 10 global public health threats facing humanity. AMR poses a serious hazard to public health on a global scale, killing at least 1.27 million people worldwide and is linked to over 5 million fatalities in 2019. In the U.S., more than 2.8 million antimicrobial-resistant infections occur each year. Antimicrobial resistance has the potential to have an impact on humans at any stage of life as well as the medical, veterinary, and agricultural sectors. This makes it one of the most important public health issues in the world. The cost of AMR to the economy is significant. In addition to death and disability, prolonged illness results in longer hospital stays, the need for more expensive medicines, and financial challenges for those impacted. Without effective antimicrobials, the success of modern medicine in treating infections, including during major surgery and cancer chemotherapy, would be at increased risk.

Antimicrobial resistance is placing the benefits of the Millennium Development Goals at risk and threatens the achievement of the Sustainable Development Goals. It is therefore a significant problem that needs to be tackled by all realms of government and society as it offers a threat to public health globally. Furthermore, a lack of access to quality antimicrobials remains a major issue. Shortage of antimicrobial drugs is affecting countries of all levels of development, especially in healthcare systems. The COVID-19 pandemic has unarguably made the entire world realize the importance of availability of adequate therapeutic agents and the preparedness of any microbial invasion.

Although antimicrobial resistance occurs naturally over time, usually through genetic changes, misuse and overuse of antimicrobials further accelerate and aggravate this process by applying evolutionary stress on the population. Antimicrobial resistance is also of great concern when dealing with deadly or immunocompromised diseases like cancer and HIV/AIDS. The development of new antimicrobials is time consuming and at the same time requires enormous economic and labor investment. There are no novel antimicrobials in the clinical pipeline. Only six of the 32 antibiotics that address the WHO list of priority infections that were listed by WHO as being in clinical development in 2021 were considered novel. Access to high-quality antimicrobials also continues to be a big problem. The shortage of antimicrobials is impacting all levels of development, particularly in healthcare systems. In this alarming situation, alternative strategies and approaches are required to fight the infections caused by multidrug resistant microbes.

Recent advancements in nanotechnology to create nanoparticles with desired physicochemical properties have been projected as a new line of defense against multidrug resistant infections. Traditional antimicrobial agents are designed to inhibit certain proteins, enzymes, DNA, or RNA, which are indispensable for the growth of the microbes and in turn the microbes switch on their defense systems that include efflux pumps, target modification, genetic changes, and many others for their survival. This evolutionary pressure at some point evolves drug-resistant microbes. Nanoparticles may offer a promising solution to deal with multidrug-resistant microbes as they can not only combat the disease-causing microbes but also act as carriers for antibiotics and other antimicrobial compounds. Due to their unique physical and chemical properties, nanoparticles have exhibited promising therapeutics applications. They exhibit antimicrobial activities by targeting disease-causing microbes using multiple pathways, which renders them the potential to either minimize or eliminate the evolution of resistance. Nanoparticle-based drug delivery systems introduce a wide range of drugs to the target sites of infection by either conjugated to their surfaces or encapsulated within their structure. Such nanoparticle-based drug delivery is considered to be safe which involves the controlled release of drugs at the targeted site without inducing systemic toxicity. Encapsulation of drugs improve their pharmacokinetic profile and therapeutic index as compared to the free drug equivalents, which drastically reduces the dose required to achieve the clinical effects as well as reduce the toxicity and undesirable side effects linked with high drug concentration and repeated dosing. Nanotechnology-based approaches could also be used to break extracellular polymeric substances (EPS) that are used by microbes to form biofilms that are hard to treat.

While an abundance of research articles and reviews currently address the issue of antimicrobial resistance, there exists a notable gap in literature—a comprehensive exploration of the potential of Nanotechnology as an alternative and efficient strategy in combating this problem, in the form of a book. Our anticipation is that this book will meet the expectations of readers by providing thorough coverage of the latest developments in nanotechnology as a novel approach to addressing microbial resistance. Designed to enlighten the audience on the backdrop of antimicrobial resistance and the transformative potential of Nanotechnology, this book aims to present alternative and efficient tools for combatting microbial resistance, elucidating the underlying antimicrobial mechanisms. Intended as a valuable resource, this comprehensive literary guide is poised to assist beginner researchers and serve as a reference for the academic and research community actively engaged in addressing the challenges posed by antimicrobial resistance.

Jeddah, Saudi Arabia — Mohmmad Younus Wani
Srinagar, India — Irshad Ahmad Wani
Coimbra, Portugal — Akhilesh Rai

Contents

Editors and Contributors

About the Editors

Mohmmad Younus Wani is an associate professor of Organic and Medicinal Chemistry in the College of Science, University of Jeddah, Jeddah, Saudi Arabia. He received his Ph.D. in Organic and Medicinal Chemistry from Jamia Millia Islamia, New Delhi, India, and Postdoctoral training from the University of Coimbra, Portugal, and Texas Therapeutics Institute, UTHealth, Houston, Texas, USA, on the development of new treatment strategies and nontraditional approaches to combat antimicrobial resistance (AMR). With extensive experience in the field of medicinal chemistry and drug discovery, he has authored/coauthored over 60 peer-reviewed research articles, 05 books, and several US patents, and is the recipient of many prestigious honors, awards, and grants. He is a professional member of many scientific societies and is on the editorial board of many reputed scientific journals. Dr. Wani is working at the interface of chemistry and biology to develop novel small molecule inhibitors, new treatment strategies, and novel approaches that could circumvent AMR, which is within the ambit of WHO's global action plan on antimicrobial resistance.

Irshad Ahmad Wani is an assistant professor of chemistry in the Postgraduate Department of Chemistry at the Govt. Postgraduate Degree College Anantnag, University of Kashmir, India. He formerly worked as a Lecturer in Chemistry at the J&K School Education Department from 2010 to 2017. Dr. Wani has been teaching Chemistry at the Undergraduate and Postgraduate levels in several colleges of the J&K Higher Education Department since 2017. Dr. Irshad A. Wani earned his master's degree in chemistry in 2007 and his Ph.D. in Nanochemistry in 2012 from Jamia Millia Islamia in New Delhi, India. He has received various prestigious awards notably ICTSGS Services Award (2021) conferred by SPAST foundation & B N Kailoo Memorial ISCAS Medal (2009) conferred by ISCAS, India. He has a good track record of publications and has published research papers in various esteemed international journals. He has also authored several book chapters and is also a co-editor of some edited books. Dr. Wani is also a life member of various prestigious scientific societies like Materials Research Society of India, India Association of Solid-State Chemists & Allied Scientists, Society for Materials Chemistry, etc. His current research interests include green synthesis of inorganic nanoparticles, inorganic-organic hybrid nanoparticles, and conjugation of nanoparticles with various significant molecules for biomedical, therapeutic, and sensing applications.

Akhilesh Rai is assistant investigator (equivalent to assistant professor) at the University of Coimbra, Portugal. He has done his postdoctoral work at Nottingham Trent University UK, Universite Catholique de Louvain Belgium, and Center of Neuroscience and Cell Biology, University of Coimbra, on various European projects. His research interests include nanotechnology-based approaches for the regenerative medicine applications with specialization in nanomaterials, antimicrobial peptides, antimicrobial agents, nanomaterial-cell interactions, skin regeneration, and drug delivery. He has been conferred with various prestigious awards notably Portuguese Science and Technology Foundation Investigator award, BluePharma-University of Coimbra award, and CSIR-India research fellowship. He is an active reviewer and is on the editorial board of various international journals. He has published more than 40 research papers in peer-reviewed international journals, 4 book chapters, and 3 patents with more than 7000 citations. He was a cofounder of the spinoff company (CureMat Technology), dedicated to developing innovative and skin-regenerative wound dressing.

Contributors

Vikrant Abbot Department of Pharmaceutical Sciences, Chandigarh Group of Colleges, Jhanjeri, Mohali, Punjab, India

Nuno M. F. Alves CDRSP—Centro para o Desenvolvimento Rápido e Sustentado do Produto, Instituto Politécnico de Leiria, Marinha Grande, Leiria, Portugal

Balaprasad G. Ankamwar Bio-Inspired Materials Research Laboratory, Department of Chemistry, Savitribai Phule Pune University (Formerly University of Pune), Ganeshkhind, Pune, India

Manoj Balachandran Department of Physics and Electronics, CHRIST (Deemed to be University), Bengaluru, Karnataka, India

Pranjal K. Baruah Department of Applied Sciences, GUIST, Gauhati University, Guwahati, Assam, India

Khalid Mujasam Batoo King Abdullah Institute for Nanotechnology, King Saud University, Riyadh, Saudi Arabia

Smritilekha Bera School of Chemical Sciences, Central University of Gujarat, Gandhinagar, India

Mudasir Ahmad Bhat School of Biosciences and Biotechnology, Baba Ghulam Shah Badshah University, Rajouri, India

Mujtaba Aamir Bhat School of Biosciences and Biotechnology, Baba Ghulam Shah Badshah University, Rajouri, India

Rita Chandel Department of Chemistry, Government Degree College, Sanjauli, Shimla, Himachal Pradesh, India

Chung-Ming Chang Master & Ph.D. Program in Biotechnology Industry, Chang Gung University, Taoyuan City, Taiwan

Department of Medical Biotechnology and Laboratory Science, Chang Gung University, Taoyuan City, Taiwan

Susana P. G. Costa Centre of Chemistry, University of Minho, Braga, Portugal

Darshana S. Dhokane Bio-Inspired Materials Research Laboratory, Department of Chemistry, Savitribai Phule Pune University (Formerly University of Pune), Ganeshkhind, Pune, India

Juliana R. Dias CDRSP—Centro para o Desenvolvimento Rápido e Sustentado do Produto, Instituto Politécnico de Leiria, Marinha Grande, Leiria, Portugal

Jayati Ray Dutta Department of Biological Sciences, Birla Institute of Technology and Science (BITS), Pilani, Hyderabad Campus, Hyderabad, Telangana, India

Helena P. Felgueiras Centre for Textile Science and Technology (2C2T), University of Minho, Guimarães, Portugal

Carolina A. M. Ferreira CDRSP—Centro para o Desenvolvimento Rápido e Sustentado do Produto, Instituto Politécnico de Leiria, Marinha Grande, Leiria, Portugal

ICBAS—Instituto de Ciências Biomédicas Abel Salazar, Universidade do Porto, Porto, Portugal

CECA—Centro de Estudos de Ciência Animal, Instituto de Ciências, Tecnologias e Agroambiente (ICETA) da Universidade do Porto, Praça Gomes Teixeira, Apartado, Porto, Portugal

MARE—Centro de Ciências do Mar e do Ambiente, Instituto Politécnico de Leiria, Peniche, Portugal

Ramakrishnan Ganesan Department of Chemistry, Birla Institute of Technology and Science (BITS), Pilani, Hyderabad Campus, Hyderabad, Telangana, India

Sara F. C. Guerreiro CDRSP—Centro para o Desenvolvimento Rápido e Sustentado do Produto, Instituto Politécnico de Leiria, Marinha Grande, Leiria, Portugal

i3S—Instituto de Investigação e Inovação em Saúde, Universidade do Porto, Porto, Portugal

Departamento de Física Médica, Instituto Português de Oncologia (IPO-Porto), Porto, Portugal

FEUP—Faculdade de Engenharia, Universidade do Porto, Porto, Portugal

Bhanushree Gupta Department of Chemistry, Centre for Basic Sciences, Pt. Ravishankar Shukla University, Raipur, Chhattisgarh, India

Monika Gupta Department of Chemistry, University of Jammu, Jammu, India

Saif Hameed Amity Institute of Biotechnology, Amity University Haryana, Gurugram (Manesar), India

Arif Tasleem Jan School of Biosciences and Biotechnology, Baba Ghulam Shah Badshah University, Rajouri, India

B. M. Kavya Bowring and Lady Curzon Hospital, Bengaluru, Karnataka, India

Mobina Kouser Department of Chemistry, University of Jammu, Jammu, India

Kamil Kuca Department of Chemistry, Faculty of Science, University of Hradec Kralove, Hradec Kralove, Czech Republic

Aumed, a.s, Praha, Czech Republic

Ashok Kumar School of Chemical Sciences, Devi Ahilya Vishwavidyalaya, Indore, MP, India

Ashwani Kumar Department of Chemistry, University of Jammu, Jammu, Jammu and Kashmir, India

K. Vineeth Kumar Department of Oral and Maxillo-Facial Surgery, Faculty of Dental Sciences, M.S. Ramaiah University of Applied Sciences, Bengaluru, Karnataka, India

Santosh Kumar Department of Chemistry, School of Basic and Applied Sciences, Harcourt Butler Technical University, Kanpur, Uttar Pradesh, India

Karishma Mahajan Department of Biotechnology and Bioinformatics, Jaypee University of Information Technology, Waknaghat, Solan, Himachal Pradesh, India

Musa Marimani Anatomical Pathology, School of Pathology, Health Sciences, University of the Witwatersrand, Johannesburg, South Africa

Elina Marinho Centre for Textile Science and Technology (2C2T), University of Minho, Guimarães, Portugal

Aleena Ann Mathew Department of Physics and Electronics, CHRIST (Deemed to be University), Bengaluru, Karnataka, India

Nigar Mehboob School of Pharmacy, DY Patil Deemed to be University, Navi Mumbai, Maharashtra, India

Riya Mukherjee Graduate Institute of Biomedical Sciences, Chang Gung University, Taoyuan City, Taiwan

Master & Ph.D. Program in Biotechnology Industry, Chang Gung University, Taoyuan City, Taiwan

Shruthi Nagaraja Department of Conservative Dentistry and Endodontics, Faculty of Dental Sciences, M.S. Ramaiah University of Applied Sciences, Bengaluru, Karnataka, India

Tatiana Padrão CDRSP—Centro para o Desenvolvimento Rápido e Sustentado do Produto, Instituto Politécnico de Leiria, Marinha Grande, Leiria, Portugal i3S—Instituto de Investigação e Inovação em Saúde, Universidade do Porto, Porto, Portugal

FEUP—Faculdade de Engenharia, Universidade do Porto, Porto, Portugal

INEB—Instituto de Engenharia Biomédica, Universidade do Porto, Porto, Portugal

Rutuja S. Palaskar Bio-Inspired Materials Research Laboratory, Department of Chemistry, Savitribai Phule Pune University (Formerly University of Pune), Ganeshkhind, Pune, India

Ramendra Pati Pandey School of Health Sciences and Technology (SoHST), UPES, Dehradun, Uttarakhand, India

C. Pushpalatha Department of Pedodontics and Preventive Dentistry, Faculty of Dental Sciences, M.S. Ramaiah University of Applied Sciences, Bengaluru, Karnataka, India

Poornima Ramesh Department of Conservative Dentistry and Endodontics, Faculty of Dental Sciences, M.S. Ramaiah University of Applied Sciences, Bengaluru, Karnataka, India

Gagandeep Kour Reen Department of Chemistry, Jawaharlal Nehru Govt. Degree College, Barwaha, MP, India

Suriya Rehman Department of Epidemic Disease Research, Institute of Research and Medical Consultations (I.R.M.C.), Imam Abdulrahman Bin Faisal University, Dammam, Saudi Arabia

Ana R. M. Ribeiro Centre for Textile Science and Technology (2C2T), University of Minho, Guimarães, Portugal

Richu Department of Chemistry, University of Jammu, Jammu, Jammu and Kashmir, India

Iram Saba Research and Scientific Center, Sultan Bin Abdulaziz Humanitarian City, Riyadh, Saudi Arabia

Amity Institute of Biotechnology, Amity University Haryana, Gurugram (Manesar), India

Partha Pratim Sarma Department of Applied Sciences, GUIST, Gauhati University, Guwahati, Assam, India

Archi Sawant School of Pharmacy, DY Patil Deemed to be University, Navi Mumbai, Maharashtra, India

Sheezma Nazir Shah School of Biosciences and Biotechnology, Baba Ghulam Shah Badshah University, Rajouri, India

Neha Sharma Department of Chemistry, University of Jammu, Jammu, India

Poonam Sharma Department of Biotechnology and Bioinformatics, Jaypee University of Information Technology, Waknaghat, Solan, Himachal Pradesh, India

Pratibha Sharma School of Chemical Sciences, Devi Ahilya Vishwavidyalaya, Indore, MP, India

Rajeshree Shinde Department of Science, Sir Vithaldas Thackersey College of Home Science, SNDTWU, Mumbai, India

A. Francisca G. Silva Centre of Chemistry, University of Minho, Braga, Portugal

Himani Singh Department of Chemistry, University of Jammu, Jammu, Jammu and Kashmir, India

Ira Singh Department of Chemistry, School of Basic and Applied Sciences, Harcourt Butler Technical University, Kanpur, Uttar Pradesh, India

Namrata Singh Department of Engineering Sciences, Ramrao Adik Institute of Technology, DY Patil University, Navi Mumbai, India

Una Stamenović Vinča Institute of Nuclear Sciences—National Institute of the Republic of Serbia, University of Belgrade, Belgrade, Serbia

Marta O. Teixeira Centre for Textile Science and Technology (2C2T), University of Minho, Guimarães, Portugal

Anindita Thakur Department of Biological Sciences, Birla Institute of Technology and Science (BITS), Pilani, Hyderabad Campus, Hyderabad, Telangana, India

Sichao Tian Institute of Medicinal Plant Development, Chinese Academy of Medical Sciences and Peking Union Medical College, Beijing, China

Meera Varghese Department of Physics and Electronics, CHRIST (Deemed to be University), Bengaluru, Karnataka, India

Shwetha Venkataramana Faculty of Dental Sciences, Department of Oral Medicine and Radiology, M.S. Ramaiah University of Applied Sciences, Bengaluru, Karnataka, India

Vesna Vodnik Vinča Institute of Nuclear Sciences—National Institute of the Republic of Serbia, University of Belgrade, Belgrade, Serbia

Ivana Vukoje Vinča Institute of Nuclear Sciences—National Institute of the Republic of Serbia, University of Belgrade, Belgrade, Serbia

Kaiser Wani Department of Biochemistry, College of Science, King Saud University, Riyadh, Saudi Arabia

University Institute of Biotechnology, Chandigarh University, Mohali, India

Qing-Hua Xu Department of Chemistry, National University of Singapore, Singapore and National University of Singapore (Suzhou) Research Institute, Suzhou, China

Peiyan Yuan School of Pharmaceutical Sciences, Shenzhen Campus of Sun Yat-sen University, Shenzhen, China

1 Antimicrobial Resistance: An Overview

Sheezma Nazir Shah, Mujtaba Aamir Bhat,
Mudasir Ahmad Bhat, and Arif Tasleem Jan

Abstract

Recounting the struggle against infectious diseases reconciles us with the failures. Following discovery, clinical implementation of different classes of antibiotics resulted in the enhancement in the occurrence of resistance among bacteria that plagued public health for a longer time. The relentless threat posed by microbial drug resistance has achieved new dimensions of pandemic scale as it has compromised the therapeutic potential of available drugs. Associated with the excessive and imprudent use of different classes of antibiotics, the opportunistic organisms come up with new strategies and mechanisms that not only thwart the treatment of diseases with available drugs but also worsen the situation by making infections impossible to treat in clinical settings. Representing a problem of global magnitude, it is imperative to design strategies for their detection, take steps to implement the appropriate use of antibiotics, monitor their impact on resistance development as part of the evaluation, and develop combative strategies to restrict the emergence and transmission of microbial-based infections among humans. This chapter delves into the widespread use of antibiotics and the emergence of resistance across various antibiotic classes. It explores the mechanistic aspects of resistance development in ESKAPE pathogens, providing a summary of factors contributing to this phenomenon and emphasizing the grave implications for human health.

Keywords

Antibiotics · Infectious diseases · Resistance mechanisms · Public health · Therapeutics

---

S. N. Shah · M. A. Bhat · M. A. Bhat · A. T. Jan (✉)
School of Biosciences and Biotechnology, Baba Ghulam Shah Badshah University, Rajouri, India
e-mail: atasleem@bgsbu.ac.in

M. Y. Wani et al. (eds.), *Nanotechnology Based Strategies for Combating Antimicrobial Resistance*, https://doi.org/10.1007/978-981-97-2023-1_1

1.1 Introduction

Antimicrobial resistance refers to the ability of microorganisms to withstand the adverse effects of different classes of antibiotics either singly or in combination, to which they were susceptible. It represents a global concern as it imposes a serious threat to the effective management of infectious diseases (O'Neill 2016; Wencewicz 2019). Although antibiotics were celebrated as the ultimate defense in the battle against different diseases, they are losing their efficacy owing to the rise of resistance among disease-causing pathogens (Hamad 2010; Simpkin et al. 2017). It has sparked profound concerns within the scientific and medical communities due to its potential to undermine the effectiveness of vital medications. The development of resistance among microbial pathogens is a foremost concern in the health-care sector, as emphasized by prominent organizations such as the Centre for Disease Prevention and Control (CDC) and the World Health Organization (WHO), which listed it as one of the three top dangers to global health of the twenty-first century (Youde 2012; Prestinaci et al. 2015). The CDC estimates that at least 23,000 deaths occur annually due to antibiotic-resistant infections in the United States alone. Moreover, a recent report suggests that by 2050, antibiotic resistance could result in premature deaths of approximately 300 million worldwide and cause economic losses of up to $100 trillion globally (Dadgostar 2019).

The development of antimicrobial resistance among microbial pathogens requires consideration of some key points. First, intrinsic resistance is the natural occurrence of resistance mechanisms to withstand the action of antimicrobial compounds. These organisms do not represent the primary concern in clinical settings as they were already resistant before exposure to the antimicrobials. The focus is primarily on acquired resistance, where a bacterial population initially susceptible to an antimicrobial develops resistance over a defined span of time. Second, assessing antimicrobial resistance or susceptibility in clinical practice is a complex process. As such, clinical susceptibility breakpoints are established based on in vitro activities of an antimicrobial drug against different pathogens by considering the pharmacological factors such as drug concentrations at the site of infection (Martinez 2014). Although, the breakpoints categorize bacteria as susceptible, intermediate, or resistant, the interpretation of susceptibility patterns changes based on the concentration of drug at the infection site and the specific type of infection being treated in clinical settings (Munita and Arias 2016; Torumkuney et al. 2022).

The challenge posed by antimicrobial resistance is intricate, with multiple factors contributing to its development. The limited development of new antibiotics exacerbates the problem, leading to potentially untreatable infections. Understanding the concepts of intrinsic and acquired resistance, as well as the complexities of interpreting susceptibility patterns, is crucial in addressing the challenges posed by antimicrobial resistance. Efforts are required to consolidate and synthesize the available information to provide a clearer understanding of antimicrobial resistance and its associated mechanisms. Effective management of the issue requires a comprehensive and global approach involving various disciplines that could strategize action plans to advocate the issue through the involvement of both national and international organizations. The programs should include prompt and accurate identification of the pathogen, comprehend the mechanisms behind its resistance, and recognize the modes of transmission (Santajit and Indrawattana 2016; Xue et al. 2019). As part of the endeavor, it is essential to improve current knowledge and understanding of the early identification of pathogens and prioritize efficient prevention and control tactics capable of restricting their spread. By doing so, it will be possible to facilitate the development of effective prevention and control strategies as well as enhance the dissemination of knowledge among health-care professionals and researchers. A comprehensive and organized approach to understanding antimicrobial resistance and exploration of the variations of known mechanisms, operation of the combinatorial mechanisms, and the emergence of novel mechanisms will be crucial in combating this pressing global health challenge. The present study will concentrate mostly on aspects that can help us to re-evaluate our assumptions about antibiotics, resistance development against a broad range of antibiotics, functional aspects of resistance among ESKAPE pathogens, and evaluation of the effects on human health due to the development of resistance.

1.2 Antibiotics and Its Resistance

The term "antibiotic" means compounds of microbial origin capable of restricting or inhibiting a broad range of bacteria and fungi. The term encompasses both natural and engineered molecules with bacteriostatic or bactericidal properties (Nicolaou and Rigol 2018). Antibiotics constitute a potent category of pharmaceutical compounds that play a pivotal role in facilitating the hosts' immune system in its efforts to counteract and resolve the infectious condition; thereby increasing human life expectancy. By effectively combating bacterial infections, it offers protection for improvement in surgical methods, organ transplants, and other disease treatments. The incredible adaptability of bacteria that allowed bacteria to evolve or acquire a variety of mechanisms to withstand antibiotic exposure quickly put this remarkable advancement in jeopardy (Weatherall et al. 2006). With the imposition of a substantial economic strain, infections resulting from antibiotic-resistant bacteria have considerable consequences on patients' health outcomes and survival rates (Cassini et al. 2019). The proliferation of extensive resistance (XDR) and bacteria with multiple resistance (MDR), combined with the slow progress and limited availability of new antibiotics, severely curtails the range of treatment options for infections caused by these resistant strains (Fig. 1.1). Consequently, health-care providers encounter significant challenges in effectively combating infections caused by MDR and XDR bacteria (Mi et al. 2018).

Antibiotic resistance is a phenomenon where bacteria become less susceptible to antibiotics, thereby making infections difficult to control. Bacteria possess various mechanistic approaches against antibiotics, including horizontal gene transfer, plasmids, mutations, and evolutionary adaptations driven by selective pressure from antibiotics (Dadgostar 2019). Acquisition of antibiotic resistance genes (ARGs) plays a crucial role in their ability to withstand the effect of a broad range of

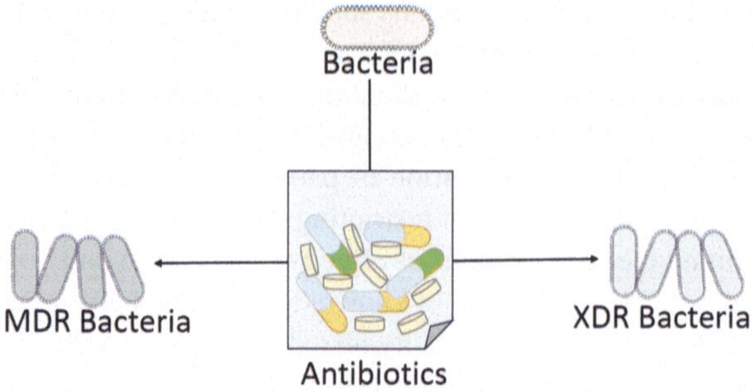

Fig. 1.1 Antibiotic over-usage leading to extremely resistant phenotypes

antibiotics (Wright 2007). In the bacterial genome, ARGs are either present inherently or acquired from other coexisting bacteria through horizontal gene transfer. Bacteria having ARGs naturally within their genome typically exhibit reduced susceptibility to antibiotics, even if they do not inhabit environments with consistently high antibiotic exposure. These bacteria are often opportunistic pathogens with versatile genetic profiles, encompassing a wide range of metabolic functions. This adaptive ability allows them to thrive and persist in various ecological niches (Martinez 2008). Bacteria have evolved inherent resistance by encompassing mechanisms that can render the drugs ineffective (e.g., β-lactamases), altering critical biosynthetic pathways, modifying target sites like protein biosynthesis, folic acid biosynthesis, and cell wall construction, as well as expelling drugs from bacterial cells (efflux). These genetic modifications can be disseminated among bacteria through mechanisms of gene transfer like transduction, transformation, or conjugation. The once-unquestioned efficacy of antibiotics now faces a formidable challenge, necessitating a comprehensive investigation into the mechanisms underpinning antibiotic resistance and an urgent search for innovative solutions to preserve their continued effectiveness in combating bacterial infections while ensuring originality and integrity in scientific discourse (Agyare et al. 2019).

The effectiveness of new antibiotics in treating common infections is diminishing rapidly because microorganisms are evolving resistance to antibiotics at an alarming rate. A significant factor contributing to the problem is the limited number of antibiotics being developed (Nicolaou and Rigol 2018; Hays et al. 2019). An antibiotic pipeline is relatively sparse, and the finding of new antimicrobial agents has slowed down in the private sector. Additionally, excessive and inappropriate antibiotic use in different countries, particularly in hospitals, contributes to genetic mutations and selection pressures that fuel the rise of resistance (Aslam et al. 2018). This trend has even led to the rise of untreatable strains of bacteria, such as *Enterobacteriaceae* having carbapenem resistance. The spread of resistance is exacerbated by factors like poor hygiene and inadequate sanitation, and global factors like travel, transmission, and trade (Laxminarayan et al. 2013). Particularly perilous are ESKAPE pathogens, which encompass vancomycin-resistant *Enterococci*, methicillin-resistant *Staphylococcus aureus* (MRSA), *Acinetobacter baumannii*,

Klebsiella pneumoniae, *Pseudomonas aeruginosa*, and *Enterobacter* spp. These strains of bacteria develop resistance to drugs and are responsible for challenging to treat infections acquired in health-care settings. Bacterial strains develop resistance to nearly all authorized antibacterial drugs, including ceftaroline, ceftazidime, daptomycin, erythromycins, gentamicin, imipenem, levofloxacin, linezolid, methicillin, tetracyclines, and vancomycin.

1.3 Resistance among ESKAPE Pathogens

The term ESKAPE was coined to describe a group of nosocomial pathogens with the ability to 'escape' the effects of the broad range of antibiotics (Ma et al. 2020). These pathogens including both gram-positive and gram-negative bacteria pose a serious threat to patient health and health-care systems worldwide. The gram-positive bacteria in this group are *Staphylococcus aureus* and *Enterococcus faecium*, while their counterparts including *Acinetobacter baumannii*, *Klebsiella pneumoniae*, *Pseudomonas aeruginosa*, and *Enterobacter* spp are gram-negative (Santajit and Indrawattana 2016). These bacteria, which are frequently linked to hospital-acquired infections, have the potential to be deadly, especially among critically ill and immunocompromised patients. They are well recognized for having the capacity to acquire resistance to numerous antibiotics, which makes treatment difficult and occasionally futile (Rice 2010).

Recognizing the urgent need for new treatment options, the WHO has highlighted 12 species of bacteria for which novel chemotherapeutics are urgently required, including all ESKAPE pathogens. This highlights the importance of developing innovative strategies to combat the multidrug resistance and virulence exhibited by these pathogens and protect patient health (Tacconelli et al. 2018).

1.3.1 *Enterococcus faecium*

It is a gram-positive bacteria that is thought to be a commensal member of the human gut microbiota. This bacteria is problematic in hospital settings because it can cause nosocomial infections and epidemics (Elsner et al. 2000). Drug-resistant enterococcal infections, in particular vancomycin-resistant Enterococci (VRE), have become more common in recent decades. In 2017, the prevalence of VRE reached 14.9% (European Union) and 30% (United States) (Mitchell et al. 2022). Infections caused by these multidrug-resistant strains of enterococci are extremely challenging and, in some cases, even impossible to treat.

Staphylococcus aureus is a gram-positive bacterium inhabiting the nose, besides serving as a normal component of the skin's microbiota. Many individuals intermittently or permanently carry *Staphylococcus aureus*, with carriage rates reaching around 50% of the population (Lowy 1998). It is a commensal and can also cause a series of infections, like infective endocarditis, osteomyelitis, soft and skin tissue infections, and bacteremia (Skinner and Keefer 1941). Before the antibiotic era,

Staphylococcus aureus infections had high mortality rates; however, things changed when penicillin was discovered in the 1940s and the prognosis improved. However, its extensive use proceeds with the evolution of the penicillin-resistant strains that produce β-lactamase enzymes (Lobanovska and Pilla 2017). Shortly after semi-synthetic penicillins were introduced in 1961, MRSA strains appeared; nevertheless, evidence suggests that methicillin-resistant existed before clinical usage (Turner et al. 2019). MRSA strains are now prevalent worldwide, with varying levels of prevalence across different countries in Europe. The methicillin resistance of *Staphylococcus aureus* is caused by the gene *mec*A encoding a modified PBP2a (penicillin-binding protein). With its location on the mobile genetic element, SCC*mec* (staphylococcal cassette chromosome *mec*), it decreases the microbial affinity of the β-lactam antibiotics, thereby rendering them ineffective (Hartman and Tomasz 1984; Harkins et al. 2017; Katayama et al. 2000).

MRSA transmission commonly occurs in health-care settings and can be facilitated by contact with contaminated surfaces, such as hospital coats, mobile phones, and tablets (Frey et al. 2019). Antibiotics of the glycopeptide category, such as vancomycin and teicoplanin, are commonly available therapies for MRSA infections in clinics. However, decreased sensitivity to teicoplanin (Kaatz et al. 1990) and the emergence of vancomycin resistance in *Staphylococcus aureus* (VRSA) has led to the search for effective alternatives to them (Chang et al. 2003; Melo-Cristino et al. 2013). The effective alternatives for vancomycin include linezolid, daptomycin, tigecycline, trimethoprim-sulfamethoxazole (TMP-SMX), and certain fifth-generation β-lactams. If reduced susceptibility to both vancomycin and daptomycin is detected, combinations or alternatives such as streptogramin antibiotics, TMP-SMX, linezolid, or telavancin may be recommended (Liu et al. 2011). The rise and spread of multidrug-resistant strains of *Enterococcus faecium* and *Staphylococcus aureus* pose significant challenges in health-care settings, highlighting the urgent need for effective treatment strategies and plans to implement effective infection control measures.

1.3.2 *Klebsiella pneumoniae*

Nosocomial infections prevalent in hospital settings are often linked to gram-negative bacteria, *Klebsiella pneumoniae*. Different strains of *K. pneumoniae* have acquired resistance via the production of the β-lactamases that attribute their strength to withstand the effect of carbapenems, cephalosporins, and penicillin. With such attributes, the control of *K. pneumoniae* strains with *the bla*KPC gene (KPC enzymes) offers strong challenges in combatting the infections caused by carbapenemase-producing *K. pneumoniae* strains (Reyes et al. 2019).

Carbapenems are considered the last-resort antibiotics for severe gram-negative infections, but the spread of carbapenem resistance in *K. pneumoniae* strains has reduced the effectiveness of these drugs. These resistant strains are often associated with health-care settings and are commonly referred to as carbapenem-resistant Enterobacteriaceae (CRE). The presence of KPC enzymes, encoded by *bla*KPC, in

K. pneumoniae isolates confers resistance to carbapenems, severely limiting treatment options. The rise in carbapenem resistance among *K. pneumoniae* strains is a significant concern for health-care providers and public health officials (Santajit and Indrawattana 2016; Queenan and Bush 2007). In addition to carbapenem resistance mediated by KPC enzymes, some strains of *K. pneumoniae* also produce a metallo-β-lactamase called New Delhi metallo-β-lactamase (NDM-1), encoded by *bla*NDM1 (Yong et al. 2009). As NDM-1 hydrolyzes carbapenems and several other classes of β-lactam antibiotics, *K. pneumoniae* strains harboring NDM-1 further contribute to carbapenem resistance and pose a serious threat to treatment options.

Considering the rise in carbapenem resistance, antibiotics preferably aminoglycosides and fluoroquinolones are used for the treatment of infections brought on by *K. pneumoniae* (Tilahun et al. 2021). However, regular use of these antibiotics led to the development of resistance against them, making it more challenging for the management of *K. pneumoniae* infections. The emergence of multidrug-resistant *K. pneumoniae* strains, such as those that produce KPC enzymes and NDM-1, highlights the urgent requirement for efficient infection control methods, prudent antibiotic administration, and the development of novel treatment options to address these challenging infections (Kumarasamy et al. 2010).

1.3.3 *Acinetobacter baumannii*

A. baumannii is an opportunistic pathogen (gram-negative bacillus) that affects immunocompromised patients who have spent an extended period in the hospital setting. It is associated with nosocomial infections, meaning infections acquired within health-care facilities (Montefour et al. 2008). The ability of *A. baumannii* to persist in hospital environments and its high resistance to varied classes of antibiotics make it a significant threat, particularly in the intensive care units where vulnerable patients are often found (Nocera et al. 2021). Carbapenem-resistant *A. baumannii* (CRAB) strains have become increasingly common, and they possess various resistance mechanisms. Two notable types of resistance genes found in CRAB strains are imipenem metal-β-lactamases (*bla*IMP) and oxacillinase serine β-lactamases (*bla*OXA). These enzymes enable *A. baumannii* to evade the action of many conventional antibiotics, such as carbapenems (Vila et al. 2007; Boucher et al. 2009). The presence of these resistance genes significantly limits treatment options and increases the difficulty of managing infections of *A. baumannii.*

The rate of multidrug-resistant *A. baumannii* strains has been on the rise globally, highlighting the pressing need for novel antimicrobial agents. Its emergence as a possible threat to the health-care system made WHO designate it in Group 1 for the development of novel antimicrobial agents (World Health Organization 2017). This classification underscores the importance of addressing the growing challenge of *A. baumannii* due to multidrug resistance and the requirement of innovative strategies to combat this pathogen. Strict infection control measures, including isolation precautions, hand hygiene, and environmental cleaning, are used to fight against the infections of *A. baumannii*. Keeping this in view, it is essential to have effective management plants for controlling *A. baumannii* infections and mitigate their impact on patient outcomes (Weber et al. 2010).

1.3.4 *Pseudomonas aeruginosa*

Pseudomonas aeruginosa is a gram-negative bacteria listed in WHO priority group 1 for infections that need innovative medical interventions. It naturally has minimal susceptibility to several antibiotics and a strong tendency to develop resistance. The resistance to carbapenem antibiotics is increased due to a reduction in the permeability of porin and the overproduction of *Amp*C β-lactamase (Fukuoka et al. 1993). Additionally, *Pseudomonas aeruginosa* express ESBLs (extended-spectrum β-lactamases), imipenem metallo-β-lactamases, and KPCs, leading to extensive resistance against carbapenems. The rising occurrence of MDR *Pseudomonas aeruginosa* strains poses significant challenges for effective treatments. While colistin, a last-resort antibiotic, is still mostly effective, despite resistance reported during extensive usage in the farming of poultry and pig (Pedersen et al. 2018; Hameed et al. 2019).

1.3.5 *Enterobacter* spp

Enterobacter spp representing the gram-negative bacteria found in immunocompromised hospitalized patients are notorious for harboring multiple antibiotic resistance genes, including carbapenemases and ESBLs (Schwaber et al. 2003). In these species, *Amp*C β-lactamase, capable of hydrolyzing penicillins and broad-spectrum cephalosporins, is hyperproduced (Schwaber et al. 2003). Carbapenem and β-lactam antibiotics stimulate the overproduction of *Amp*C β-lactamase, which increases mortality rates and resistance during treatment (Huh et al. 2014). Certain strains of *Enterobacter* are only sensitive to a few number of last-resort antibiotics, including tigecycline and colistin, because of their capacity to quickly develop resistance (Pendleton et al. 2013). Recent reports of colistin and tigecycline resistance highlight the urgent need for new antimicrobial medications. These reports include either chromosomal mutations in two-component systems or plasmid-mediated mobilized colistin resistance (*mcr*) genes (Pournaras et al. 2016; Berglund 2019).

The spore-forming, gram-positive bacteria *Clostridioides difficile* (formerly known as *Clostridium difficile*) has emerged as a human pathogen in addition to the ESKAPE pathogens and is becoming increasingly resistant to metronidazole (Dingsdag and Hunter 2018). Since 2016, the ECDC has been tracking *C. difficile* infections (CDIs), and it has found that a sizable fraction of these infections exhibit metronidazole resistance (Huemer et al. 2020). The rise of metronidazole resistance is particularly concerning, as *C. difficile* isolates from both humans and animals have been shown to carry a novel plasmid (pCD-METRO) that confers resistance (Boekhoud et al. 2020). To lessen the occurrence and spread of drug-resistant *C.*

difficile, CDIs need to be thoroughly monitored internationally, and broad-spectrum antibiotic overuse must be avoided.

1.4 Antibiotic Resistance

1.4.1 Resistance to β-Lactam Antibiotics

1.4.1.1 Methicillin

Staphylococcus aureus infections are more challenging to treat due to escalating rates of resistance, particularly to β-lactam antibiotics. The presence of the genes *mec*C or *mec*A, which encode different proteins PBP2c or PBP2a (penicillin-binding proteins), is primarily responsible for this resistance (Zhu et al. 2022). By imparting resistance to oxacillin, these proteins provide resistance to nearly all β-lactam antibiotics. The SCC*mec* (staphylococcal chromosomal cassette), which contains the *mec*C and *mec*A genes, is available in several forms. Some SCC*mec* types also possess additional genes that code for non-β-lactam antibiotic resistance, in addition to the *mec*A gene. These multidrug-resistant SCC*mec* types are commonly associated with hospital strains of *Staphylococcus aureus* (Cercenado and Ruiz de Gopegui 2008). In 2018, a strain of *Staphylococcus aureus* was discovered during routine MRSA screening carrying the *mec*B gene, which was found on a plasmid (Mlynarczyk-Bonikowska et al. 2022). Previously, this gene had only been reported in *Macrococcus caseolyticus* (Schwendener and Perreten 2022). The plasmid-mediated methicillin resistance mechanism, transferable between bacteria, represents a new risk level for the widespread dissemination in *staphylococci* of broad β-lactam resistance (Urban-Chmiel et al. 2022). In a scientific context, it is imperative to investigate the occurrence of mecB-mediated methicillin resistance within the populations of MRSA and methicillin-resistant coagulase-negative staphylococci (MRCoNS) among human subjects (Becker et al. 2018). To fully grasp the significance of this novel resistance mechanism and its potential implications for the treatment and management of staphylococcal infections resistant to antibiotics, further research is needed.

1.4.1.2 Cephalosporins

While maintaining the susceptibility to carbapenems, Enterobacteria were found to exhibit resistance to third-generation cephalosporins often attributed to ESBLs or *Amp*C type β-lactamases production (Meini et al. 2019). ESBLs can hydrolyze a wide variety of β-lactam antibiotics including cephalosporins of the third-generation. Similarly, *Amp*C β-lactamases can also provide resistance to these cephalosporins. *Amp* C β-lactamases can exist in two forms: chromosomal (*cAmp*C) or plasmid-mediated (*pAmp*C) (Asenjo et al. 2021). Besides ESBLs and *Amp*C β-lactamases, other factors can contribute to resistance in Enterobacteria. These include reduced permeability of the bacterial cell wall and the existence of efflux pumps capable of removing antibiotics at the cell surface. In some cases, there have been reports of strains carrying modified versions of chromosomal *Amp*C (*cAmp*C) enzymes that hydrolyze cefepime, a fourth-generation cephalosporin (Varela et al. 2021). These variants are known as extended-spectrum *cAmp*C (cESAC). Collectively, these diverse resistance mechanisms in Enterobacteria can lead to decreased susceptibility or resistance to cephalosporins, particularly third-generation ones, while still retaining susceptibility to carbapenems (Iredell et al. 2016). The investigation of extended-spectrum *cAmp*C (cESAC) in *Pseudomonas aeruginosa* has been somewhat restricted compared to other bacterial species. This limitation is attributed to the considerable genetic diversity and polymorphism observed in the sequence of *cAmp*C within this bacterium. However, a study has offered insightful information on these variants (Berrazeg et al. 2015). The work brought attention to the existence of cESAC variants, which result from changes in the β-loop of the β-lactamase enzyme. These variants have significant therapeutic implications and are crucial to comprehend their impact on antibiotic resistance in *Pseudomonas aeruginosa.*

Lately, significant improvements in the development of novel antibiotics and combinations with β-lactamase inhibitors have been made. These innovative treatments, like ceftolozane/tazobactam, ceftazidime/avibactam, and meropenem/vaborbactam, have started to address the limited options for combating gram-negative drug-resistant bacteria (Voulgaris et al. 2019). Furthermore, other drugs like aztreonam/avibactam, imipenem/relebactam, and cefiderocol are currently undergoing the process of obtaining marketing authorization (Tilahun et al. 2021). These emerging treatment alternatives provide hope in tackling the challenges posed by gram-negative pathogens that demonstrate antibiotic resistance. The development of these new therapies offers a promising prospect in combatting the infections caused by gram-negative bacteria (Ma et al. 2020). Fraile-Ribot et al. (2017) made a significant discovery of OXA-539, a new extended-spectrum β-lactamase originating from OXA-2. During their research, they identified a duplication of a crucial aspartate residue in OXA-539. In transformation studies conducted on the PAO1 strain, it was observed that the MICs (minimum inhibitory concentrations) of cefepime, ceftazidime, ceftolozane/tazobactam, and ceftazidime/avibactam were higher in the transformant having OXA-539 compared to isolates having OXA-2. Interestingly, the opposite trend was observed for meropenem, where the MICs were lower in the transformant carrying OXA-539 compared to isolates carrying OXA-2. This finding indicates the distinct impact of OXA-539 on different antibiotics, highlighting the importance of understanding its implications for antibiotic resistance. The development of resistance in *Pseudomonas aeruginosa* after ceftolozane/tazobactam therapy has only been briefly mentioned in published papers (Fraile-Ribot et al. 2017). According to whole-genome sequencing, the emergence of resistance strains frequently involves a number of changes, such as those that result in *Amp*C structural modification and overexpression, such as substitutions or deletions of amino acids (Fraile-Ribot et al. 2018). It has been shown that isolates that exhibit cross-resistance to ceftazidime/avibactam also exhibit resistance to ceftolozane/tazobactam. The MICs of carbapenems and piperacillin/tazobactam for these strains have, nevertheless, slightly decreased (Fraile-Ribot et al. 2018).

The widespread usage of cephalosporins and ESBLs has emerged as a diverse group of class A β-Lactamases. These enzymes, exemplified by over 172 distinct *bla*CTX-M variants and 220 variants of *bla*TEM documented thus far, constitute the predominant β-lactamases among Enterobacteriaceae. Owing to their higher hydrolytic activity, ESBLs were found to confer resistance to a range of antibiotics including aminopenicillins (like amoxicillin or ampicillin), carboxypenicillins (such as tricarcillin or carbenicillin), ureidopenicillins (like piperacillin), cephalosporins (including cephaloridine, cephalothin, and cefuroxime), oxi-imino cephalosporins (such as ceftriaxone and cefotaxime), cefpirome, and cefepime (Thomas and Nielsen 2005). Due to their extensive host compatibility and effective conjugation, the *bla*TEM and *bla*CTX-M gene have undergone multiple horizontal gene transfer events, becoming integrated into plasmids across diverse bacterial species (Barlow et al. 2008; Woodford et al. 2009).

The study on ESBL genes (specifically *bla*TEM and *bla*CTX-M) of indigenous organisms of the Yamuna River in India by Azam et al. (2016) revealed that approximately 40% of the isolated organisms exhibited concurrent mercury tolerance genes. The computational assessment of the new $bla_{\text{CTX-M-152}}$ variant involved a multi-step process, including structure generation through modeling and subsequent docking simulations. These simulations provided valuable information about the catalytic behavior of CTX-M-152. CTX-M-152 has been identified as an outlier within the CTX-M group 25 due to distinctive substitutions, including G89D, Q26H, P99S, T154A, and D146G. It is noteworthy that this particular variant relies on the presence of five specific amino acid residues (Asn132, Ser70, Gly238, Ser237, and Arg273) for its role in positioning cefotaxime within the active site. This positioning is facilitated by the formation of seven hydrogen bonds, resulting in a binding energy of approximately −7.6 kcal/mol (Azam et al. 2016).

1.4.1.2.1 Ceftobiprole and Ceftaroline

For gram-positive bacteria that exhibit resistance to β-lactam, two new cephalosporins, ceftobiprole and ceftaroline, have been developed. These antibiotics exhibit a strong affinity for PBP2a, a critical protein associated with resistance (King et al. 2016). However, there have been recent findings of ceftaroline-resistant clinical strains of *Streptococcus pneumoniae*. These strains showed mutations in various PBPs, particularly PBP2x, which is strongly linked to resistance against this cephalosporin (Pfaller et al. 2017). Studies on ceftobiprole resistance are few and mostly concentrate on laboratory mutants compared to clinical isolates. However, a triple mutation (N146K-N204K-G246E) in PBP2a has been found, resulting in considerable resistance in African MRSA strains to both ceftaroline and ceftobiprole (Schaumburg et al. 2016). A novel insertion in PBP2a and amino acid alterations in all studied PBPs (PBP1–PBP4) were discovered through the examination of clinical MRSA samples, and these changes may help explain resistance to these new cephalosporins without harming other β-lactams. Treatment of these infections is made more difficult by the unavailability of therapeutically useful β-lactams against MRSA as a result of the advent of antibiotic resistance (Morroni et al. 2018).

1.4.1.3 Carbapenem

Resistance to carbapenems in gram-negative bacteria primarily involves two main mechanisms i.e., the production of carbapenemases, which are hydrolytic enzymes, and increased expression of β-lactamases (ESBL's and cephalosporinases) with limited or no activity against carbapenems. Additionally, reduced permeability due to alterations in porins can contribute to resistance (Tooke et al. 2019). Clinically relevant carbapenemases are classified into two classes: A and B. Class A carbapenemases encompass NmcA/IMI, KPC, GES, and SME, each with different variants. Recently, two novel plasmid-mediated class A carbapenemases, FRI-1 in an *Enterobacter cloacae* strain and BKC-1 in *K. pneumonia* a Brazilian strain, were identified (Dortet et al. 2015). BKC-1 exhibits weak imipenem hydrolysis but confers resistance to β-lactams (monobactams, cephalosporins, penicillins, and carbapenems), while FRI-1 imparts high resistance to penicillins, narrow-spectrum aztreonam, carbapenems, and cephalosporins (and limited activity against broad-spectrum cephalosporins, such as cefotaxime, ceftazidime, and cefepime) (Dortet et al. 2015). Class B carbapenemases are primarily categorized into three types: IMP, NDM, and VIM, with several variants described for each type. A *Pseudomonas aeruginosa* strain carrying a novel class B carbapenemase called HMB-1 (Pfennigwerth et al. 2017). The gene *bla*HMB-1 was found on the chromosome and showed 74.3% genetic similarity to the closest *bla*KHM-1gene. Transformation studies demonstrated a six-fold increase in meropenem minimum inhibitory concentration in an *E. coli* recipient, but no change in aztreonam susceptibility was observed (Pfennigwerth et al. 2017).

Through the existence of efflux pumps, gram-negative bacteria exhibit resistance to different classes of antibiotics. Of them, RND (resistance-nodulation-cell division) family member AcrAB-TolC is actively engaged in imparting resistance to a wide variety of antibiotics (Schuster et al. 2017; Atac et al. 2018). Although the efflux pumps in carbapenem resistance have received limited research attention, recent clinical observations have highlighted their clinical implications. The minimum inhibitory concentrations for carbapenems were found higher in mutant carbapenemase-producing enterobacteria (with changes in the AcrAB-TolC efflux pump) than in control strains (Saw et al. 2016). The isolate showed intermediate susceptibility to imipenem, resistance to meropenem, ertapenem, and piperacillin-tazobactam, and susceptibility to aztreonam as well as many third- and fourth-generation cephalosporins (cefepime, ceftazidime, and ceftriaxone) (Yang et al. 2018). Genome sequencing revealed that the porins OmpK36 and OmpK35, which have previously been linked to carbapenem resistance, are deficient, and the efflux pump is overexpressed as a result of mutations in the acrR regulator gene, which was discovered using genome sequencing (Yang et al. 2018). These findings emphasize the role of efflux pumps in attributing resistance to broader classes of antibiotics among gram-negative bacteria.

1.4.2 Resistance to Antibiotics of Non-β-Lactams Category

1.4.2.1 Aminoglycosides

The growing prevalence of 16S rRNA methyltransferases is becoming a cause for heightened concern, primarily because these enzymes possess the capability to provide substantial resistance to clinically important aminoglycoside antibiotics. Even though the majority of these methyltransferases were first identified before 2014, their spread has experienced considerable advancements since that time (Fournier et al. 2022). Nine distinct 16S rRNA methyltransferases, including ArmA and its variations RmtA through RmtH (Doi et al. 2016), have been discovered. Notably, a fresh variety known as RmtB4 has just lately been reported (Costello et al. 2019). Recent research suggests that methyltransferases are largely proliferating among *Enterobacteriaceae* that produce carbapenemase, thereby providing a risk in treating MDR gram-negative infections (Doi et al. 2016; Taylor et al. 2018). ArmA, RmtB, RmtC, and RmtF are the most often reported 16S rRNA methyltransferases, and they are frequently discovered along with isolates that produce NDM or KPC carbapenemase (Doi et al. 2016; Costello et al. 2019; Taylor et al. 2018; Poirel et al. 2018; Galani et al. 2019).

1.4.2.1.1 Plazomicin

Next-generation aminoglycoside plazomicin was created to combat enzymes that inactivate aminoglycosides, a frequent route of resistance to aminoglycosides. It is used to treat patients with severe infections brought on by multidrug-resistant *Enterobacteriaceae*, such as strains that are carbapenem-resistant and produce ESBLs. Plazomicin showed sustained activity against the majority of carbapenemase-producing *K. pneumoniae* in a recent study conducted in Greek hospitals (Galani et al. 2019), with consistently lower MICs in contrast to other aminoglycosides, even when aminoglycoside-modifying enzymes are present, and 7.7% of the isolates (1 KPC plus OXA-48 producer, 6 VIM, and 16 KPC) carried the *rmt*B gene (22 isolates) or the *arm*A gene (1 isolate), making them resistant to plazomicin (Galani et al. 2019).

1.4.2.2 Tigecycline

Tigecycline, a member of the glycylcycline family, is a derivative of minocycline with broad-spectrum antimicrobial activity (Olson et al. 2006). The expression of the MepA efflux pump is often correlated with incidences of tigecycline resistance among *Staphylococcus aureus*. The mechanisms of tigecycline resistance in enterococci have recently been reported to Niebel et al. (2015), who highlighted the possibility of a link between tigecycline resistance among clinical strains of *Enterococcus faecium* and mutations in the gene *rps*J encoding S10 structural protein of the smaller subunit of the ribosome. Despite being a strong alternative in treating infections caused by gram-negative bacteria that are extensively drug-resistant (XDR) and multidrug-resistant (MDR), it has lately been observed that resistance against tigecycline is on the rise (Karaiskos and Giamarellou 2014). Overexpression of the efflux pump (*Acr*AB-TolC) is the most well-known resistance mechanism, and it results from mutations in the regulating gene's coding regions (Sheng et al. 2014). This mechanism has been seen in other enterobacteria, like *Enterobacter cloacae*, and has been investigated extensively in *K. pneumoniae* (Keeney et al. 2007; Ye et al. 2017). They observed that the ribosomal binding site sequence is suppressed, *ram*R's translation is impaired, and levels of *ram*A and *acr*AB are elevated, leading to its rapid appearance after 41 days of treatment (Ye et al. 2017). Additionally, tigecycline resistance is also linked to overexpression of the *Oqx*AB pump (Zhong et al. 2014). Recently, it has been revealed that tigecycline resistance can occur through mechanisms, which are independent of the *Acr*AB system. In 2014, tigecycline-resistant *K. pneumoniae* isolates from the high-risk clone ST512 that produces KPC-3-carbapenemase were found to be tigecycline-resistant (Zhong et al. 2014; Mendes et al. 2022). It is interesting that this resistance was not linked to changes in the gene *ram*R. Instead, a mutation in the gene *rps*J, which codes for the ribosomal protein S10, led to an alteration in one of the protein's essential amino acids. The S10 protein functions as tigecycline's binding target and is a component of the ribosomal RNA 30S subunit (Villa et al. 2014). The insertion of an element that resembled IS5 in the promoter region was linked to this increased expression. This discovery established a connection between the lack of tigecycline susceptibility seen in vivo and the existence of this genetic component in *K. pneumoniae* isolates (Nielsen et al. 2014).

1.4.2.2.1 Eravacycline

The first fluorocycline produced from minocycline, eravacycline, exhibits greater efficacy against both gram-positive and gram-negative bacteria (Bassetti and Righi 2014). The *Oqx*AB pump is expressed more frequently in resistant and heteroresistant bacteria than the *Acr*AB-TolC pump. Notably, *ram*A regulatory gene overexpression is large, but it has little impact on the *Acr*AB-TolC pump. Additionally, the MacAB-TolC pump is crucial in the emergence of eravacycline resistance (Zheng et al. 2018).

1.4.2.3 Colistin

The escalation in the prevalence of carbapenemase-producing enterobacteria has resulted in an elevated reliance on colistin, an older antibiotic known for its notable toxicity (Bialvaei and Samadi Kafil 2015). Colistin emerged as one of the limited therapeutic options available in the treatment of infections caused by extensively drug-resistant gram-negative bacteria. In recent years, novel chromosomal and plasmid-mediated mechanisms were found imparting resistance to polymyxins, the antibiotic class to which colistin belongs (El-Sayed Ahmed et al. 2020).

Addition of the cationic groups like L-Ara4n and pEtN that cause modification of the lipopolysaccharide lipid A (LPS) is believed to be a promising mechanism involved in the acquisition of resistance to polymyxins. These sugars are produced by the gene *pmr*E and the *pmr*HFIJKLM operon, whereas the synthesis of pEtN is encoded by the *pmr*C gene. As a result of these alterations, positively charged LPS is produced, which decreases the LPS's attraction for positively charged polymyxins. Overexpression of LPS modification-controlling transcriptional regulatory mechanisms, which is frequently brought on by mutations, is the most prevalent method of chromosomal resistance (Gogry et al. 2021). *K. pneumoniae* can develop resistance to colistin through genetic changes in several regulatory systems. These include alterations in *mgr*B, (Cannatelli et al. 2013; Poirel et al. 2015), *Pmr*B (*Pmr*AB system), *pho*Q, and *pho*P (*Pho*PQ system) (Cannatelli et al. 2014; Jayol et al. 2014; Jayol et al. 2015) as well as the presence of specific amino acid substitutions in *Crr*AB (Wright et al. 2015; Cheng et al. 2016) as depicted in Fig. 1.2. A partial nucleotide loss in the *pho*P gene allows heteroresistant *K. pneumoniae* strains to return to susceptibility (Wright et al. 2015). The upregulation of the *Acr*AB-TolC efflux pump among *Enterobacter asburiae* strains is correlated with colistin resistance via enhanced expression of the *sox*RS stress response regulatory system (Telke et al. 2017).

By the end of 2015, it was thought that chromosomal changes were the sole way for colistin resistance to develop. But that same year, *mcr*-1, a horizontally and plasmid-mediated transmissible mechanism, appeared in China (Liu et al. 2016). The phosphoethanolamine transferase that the *mcr*-1 gene encodes changes lipid A, resulting in colistin resistance in a variety of enterobacteria. The MIC of polymyxin is increased by four- to eightfold in bacteria containing this gene, which is sufficient to give colistin resistance (Liu et al. 2016). Since its discovery, attempts have been undertaken to locate the *mcr*-1 gene in strain collections already in existence. Isolates containing this gene have been discovered before it was first described (Rolain et al. 2016; Prim et al. 2016; Hasman et al. 2015). Human strains of patients who had never received colistin treatment before have primarily shown transferable resistance to it. The consumption of tainted food, which leads to intestinal colonization, has been linked to the acquisition of mcr-1 (Hasman et al. 2015). It has been documented that mcr-1 can co-occur with other resistance mechanisms including ESBLs (Haenni et al. 2016), or carbapenemases such as NDM-5 (Mediavilla et al. 2016), VIM-1 (Poirel et al. 2016), or KPC (Di Pilato et al. 2016) have also been described. There are enterobacterial variants of the mcr gene family, ranging from

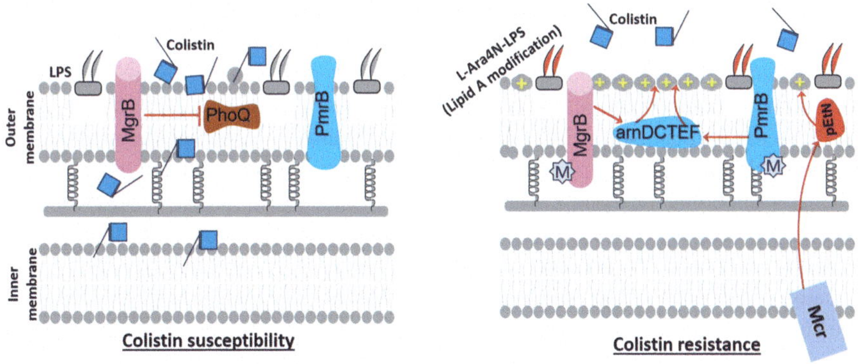

Fig. 1.2 Colistin resistance in bacteria. The figure shows the basic assembly of the membrane structure with associated protein and their transition on mutation, leading to modification of lipid A that attributes resistance to colistin

mcr-1 to mcr-9, which exhibit notable genetic differences (Wang et al. 2018; Carroll et al. 2019). Additionally, nucleotide variants within the same cluster of genes (such as *mcr*-1.1 and *mcr*-1.2) have been described (Di Pilato et al. 2016).

1.4.2.4 Macrolides

Macrolide resistance among bacteria is known to occur by 03 mechanisms: (1) target change by mutation or methylation, (2) antibiotic inactivation, and (3) active efflux pumps. One important method granting macrolide resistance is the methylation of ribosomes by erm genes, especially in specific gram-positive bacteria. The two mechanisms that have historically caused Campylobacter spp. to exhibit low levels of macrolide resistance are (1) mutations in the target location (23S rRNA or genes (*rpl*D and *rpl*V) producing ribosomal proteins (L4 and L22)) and (2) efflux pumps (*Cme*ABC). However, a recent discovery of the *erm*(B) gene in *Campylobacter coli* and *Campylobacter jejuni* strains on the bacterial chromosome and plasmids marks a significant advancement (Qin et al. 2014; Wang et al. 2014).

1.4.2.5 Fluoroquinolones

Fluoroquinolone resistance in clinically relevant gram-positive bacteria is attributed to mutations in the gene sequence of *par*C and *gyr*A genes (Poole 2000). In addition to this, resistance to fluoroquinolone among gram-positive bacteria, such as *Listeria monocytogenes*, is also attributed to the presence of efflux pumps within the major facilitator superfamily (MFS). Lately, it was found that FepA (a member of the multidrug and toxic compound extrusion (MATE) family) is responsible for conferring resistance to ciprofloxacin and norfloxacin in *Listeria monocytogenes* (Guerin et al. 2014).

1.4.2.6 Vancomycin

The glycopeptide antibiotic vancomycin is used to treat infections of the gram-positive bacterial origin (Tarai et al. 2013). The glycopeptide resistance is attributed to the thickening of the cell wall in *Staphylococcus aureus* and the presence of genes (*van*A and *van*B) in enterococci. Despite reports of *S. aureus* strains with high vancomycin resistance (VRSA) and intermediate susceptibility (VISA), VRSA isolates are still rare. The impact of glycopeptide resistance is more significant within the Enterococcus genus, (Ruef 2004).Yamaguchi et al. (2015) during their study on the clinical isolate MRSA found its resistance to vancomycin attributed to single nucleotide polymorphisms (SNPs) in two previously unrelated genes: *cap*B and *lyt*N encoding tyrosine kinase (role in capsular polysaccharide biosynthesis) and another coding for peptidoglycan hydrolase (N-acetylmuramyl-l-alanine amidase), respectively. The presence of these substitutions in *cap*B and *lyt*N was found to cause resistance to vancomycin, even in the absence of typical van genes. Notably, the strain also exhibited resistance to rifampicin and daptomycin. Thus, it has been concluded that these new substitutions were responsible for the increased vancomycin MIC and subsequent resistance (Yamaguchi et al. 2015).

1.4.2.7 Linezolid

The initial oxazolidinone compound introduced for clinical use, namely linezolid, is experiencing a growing utilization in the control of enterococci and MRSA infections. Until the year 2015, the primary determinant associated with resistance to linezolid in enterococci was characterized by chromosomal mutations occurring within the 23S ribosomal RNA subunit, with specific attention given to the G2576T alteration (Rybak and Roberts 2015). *Enterococcus faecalis* and *Enterococcus faecium* strains obtained in China were the first to contain the novel transferable gene *optr*A, which was discovered in 2015. This gene provides resistance to the phenicol class of antibiotics (chloramphenicol) as well as the oxazolidinones (linezolid and tedizolid) (Wang et al. 2015). Since then, enterococci have mostly been associated with the *optr*A gene, but certain strains of *Staphylococcus suis* with a porcine origin have also been identified. The discovery of similar *optr*A variations on presumably identical plasmids in different strains raises the possibility that these plasmids harboring this gene have disseminated (Bender et al. 2018). The *optr*A gene has been linked to particular *E. faecalis* clones, including ST480 and ST585 (Argudín et al. 2019; Càmara et al. 2019).

Genome sequencing of a clinical MRSA strain revealed *poxt*A gene conferring resistance to phenicols, tetracyclines, and oxazolidinones (Antonelli et al. 2018). Due to the gene's location on a plasmid that promotes interspecies transfer, it was recently discovered in an *E. faecium* clinical strain as well (Papagiannitsis et al. 2019). The ATP-binding cassette F (ABC-F) family of proteins, known to be encoded by the *optr*A and *poxt*A genes, function by modifying the bacterial ribosome, thereby providing resistance against the antimicrobial properties of these antibiotics. The appearance of new plasmid-mediated, transferable genes raises apprehensions about inter- and intraspecies dissemination among different bacterial genera that share a common ecological niche.

1.4.2.8 Fosfomycin

It is a broad-spectrum antibiotic, which prevents the formation of cell walls. It has long been understood that fosfomycin resistance results from the development of fosfomycin-inactivating enzymes that are encoded by chromosome or plasmid. The Enterobacterales, *Pseudomonas* species, and *Acinetobacter* species are frequently reported to contain the *Fos*A genes. New subtypes of plasmid-mediated genes, including *fos*A5 (Ma et al. 2015), *fos*A6 (Guo et al. 2016), *fos*A7 (Rehman et al. 2017), and *fos*A8 (Poirel et al. 2019), have been discovered. Furthermore, it has been discovered that these genes coexist on plasmids with other resistance genes that give resistance to β-lactams, aminoglycosides, and fluoroquinolones (Guo et al. 2016).

1.4.2.9 Daptomycin

Daptomycin, a lipopeptide antibiotic, is used to treat infections caused by *Staphylococcus aureus* and other gram-positive bacteria. It interacts with the phosphatidylglycerol (PG) of the bacterial cell membrane, reflecting its mechanism of action (Steenbergen et al. 2005). The *mpr*F gene, which codes for the production of the dual-purpose enzyme MprF, was found to impart resistance to daptomycin among *Staphylococcus aureus* (Kotsogianni et al. 2021). These changes make it easier for the enzyme to add a positive charge to the bacterial peptidoglycan, which repels the daptomycin-Ca^{2+} combination electrostatically and prevents it from attaching to the target. *Staphylococcus aureus* strains with increased daptomycin MIC values outside of the sensitive range have also been found to contain mutations in other genes, such as *yyc*G (encode histidine kinase), *rpo*B, and *rpo*C (RNA polymerase subunits) (Kakoullis et al. 2021). Although not yet completely understood, the mechanisms of daptomycin resistance in enterococci appear to be distinct from those seen in *Staphylococcus aureus* (Montero et al. 2008). Other therapeutically significant bacteria including *Corynebacterium striatum* and *Streptococcus mitis* have also developed daptomycin resistance. Mutations in the *pgs*A2 gene, which codes for a peptidoglycan synthase, cause a considerable loss of peptidoglycan and give high-level resistance in *C. striatum*, a strain that developed high resistance to daptomycin throughout treatment (Asenjo et al. 2021). *S.mitis/oralis* exhibits decreased susceptibility to daptomycin compared to other streptococci of the viridans group and has the propensity to establish high levels of resistance quickly. *Staphylococcus mitis/oralis* has been found to have mutations in the gene *cds*A encoding a phosphatidate cytidyltransferase with its role in the synthesis of membrane phospholipids. These mutations cause peptidoglycan and cardiolipin to disappear, which increases daptomycin resistance (Mishra et al. 2017).

1.5 Mechanistic Basis of Antibiotic Resistance

Antibiotic resistance mechanisms encompass various genetic changes in microorganisms that allow them to withstand the effects of antibiotics. These mechanisms involve genetic mutations that modify antibiotic targets or metabolic pathways, horizontal gene transfer for acquiring resistance genes, efflux pumps that eliminate antibiotics from the cell, enzymatic inactivation of antibiotics, and the formation of protective biofilms as shown in Fig. 1.3, collectively contributing to the emergence and dissemination of resistance within bacterial populations (Munita and Arias 2016). Genetic strategies are used by bacteria to evolve and fight against antibiotics and the strategies include resistance genes being acquired by horizontal gene transfer from other bacteria as well as changes in the genes involved in the antibiotic mechanism of action (Dantas and Sommer 2014). Bacteria are able to thrive and endure in the presence of antibiotics by using these processes, demonstrating their remarkable genetic plasticity and evolutionary responsiveness to environmental difficulties. In mutational resistance, a subset of bacteria originating from a previously susceptible population undergoes genetic changes impacting the effectiveness of antimicrobial agents, thereby enabling bacteria to thrive and reproduce even in the presence of the drug (Reygaert 2018). The entire cell homeostasis frequently suffers as a result of these mutational changes that impart resistance, which reduces fitness. As a result, these alterations are only kept when absolutely necessary when the antibiotic is present (Munita and Arias 2016). Antimicrobial resistance mutations

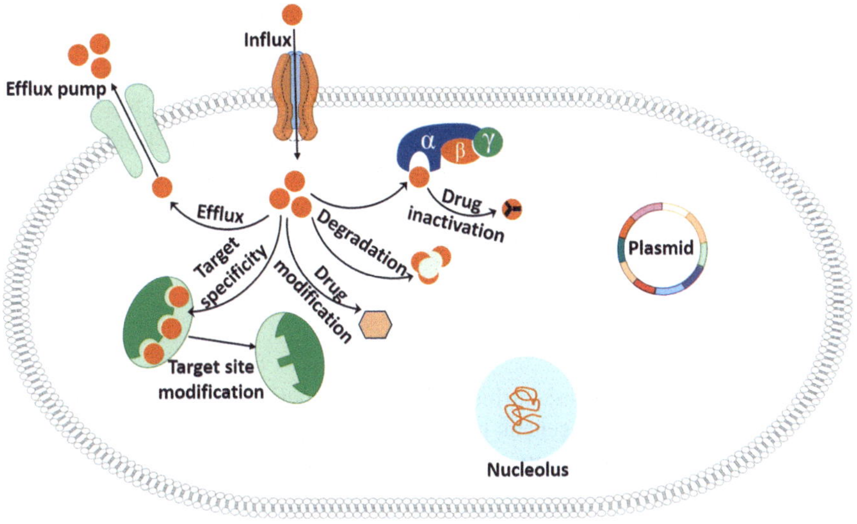

Fig. 1.3 Operating resistance mechanisms in bacteria

typically have a variety of effects on how an antibiotic works. These include altering the medication affinity for the antimicrobial target, decreasing drug uptake into the cell, activating efflux mechanisms to expel the toxic chemical, or altering regulatory networks to have a broad impact on crucial metabolic pathways. As a result, the diversity and complexity of resistance arising from acquired mutational alterations can differ (Martinez and Baquero 2000).

For millions of years, antimicrobial compounds have existed, leading bacteria to develop intricate mechanisms of drug resistance in response to them (Uddin et al. 2021). A single bacterial cell can use a number of resistance mechanisms simultaneously to fend off an antibiotic. For instance, fluoroquinolone (FQ) resistance can arise through three distinct biochemical pathways, all of which might concurrently exist within a single bacterial strain, resulting in a cumulative impact that frequently escalates resistance levels: (1) genetic mutations occurring in genes responsible for encoding the FQ target sites, namely topoisomerase IV and DNA gyrase, (2) elevated expression of efflux pumps pertaining to FQ molecules, and (3) safeguarding the FQ target site by Qnr protein (Aldred et al. 2014).

On the other hand, it seems that specific resistance mechanisms are preferred by particular bacterial species. While the production of lactamases is the main mechanism of lactam resistance among gram-negative bacteria, alteration of the penicillin-binding protein's (PBPs) target location is associated with conferring resistance to gram-positive bacteria (Munita and Arias 2016). Major variations in the cell envelope structure among different bacterial groups are thought to have an impact on this preference for particular pathways. The existence of an outer membrane in gram-negative bacteria enables regulation of the influx of molecules into the periplasmic region. In order to reach the PBPs, which are found in the inner membrane, many lactams need certain porins. By controlling these molecules' access to the periplasmic region, the bacterial cell is able to produce enough lactamases to favor the breakdown of the antibiotic (Yamashita and Buchanan 2010). Gram-positive bacteria, on the other hand, do not have this 'compartmentalization' edge since they lack a membrane on the outside. However, some situations show that β-lactamases can still be effective in gram-positive organisms even in the absence of compartmentalization, as shown by the formation of *staphylococcal penicillinase* (Clark and Pazdernik 2012).

1.5.1 Intrinsic Resistome

Bacterial species often possess intrinsic resistance (inherited or natural) against several antibiotics due to their genetic makeup or characteristics. 'Intrinsic resistome' is the name given to the group of genes that cause intrinsic resistance. Whole-genome sequencing, antibiotic selection based on experimental evolution, extensive transposon-tagged mutant libraries, and other techniques have all been used in the studies of intrinsic resistance. According to the findings, a bacterial genome's intrinsic resistome may make up as much as 3% of its total size. Thus, natural antibiotic resistance is a complicated phenomenon involving not only a small number of genes that have developed in response to antibiotics but also certain genes encoding proteins as well as metabolic enzymes that control essential bacterial functions (Fajardo et al. 2008; Linares et al. 2010). Bacteria may become more prone to antibiotics due to mutations in the components causing intrinsic resistance, making them attractive targets for new medicines when combined with existing antimicrobial treatments (Martínez 2012). It has been found that inhibiting the intrinsic resistance to antibiotics may broaden the mutant selection, raising several questions about the efficacy of such mixtures. If this growth occurs in systems other than those already studied, more research is needed to confirm this (García-León et al. 2012).

If all strains of a certain bacterial species consistently have MIC above a predetermined breakpoint, that species is thought to be intrinsically resistant to that antibiotic from a therapeutic standpoint. As a result of pre-antibiotic factors like lack of a susceptible target or decreased antibiotic permeability, intrinsic resistance is a result of these circumstances. For this kind of resistance, the bacterium and the antibiotic do not need to have any sort of evolutionary connection (Martínez et al. 2015). Comprehensive research on bacteria like *Pseudomonas aeruginosa* and *E. coli* has revealed that target absence or reduced permeability alone is insufficient to explain intrinsic resistance (Fajardo et al. 2008). These investigations employ an ecological definition of resistance, according to which any mutant exhibiting decreased susceptibility to an antibiotic compared to its parental strain is considered resistant, regardless of whether the MIC surpasses the clinically established breakpoint. This mutant helps to identify elements that play a vital role in increasing resistance levels (acquired resistance). The inactivation of the protein that the altered gene encodes diminishes the level of natural resistance, as seen when a mutant is more sensitive than its original strain. These mutants help identify elements contributing to intrinsic resistance (Alvarez-Ortega et al. 2010, 2011; Liu et al. 2010; Fernández et al. 2013).

1.5.2 Extrinsic Resistome

Extrinsic resistance refers to resistance mechanisms that bacteria acquire from external sources, often through horizontal gene transfer. Antibiotic resistance genes may not always portend an immediate risk of their expansion in a specific ecosystem. It is essential to differentiate between resistance genes already integrated into the systems and widely distributed throughout human pathogens and solely residing within the chromosomes of environmental bacteria. While the presence of the latter requires cautious investigation before making conclusions, the former's presence in an ecosystem should be regarded as a potential threat to dissemination. For instance, the detection of housekeeping genes (like *E. coli ampC*), which are chromosomally encoded, in gut metagenome can serve as an indicator of *E. coli* presence but does not always signify an imminent risk of resistance spread (Martínez et al. 2015). Due to the scarcity of documented origins of resistance genes acquired through horizontal gene transfer, it remains challenging to estimate the probability that a unique resistance gene within a specific metagenome will be acquired and disseminated among human pathogens (Lupo et al. 2012).

Nevertheless, it is still possible to assess the risks posed by these genes in fostering resistance emergence. The identification of novel resistance mechanisms against existing or novel drugs may hold greater significance than identifying inactivating enzymes associated with known resistance gene categories (Martinez 2014). When investigating the processes of resistance acquisition and dissemination, it is essential to consider the impact of human antibiotic usage on the emergence of resistance in natural environments. Evidence related to resistance-conferring genes suggests that antibiotic use has disproportionately enriched a limited set of genes, particularly those currently residing in gene-transfer sites within human pathogens (Forsberg et al. 2012; Martinez 2008). It might be considered biological pollution when these genes are released into natural ecosystems from farms or hospital settings (Martinez 2009). Antibiotic resistance components are self-replicative, resulting in their ubiquitous existence, even in healthy environments where antibiotic selection pressure has not occurred, unlike other contaminants whose concentration decreases with time and distance from the source (Jian et al. 2021). The farms and wastewater treatment facilities provide a high risk for the development, upkeep, and spread of antibiotic resistance. Pathogenic bacteria may come in contact with potential providers of genes related to resistance in these circumstances (Baquero et al. 2008). Chemical pollutants like heavy metals or biocides, which can potentially influence the selection of microbes that are resistant to antibiotics, can increase this risk (Sanchez et al. 2005).

1.5.3 HGT and Antibiotic Resistance

HGT (horizontal gene transfer) as depicted in Fig. 1.4, involving the acquisition of foreign DNA, plays a significant role in bacterial evolution and is a key factor responsible for the creation of antibiotic resistance in bacteria (Emamalipour et al. 2020). Natural compounds found in the environment, particularly in soil, are the source of many clinical antibacterial agents. Bacteria that cohabit with these environmental compounds often carry inherent genetic elements of resistance. This collection of resistant genes, known as the "environmental resistome," serves as a rich reservoir of antibiotic resistance determinants that can be acquired by clinically important bacteria. This genetic exchange contributes to the widespread dissemination of resistance against commonly used antibiotics (Mutuku et al. 2022). Bacteria traditionally take up foreign genetic material through three primary mechanisms: transformation (naked DNA uptake), transduction (facilitated by bacteriophages), and conjugation (also known as bacterial "sex"). The transformation is one of the simplest forms of HGT, but only a limited number of species of bacteria that are clinically relevant have the inherent ability to uptake naked DNA for resistance development (Thomas and Nielsen 2005).

Transformation, transduction, and conjugation are the three primary methods by which bacteria obtain external genetic material (Lerminiaux and Cameron 2019). Only a small subset of clinically significant bacterial species may spontaneously integrate such DNA to evolve resistance, and transformation entails the reception of bare DNA. Conjugation is a very effective way of gene transfer, and it frequently occurs in the hospital environment when resistance emerges. Cells interact with one another to form conjugates, which is especially common in the gastrointestinal tract of people taking antibiotics. Mobile genetic elements (MGEs) are frequently used in conjugation as a means of transferring important genetic data (Manson et al. 2010). The two MGEs that are crucial for the emergence and spread of antibiotic resistance in species that are relevant for therapeutic purposes are transposons and plasmids. Integrons also serve as an effective mechanism for the accumulation of genes that are related to antibiotic resistance. Integrons are site-specific

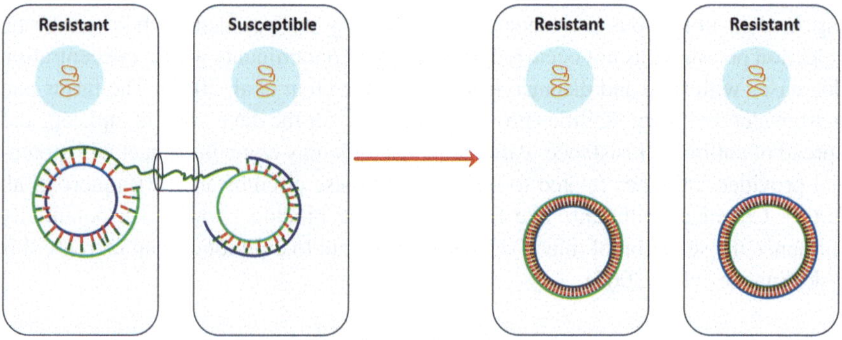

Fig. 1.4 Horizontal transfer of plasmid that leads to development of resistance

recombination mechanisms that can insert portable gene cassettes into the chromosomes of bacteria. They offer a streamlined procedure for adding new genes and the tools required to guarantee their expression (Thomas and Nielsen 2005). Among the primary drivers of bacterial evolution is the use of integrons, a potent method for genetic exchange (Stokes and Gillings 2011).

1.5.4 Plasmid-Mediated Antibiotic Resistance

Bacteria acquire antibiotic resistance through two primary mechanisms: chromosomal mutations and the uptake of transposable genetic elements, like plasmids, through HGT (Bacilli 2018). Through conjugation, plasmids may be horizontally transmitted between bacteria and are capable of independent replication outside of bacterial chromosomes. They play a critical role in the evolutionary process of antibiotic resistance by facilitating the dissemination of resistance genes among clinically significant pathogens (Bacilli 2018; Carattoli 2013). Plasmids are essential in enabling the spread of genes related to antibiotic resistance among bacterial pathogens. However, the fitness cost of introducing an antibiotic-resistant plasmid into a new bacterial host is frequently high, making the plasmid-bearing bacterium less competitive in the absence of antibiotics. But the plasmid gives the bacteria a chance to survive in the presence of antibiotics (Lenski 1998; San Millan and MacLean 2017). The plasmid and bacterial modifications that take place during their early cohabitation with antibiotics are crucial for infection development and the maintenance of plasmid-mediated resistance even after the treatment is stopped (Rodríguez-Beltrán et al. 2021).

Conjugative plasmids, especially in bacterial families like Enterobacteriaceae and Enterococcaceae, had a major role in the spread of antibiotic resistance (Boucher et al. 2009; Vincent 2003; Rozwandowicz et al. 2018). Some plasmid-bacterium associations become successful, leading to the emergence of "superbugs" that rapidly disseminate within the clinical environment (Boucher et al. 2009; Dimou et al. 2012). There is broad evidence of these particular relationships between plasmids and bacterial clones, with some plasmids that confer antibiotic resistance being intimately associated with particular bacterial lineages. Notable examples include the association between sequence types 11 (ST11) or ST405 of *Klebsiella pneumoniae* and the plasmid pOXA-48 (carrying the carbapenemase gene *bla*OXA-48) (Oteo et al. 2015; Brañas et al. 2015), and the association between ST131 clade C2/H30Rx of *Escherichia coli* and IncFII plasmids carrying *bla*CTX-M (Stoesser et al. 2016). However, much remains to be discovered regarding the molecular and evolutionary mechanisms that contribute to the success of these associations in the clinical context.

1.6 Bioflim and Antibiotic Resistance

Complex populations of bacteria called biofilms attach to both surfaces (living and non-living) and are enclosed in a self-made matrix. As they have been related to the emergence of several bacterial illnesses that are difficult to properly treat with medicines, these biofilm formations have major medical value. Bacteria's resistance to the effects of antimicrobial agents can be improved by the creation of biofilms. This characteristic of biofilms presents a significant obstacle in the fight against illnesses brought on by bacteria found in biofilms. The 1999 study by Costerton, Stewart, and Greenberg highlighted the role that biofilms play in the development of these illnesses (Costerton et al. 1999).

1.6.1 Biofilm Resistance and Tolerance

Even in the presence of concentrations that would typically prevent their development, microorganisms that are resistant to antimicrobial treatments can flourish. The MIC, or the lowest concentration at which an antimicrobial drug can stop a bacterium from growing, is commonly used to assess resistance. Often, genetic mutations or HGT-acquired antibiotic-resistance genes are blamed for this resistance. However, wild-type genes and cellular features can also come with a built-in resistance mechanism (Cox and Wright 2013; Blair et al. 2015). Because their outer membranes are impermeable, gram-negative bacteria, for instance, have inherent resistance to several medications, including vancomycin. Contrarily, tolerance describes a microorganism's capacity to endure the presence of bactericidal antimicrobial substances without developing or perishing. Tolerance is measured using the MBC (minimum bactericidal concentration), which exhibits the lowest concentration of a bactericidal antibiotic that can kill at least 99.9% of the cells in a culture. Even after a bactericidal drug has bonded to its target, tolerance mechanisms prevent harmful effects from occurring (Lewis 2008). Reduced growth rate, persister cells, and methods to combat antibiotic-induced oxidative stress are a few examples of biofilm tolerance mechanisms. Several authors have used the term "recalcitrance" while others have used the terms "resistance" and "tolerance" interchangeably to describe the various strategies utilized by biofilms to withstand antimicrobial therapy (Mah and O'Toole 2001). Instead, a neutral position would be to say that biofilms show decreased antibiotic sensitivity or susceptibility via a variety of ways (Gilbert et al. 2002; Lebeaux et al. 2014).

1.6.2 Biofilm Susceptibility Testing

Efforts have been made to establish standardized biofilm protocols to address the differences in susceptibility testing procedures and biofilm growth across laboratories. ASTM International has developed several approved standards for biofilm growth and susceptibility testing, primarily focusing on *Pseudomonas aeruginosa*.

These standards include ASTM E2647 for growing *P. aeruginosa* biofilms in drip flow reactors with low shear and continuous flow conditions (Hotterbeekx et al. 2016), ASTM E2196 for moderate shear conditions using rotating disk reactors, ASTM E2562 for high shear conditions using CDC biofilm reactors (Hotterbeekx et al. 2016), and ASTM E2871 for treating biofilms grown on coupons in CDC biofilm reactors (Ausbacher et al. 2018). However, access to ASTM standards requires a subscription fee, which may limit their widespread implementation. Furthermore, the current ASTM-approved methods related *P. aeruginosa* are limited, and creating standard operating procedures for other microbes would help advance the field of biofilm susceptibility testing.

In addition to ASTM standards, other methods have been employed for biofilm susceptibility testing. Viable colony counts, where biofilm biomass is harvested and plated to quantify the number of cells, can be used by comparing the relative susceptibilities of mutant and wild-type strains to antimicrobial agents (Hall and Mah 2017). The effects of antimicrobial therapy may be evaluated, as well as the function of biofilm tolerance or resistance mechanisms, using microscopy and staining methods. High-throughput screening has been made possible by microscopic techniques (Müsken et al. 2010) described to evaluate the antibiotic susceptibilities of *P. aeruginosa* biofilms generated in microtiter plates (Müsken et al. 2010). BioFILM PA for *P. aeruginosa* is one of the biofilm antimicrobial susceptibility tests created by Innovotech Inc. and authorized for clinical use in Canada (Waters and Ratjen 2017). The MBEC (minimum biofilm eradication concentration) Assay, described in ASTM E2799 (ASTM International 2012), offers versatility in assessing biofilm susceptibility. On pegs fastened to the lid of a microtiter plate, biofilms are produced, and antimicrobial challenges, recovery, and assessment of viability can be performed. MBEC Assay System can be modified for a variety of uses, including biofilm imaging and high-throughput genetic screens (Harrison et al. 2006; Harrison et al. 2010). However, it should be noted that the MBEC Assay System may be relatively more expensive compared to other susceptibility testing methods. While there are standardized methods available, further development of standard protocols for different microorganisms and accessibility of these methods remain areas for improvement in the field of biofilm susceptibility testing.

1.7 Resistance in the Environment

The dissemination of bacteria and genetic elements among animals, humans, and the surrounding environment represents a significant global health concern due to the potential emergence and spread of antibiotic resistance. Pathogens routinely pick up novel resistance components from other species, despite the fact that there are many barriers in place to prevent the transfer of bacteria and genes. This constant acquisition of resistance genes undermines our ability to the prevention and treatment of bacterial infections effectively (Larsson and Flach 2022). Therefore, it is crucial to implement comprehensive strategies, including enhanced surveillance, responsible antibiotic usage, control measures, and infection prevention, and the development of new antimicrobial treatments to combat the growing threat of antibiotic resistance. The two basic ways by which antibiotic resistance might develop are the acquisition of foreign DNA and changes in the bacterial genome already present. Antibiotics put a lot of strain on bacteria, which results in frequent mutations that might emerge in humans or animals receiving treatment (Larsson and Flach 2022). However, these selection pressures are less frequent in other contexts, and in general, external factors have little to no influence on the mutation-based evolution of resistance in the majority of infections (Reygaert 2018).

On the other hand, ecosystems like water, soil, and others that have a variety of ecological niches are more likely to see the absorption of novel resistance factors (Rinke et al. 2013). The variety seen in the microbiota of people and domestic animals is surpassed by that found in these habitats, which act as a huge gene pool (Schulz et al. 2017). This raises the possibility of microorganisms acquiring novel antibiotic resistance genes (ARGs). The resistance amount, co-selection possibilities, virulence, and potential of transmission can all be affected by changes in the genomic background around ARGs, further complicating the resistance problem (Ahmad et al. 2021).

When compared to the spread of genotypes that are already extensively circulated, the effects of evolutionary processes that produce infections with unique and potent resistance genotypes are very different. A new genotype that is harder to treat can spread globally as a result of even a single occurrence, having irreversible effects (Andersson and Hughes 2011). These crucial evolutionary events are uncommon and singular in character, which makes it challenging to forecast them. However, the capacity to postpone or stop their emergence can be extremely advantageous in the fight against antibiotic resistance.

1.7.1 Pollution as a Driver

Antibiotics can enter the environment through various pathways, including human and animal excretions (Wang et al. 2020; Tran et al. 2018; Szymańska et al. 2019), improper disposal of unused drugs (Anwar et al. 2020), direct contamination in aquaculture (Cabello et al. 2013; Cabello et al. 2016), or plant production and waste streams from antibiotic production (Larsson 2014; Larsson et al. 2007; Milaković et al. 2019; Bielen et al. 2017; Fick et al. 2009). However, in the environment, the concentrations of antibiotics are generally low, often below the levels required to select resistant strains (Bengtsson-Palme and Larsson 2016; Gullberg et al. 2011). While there is a correlation between antibiotic concentrations and the abundance of antibiotic ARGs in environmental samples, this is mainly attributed to varying pollution levels from human excreta rather than on-site selection of resistant bacteria by antibiotic residues (Larsson and Flach 2022).

Nonetheless, there are instances where antibiotic concentrations in certain environments, such as sewage treatment plants (Chow et al. 2021; Bengtsson-Palme et al. 2016), exceed the suspected levels that could promote resistance, suggesting a potential role of antibiotic pollution in resistance evolution. Some studies have reported an increase in the relative abundance of specific ARGs in these environments. However, determining whether these changes are solely due to taxonomic shifts unrelated to the direct selection of resistant strains within species poses challenges or antibiotic selection pressures (Yang et al. 2014; Bengtsson-Palme et al. 2016; Manaia et al. 2018). It is conceivable that ARGs found in fecal bacteria introduced into the environment may contribute to the emergence of clinically relevant resistance through horizontal transfer to pathogens capable of infecting humans. However, the probability of such occurrences is greater within the human or animal microbiota, where frequent selection pressures and interactions between commensal organisms and pathogens take place, and where no environmental transmission barriers exist (Crits-Christoph et al. 2022). Therefore, it is contended that the clinical significance of resistance evolution resulting from the environmental dissemination of ARGs already commonly encountered in the human microbiota is likely to be limited.

1.8 Health Assessment

Numerous health problems, such as gastrointestinal (GI) disease, urinary tract infections (UTIs), respiratory disorders, and infections of the eyes, ears, and skin, can result from exposure to fecal pollution in ambient waters. Since GI illness is generally seen as the most delicate and readily noticed health result, it is widely employed as a major health indicator in assessing the microbiological hazards connected to recreational water activities (US EPA 2012; Wade et al. 2008, 2010). The majority of GI illnesses connected to exposure to recreational water are typically caused by viruses (Korzeniewska et al. 2013). When drug-resistant strains like MRSA are present, bacterial pathogens like *Staphylococcus aureus* might offer more serious health hazards.

1.8.1 Environmental Transmission

Certain resistant bacteria may be able to colonize or infect hosts through environmental transmission events (Leonard et al. 2018; Manaia et al. 2018; Schijven et al. 2015). These events entail the transmission of pathogens that are resistant from one person to another, whereas evolutionary events speak about modifications to their DNA sequence, such as genetic transmission between bacterial species. In comparison to an evolutionary event that results in the creation of a novel and effective resistance genotype in pathogens, which has the potential to have global repercussions, the effects of a transmission event of a widely circulating resistant pathogen among people are relatively limited. However, environmental transmission can have a major impact on overall resistance patterns in areas with limited infrastructure for processing fecal waste, such as many low and middle-income countries (Collignon et al. 2018).

Raw veggies are one example of a food source that could expose you to bacteria that are resistant. Fresh food contamination frequently results in infections (Uyttendaele et al. 2015), however, it is not yet known whether eating infected produce increases the risk of contracting resistant strains. However, research has demonstrated that a variety of resistant bacteria with varied mobile components are frequently present in fresh produce (Uyttendaele et al. 2015; Blau et al. 2018; Reid et al. 2020). Therefore, appropriate practices for the use of human and animal waste in agriculture are required to prevent the spread of disease and the emergence of resistance (Zhu et al. 2019). Although a wide variety of ARGs may be present in urban air (Pal et al. 2016), there is little evidence to support long-distance aerial transmission of bacterial pathogens (Kozajda et al. 2019).

Based on information on environmental emissions, destiny, and exposure, the hazards connected with the spread of resistant bacteria can theoretically be assessed and modeled. However, reliable models that outline the environmental transmission pathway are few and far between (Ashbolt et al. 2013; Franz et al. 2014). ARG-based models are less likely to be predictive than those that focus on a particular resistant bacterium while taking into account the bacterial host and the genetic background of mobile ARGs. Lack of knowledge regarding the necessary doses for widespread bacterial colonization is one such knowledge gap (Larsson and Flach 2022). It is still unclear how selecting agents contributes to environmental transmission. Despite the fact that some infections survive there, for the majority of them, the environment is a more hostile place than a human or animal host, which frequently restricts their growth. The possibility that environmental exposure will become sufficient for colonization or infection of a human or animal host may, therefore, be slightly influenced by small growth differences between resistant and non-resistant strains, caused by exposure to sub-MICs of antibiotics (Bengtsson-Palme et al. 2018).

1.8.2 Exposure Estimates and Risk Assessments

AMR exposure in humans during recreational water activities was evaluated in a study by Leonard et al. (2015) that was carried out in England and Wales. They evaluated the *E. coli* density and water ingestion volumes related to various water sports (such as swimming, surfing, diving, boating, and wading/splashing), as well as the existence of third-generation cephalosporin-resistant ESBL-producing *E. coli* in coastal water samples. By merging these data, they calculated the typical quantity of third-generation cephalosporin-resistant ESBL-producing *E. coli* consumed during each water sport. The study found that there is still a noticeable danger of human exposure to antibiotic-resistant bacteria among recreational water users, despite the low prevalence of such resistance (0.12%). According to their data, at least one third-generation cephalosporin-resistant *E. coli* was likely consumed during over 6.3 million water sports activities in 2012. It should be highlighted that this study may have underestimated the true degree of recreational exposure to AMR bacteria because it only looked at third-generation cephalosporin resistance.

In a separate investigation conducted by O'Flaherty et al. in 2019, the study examined potential human exposures to antibiotic-resistant *E. coli* bacteria in the vicinity of wastewater treatment plants (WWTPs) and tourist beaches in central Italy. To gauge the volume of water consumed by beachgoers, the researchers conducted a field study and developed a quantitative risk assessment model. This model incorporated factors such as incidental water ingestion during swimming, the dispersion and reduction of antibiotic-resistant *E. coli* from WWTP effluent into the river, and the dilution of these bacteria from the river into the swimming areas. The model's projections indicated that the expected levels of human exposure to antibiotic-resistant *E. coli* during recreational water activities ranged from 0 to 345 colony-forming units per 100 ml of water. These results suggest that engaging in recreational water activities can potentially result in human exposure to antibiotic resistance.

1.8.3 Mortality (Direct and Indirect Attribute to Antibiotic Rresistance)

1.8.3.1 Direct Mortality Attributed to Antibiotic Resistance

Antibiotic resistance-related deaths that could have been avoided if the causing organism had not been resistant to antibiotics are referred to as "direct mortality." It can be difficult to determine the specific number of deaths that can be directly linked to antibiotic resistance; frequently, estimates are made using statistics that are frequently used in grant applications and reports. For example, figures such as "at least" 25,000–33,000 deaths per year in Europe or 23,000 deaths per year in the United States are frequently referenced (de Kraker et al. 2016; Boolchandani et al. 2019; Cassini et al. 2019). The need to distinguish between deaths inflicted by people infected with antibiotic-resistant bacteria and deaths particularly brought on by the infection's resistance to conventional therapy makes it challenging to get reliable statistics.

Age and underlying disorders are only two examples of variables that affect mortality rates. It is conceivable for critically sick individuals who have germs that are multi-drug resistant to be mistakenly labeled as "deaths due to resistance," even if their cause of death was unrelated to antibiotic resistance and they were being exposed to antibiotics for prophylactic or therapeutic purposes. In order to attain precise and reliable data, it is essential to conduct rigorous case-control studies. These studies should involve a comparison of mortality rates among patients afflicted with the same infection, where one group is affected by antibiotic-resistant bacteria and the other by non-resistant strains. Both groups should receive identical treatment, belong to the same age bracket, possess similar underlying medical conditions, and be situated in identical health-care settings (Baquero 2021).

It is of paramount importance to bear in mind that in low-income countries, mortality rates due to infectious diseases are notably elevated. These nations grapple with challenges such as malnutrition, inadequate sanitation, and limited access to appropriate health-care resources, including intensive care facilities, which are reminiscent of the health-care landscape in Europe two centuries ago before the advent of antibiotics. Remarkably, these countries, until recently, were effectively in a state akin to the pre-antibiotic era and have experienced a demonstrable positive impact from the introduction of antibiotics in reducing overall mortality (Abat et al. 2018). As a consequence, the burden of mortality associated with antibiotic resistance, encompassing resistance to antibiotics that are accessible and affordable, as well as resistance to advanced antibiotics like third-generation cephalosporins and carbapenems, may be considerably more pronounced in low-income countries when juxtaposed with the United States or the European Union (Huynh et al. 2015; Founou et al. 2017).

1.8.3.2 Indirect Mortality Mediated by Antibiotic Resistance

The impact of antibiotics on the composition of the healthy human microbiota can lead to changes in the proportions of microorganisms. Antibiotic resistance in the dominant commensal species, which are typically non-pathogenic, may actually serve to protect against the overgrowth of more harmful pathogens, thus enhancing the resilience of the microbiota. Bacteremic infections caused by deadly organisms such as *E. coli*, *Enterococcus faecium*, *Enterococcus faecalis*, and *K. pneumoniae* are relatively rare compared to the overall microbial population in the intestines (Baquero 2021). Intensive antibiotic exposure, whether for prevention or treatment, can result in an increased abundance of antibiotic-resistant pathogenic organisms, particularly in elderly individuals where bacteremia often originates from urinary tract infections. The accumulation of antibiotic-resistant bacteria in the microbiota may be influenced by an individual's history of antibiotic use throughout their life (Baquero 2007). Similar to decreasing risk factors like cigarette use, a high salt diet, or a diet rich in cholesterol for cardiovascular diseases, making efforts to reduce antibiotic exposure might be a preventative intervention for mortality. At least in high-income nations, the risk of death seems to be comparable for populations of multi-resistant and susceptible *E. coli*, even if a larger frequency of bacteremic episodes is linked to a rise in mortality (de Lastours et al. 2020). However, the effect of antibiotic resistance on mortality is probably more important in low-income nations with limited access to cutting-edge medical treatment (Becker et al. 2009).

1.9 Conclusions and Future Prospective

The emergence of resistance due to unchecked use of antibiotics has impressed upon a great threat to the global health-care system. Although there has been a slight decline in global antibiotic consumption, overall usage remains substantial and has led to the emergence of drug-resistant pathogens. The escalating burden of antibiotic resistance genes (ARGs) among the pathogenic organisms, and human-induced pollutants in environmental settings serve as a reservoir, particularly concerning their anthropogenic dissemination. Significant enrichment of antibiotic-resistant bacteria (ARB) and ARGs in environmental settings, such as hospitals, farms, wastewater treatment plants (WWTPs), and agricultural settings contributes significantly to the widespread dissemination of antibiotic resistance genes (ARGs) and presents a significant threat to public health. As infectious diseases are linked to resistant pathogens, their health risks significantly impact public health and impose economic burdens on nations. Having adverse repercussions on environmental health, there is a pressing need for the advancement of rapid, accurate, and cost-effective environmental monitoring tools and advanced techniques for real-time identification of the resistance determinants.

A "One Health" perspective, considering human, animal, and environmental factors, is vital for comprehending the intricate gene flow processes that contribute to overall resistome. This approach aims to identify priority factors for mitigating ARG transmission in clinical settings. The primary objective is to identify health-risk ARGs in different contexts and quantify the resistome's flow between humans and the environment including exposure pathways and doses. Vigilant control of factors capable of generating or exacerbating antibiotic resistance transmission can alleviate the burden of antibiotic resistance on humans. It concentrates on minimizing the development and transmission of resistance determinants, highlighting the potential need to prevent the infiltration of external ARB and halt their transmission. Lastly, antibiotic stewardship, endorsed by global health organizations, should be comprehensively implemented through national policies to curb the emergence of both environmental and clinical resistomes. Conquering the universal threat posed by antibiotic resistance, coupled with its potential impact on public health, necessitates cumulative and sustained efforts from the global community.

Acknowledgments Authors would like to thank their colleague Ms. Saima Jan, whose suggestions helped in improving the contents of the chapter.

Author Contributions All authors contributed equally in generating the draft of the different sections and in the repeated editing of the contents present in the finalized version of the chapter. All authors have read and agreed to the publication of the final version of the chapter.

Funding The work was supported by J&K Science Technology and Innovation Council (J&K ST&IC), India grant (JKST&IC/SRE/996-998), and DST-SERB Core Grant (CRG/2019/004106).

Conflicts of Interest The authors declare no conflict of interest.

References

Abat C, Gautret P, Raoult D (2018) Benefits of antibiotics burden in low-income countries. Proc Natl Acad Sci 115(35):E8109–E8110

Agyare C, Etsiapa Boamah V, Ngofi Zumbi C, Boateng Osei F (2019) Antibiotic use in poultry production and its effects on bacterial resistance. In: Antimicrobial resistance - a global threat. IntechOpen, London

Ahmad I, Malak HA, Abulreesh HH (2021) Environmental antimicrobial resistance and its drivers: a potential threat to public health. J Glob Antimicrob Resistance 27:101–111

Aldred KJ, Kerns RJ, Osheroff N (2014) Mechanism of quinolone action and resistance. Biochemistry 53(10):1565–1574

Alvarez-Ortega C, Wiegand I, Olivares J, Hancock RE, Martínez JL (2010) Genetic determinants involved in the susceptibility of Pseudomonas aeruginosa to β-lactam antibiotics. Antimicrob Agents Chemother 54(10):4159–4167

Alvarez-Ortega C, Wiegand I, Olivares J, Hancock RE, Martínez JL (2011) The intrinsic resistome of Pseudomonas aeruginosa to β-lactams. Virulence 2(2):144–146

Andersson DI, Hughes D (2011) Persistence of antibiotic resistance in bacterial populations. FEMS Microbiol Rev 35(5):901–911

Antonelli A, D'Andrea MM, Brenciani A, Galeotti CL, Morroni G, Pollini S et al (2018) Characterization of poxtA, a novel phenicol–oxazolidinone–tetracycline resistance gene from an MRSA of clinical origin. J Antimicrob Chemother 73(7):1763–1769

Anwar M, Iqbal Q, Saleem F (2020) Improper disposal of unused antibiotics: an often-overlooked driver of antimicrobial resistance. Expert Rev Anti-Infect Ther 18(8):697–699

Argudín MA, Youzaga S, Dodémont M, Heinrichs A, Roisin S, Deplano A et al (2019) Detection of optrA-positive enterococci clinical isolates in Belgium. Eur J Clin Microbiol Infect Dis 38:985–987

Asenjo A, Oteo-Iglesias J, Alós JI (2021) What's new in mechanisms of antibiotic resistance in bacteria of clinical origin? Enferm Infecc Microbiol Clin 39(6):291–299

Ashbolt NJ, Amézquita A, Backhaus T, Borriello P, Brandt KK, Collignon P et al (2013) Human health risk assessment (HHRA) for environmental development and transfer of antibiotic resistance. Environ Health Perspect 121(9):993–1001

Aslam B, Wang W, Arshad MI, Khurshid M, Muzammil S, Rasool MH et al (2018) Antibiotic resistance: a rundown of a global crisis. Infect Drug Resistance 11:1645–1658

ASTM International (2012) ASTM E2799-12 standard test method for testing disinfectant efficacy against Pseudomonas aeruginosa biofilm using the MBEC assay. ASTM International, West Conshohocken

Atac N, Kurt-Azap O, Dolapci I, Yesilkaya A, Ergonul O, Gonen M, Can F (2018) The role of AcrAB–TolC efflux pumps on quinolone resistance of E. coli ST131. Curr Microbiol 75:1661–1666

Ausbacher D, Lorenz L, Pitts B, Stewart PS, Goeres DM (2018) Paired methods to measure biofilm killing and removal: a case study with Penicillin G treatment of Staphylococcus aureus biofilm. Lett Appl Microbiol 66(3):231–237

Azam M, Jan AT, Haq QM (2016) Bla CTX-M-152, a novel variant of CTX-M-group-25, identified in a study performed on the prevalence of multidrug resistance among natural inhabitants of river Yamuna, India. Front Microbiol 7:176

Bacilli GN (2018) Resistance of gram-negative bacilli to antimicrobials. In: Antimicrobial resistance in the 21st century. Springer, Cham, p 71

Baquero F (2007) Topical debates evaluation of risks and benefits of consumption of antibiotics: from individual to public health. In: Encyclopedia of infectious diseases: modern methodologies. Elsevier, Amsterdam, pp 509–516

Baquero F (2021) Threats of antibiotic resistance: an obliged reappraisal. Int Microbiol 24(4):499–506

Baquero F, Martínez JL, Cantón R (2008) Antibiotics and antibiotic resistance in water environments. Curr Opin Biotechnol 19(3):260–265

Barlow M, Reik RA, Jacobs SD, Medina M, Meyer MP, McGowan JE Jr, Tenover FC (2008) High rate of mobilization for blaCTX-Ms. Emerg Infect Dis 14(3):423

Bassetti M, Righi E (2014) Eravacycline for the treatment of intra-abdominal infections. Expert Opin Investig Drugs 23(11):1575–1584

Becker JU, Theodosis C, Jacob ST, Wira CR, Groce NE (2009) Surviving sepsis in low-income and middle-income countries: new directions for care and research. Lancet Infect Dis 9(9):577–582

Becker K, van Alen S, Idelevich EA, Schleimer N, Seggewiß J, Mellmann A et al (2018) Plasmid-encoded transferable mecB-mediated methicillin resistance in Staphylococcus aureus. Emerg Infect Dis 24(2):242

Bender JK, Fleige C, Lange D, Klare I, Werner G (2018) Rapid emergence of highly variable and transferable oxazolidinone and phenicol resistance gene optrA in German Enterococcus spp. clinical isolates. Int J Antimicrob Agents 52(6):819–827

Bengtsson-Palme J, Larsson DJ (2016) Concentrations of antibiotics predicted to select for resistant bacteria: proposed limits for environmental regulation. Environ Int 86:140–149

Bengtsson-Palme J, Hammaren R, Pal C, Östman M, Björlenius B, Flach CF et al (2016) Elucidating selection processes for antibiotic resistance in sewage treatment plants using metagenomics. Sci Total Environ 572:697–712

Bengtsson-Palme J, Kristiansson E, Larsson DJ (2018) Environmental factors influencing the development and spread of antibiotic resistance. FEMS Microbiol Rev 42(1):fux053

Berglund B (2019) Acquired resistance to colistin via chromosomal and plasmid-mediated mechanisms in Klebsiella pneumoniae. Infect Microb Dis 1(1):10–19

Berrazeg M, Jeannot K, Ntsogo Enguéné VY, Broutin I, Loeffert S, Fournier D, Plésiat P (2015) Mutations in β-lactamase AmpC increase resistance of Pseudomonas aeruginosa isolates to antipseudomonal cephalosporins. J Antimicrob Agents 59(10):6248–6255

Bialvaei AZ, Samadi Kafil H (2015) Colistin, mechanisms and prevalence of resistance. Curr Med Res Opin 31(4):707–721

Bielen A, Šimatović A, Kosić-Vukšić J, Senta I, Ahel M, Babić S et al (2017) Negative environmental impacts of antibiotic-contaminated effluents from pharmaceutical industries. Water Res 126:79–87

Blair JM, Bavro VN, Ricci V, Modi N, Cacciotto P, Kleinekathöfer U et al (2015) AcrB drug-binding pocket substitution confers clinically relevant resistance and altered substrate specificity. Proc Natl Acad Sci 112(11):3511–3516

Blau K, Bettermann A, Jechalke S, Fornefeld E, Vanrobaeys Y, Stalder T et al (2018) The transferable resistome of produce. MBio 9(6):10–1128

Boekhoud IM, Hornung BV, Sevilla E, Harmanus C, Bos-Sanders IM, Terveer EM et al (2020) Plasmid-mediated metronidazole resistance in Clostridioides difficile. Nat Commun 11(1):598

Boolchandani M, D'Souza AW, Dantas G (2019) Sequencing-based methods and resources to study antimicrobial resistance. Nat Rev Genet 20(6):356–370

Boucher HW, Talbot GH, Bradley JS, Edwards JE, Gilbert D, Rice LB et al (2009) Bad bugs, no drugs: no ESKAPE! An update from the Infectious Diseases Society of America. Clin Infect Dis 48(1):1–12

Brañas P, Villa J, Viedma E, Mingorance J, Orellana MA, Chaves F (2015) Molecular epidemiology of carbapenemase-producing Klebsiella pneumoniae in a hospital in Madrid: successful establishment of an OXA-48 ST11 clone. Int J Antimicrob Agents 46(1):111–116

Cabello FC, Godfrey HP, Tomova A, Ivanova L, Dölz H, Millanao A, Buschmann AH (2013) Antimicrobial use in aquaculture re-examined: its relevance to antimicrobial resistance and to animal and human health. Environ Microbiol 15(7):1917–1942

Cabello FC, Godfrey HP, Buschmann AH, Dölz HJ (2016) Aquaculture as yet another environmental gateway to the development and globalisation of antimicrobial resistance. Lancet Infect Dis 16(7):e127–e133

Càmara J, Camoez M, Tubau F, Pujol M, Ayats J, Ardanuy C, Domínguez MÁ (2019) Detection of the novel optrA gene among linezolid-resistant enterococci in Barcelona, Spain. Microb Drug Resist 25(1):87–93

Cannatelli A, D'Andrea MM, Giani T, Di Pilato V, Arena F, Ambretti S et al (2013) In vivo emergence of colistin resistance in Klebsiella pneumoniae producing KPC-type carbapenemases mediated by insertional inactivation of the PhoQ/PhoP mgrB regulator. Antimicrob Agents Chemother 57(11):5521–5526

Cannatelli A, Di Pilato V, Giani T, Arena F, Ambretti S, Gaibani P et al (2014) In vivo evolution to colistin resistance by PmrB sensor kinase mutation in KPC-producing Klebsiella pneumoniae is associated with low-dosage colistin treatment. Antimicrob Agents Chemother 58(8):4399–4403

Carattoli A (2013) Plasmids and the spread of resistance. IJMM 303:298–304

Carroll LM, Gaballa A, Guldimann C, Sullivan G, Henderson LO, Wiedmann M (2019) Identification of novel mobilized colistin resistance gene mcr-9 in a multidrug-resistant, colistin-susceptible Salmonella enterica serotype Typhimurium isolate. MBio 10(3):10–1128

Cassini A, Högberg LD, Plachouras D, Quattrocchi A, Hoxha A, Simonsen GS et al (2019) Attributable deaths and disability-adjusted life-years caused by infections with antibiotic-resistant bacteria in the EU and the European Economic Area in 2015: a population-level modelling analysis. Lancet Infect Dis 19(1):56–66

Cercenado E, Ruiz de Gopegui E (2008) Community-acquired methicillin-resistant Staphylococcus aureus. Enferm Infecc Microbiol Clin 26:19–24

Chang S, Sievert DM, Hageman JC, Boulton ML, Tenover FC, Downes FP et al (2003) Infection with vancomycin-resistant Staphylococcus aureus containing the vanA resistance gene. N Engl J Med 348(14):1342–1347

Cheng YH, Lin TL, Lin YT, Wang JT (2016) Amino acid substitutions of CrrB responsible for resistance to colistin through CrrC in Klebsiella pneumoniae. Antimicrob Agents Chemother 60(6):3709–3716

Chow LK, Ghaly TM, Gillings MR (2021) A survey of sub-inhibitory concentrations of antibiotics in the environment. J Environ Sci 99:21–27

Clark DP, Pazdernik NJ (2012) Molecular biology. Elsevier, Amsterdam

Collignon P, Beggs JJ, Walsh TR, Gandra S, Laxminarayan R (2018) Anthropological and socio-economic factors contributing to global antimicrobial resistance: a univariate and multivariable analysis. Lancet Planetary Health 2(9):e398–e405

Costello SE, Deshpande LM, Davis AP, Mendes RE, Castanheira M (2019) Aminoglycoside-modifying enzyme and 16S ribosomal RNA methyltransferase genes among a global collection of Gram-negative isolates. J Glob Antimicrob Resistance 16:278–285

Costerton JW, Stewart PS, Greenberg EP (1999) Bacterial biofilms: a common cause of persistent infections. Science 284(5418):1318–1322

Cox G, Wright GD (2013) Intrinsic antibiotic resistance: mechanisms, origins, challenges and solutions. Int J Med Microbiol 303(6-7):287–292

Crits-Christoph A, Hallowell HA, Koutouvalis K, Suez J (2022) Good microbes, bad genes? The dissemination of antimicrobial resistance in the human microbiome. Gut Microbes 14(1):2055944

Dadgostar P (2019) Antimicrobial resistance: implications and costs. Infect Drug Resistance 12:3903–3910

Dantas G, Sommer MO (2014) How to fight back against antibiotic resistance. Am Sci 102(1):42–51

de Kraker ME, Stewardson AJ, Harbarth S (2016) Will 10 million people die a year due to antimicrobial resistance by 2050? PLoS Med 13(11):e1002184

de Lastours V, Laouénan C, Royer G, Carbonnelle E, Lepeule R, Esposito-Farèse M et al (2020) Mortality in Escherichia coli bloodstream infections: antibiotic resistance still does not make it. J Antimicrob Chemother 75(8):2334–2343

Di Pilato V, Arena F, Tascini C, Cannatelli A, Henrici De Angelis L, Fortunato S et al (2016) mcr-1.2, a new mcr variant carried on a transferable plasmid from a colistin-resistant KPC carbapenemase-producing Klebsiella pneumoniae strain of sequence type 512. Antimicrob Agents Chemother 60(9):5612–5615

Dimou V, Dhanji H, Pike R, Livermore DM, Woodford N (2012) Characterization of Enterobacteriaceae producing OXA-48-like carbapenemases in the UK. J Antimicrob Chemother 67(7):1660–1665

Dingsdag S, Hunter N (2018) Metronidazole: an update on metabolism, structure-cytotoxicity and resistance mechanisms. J Antimicrob Chemother 73:265–279

Doi Y, Wachino JI, Arakawa Y (2016) Aminoglycoside resistance: the emergence of acquired 16S ribosomal RNA methyltransferases. Infect Dis Clin 30(2):523–537

Dortet L, Poirel L, Abbas S, Oueslati S, Nordmann P (2015) Genetic and biochemical characterization of FRI-1, a carbapenem-hydrolyzing class A β-lactamase from Enterobacter cloacae. Antimicrob Agents Chemother 59(12):7420–7425

El-Sayed Ahmed MAEG, Zhong LL, Shen C, Yang Y, Doi Y, Tian GB (2020) Colistin and its role in the Era of antibiotic resistance: an extended review (2000–2019). Emerg Microb Infect 9(1):868–885

Elsner HA, Sobottka I, Mack D, Laufs R, Claussen M, Wirth R (2000) Virulence factors of Enterococcus faecalis and Enterococcus faecium blood culture isolates. Eur J Clin Microbiol Infect Dis 19:39–42

Emamalipour M, Seidi K, Zununi Vahed S, Jahanban-Esfahlan A, Jaymand M, Majdi H et al (2020) Horizontal gene transfer: from evolutionary flexibility to disease progression. Front Cell Dev Biol 8:229

Fajardo A, Martínez-Martín N, Mercadillo M, Galan JC, Ghysels B, Matthijs S et al (2008) The neglected intrinsic resistome of bacterial pathogens. PLoS One 3(2):e1619

Fernández L, Álvarez-Ortega C, Wiegand I, Olivares J, Kocíncová D, Lam JS et al (2013) Characterization of the polymyxin B resistome of Pseudomonas aeruginosa. Antimicrob Agents Chemother 57(1):110–119

Fick J, Söderström H, Lindberg RH, Phan C, Tysklind M, Larsson DJ (2009) Contamination of surface, ground, and drinking water from pharmaceutical production. Environ Toxicol Chem 28(12):2522–2527

Forsberg KJ, Reyes A, Wang B, Selleck EM, Sommer MO, Dantas G (2012) The shared antibiotic resistome of soil bacteria and human pathogens. Science 337(6098):1107–1111

Founou RC, Founou LL, Essack SY (2017) Clinical and economic impact of antibiotic resistance in developing countries: a systematic review and meta-analysis. PLoS One 12(12):e0189621

Fournier C, Poirel L, Despont S, Kessler J, Nordmann P (2022) Increasing trends of association of 16S rRNA methylases and carbapenemases in Enterobacterales clinical isolates from Switzerland, 2017–2020. Microorganisms 10(3):615

Fraile-Ribot PA, Mulet X, Cabot G, del Barrio-Tofiño E, Juan C, Pérez JL, Oliver A (2017) In vivo emergence of resistance to novel cephalosporin–β-lactamase inhibitor combinations through the duplication of amino acid D149 from OXA-2 β-lactamase (OXA-539) in sequence type 235 Pseudomonas aeruginosa. J Antimicrob Agents 61(9):10–1128

Fraile-Ribot PA, Cabot G, Mulet X, Periañez L, Martín-Pena ML, Juan C et al (2018) Mechanisms leading to in vivo ceftolozane/tazobactam resistance development during the treatment of infections caused by MDR Pseudomonas aeruginosa. J Antimicrob Chemother 73(3):658–663

Franz E, Schijven J, de Roda Husman AM, Blaak H (2014) Meta-regression analysis of commensal and pathogenic Escherichia coli survival in soil and water. Environ Sci Technol 48(12):6763–6771

Frey PM, Marti GR, Droz S, de Roche M, Suter-Riniker F, Aujesky D, Brugger SD (2019) Bacterial colonization of handheld devices in a tertiary care setting: a hygiene intervention study. Antimicrob Resist Infect Control 8(1):1–4

Fukuoka T, Ohya S, Narita T, Katsuta M, Iijima M, Masuda N et al (1993) Activity of the carbapenem panipenem and role of the OprD (D2) protein in its diffusion through the Pseudomonas aeruginosa outer membrane. Antimicrob Agents Chemother 37(2):322–327

Galani I, Nafplioti K, Adamou P, Karaiskos I, Giamarellou H, Souli M (2019) Nationwide epidemiology of carbapenem resistant Klebsiella pneumoniae isolates from Greek hospitals, with regards to plazomicin and aminoglycoside resistance. BMC Infect Dis 19(1):1–12

García-León G, Sánchez MB, Martínez JL (2012) The inactivation of intrinsic antibiotic resistance determinants widens the mutant selection window for quinolones in Stenotrophomonas maltophilia. Antimicrob Agents Chemother 56(12):6397–6399

Gilbert P, Maira-Litran T, McBain AJ, Rickard AH, Whyte FW (2002) The physiology and collective recalcitrance of microbial biofilm communities. Adv Microb Physiol 46:202–256

Gogry FA, Siddiqui MT, Sultan I, Haq QMR (2021) Current update on intrinsic and acquired colistin resistance mechanisms in bacteria. Front Med 8:677720

Guerin F, Galimand M, Tuambilangana F, Courvalin P, Cattoir V (2014) Overexpression of the novel MATE fluoroquinolone efflux pump FepA in Listeria monocytogenes is driven by inactivation of its local repressor FepR. PLoS One 9(9):e106340

Gullberg E, Cao S, Berg OG, Ilbäck C, Sandegren L, Hughes D, Andersson DI (2011) Selection of resistant bacteria at very low antibiotic concentrations. PLoS Pathog 7(7):e1002158

Guo Q, Tomich AD, McElheny CL, Cooper VS, Stoesser N, Wang M et al (2016) Glutathione-S-transferase FosA6 of Klebsiella pneumoniae origin conferring fosfomycin resistance in ESBL-producing Escherichia coli. J Antimicrob Chemother 71(9):2460–2465

Haenni M, Poirel L, Kieffer N, Châtre P, Saras E, Métayer V et al (2016) Co-occurrence of extended spectrum β lactamase and MCR-1 encoding genes on plasmids. Lancet Infect Dis 16(3):281–282

Hall CW, Mah TF (2017) Molecular mechanisms of biofilm-based antibiotic resistance and tolerance in pathogenic bacteria. FEMS Microbiol Rev 41(3):276–301

Hamad B (2010) The antibiotics market. Nat Rev Drug Discov 9(9):675

Hameed F, Khan MA, Muhammad H, Sarwar T, Bilal H, Rehman TU (2019) Plasmid-mediated mcr-1 gene in Acinetobacter baumannii and Pseudomonas aeruginosa: first report from Pakistan. Rev Soc Bras Med Trop 52:e20190237

Harkins CP, Pichon B, Doumith M, Parkhill J, Westh H, Tomasz A et al (2017) Methicillin-resistant Staphylococcus aureus emerged long before the introduction of methicillin into clinical practice. Genome Biol 18(1):1–11

Harrison JJ, Ceri H, Yerly J, Stremick CA, Hu Y, Martinuzzi R, Turner RJ (2006) The use of microscopy and three-dimensional visualization to evaluate the structure of microbial biofilms cultivated in the Calgary Biofilm Device. Biol Procedures Online 8:194–215

Harrison JJ, Stremick CA, Turner RJ, Allan ND, Olson ME, Ceri H (2010) Microtiter susceptibility testing of microbes growing on peg lids: a miniaturized biofilm model for high-throughput screening. Nat Protoc 5(7):1236–1254

Hartman BJ, Tomasz A (1984) Low-affinity penicillin-binding protein associated with beta-lactam resistance in Staphylococcus aureus. J Bacteriol 158(2):513–516

Hasman H, Hammerum AM, Hansen F, Hendriksen RS, Olesen B, Agersø Y et al (2015) Detection of mcr-1 encoding plasmid-mediated colistin-resistant Escherichia coli isolates from human bloodstream infection and imported chicken meat, Denmark 2015. Eur Secur 20(49):30085

Hays JP, Mitsakakis K, Luz S, van Belkum A, Becker K, van den Bruel A et al (2019) The successful uptake and sustainability of rapid infectious disease and antimicrobial resistance point-of-care testing requires a complex 'mix-and-match' implementation package. Eur J Clin Microbiol Infect Dis 38:1015–1022

Hotterbeekx A, Lanckacker E, Moons P, Lammens C, Kerstens M, Ieven M et al (2016) Characterizing the in vitro biofilm phenotype of Staphylococcus epidermidis isolates from central venous catheters. J Microbiol Methods 127:95–101

Huemer M, Mairpady Shambat S, Brugger SD, Zinkernagel AS (2020) Antibiotic resistance and persistence-Implications for human health and treatment perspectives. EMBO Rep 21(12):e51034

Huh K, Kang CI, Kim J, Cho SY, Ha YE, Joo EJ et al (2014) Risk factors and treatment outcomes of bloodstream infection caused by extended-spectrum cephalosporin-resistant Enterobacter species in adults with cancer. Diagn Microbiol Infect Dis 78(2):172–177

Huynh BT, Padget M, Garin B, Herindrainy P, Kermorvant-Duchemin E, Watier L et al (2015) Burden of bacterial resistance among neonatal infections in low income countries: how convincing is the epidemiological evidence? BMC Infect Dis 15(1):1–13

Iredell J, Brown J, Tagg K (2016) Antibiotic resistance in Enterobacteriaceae: mechanisms and clinical implications. BMJ 352:h6420

Jayol A, Poirel L, Brink A, Villegas MV, Yilmaz M, Nordmann P (2014) Resistance to colistin associated with a single amino acid change in protein PmrB among Klebsiella pneumoniae isolates of worldwide origin. Antimicrob Agents Chemother 58(8):4762–4766

Jayol A, Nordmann P, Brink A, Poirel L (2015) Heteroresistance to colistin in Klebsiella pneumoniae associated with alterations in the PhoPQ regulatory system. Antimicrob Agents Chemother 59(5):2780–2784

Jian Z, Zeng L, Xu T, Sun S, Yan S, Yang L et al (2021) Antibiotic resistance genes in bacteria: occurrence, spread, and control. J Basic Microbiol 61(12):1049–1070

Kaatz GW, Seo SM, Dorman NJ, Lerner SA (1990) Emergence of teicoplanin resistance during therapy of Staphylococcus aureus endocarditis. J Infect Dis 162(1):103–108
Kakoullis L, Papachristodoulou E, Chra P, Panos G (2021) Mechanisms of antibiotic resistance in important gram-positive and gram-negative pathogens and novel antibiotic solutions. Antibiotics 10(4):415
Karaiskos I, Giamarellou H (2014) Multidrug-resistant and extensively drug-resistant Gram-negative pathogens: current and emerging therapeutic approaches. Expert Opin Pharmacother 15(10):1351–1370
Katayama Y, Ito T, Hiramatsu K (2000) A new class of genetic element, staphylococcus cassette chromosome mec, encodes methicillin resistance in Staphylococcus aureus. Antimicrob Agents Chemother 44(6):1549–1555
Keeney D, Ruzin A, Bradford PA (2007) RamA, a transcriptional regulator, and AcrAB, an RND-type efflux pump, are associated with decreased susceptibility to tigecycline in Enterobacter cloacae. Microb Drug Resist 13(1):1–6
King DT, Sobhanifar S, Strynadka NC (2016) One ring to rule them all: Current trends in combating bacterial resistance to the β-lactams. Protein Sci 25(4):787–803
Korzeniewska E, Korzeniewska A, Harnisz M (2013) Antibiotic resistant Escherichia coli in hospital and municipal sewage and their emission to the environment. Ecotoxicol Environ Saf 91:96–102
Kotsogianni I, Wood TM, Alexander FM, Cochrane SA, Martin NI (2021) Binding studies reveal phospholipid specificity and its role in the calcium-dependent mechanism of action of daptomycin. ACS Infect Dis 7(9):2612–2619
Kozajda A, Jeżak K, Kapsa A (2019) Airborne Staphylococcus aureus in different environments-a review. Environ Sci Pollut Res 26:34741–34753
Kumarasamy KK, Toleman MA, Walsh TR, Bagaria J, Butt F, Balakrishnan R et al (2010) Emergence of a new antibiotic resistance mechanism in India, Pakistan, and the UK: a molecular, biological, and epidemiological study. Lancet Infect Dis 10(9):597–602
Larsson DJ (2014) Pollution from drug manufacturing: review and perspectives. Philos Trans R Soc B Biol Sci 369(1656):20130571
Larsson DJ, Flach CF (2022) Antibiotic resistance in the environment. Nat Rev Microbiol 20(5):257–269
Larsson DJ, de Pedro C, Paxeus N (2007) Effluent from drug manufactures contains extremely high levels of pharmaceuticals. J Hazard Mater 148(3):751–755
Laxminarayan R, Duse A, Wattal C, Zaidi AK, Wertheim HF, Sumpradit N et al (2013) Antibiotic resistance-the need for global solutions. Lancet Infect Dis 13(12):1057–1098
Lebeaux D, Ghigo JM, Beloin C (2014) Biofilm-related infections: bridging the gap between clinical management and fundamental aspects of recalcitrance toward antibiotics. Microbiol Mol Biol Rev 78(3):510–543
Lenski RE (1998) Bacterial evolution and the cost of antibiotic resistance. Int Microbiol 1(4):265–270
Leonard AF, Zhang L, Balfour AJ, Garside R, Gaze WH (2015) Human recreational exposure to antibiotic resistant bacteria in coastal bathing waters. Environ Int 82:92–100
Leonard AF, Zhang L, Balfour AJ, Garside R, Hawkey PM, Murray AK et al (2018) Exposure to and colonisation by antibiotic-resistant *E. coli* in UK coastal water users: environmental surveillance, exposure assessment, and epidemiological study (Beach Bum Survey). Environ Int 114:326–333
Lerminiaux NA, Cameron AD (2019) Horizontal transfer of antibiotic resistance genes in clinical environments. Can J Microbiol 65(1):34–44
Lewis K (2008) Multidrug tolerance of biofilms and persister cells. In: Bacterial biofilms. Springer, Cham, pp 107–131
Linares JF, Moreno R, Fajardo A, Martínez-Solano L, Escalante R, Rojo F, Martínez JL (2010) The global regulator Crc modulates metabolism, susceptibility to antibiotics and virulence in Pseudomonas aeruginosa. Environ Microbiol 12(12):3196–3212

Liu A, Tran L, Becket E, Lee K, Chinn L, Park E et al (2010) Antibiotic sensitivity profiles determined with an Escherichia coli gene knockout collection: generating an antibiotic bar code. Antimicrob Agents Chemother 54(4):1393–1403
Liu C, Bayer A, Cosgrove SE, Daum RS, Fridkin SK, Gorwitz RJ et al (2011) Clinical practice guidelines by the Infectious Diseases Society of America for the treatment of methicillin-resistant Staphylococcus aureus infections in adults and children. Clin Infect Dis 52(3):e18–e55
Liu YY, Wang Y, Walsh TR, Yi LX, Zhang R, Spencer J et al (2016) Emergence of plasmid-mediated colistin resistance mechanism MCR-1 in animals and human beings in China: a microbiological and molecular biological study. Lancet Infect Dis 16(2):161–168
Lobanovska M, Pilla G (2017) Focus: drug development: Penicillin's discovery and antibiotic resistance: lessons for the future? Yale J Biol Med 90(1):135
Lowy FD (1998) Staphylococcus aureus infections. N Engl J Med 339(8):520–532
Lupo A, Coyne S, Berendonk TU (2012) Origin and evolution of antibiotic resistance: the common mechanisms of emergence and spread in water bodies. Front Microbiol 3:18
Ma Y, Xu X, Guo Q, Wang P, Wang W, Wang M (2015) Characterization of fosA5, a new plasmid-mediated fosfomycin resistance gene in Escherichia coli. Lett Appl Microbiol 60(3):259–264
Ma YX, Wang CY, Li YY, Li J, Wan QQ, Chen JH et al (2020) Considerations and caveats in combating ESKAPE pathogens against nosocomial infections. Adv Sci 7(1):1901872
Mah TFC, O'Toole GA (2001) Mechanisms of biofilm resistance to antimicrobial agents. Trends Microbiol 9(1):34–39
Manaia CM, Rocha J, Scaccia N, Marano R, Radu E, Biancullo F et al (2018) Antibiotic resistance in wastewater treatment plants: Tackling the black box. Environ Int 115:312–324
Manson JM, Hancock LE, Gilmore MS (2010) Mechanism of chromosomal transfer of Enterococcus faecalis pathogenicity island, capsule, antimicrobial resistance, and other traits. Proc Natl Acad Sci 107(27):12269–12274
Martinez JL (2008) Antibiotics and antibiotic resistance genes in natural environments. Science 321(5887):365–367
Martinez JL (2009) Environmental pollution by antibiotics and by antibiotic resistance determinants. Environ Pollut 157(11):2893–2902
Martínez JL (2012) The antibiotic resistome: challenge and opportunity for therapeutic intervention. Future Med Chem 4(3):347–359
Martinez JL (2014) General principles of antibiotic resistance in bacteria. Drug Discov Today Technol 11:33–39
Martinez JL, Baquero F (2000) Mutation frequencies and antibiotic resistance. Antimicrob Agents Chemother 44(7):1771–1777
Martínez JL, Coque TM, Baquero F (2015) What is a resistance gene? Ranking risk in resistomes. Nat Rev Microbiol 13(2):116–123
Mediavilla JR, Patrawalla A, Chen L, Chavda KD, Mathema B, Vinnard C et al (2016) Colistin- and carbapenem-resistant Escherichia coli harboring mcr-1 and bla NDM-5, causing a complicated urinary tract infection in a patient from the United States. MBio 7(4):10–1128
Meini S, Tascini C, Cei M, Sozio E, Rossolini GM (2019) AmpC β-lactamase-producing Enterobacterales: what a clinician should know. Infection 47:363–375
Melo-Cristino J, Resina C, Manuel V, Lito L, Ramirez M (2013) First case of infection with vancomycin-resistant Staphylococcus aureus in Europe. Lancet 382(9888):205
Mendes G, Ramalho JF, Bruschy-Fonseca A, Lito L, Duarte A, Melo-Cristino J, Caneiras C (2022) Whole-genome sequencing enables molecular characterization of non-clonal group 258 high-risk clones (ST13, ST17, ST147 and ST307) among carbapenem-resistant Klebsiella pneumoniae from a tertiary university hospital centre in Portugal. Microorganisms 10(2):416
Mi G, Shi D, Wang M, Webster TJ (2018) Reducing bacterial infections and biofilm formation using nanoparticles and nanostructured antibacterial surfaces. Adv Healthc Mater 7(13):1800103
Milaković M, Vestergaard G, González-Plaza JJ, Petrić I, Šimatović A, Senta I et al (2019) Pollution from azithromycin-manufacturing promotes macrolide-resistance gene propagation and induces spatial and seasonal bacterial community shifts in receiving river sediments. Environ Int 123:501–511

Mishra NN, Tran TT, Seepersaud R, García-De-La-Mària C, Faull K, Yoon A et al (2017) Perturbations of phosphatidate cytidylyltransferase (CdsA) mediate daptomycin resistance in *Streptococcus mitis/oralis* by a novel mechanism. Antimicrob Agents Chemother 61(4):10–1128

Mitchell M, Nguyen SV, Macori G, Bolton D, McMullan G, Drudy D, Fanning S (2022) *Clostridioides difficile* as a potential pathogen of importance to one health: a review. Foodborne Pathog Dis 19(12):806–816

Mlynarczyk-Bonikowska B, Kowalewski C, Krolak-Ulinska A, Marusza W (2022) Molecular mechanisms of drug resistance in Staphylococcus aureus. Int J Mol Sci *23*(15):8088

Montefour K, Frieden J, Hurst S, Helmich C, Headley D, Martin M, Boyle DA (2008) *Acinetobacter baumannii*: an emerging multidrug-resistant pathogen in critical care. Crit Care Nurse 28(1):15–25

Montero CI, Stock F, Murray PR (2008) Mechanisms of resistance to daptomycin in *Enterococcus faecium*. Antimicrob Agents Chemother 52(3):1167–1170

Morroni G, Brenciani A, Brescini L, Fioriti S, Simoni S, Pocognoli A et al (2018) High rate of ceftobiprole resistance among clinical *methicillin-resistant Staphylococcus aureus* isolates from a hospital in central Italy. J Antimicrob Agents 62(12):10–1128

Munita JM, Arias CA (2016) Mechanisms of antibiotic resistance. In: Virulence mechanisms of bacterial pathogens. Wiley, Hoboken, pp 481–511

Müsken M, Di Fiore S, Römling U, Häussler S (2010) A 96-well-plate–based optical method for the quantitative and qualitative evaluation of *Pseudomonas aeruginosa* biofilm formation and its application to susceptibility testing. Nat Protoc *5*(8):1460–1469

Mutuku C, Gazdag Z, Melegh S (2022) Occurrence of antibiotics and bacterial resistance genes in wastewater: resistance mechanisms and antimicrobial resistance control approaches. World J Microbiol Biotechnol 38(9):152

Nicolaou KC, Rigol S (2018) A brief history of antibiotics and select advances in their synthesis. J Antibiot 71(2):153–184

Niebel M, Quick J, Prieto AMG, Hill RL, Pike R, Huber D et al (2015) Deletions in a ribosomal protein-coding gene are associated with tigecycline resistance in *Enterococcus faecium*. Int J Antimicrob Agents 46(5):572–575

Nielsen LE, Snesrud EC, Onmus-Leone F, Kwak YI, Avilés R, Steele ED et al (2014) IS 5 element integration, a novel mechanism for rapid in vivo emergence of tigecycline nonsusceptibility in *Klebsiella pneumoniae*. Antimicrob Agents Chemother 58(10):6151–6156

Nocera FP, Attili AR, De Martino L (2021) *Acinetobacter baumannii*: its clinical significance in human and veterinary medicine. Pathogens 10(2):127

O'Neill J (2016) Tackling drug-resistant infections globally: final report and recommendations

Olson MW, Ruzin A, Feyfant E, Rush TS III, O'Connell J, Bradford PA (2006) Functional, biophysical, and structural bases for antibacterial activity of tigecycline. Antimicrob Agents Chemother 50(6):2156–2166

Oteo J, Ortega A, Bartolomé R, Bou G, Conejo C, Fernández-Martínez M et al (2015) Prospective multicenter study of carbapenemase-producing Enterobacteriaceae from 83 hospitals in Spain reveals high in vitro susceptibility to colistin and meropenem. Antimicrob Agents Chemother 59(6):3406–3412

Pal C, Bengtsson-Palme J, Kristiansson E, Larsson DG (2016) The structure and diversity of human, animal and environmental resistomes. Microbiome 4(1):1–15

Papagiannitsis CC, Tsilipounidaki K, Malli E, Petinaki E (2019) Detection in Greece of a clinical Enterococcus faecium isolate carrying the novel oxazolidinone resistance gene poxtA. J Antimicrob Chemother 74(8):2461–2462

Pedersen MG, Olesen HV, Jensen-Fangel S, Nørskov-Lauritsen N, Wang M (2018) Colistin resistance in Pseudomonas aeruginosa and Achromobacter spp. cultured from Danish cystic fibrosis patients is not related to plasmid-mediated expression of mcr-1. J Cyst Fibros 17(2):e22–e23

Pendleton JN, Gorman SP, Gilmore BF (2013) Clinical relevance of the ESKAPE pathogens. Expert Rev Anti-Infect Ther 11(3):297–308

Pfaller MA, Mendes RE, Flamm RK, Jones RN, Sader HS (2017) Ceftaroline activity against multidrug-resistant Streptococcus pneumoniae from US medical centers (2014) and molecular characterization of a single ceftaroline nonsusceptible isolate. Microb Drug Resist 23(5):571–579

Pfennigwerth N, Lange F, Belmar Campos C, Hentschke M, Gatermann SG, Kaase M (2017) Genetic and biochemical characterization of HMB-1, a novel subclass B1 metallo-β-lactamase found in a Pseudomonas aeruginosa clinical isolate. J Antimicrob Chemother 72(4):1068–1073

Poirel L, Jayol A, Bontron S, Villegas MV, Ozdamar M, Türkoglu S, Nordmann P (2015) The mgrB gene as a key target for acquired resistance to colistin in Klebsiella pneumoniae. J Antimicrob Chemother 70(1):75–80

Poirel L, Kieffer N, Liassine N, Thanh D, Nordmann P (2016) Plasmid-mediated carbapenem and colistin resistance in a clinical isolate of Escherichia coli. Lancet Infect Dis 16(3):281

Poirel L, Goutines J, Aires-de-Sousa M, Nordmann P (2018) High rate of association of 16S rRNA methylases and carbapenemases in Enterobacteriaceae recovered from hospitalized children in Angola. J Antimicrob Agents 62(4):10–1128

Poirel L, Vuillemin X, Kieffer N, Mueller L, Descombes MC, Nordmann P (2019) Identification of FosA8, a plasmid-encoded fosfomycin resistance determinant from Escherichia coli, and its origin in Leclercia adecarboxylata. Antimicrob Agents Chemother 63(11):10–1128

Poole K (2000) Efflux-mediated resistance to fluoroquinolones in gram-positive bacteria and the mycobacteria. Antimicrob Agents Chemother 44(10):2595–2599

Pournaras S, Koumaki V, Spanakis N, Gennimata V, Tsakris A (2016) Current perspectives on tigecycline resistance in Enterobacteriaceae: susceptibility testing issues and mechanisms of resistance. Int J Antimicrob Agents 48(1):11–18

Prestinaci F, Pezzotti P, Pantosti A (2015) Antimicrobial resistance: a global multifaceted phenomenon. Pathogens Global Health 109(7):309–318

Prim N, Rivera A, Rodríguez-Navarro J, Español M, Turbau M, Coll P, Mirelis B (2016) Detection of mcr-1 colistin resistance gene in polyclonal Escherichia coli isolates in Barcelona, Spain, 2012 to 2015. Eur Secur 21(13):30183

Qin S, Wang Y, Zhang Q, Zhang M, Deng F, Shen Z et al (2014) Report of ribosomal RNA methylase gene erm (B) in multidrug-resistant Campylobacter coli. J Antimicrob Chemother 69(4):964–968

Queenan AM, Bush K (2007) Carbapenemases: the versatile β-lactamases. Clin Microbiol Rev 20(3):440–458

Rehman MA, Yin X, Persaud-Lachhman MG, Diarra MS (2017) First detection of a fosfomycin resistance gene, fosA7, in Salmonella enterica serovar Heidelberg isolated from broiler chickens. Antimicrob Agents Chemother 61(8):10–1128

Reid CJ, Blau K, Jechalke S, Smalla K, Djordjevic SP (2020) Whole genome sequencing of Escherichia coli from store-bought produce. Front Microbiol 10:3050

Reyes J, Aguilar AC, Caicedo A (2019) Carbapenem-resistant Klebsiella pneumoniae: microbiology key points for clinical practice. Int J Gen Med 12:437–446

Reygaert WC (2018) An overview of the antimicrobial resistance mechanisms of bacteria. AIMS Microbiol 4(3):482

Rice LB (2010) Progress and challenges in implementing the research on ESKAPE pathogens. Infect Control Hosp Epidemiol 31(S1):S7–S10

Rinke C, Schwientek P, Sczyrba A, Ivanova NN, Anderson IJ, Cheng JF et al (2013) Insights into the phylogeny and coding potential of microbial dark matter. Nature 499(7459):431–437

Rodríguez-Beltrán J, DelaFuente J, Leon-Sampedro R, MacLean RC, San Millan A (2021) Beyond horizontal gene transfer: the role of plasmids in bacterial evolution. Nat Rev Microbiol 19(6):347–359

Rolain JM, Kempf M, Leangapichart T, Chabou S, Olaitan AO, Le Page S et al (2016) Plasmid-mediated mcr-1 gene in colistin-resistant clinical isolates of Klebsiella pneumoniae in France and Laos. Antimicrob Agents Chemother 60(11):6994

Rozwandowicz M, Brouwer MSM, Fischer J, Wagenaar JA, Gonzalez-Zorn B, Guerra B et al (2018) Plasmids carrying antimicrobial resistance genes in Enterobacteriaceae. J Antimicrob Chemother 73(5):1121–1137
Ruef C (2004) Epidemiology and clinical impact of glycopeptide resistance in Staphylococcus aureus. Infection 32(6):315–327
Rybak JM, Roberts K (2015) Tedizolid phosphate: a next-generation oxazolidinone. Infect Dis Ther 4:1–14
San Millan A, MacLean RC (2017) Fitness costs of plasmids: a limit to plasmid transmission. Microbiol Spectr 5(5):10–1128
Sanchez P, Moreno E, Martinez JL (2005) The biocide triclosan selects Stenotrophomonas maltophilia mutants that overproduce the SmeDEF multidrug efflux pump. Antimicrob Agents Chemother 49(2):781–782
Santajit S, Indrawattana N (2016) Mechanisms of antimicrobial resistance in ESKAPE pathogens. Biomed Res Int 2016:2475067
Saw HT, Webber MA, Mushtaq S, Woodford N, Piddock LJ (2016) Inactivation or inhibition of AcrAB-TolC increases resistance of carbapenemase-producing Enterobacteriaceae to carbapenems. J Antimicrob Chemother 71(6):1510–1519
Schaumburg F, Peters G, Alabi A, Becker K, Idelevich EA (2016) Missense mutations of PBP2a are associated with reduced susceptibility to ceftaroline and ceftobiprole in African MRSA. J Antimicrob Chemother 71(1):41–44
Schijven JF, Blaak H, Schets FM, de Roda Husman AM (2015) Fate of extended-spectrum β-lactamase-producing Escherichia coli from faecal sources in surface water and probability of human exposure through swimming. Environ Sci Technol 49(19):11825–11833
Schulz F, Eloe-Fadrosh EA, Bowers RM, Jarett J, Nielsen T, Ivanova NN et al (2017) Towards a balanced view of the bacterial tree of life. Microbiome 5:1–6
Schuster S, Vavra M, Schweigger TM, Rossen JW, Matsumura Y, Kern WV (2017) Contribution of AcrAB-TolC to multidrug resistance in an Escherichia coli sequence type 131 isolate. Int J Antimicrob Agents 50(3):477–481
Schwaber MJ, Graham CS, Sands BE, Gold HS, Carmeli Y (2003) Treatment with a broad-spectrum cephalosporin versus piperacillin-tazobactam and the risk for isolation of broad-spectrum cephalosporin-resistant Enterobacter species. Antimicrob Agents Chemother 47(6):1882–1886
Schwendener S, Perreten V (2022) The bla and mec families of β-lactam resistance genes in the genera Macrococcus, Mammaliicoccus and Staphylococcus: an in-depth analysis with emphasis on Macrococcus. J Antimicrob Chemother 77(7):1796–1827
Sheng ZK, Hu F, Wang W, Guo Q, Chen Z, Xu X et al (2014) Mechanisms of tigecycline resistance among Klebsiella pneumoniae clinical isolates. Antimicrob Agents Chemother 58(11):6982–6985
Simpkin VL, Renwick MJ, Kelly R, Mossialos E (2017) Incentivising innovation in antibiotic drug discovery and development: progress, challenges and next steps. J Antibiot 70(12):1087–1096
Skinner D, Keefer CS (1941) Significance of bacteremia caused by Staphylococcus aureus: a study of one hundred and twenty-two cases and a review of the literature concerned with experimental infection in animals. Arch Intern Med 68(5):851–875
Steenbergen JN, Alder J, Thorne GM, Tally FP (2005) Daptomycin: a lipopeptide antibiotic for the treatment of serious Gram-positive infections. J Antimicrob Chemother 55(3):283–288
Stoesser N, Sheppard AE, Pankhurst L, De Maio N, Moore CE, Sebra R et al (2016) Evolutionary history of the global emergence of the Escherichia coli epidemic clone ST131. MBio 7(2):10–1128
Stokes HW, Gillings MR (2011) Gene flow, mobile genetic elements and the recruitment of antibiotic resistance genes into Gram-negative pathogens. FEMS Microbiol Rev 35(5):790–819
Szymańska U, Wiergowski M, Sołtyszewski I, Kuzemko J, Wiergowska G, Woźniak MK (2019) Presence of antibiotics in the aquatic environment in Europe and their analytical monitoring: recent trends and perspectives. Microchem J 147:729–740

Tacconelli E, Carrara E, Savoldi A, Harbarth S, Mendelson M, Monnet DL et al (2018) Discovery, research, and development of new antibiotics: the WHO priority list of antibiotic-resistant bacteria and tuberculosis. Lancet Infect Dis 18(3):318–327

Tarai B, Das P, Kumar D (2013) Recurrent challenges for clinicians: emergence of methicillin-resistant Staphylococcus aureus, vancomycin resistance, and current treatment options. J Lab Phys 5(2):71–78

Taylor E, Sriskandan S, Woodford N, Hopkins KL (2018) High prevalence of 16S rRNA methyltransferases among carbapenemase-producing Enterobacteriaceae in the UK and Ireland. Int J Antimicrob Agents 52(2):278–282

Telke AA, Olaitan AO, Morand S, Rolain JM (2017) soxRS induces colistin hetero-resistance in Enterobacter asburiae and Enterobacter cloacae by regulating the acrAB-tolC efflux pump. J Antimicrob Chemother 72(10):2715–2721

Thomas CM, Nielsen KM (2005) Mechanisms of, and barriers to, horizontal gene transfer between bacteria. Nat Rev Microbiol 3(9):711–721

Tilahun M, Kassa Y, Gedefie A, Ashagire M (2021) Emerging carbapenem-resistant Enterobacteriaceae infection, its epidemiology and novel treatment options: a review. Infect Drug Resistance 14:4363–4374

Tooke CL, Hinchliffe P, Bragginton EC, Colenso CK, Hirvonen VH, Takebayashi Y, Spencer J (2019) β-Lactamases and β-lactamase inhibitors in the 21st century. J Mol Biol 431(18):3472–3500

Torumkuney D, Kundu S, Vu GV, Nguyen HA, Pham HV, Kamble P et al (2022) Country data on AMR in Vietnam in the context of community-acquired respiratory tract infections: links between antibiotic susceptibility, local and international antibiotic prescribing guidelines, access to medicines and clinical outcome. J Antimicrob Chemother 77(1):26–34

Tran NH, Reinhard M, Gin KYH (2018) Occurrence and fate of emerging contaminants in municipal wastewater treatment plants from different geographical regions-a review. Water Res 133:182–207

Turner NA, Sharma-Kuinkel BK, Maskarinec SA, Eichenberger EM, Shah PP, Carugati M et al (2019) Methicillin-resistant Staphylococcus aureus: an overview of basic and clinical research. Nat Rev Microbiol 17(4):203–218

Uddin TM, Chakraborty AJ, Khusro A, Zidan BRM, Mitra S, Emran TB et al (2021) Antibiotic resistance in microbes: history, mechanisms, therapeutic strategies and future prospects. J Infect Public Health 14(12):1750–1766

Urban-Chmiel R, Marek A, Stępień-Pyśniak D, Wieczorek K, Dec M, Nowaczek A, Osek J (2022) Antibiotic resistance in bacteria-A review. Antibiotics 11(8):1079

US EPA (2012) Recreational water quality criteria. Health and Ecological Criteria Division OoSaT, 42

Uyttendaele M, Jaykus LA, Amoah P, Chiodini A, Cunliffe D, Jacxsens L et al (2015) Microbial hazards in irrigation water: standards, norms, and testing to manage use of water in fresh produce primary production. Compr Rev Food Sci Food Saf 14(4):336–356

Varela MF, Stephen J, Lekshmi M, Ojha M, Wenzel N, Sanford LM et al (2021) Bacterial resistance to antimicrobial agents. Antibiotics 10(5):593

Vila J, Martí S, Sanchez-Céspedes J (2007) Porins, efflux pumps and multidrug resistance in Acinetobacter baumannii. J Antimicrob Chemother 59(6):1210–1215

Villa L, Feudi C, Fortini D, García-Fernández A, Carattoli A (2014) Genomics of KPC-producing Klebsiella pneumoniae sequence type 512 clone highlights the role of RamR and ribosomal S10 protein mutations in conferring tigecycline resistance. Antimicrob Agents Chemother 58(3):1707–1712

Vincent JL (2003) Nosocomial infections in adult intensive-care units. Lancet 361(9374):2068–2077

Voulgaris GL, Voulgari ML, Falagas ME (2019) Developments on antibiotics for multidrug resistant bacterial Gram-negative infections. Expert Rev Anti-Infect Ther 17(6):387–401

Wade TJ, Calderon RL, Brenner KP, Sams E, Beach M, Haugland R et al (2008) High sensitivity of children to swimming-associated gastrointestinal illness: results using a rapid assay of recreational water quality. Epidemiology 19:375–383

Wade TJ, Sams E, Brenner KP, Haugland R, Chern E, Beach M et al (2010) Rapidly measured indicators of recreational water quality and swimming-associated illness at marine beaches: a prospective cohort study. Environ Health 9(1):1–14
Wang Y, Zhang M, Deng F, Shen Z, Wu C, Zhang J et al (2014) Emergence of multidrug-resistant Campylobacter species isolates with a horizontally acquired rRNA methylase. Antimicrob Agents Chemother 58(9):5405–5412
Wang Y, Lv Y, Cai J, Schwarz S, Cui L, Hu Z et al (2015) A novel gene, optrA, that confers transferable resistance to oxazolidinones and phenicols and its presence in Enterococcus faecalis and Enterococcus faecium of human and animal origin. J Antimicrob Chemother 70(8):2182–2190
Wang X, Wang Y, Zhou Y, Li J, Yin W, Wang S et al (2018) Emergence of a novel mobile colistin resistance gene, mcr-8, in NDM-producing Klebsiella pneumoniae. Emerg Microb Infect 7(1):1–9
Wang J, Chu L, Wojnárovits L, Takács E (2020) Occurrence and fate of antibiotics, antibiotic resistant genes (ARGs) and antibiotic resistant bacteria (ARB) in municipal wastewater treatment plant: an overview. Sci Total Environ 744:140997
Waters V, Ratjen F (2017) Standard versus biofilm antimicrobial susceptibility testing to guide antibiotic therapy in cystic fibrosis. Cochrane Database Syst Rev 6:CD009528
Weatherall D, Greenwood B, Chee HL, Wasi P (2006) Control: past, present, and future. Disease control priorities in developing countries, 119
Weber DJ, Rutala WA, Miller MB, Huslage K, Sickbert-Bennett E (2010) Role of hospital surfaces in the transmission of emerging health care-associated pathogens: norovirus, Clostridium difficile, and Acinetobacter species. Am J Infect Control 38(5):S25–S33
Wencewicz TA (2019) Crossroads of antibiotic resistance and biosynthesis. J Mol Biol 431(18):3370–3399
Woodford N, Carattoli A, Karisik E, Underwood A, Ellington MJ, Livermore DM (2009) Complete nucleotide sequences of plasmids pEK204, pEK499, and pEK516, encoding CTX-M enzymes in three major Escherichia coli lineages from the United Kingdom, all belonging to the international O25: H4-ST131 clone. Antimicrob Agents Chemother 53(10):4472–4482
World Health Organization (2017) Prioritization of pathogens to guide discovery, research and development of new antibiotics for drug-resistant bacterial infections, including tuberculosis (No. WHO/EMP/IAU/2017.12). World Health Organization
Wright GD (2007) The antibiotic resistome: the nexus of chemical and genetic diversity. Nat Rev Microbiol 5(3):175–186
Wright MS, Suzuki Y, Jones MB, Marshall SH, Rudin SD, van Duin D et al (2015) Genomic and transcriptomic analyses of colistin-resistant clinical isolates of Klebsiella pneumoniae reveal multiple pathways of resistance. Antimicrob Agents Chemother 59(1):536–543
Xue W, Zhou Q, Li F (2019) Bacterial community changes and antibiotic resistance gene quantification in microbial electrolysis cells during long-term sulfamethoxazole treatment. Bioresour Technol 294:122170
Yamaguchi T, Suzuki S, Okamura S, Miura Y, Tsukimori A, Nakamura I et al (2015) Evolution and single-nucleotide polymorphisms in methicillin-resistant Staphylococcus aureus strains with reduced susceptibility to vancomycin and daptomycin, based on determination of the complete genome. Antimicrob Agents Chemother 59(6):3585–3587
Yamashita S, Buchanan SK (2010) Solute and ion transport: outer membrane pores and receptors. EcoSal Plus 4(1):10–1128
Yang Y, Li B, Zou S, Fang HH, Zhang T (2014) Fate of antibiotic resistance genes in sewage treatment plant revealed by metagenomic approach. Water Res 62:97–106
Yang S, Hemarajata P, Shevy L, Maciariello M, Culbreath K, Bush K, Humphries R (2018) Unusual carbapenem resistant but ceftriaxone and cefepime susceptible Klebsiella oxytoca isolated from a blood culture: case report and whole-genome sequencing investigation. IDCases 11:9–11
Ye M, Ding B, Qian H, Xu Q, Jiang J, Huang J et al (2017) In vivo development of tigecycline resistance in Klebsiella pneumoniae owing to deletion of the ramR ribosomal binding site. Int J Antimicrob Agents 50(4):523–528

Yong D, Toleman MA, Giske CG, Cho HS, Sundman K, Lee K, Walsh TR (2009) Characterization of a new metallo-β-lactamase gene, bla NDM-1, and a novel erythromycin esterase gene carried on a unique genetic structure in Klebsiella pneumoniae sequence type 14 from India. Antimicrob Agents Chemother 53(12):5046–5054

Youde J (2012) Global health governance. Polity Press, Malden

Zheng JX, Lin ZW, Sun X, Lin WH, Chen Z, Wu Y, Yu ZJ (2018) Overexpression of OqxAB and MacAB efflux pumps contributes to eravacycline resistance and heteroresistance in clinical isolates of Klebsiella pneumoniae. Emerg Microb Infect 7(1):1–11

Zhong X, Xu H, Chen D, Zhou H, Hu X, Cheng G (2014) First emergence of acrAB and oqxAB mediated tigecycline resistance in clinical isolates of Klebsiella pneumoniae pre-dating the use of tigecycline in a Chinese hospital. PLoS One 9(12):e115185

Zhu YG, Zhao Y, Zhu D, Gillings M, Penuelas J, Ok YS et al (2019) Soil biota, antimicrobial resistance and planetary health. Environ Int 131:105059

Zhu Y, Huang WE, Yang Q (2022) Clinical perspective of antimicrobial resistance in bacteria. Infect Drug Resistance 15:735–746

Role of Nanomedicine in Overcoming Antimicrobial Resistance: Challenges and Opportunities

2

Riya Mukherjee, Chung-Ming Chang, Ramendra Pati Pandey, and Saif Hameed

Abstract

In the face of escalating threats posed by antimicrobial resistance (AMR) to global health, this chapter explores the pivotal role of nanomedicine as a promising solution. Delving into the dynamic landscape of nanoscale interventions, the chapter offers a comprehensive analysis of the challenges and opportunities inherent in leveraging nanomedicine to combat AMR. With its unique capability to manipulate matter at the nanoscale, nanomedicine opens unprecedented avenues for creative responses to tackle the complexities of AMR. The chapter meticulously examines various classifications of nanosystems designed to address AMR, encompassing nanomaterials with intrinsic antimicrobial capabilities, nanocarriers for precise drug administration, and the synergistic effects of nanosystems in tandem with antibiotics. Acknowledging the intricacies of the battle against AMR, the chapter candidly discusses the hurdles faced by nanomedicine in this endeavor. Despite these challenges, an optimistic outlook per-

R. Mukherjee
Graduate Institute of Biomedical Sciences, Chang Gung University, Taoyuan City, Taiwan

Master & Ph.D. Program in Biotechnology Industry, Chang Gung University, Taoyuan City, Taiwan

C.-M. Chang
Master & Ph.D. Program in Biotechnology Industry, Chang Gung University, Taoyuan City, Taiwan

Department of Medical Biotechnology and Laboratory Science, Chang Gung University, Taoyuan City, Taiwan

R. P. Pandey
School of Health Sciences and Technology (SoHST), UPES, Dehradun, Uttarakhand, India

S. Hameed (✉)
Amity Institute of Biotechnology, Amity University Haryana, Gurugram (Manesar), India

M. Y. Wani et al. (eds.), *Nanotechnology Based Strategies for Combating Antimicrobial Resistance*, https://doi.org/10.1007/978-981-97-2023-1_2

vades the narrative, underscoring the potential of nanomedicine to contribute significantly to the fight against AMR. This chapter not only provides a compelling overview of the evolving role of nanomedicine in overcoming AMR but also invites readers to consider the fascinating opportunities presented by this transformative field. It encourages active engagement in the global effort to combat AMR, painting a vivid picture of a future where nanomedicine serves as a powerful ally in safeguarding lives against drug-resistant infections.

Keywords

Nanomedicine · Antimicrobial resistance · Global health · Drug-resistant infections

2.1 Introduction

Antimicrobial resistance (AMR) occurs when microbes evolve mechanisms to protect themselves from the impact of antimicrobials. This phenomenon arises when bacteria gain the ability to withstand exposure to antibiotics. Improper and often excessive use of antibiotics in medical procedures and intensive animal farming have led to the emergence of superbugs that are resistant to antimicrobials. Treating resistant microbes becomes challenging, as it necessitates higher doses, alternative medications that may be more toxic and expensive. Antimicrobial resistance (AMR) is an escalating crisis with far-reaching consequences. The rising prevalence of multidrug-resistant (MDR) organisms poses a formidable challenge in treating healthcare-associated illnesses (Chakraborty et al. 2022). These organisms, classified as extensively drug-resistant (XDR) or totally drug resistant (TDR), commonly known as superbugs, are rendering current antibiotics less effective. The pathogenesis of AMR involves critical factors such as genetic mutations, where microorganisms develop resistance through genetic changes; the perilous overuse and misuse of antimicrobials; horizontal gene transfer, allowing bacteria to share resistance genes; biofilm formation; and the alarming scarcity of new antibiotics (Blume et al. 2012). Urgent action is imperative to curb this growing threat and some multidrug-resistant organisms. Understanding the primary pathogen–drug combinations driving bacterial antimicrobial resistance (AMR) and global trends is crucial. The Lancet provides a comprehensive overview of the current state of AMR worldwide, highlighting its emergence as a leading global cause of mortality across 204 countries and territories. Recent estimates reveal that drug-resistant infections led to 1.27 million deaths globally in 2019 (Fig. 2.1), with a potential staggering rise to 10 million annual deaths by 2050 if unchecked (Collaborators 2022). Uncontrolled AMR could result in a yearly GDP reduction of US$3.4 trillion and push an additional 24 million people into extreme poverty over the next decade (UNEP 2023). Inappropriate antibiotic use has given rise to more aggressive, drug-resistant strains, posing significant challenges to conventional antibacterial medications. Issues such as poor water solubility, reduced stability, and challenging targeting further complicate the use of these drugs. Despite the urgent need to combat AMR, investment in the

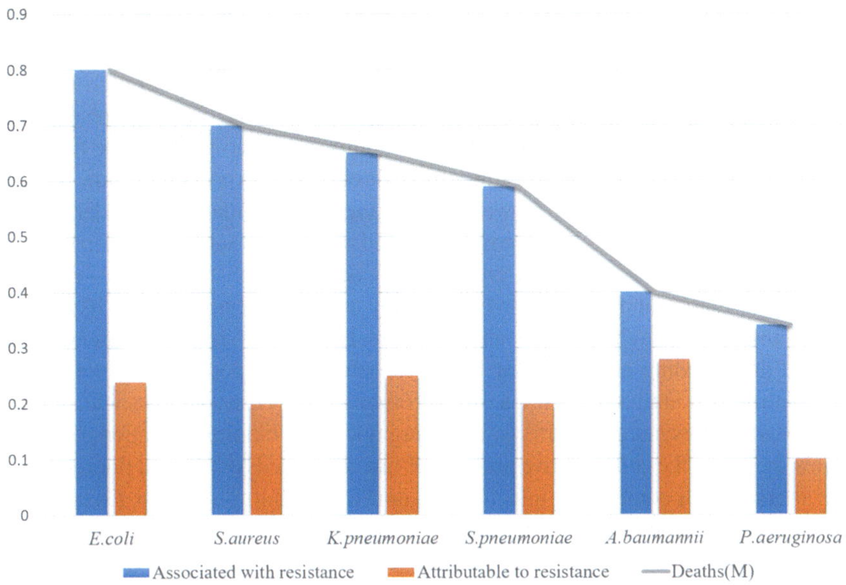

Fig. 2.1 Six pathogens responsible for around 80% of the 1.27 million deaths attributed directly to antimicrobial resistance in 2019

discovery of new antimicrobial medications has significantly declined, resulting in fewer new antibiotics being introduced to the market in recent years. Immediate and concerted efforts are necessary to address this critical issue. Boosting the effectiveness of existing antibiotics presents a pragmatic and expeditious strategy in the battle against drug resistance. In the face of the severe human and financial consequences linked to antibiotic resistance, the urgency for innovative approaches is paramount. Nanotechnology emerges as a pivotal player in this scenario. With engineered structures ranging from 1 to 100 nm, nanoparticles (NPs) offer a promising avenue to tackle infections and counter the colossal challenge of antimicrobial resistance (Eleraky et al. 2020). The distinctive physical and chemical attributes of NPs have demonstrated therapeutic potential, making them a compelling solution. This chapter delves into a compelling exploration of the possibilities and challenges inherent in leveraging nanomedicines to combat the escalating threat of antimicrobial resistance, underscoring the imperative need for transformative advancements in the field.

2.2 Mechanism of Nanosystems to Combat Antimicrobial Resistance

At the core of microbial existence lies an unwavering drive to grow, survive, and proliferate swiftly within their environment. Armed with remarkable adaptability, microbes deploy an array of survival strategies, making them formidable adversaries.

Yet, the arsenal of antibiotics wielded against them possesses the potential to disrupt their growth, alter genetic makeup, induce drug resistance, and ensure their survival. The molecular correlation between bacteria and antibiotics unfolds through various intricate mechanisms (Fig. 2.2). Microbes, in their relentless quest for resistance, employ strategies such as reducing cell permeability, inactivating enzymes, modifying target sites/enzymes, and boosting efflux through overexpressing efflux pumps (Horikoshi and Serpone 2013). However, emerging as a formidable contender in this microbial battleground are nanoparticles, armed with their antimicrobial prowess. Within the realm of nanosystems, crucial physicochemical characteristics—particle size, surface charge, and solubility—hold the key to orchestrating activities like intracellular absorption, biodistribution, and clearance. Nanometer-sized particles not only enhance drug-loading efficiency for both hydrophilic and lipophilic antibiotics but also amplify their antibacterial impact. The combination of nanoparticles with antibiotics gives rise to synergistic effects, thwarting microbial biofilm formation. Notable studies showcase the collaborative prowess of functionalized AuNPs and fluoroquinolone antibiotics in combatting infections caused by multidrug-resistant *Escherichia coli*, both in vitro and in vivo (Pelgrift and Friedman 2013).

We are currently on the brink of a new era, nanoparticles emerge as the heralds of the next generation of antibiotics, promising innovative and potent solutions to counter the evolving challenges posed by microbial resistance. Nanoparticles deploy a spectrum of mechanisms to halt bacterial growth, ranging from inducing intracellular effects and directly interacting with bacteria to disrupting membrane potential and integrity.

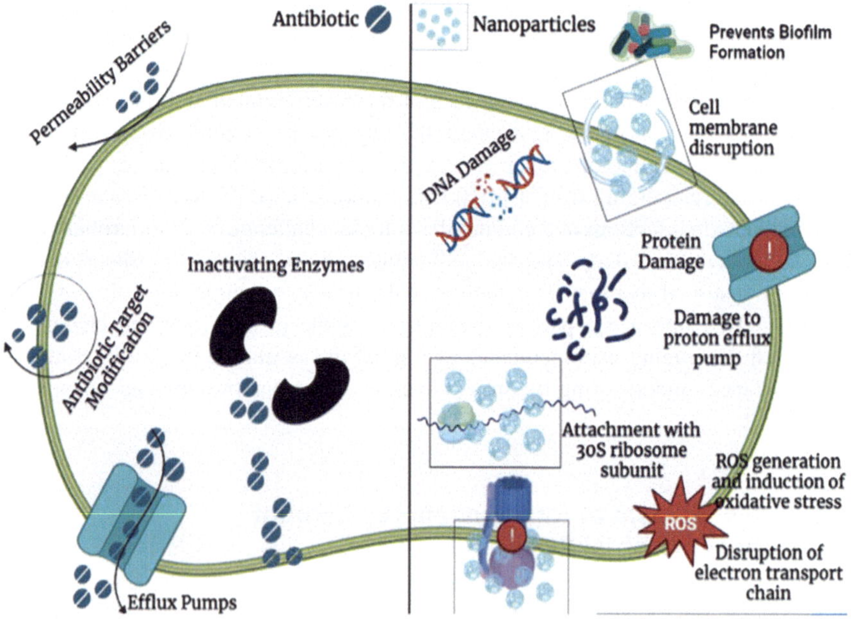

Fig. 2.2 Mechanisms of antimicrobial resistance (AMR) in microbes (left) and the actions of nanoparticles against AMR (right)

Additionally, they stimulate host immune responses, inhibit biofilm formation, generate reactive oxygen species (ROS), and impede RNA and protein synthesis (Xie et al. 2023). Intriguingly, nanosized systems showcase a unique ability to curtail the stimulation of resistance, effectively thwarting bacteria from developing resistance strategies.

2.3 Classification of Nanosystems

Antimicrobial resistance can arise through various pathways, such as enzyme activation, reduced cell permeability, alterations in target sites, and increased efflux caused by the overexpression of efflux pumps. Nanoparticles (NPs) can counteract all these resistance mechanisms employed by organisms. Furthermore, NPs demonstrate synergistic effects against bacteria, hinder biofilm production, and effectively address multidrug resistance when combined with antibiotics. Nanosystems are commonly categorized broadly based on their structural properties and the materials used in their preparation. Based on this, there are primarily two types of nanosystems: organic nanosystems and inorganic nanosystems (Fig. 2.3).

2.3.1 Inorganic Nanosystems

Silver Nanoparticles (AgNPs) The antibacterial capabilities of silver nanoparticles (AgNPs) have been the subject of substantial research (Table 2.1). Numerous microorganisms, including bacteria, fungi, and viruses, can grow more slowly as a result. Microorganisms' cell membranes can be damaged by AgNPs, which causes cell lysis and death. They can also interact with microbial biomolecules and enzymes, causing them to behave abnormally. In order to stop microbial colonization and the development of biofilms, AgNPs have been added to a variety of materials, including coatings, films, and dressings (Baptista et al. 2018).

Zinc Oxide Nanoparticles (ZnONPs) Significant antibacterial action has been demonstrated by zinc oxide nanoparticles (ZnONPs) against a range of microorganisms, including bacteria, fungi, and viruses. ZnONPs have the ability to release zinc ions, which can inhibit or kill bacteria by interfering with their biological functions. ZnONPs' antibacterial properties depend on their capacity to disrupt bacterial cell membrane integrity, reduce the hydrophobicity of the cell surface, and downregulate specific gene expression (Table 2.1). Additionally, the ZnONPs treatment increased the bacterial damage within the cell by generating reactive oxygen species. Furthermore, ZnONPs have been found to inhibit the formation of biofilms and the activity of hemolysin toxin produced by *Staphylococcus aureus*. This means that ZnONPs have the ability to hinder the formation of biofilms and the damaging effects of the hemolysin toxin, potentially offering a way to combat the harmful activities of *Staphylococcus aureus*. ZnONPs have been employed for topical treatments against microbial infections in a variety of formulations, including creams, ointments, and hydrogels (Gupta et al. 2017).

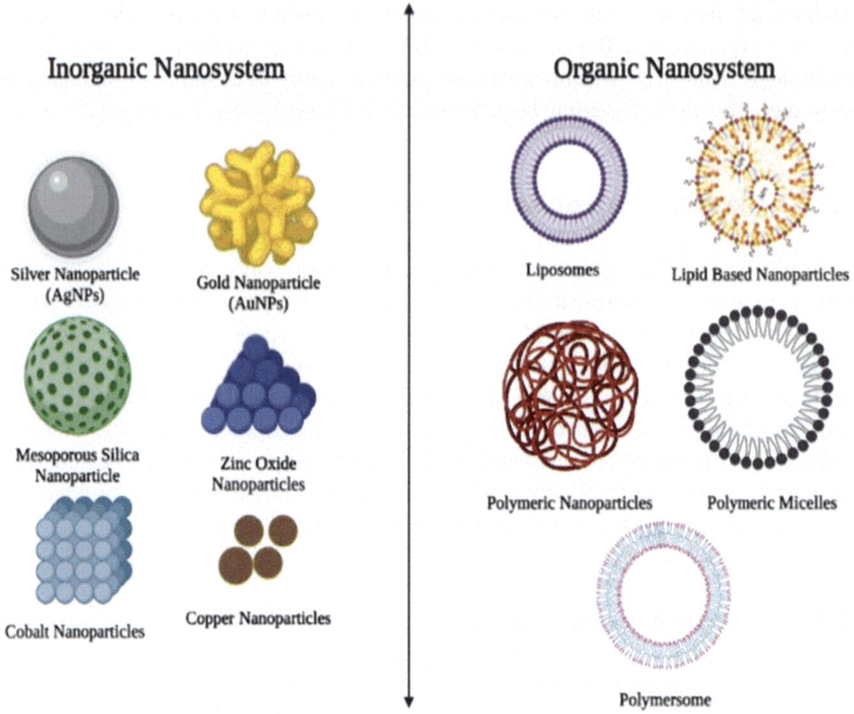

Fig. 2.3 The two types of nanosystems: inorganic nanosystem and organic nanosystem

Copper Nanoparticles (CuNPs) CuNPs, copper nanoparticles, have also proven to have antibacterial qualities. They can damage and kill cells by interfering with DNA and microbial enzymes (Zazo et al. 2016). CuNPs have been used in coatings, clothing, and medical equipment to stop the growth of biofilms and microorganisms, lowering the possibility of AMR (Table 2.1).

Silica Nanoparticles Silica nanoparticles (SiNPs) have emerged as a promising option for various biomedical applications, particularly in cancer and antimicrobial therapeutics (Table 2.1). SiNPs offer versatility, especially in the realm of antimicrobial treatments, owing to silica's low chemical reactivity, biocompatibility, and thermal stability, which aid in efficient drug delivery. The unique structure of silica allows for optimal drug loading and controlled delivery to the targeted site (Tan et al. 2021). Mesoporous silica nanoparticles have shown positive outcomes in multiple treatment phases, addressing aspects such as biofilm targeting, adjuvant capacity, and controlled drug release. The tunable pore properties of SiNPs enable the loading of diverse payloads, including drugs, dyes, and metals, leading to effective therapeutic outcomes and tailored interactions with the payload for well-controlled release and functional activity (Hetta et al. 2023).

Table 2.1 Inorganic nanoparticles mechanism against the targeted bacteria and its properties

Nanoparticles (NPs)	Antimicrobial activity mechanisms	Targeted microorganisms	Size of the nanoparticles (nm)
Silver nanoparticles (AgNPs)	AgNPs generate reactive oxygen species (ROS), disintegrate the cell membrane, inhibit cell wall synthesis, increase membrane permeability, and cause ribosome destabilization and intercalation between DNA bases	*Staphylococcus epidermidis*, MRSA, vancomycin-resistant *Enterococcus* (VRE), extended-spectrum beta-lactamase (ESBL)-producing organisms, MDR *Escherichia coli, Pseudomonas aeruginosa, Klebsiella pneumoniae*, carbapenem- and polymyxin B-resistant *A. baumannii*, carbapenem-resistant *P. aeruginosa*, and carbapenem-resistant *Enterobacteriaceae* (CRE)	1–100
Zinc oxide nanoparticles (ZnONPs)	ZnONPs generate ROS, cause the disruption of the membrane, cause lipid and protein damage, and also cause adsorption to the cell surface	ESBL-producing *E. coli, K. pneumoniae, Enterobacter aerogenes, E. coli, Klebsiella oxytoca, K. pneumoniae*, MRSA	10–100
Copper nanoparticles (CuNPs)	Disappearance of cell membrane, DNA degradation, protein oxidation, ROS generation, and lipid peroxidation	MDR *E. coli, A. baumannii*	2–350
Silica nanoparticles	Disruption of bacterial cell membrane, generation of ROS, inhibition of bacterial enzymes, and induction of stress responses	*S. aureus, E. coli, P. aeruginosa, S. typhimurium, S. mutans*	10–500
Cobalt nanoparticles (CoNps)	Disruption of the bacterial cell membrane, generation of ROS, inhibition of bacterial enzymes, and disruption of bacterial biofilms	*S. aureus, E. coli, P. aeruginosa, S. enteritidis, C. albicans*	10–30

Cobalt Nanoparticles Previous research has suggested that cobalt nanoparticles may have antibacterial properties against *E. coli* strains. Further study has demonstrated cobalt oxide nanoparticles' anticancer and antibacterial effects on bacteria and colon cancer cells (Table 2.1). Cobalt nanosuspension exhibits the strongest antibacterial effects against *S. aureus*, a multidrug-resistant organism. Additionally, seen were cell wall rupturing, cell wall creation of holes, and cell shrinkage (Lee et al. 2019).

2.3.2 Organic Nanosystems

2.3.2.1 Liposomes

Considering their distinct characteristics, such as biocompatible, biodegradable, and nontoxic vesicles, liposomes are the first nanoscale drug that has been approved for medical use. A lipid bilayer structure that encloses an aqueous compartment makes up liposomes. In contrast to traditional antibiotics, liposomes have the unique ability to enhance the pharmacokinetics and pharmacodynamics of antibiotics and sustain drug release (Table 2.2). This helps to maintain the proper antibiotic level for a long period. Dicloxacillin-loaded liposome administration was tested for effectiveness against MRSA infections (Power of Particles 2023). The improved antibacterial activity was confirmed by the drug-loaded liposomes. As a result, liposomes as drug delivery methods offer improved protection and biodistribution of antibiotics. The liposomes' structure is particular to the targeted bacterial biofilm. Positively charged liposomes interact most strongly with negatively charged bacterial biofilms. The effectiveness of neutral and negatively charged gentamicin-loaded liposomes in targeting the pathogenic strains of *P. aeruginosa* (*Pseudomonas aeruginosa*) and *K. oxytoca* (*Klebsiella oxytoca*) was examined by Solleti et al. (2015). Unexpectedly, anionic liposomes outperformed neutral antibiotic-loaded liposomes and free gentamicin in terms of drug encapsulation and gentamicin's ability to target bacterial biofilms. This could be better understood in terms of the improved delivery of antibiotic-loaded negatively charged liposomes through a fusion mechanism that permits the direct injection of liposome-entrapped antibiotics into the cytoplasm of bacteria in spite of the repulsive forces. A positively charged antibiotic (gentamicin) is more effectively encapsulated within negatively charged liposomes based on electrostatic interaction (Author 2021).

Usually, liposomes are categorized on the basis of their physicochemical characteristics, composition, preparation technique, vesicle size, and bilayers. Table 2.2 depicts various classifications of liposomes used against bacterial biofilm infections. Nanostructured lipid carriers (NLCs) are considered the advanced type of SLNs. NLCs are composed of a rigid matrix blended with liquid oil to form an unstructured matrix (Author 2021). Unlike SLNs, they form an imperfect core and an amorphous matrix for better drug-loading ability and minimized drug escape from the matrix during storage.

2.3.2.2 Polymeric Micelles

Antimicrobial resistance (AMR) has been researched using polymeric micelles as prospective medication delivery methods. Amphiphilic block copolymers make up polymeric micelles, which self-assemble into nanoparticles with a hydrophobic core and a hydrophilic shell. Drugs, including antibiotics, can be encapsulated in these micelles, which enhances the pharmacokinetics and pharmacodynamics of the drug. Polymeric micelles have been shown in numerous studies to increase the effectiveness of antibiotics against AMR bacteria. For instance, ciprofloxacin-loaded polymeric micelles have demonstrated efficacy against methicillin-resistant *Staphylococcus aureus* (MRSA) and *Pseudomonas aeruginosa*, two bacteria that

Table 2.2 Classification of liposomes used against bacterial infections

Type of liposome	Composition of lipid	Antibiotic used/ neutraceutical agents	Targeted bacterial biofilm strain	Outcomes of the type of liposome
Conventional lipososmes	Egg yolk soyabean L-alpha phosphatidylcholine Dipalmitoylphosphatidylcholine pH-sensitive lipids: Phosphatidylcholine: cholesterol Egg yolk lecithin	Ofloxacin Ampicillin Calcifediol Vancomycin Cinnamaldehyde	Any ocular postsurgery infections *P. aeruginosa* MRSA *S. aureus*	• More effective localized liposomal antibiotic delivery to the infection location • High capacity for loading and encapsulating • Very effective targeting to the bronchial cells • High bacterial killing property • Enhanced in vitro antimicrobial activities • Improved stability and sustained antimicrobial action
Fusogenic lipososmes	Dope/CHe MS	Fusidic acid	*S. epidermis*, *A. baumanii*	• Liposomal preparations have excellent antibacterial action against both Gram-positive and Gram-negative pathogens
Surface-modified liposomes	Phospholipid, cholesterol, tween 80, Vitamin E	Gallic acid	*E. coli* *S. aureus*	• Lactoferrin gallic acid liposomes had better storage stability and more antibacterial power than gallic acid liposomes. Enhanced potential for lactoferrin liposomes as an efficient method of delivering nutraceuticals in food

are frequently linked to AMR (Rancan et al. 2023). Similar to this, vancomycin-loaded polymeric micelles have been demonstrated to be effective against vancomycin-resistant *Enterococcus faecalis*. One benefit of employing polymeric micelles as drug delivery systems to combat AMR is that they can enhance the solubility and bioavailability of medications that are frequently insufficiently soluble in water, which can boost the effectiveness of those medications. Additionally, polymeric micelles can promote drug absorption into bacterial biofilms, a frequent source of AMR, and drug accumulation at the site of infection (Rancan et al. 2023).

2.3.2.3 Lipid-Based Nanoparticles

The two main categories of lipid-based nanoparticles are solid lipid nanoparticles (SLNs) and nanostructured lipid carriers (NLCs). To overcome the instability issues with liposomes, high melting-stable lipids were combined to create SLNs, which are drug delivery devices in colloid form. Better targeting and the ability for controlled release are provided by SLNs. SLNs have shown the ability to lessen AMR processes (Abo-Shama et al. 2020). SLNs can enhance cell uptake to treat intracellular infections, reduce biofilm formation or the survival of biofilm-forming bacteria, reduce efflux pump-mediated drug expulsion, avoid the effects of antibiotic-modifying enzymes, and reduce efflux pump-mediated drug expulsion.

2.4 Synergistic Effects of Nanosystems with Antibiotics

When nanoparticles are covered with other antibiotics, their activity typically increases significantly. If the nanoparticles can be employed in addition to the antibiotics, the bacterial resistance can be somewhat reduced. Antibiotics and nanoparticles together have been found to have synergistic effects in treating bacterial infections. It can be demonstrated in a specific way that the synergistic effects of nanosystems and antibiotics play a significant role in the fight against resistance (Fig. 2.4). Antibiotics can be added to nanoparticles or attached to their surfaces, shielding them from the chemicals and enzymes that would otherwise break them down.

2.4.1 Enhanced Antibacterial Activity

By enhancing the solubility, stability, and bioavailability of antibiotics, nanosystems can enhance their antibacterial activity. For instance, nanoparticle-encapsulating antibiotics can prevent them from degrading, improve their absorption by bacteria, and enhance their release at the infection site. Higher antibiotic concentrations at the infection site may result from this, which would increase bacterial eradication (Abo-Shama et al. 2020).

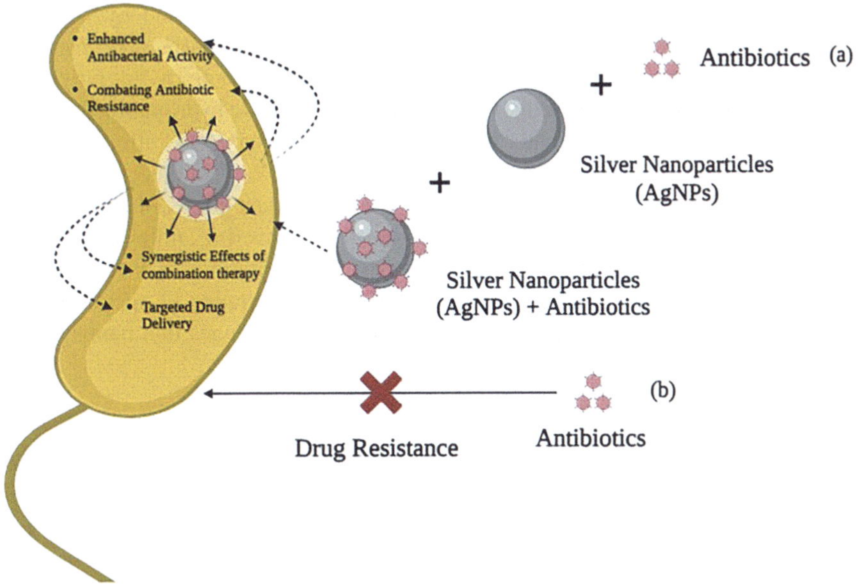

Fig. 2.4 Synergistic effects of nanosystems with antibiotics. (**a**) Silver nanoparticles (AgNPs) with antibiotics have synergistic effects in treating bacterial infections and (**b**) only conventional antibiotics emerging resistance

2.4.2 Combating Antibiotic Resistance

Addressing bacterial illnesses poses a considerable challenge due to the rise of antibiotic resistance. Nanosystems offer a viable solution to combat antibiotic resistance, with nanoparticles playing a crucial role. Efflux pump inhibitors, which are proteins bacteria use to expel drugs and reduce intracellular drug concentration, can be synthesized by nanoparticles. These nanoparticles enhance the intracellular accumulation of antibiotics, boosting their effectiveness against resistant bacteria by inhibiting efflux pumps (Balderrama-González et al. 2021). Nanoparticles primarily employ two mechanisms as promising antimicrobial agents against bacteria: (1) disrupting membrane potential and integrity and (2) inducing oxidative stress through the generation of reactive oxygen species (ROS). These mechanisms can act independently or simultaneously. Inorganic nanoparticles, known for their high surface-area-to-volume ratio and unique physical and chemical properties, exhibit enhanced antibacterial activity (Balderrama-González et al. 2021). Silver nanoparticles (AgNPs), renowned for their potent bactericidal effect surpassing other metal oxides, are widely used as effective antimicrobials with a reduced risk of inducing resistance. Gold nanoparticles (AuNPs), another category of inorganic nanoparticles, are extensively explored as antimicrobial agents due to their photothermal activity, biocompatibility, and ease of modification with small antimicrobial drugs. Polymeric nanoparticles can directly eliminate microbes by interacting with negatively charged bacterial cell walls. Incorporating cationic and hydrophobic moieties into polymer chains, such as quaternary ammonium groups, alkyl pyridiniums, and phosphonium, imparts intrinsic antimicrobial activity to polymeric nanoparticles. Cationic groups disrupt cell membranes, while hydrophobic moieties aid in membrane penetration and rupture. Increasing the density of cationic charges has been shown to enhance electrostatic interactions with anionic membranes, thereby improving the antibacterial effect. Beyond direct bacterial elimination, polymeric nanoparticles serve as effective carriers for delivering antibiotics, antimicrobial peptides, and agents to target body parts. The limitations of clinical antibiotic therapy, including low drug bioavailability, poor penetration to infection sites, antibiotic side effects, and resistance, are significant challenges that nanosystems aim to overcome (Chakraborty et al. 2022).

2.4.3 Synergistic Effects of Combination Therapy

Nanosystems can be utilized in combination with antibiotics to produce a synergistic impact in combination therapy. For instance, combining several antibiotics or other antimicrobial compounds into nanoparticles can have a synergistic effect against bacteria. In comparison to taking antibiotics alone, the combined action of many antibiotics can damage bacterial cell walls, prevent protein synthesis, and interfere with bacterial metabolism. This results in increased antibacterial activity (Adeniji et al. 2022).

2.4.4 Targeted Drug Delivery

Antibiotics can be delivered to the site of infection using nanosystems that are specifically engineered to target bacteria. This can prevent harm to healthy cells and lessen the systemic toxicity of antibiotics. For instance, functionalizing nanoparticles with ligands that only attach to the surfaces of bacterial cells enables the targeted delivery of antibiotics to bacteria. This method of tailored medication delivery can lessen the negative effects of antibiotics while increasing their efficacy (Abo-Shama et al. 2020).

Bacterial biofilms are intricate structures that can shield germs from the effects of antibiotics. Biofilms can be broken up by nanosystems, which makes bacteria more sensitive to antibiotics. For instance, antibiotics can enter biofilms and kill bacteria thereby disrupting the extracellular matrix caused by nanoparticle penetration.

2.5 Novel Approaches for Combatting Antimicrobial Resistance with Nanosystems

The details in the studies bring down to the novel approaches that are utilized for combating antibacterial resistance.

2.5.1 Lipid Polymer Nanoparticles (LPNs)

These are a particular class of nanocarriers that can encapsulate medications or other therapeutic agents, such as antimicrobial peptides, to increase their efficacy and lessen their toxicity. LPNs have drawn interest as a viable remedy for the escalating issue of antibiotic resistance in recent years. In comparison to traditional antibiotics, LPNs have better absorption, less toxicity, and more selectivity. Urinary tract infections, biliary tract infections, and endocarditis have all been linked to *Enterococcus faecalis* (*E. faecalis*) biofilms (Wan et al. 2021). Due to its opportunistic nature and ability to internalize into host cells, *E. faecalis* offers significant difficulties for standard antibiotic therapy. Ampicillin's (Amp) effectiveness against challenging *E. faecalis* infections is significantly diminished by its incapacity to eliminate bacteria concealed in biofilms and intracellular niches (Mubeen et al. 2021). Amp was loaded into Lipid-Polymer Hybrid Nanoparticles (LPNs), a highly effective nano delivery platform made up of a special mix of DOTAP lipid shell and PLGA polymeric core, to increase the effectiveness of Amp against various kinds of *E. faecalis* infections. These nanoparticles (Amp-LPNs) were studied for their antibacterial efficacy in a protozoa infection model, resulting in a significantly increased multiplicity of infections (Muteeb 2023).

2.5.2 Phage Therapy

Using bacteriophages, which are viruses that only infect and kill bacteria, is known as phage therapy. Phages can be used to target particular bacterial strains that have developed antibiotic resistance since they are very specific to the type of bacteria they infect. Phages are the most prevalent creatures in our environment and are found everywhere in nature. They play a critical role in the creation and regulation of microbial communities. There has been a resurgence in research into the use of bacteriophages in the treatment of bacterial illnesses due to the advent and spread of global antibiotic resistance and the need to find safer alternatives. However, high specificity and poor pharmacokinetic qualities are the main limitations that prevent the use of phage therapy.

By creating a phage cocktail loaded with liposomes rather than employing monophage therapy to treat *K. pneumoniae*, researches have shown that the use of nano drug delivery systems can increase the potential of phage therapy. The study's results showed that liposomal-entrapped phage cocktail treatment significantly reduced the number of bacteria in the main organs and blood of infected mice when compared to nonliposomal-free phage cocktail treatment (Muteeb 2023).

2.5.3 Nanoantibiotic

The transformation of therapeutic substances into nanosized assemblies is possible with the help of the promising nanoantibiotic technique, which is regarded as a carrier-free drug delivery method. The ability of this strategy to alter the physical characteristics of antibiotics, boost their rate of dissolution, increase drug bioavailability, lessen side effects, improve contact with microorganisms, improve interaction with and penetration within bacterial membranes, and thus perform better against antibiotic-resistant strains, makes it very interesting (Mubeen et al. 2021).

2.6 Challenges of Nanomedicine in Combatting Antimicrobial Resistance

Despite all the benefits that nanomedicines have in the fight against AMR, the clinical translation of nanobased formulations is never without its difficulties (Table 2.3). The design of the nanoformulation and the best method of administration should be the top priorities. It is necessary to change the emphasis placed on removing various obstacles caused by production, costing, and quality control tests. Preclinical evaluation is another important activity that needs to be considered, including toxicity detection and in vivo estimation. The establishment and development of advanced toxicity assessments for nanomedicines are necessary. It takes a long time and money to get an idea to market, hence attempts should be made to speed up this process. For sensitive, verified, and standardizable examinations containing techniques to correctly evaluate the nanotoxicology of nanoparticulates during the early stages of clinical development, clear regulatory criteria tailored for nanoparticulates should be defined. The development of regulatory protocols, methodologies, and tools to ensure uniform good manufacturing practice (GMP) production and

Table 2.3 Disadvantages and advantages of nano antimicrobials

Advantages	Disadvantages
Targeted drug delivery via specific accumulation	Nanomaterial accumulation in tissues and organs following intravenous injection
Fewer side effects of chemical antimicrobials	High systemic exposure to medicines delivered locally in the right quantities for desired therapeutic use
Controlled drug release and broad therapeutic index	Nanotoxicity
Can cross tissue barriers, low immunosuppression, and improved solubility	Lack of methods for characterization that are unaffected by the characteristics of nanoparticles (NPs)

characterization, safety, and economical clinical trial design is challenging since nanoparticulates represent different kinds of nanostructures. To make the process from creation to commercialization simpler and to cut down on the time and money spent on negotiating licensing and partnership agreements, nanoparticulates' patents and intellectual properties (IP) must be made simpler.

2.7 Conclusion and Future Perspectives

Over the next few years, there will likely be a major increase in the need for nanomedicine. The global market for nanomedicine, according to a study by Grand View Research, was estimated to be worth USD 111.4 billion in 2020 and is projected to increase at a CAGR of 13.3% from 2021 to 2028 (Grand View Research 2023). One of the main factors propelling market expansion is the rising incidence of chronic diseases like cancer, neurological disorders, and cardiovascular diseases. The report also identifies the growing utilization of nanotechnology as a major market growth driver in medication delivery systems, diagnostic imaging, and illness monitoring. AMR poses a serious risk to world health and has the potential to reduce the efficacy of contemporary therapy. Antibiotic misuse and overuse have contributed to the emergence of resistant bacterial strains, making it challenging to treat diseases that were once easily treatable. AMR can be prevented with the help of nanomedicine, and many methods are being investigated.

References

Abo-Shama UH, El-Gendy H, Mousa WS, Hamouda RA, Yousuf WE, Hetta HF, Abdeen EE (2020) Synergistic and antagonistic effects of metal nanoparticles in combination with antibiotics against some reference strains of pathogenic microorganisms. Inf Drug Resistance 13:351–362. https://doi.org/10.2147/IDR.S234425

Adeniji OO, Nontongana N, Okoh JC, Okoh AI (2022) The potential of antibiotics and nanomaterial combinations as therapeutic strategies in the management of multidrug-resistant infections: a review. Int J Mol Sci 23(23):15038. https://doi.org/10.3390/ijms232315038

Author (2021) Let's talk about lipid nanoparticles. Nat Rev Mater 6:99. https://doi.org/10.1038/s41578-021-00281-4

Balderrama-González A-S, Piñón-Castillo H-A, Ramírez-Valdespino C-A, Landeros-Martínez L-L, Orrantia-Borunda E, Esparza-Ponce H-E (2021) Antimicrobial resistance and inorganic nanoparticles. Int J Mol Sci 22(23):12890. https://doi.org/10.3390/ijms222312890

Baptista PV et al (2018) Nano-strategies to fight multidrug resistant bacteria—"a battle of the titans". Front Microbiol 9:1441

Blume OR, Yost SE, Kaplan B (2012) Antibody-mediated rejection: pathogenesis, prevention, treatment, and outcomes. J Transpl 2012:201754. https://doi.org/10.1155/2012/201754

Chakraborty N, Jha D, Roy I et al (2022) Nanobiotics against antimicrobial resistance: harnessing the power of nanoscale materials and technologies. J Nanobiotechnol 20:375. https://doi.org/10.1186/s12951-022-01573-9

Collaborators (2022) Global burden of bacterial antimicrobial resistance in 2019: a systematic analysis. Lancet 399(10325):629–655

Eleraky NE, Hassan SB, Omar MM (2020) Nanomedicine fight against antibacterial resistance: an overview of the recent pharmaceutical innovations. Pharmaceutics 12(2):142

Grand View Research (2023) Nanomedicine market size, share & trends analysis report by application (drug delivery), by indication (clinical oncology, infectious diseases), by molecule type, by region, and segment forecasts, 2023-2030. https://www.grandviewresearch.com/industry-analysis/nanomedicine-market

Gupta A et al (2017) Synergistic antimicrobial therapy using nanoparticles and antibiotics for the treatment of multidrug-resistant bacterial infection. Nano Futures 1(1):015004

Hetta HF et al (2023) Nanotechnology as a promising approach to combat multidrug resistant bacteria: a comprehensive review and future perspectives. Biomedicine 11(2):413

Horikoshi S, Serpone N (2013) Microwaves in nanoparticle synthesis: fundamentals and applications. Wiley, Hoboken

Lee N-Y, Ko W-C, Hsueh P-R (2019) Nanoparticles in the treatment of infections caused by multidrug-resistant organisms. Front Pharmacol 10:1153

Mubeen B, Ansar AN, Rasool R, Ullah I, Imam SS, Alshehri S, Ghoneim MM, Alzarea SI, Nadeem MS, Kazmi I (2021) Nanotechnology as a novel approach in combating microbes providing an alternative to antibiotics. Antibiotics 10(12):1473. https://doi.org/10.3390/antibiotics10121473

Muteeb G (2023) Nanotechnology-A light of hope for combating antibiotic resistance. Microorganisms 11(6):1489. https://doi.org/10.3390/microorganisms11061489

Pelgrift RY, Friedman AJ (2013) Nanotechnology as a therapeutic tool to combat microbial resistance. Adv Drug Deliv Rev 65:1803–1815

Power of Particles (2023) Nanoantibiotics to fight antimicrobial resistance: teaching an old drug new tricks. Available from: https://www.powerofparticles.com/using-nanoantibiotics-to-combat-antimicrobial-resistance

Rancan F, Jurisch J, Hadam S, Vogt A, Blume-Peytavi U, Bayer IS, Contardi M, Schaudinn C (2023) Ciprofloxacin-loaded polyvinylpyrrolidone foils for the topical treatment of wound infections with methicillin-resistant Staphylococcus aureus (MRSA). Pharmaceutics 15(7):1876. https://doi.org/10.3390/pharmaceutics15071876. PMID: 37514062; PMCID: PMC10385417.

Solleti VS, Alhariri M, Halwani M, Omri A (2015) Antimicrobial properties of liposomal azithromycin for *Pseudomonas* infections in cystic fibrosis patients. J Antimicrob Chemother 70(3):784–796. https://doi.org/10.1093/jac/dku452

Tan CH et al (2021) Lipid-polymer hybrid nanoparticles enhance the potency of ampicillin against Enterococcus faecalis in a protozoa infection model. ACS Inf Dis 7(6):1607–1618

UNEP (2023) Antimicrobial resisitance: a global threat. Available from: https://www.unep.org/explore-topics/chemicals-waste/what-we-do/emerging-issues/antimicrobial-resistance-global-threat

Wan F, Draz MS, Gu M, Yu W, Ruan Z, Luo Q (2021) Novel strategy to combat antibiotic resistance: a sight into the combination of CRISPR/Cas9 and nanoparticles. Pharmaceutics 13(3):352. https://doi.org/10.3390/pharmaceutics13030352

Xie M et al (2023) Antibacterial nanomaterials: mechanisms, impacts on antimicrobial resistance and design principles. Angew Chem Int Ed 62(17):e202217345

Zazo H, Colino CI, Lanao JM (2016) Current applications of nanoparticles in infectious diseases. J Control Release 224:86–102

Nanotechnology-Based Tools to Overcome Antimicrobial Resistance

3

Karishma Mahajan, Rita Chandel, Poonam Sharma, and Vikrant Abbot

Abstract

The World Health Organisation (WHO) has called for the rapid development of effective technology to tackle antimicrobial resistance (AMR). Poor diagnosis and overuse of medications induce antimicrobial resistance (AMR). This chapter is focused on therapeutic modalities that have been formulated and suggested to address this particular issue. Recent studies have shown the efficacy of nanotechnology in addressing the issue of antimicrobial resistance. The present chapter provides an overview of recent advancements in nanotechnology that exhibit potential for combating resistant pathogens. The feasibility and potential of these technologies for novel antimicrobial therapies are also evaluated. In this chapter, various nanotechnology approaches that are being rapidly developed for the management of antimicrobial resistance (AMR) have been discussed. Furthermore, the impact of inorganic carriers, specifically metal nanoparticles such as copper, zinc, and silver on biofilm populations, has been examined. This chapter provides an extensive overview of the future prospects of nanotechnology for overcoming antimicrobial resistance.

K. Mahajan · P. Sharma (✉)
Department of Biotechnology and Bioinformatics, Jaypee University of Information Technology, Waknaghat, Solan, Himachal Pradesh, India
e-mail: poonam.sharma@juit.ac.in

R. Chandel
Department of Chemistry, Government Degree College, Sanjauli, Shimla, Himachal Pradesh, India

V. Abbot
Department of Pharmaceutical Sciences, Chandigarh Group of Colleges, Jhanjeri, Mohali, Punjab, India

M. Y. Wani et al. (eds.), *Nanotechnology Based Strategies for Combating Antimicrobial Resistance*, https://doi.org/10.1007/978-981-97-2023-1_3

Keywords

Antimicrobial resistance · Nanoparticles · Biofilms · Nanosphere · Nanotechnology

3.1 Introduction

Antimicrobials are one of the most successful forms of chemotherapy and have helped in controlling the infectious diseases that endanger human civilization (Aminov 2010). Agents that kill bacteria or prevent their growth are known as antimicrobials, which come from the Greek words anti (against), mikros (small), and bios (life). Bacterial antibiotics are naturally occurring chemicals with antimicrobial properties. On the other hand, an antimicrobial is a chemical that kills or suppresses microbial development without harming the host in any significant way. Antimicrobials can be broken down into subcategories based on the types of microorganisms they are effective against. It is well known that antibiotics kill bacteria, antivirals kill viruses, antifungals kill fungi, and antiprotozoals kill protozoa (Das and Patra 2017). Microbicidal antimicrobials are distinguished from biostatic antimicrobials by their ability to kill bacteria. Antibiotics have been used by humans since long before the so-called "antibiotic era" of the twentieth century. An ancient Indian medical practice called Ayurveda dates back some 5000 years and emphasizes the need to eliminate disease-causing germs. Suvarna Bhasma, also known as gold calyx, was used to treat tuberculosis, whereas Mahasudarshan Kwath was used to treat malaria, and Triphala Guggulu was used for surgical prophylaxis (Vadhana et al. 2015). Establishing the critical role of germs in infectious illness causation, the germ theory of disease paved the way for the "modern antibiotic era." When Alexander Fleming isolated penicillin from *Penicillium rubens* in 1928, it was a huge breakthrough in the field of antibiotics. Many potentially fatal infectious diseases can now be treated successfully with penicillin, thanks to its clinical applications. Meningitis, infant bed fever, and pneumonia were all effectively treated using the first synthetic antibacterial sulphonamide created by Gerhard Domagk in 1935. Antibiotics like these serve as a model for future antibiotic discovery and applications in clinical settings. The emergence of resistance in microorganisms is a major threat to the antimicrobial's clinical utility. Penicillinase, an enzyme found in bacteria, was discovered by Alexander Fleming's team long before antibiotics were used clinically. Due to the rise of penicillin-resistant bacteria, especially among *Staphylococci*, most nations have made the drug available only "by prescription." Methicillin, a semisynthetic penicillin variant that is resistant to the enzyme penicillinase, was created as a countermeasure. However, methicillin-resistant *Staphylococcus aureus* (MRSA) emerged after only a few years of clinical use and is now a major challenge for antimicrobial therapy everywhere. Antibiotic resistance has emerged predictably, and this trend holds true for all antibiotics that entered widespread use after the middle of the twentieth century. The EU and the United States have recorded significant mortality rates due to multidrug-resistant bacterial infections at 25,000 and 63,000 individuals annually, respectively. More than 20,000 antibiotic-resistant genes of 400 types have been predicted from available genome sequences, leading scientists to warn that the world will return to a pre-antibiotic era plagued by life-threatening microbial infections (Liu and Pop 2009). One of the primary medical issues of the twenty-first century is therefore the identification of innovative antimicrobial medicines to which microorganisms cannot quickly build resistance. In order for antimicrobial therapy to be successful, the development of efficient drug delivery techniques is necessary, in addition to the discovery of new antimicrobials. Local and systemic adverse effects of antimicrobials, as well as a lower therapeutic index, would result from inefficient drug distribution. Despite the drawbacks of traditional antibiotic therapy, the current therapeutic application of nanotechnology has the potential to completely transform the field. There is less likelihood of nanoparticles (NPs) causing microbial resistance than other potential antibacterial agents (Huh and Kwon 2011). Loading medicines on NP-based carriers can increase the therapeutic index of antimicrobials compared to their free drug counterparts. By encasing antimicrobial drugs in NPs, their serum solubility can be improved, circulation time in the body can be increased, drugs can be kept in specific tissues, and can be used in combination therapies where multiple drugs are delivered to the same cell (Zhang et al. 2010). Given the gravity of the problems in antimicrobial therapy, it is important to examine recent scientific developments in our knowledge of the mechanisms of microbial antibiotic resistance, the counter strategies against antibiotic-resistant pathogenic strains, and antimicrobial drug delivery. The development of antibiotic-resistant organisms, methods to combat this problem, and the administration of antimicrobials via nanostructured biomaterials have been elaborately discussed in this chapter.

3.2 Classifications of Nanoparticles

Nanoparticles may be broadly classified into two distinct categories: organic and inorganic (Fig. 3.1). The first category includes polymeric micelles, liposomes, solid lipid nanoparticles, nanostructured lipid carriers, and compact polymeric nanoparticles. The second category comprises magnetic nanoparticles, mesoporous silica nanoparticles, and metal nanoparticles. Nanoparticles may be classified using a variety of criteria, including their form, size, and chemical characteristics. Nanoparticles may be classified into many significant categories based on their distinctive physical and chemical properties. These categories include carbon-based, lipid, polymer, ceramic, semiconductor, metal, and others (Lisziewicz et al. 2012).

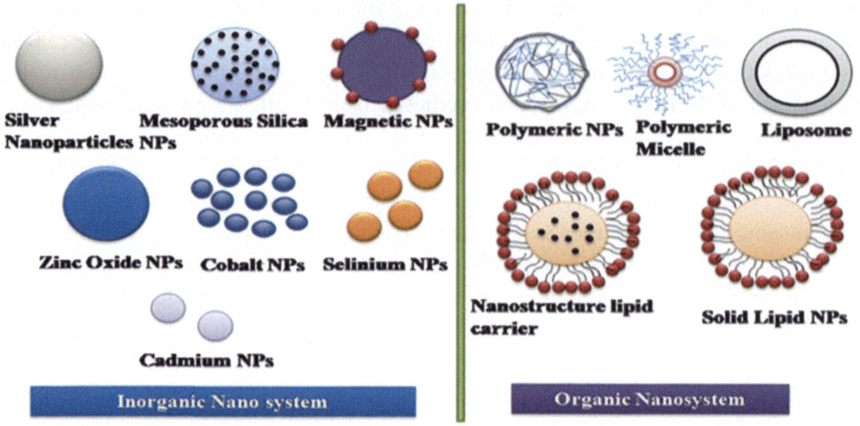

Fig. 3.1 Classification of nanosystems

3.3 Mechanical Properties of Nanomaterials

The term "mechanical properties of nanomaterials" refers to the characteristics that nanoparticles exhibit when subjected to varied environments and a wide variety of external stresses. Extensive scholarly research has been conducted on the mechanical characteristics of nanomaterials. Nevertheless, the primary emphasis of these investigations centres on enhancing the mechanical characteristics of nanomaterials through the incorporation of nanoparticles into the matrix, which often does not possess nanoscale features. Limited studies have been conducted on the mechanical characteristics of pure nanomaterials. Additional information about the mechanical characteristics of metal nanoparticles has been acquired, including Vickers hardness, fracture toughness, fracture strength, ultimate tensile strength, and impact strength (Wu et al. 2020).

3.4 Current Antimicrobial Resistance Threats

The use of antibiotics for the treatment of microbial infections in both humans and animals poses a significant risk for the dissemination of antibiotic resistance (Austin et al. 1999). The inappropriate use of antibiotics has resulted in the transformation of pathogenic microorganisms that cause epidemics into forms that are resistant to multiple drugs (Davies 1996). Despite the significant worldwide risks associated with antimicrobial resistance, it has received insufficient attention in both developing and developed countries. It is imperative for every nation to devise a customized approach to combating antimicrobial resistance that is suited to its unique circumstances. Antimicrobial resistance is a more severe issue in developing countries due to the widespread availability and misuse of antibiotics in excessive dosages, as well as financial limitations that impede the replacement of older antibiotics with newer, more costly alternatives. This phenomenon is likely to result in the emergence of antimicrobial-resistant strains (Ganesh Kumar et al. 2013). New Delhi metallo-beta-lactamase-1 (NDM-1) is a commonly recognized type of antibiotic resistance that is generated by the gene blaNDM-1. This gene is carried on a plasmid and has the potential to be transmitted between bacterial species such as *Escherichia coli* and *Klebsiella pneumoniae*. As a consequence, bacterial strains have emerged exhibiting extensive resistance to antibiotics, including carbapenems. *Mycobacterium tuberculosis*, a pathogen with a long history, has re-emerged in the twentieth century as a multidrug-resistant strain. This variant has affected a significant proportion of the global population, including both developed and developing nations. The use of nanostructures for antimicrobial therapy is being explored as a potential solution. The efficacy of antituberculosis medications has been impaired due to the fast appearance and dissemination of strains that are resistant to four or more frontline therapies for the infection. These strains, known as highly drug-resistant and entirely drug-resistant strains, provide a significant challenge in the management and control of the disease (Davies 1996). The microorganisms responsible for hospital-acquired infections (HAIs) have emerged as a global concern. Methicillin-resistant *Staphylococcus aureus* (MRSA) is widely recognized as the most infamous multidrug-resistant pathogen responsible for nosocomial infections. MRSA has transitioned from being a causative agent of healthcare-associated infections (HAIs) to a significant pathogen in the community. This is attributed to the acquisition of pathogenic genes that encode cytotoxic Pantone-Valentine leukocidin, which enhances its virulence (Davies 1996). *Pseudomonas aeruginosa* was initially responsible for burn wound infections and subsequently evolved into a significant nosocomial infection. *Pseudomonas aeruginosa*, a bacterial species known for its production of metallo-beta-lactamase, has shown resistance to carbapenem antibiotics. The rise of multidrug-resistant Enterobacteriaceae strains that produce specifically ESBL-producing *K. pneumoniae*, extended spectrum β-lactamases (ESBLs), and *Salmonella enterica* strains resistant to ciprofloxacin, as well as vancomycin-intermediate Staphylococci, fluoroquinolone-resistant Salmonella, and *Acinetobacter baumannii*, *P. aeruginosa* strains resistant to ciprofloxacin, cefepime, and ceftazidime pose substantial challenges in the field of antimicrobial therapy (Ganesh Kumar et al. 2013). The genome analysis of *A. baumanni* has revealed the existence of a minimum of 28 genomic islands that carry genes responsible for antibiotic resistance. This finding has led to significant apprehension due to the swift emergence of antibiotic-resistant strains of *A. baumanni*. The extensive use of antibiotics such as fluoroquinolones, newer penicillins, and expanded-spectrum cephalosporins, has led to the emergence of a hyper-virulent and toxic HAI strain of clostridium. The avian influenza outbreak that occurred globally was caused by the influenza A (H1N1) virus. It was discovered that this virus was vulnerable to neuraminidase inhibitors, specifically oseltamivir and zanamivir. In June 2009, reports emerged regarding resistance to oseltamivir, and subsequently, 570 cases of oseltamivir-resistant infections have been documented globally (Potdar et al. 2013). Due to the urgency of the issue of antibiotic resistance, it is imperative to increase awareness regarding this matter and devise effective approaches to address it.

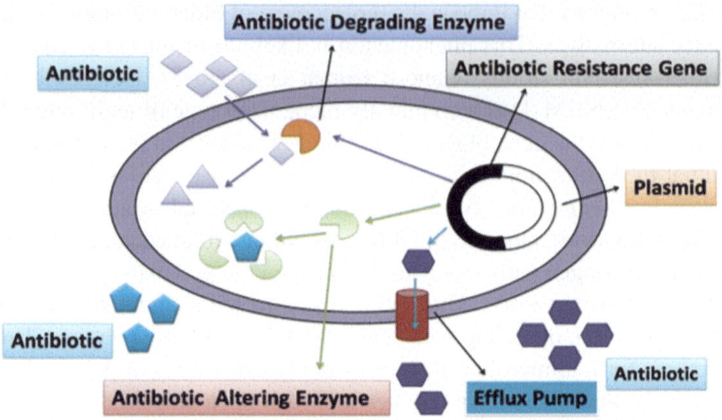

Fig. 3.2 Mechanism of antimicrobial resistance

3.5 Microbial Resistance Mechanism to Antimicrobials

When antimicrobial therapies began to show widespread success in the late 1960s, US Surgeon General William H. Stewart famously declared, "It is time to close the book on infectious diseases and declare the war against pestilence won." This once-held belief among the medical profession has since been disproven. Microbial infections are the second-greatest cause of death worldwide, and antibiotic resistance has contributed to this crisis (Spellberg et al. 2008). Therefore, it is of paramount biochemical importance to determine the mechanism of biochemical resistance. Biochemical and genetic features of the strain contribute to the microorganisms' ability to withstand antimicrobials. The mechanisms through which microorganisms develop resistance to antibiotics vary widely depending on factors such as the kind of microbe, drug, target site, and whether or not the resistance is transmitted by a plasmid or a chromosomal mutation (FTB n.d.). While the genetic and biochemical versatility of bacteria is a boon to medicine, antibiotics are becoming less effective because resistance mechanisms are so pervasive (Davies and Davies 2010). The mechanism of antimicrobial resistance is shown in Fig. 3.2.

3.6 Role of Biofilms in Antimicrobial Resistance

Due to the inherent antibiotic resistance that pathogenic microbial biofilm provides through its lifestyle, it is considered a global threat. When microorganisms form communities in a clinical setting, they can cause life-threatening infections. Antibiotics at large doses for extended periods of time are generally ineffective against this cellular organization, which contributes to the persistence of the infection. When biofilms form on medical equipment, they not only hinder treatment but also pose a threat of infection. Researchers from all over the world have responded to the issue posed by biofilms by proposing or developing solutions to biofilm control. The newly developed drugs have the potential to be implemented in clinical settings to prevent or eliminate pathogenic biofilms (Ribeiro et al. 2016). Despite advances in biofilm research, figuring out how to treat biofilm infections with antibiotics is still a mystery. The history of an antimicrobial agent's pharmacokinetic (PK) and pharmacodynamic (PD) characteristics is crucial for developing an effective dosage schedule and preventing the emergence of antimicrobial tolerance and resistance in biofilm infections. Because bacterial biofilms differ from plankton-grown cells in growth rate, gene expression, and metabolism, it is difficult to extrapolate the results from PK/PD studies of antibiotics performed on planktonic cells to biofilms (Donelli 2014). Multiple procedures have been established for in vitro and in vivo studies of antibiotic pharmacokinetics and pharmacodynamics in *Pseudomonas aeruginosa* biofilm infections. It is important to stress that no biofilm procedures have been approved for clinical usage or shown to be helpful in directing antibiotic therapy (Frieri et al. 2017).

3.6.1 Biofilm Formation

In most cases, there are five primary steps involved in the production of biofilm by bacterial pathogens on any substratum or layer. During the first stage of development, free-swimming planktonic cells temporarily and reversibly attach themselves to biotic or abiotic surfaces through weak interactions such as acid–base, hydrophobic, Van der Waals, and electrostatic forces. Secondly, bacterial pathogens permanently attach to the surface using stronger connections such as collagen-binding adhesive proteins, lipopolysaccharides, flagella, and pili. This process is known as colonization. The proliferation stage is characterized by the substantial accumulation of multilayered bacterial cells as well as the production of huge quantities of EPS. This is followed by the maturation stage in which the attached multilayered bacterial cells develop into a biofilm that has the typical three-dimensional structure of a biofilm. Finally, the dispersion step occurs after the full growth of the biofilm and involves the use of both mechanical and active methods to break up the biofilm into smaller pieces (Fig. 3.3) (Koválóvá et al. 2014; Srinivasan et al. 2021; Yin et al. 2019).

3.6.2 Characteristics of Biofilm Formation

The process of biofilm formation is characterized by a gradual progression. Bacterial cells exhibit an initial migration onto a surface, followed by a reversible adhesion to said surface. The irreversible adhesion resulting from the expansion of micro-colonies during the second step induces the formation of an extracellular polymeric substance (EPS) matrix. The manifestation of the fully formed three-dimensional biofilm structure subsequently becomes evident. Mature biofilms have been observed to exhibit reduced susceptibility to antibacterial drugs and the immune system's defences. During the process of biofilm dispersal, the cells are able to

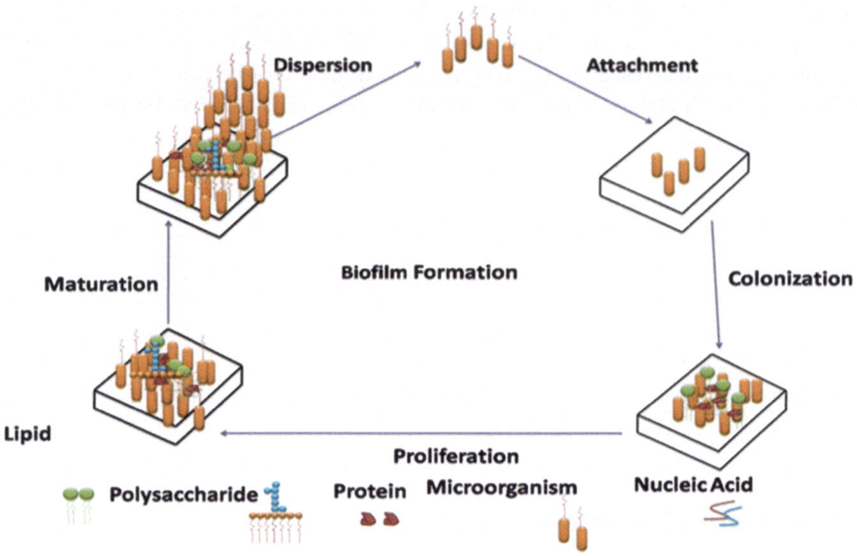

Fig. 3.3 Mechanism of biofilm formation

withstand lysis and are subsequently discharged from the biofilm community. Bacterial organisms have the ability to generate a biofilm on either a biotic or abiotic substrate within the host organism. Typically, abiotic surfaces undergo habituation coating by proteins or other biological molecules, which subsequently alter cell adherence. The involvement of host cells in the production of biofilms is significant, as their constituent components can be assimilated into the matrix of the biofilm (Lynch and Robertson 2008; Römling and Balsalobre 2012).

3.7 Nanotechnology for Antimicrobial Resistance

Antibiotic resistance occurs when bacteria develop the ability to evade the medications designed to kill them. As a result of its rapid expansion, it is becoming a major problem for public health around the world. At the very least, it was directly or indirectly responsible for 1.27 million deaths in 2019. The United States experiences about three million cases of illnesses resistant to antibiotics each year, and the cost of treating the six most prevalent cases is more than $4.6 billion. The annual cost to the EU is estimated at €1.5 billion, with around 700,000 new cases. Culture methods that expose bacteria to antibiotics are used in antibiotic sensitivity testing (AST), whereas genetic studies can reveal whether or not bacteria carry genes that confer resistance. In a medical context, a 24-hour AST or longer is not uncommon, and it can literally mean the difference between life and death for bacteria that develop slowly. Some more rapid ASTs have been created in recent years, although they are typically complicated and expensive (Villalba et al. 2023). In order to address these concerns, it is essential to develop novel antimicrobial compounds that exhibit both non-resistance and minimal adverse effects. However, the process of synthesizing novel antimicrobial medicines remains a challenging endeavour. In light of this difficulty, there is a pressing need for the creation of an innovative and effective medication delivery system. This system should aim to enhance the therapeutic index of existing antimicrobial drugs, while concurrently minimizing local and systemic adverse effects. Additionally, it is essential that this delivery method does not contribute to the emergence of antimicrobial resistance among microorganisms. In the present day, the use of nanosized vehicles (NVs) to transport antimicrobial medications, known as nanomedicine, has promise in addressing the difficulties often encountered with traditional therapeutic approaches and delivery methods. Nanosized vehicles (NVs) of varied dimensions are used in the field of nanomedicine for the purpose of encapsulating and transporting antimicrobial medicines. The novel nanovectors (NVs) possess distinctive characteristics, such as a decreased likelihood of resistance development by microorganisms, notably improved therapeutic effectiveness, higher solubility in serum compared to free drugs, extended systemic drug circulation durations, prolonged therapeutic impacts, diminished adverse effects on healthy tissues or cells, and the ability to facilitate combination therapy by delivering multiple drugs to a specific cell site (Zhang et al. 2010). Nanovesicles (NVs) are small particles or vesicles with dimensions ranging from 1 to 100 nm. NVs possess distinct and sophisticated physicochemical characteristics that differentiate them from bulk materials, individual atoms, or molecules. These properties primarily stem from quantum effects and result in a significantly increased surface area to volume ratio. In the year 2000, the National Institutes of Health in the United States initiated the National Nanotechnology Initiative. This programme aims to provide support, coordination, and advancement for research and development endeavours focused on nanoscale projects. Nanotechnology involves more than mere reduction in size; rather, it entails the exploitation of distinct physical, chemical, mechanical, and optical characteristics inherent to materials at the nanoscale (Weissig and Guzman-Villanueva 2015). There are two approaches for producing the NVs (a) bottom-up approaches and (b) top-down approaches (Wang and Xia 2004). Each approach has several techniques for synthesizing NVs. The selection of a technique for synthesizing NVs depends on the nature of the material, the type of NVs, and its application in different fields. Currently, nanotechnology is applied in almost all fields including medicine, energy, electronics, chemistry, physics, materials science, and defense.

3.7.1 Nanomaterials Against Bacteria

Nanomaterials possess distinct physical and chemical properties that are significantly divergent from those of bulk materials, owing to their nanometre scale range (1–100 nm) in at least one dimension. Nanoparticles constitute a fundamental component of nanomaterials. Numerous types of nanoparticles are used for antimicrobial resistance (AMR) purposes. Nanoparticles (NPs) have been found to exhibit antibacterial properties through various mechanisms, including direct interaction with the bacterial cell wall, inhibition of biofilm formation, stimulation of innate and adaptive host immune responses, production of reactive oxygen species (ROS), and induction of intracellular effects such as interactions with DNA and/or proteins (Baptista et al. 2018). Different types of nanoparticles, their therapeutic uses, and mechanism of action are shown in Table 3.1 (Das and Patra 2017).

Table 3.1 Types of nanoparticles, therapeutic uses, and mechanism of action

Nanoparticle type	Therapeutic uses	Mechanism of action
Silver nanoparticles	Highly efficient against bacteria that are resistant to many drugs, such as methicillin-resistant coagulase-negative staphylococci (e.g. *Staphylococcus epidermidis*), methicillin-resistant *Staphylococcus aureus* (MRSA), ESBL-positive *Klebsiella pneumoniae,* and vancomycin-resistant *Enterococcus faecium* (Nanda and Saravanan 2009)	Increased permeability of membranes, suppression of DNA replication, stimulation of free radical production, and suppression of signal transmission (Matsumura et al. 2003)
Zinc oxide	Effective against MRSA, *Salmonella typhimurium*, *K. pneumoniae*, Listeria monocytogenes, *Salmonella enteritidis*, Lactobacillus, Streptococcus mutants, and methicillin-resistant *S. epidermidis* (MRSE) (Rai and Bai 2011)	The breakdown of bacterial cell membrane lipids and proteins and the generation of reactive oxygen species (ROS) (Jin et al. 2009)
Titanium dioxide	Antibacterial activity against *Bacteroides fragilis, Pseudomonas aeruginosa*, *Enterococcus hirae*, *S. typhimurium E. coli*, *S. aureus*, and *Lactobacillus acidophilus*. Generation of hydroxyl radicals leads to an initial oxidative attack on the microbial cell membrane (Hajipour et al. 2012)	The deterioration of the microbial membrane restricts essential biological processes such as oxidative phosphorylation, respiration, and semi-permeability (Maness et al. 1999)
Gold nanoparticles	This treatment demonstrates efficacy in targeting and inhibiting the growth of cancerous cells. The action of vancomycin-coated gold nanoparticles (Au NPs) is increased against vancomycin-resistant enterococci. In contrast, ceflacor-reduced Au NPs exhibit a wide specificity, unlike ceflacor or Au NPs used individually. The antibacterial activity of chitosan-capped Au NPs combined with free ampicillin exhibits a twofold increase (Rai et al. 2010)	The cell membrane's bilayer is subject to strong electrostatic attractions from negatively charged entities, which in turn stimulate the creation of reactive oxygen species (ROS) (Beyth et al. 2015)
Copper nanoparticles	Nanocomposites have been used as a promising strategy to enhance the efficacy of antibiotics, particularly in combating hospital-acquired conditions (Ruparelia et al. 2008)	The production of reactive oxygen species (ROS) may induce deleterious consequences by disrupting the biosynthesis of amino acids and DNA (Esteban-Tejeda et al. 2009)

(continued)

Table 3.1 (continued)

Nanoparticle type	Therapeutic uses	Mechanism of action
Chitosan	The use of chitosan conjugated with sulphamethoxazole has been shown to be effective in the treatment of drug-resistant strains of *Pseudomonas aeruginosa* (Tin et al. 2009)	The interaction between chitosan and the negatively charged surface of bacterial cells leads to an elevation in the permeability of the microbial cell wall. This binding process also results in the chelation of trace metals, which subsequently hinders the activities of enzymes. Additionally, the development of microorganisms is affected by this phenomenon (Qi et al. 2004)
Nitric oxide-releasing nanoparticles	In this study, antibiotic-resistant clinical isolates, including *Escherichia coli*, *Streptococcus pyogenes*, *Klebsiella pneumoniae*, *Enterococcus faecalis*, and *Pseudomonas aeruginosa* were evaluated (Hajipour et al. 2012)	The alteration of important bacterial proteins is a consequence of changes in bacterial membrane structure and the production of reactive nitrogen species (Hajipour et al. 2012)
Carbon nanotubes (CNTs)	Enhanced antibacterial efficacy against drug-resistant strains of both Gram-negative bacteria and Gram-positive (Yang et al. 2010)	The aggregation of cells with carbon nanotubes (CNTs) results in detrimental effects on the bacterial cell wall, leading to the subsequent release of their DNA content (Yang et al. 2010)

3.7.2 Nanomedicines for Antimicrobial Resistance

New antimicrobial medication compounds that do not cause resistance or negative effects are needed to fix these problems. However, developing novel antimicrobial medications is not a simple process, because of this difficulty, a novel and effective drug delivery system is required that can improve the therapeutic index of commonly used antimicrobial drugs without causing resistance in microorganisms and with minimal adverse effects on the human body. These problems are typically connected with traditional therapy and delivery systems; however, they may be solved by the use of antimicrobial medicines placed in nanosized vehicles (nanomedicine). Nanomedicine makes use of nanosize vehicles (NVs) of varying sizes to transport antimicrobial medicines. There is less of a chance of drug resistance developing in microorganisms when using these novel NVs because they are more soluble in serum than free drugs, their therapeutic effects last longer, and they have less of an impact on healthy tissues and cells (Bhakay et al. 2018).

3.8 Recent Advancements in Combating Antimicrobial Resistance

In today's world, antimicrobial resistance is a serious problem; hence, finding a solution to this problem using any new technology is extremely vital. Nanotechnology is a relatively recent field of study that is currently being applied to the reduction of antibiotic resistance. Over the course of the past year, researchers have developed a variety of solutions to combat this issue. When it comes to delivering antibiotics and other antibacterial medications, chitosan is the natural polymer of choice as a nanocarrier (Spirescu et al. 2021). Qiu et al. created nanoparticles of gentamycin by phosphatidylcholine-chitosan hybrid coated. The lipid component was included because lipid medication carriers have the ability to merge with the phospholipid membrane of bacteria. The results showed that the synthesized system could prevent the growth and biofilm formation of both Gram-positive and Gram-negative bacteria (Qiu et al. 2020). Alruwaili et al. created hydrogel systems containing gentamycin for the purpose of ocular administration. The use of pH-sensitive carbopol polymer solutions in the formulation of gentamycin nanoparticles composed of chitosan has been shown to augment the antibacterial properties of chitosan nanoparticles containing gentamycin (Alruwaili et al. 2020). Evangelista et al. synthesized supramolecular polyelectrolyte complexes for antibacterial applications by exploiting interactions between the positively charged $-NH_3$ + groups of the cyclodextrin-grafted chitosan and the negatively charged $-SO_3$ groups of the carrageenan. Because of its ability to form host–guest inclusion complexes with silver sulphadiazine molecules, cyclodextrin was employed to deliver silver ions to kill bacteria (Evangelista et al. 2020). The study conducted by Walvekar et al. investigated the effectiveness of different levels of hyaluronic acid-oleylamine conjugation as nanocarriers for medication delivery in combating methicillin-resistant *Staphylococcus aureus*. Polymersomes, which are nanocapsules composed of hydrophilic polymers that are grafted with long fatty acids, have the ability to undergo self-assembly and form spherical drug carriers. In this study, these polymersomes were used for the encapsulation of vancomycin antibiotic (Walvekar et al. 2019). Oliveira et al. developed a bilayer biomembrane for wound management. The first layer of the biomembrane consisted of sodium alginate nanoparticles carrying the antibiotic polymyxin B sulphate, while the second layer included chitosan, hydroxypropyl methylcellulose, and lidocaine chloride, serving as an anaesthetic medication (Oliveira et al. 2020). Various nanoformulations used for antimicrobial resistance are shown in Table 3.2.

Table 3.2 List of several nanoformulations for antibacterial resistance

S. No.	Formulation	Drug	Types of nanosystems	Therapeutic uses	References
1.	Presenting vinomycine on nanoparticles to enhance antimicrobial activities	Vinomycine	Au nanoparticles	Increase in vitro antibacterial activity	Gu et al. (2003)
2.	One-step synthesis of positively charged gold nanoclusters as effective antimicrobial nanoagents against multidrug-resistant bacteria and biofilms	(11-mercaptoundecyl)-N,N,N-trimethylammonium bromide (MTAB)	Gold nanoclusters	Antibacterial activity against both Gram-negative bacteria and Gram-positive bacteria	Li et al. (2020)
3.	Chitosan nanoparticles efficiently enhance the dispersibility, stability, and selective antibacterial activity of insoluble isoflavonoids	Isoflavonoids	Chitosan nanoparticles	Increase dispersibility and antibacterial activity	Wu et al. (2023)
4.	Collagen hydrogel with multiple antimicrobial mechanisms as antibacterial wound dressing	Cypate-conjugated antimicrobial peptides	Liposomes, hydrogel	Chronic wounds, with necrosis, persistent inflammation	Wang et al. (2023)
6.	Inhibition of polymicrobial biofilms of *Candida albicans- Staphylococcus aureus/Streptococcus mutans* by Fucoidan-gold nanoparticles	Fucoidan	Gold nanoparticles	Inhibit the growth of bacterial and fungal pathogens	Tabassum et al. (2023)
7.	Inhibition of mixed biofilms of *Candida albicans* and *Staphylococcus aureus* by β-caryophyllene-gold nanoparticles	β-Caryophyllene	Gold nanoparticles	Decrease polymicrobial infections.	Khan et al. (2023)
8.	Preparation and characterization of chitosan-silver nanocomposite films and their antibacterial activity against *Staphylococcus aureus*	Chitosan	Silver nanocomposite films	Strong bactericidal action	Regiel et al. (2013)

(continued)

Table 3.2 (continued)

S. No.	Formulation	Drug	Types of nanosystems	Therapeutic uses	References
9.	Inhibitory activities of phloroglucinol-chitosan nanoparticles on mono- and dual-species biofilms of Candida albicans and bacteria	Phloroglucinol	Chitosan nanoparticles	For treating infections caused by biofilm-forming pathogens	Khan et al. (2022)
10.	Nanochitosan antimicrobial activity against *Streptococcus mutans* and *Candida albicans* dual-species biofilms	Chitosan	Nanoparticles	Broad antimicrobial spectrum	Ikono et al. (2019)
11.	Gold nanoclusters for targeting methicillin-resistant *Staphylococcus aureus* in vivo	Methicillin	Gold nanoclusters	Increase antibacterial activity	Xie et al. (2018)
12	Mercaptopyrimidine-conjugated gold nanoclusters as nanoantibiotics for combating multidrug-resistant superbugs	Mercaptopyrimidine	Gold nanoclusters		Zheng et al. (2018)
13.	Biological synthesis and characterization of silver nanoparticles with *Capparis zeylanica* L. leaf extract for potent antimicrobial and anti-proliferation efficiency	*Capparis zeylanica* leaves	Silver nanoparticles	Antimicrobial and antiproliferative potency	Nilavukkarasi et al. (2020)

3.9 Future Perspectives

Nanotechnology is incredibly flexible and can be applied to a wide variety of fields at the moment. Due to the potential for early diagnosis of AMR infections and, thus, successful patient treatment, the application of this field to the development of antimicrobial resistance is highly promising (Saxena et al. 2022). Antimicrobial resistance can be greatly lowered by using nanotechnology. The Food and Drug Administration (FDA) or European Medicines Agency (EMA) has given their stamp of approval to nanotechnologies for antimicrobial resistance. Advances in the clinical use of nanomedicines as antimicrobials' development of commercially viable, clinically proven antibacterial nanomedicines are still in its infancy. Multiple antibacterial medication formulations based on nanotechnology are currently in clinical development at various pharmaceutical companies. Table 3.3 provides a list of antibacterial nanotherapeutics that have been approved for use in clinical settings. Genetic immunity funded the first clinical trial of an antibacterial nanomedicine in 2005 (Lisziewicz et al. 2012). The idea of using nanotechnology in drug development has been quite successful. While there has been significant progress over the past decade, only a small number of nanomedicines have made it to the market. The difficulty in gaining access to information and researching prior nanomedicine trends is one of the key causes of the low prevalence of nanomedicine in clinical applications. The length of time it takes to complete a clinical study is a key drawback (Hawthorne et al. 2017). The Food and Drug Administration (FDA) and the European Medicines Agency (EMA) have both given their stamp of approval to numerous nanoparticles and nanomedicines (drug-loaded nanoparticles) designed to treat a wide range of microbial infections (Clancy et al. 2013; Henry et al. 2014). Only a select few antibacterial nanomedicines have been cleared for use in humans. For instance, the treatment of pulmonary illness caused by *Mycobacterium avium* complex with the amikacin liposomal nanosuspension Arikace® has been licenced for clinical usage (Clancy et al. 2013). Clinical trials have shown that Silvasorb®, a nanoformulation containing silver, can improve the skin's absorption of biologics when applied topically for the treatment of microbial infections. Amphotericin-B is a lipid-based nanoformulation available under the brand name Amphotec® (Clemons and Stevens 1998). Because of its quick and uniform distribution throughout the body, nanomedicine administered systemically has greater potential to impact clinical care. Overall, several types of NCs are being developed to deliver different antimicrobial drugs, but lipid-based nanoformulations are majorly approved by the US FDA because they are highly biocompatible, biodegradable, and non-toxic in nature. Nanotechnology has the potential to become a game-changing tool for fixing this issue in the not-too-distant future.

Table 3.3 List of FDA/EMA approved formulations

Trade name (manufacturer)	Antimicrobial drug	Formulation type	Targeted pathogen	Indications	Route of administration	Clinical trial phase/ approval
Arikace®/ARIKAYCE (EMA)	Amikacin	Liposome suspension	*P. aeruginosa*	NTM lung disease	Inhalation	Approved in 2018
Lipoquin® (Aradigm)	Ciprofloxacin	Liposome	*P. aeruginosa*	Chronic lung infection	Inhalation	II
Pulmaquin® (Aradigm)	Ciprofloxacine	Liposome	*P. aeruginosa*	Chronic lung infection	Inhalation	II–III
SilvaSorb Gel (MedLine)		Hydrogels	–	Wound healing	Topical	III

3.10 Conclusion

Antimicrobial resistance (AMR) is a growing public health emergency that necessitates swift and coordinated action. Drug-resistant bacteria have emerged as a result of the overuse and abuse of antibiotics, making once-effective therapies worthless. This situation offers serious economic and societal consequences in addition to endangering our ability to fight infectious diseases. A multimodal strategy is needed to combat AMR, including better infection prevention and control procedures, responsible antibiotic stewardship, and the creation of novel antimicrobial medicines. Global cooperation is also necessary to address this expanding threat because AMR does not consider national boundaries and affects humans, animals, and the environment equally. The preservation of antibiotics and the development of sustainable practices must be prioritized in society's knowledge, healthcare practices, and governmental policy. Antibiotics must continue to be powerful weapons in our fight against infectious diseases, and this requires our collaborative commitment to stop antimicrobial resistance.

References

Alruwaili NK, Zafar A, Imam SS, Alharbi KS, Alotaibi NH, Alshehri S, Alhakamy NA, Alzarea AI, Afzal M, Elmowafy M (2020) Stimulus responsive ocular gentamycin-ferrying chitosan nanoparticles hydrogel: formulation optimization, ocular safety and antibacterial assessment. Int J Nanomedicine 15:4717–4737. https://doi.org/10.2147/IJN.S254763

Aminov RI (2010) A brief history of the antibiotic era: lessons learned and challenges for the future. Front Microbiol 1:134. https://doi.org/10.3389/FMICB.2010.00134

Austin DJ, Kristinsson KG, Anderson RM (1999) The relationship between the volume of antimicrobial consumption in human communities and the frequency of resistance. Proc Natl Acad Sci USA 96(3):1152–1156. https://doi.org/10.1073/PNAS.96.3.1152

Baptista PV, McCusker MP, Carvalho A, Ferreira DA, Mohan NM, Martins M, Fernandes AR (2018) Nano-strategies to fight multidrug resistant bacteria "A battle of the titans". Front Microbiol 9:1441. https://doi.org/10.3389/FMICB.2018.01441/BIBTEX

Beyth N, Houri-Haddad Y, Domb A, Khan W, Hazan R (2015) Alternative antimicrobial approach: nano-antimicrobial materials. Evid Based Complement Alternat Med 2015:246012. https://doi.org/10.1155/2015/246012

Bhakay A, Rahman M, Dave RN, Bilgili E (2018) Bioavailability enhancement of poorly water-soluble drugs via nanocomposites: formulation–processing aspects and challenges. Pharmaceutics 10(3):86. https://doi.org/10.3390/PHARMACEUTICS10030086

Clancy JP, Dupont L, Konstan MW, Billings J, Fustik S, Goss CH, Lymp J, Minic P, Quittner AL, Rubenstein RC, Young KR, Saiman L, Burns JL, Govan JRW, Ramsey B, Gupta R (2013) Phase II studies of nebulised Arikace in CF patients with Pseudomonas aeruginosa infection. Thorax 68(9):818–825. https://doi.org/10.1136/THORAXJNL-2012-202230/-/DC1

Clemons KV, Stevens DA (1998) Comparison of fungizone, amphotec, ambisome, and abelcet for treatment of systemic murine cryptococcosis. Antimicrob Agents Chemother 42(4):899–902. https://doi.org/10.1128/AAC.42.4.899

Das B, Patra S (2017) Antimicrobials: meeting the challenges of antibiotic resistance through nanotechnology. In: Nanostructures for antimicrobial therapy: nanostructures in therapeutic medicine series. Elsevier, Amsterdam. https://doi.org/10.1016/B978-0-323-46152-8.00001-9

Davies J (1996) Origins and evolution of antibiotic resistance. Microbiologia 12(1):9–16. https://doi.org/10.1128/MMBR.00016-10

Davies J, Davies D (2010) Origins and evolution of antibiotic resistance. Microbiol Mol Biol Rev 74(3):9–16. https://doi.org/10.1128/MMBR.00016-10

Donelli G (2014) Microbial biofilms, methods and protocols edited. Springer, Cham, p 1147. https://doi.org/10.1007/978-1-4939-0467-9

Esteban-Tejeda L, Malpartida F, Esteban-Cubillo A, Pecharromn C, Moya JS (2009) Antibacterial and antifungal activity of a soda-lime glass containing copper nanoparticles. Nanotechnology 20(50):505701. https://doi.org/10.1088/0957-4484/20/50/505701

Evangelista TFS, Andrade GRS, Nascimento KNS, dos Santos SB, de Fátima Costa Santos M, Da Ros Montes Doca C, dos Estevam C, Gimenez IF, Almeida LE (2020) Supramolecular polyelectrolyte complexes based on cyclodextrin-grafted chitosan and carrageenan for controlled drug release. Carbohydr Polym 245:116592. https://doi.org/10.1016/J.CARBPOL.2020.116592

Frieri M, Kumar K, Boutin A (2017) Antibiotic resistance. J Infect Public Health 10(4):369–378. https://doi.org/10.1016/j.jiph.2016.08.007

FTB (n.d.) Antibiotic resistance mechanisms in bacteria: biochemical and genetic aspects. https://www.ftb.com.hr/about-ftb/70-volume-46-issue-no-1/291-antibiotic-resistance-mechanisms-in-bacteria-biochemical-and-genetic-aspects. Accessed 3 May 2023

Ganesh Kumar S, Adithan C, Harish BN, Sujatha S, Roy G, Malini A (2013) Antimicrobial resistance in India: a review. J Nat Sci Biol Med 4(2):286–291. https://doi.org/10.4103/0976-9668.116970

Gu H, Ho PL, Tong E, Wang L, Xu B (2003) Presenting vancomycin on nanoparticles to enhance antimicrobial activities. Nano Lett 3(9):1261–1263. https://doi.org/10.1021/NL034396z

Hajipour MJ, Fromm KM, Akbar Ashkarran A, Jimenez de Aberasturi D, de Larramendi IR, Rojo T, Serpooshan V, Parak WJ, Mahmoudi M (2012) Antibacterial properties of nanoparticles. Trends Biotechnol 30(10):499–511. https://doi.org/10.1016/J.TIBTECH.2012.06.004

Hawthorne GH, Bernuci MP, Bortolanza M, Issy AC, Del-Bel E (2017) Clinical developments in antimicrobial nanomedicine: toward novel solutions. In: Nanostructures for antimicrobial therapy: nanostructures in therapeutic medicine series. Elsevier, Amsterdam, pp 653–668. https://doi.org/10.1016/B978-0-323-46152-8.00029-9

Henry BD, Neill DR, Becker KA, Gore S, Bricio-Moreno L, Ziobro R, Edwards MJ, Mühlemann K, Steinmann J, Kleuser B, Japtok L, Luginbühl M, Wolfmeier H, Scherag A, Gulbins E, Kadioglu A, Draeger A, Babiychuk EB (2014) Engineered liposomes sequester bacterial exotoxins and protect from severe invasive infections in mice. Nat Biotechnol 33(1):81–88. https://doi.org/10.1038/nbt.3037

Huh AJ, Kwon YJ (2011) "Nanoantibiotics": a new paradigm for treating infectious diseases using nanomaterials in the antibiotics resistant era. J Control Release 156(2):128–145. https://doi.org/10.1016/J.JCONREL.2011.07.002

Ikono R, Vibriani A, Wibowo I, Saputro KE, Muliawan W, Bachtiar BM, Mardliyati E, Bachtiar EW, Rochman NT, Kagami H, Xianqi L, Nagamura-Inoue T, Tojo A (2019) Nanochitosan antimicrobial activity against Streptococcus mutans and Candida albicans dual-species biofilms. BMC Res Notes 12(1):383. https://doi.org/10.1186/S13104-019-4422-X

Jin T, Sun D, Su JY, Zhang H, Sue HJ (2009) Antimicrobial efficacy of zinc oxide quantum dots against Listeria monocytogenes, Salmonella Enteritidis, and Escherichia coli O157:H7. J Food Sci 74(1):46. https://doi.org/10.1111/J.1750-3841.2008.01013.X

Khan F, Oh D, Chandika P, Jo DM, Bamunarachchi NI, Jung WK, Kim YM (2022) Inhibitory activities of phloroglucinol-chitosan nanoparticles on mono- and dual-species biofilms of Candida albicans and bacteria. Colloids Surf B Biointerfaces 211:112307. https://doi.org/10.1016/J.COLSURFB.2021.112307

Khan F, Tabassum N, Jeong G-J, Jung W-K, Kim Y-M (2023) Inhibition of mixed biofilms of Candida albicans and Staphylococcus aureus by β-caryophyllene-gold nanoparticles. Antibiotics 12(4):726. https://doi.org/10.3390/antibiotics12040726

Kovalóvá Z, Zahoran M, Zahoranová A, Machala Z (2014) Streptococci biofilm decontamination on teeth by low-temperature air plasma of dc corona discharges. J Phys D Appl Phys 47(22):224014. https://doi.org/10.1088/0022-3727/47/22/224014

Li Y, Zhen J, Tian Q, Shen C, Zhang L, Yang K, Shang L (2020) One step synthesis of positively charged gold nanoclusters as effective antimicrobial nanoagents against multidrug-resistant bacteria and biofilms. J Colloid Interface Sci 569:235–243. https://doi.org/10.1016/J.JCIS.2020.02.084

Lisziewicz J, Bakare N, Calarota SA, Bánhegyi D, Szlávik J, Újhelyi E, Toke ER, Molnár L, Lisziewicz Z, Autran B, Lori F (2012) Single DermaVir immunization: dose-dependent expansion of precursor/memory T cells against all HIV antigens in HIV-1 infected individuals. PLoS ONE 7(5):e35416. https://doi.org/10.1371/JOURNAL.PONE.0035416

Liu B, Pop M (2009) ARDB–antibiotic resistance genes database. Nucleic Acids Res 37:656. https://doi.org/10.1093/NAR/GKN656

Lynch AS, Robertson GT (2008) Bacterial and fungal biofilm infections. Annu Rev Med 59:415–428. https://doi.org/10.1146/ANNUREV.MED.59.110106.132000

Maness PC, Smolinski S, Blake DM, Huang Z, Wolfrum EJ, Jacoby WA (1999) Bactericidal activity of photocatalytic TiO_2 reaction: toward an understanding of its killing mechanism. Appl Environ Microbiol 65(9):4094–4098. https://doi.org/10.1128/AEM.65.9.4094-4098.1999

Matsumura Y, Yoshikata K, Kunisaki S, ichi, & Tsuchido, T. (2003) Mode of bactericidal action of silver zeolite and its comparison with that of silver nitrate. Appl Environ Microbiol 69(7):4278–4281. https://doi.org/10.1128/AEM.69.7.4278-4281.2003

Nanda A, Saravanan M (2009) Biosynthesis of silver nanoparticles from Staphylococcus aureus and its antimicrobial activity against MRSA and MRSE. Nanomedicine 5(4):452–456. https://doi.org/10.1016/J.NANO.2009.01.012

Nilavukkarasi M, Vijayakumar S, Prathip Kumar S (2020) Biological synthesis and characterization of silver nanoparticles with Capparis zeylanica L. leaf extract for potent antimicrobial and anti proliferation efficiency. Mater Sci Energy Technol 3:371–376. https://doi.org/10.1016/J.MSET.2020.02.008

Oliveira DML, Rezende PS, Barbosa TC, Andrade LN, Bani C, Tavares DS, da Silva CF, Chaud MV, Padilha F, Cano A, de Albuquerque Júnior RLC, Souto EB, Severino P (2020) Double membrane based on lidocaine-coated polymyxin-alginate nanoparticles for wound healing: In vitro characterization and in vivo tissue repair. Int J Pharm 591:120001. https://doi.org/10.1016/J.IJPHARM.2020.120001

Potdar VA, Padbidri VV, Chadha MS (2013) Oseltamivir-resistant influenza A(H1N1) pdm09 virus: first reported case from India. J Public Health 2:3–4. https://doi.org/10.4103/2224-3151.115828

Qi L, Xu Z, Jiang X, Hu C, Zou X (2004) Preparation and antibacterial activity of chitosan nanoparticles. Carbohydr Res 339(16):2693–2700. https://doi.org/10.1016/J.CARRES.2004.09.007

Qiu Y, Xu D, Sui G, Wang D, Wu M, Han L, Mu H, Duan J (2020) Gentamicin decorated phosphatidylcholine-chitosan nanoparticles against biofilms and intracellular bacteria. Int J Biol Macromol 156:640–647. https://doi.org/10.1016/J.IJBIOMAC.2020.04.090

Rai RV, Bai JA (2011) Nanoparticles and their potential application as antimicrobials. Formatex 2011:197–209

Rai A, Prabhune A, Perry CC (2010) Antibiotic mediated synthesis of gold nanoparticles with potent antimicrobial activity and their application in antimicrobial coatings. J Mater Chem 20(32):6789–6798. https://doi.org/10.1039/C0JM00817F

Regiel A, Irusta S, Kyzioł A, Arruebo M, Santamaria J (2013) Preparation and characterization of chitosan-silver nanocomposite films and their antibacterial activity against Staphylococcus aureus. Nanotechnology 24(1):15101. https://doi.org/10.1088/0957-4484/24/1/015101

Ribeiro SM, Felício MR, Boas EV, Gonçalves S, Costa FF, Samy RP, Santos NC, Franco OL (2016) New frontiers for anti-biofilm drug development. Pharmacol Ther 160:133–144. https://doi.org/10.1016/j.pharmthera.2016.02.006

Römling U, Balsalobre C (2012) Biofilm infections, their resilience to therapy and innovative treatment strategies. J Intern Med 272(6):541–561. https://doi.org/10.1111/JOIM.12004

Ruparelia JP, Chatterjee AK, Duttagupta SP, Mukherji S (2008) Strain specificity in antimicrobial activity of silver and copper nanoparticles. Acta Biomater 4(3):707–716. https://doi.org/10.1016/J.ACTBIO.2007.11.006

Saxena S, Punjabi K, Ahamad N, Singh S, Bendale P, Banerjee R (2022) Nanotechnology approaches for rapid detection and theranostics of antimicrobial resistant bacterial infections. ACS Biomater Sci Eng 8(6):2232–2257. https://doi.org/10.1021/ACSBIOMATERIALS.1C01516

Spellberg B, Guidos R, Gilbert D, Bradley J, Boucher HW, Scheld WM, Bartlett JG, Edwards J (2008) The epidemic of antibiotic-resistant infections: a call to action for the medical community from the Infectious Diseases Society of America. Clin Infect Dis 46(2):155–164. https://doi.org/10.1086/524891

Spirescu VA, Chircov C, Grumezescu AM, Andronescu E (2021) Polymeric nanoparticles for antimicrobial therapies: an up-to-date overview. Polymers. https://doi.org/10.3390/polym13050724

Srinivasan R, Santhakumari S, Poonguzhali P, Geetha M, Dyavaiah M, Xiangmin L (2021) Bacterial biofilm inhibition: a focused review on recent therapeutic strategies for combating the biofilm mediated infections. Front Microbiol 12(May):1–19. https://doi.org/10.3389/fmicb.2021.676458

Tabassum N, Khan F, Kang MG, Jo DM, Cho KJ, Kim YM (2023) Inhibition of polymicrobial biofilms of Candida albicans- Staphylococcus aureus/Streptococcus mutans by fucoidan-gold nanoparticles. Mar Drugs 21(2):123. https://doi.org/10.3390/MD21020123

Tin S, Sakharkar KR, Lim CS, Sakharkar MK (2009) Activity of chitosans in combination with antibiotics in Pseudomonas aeruginosa. Int J Biol Sci 5(2):153–160. https://doi.org/10.7150/IJBS.5.153

Vadhana P, Singh BR, Bharadwaj M (2015) Emergence of herbal antimicrobial drug resistance in clinical bacterial isolates. Pharm Anal Acta 6(10):1000434. https://doi.org/10.4172/2153-2435.1000434

Villalba MI, Rossetti E, Bonvallat A, Yvanoff C, Radonicic V, Willaert RG, Kasas S (2023) Simple optical nanomotion method for single-bacterium viability and antibiotic response testing. Proc Natl Acad Sci 120(18):e2221284120. https://doi.org/10.1073/PNAS.2221284120

Walvekar P, Gannimani R, Salih M, Makhathini S, Mocktar C, Govender T (2019) Self-assembled oleylamine grafted hyaluronic acid polymersomes for delivery of vancomycin against methicillin resistant Staphylococcus aureus (MRSA). Colloids Surf B Biointerfaces 182:110388. https://doi.org/10.1016/J.COLSURFB.2019.110388

Wang Y, Xia Y (2004) Bottom-up and top-down approaches to the synthesis of monodispersed spherical colloids of low melting-point metals. Nano Lett 4(10):2047–2050. https://doi.org/10.1021/nl048689j

Wang X, Zhao D, Li Y, Zhou X, Hui Z, Lei X, Qiu L, Bai Y, Wang C, Xia J, Xuan Y, Jiang P, Wang J (2023) Collagen hydrogel with multiple antimicrobial mechanisms as anti-bacterial wound dressing. Int J Biol Macromol 232:123413. https://doi.org/10.1016/j.ijbiomac.2023.123413

Weissig V, Guzman-Villanueva D (2015) Nanopharmaceuticals (part 2): products in the pipeline. Int J Nanomedicine 10(1):1245–1257. https://doi.org/10.2147/IJN.S65526

Wu Q, Miao WS, Du Zhang Y, Gao HJ, Hui D (2020) Mechanical properties of nanomaterials: a review. Nanotechnol Rev 9(1):259–273. https://doi.org/10.1515/ntrev-2020-0021

Wu Y, Gao H, Liu J, Liang H (2023) Chitosan nanoparticles efficiently enhance the dispersibility, stability and selective antibacterial activity of insoluble isoflavonoids. Int J Biol Macromol 232:123420. https://doi.org/10.1016/J.IJBIOMAC.2023.123420

Xie Y, Liu Y, Yang J, Liu Y, Hu F, Zhu K, Jiang X (2018) Gold nanoclusters for targeting methicillin-resistant staphylococcus aureus in vivo. Angew Chem 57(15):3958–3962. https://doi.org/10.1002/ANIE.201712878

Yang C, Mamouni J, Tang Y, Yang L (2010) Antimicrobial activity of single-walled carbon nanotubes: length effect. Langmuir 26(20):16013–16019. https://doi.org/10.1021/LA103110G

Yin W, Wang Y, Liu L, He J (2019) Biofilms: the microbial "protective clothing" in extreme environments. Int J Mol Sci 20(14):3423. https://doi.org/10.3390/IJMS20143423

Zhang L, Pornpattananangkul D, Hu C-M, Huang C-M (2010) Development of nanoparticles for antimicrobial drug delivery. Curr Med Chem 17(6):585–594. https://doi.org/10.2174/092986710790416290

Zheng Y, Liu W, Qin Z, Chen Y, Jiang H, Wang X (2018) Mercaptopyrimidine-conjugated gold nanoclusters as nanoantibiotics for combating multidrug-resistant superbugs. Bioconjug Chem 29(9):3094–3103. https://doi.org/10.1021/acs.bioconjchem.8b00452

4 Metal Nanoparticles As Alternative Antimicrobial Agents to Combat Multidrug Resistance Bacteria

Sichao Tian, Peiyan Yuan, and Qing-Hua Xu

Abstract

The rise of multidrug-resistant (MDR) bacteria poses a significant global health concern. The development of novel and effective antimicrobial agents is a pressing challenge. Metal nanoparticles (MNPs) and their bio-conjugates have attracted considerable attention as potential antimicrobial agents due to their distinct properties. They hold promise for selective detection and treatment of bacteria. This chapter presents a summary of recent endeavors in creating diverse antimicrobial agents based on MNPs and their conjugates to address the issue of MDR bacteria. The discussion commences with an introduction to fundamental concepts and mechanisms involving the detection of bacteria through optical, magnetic, and electrochemical methods, as well as therapeutic approaches including photothermal, photodynamic, and chemotherapy methods. Representative examples are examined to showcase both in vitro and in vivo proof-of-concept applications. Finally, challenges and a prospective outlook of this research area were summarized.

S. Tian
Institute of Medicinal Plant Development, Chinese Academy of Medical Sciences and Peking Union Medical College, Beijing, China

P. Yuan
School of Pharmaceutical Sciences, Shenzhen Campus of Sun Yat-sen University, Shenzhen, China

Q.-H. Xu (✉)
Department of Chemistry, National University of Singapore, Singapore, and National University of Singapore (Suzhou) Research Institute, Suzhou, China
e-mail: chmxqh@nus.edu.sg

M. Y. Wani et al. (eds.), *Nanotechnology Based Strategies for Combating Antimicrobial Resistance*, https://doi.org/10.1007/978-981-97-2023-1_4

Keywords

Multidrug-resistant (MDR) bacteria · Antimicrobial agents · Nanoparticles · Therapeutic approaches

4.1 Introduction

The widespread occurrence of bacteria that are resistant to multiple drugs, known as multidrug-resistant (MDR) bacteria, is a major and alarming issue for public health systems around the world (Boucher et al. 2009; Yang et al. 2017). Traditional antibiotic treatments, which have been widely used to manage bacterial infections, are no longer effective against the increasing occurrence of MDR bacteria (Olaitan et al. 2014; Wang et al. 2020a). The scarcity of potent antibacterial medications and the widespread dissemination of MDR bacteria have led to a crisis of MDR resistance (Brown and Wright 2016). This necessitates the creation of innovative therapeutic approaches to serve as alternative antimicrobial agents (Wei et al. 2020; Huo et al. 2021). Efficient identification and management of MDR bacteria are crucial in the fight against them (Deusenbery et al. 2021; Huang et al. 2023). Prompt and precise identification of bacterial infections is essential for halting the spreading of bacterial infection and developing suitable intervention measures (Zhao et al. 2021; Kuang et al. 2023; Peukert et al. 2022). The development of therapeutic interventions capable of effectively combating bacterial infections is highly demanded, especially for those infections that have developed drug resistance (Pan et al. 2022; Feng et al. 2021).

Metal nanoparticles (MNPs) and their bio-conjugates have garnered significant interest as potential antimicrobial agents (Sanchez-Lopez et al. 2020; Wu et al. 2021). MNPs have distinct advantages that make them suitable for both detecting and treating bacterial infections (Peng et al. 2018). MNPs possess inherent benefits including a remarkably large surface area, precise control over size and morphology, easy surface modification, as well as distinctive optical, mechanical, electromagnetic, and chemical properties (Wang et al. 2020b; Saha et al. 2012). Extensive research has been conducted in the past 20 years to investigate the antimicrobial properties of metal-based nanoparticles (NPs). Silver (Ag) and copper (Cu) have been utilized for thousands of years as antimicrobial agents for the purpose of preserving food and disinfecting water (Lemire et al. 2013). The antimicrobial properties and range of effectiveness of MNPs have greatly improved with the recent advancements in nanotechnology. The innate antimicrobial properties of MNPs can be intensified by the extremely small size of MNPs, enabling more robust physical interactions with the surface of bacteria (Stoimenov et al. 2002). The effectiveness of these nanoparticles as antimicrobial agents is further enhanced by the unique properties of MNPs, which include the photothermal and photodynamic effects. The antimicrobial effect of MNPs is derived from several distinct mechanisms, including the release of metal ions, the production of reactive oxygen species (ROS), direct disruption of bacterial cell membranes, or a combination of these actions (Stoimenov et al. 2002; Hajipour et al. 2012). MNPs possess not only improved antimicrobial properties but also have the potential to be used as detection/imaging agents for bacterial pathogens. This is because they exhibit a distinctive characteristic known as local surface plasmon resonance (LSPR). The distinctive characteristic of LSPR in MNPs enables the development of diverse optical techniques, such as colorimetric, fluorescence, and nonlinear optical methods, for the identification of bacterial pathogens. MNPs possessing magnetic and electrochemical characteristics have also been investigated for the purpose of identifying bacterial pathogens (Chen et al. 2017).

The efficiency of bacterial detection and therapy of MNPs has been enhanced through the modification with appropriate conjugates (Zhou et al. 2015; Chang et al. 2019a). MNP conjugates demonstrate exceptional properties that go beyond the capabilities of their individual components through conjugation with biomolecules (Pandey et al. 2021), photosensitizers (Rossi et al. 2020; Elashnikov et al. 2019), and fluorescent dyes (Panda et al. 2008). They present a comprehensive strategy for combating infections by combining the specificity of biomolecules with the strengths of MNPs. For example, conjugated MNPs with photosensitizers enabled light activation for localized cytotoxicity and targeted bacterial recognition (Zhou et al. 2018). When MNPs are conjugated with targeting ligands, real-time imaging is made possible and in vivo navigation accuracy is increased (Lu et al. 2020). By combining the optical, magnetic, and electrical characteristics of MNPs with conjugated molecules, these MNP conjugates enable the development of platforms with enhanced therapeutic efficacy, specificity, and sensitivity (Sanchez-Lopez et al. 2020).

This chapter aims to provide a comprehensive overview of recent progress in developing antimicrobial agent based on MNPs and their conjugates to address the challenges posed by the MDR bacteria (Scheme 4.1). The investigation includes both therapeutic modalities that take advantage of MNPs' diverse properties to circumvent bacterial resistance mechanisms and detection strategies that use the peculiar properties of MNPs for precise identification. This chapter highlights several creative accomplishments and emphasizes the reasoning behind the selection of MNPs and their conjugates. By utilizing the extraordinary potential of MNPs and their bio-conjugates, we hope to show how to identify and treat bacterial infections.

4.2 MNPs and Their Conjugates for Bacterial Detection

The unique physicochemical properties of MNPs and their conjugates make them appealing candidates for developing bacterial detection methods (Peng et al. 2020; Lee et al. 2021). When MNPs interact with incident light, a unique LSPR property results from the collective oscillations of conduction band electrons (Chen et al. 2013). The LSPR spectrum can be tuned over a wide range of wavelengths from ultraviolet to near-infrared (NIR), enabling precise manipulation and customized detection (Zhang et al. 2019). In particular, when the LSPR mode is excited, a significant amplification of the local electromagnetic field occurs in the vicinity of MNPs. This phenomenon has been strategically utilized to improve a range of optical signals, such as surface-enhanced Raman scattering (Lim et al. 2010),

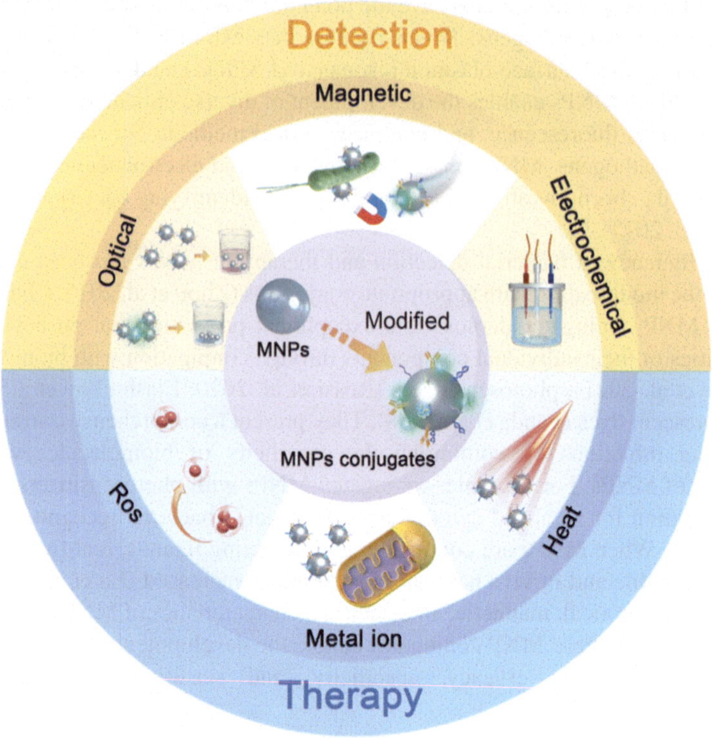

Scheme 4.1 Illustrative summary of various applications of metal nanoparticles in detecting and treating bacterial pathogens

fluorescence (Kinkhabwala et al. 2009), and two-photon photoluminescence (2PPL) (Yuan et al. 2012; Jiang et al. 2013a). All of these characteristics highlight how versatile MNPs are, making them ideal for supporting the creation of cutting-edge biological detection and imaging platforms, particularly in the field of bacterial detection.

4.2.1 MNPs

In 1996, Mirkin et al. used gold (Au) NPs to detect target DNA to demonstrate the use of MNPs in colorimetric detection (Mirkin et al. 1996). The basis of colorimetric detection is a target-induced visual color change that can be seen with the unaided eye for qualitative detection or quantitatively analyzed using UV–vis spectrometry. Aggregation of MNPs led to the phenomenon of interparticle plasmonic coupling, which caused a shift toward longer wavelengths in the extinction spectrum and subsequently resulted in a color change (Halas et al. 2011). The color change is easily visible to the unaided eye because MNPs have an extinction coefficient of several orders of magnitude larger than that of conventional organic dye molecules (Mayer and Hafner 2011). Upon the formation of MNP aggregates, the solution undergoes a color change. Specifically, Au NPs change from red to purple, whereas Ag NPs change from yellow to brown. Due to the absence of intricate apparatus, this detection method is highly convenient. A wide range of techniques, encompassing nucleic acids, tiny molecules, enzymes, and infections, have been devised for the identification of different biological entities (Huang et al. 2017; Ray et al. 2012).

The benefits of controlled synthesis, simple surface modification, high solubility, and exceptional biocompatibility make Au NPs the most popular MNPs in colorimetric detection techniques. The aggregation of Au NPs is typically the basis for the colorimetric detection of bacteria. For instance, unmodified Au NPs and an asymmetric polymerase chain reaction (asPCR) with propidium monoazide (PMA) were utilized to find viable emetic *B. cereus* (Li et al. 2018). Long genomic DNA fragments produced by asPCR in the presence of viable target bacteria could be absorbed onto the surface of Au NPs through the coordination of the exposed bases' nitrogen and gold atoms. The addition of NaCl solution caused a color shift that was visible to the unaided eye in a matter of minutes. Chen et al. used salt-induced aggregation of unmodified Au NPs as a colorimetric signal output to successfully detect *Salmonella typhimurium* (Chen et al. 2021). In the presence of *Salmonella typhimurium*, the hairpin structure on the surface of Au NPs desorbs to form a significant amount of double-stranded DNA (dsDNA). As a result, the well-dispersed Au NPs formed aggregates, changing the solution color from red to blue-gray. The signal output of *Salmonella typhimurium* was linear over the concentration range of 10^2 to 10^6 colony-forming units (CFU)/mL, with a sensitivity of 2.4×10^2 CFU/mL under optimal conditions.

The investigation of more complex nanomaterials, such as core–shell structured MNPs, has gained attention due to the remarkable success of colorimetric bacterial detection using MNPs (Ding et al. 2017). MNPs can have different colors depending on the thickness of their outer shell. Zhang et al. created a plasmonic time-temperature indicator (TTI) that uses a sharp color shift to detect the growth of bacteria and perishable food items on Au nanorods (NRs) (Zhang et al. 2013). The concentration of the relevant reagents was adjusted, resulting in the synchronized self-evolution of TTI and the growth of *Escherichia coli* (*E. coli*) at any temperature within the normal range. This TTI can be used to monitor the growth of bacteria and perishables by using a sharp color change from red to green.

The primary use of unmodified MNPs is in colorimetric detection. MNP conjugates have been the subject of most research, which will be discussed in the following section. More localization accuracy, increased selectivity, and a variety of detection techniques are made possible by MNP conjugates. Moreover, the enhanced signal amplification, durability, and biocompatibility characteristics of the MNP conjugates contribute to the overall efficacy of MNP-based assays.

4.2.2 MNP Conjugates

4.2.2.1 Colorimetric Detection

Another enzyme-nanoparticle conjugation system-based colorimetric sensing scheme for *E. coli* bacteria was reported by Miranda et al. (2011). This approach entailed the utilization of Au NPs containing quaternary amine head groups with a positive charge to interact with β-galactosidase (β-Gal) by electrostatic binding, thereby impeding the enzyme's activity. An enzyme-amplified colorimetric signal was produced when Au NPs bound to *E. coli* bacteria and released β-Gal, which then restored the enzyme's activity. Because of their specific design, Au NPs can function as important regulators during bacterial interactions, controlling the enzyme's active state and thereby amplifying the detection signal. By combining the interactions of Au NPs with both bacteria and enzymes, this technique offers a sensitive and focused approach to the detection of bacteria. The colorimetric assay successfully identified the presence of 10^2 bacteria mL^{-1} in solution and 10^4 bacteria mL^{-1} in a field-friendly test strip.

Antibodies can function as conjugates to identify the target bacteria with selectivity. Wang and colleagues devised a colorimetric method that employs Au NPs modified with antibodies to identify the intended bacteria with precision (Wang et al. 2010). The Au NPs, which have been modified with antibodies, specifically aggregate on the *Salmonella typhimurium* bacteria target, while only a small number of Au NPs adhere to other bacteria. This is because the anti-Salmonella antibody has a strong affinity for the *Salmonella typhimurium* bacteria. The color of the solution clearly changed from red to blue as Au NPs aggregated. MDR *Salmonella typhimurium* has also been selectively detected using antibody-conjugated Au NPs (Khan et al. 2011). The MDR bacteria that were being targeted were successfully detected in both the solution and the food sample, with the detection limit being relatively low.

4.2.2.2 Fluorescence-Based Detection

Compared to colorimetric techniques, fluorescence-based methods have much higher sensitivity and better detection limits, and they even make single-molecule detection possible. The optical characteristics of neighboring fluorophores can be changed by MNP interaction, leading to either fluorescence enhancement or quenching depending on the proximity and spectral properties of the fluorophores (Yuan et al. 2012; Huang et al. 2017; Lee et al. 2011).

Fluorescence quenching induced by fluorescence resonance energy transfer (FRET) is a common method. Because MNPs have tunable extinction spectra and high molar extinction coefficients, they can effectively act as excellent energy acceptors to quench the fluorescence of nearby fluorophores. Au NPs have been known to be excellent fluorescence quenchers (Swierczewska et al. 2011). MNPs-induced quenching has found widespread application in biosensing techniques using two primary tactics: target-induced fluorescence quenching and target-induced fluorescence recovery. Both of these strategies are focused on the quenching of fluorescence.

Fluorophores with a high emission quantum yield (QY) are typically employed in target-induced fluorescence quenching strategies, and they are typically far from Au NPs in the absence of a target. The distance between fluorophores and Au NPs decreases in the presence of the target analyst, leading to an efficient energy transfer from fluorophores to Au NPs, thereby effectively quenching the fluorescence of fluorophores (Yu et al. 2017). On the other hand, the target-induced fluorescence recovery method involves the restoration of fluorescence in fluorophores when they detach from the surface of Au NPs in the presence of the target. Initially, the fluorescence of the fluorophores is suppressed by the Au NPs, which are in the form of preconjugated fluorophores on the Au NPs. A highly amplified fluorescence assay was developed by Zhang et al. to quickly detect *Salmonella enteritidis* (Zhang et al. 2009). Au NPs were modified on target-specific DNA probes and fluorescein-labeled barcode DNA to serve as the transducer and fluorescence quencher in this assay. The preconcentrator and separator were the second target-specific DNA probe-modified magnetic NPs. Magnetic NP-Au NP conjugates were created in the presence of the target and were distinguished from the unreacted NPs by a magnetic field. Following the release of the fluorescence barcode DNA probes, the fluorescence signal—which is an exponential function of the target concentration—was recovered. By combining cationic Au NPs with conjugated polymers (PPE), Phillips et al. reported a fluorescence-based biosensor for bacteria detection. Au NPs and PPE functioned as the quenchers and flares, respectively (Fig. 4.1a) (Phillips et al. 2008). Fluorescence quenching was initially caused by the anionic PPE's electrostatic interactions with positively charged Au NPs. When the target bacteria were added to the Au-PPE conjugates, the positively charged Au NPs and negatively charged bacteria interacted to release PPE, which led to a notable recovery of fluorescence. This Au-PPE-conjugated system was successfully used to differentiate 12 different bacteria. Six biofilm strains and bacterial species could be quickly distinguished using the ratio-metric response of the readout. By using vancomycin-functionalized Au nanoclusters (NCs) and aptamer-modified Au NPs as the energy donor and acceptor, respectively (Fig. 4.1b), Yu et al. demonstrated a linear variation of fluorescence signals with the logarithm concentration of *S. aureus* in the range of 20 to 10^8 CFU mL^{-1} with a LOD of 10 CFU mL^{-1} (Yu et al. 2017). When bacteria were added to the system, the connections between Au NCs and Au NPs were immediately destroyed, causing Au NCs to be released and fluorescence to recover. By utilizing upconversion nanoparticles (UCNPs) functionalized with the corresponding complementary DNA (cDNA) and aptamers conjugated Au NPs, Jin et al. (2017) developed a novel FRET-based detection platform for fast, ultrasensitive, and specific bacteria detection. When hybridizing their surface-modifying aptamers and cDNA, FRET can occur to quench the upconversion fluorescence due to the spectral overlap between the emission spectra of UCNPs and the absorption spectra of Au NPs (Fig. 4.1c). Upconversion fluorescence was recovered when the aptamers preferentially bound to the target bacteria, causing them to separate from Au NPs and release UCNPs.

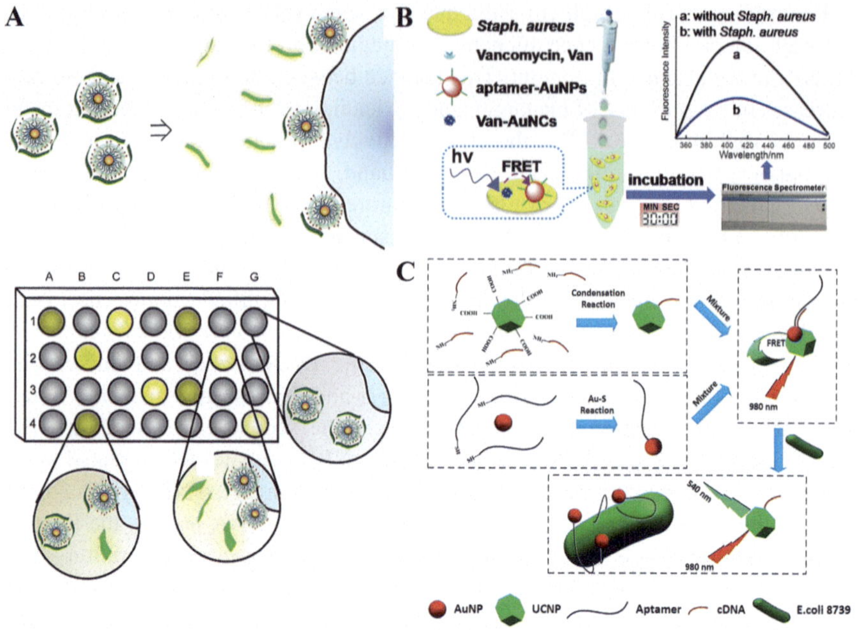

Fig. 4.1 MNP conjugates for bacterial detection based on the fluorescent method. (**a**) Target-induced fluorescence change of Au-PPE conjugates for different bacteria detection. Reproduced with permission. Copyright 2008, Wiley-VCH. (**b**) Dual-recognition FRET-based platform for one-step sensitive detection of pathogenic bacteria using fluorescent vancomycin-Au NCs and aptamer-Au NPs. Copyright 2017, ACS Publications. (**c**) UCNPs-based FRET apta-sensor for rapid and ultrasensitive bacteria detection. Copyright 2017, Elsevier

4.2.2.3 Nonlinear Optical (NLO) Response-based Detection

Plasmon coupling-induced strong local electromagnetic field amplification can also be employed to enhance NLO responses such as 2PPL of MNPs (Jiang et al. 2013a; Han et al. 2012). Because of their unique advantages of confined excitation and deep penetration, the two-photon excitation techniques are especially well-suited for biological applications (He et al. 2008a). Au NRs have been known to display strong 2PPL (Li et al. 2019) and aggregated MNPs have been demonstrated to display much stronger 2PPL compared to well-dispersed MNPs (Guan et al. 2010, 2013; Jiang et al. 2013b; Yuan et al. 2014). Using a high-affinity peptide sequence specific for the spores of *Bacillus subtilis*, an established and commonly used simulant species of Bacillus anthracis, the causative agent of anthrax, He et al. show that Au NRs can be used in bacterial detection based on 2PPL imaging (Fig. 4.2a) (He et al. 2008b). Using this technique, a homing peptide with a cysteine at the end was used to functionalize Au NRs. Even at 30 mW of output power, Bacillus spores incubated with Au NRs conjugated to the control peptide did not produce detectable signals, in contrast to the spores incubated with peptide-functionalized Au NRs, which produced clear 2PPL images. Au NRs can be used to detect spores with high

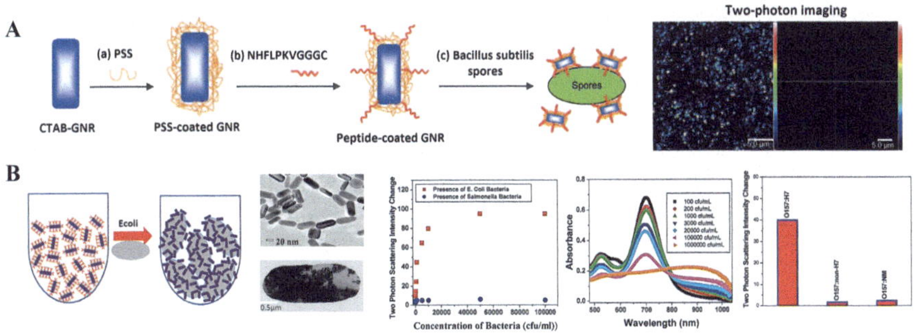

Fig. 4.2 MNP conjugates for bacterial detection based on the nonlinear optical method. (**a**) Two-photon luminescence imaging of bacillus spores using peptide-functionalized Au NRs. Copyright 2008, Tsinghua Press and Springer-Verlag. (**b**) Aggregation of antibody-conjugated Au NR for *E. coli* bacteria based on two-photon scattering technique. Copyright 2009, ACS Publications

fidelity at femtomolar concentrations due to the high affinity of the peptide labels and the high sensitivity of 2PPL.

The aggregation of Au NRs has also been used for NLO response-based detection (Polavarapu et al. 2014). Aggregation of Au NRs resulted in the loss of symmetry centers, which increased the significance of electric dipoles' contribution to two-photon scattering (Olesiak-Banska et al. 2019). The intensity of two-photon scattering will increase with Au NR aggregation because the contribution of electric dipoles is many times greater than that of multipole moments. Singh and colleagues have demonstrated the use of an antibody-conjugated Au NR-based two-photon scattering technique to produce a rapid and highly sensitive assay for the detection of *E. coli* bacteria (Fig. 4.2b) (Singh et al. 2009). About 40 times more two-photon scattering intensity is observed when different concentrations of *E. coli* O157:H7 bacteria are mixed with anti-*E. coli* antibody-conjugated nanorods. At a 50 CFU/mL level, *E. coli* bacteria can be detected fast, precisely, and with good discrimination against other bacteria without the need for amplification or enrichment. The entire bioassay process, from bacteria binding to detection and analysis, took less than 15 min.

4.2.2.4 Magnetic Methods

Magnetic NPs have been widely used in several biological applications over the past few decades, including drug delivery, biosensing, and magnetic resonance imaging (Colombo et al. 2012). In order to create a variety of schemes for the detection of proteins, bacteria, and biomarkers for cancer cells, magnetic NPs have been integrated with various analytical techniques, such as optical and electrochemical techniques (Gao et al. 2006, 2009). Magnetic NPs have been widely used as agents for separation, concentration, and target identification because they make it simple to control their location and motion through the application of external magnetic fields (Gu et al. 2006). Because biological samples typically lack magnetic properties, the use of magnetic NPs enables highly sensitive detections with minimal background signal and exceptional signal-to-noise ratio. The magnetic biosensors function by means of target-induced alterations in the magnetic and/or optical signals of conjugates of metal–magnetic NPs when they are subjected to magnetic agitation (Lee et al. 2015).

Efforts have been made to simultaneously detect two types of pathogens rather than one at a time to increase the detection efficiency. As the capture probe, Fe_3O_4 magnetic Au NPs coated with aptamers of *Salmonella typhimurium* and *Staphylococcus aureus* were employed because they can bind with two pathogens. *Salmonella typhimurium* and *Staphylococcus aureus* were among the bacteria that the magnetic force assisted in separating from others (Fig. 4.3a) (Zhang et al. 2015a). As particular Raman tags, 5,5′-dithiobis, and mercaptobenzoic acid were used to functionalize the shell Au probes. Raman signals could be used to sense the presence of *Staphylococcus aureus* and *Salmonella typhimurium* sandwiched together. The limit of detection (LOD) was 35 CFU mL^{-1} for *Staphylococcus aureus* and 15 CFU mL^{-1} for *Salmonella typhimurium.*

Using a plasmonic sensor array, magnetic NPs have been combined with optical colorimetric analytical techniques to detect both urease-positive and urease-negative bacteria (Fig. 4.3b) (Santopolo et al. 2019). The positively charged polymer-coated magnetic beads in this sensor array mediated the electrostatic interaction between negatively charged bacteria. At an acidic pH, the magnetic bead–bacteria complex quickly accumulated when bovine serum albumin was present. The solution changed the color from red to blue due to the aggregation of magnetic beads since the urease-negative bacteria attached to them were unable to produce urease or alkalize the mixture. This plasmonic nano-sensor can detect a low concentration (LOD of 10 cells mL^{-1}) of urease-positive/-negative bacteria in a short period of time (40 min).

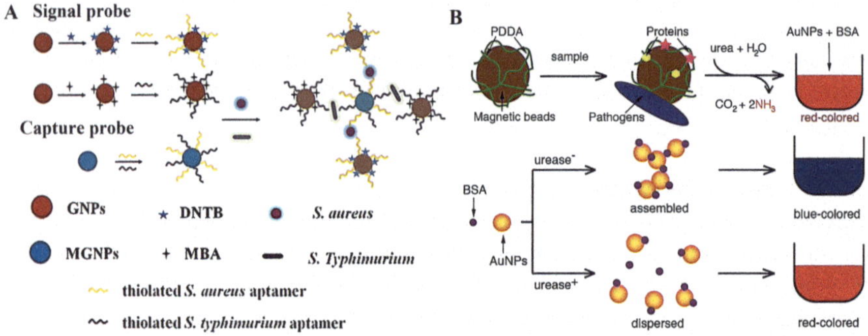

Fig. 4.3 MNP conjugates for bacterial detection based on the magnetic method. (**a**) Schematic illustration of the aptasensor for simultaneous detection of *S. aureus* and *S. typhimurium* based on Au NPs-enhanced Raman intensity. Copyright 2015, Elsevier. (**b**) Ultrafast and ultrasensitive naked-eye detection of urease-positive bacteria with plasmonic nano-sensors. Copyright 2019, ACS Publications

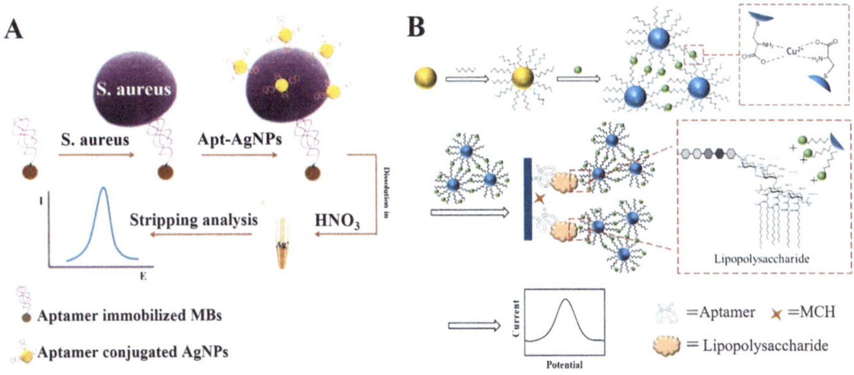

Fig. 4.4 MNP conjugates for bacterial detection based on the electronic method. (**a**) Aptamer conjugated Ag NP-based electrochemical biosensor for *S. aureus* detection. Reproduced with permission. Copyright 2015, Elsevier. (**b**) Copper ion-assisted Au NPs aggregates for electrochemical signal amplification of lipopolysaccharide sensing. Copyright 2019, Elsevier

4.2.2.5 Electrochemical Methods

Noble MNPs' large surface area and strong conductivity make them excellent materials for a variety of target analytes' electrochemical detection (Pallares et al. 2019; Su et al. 2016). Electrochemical biosensors primarily consist of biological recognition components capable of selective binding to the analyte, as well as an electrochemical transducer that can generate electrical signals to which the analyte responds. Electrochemical detections have many advantages over other conventional methods, including high sensitivity, quick response, affordability, and ease of use (Afonso et al. 2013; Manickam et al. 2020). Abbaspour et al. created an intriguing electrochemical immunosensor for *S. aureus* detection (Fig. 4.4a) (Abbaspour et al. 2015). For quick magnetic separation in this experiment, streptavidin-conjugated magnetic NPs modified with biotinylated primary anti-*S. aureus* aptamers served as capture probes, and Ag NPs modified with secondary anti-*S. aureus* aptamers served as electrochemical signal amplifiers. The detection specificity was increased by the dual-aptamer design. Sandwich complexes containing *S. aureus* bacteria were created, and subsequent separation of the complexes under the influence of an external magnet produced the differential pulse stripping voltammetric readout. Wan et al. reported using copper ion-assisted aptamer-functionalized Au NP aggregation to detect lipopolysaccharide (Fig. 4.4b) (Wang et al. 2019). Differential pulse voltammetry was utilized to test the sensing performance of the electrochemical sensor, and the stripping peak currents of Cu reoxidized to Cu^{2+} were employed to monitor the level of lipopolysaccharide. The electrochemical aptasensor was very sensitive to lipopolysaccharide, with a detection limit of 0.033 pg/mL.

4.2.2.6 Other Methods

Dynamic light scattering (DLS) is another crucial test technique to identify bacteria using the size and size distribution of MNPs because aggregation will alter the size of NPs. DLS is a technique that relies on how light interacts with particles. Measurements of narrow particle size distributions, particularly in the 2–500 nm range, can be performed with this method (Tomaszewska et al. 2013). The primary benefits of DLS detection are its affordability, fast, and ease of use (Tomaszewska et al. 2013). A food-borne pathogen that is Gram-positive and has an infectious dose of less than 1000 cells is *Listeria monocytogenes*. Au NPs coated with a monoclonal antibody against *Listeria monocytogenes* may serve as effective probes for the detection of bacteria (Fig. 4.5a) (Huang et al. 2015). Several antibody-coated Au NPs were able to capture *Listeria monocytogenes* in the presence of the bacteria, resulting in the formation of dimers, trimers, or multiaggregates of Au NPs. The level of *Listeria monocytogenes* could be determined through DLS, with an LOD of 35 CFU mL^{-1} in PBS and 22 CFU g^{-1} in lettuce samples.

Because the light scattering from larger NPs is so strong that the light scattering from smaller NPs is hidden, in certain cases the detection limit of DLS was lower than that of UV–vis measurement (Tomaszewska et al. 2013). Gram-negative food-borne illnesses can be caused by *Campylobacter jejuni*. Using complementary RNA-functionalized Au NPs as indicators, specific *Campylobacter jejuni* DNA fragments were found in food (Fig. 4.5b) (McVeya et al. 2017). RNA-functionalized Au NPs bound to target DNA in the presence of DNA fragments from *Campylobacter jejuni*. RNase H was able to cleave and remove all the RNA from Au NPs, which caused bare Au NPs to aggregate and turn blue when exposed to NaCl. RNA-functionalized Au NPs would monodisperse and turn red when RNA fragments repelled one another in the absence of target DNA. This method used the complementation between RNA and DNA to selectively detect the microbes, showing a sensitivity of 40.7 fM by UV–vis and 2.45 fM by DLS.

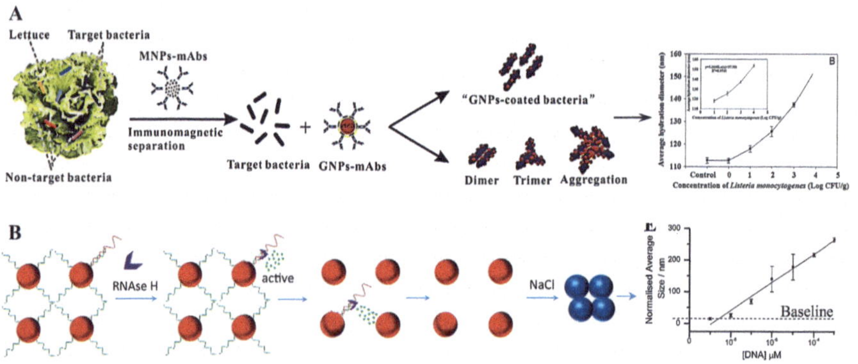

Fig. 4.5 MNP conjugates for bacterial detection based on the DLS method. (**a**) Antibody-coated Au NPs act as useful probes for bacterial detection based on DLS. Copyright 2015, Elsevier. (**b**) Endonuclease-controlled aggregation of Au NPs for the ultrasensitive detection of pathogenic bacterial DNA. Copyright 2017, Elsevier

4.3 MNPs for Bacterial Treatment

Strong broad-spectrum antimicrobial effects have also been reported for noble MNPs such as Au, Ag, and Cu NPs and their conjugates (Zhang et al. 2015b). These MNPs and MNP conjugates are typically toxic due to their ability to release metal ions, produce ROS, or undergo photothermal/photodynamic effects when exposed to light (Sanchez-Lopez et al. 2020; Usman et al. 2013). In the following section, typical instances of MNPs with these various mechanisms will be covered in detail.

4.3.1 MNPs

4.3.1.1 Mechano-bactericidal

It is generally believed that Au NPs are physiologically inert and do not possess antibacterial properties (Zhang et al. 2015b). In an investigation comparing the minimum inhibitory concentrations (MIC) of Ag NPs and Au NPs against *S. aureus*, Ag NPs demonstrated an inhibitory effect only at extremely high concentrations of 197 g/mL, with an MIC value of 4.86±2.71 g/mL and a minimum bactericidal concentration of 6.25 g/mL. Despite this, recent studies have demonstrated the antibacterial properties of pure Au NPs in a variety of forms. Penders et al. demonstrated that Au nanoflowers (NFs) and nanostars (NSs) were demonstrated to have significant antibacterial activity against *S. aureus* (Penders et al. 2017), and Au NFs were superior at concentrations of 250–500 g/mL. Despite the relatively high concentration of Au NPs used, the particles exhibit notable cell cytotoxicity when tested against human dermal fibroblasts. The reason for the antibacterial activity of Au NFs was identified as the high-aspect ratio spikes and pillars, which led to membrane rupture and increased local stress on the bacterial cell wall.

The local stress that Au NPs apply to bacterial membranes can result in membrane rupture, rendering them the antibacterial effect. The local stress can result from the attraction forces between Au NPs and bacterial membranes due to electrostatic interactions of their opposing surface charges. It needs to be noted that a recently developed research field called mechano-bactericidal activity of nanomaterials holds the potential to be effective in preventing infections, biofilm formation, and bacterial adhesion as traditional antibacterial strategies (Ivanova et al. 2012, 2013; Linklater et al. 2017). The goal of this approach is to produce antibacterial nanomaterials with a specific nanostructure shape that, when in contact with bacterial cells, can apply deadly mechanical forces. A recent study for star-shaped and quasi-spherical Au NPs employed this technique (Linklater et al. 2020). Evidence shows that the adsorption of NPs results in an elevation in membrane tension, leading to mechanical deformation, cell rupture, and subsequent cell death. According to a biophysical model, the interaction between NPs and membranes might lead to the constriction and stretching of the membranes. This widespread occurrence was experimentally confirmed using model membranes, Gram-positive and Gram-negative bacteria, *Pseudomonas aeruginosa* and *Staphylococcus aureus*, and model membranes. The process by which nontranslocating NPs cause the mechanical death of bacterial cells through membrane tension is supported by both theoretical and experimental evidence.

4.3.1.2 Releasing Metal Ions

Strong coordination bonds can be formed between metal ions and the N, O, or S atoms of biomolecules (Ma et al. 2009). Ag ions, for instance, have high affinity to thiols, phosphates, and amines of proteins and DNA (Le Ouay and Stellacci 2015). These biomolecules' functionality may be affected by the binding of metal ions, which may be used for antimicrobial purposes. Since the binding interaction between biomolecules and metal ions is typically nonspecific, antimicrobial agents associated with metal ions typically exhibit broad-spectrum activity. It is common for clinical applications to raise concerns regarding the biocompatibility of metal ion-associated antimicrobial agents. This is due to the fact that metal ions have the ability to target both normal mammalian cells and antibiotic-resistant bacteria.

Ag^+ ions are released by Ag NPs, which is the main source of their potent antibacterial activity. The relationship between Ag NPs and bacterial cells was investigated by Xiu et al. (2012) (Fig. 4.6a). In this study, bacterial cells were treated under various oxygen conditions using Ag NPs of varying sizes and surface moieties. Under aerobic conditions, it was discovered that Ag NPs with various coatings and sizes released Ag^+, which significantly reduced the number of bacteria. Ag NPs, on the other hand, demonstrated no bactericidal activity in anaerobic conditions, suggesting that Ag^+ is linked to antimicrobial activity. Another study that used Ag NPs to reduce bacteria more when aerobic conditions were met provided additional evidence for this conclusion (Lemire et al. 2013). Park et al., however, showed that bactericidal activity was also impacted by Ag^+-mediated ROS generation. They

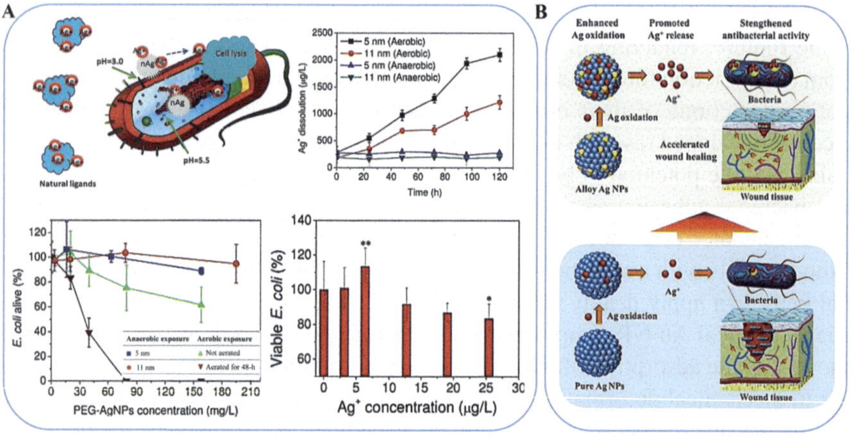

Fig. 4.6 MNPs for bacterial therapy based on ion release. (**a**) Ag NPs exert toxicity to bacterial cells through Ag^+ ion release. Copyright 2012, ACS Publication. (**b**) Upshift of the d-band center toward the Fermi level for promoting the silver ion release, bacteria inactivation, and wound healing of alloy Ag NPs. Copyright 2019, ACS Publication

demonstrated that ROS production accounted for nearly half of the antibacterial performance of Ag NPs (Marambio-Jones and Hoek 2010). According to this work, at lower concentrations, ROS appears to be the main antimicrobial mechanism; at higher concentrations, however, Ag^+ release becomes more significant. When exposed to UV radiation, ROS serves as the main antibacterial mechanism (Li et al. 2013).

Chang et. al studied the bactericidal ability of pure and alloy Ag NPs (Chang et al. 2019b). When Ag NPs were doped with Au, Pt, and Pd, their d-band centers were tuned to shift upward toward the Fermi level. The resulting Pd–Ag alloy NPs exhibited the largest shift, followed by Pt–Ag and Au–Ag NPs (Fig. 4.6b). When alloy Au–Ag, Pt–Ag, and Pd–Ag NPs are formed, the d-band center of Ag NPs can be changed to move upward toward the Fermi level. The largest shift is seen in Pd–Ag NPs, which are followed by Pt–Ag and Au–Ag NPs. Moving the d-band center upward is found to be an effective way to increase antibacterial activity, as it can promote Ag oxidation and subsequent Ag^+ release, enhance antibacterial activity, and hasten wound healing.

4.3.1.3 Photothermal

MNPs have special optical qualities that allow light irradiation to further increase their antibacterial activity. Noble MNPs, depending on their size, shape, and level of aggregation, exhibit strong light absorption in the visible to near-infrared range (Hartland 2011). A higher local temperature results from the high conversion efficiency of the absorbed light into heat. Because the photothermal effect disrupts membranes and causes the denaturalization of proteins, it has been used to eradicate bacteria. As a result, several Au nanostructures, including Au nanocages, Au nanostars, Au nanoshells, Au NRs, and Au nanoflowers, have been utilized for the photothermal (PTT) treatment of infectious diseases.

Zharov et al. (2006) reported the first studies on the photothermal activity of Au NPs under NIR irradiation. They discovered that Au NPs demonstrated bactericidal activity against *S. aureus* under laser irradiation, while laser irradiation by itself did not show any bactericidal activity. Au nanostars possess distinct optical characteristics due to their sharp tips, including a narrow and adjustable LSPR band. On a glass surface, Taglietti and colleagues grafted an Au NS monolayer, which was then further functionalized with thiols to provide wettability (Rovati et al. 2019). Au NSs were found to eliminate almost all typical bacterial strains as a result of NIR irradiation (808 nm, 0.264 W cm^{-2}) in 30 min. Yang et al. immobilized NIR light-excited Au NRs on the Ti surface by the electrostatic surface self-assembly technique (Fig. 4.7a) (Yang et al. 2019). Under 808 nm irradiation, the generated Au NRs modified surface was found to be resistant to four strains of Gram-positive cocci (*Staphylococcus aureus* and *Staphylococcus epidermidis*) and Gram-negative bacilli (*E. coli* and *Pseudomonas aeruginosa*). Five nanometer Pt nanodots were modified and applied to the surface of Au NRs to create Au@Pt NRs. Compared to Au NRs, Au@Pt NRs' LSPR absorbance band could be adjusted from 755 nm to 845 nm by altering the Pt loading from 0.05 to 0.2. The Au@$Pt_{0.1}$ NRs demonstrated the best photothermal therapy efficacy against both Gram-positive and Gram-negative

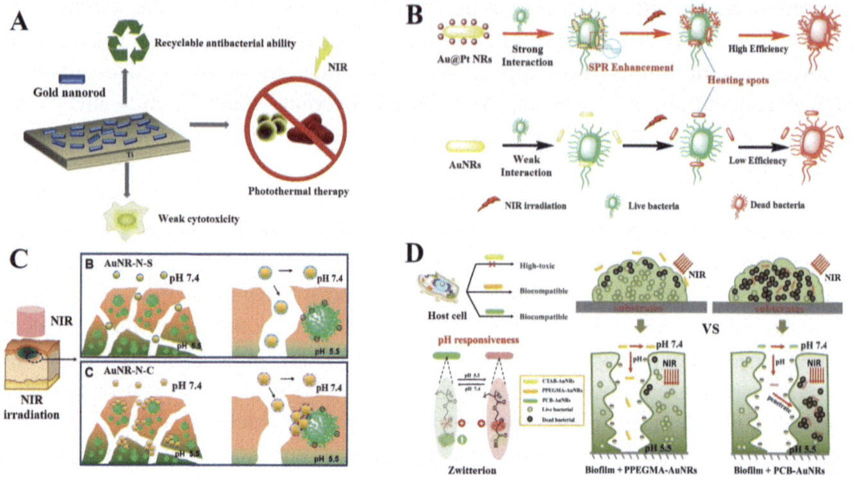

Fig. 4.7 MNPs for bacterial therapy based on photothermal. (**a**) Au NRs for the photothermal killing of bacteria. Copyright 2018, Elsevier. (**b**) Photothermal lysis of pathogenic bacteria by platinum nanodots decorated Au NRs under near-infrared irradiation. Copyright 2018, Elsevier. (**c**) Au NPs with effective adherence and enhanced photothermal ablation of MRSA biofilm. Copyright 2017, ACS Publication. (**d**) Au NRs with surface charge switchable activities for enhanced photothermal killing of bacteria and eradication of biofilm. Copyright 2020, the Royal Society of Chemistry

bacteria under NIR irradiation. Pt nanodots added to the surface of Au@Pt NRs not only increased their photothermal conversion efficiency but also increased their affinity for bacteria, greatly reducing their cytotoxicity. Bacterial cell walls were damaged as a result of photothermal lysis over Au@Pt NRs, suggesting that the thermal ablation mechanism was most likely the basis for bacterial death (Fig. 4.7b) (Zhang et al. 2018).

It has been reported that aggregated Au NPs enhance photothermal effects (Guan et al. 2019; Tian et al. 2022). Surface-adaptive Au NPs with improved photothermal ablation of methicillin-resistant *Staphylococcus aureus* (MRSA) biofilm and effective adherence were reported by Hu et al. (2017) (Fig. 4.7c). It was discovered that the surface-adaptive mixed charged zwitterionic 14 nm Au NPs could rapidly assemble on an acidic biofilm and successfully adhere to bacteria while remaining steadily distributed in healthy tissue. The aggregated Au NPs with enhanced NIR absorbance can effectively convert NIR light energy into localized heat, leading to thermal ablation of MRSA biofilm, whereas the dispersed Au NPs demonstrated negligible toxicity to healthy tissues.

There is great potential for treating bacterial infections with pH-responsive Au NPs, especially those involving drug-resistant bacteria and related biofilms. By combining polyacrylate with pendant carboxyl betaine (PCB) groups, Au NRs with pH-induced surface charge switchable activities were created (Fig. 4.7d) (Qiao et al. 2020). In acidic solutions, PCB-Au NRs' negatively charged changed to positively charged, which is indicative of bacterial infections. Because of the enhanced adherence of Au NRs to the bacteria's membrane and the deep penetration of the particles within the bacterial biofilm, the positively charged PCB-Au NRs demonstrated improved photothermal disinfection efficiency against bacteria in both planktonic and biofilm phenotypes.

4.3.1.4 MNPs for Combination Therapy

Combination therapy is attractive owing to its advantages of resistance emergence, polymicrobial infections, broadening antibacterial spectrum, and synergy (Baronia and Ahmed 2014). Ag^{+}, Zn^{2+}, and Cu^{2+} are examples of germicidal metal ions that exhibit strong germicidal effects by rupturing the bacterial membrane and inactivating the intracellular proteins directly (Yuan et al. 2018). In conjunction with PTT, these metal ions could achieve noninvasive, on-demand release while continuing to exhibit remarkable germicidal activity at low concentrations without causing adverse effects. As a result, they have been thoroughly investigated and used in the synergistic treatment of PTT and chemotherapy.

Due to their porous walls and hollow structure, gold nanocages (Au NCs) are excellent choices for the encapsulation and delivery of antibacterial medications. Based on the galvanic replacement reaction, Zhao et al. synthesized Au–Ag NCs, in which Ag NPs function as broad-spectrum antibacterial agents and Au NCs as NIR light-activated photothermal agents (Fig. 4.8a) (Wu et al. 2019). Moreover, the prolonged sustained release of Ag^{+} was made possible by coating a mesoporous silica layer on Au–Ag NCs. The temperature of a 50 mg mL^{-1} as-prepared Au–Ag@SiO_2 suspension increased from 20.7 to 57.4 °C after 10 min NIR irradiation (808 nm, 1 W cm^{-2}), which was effective in killing bacteria. Their inherent photothermal properties and long-lasting Ag ion release were fully utilized for combined anti-infective therapy induced by NIR. The as-prepared Au–Ag@SiO_2 NCs' broad-spectrum antibacterial ability was validated in vitro against the Gram-positive and Gram-negative bacteria *Staphylococcus aureus* and *E. coli*. Furthermore, with the

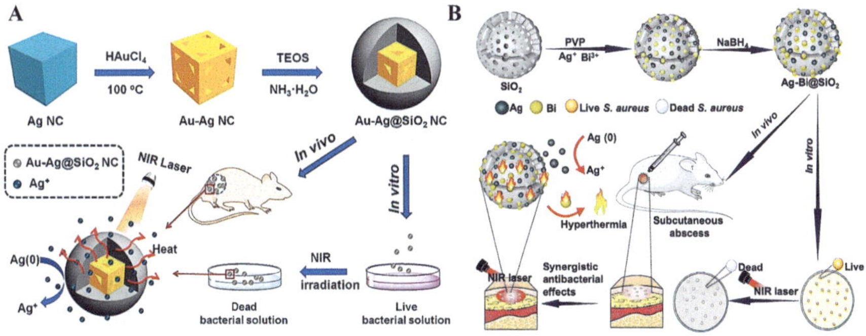

Fig. 4.8 MNPs for bacterial therapy based on combination therapy. (**a**) Silica-coated gold–silver nanocages as photothermal antibacterial agents for combined anti-infective therapy. Copyright 2019, ACS Publication. (**b**) Mesoporous silica-supported silver-bismuth nanoparticles as photothermal agents for skin infection synergistic antibacterial therapy. Copyright 2020, Wiley-VCH

aid of NIR irradiation, Au–Ag@SiO_2 NCs demonstrate efficient treatment of the *S. aureus* biofilm. Because of their exceptional near-infrared absorption capability, bismuth nanoparticles, or Bi NPs, have also been used as a perfect photothermal agent. Dong's group prepared Ag and Bi NPs coated by mesoporous silica (Fig. 4.8b) (Cao et al. 2020). By inducing hyperthermia through 808 nm laser irradiation, Bi NPs have the potential to disrupt biofilm structure, accelerate Ag^+ penetration, and improve antibacterial efficacy against MRSA and biofilms.

4.3.2 MNP Conjugates

4.3.2.1 Photothermal Therapy

It is usually not precise enough to target bacteria just by using the charge of stabilizers. Photothermal sterilization can be accomplished more precisely by targeting bacteria with antibody-conjugated MNPs. Additionally, photothermal reagents for the targeted destruction of bacteria were explored using popcorn-shaped NPs and antibody-conjugated Au NRs (Norman et al. 2008). Salmonella was developed by conjugating oven-shaped Au NPs with anti-Salmonella antibody, which had a killing efficiency of 97% when exposed to light irradiation at 670 nm for 15 minutes. This was done while *E. coli* bacteria were still alive under the same conditions. In addition, the antibiotic vancomycin has been utilized as a target ligand in the process of conjugation with polygonal Au NPs (Huang et al. 2007). Upon exposure to NIR light at 808 nm for 5 min, vancomycin-conjugated Au NPs were found to eradicate over 99% of bacteria, including pandrug-resistant *A. baumannii*, MRSA, *E. coli*, and *S. aureus*.

Zhao et al. designed a biomimetic approach combining photothermal methods to eradicate drug-resistant infections (Zhao et al. 2019). This novel blocking strategy mimics the natural carbohydrate-lectin interaction. The developed antibacterial nanoconjugate formulation virtually eliminates bacteria and inhibits biofilms, with a 90% capacity, against *P. aeruginosa* infections. The remarkable antibacterial effect was ascribed to the hetero-multivalent and three-dimensional presentation of two varieties of glycomimetic polymers on nanostructures ranging from 50 to 100 nm in size, which were specifically designed to target the LecA and LecB lectins of *P. aeruginosa*. The remarkable antibacterial effect was ascribed to the capacity of Au NRs to transform near-infrared light into thermal energy, leading to the demise of bacterial cells and additional eradication of the bacteria. This allowed the host to recuperate from the acute inflammation triggered by a drug-resistant strain of *P. aeruginosa* that was clinically isolated.

4.3.2.2 Photodynamic and Photothermal Combination Therapy

Light irradiation can boost the ROS generation of MNP conjugates in addition to the photothermal effect, increasing their bactericidal activity. The principal mechanism of ROS generation is that absorption of light promotes the molecules onto higher energy states where they can readily react with oxygen or water to produce ROS. Photodynamic therapy and photothermal therapy can work in synergy because of MNPs' intrinsic photothermal properties. This section will cover the latest advancements in antibacterial MNPs that exhibit photothermal and photodynamic effects.

As was mentioned in the previous section, one significant antibacterial mechanism is oxidative stress brought on by the production of ROS. Photodynamic effects of MNPs can further strengthen their antibacterial action. The distance dependence of plasmon-enhanced $^1O_2$ production has been demonstrated for controlled systems by varying the separation distance between the $^1O_2$ source and the metallic core Knowledge of such distance dependence would allow the engineering of photosensitizing plasmonic nanomedicines with optimized photodynamic efficiency. Zhao et al. developed a core–shell nanostructure (AuNR/SiO_2-T790) consisting of an Au NR core, a silica shell with variable thickness, and a photosensitizer molecule layer (T790), to investigate distance-dependent one- and two-photon excitation fluorescence enhancement (Fig. 4.9a) (Zhao et al. 2014). The separation distance between Au NRs and T790 was varied by changing the thickness of the silica shell spacer from 13 to 42 nm. The optimum structure for $^1O_2$ generation and photodynamic activity was found to occur at shell thickness of 20 and 35 nm under one- and two-photon excitation, respectively. Planas et al. synthesized a similar series of core–shell silver-silica NPs with a photosensitizer, Rose Bengal, tethered to their surface

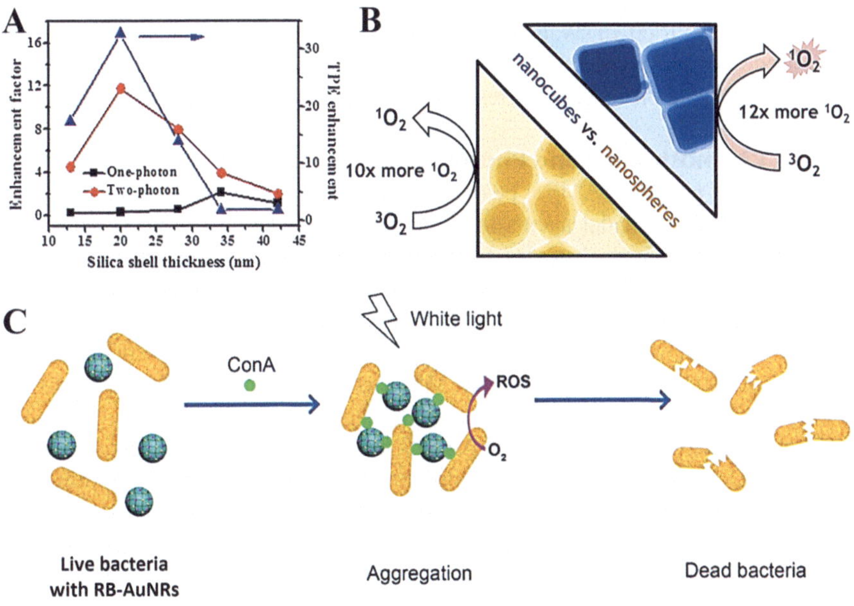

Fig. 4.9 MNP conjugates for bacterial treatment based on the photodynamic and photothermal combination therapy. (**a**) Distance-dependent plasmon-enhanced $^1O_2$ production and emission for bacterial inactivation. Copyright 2016, ACS Publication. (**b**) Hybrid silver nanocubes for improved plasmon-enhanced singlet oxygen production and inactivation of bacteria. Copyright 2019, ACS Publication. (**c**) Gold/polymer nanocomposites and their application in a controlled antibacterial assay. Copyright 2016, ACS Publication

(Planas et al. 2016). By varying the silica shell thickness from 5 to 100 nm, significant enhancement in $^1O_2$ generation for silica shell thicknesses was found to occur between 10 and 20 nm, which can be utilized to boost both photo-antibacterial activity and detection limit of $^1O_2$ in cells.

In addition to the separation distance between the molecule and the metal surface, the factors affecting the extent of such plasmonic coupling include, but are not limited to, the degree of spectral overlap between the surface plasmon band and the absorption profile of the molecule of interest (Lovell et al. 2010), its triplet and $^1O_2$ production quantum yield (Zhang et al. 2008), and the metal nanostructure size, shape, and composition (Asselin et al. 2016; Khaing et al. 2012; Camacho et al. 2015). A highly homogenous form of silver nanocubes (Ag@SiO_2-RB NCs) with silica shell thicknesses ranging from 5 to 50 nm was prepared through the fabrication of Rose Bengal-decorated silica-coated silver nanocubes (Fig. 4.9b) (Macia et al. 2019). Nanoparticles with 10 nm silica shell displayed an optimum enhancement of $^1O_2$ generation by 12-fold. The Ag@SiO_2-RB NCs exhibited superior $^1O_2$ generation, resulting in increased antibacterial activity during photodynamic inactivation tests including both Gram-positive and Gram-negative bacterial strains.

Due to their ability to regulate the activity of nearby photosensitizers, Au NPs are frequently utilized in sterilization processes. RB molecules in nanocomposites showed improved phosphorescence, longer triplet state lifetimes, reduced fluorescence, and increased generation of ROS because of increased intersystem crossing efficiency and SPR effect (Wang et al. 2016). It was found that concanavalin A (Con A) caused the Au NPs to clump together, which shortened the distance between the bacteria and the photosensitizers, resulting in enhanced sterilization (Fig. 4.9c).

4.3.2.3 Chemotherapy and Photothermal Combination Therapy

Antimicrobial peptides (AMPs) possess strong antibacterial properties and can be used as novel bactericidal agents in chemotherapy (Xuan et al. 2023). However, there are technological obstacles to the therapeutic applications because of the inherent limitations of peptides such as cytotoxicity, bioavailability, and instability (Frei et al. 2023; Wang et al. 2021). MNPs are not likely to develop drug resistance as their antibacterial properties have completely different mechanisms of action (Wang et al. 2021; Chen et al. 2020a). MNPs are regarded as the better options to combat bacterial infections in combination therapy with AMPs.

Conjugation of AMP with Ag and Au NPs has been developed because functional groups of AMP, such as sulfhydryl groups, amino groups, and carboxyl groups, can readily chelate with noble metals (Ghosh et al. 2008). Chen modified the surface of Au NRs using buforin 2b (BF2b), an artificial AMP optimized from buforin, in order to create a specific drug delivery method for killing MRSA (Fig. 4.10a) (Chen et al. 2020b). In contrast to traditional antibiotics, BF2b exhibits a unique antibacterial profile and a reduced potential to induce drug resistance in bacteria. Conversely, Au NRs possess several special qualities, such as the ability to convert light into heat to allow photothermal sterilization. DNA aptamers combined with Au NPs (Au NP-Apt;

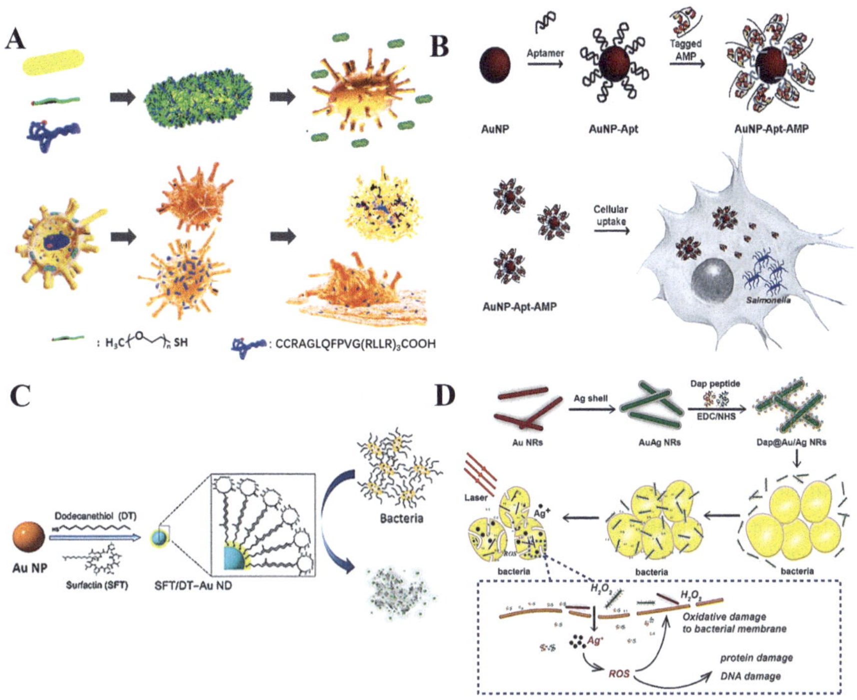

Fig. 4.10 MNP conjugates for bacterial treatment based on the chemotherapy and photothermal combination therapy. (**a**) Integration of AMPs and Au NRs for bimodal antibacterial applications. Copyright 2020, The Royal Society of Chemistry. (**b**) Au NP-DNA aptamer conjugate-assisted delivery of AMPs effectively eliminates intracellular *Salmonella enterica* serovar *Typhimurium*. Copyright 2016, Elsevier. (**c**) Self-assembly of AMPs on Au nanodots against MDR bacteria. Copyright 2015, WILEY-VCH. (**d**) Intelligent peptide nanorods against drug-resistant bacterial infection and promote wound healing by photothermal therapy. Copyright 2021, Elsevier

Fig. 4.10b) effectively delivered AMPs into mammalian living systems (Yeom et al. 2016). This system was introduced into HeLa cells that were infected with *S. Typhimurium* (*Salmonella enterica* serovar *Typhimurium*). The viability of the host cells was improved as a result of the removal of intracellular *S. Typhimurium* cells. Surfactin (SFT), an AMP, and 1-dodecanethiol (DT) were coupled onto Au NPs (SFT/DT-Au NDs) (Fig. 4.10c) (Chen et al. 2015). Using the complex antibacterial mechanisms of MNPs, Au NPs not only act as carriers but also significantly reduce the possibility that bacteria will become resistant to drugs. When it comes to both drug-resistant and nondrug-resistant bacteria, SFT/DT-Au NDs exhibit superior antibacterial activity compared to SFT. The threshold inhibitory concentration values of SFT/DT-Au NDs are significantly lower (>80-fold) than those of SFT.

Ag NPs are known to display potent antibacterial properties (Xiu et al. 2012) and can combine with other antimicrobial molecules to achieve the synergistic effect of bacterial inhibition. Chemical conjugation of a class of AMPs known as Dap with Au/Ag NRs resulted in the development of a novel photothermal antibacterial system known as Dap@Au/Ag NRs. This system exhibits broad-spectrum antibacterial activity and is resistant to bacterial resistance (Fig. 4.10d) (Dong et al. 2022). Low concentrations of H_2O_2 have been found to readily etch Dap@Au/Ag NRs, releasing silver ions with antibacterial toxicity that are subsequently used in combination therapy with AMPs. Additionally, etched Au NRs exhibit outstanding photothermal activity and good biocompatibility, making them useful in the fight against antibiotic-resistant bacteria. The high toxicity and low antibacterial efficiency of conventional antibiotics and can be overcome by this synergistic photothermal antibacterial system. It can also deal with the problem of drug-resistant strains of bacteria causing bacterial infections in the interim.

4.4 MNPs for Bacterial Theranostics

Theranostics provides opportunities for simultaneous, accurate diagnosis, and treatment of bacterial infections (He et al. 2019). MNPs can be used as imaging and therapeutic agents on their own, as well as in combination with fluorescent dyes, photosensitizers, biomolecules, and other substances to facilitate the diagnosis and treatment of bacteria (Yang et al. 2017; Li et al. 2014; Pornpattananangkul et al. 2011). Here, we go over the various approaches to treating and diagnosing bacteria using MNPs and their conjugates.

4.4.1 MNPs

Cationic Au NPs were utilized by Yuan et al. as two-photon imaging and antibacterial agents (Fig. 4.11a) (Yuan et al. 2015). This was accomplished by taking advantage of the unique optical properties and antibacterial activity of these particles. Cationic Au NPs (approximately 42 nm) aggregated on the surfaces of both Gram-positive and Gram-negative bacteria, specifically *S. aureus* and *E. coli*, due to electrostatic interactions. The utilization of positively charged Au NPs at very low levels allows for the enhancement of aggregation in 2PPL, enabling the successful imaging of bacteria using near-infrared irradiation. In addition, there was no noticeable harm to human dermal fibroblasts when the surfaces of bacteria treated with aggregated Au NPs exhibited photothermal/photodynamic effects, resulting in ~90% killing efficiency against *S. aureus* and *E. coli*. Because mammalian cells are less negatively charged than bacterial cells, there are likely fewer Au NP aggregates formed in mammalian cells, resulting in the selectivity of Au NPs. Afterward, core–shell Au@Ag NPs were made to obtain even more potent antibacterial activity (Fig. 4.11b) (Ding et al. 2017). Under near-infrared irradiation, Au@Ag NPs displayed much stronger photo-induced antibacterial activity than Au NPs of similar sizes owing to higher photothermal effects and simultaneous release of Ag^+ and ROS generation. Biofilm is a serious health risk that is typically very difficult to eradicate by standard antibiotic treatments. These Au@Ag NPs can effectively eradicate bacteria inside *S. aureus* biofilm when exposed to near-infrared irradiation.

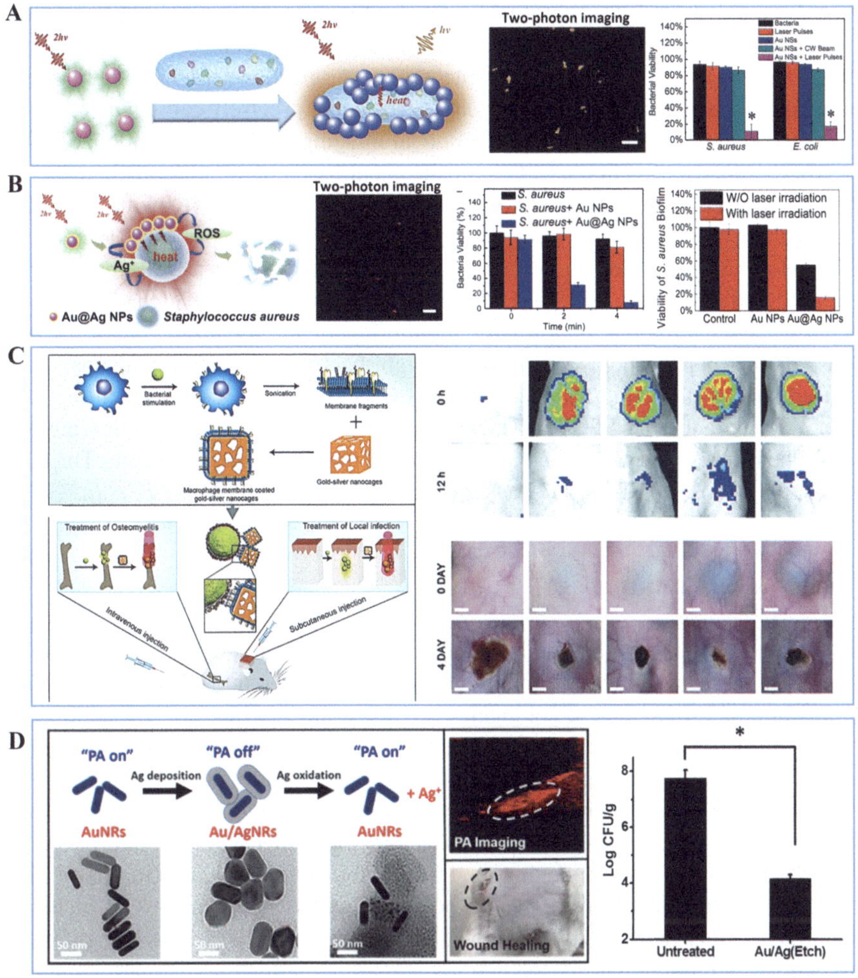

Fig. 4.11 MNPs for theranostics. (**a**) Two-photon pictures of *E. coli* and the viability of *S. aureus* and *E. coli* treated with Au NPs demonstrate the dual-functionality of Au NPs. Reproduced with permission. Copyright 2015, Wiley-VCH. (**b**) Au@Ag NPs formed aggregates on bacterial surface allowing simultaneous two-photon excitation imaging and photo-enhanced antibacterial effect against *S. aureus*. Reproduced with permission. Copyright 2017, Elsevier. (**c**) Pretreated macrophage-membrane-coated Au–Ag nanocages for treatment of bacterial infections in vivo. Copyright 2018, Wiley-VCH. (**d**) A gold/silver hybrid nanoparticle for treatment and photoacoustic imaging of bacterial infection. Copyright 2018, ACS Publication

It is critical to identify the proper biomolecule or protein on the bacterial membrane to serve as a target for multifunctional nanoparticle bioconjugate in photothermal therapy-based selective antibacterial therapy. To accomplish precise targeting in vivo, MNPs can be coated with cell membranes. Macrophages were preincubated with bacteria by Zhang et al. to stimulate the expression of toll-like receptors (also known as microbial molecular binding receptors, or TLRs) on the membranes (Wang et al. 2018). Au–Ag nanocages were coated with membranes that were obtained from pretreated macrophages, as shown in Fig. 4.11c. After attaching themselves to bacteria, the nanocages could precisely cause hyperthermia in response to near-infrared light with optical imaging capabilities simultaneously.

MNPs can also function as a theranostic agent based on photoacoustic imaging. Au/Ag hybrid NPs were prepared by Kim et al. by covering Au NRs with Ag, which resulted in a reduction in the photoacoustic signal of the microparticles (Fig. 4.11d) (Kim et al. 2018). Following the oxidative etching of the silver shell, the silver is released while the photoacoustic contrast is simultaneously recovered. This photoacoustic signal provides noninvasive monitoring of the localized release of Ag^+ ions. These nanoparticles exhibited a high bactericidal activity against both methicillin-resistant *Staphylococcus aureus* (MRSA) and *E. coli*. The killing efficiency was greater than 99.99%. These nanoparticles achieved this by releasing Ag^+ ions when exposed to ROS such as hydrogen peroxide and peroxynitrite. The concentration of Ag^+ ions released was 32 μM for MRSA and 8 μM for *E. coli*.

4.4.2 MNP Conjugates

The numerous benefits of using MNP conjugates in bacterial diagnostics and treatments are ascribed to the multifunctional capability of the integrated biomolecules (Nguyen et al. 2022). These conjugates provide a flexible platform for addressing bacterial-related problems by utilizing the multifunctionality of MNPs to fulfill a variety of roles. For instance, the attached targeted molecules can allow for the precise identification and management of various bacterial strains (Wang et al. 2010). Transferrin and Au NRs were coupled to enable improved cancer cell microsurgery, two-photon luminescence imaging, and effective targeting (Li and Gu 2010). This multimodal approach demonstrates the potential of MNP conjugates as all-inclusive solutions in this field by highlighting their effectiveness in addressing different aspects of bacterial diagnosis and therapy.

The aggregation of MNPs for bacterial detection and antibacterial treatments in vivo has also attracted more attention. Although Au NPs do not aggregate on bacterial membranes, Yang et al. showed that Au NPs can form aggregates under light stimulation after being swallowed by bacteria (Fig. 4.12a) (Yang et al. 2022). The photoactive crosslinkers in diazirine cause the nanoprobes to aggregate under 405 nm laser illumination, which significantly improves the photoacoustic signals. Ce6-modified Au NPs, diazirine, and glucose polymer (GP) make up the nanoprobes. Distinguishing between bacterial infections and other inflammation-related symptoms is difficult because most published nanoagents exhibit low selectivity toward bacterial cells when compared to mammalian cells. Consequently, this method provides higher antibacterial rates during a short treatment and enhanced imaging sensitivity. As few as 10^5 CFU of bacteria, or about two orders of magnitude less than most optical contrast agents, can be distinguished in vivo. For many

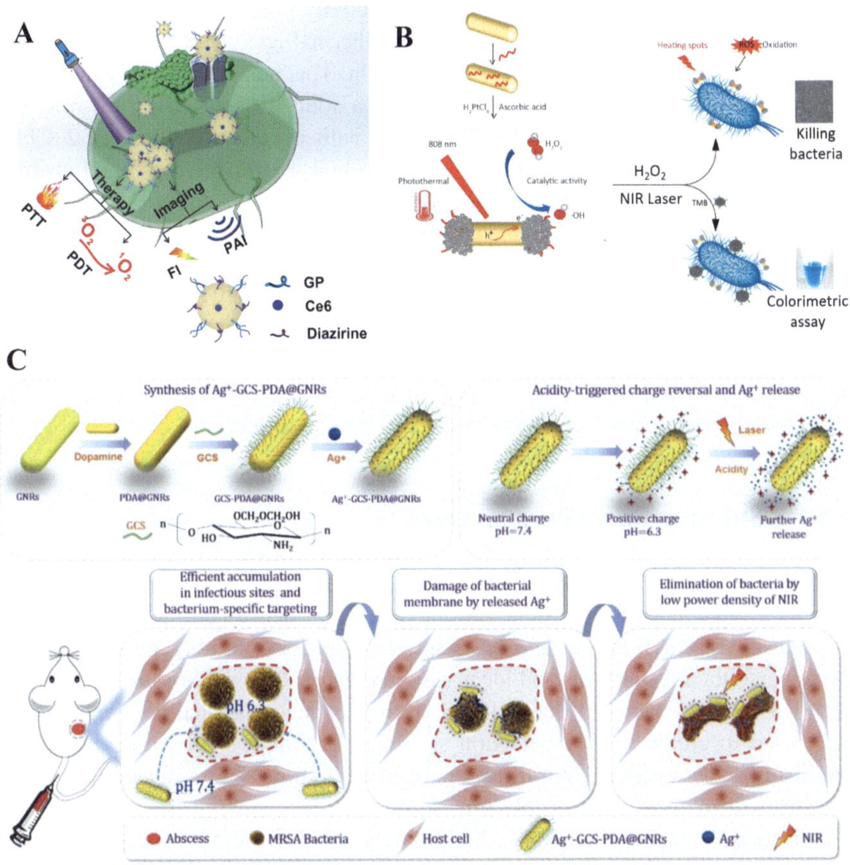

Fig. 4.12 MNPs-conjugates for theranostics. (**a**) Glucose polymer-modified Au NPs aggregate in the bacterial cells for fluorescence imaging, enhanced photoacoustic signal, and antibacterial activity, enabling sensitive imaging of bacteria in vivo. (**b**) DNA-encoded bimetallic Au–Pt dumbbell nanozyme for high-performance detection and eradication of *Escherichia coli* O157:H7. (**c**) Ag^+-GCS-PDA@GNRs is a robust and practical platform for use in synergistic bacteria ablating

in vivo circumstances, such high sensitivity is sufficient. The method of in situ aggregation for sterilization in the body has gradually emerged in recent years. A photoacoustic contrast agent based on functional peptide-modified Au NPs (Au NPs@ P1) was developed by Lu et al. (2020). By using a bacterial enzyme that is overexpressed, these nanoparticles can be made with precisely customized surface peptides that cause the Au NPs to aggregate in situ. Au NPs accumulation onto bacterial cells was significantly enhanced by the active targeting and in situ aggregation of Au NPs@P1, showing over three times higher levels than those of the active targeting candidate.

DNA-modified Au–Pt NPs were applied in the simultaneous detection and eradication of pathogenic bacteria (Fig. 4.12b) (Lu et al. 2021). The combined action of

Au and Pt results in high nanozyme catalytic activity in the as-prepared nanoparticles. The Au–Pt NPs exhibit strong photothermal conversion efficiency and increased catalytic activity when exposed to light. This means that they can be used to eradicate and detect *E. coli* O157:H7 with a solid 95% efficacy in 5 min and excellent linear detection (10–10^7 CFU/mL) with a detection limit of 2 CFU/mL. Liu et al. developed an efficient antimicrobial hybrid for combined chemo-photothermal therapy based on polydopamine (PDA)-coated Au NRs (Fig. 4.12c) (Liu et al. 2018). Ag^+-glycol chitosan-PDA@Au NRs and high silver ions loading efficiency were both attained by the PDA coating. Furthermore, grafting imaging agents was done by using the free amines present on the glycol chitosan. In addition to demonstrating an exceptional chemo-photothermal synergistic therapeutic effect on abscesses, the designed antimicrobial hybrid could specifically target and accumulate the focal infection sites, resulting in adequate bacterial eradication, accelerated wound healing, and decreased damage to normal tissue.

4.5 Summary and Perspectives

The unique optical, magnetic, and electrical properties of MNP conjugates, along with their large surface-to-volume ratio and ease of surface modification, have led to extensive research in the field of bacterial diagnostics and therapy. This review has covered the mechanisms and ideas behind creating an assay for the detection of MNPs-associated bacteria. Benefiting from their exceptional qualities, MNPs have been used to create several detection schemes for simple signal transduction or promising detection signal amplification. Numerous strategies for detecting bacteria have been developed, such as colorimetric, fluorescence, nonlinear optical, electrochemical, and magnetic methods. Additionally, MNPs' antibacterial mechanisms and uses in treating bacterial infections are examined. Using metal ion release, ROS generation, or photothermal/photodynamic effects, MNPs with exceptional antibacterial properties have been widely used in vitro to treat bacteria. The antibacterial effectiveness and action spectrum of metal-based NPs were found to be influenced by their size, shape, and chemical composition. These discoveries might provide insight into creating effective and all-encompassing metal-based nanomedicine to treat bacterial infections.

This field has advanced beyond standalone MNPs. MNP conjugates expand the scope of bacterial diagnosis and treatment. These finely tuned conjugates transform the way we treat drug-resistant bacteria by integrating MNPs with functional biomolecules. Targeted therapy, for example, is made possible by metal nanoparticle-antibody conjugates to interact with specific bacterial markers. When MNPs and nucleic acid aptamers are combined, novel detection platforms are created that provide sensitive and quick results. In addition, controlled treatment is ensured by polymer-metal nanoparticle hybrids that release agents in response to cues. This carefully planned combination marks a paradigm change, bringing in a new age of accurate, powerful, and flexible antibacterial treatments.

The effectiveness of metal-based NPs as agents for bacterial detection and treatment has been established. Before these NPs can be used clinically as therapeutic and diagnostic agents, there are still a few obstacles to be addressed. By using different surface modification techniques, the material stability and detection selectivity need to be further enhanced. High-affinity recognition agents produce strong anti-interference ability in NPs, while effective capping agents lead to high stability in NPs. To create quick, accurate, and focused biosensors to use in intricate biological settings, both modification techniques are crucial. Currently, most documented methods for detecting bacteria using metal-based NPs are still in the proof-of-principle phase and were examined on simplified specimens rather than bodily fluids. To be used in practical applications, extensive experiments involving various complex biological samples ought to be carried out.

While this review summarizes the strong antibacterial activity of metal-based NPs, the biocompatibility of these nanomaterials is an equally important and frequently overlooked criterion for clinical applications. It is extremely difficult to reduce the toxicity of metal-based NPs to normal eukaryotic cells without sacrificing their ability to fight bacteria. The broad action spectrum of MNPs makes them attractive for treating most infectious diseases. As bacterial resistance is less likely to emerge for nanomaterials with multiple action mechanisms, they are more likely to be effective against a variety of pathogenic bacteria. While photodynamic and photothermal therapies are useful in the treatment of drug-resistant bacteria, it should be noted that most reported antibacterial studies use optical densities higher than the skin tolerance threshold (Kalluru et al. 2016). The risk of injury to healthy tissues exists when using high-power density in the fight against infection. It is expected that future research on photothermal derivatives for the treatment of bacterial illnesses will use very low light intensities. Notwithstanding the encouraging outcomes these metal-based structures have shown in the treatment of infectious diseases, several crucial fundamental questions regarding nanomaterials still need to be thoroughly investigated. These include the mechanisms of synergistic action, lesion accumulation, metabolic pathways, and toxicities toward critical organs. These multifunctional MNPs can be used for concurrent detection and treatment to enable imaging-guided therapy, opening a new avenue for the advancement of nanomedicine.

This review aims to assist medical professionals and clinicians in searching for alternative therapies for infections brought on by biofilms or drug-resistant bacteria, as well as biomedical researchers interested in developing new antibacterial materials and technologies. Biomedical applications such as drug-resistant bacterial killing, persistent biofilm removal, and tumor lysis are greatly aided by suitably engineered nanocomposites with high light-to-heat/ROS conversion capacity.

References

Abbaspour A, Norouz-Sarvestani F, Noon A, Soltani N (2015) Aptamer-conjugated silver nanoparticles for electrochemical dual-aptamer-based sandwich detection of staphylococcus aureus. Biosens Bioelectron 68:149–155

Afonso AS, Perez-Lopez B, Faria RC, Mattoso LHC, Hernandez-Herrero M, Roig-Sagues AX, Maltez-da Costa M, Merkoci A (2013) Electrochemical detection of Salmonella using gold nanoparticles. Biosens Bioelectron 40(1):121–126

Asselin J, Legros P, Grégoire A, Boudreau D (2016) Correlating metal-enhanced fluorescence and structural properties in Ag@SiO_2 core-shell nanoparticles. Plasmonics 11(5):1369–1376

Baronia A, Ahmed A (2014) Current concepts in combination antibiotic therapy for critically ill patients. Indian J Crit Care Med 18(5):310–314

Boucher HW, Talbot GH, Bradley JS, Edwards JE, Gilbert D, Rice LB, Scheld M, Spellberg B, Bartlett J (2009) Bad bugs, no drugs: no ESKAPE! An update from the Infectious Diseases Society of America. Clin Infect Dis 48(1):1–12

Brown ED, Wright GD (2016) Antibacterial drug discovery in the resistance era. Nature 529(7586):336–343

Camacho SA, Aoki PHB, Albella P, Oliveira ON, Constantino CJL, Aroca RF (2015) Increasing the enhancement factor in plasmon-enhanced fluorescence with shell-isolated nanoparticles. J Phys Chem C 120(37):20530–20535

Cao C, Ge W, Yin J, Yang D, Wang W, Song X, Hu Y, Yin J, Dong X (2020) Mesoporous silica supported silver-bismuth nanoparticles as photothermal agents for skin infection synergistic antibacterial therapy. Small 16(24):2000436

Chang CC, Chen CP, Wu TH, Yang CH, Lin CW, Chen CY (2019a) Gold nanoparticle-based colorimetric strategies for chemical and biological sensing applications. Nano 9(6):861

Chang Y, Cheng Y, Feng Y, Li K, Jian H, Zhang H (2019b) Upshift of the d band center toward the fermi level for promoting silver ion release, bacteria inactivation, and wound healing of alloy silver nanoparticles. ACS Appl Mater Interfaces 11(13):12224–12231

Chen HJ, Shao L, Li Q, Wang JF (2013) Gold nanorods and their plasmonic properties. Chem Soc Rev 42(7):2679–2724

Chen W-Y, Chang H-Y, Lu J-K, Huang Y-C, Harroun SG, Tseng Y-T, Li Y-J, Huang C-C, Chang H-T (2015) Self-assembly of antimicrobial peptides on gold nanodots: against multidrug-resistant bacteria and wound-healing application. Adv Funct Mater 25(46):7189–7199

Chen J, Andler SM, Goddard JM, Nugen SR, Rotello VM (2017) Integrating recognition elements with nanomaterials for bacteria sensing. Chem Soc Rev 46(5):1272–1283

Chen Y, Gao Y, Chen Y, Liu L, Mo A, Peng Q (2020a) Nanomaterials-based photothermal therapy and its potentials in antibacterial treatment. J Control Release 328:251–262

Chen J, Dai T, Yu J, Dai X, Chen R, Wu J, Li N, Fan L, Mao Z, Sheng G, Li L (2020b) Integration of antimicrobial peptides and gold nanorods for bimodal antibacterial applications. Biomater Sci 8(16):4447–4457

Chen S, Zong X, Zheng J, Zhang J, Zhou M, Chen Q, Man C, Jiang Y (2021) A colorimetric strategy based on aptamer-catalyzed hairpin assembly for the on-site detection of salmonella typhimurium in milk. Food Secur 10(11):2539

Colombo M, Carregal-Romero S, Casula MF, Gutierrez L, Morales MP, Bohm IB, Heverhagen JT, Prosperi D, Parak WJ (2012) Biological applications of magnetic nanoparticles. Chem Soc Rev 41(11):4306–4334

Deusenbery C, Wang Y, Shukla A (2021) Recent innovations in bacterial infection detection and treatment. ACS Inf Dis 7(4):695–720

Ding X, Yuan P, Gao N, Zhu H, Yang YY, Xu QH (2017) Au-Ag core-shell nanoparticles for simultaneous bacterial imaging and synergistic antibacterial activity. Nanomedicine 13(1):297–305

Dong X, Ye J, Chen Y, Tanziela T, Jiang H, Wang X (2022) Intelligent peptide-nanorods against drug-resistant bacterial infection and promote wound healing by mild-temperature photothermal therapy. Chem Eng J 432:134061

Elashnikov R, Radocha M, Panov I, Rimpelova S, Ulbrich P, Michalcova A, Svorcik V, Lyutakov O (2019) Porphyrin-silver nanoparticles hybrids: Synthesis, characterization and antibacterial activity. Mater Sci Eng C Mater Biol Appl 102:192–199

Feng Y, Coradi Tonon C, Ashraf S, Hasan T (2021) Photodynamic and antibiotic therapy in combination against bacterial infections: efficacy, determinants, mechanisms, and future perspectives. Adv Drug Deliv Rev 177:113941

Frei A, Verderosa AD, Elliott AG, Zuegg J, Blaskovich MAT (2023) Metals to combat antimicrobial resistance. Nat Rev Chem 7(3):202–224

Gao JH, Li L, Ho PL, Mak GC, Gu HW, Xu B (2006) Combining fluorescent probes and biofunctional magnetic nanoparticles for rapid detection of bacteria in human blood. Adv Mater 18(23):3145

Gao JH, Gu HW, Xu B (2009) Multifunctional magnetic nanoparticles: design, synthesis, and biomedical applications. Acc Chem Res 42(8):1097–1107

Ghosh P, Han G, De M, Kim CK, Rotello VM (2008) Gold nanoparticles in delivery applications. Adv Drug Deliv Rev 60(11):1307–1315

Gu HW, Xu KM, Xu CJ, Xu B (2006) Biofunctional magnetic nanoparticles for protein separation and pathogen detection. Chem Commun 9:941–949

Guan Z, Polavarapu L, Xu QH (2010) Enhanced two-photon emission in coupled metal nanoparticles induced by conjugated polymers. Langmuir 26(23):18020–18023

Guan Z, Gao N, Jiang XF, Yuan P, Han F, Xu QH (2013) Huge enhancement in two-photon photoluminescence of Au nanoparticle clusters revealed by single-particle spectroscopy. J Am Chem Soc 135(19):7272–7727

Guan Z, Zhang T, Zhu H, Lyu D, Sarangapani S, Xu QH, Lang MJ (2019) Simultaneous imaging and selective photothermal therapy through aptamer-driven Au nanosphere clustering. J Phys Chem Lett 10(2):183–188

Hajipour MJ, Fromm KM, Ashkarran AA, de Aberasturi DJ, de Larramendi IR, Rojo T, Serpooshan V, Parak WJ, Mahmoudi M (2012) Antibacterial properties of nanoparticles. Trends Biotechnol 30(10):499–511

Halas NJ, Lal S, Chang WS, Link S, Nordlander P (2011) Plasmons in strongly coupled metallic nanostructures. Chem Rev 111(6):3913–3961

Han F, Guan Z, Tan TS, Xu QH (2012) Size-dependent two-photon excitation photoluminescence enhancement in coupled noble-metal nanoparticles. ACS Appl Mater Interfaces 4(9):4746–4751

Hartland GV (2011) Optical studies of dynamics in noble metal nanostructures. Chem Rev 111(6):3858–3887

He GS, Tan LS, Zheng Q, Prasad PN (2008a) Multiphoton absorbing materials: molecular designs, characterizations, and applications. Chem Rev 108(4):1245–1330

He W, Henne WA, Wei Q, Zhao Y, Doorneweerd DD, Cheng JX, Low PS, Wei A (2008b) Two-photon luminescence imaging of bacillus spores using peptide-functionalized gold nanorods. Nano Res 1(6):450–456

He J, Li C, Ding L, Huang Y, Yin X, Zhang J, Zhang J, Yao C, Liang M, Pirraco RP, Chen J, Lu Q, Baldridge R, Zhang Y, Wu M, Reis RL, Wang Y (2019) Tumor targeting strategies of smart fluorescent nanoparticles and their applications in cancer diagnosis and treatment. Adv Mater 31(40):e1902409

Hu D, Li H, Wang B, Ye Z, Lei W, Jia F, Jin Q, Ren KF, Ji J (2017) Surface-adaptive gold nanoparticles with effective adherence and enhanced photothermal ablation of methicillin-resistant Staphylococcus aureus biofilm. ACS Nano 11(9):9330–9339

Huang WC, Tsai PJ, Chen YC (2007) Functional gold nanoparticles as photothermal agents for selective-killing of pathogenic bacteria. Nanomedicine 2(6):777–787

Huang X, Xu Z, Mao Y, Ji Y, Xu H, Xiong Y, Li Y (2015) Gold nanoparticle-based dynamic light scattering immunoassay for ultrasensitive detection of Listeria monocytogenes in lettuces. Biosens Bioelectron 66:184–190

Huang XL, Liu YJ, Yung B, Xiong YH, Chen XY (2017) Nanotechnology-enhanced no-wash biosensors for in vitro diagnostics of cancer. ACS Nano 11(6):5238–5292

Huang X, Li L, Chen Z, Yu H, You X, Kong N, Tao W, Zhou X, Huang J (2023) Nanomedicine for the detection and treatment of ocular bacterial infections. Adv Mater 35:2302431

Huo J, Jia Q, Huang H, Zhang J, Li P, Dong X, Huang W (2021) Emerging photothermal-derived multimodal synergistic therapy in combating bacterial infections. Chem Soc Rev 50(15):8762–8789

Ivanova EP, Hasan J, Webb HK, Truong VK, Watson GS, Watson JA, Baulin VA, Pogodin S, Wang JY, Tobin MJ, Lobbe C, Crawford RJ (2012) Natural bactericidal surfaces: mechanical rupture of Pseudomonas aeruginosa cells by cicada wings. Small 8(16):2489–2494

Ivanova EP, Hasan J, Webb HK, Gervinskas G, Juodkazis S, Truong VK, Wu AH, Lamb RN, Baulin VA, Watson GS, Watson JA, Mainwaring DE, Crawford RJ (2013) Bactericidal activity of black silicon. Nat Commun 4:2838

Jiang C, Zhao T, Yuan P, Gao N, Pan Y, Guan Z, Zhou N, Xu QH (2013a) Two-photon induced photoluminescence and singlet oxygen generation from aggregated gold nanoparticles. ACS Appl Mater Interfaces 5(11):4972–4977

Jiang XF, Pan Y, Jiang C, Zhao T, Yuan P, Venkatesan T, Xu QH (2013b) Excitation nature of two-photon photoluminescence of gold nanorods and coupled gold nanoparticles studied by two-pulse emission modulation spectroscopy. J Phys Chem Lett 4(10):1634–1648

Jin B, Wang S, Lin M, Jin Y, Zhang S, Cui X, Gong Y, Li A, Xu F, Lu TJ (2017) Upconversion nanoparticles based FRET aptasensor for rapid and ultrasenstive bacteria detection. Biosens Bioelectron 90:525–533

Kalluru P, Vankayala R, Chiang CS, Hwang KC (2016) Nano-graphene oxide-mediated in vivo fluorescence imaging and bimodal photodynamic and photothermal destruction of tumors. Biomaterials 95:1–10

Khaing MK, Yang Y, Hu Y, Gomez M, Du H, Wang H (2012) Gold nanoparticle-enhanced and size-dependent generation of reactive oxygen species from protoporphyrin IX. ACS Nano 6:1939–1947

Khan SA, Singh AK, Senapati D, Fan Z, Ray PC (2011) Targeted highly sensitive detection of multi-drug resistant salmonella DT104 using gold nanoparticles. Chem Commun 47(33):9444–9446

Kim T, Zhang Q, Li J, Zhang L, Jokerst JV (2018) A gold/silver hybrid nanoparticle for treatment and photoacoustic imaging of bacterial infection. ACS Nano 12(6):5615–5625

Kinkhabwala A, Yu Z, Fan S, Avlasevich Y, Müllen K, Moerner WE (2009) Large single-molecule fluorescence enhancements produced by a bowtie nanoantenna. Nat Photonics 3(11):654–657

Kuang Y, Song M, Zhou X, Mi J, Zhang Z, Liu G, Shen Z, Liu Z, Chen C, Wu MX, Zhao Y, Yang B, Jiang Y (2023) Cefminox sodium carbon nanodots for treatment and bacterial detection of bloodstream infection. Chem Eng J 470:143988

Le Ouay B, Stellacci F (2015) Antibacterial activity of silver nanoparticles: a surface science insight. Nano Today 10(3):339–354

Lee YH, Lakshminarayana P, Jiang C, Yuan P, Xu Q-H (2011) Recent advances in metal-enhanced optical properties. Cosmos 06(02):167–195

Lee H, Shin TH, Cheon J, Weissleder R (2015) Recent developments in magnetic diagnostic systems. Chem Rev 115(19):10690–10724

Lee JW, Choi SR, Heo JH (2021) Simultaneous stabilization and functionalization of gold nanoparticles via biomolecule conjugation: progress and perspectives. ACS Appl Mater Interfaces 13(36):42311–42328

Lemire JA, Harrison JJ, Turner RJ (2013) Antimicrobial activity of metals: mechanisms, molecular targets and applications. Nat Rev Microbiol 11(6):371–384

Li J-L, Gu M (2010) Surface plasmonic gold nanorods for enhanced two-photon microscopic imaging and apoptosis induction of cancer cells. Biomaterials 31(36):9492–9498

Li Y, Zhang W, Niu JF, Chen YS (2013) Surface-coating-dependent dissolution, aggregation, and reactive oxygen species (ROS) generation of silver nanoparticles under different irradiation conditions. Environ Sci Technol 47(18):10293–10301

Li XN, Robinson SM, Gupta A, Saha K, Jiang ZW, Moyano DF, Sahar A, Riley MA, Rotello VM (2014) Functional gold nanoparticles as potent antimicrobial agents against multi-drug-resistant bacteria. ACS Nano 8(10):10682–10686

Li F, Li F, Yang G, Aguilar ZP, Lai W, Xu H (2018) Asymmetric polymerase chain assay combined with propidium monoazide treatment and unmodified gold nanoparticles for colorimetric detection of viable emetic bacillus cereus in milk. Sensors Actuators B Chem 255:1455–1461

Li S, Shen X, Xu QH, Cao Y (2019) Gold nanorod enhanced conjugated polymer/photosensitizer composite nanoparticles for simultaneous two-photon excitation fluorescence imaging and photodynamic therapy. Nanoscale 11(41):19551–19560

Lim DK, Jeon KS, Kim HM, Nam JM, Suh YD (2010) Nanogap-engineerable Raman-active nanodumbbells for single-molecule detection. Nat Mater 9(1):60–67

Linklater DP, Juodkazis S, Rubanov S, Ivanova EP (2017) Comment on "Bactericidal effects of natural nanotopography of dragonfly wing on Escherichia coli". ACS Appl Mater Interfaces 9(35):29387–29393

Linklater DP, Baulin VA, Le Guevel X, Fleury JB, Hanssen E, Nguyen THP, Juodkazis S, Bryant G, Crawford RJ, Stoodley P, Ivanova EP (2020) Antibacterial action of nanoparticles by lethal stretching of bacterial cell membranes. Adv Mater 32(52):e2005679

Liu M, He D, Yang T, Liu W, Mao L, Zhu Y, Wu J, Luo G, Deng J (2018) An efficient antimicrobial depot for infectious site-targeted chemo-photothermal therapy. J Nanobiotechnol 16(1):23

Lovell JF, Liu TWB, Chen J, Zheng G (2010) Activatable photosensitizers for imaging and therapy. Chem Rev 110:2839–2857

Lu SZ, Guo XY, Zou MS, Zheng ZQ, Li YC, Li XD, Li LL, Wang H (2020) Bacteria-instructed in situ aggregation of Au NPs with enhanced photoacoustic signal for bacterial infection bioimaging. Adv Healthc Mater 9(1):e1901229

Lu C, Tang L, Gao F, Li Y, Liu J, Zheng J (2021) DNA-encoded bimetallic Au-Pt dumbbell nanozyme for high-performance detection and eradication of Escherichia coli O157:H7. Biosens Bioelectron 187:113327

Ma Z, Jacobsen FE, Giedroc DP (2009) Coordination chemistry of bacterial metal transport and sensing. Chem Rev 109(10):4644–4681

Macia N, Bresoli-Obach R, Nonell S, Heyne B (2019) Hybrid silver nanocubes for improved plasmon-enhanced singlet oxygen production and inactivation of bacteria. J Am Chem Soc 141(1):684–692

Manickam P, Vashist A, Madhu S, Sadasivam M, Sakthivel A, Kaushik A, Nair M (2020) Gold nanocubes embedded biocompatible hybrid hydrogels for electrochemical detection of H(2) O(2). Bioelectrochemistry 131:107373

Marambio-Jones C, Hoek EMV (2010) A review of the antibacterial effects of silver nanomaterials and potential implications for human health and the environment. J Nanopart Res 12(5):1531–1551

Mayer KM, Hafner JH (2011) Localized surface plasmon resonance sensors. Chem Rev 111(6):3828–3857

McVeya C, Huangb F, Elliotta C, Cao C (2017) Endonuclease controlled aggregation of gold nanoparticles for the ultrasensitive detection of pathogenic bacterial DNA. Biosens Bioelectron 92:502–508

Miranda OR, Li XN, Garcia-Gonzalez L, Zhu ZJ, Yan B, Bunz UHF, Rotello VM (2011) Colorimetric bacteria sensing using a supramolecular enzyme-nanoparticle biosensor. J Am Chem Soc 133(25):9650–9653

Mirkin CA, Letsinger RL, Mucic RC, Storhoff JJ (1996) A DNA-based method for rationally assembling nanoparticles into macroscopic materials. Nature 382(6592):607–609

Nguyen VN, Zhao Z, Tang BZ, Yoon J (2022) Organic photosensitizers for antimicrobial phototherapy. Chem Soc Rev 51(9):3324–3340

Norman RS, Stone JW, Gole A, Murphy CJ, Sabo-Attwood TL (2008) Targeted photothermal lysis of the pathogenic bacteria, Pseudomonas aeruginosa, with gold nanorods. Nano Lett 8(1):302–306

Olaitan AO, Morand S, Rolain JM (2014) Mechanisms of polymyxin resistance: acquired and intrinsic resistance in bacteria. Front Microbiol 5(18):643

Olesiak-Banska J, Waszkielewicz M, Obstarczyk P, Samoc M (2019) Two-photon absorption and photoluminescence of colloidal gold nanoparticles and nanoclusters. Chem Soc Rev 48(15):4087–4117

Pallares RM, Thanh NTK, Su X (2019) Sensing of circulating cancer biomarkers with metal nanoparticles. Nanoscale 11(46):22152–22171

Pan X, Wu N, Tian S, Guo J, Wang C, Sun Y, Huang Z, Chen F, Wu Q, Jing Y, Yin Z, Zhao B, Xiong X, Liu H, Zhou D (2022) Inhalable MOF-derived nanoparticles for sonodynamic therapy of bacterial pneumonia. Adv Funct Mater 32(25):2112145

Panda BR, Singh AK, Ramesh A, Chattopadhyay A (2008) Rapid estimation of bacteria by a fluorescent gold nanoparticle-polythiophene composite. Langmuir 24(20):11995–12000

Pandey R, Chang D, Smieja M, Hoare T, Li Y, Soleymani L (2021) Integrating programmable DNAzymes with electrical readout for rapid and culture-free bacterial detection using a handheld platform. Nat Chem 13(9):895–901

Penders J, Stolzoff M, Hickey DJ, Andersson M, Webster TJ (2017) Shape-dependent antibacterial effects of non-cytotoxic gold nanoparticles. Int J Nanomedicine 12:2457–2468

Peng Y, Liu Y, Lu X, Wang S, Chen M, Huang W, Wu Z, Lu G, Nie L (2018) Ag-hybridized plasmonic au-triangular nanoplates: highly sensitive photoacoustic/raman evaluation and improved antibacterial/photothermal combination therapy. J Mater Chem B 6(18):2813–2820

Peng H, Borg RE, Nguyen ABN, Chen IA (2020) Chimeric phage nanoparticles for rapid characterization of bacterial pathogens: detection in complex biological samples and determination of antibiotic sensitivity. ACS Sens 5(5):1491–1499

Peukert C, Popat Gholap S, Green O, Pinkert L, van den Heuvel J, van Ham M, Shabat D, Brönstrup M (2022) Enzyme-activated, chemiluminescent siderophore-dioxetane probes enable the selective and highly sensitive detection of bacterial pathogens. Angew Chem Int Ed 61(25):e202201423

Phillips RL, Miranda OR, You CC, Rotello VM, Bunz UHF (2008) Rapid and efficient identification of bacteria using gold-nanoparticle - poly(para-phenyleneethynylene) constructs. Angew Chem Int Ed 47(14):2590–2594

Planas O, Macia N, Agut M, Nonell S, Heyne B (2016) Distance-dependent plasmon-enhanced singlet oxygen production and emission for bacterial inactivation. J Am Chem Soc 138(8):2762–2768

Polavarapu L, Pérez-Juste J, Xu Q-H, Liz-Marzán LM (2014) Optical sensing of biological, chemical and ionic species through aggregation of plasmonic nanoparticles. J Mater Chem C 2(36):7460–7476

Pornpattananangkul D, Zhang L, Olson S, Aryal S, Obonyo M, Vecchio K, Huang CM, Zhang LF (2011) Bacterial toxin-triggered drug release from gold nanoparticle-stabilized liposomes for the treatment of bacterial infection. J Am Chem Soc 133(11):4132–4139

Qiao Z, Yao Y, Song S, Yin M, Yang M, Yan D, Yang L, Luo J (2020) Gold nanorods with surface charge-switchable activities for enhanced photothermal killing of bacteria and eradication of biofilm. J Mater Chem B 8(15):3138–3149

Ray PC, Khan SA, Singh AK, Senapati D, Fan Z (2012) Nanomaterials for targeted detection and photothermal killing of bacteria. Chem Soc Rev 41(8):3193–3209

Rossi F, Khoo EH, Su X, Thanh NTK (2020) Study of the effect of anisotropic gold nanoparticles on plasmonic coupling with a photosensitizer for antimicrobial film. ACS Appl Biol Mater 3(1):315–326

Rovati D, Albini B, Galinetto P, Grisoli P, Bassi B, Pallavicini P, Dacarro G, Taglietti A (2019) High stability thiol-coated gold nanostars monolayers with photo-thermal antibacterial activity and wettability control. Nano 9(9):1288

Saha K, Agasti SS, Kim C, Li X, Rotello VM (2012) Gold nanoparticles in chemical and biological sensing. Chem Rev 112(5):2739–2779

Sanchez-Lopez E, Gomes D, Esteruelas G, Bonilla L, Lopez-Machado AL, Galindo R, Cano A, Espina M, Ettcheto M, Camins A, Silva AM, Durazzo A, Santini A, Garcia ML, Souto EB (2020) Metal-based nanoparticles as antimicrobial agents: an overview. Nano 10(2):292

Santopolo G, Domenech-Sanchez A, Russell SM, de la Rica R (2019) Ultrafast and ultrasensitive naked-eye detection of urease-positive bacteria with plasmonic nanosensors. ACS Sens 4(4):961–967

Singh AK, Senapati D, Wang S, Griffin J, Neely A, Candice P, Naylor KM, Varisli B, Kalluri JR, Ray PC (2009) Gold nanorod based selective identification of Escherichia coli bacteria using two-photon rayleigh scattering spectroscopy. ACS Nano 3:1906–1912

Stoimenov PK, Klinger RL, Marchin GL, Klabunde KJ (2002) Metal oxide nanoparticles as bactericidal agents. Langmuir 18(17):6679–6686

Su S, Sun H, Cao W, Chao J, Peng H, Zuo X, Yuwen L, Fan C, Wang L (2016) Dual-target electrochemical biosensing based on DNA structural switching on gold nanoparticle-decorated MoS2 nanosheets. ACS Appl Mater Interfaces 8(11):6826–6833

Swierczewska M, Lee S, Chen X (2011) The design and application of fluorophore-gold nanoparticle activatable probes. Phys Chem Chem Phys 13(21):9929–9941

Tian S, He J, Lyu D, Li S, Xu Q-H (2022) Aggregation enhanced photoactivity of photosensitizer conjugated metal nanoparticles for multimodal imaging and synergistic phototherapy below skin tolerance threshold. Nano Today 45:101543

Tomaszewska E, Soliwoda K, Kadziola K, Tkacz-Szczesna B, Celichowski G, Cichomski M, Szmaja W, Grobelny J (2013) Detection limits of DLS and UV-Vis spectroscopy in characterization of polydisperse nanoparticles colloids. J Nanomater 2013:1–10

Usman MS, El Zowalaty ME, Shameli K, Zainuddin N, Salama M, Ibrahim NA (2013) Synthesis, characterization, and antimicrobial properties of copper nanoparticles. Int J Nanomedicine 8:4467–4478

Wang SG, Singh AK, Senapati D, Neely A, Yu HT, Ray PC (2010) Rapid colorimetric identification and targeted photothermal lysis of salmonella bacteria by using bioconjugated oval-shaped gold nanoparticles. Chem Eur J 16(19):5600–5606

Wang C, Cui Q, Wang X, Li L (2016) Preparation of hybrid gold/polymer nanocomposites and their application in a controlled antibacterial assay. ACS Appl Mater Interfaces 8(42):29101–29109

Wang C, Wang Y, Zhang L, Miron RJ, Liang J, Shi M, Mo W, Zheng S, Zhao Y, Zhang Y (2018) Pretreated macrophage-membrane-coated gold nanocages for precise drug delivery for treatment of bacterial infections. Adv Mater 30(46):1804023

Wang N, Dai H, Sai L, Ma H, Lin M (2019) Copper ion-assisted gold nanoparticle aggregates for electrochemical signal amplification of lipopolysaccharide sensing. Biosens Bioelectron 126:529–534

Wang J, Yang Y, Xu Y, Zhao L, Wang L, Yin Z, Li H, Tan H, Liu K (2020a) A dual enhanced antibacterial strategy based on high chlorin e6-loaded polyethyleneimine functionalized graphene. RSC Adv 11(2):739–744

Wang L, Hasanzadeh Kafshgari M, Meunier M (2020b) Optical properties and applications of plasmonic-metal nanoparticles. Adv Funct Mater 30(51):2005400

Wang C, Hong T, Cui P, Wang J, Xia J (2021) Antimicrobial peptides towards clinical application: delivery and formulation. Adv Drug Deliv Rev 175:113818

Wei G, Yang G, Wang Y, Jiang H, Fu Y, Yue G, Ju R (2020) Phototherapy-based combination strategies for bacterial infection treatment. Theranostics 10(26):12241–12262

Wu S, Li A, Zhao X, Zhang C, Yu B, Zhao N, Xu F-J (2019) Silica-coated gold–silver nanocages as photothermal antibacterial agents for combined anti-infective therapy. ACS Appl Mater Interfaces 11(19):17177–17183

Wu Q, Peng R, Luo Y, Cui Q, Zhu S, Li L (2021) Antibacterial activity of porous gold nanocomposites via NIR light-triggered photothermal and photodynamic effects. ACS Appl Biol Mater 4(6):5071–5079

Xiu ZM, Zhang QB, Puppala HL, Colvin VL, Alvarez PJJ (2012) Negligible particle-specific antibacterial activity of silver nanoparticles. Nano Lett 12(8):4271–4275

Xuan J, Feng W, Wang J, Wang R, Zhang B, Bo L, Chen Z-S, Yang H, Sun L (2023) Antimicrobial peptides for combating drug-resistant bacterial infections. Drug Resist Updat 68:100954

Yang XL, Yang JC, Wang L, Ran B, Jia YX, Zhang LM, Yang G, Shao HW, Jiang XY (2017) Pharmaceutical intermediate-modified gold nanoparticles: against multidrug-resistant bacteria and wound-healing application via an electrospun scaffold. ACS Nano 11(6):5737–5745

Yang T, Wang D, Liu X (2019) Assembled gold nanorods for the photothermal killing of bacteria. Colloids Surf B Biointerfaces 173:833–841

Yang Y, Chu B, Cheng J, Tang J, Song B, Wang H, He Y (2022) Bacteria eat nanoprobes for aggregation-enhanced imaging and killing diverse microorganisms. Nat Commun 13(1):1255

Yeom J-H, Lee B, Kim D, Lee J-K, Kim S, Bae J, Park Y, Lee K (2016) Gold nanoparticle-DNA aptamer conjugate-assisted delivery of antimicrobial peptide effectively eliminates intracellular Salmonella enterica serovar Typhimurium. Biomaterials 104:43–51

Yu MQ, Wang H, Fu F, Li LY, Li J, Li G, Song Y, Swihart MT, Song EQ (2017) Dual-recognition forster resonance energy transfer based platform for one-step sensitive detection of pathogenic bacteria using fluorescent vancomycin-gold nanoclusters and aptamer-gold nanoparticles. Anal Chem 89(7):4085–4090

Yuan P, Lee YH, Gnanasammandhan MK, Guan Z, Zhang Y, Xu QH (2012) Plasmon enhanced upconversion luminescence of NaYF4:Yb,Er@SiO_2@Ag core-shell nanocomposites for cell imaging. Nanoscale 4(16):5132–5137

Yuan PY, Ma RZ, Guan ZP, Gao NY, Xu QH (2014) Tuning two-photon photoluminescence of gold nanoparticle aggregates with DNA and its application as turn-on photoluminescence probe for DNA sequence detection. ACS Appl Mater Interfaces 6(15):13149–13156

Yuan PY, Ding X, Guan ZP, Gao NY, Ma R, Jiang XF, Yang YY, Xu QH (2015) Plasmon-coupled gold nanospheres for two-photon imaging and photoantibacterial activity. Adv Healthc Mater 4(5):674–678

Yuan P, Ding X, Yang YY, Xu QH (2018) Metal nanoparticles for diagnosis and therapy of bacterial infection. Adv Healthc Mater 7(13):e1701392

Zhang Y, Aslan K, Previte MJR, Geddes CD (2008) Plasmonic engineering of singlet oxygen generation. Proc Natl Acad Sci USA 105:1798–1802

Zhang D, Carr DJ, Alocilja EC (2009) Fluorescent bio-barcode DNA assay for the detection of Salmonella enterica serovar Enteritidis. Biosens Bioelectron 24(5):1377–1381

Zhang C, Yin A-X, Jiang R, Rong J, Dong L, Zhao T, Sun L-D, Wang J, Chen X, Yan C-H (2013) Time-temperature indicator for perishable products based on kinetically programmable Ag overgrowth on Au nanorods. ACS Nano 7(5):4561–4568

Zhang H, Ma XY, Liu Y, Duan N, Wu SJ, Wang ZP, Xu BC (2015a) Gold nanoparticles enhanced SERS aptasensor for the simultaneous detection of Salmonella typhimurium and Staphylococcus aureus. Biosens Bioelectron 74:872–877

Zhang Y, Shareena Dasari TP, Deng H, Yu H (2015b) Antimicrobial activity of gold nanoparticles and ionic gold. J Environ Sci Health C Environ Carcinog Ecotoxicol Rev 33(3):286–327

Zhang J, Feng Y, Mi J, Shen Y, Tu Z, Liu L (2018) Photothermal lysis of pathogenic bacteria by platinum nanodots decorated gold nanorods under near infrared irradiation. J Hazard Mater 342:121–130

Zhang A, Pan S, Zhang Y, Chang J, Cheng J, Huang Z, Li T, Zhang C, de la Fuentea JM, Zhang Q, Cui D (2019) Carbon-gold hybrid nanoprobes for real-time imaging, photothermal/photodynamic and nanozyme oxidative therapy. Theranostics 9(12):3443–3458

Zhao T, Yu K, Li L, Zhang T, Guan Z, Gao N, Yuan P, Li S, Yao SQ, Xu QH, Xu GQ (2014) Gold nanorod enhanced two-photon excitation fluorescence of photosensitizers for two-photon imaging and photodynamic therapy. ACS Appl Mater Interfaces 6(4):2700–2708

Zhao Y, Guo Q, Dai X, Wei X, Yu Y, Chen X, Li C, Cao Z, Zhang X (2019) A biomimetic non-antibiotic approach to eradicate drug-resistant infections. Adv Mater 31(7):e1806024

Zhao W, Ding M, Zhang X, Xin Z, Song L, Cheng Z, Luan S (2021) Metabolism-driven disassembly of nanoprobes for bacterial detection, imaging, and photo-inactivation. Adv Funct Mater 32(12):2107574

Zharov VP, Mercer KE, Galitovskaya EN, Smeltzer MS (2006) Photothermal nanotherapeutics and nanodiagnostics for selective killing of bacteria targeted with gold nanoparticles. Biophys J 90(2):619–627

Zhou W, Gao X, Liu D, Chen X (2015) Gold nanoparticles for in vitro diagnostics. Chem Rev 115(19):10575–10636

Zhou Z, Peng S, Sui M, Chen S, Huang L, Xu H, Jiang T (2018) Multifunctional nanocomplex for surface-enhanced raman scattering imaging and near-infrared photodynamic antimicrobial therapy of vancomycin-resistant bacteria. Colloids Surf B Biointerfaces 161:394–402

Plant Extract-Mediated Synthesis and Antibacterial Potential of Metallic Nanoparticles

5

Gagandeep Kour Reen, Pratibha Sharma, and Ashok Kumar

Abstract

The emergence of microbial resistance against antimicrobial drugs has become a crucial and complex problem. According to the WHO statistics, the increased antibiotic-resistant strains of bacteria became one of the major causes of health-related problems. Thus, there is a constant need for research and innovation in this field to develop effective and easily available medications against microbes. In this regard, metal and metal oxide nanoparticles (MONPs) of copper, gold, silver, zinc, titanium, etc. have been analyzed for their role in diagnostic and therapeutic usage over the years. Till now, a diversity of metallic nanoparticles (NPs) has been proven to exhibit selective toxicity on bacteria, while showing a minimal impact on mammalian cells. Also, various synthetic methods have been explored for their synthesis in order to obey the green principles of chemistry. This chapter delineated the green synthetic approaches and the role of metal nanoparticles (MNPs) and metal oxide nanoparticles (MONPs) in their potential to treat bacterial infections. The use of various plant extracts for green and eco-friendly production of NPs has been discussed. Moreover, the chapter will provide a better understanding regarding the application of these NPs as antibacterial agents and the mechanism of action they follow inside the cells.

Keywords

Metals · Metal oxides · Nanoparticles · Plant Extract · Antibacterial · Green Synthesis

G. K. Reen (✉)
Department of Chemistry, Jawaharlal Nehru Government Degree College, Barwaha, MP, India
e-mail: dr.gagandeepkreen@mp.gov.in

P. Sharma · A. Kumar
School of Chemical Sciences, Devi Ahilya Vishwavidyalaya, Indore, MP, India

M. Y. Wani et al. (eds.), *Nanotechnology Based Strategies for Combating Antimicrobial Resistance*, https://doi.org/10.1007/978-981-97-2023-1_5

5.1 Introduction

The evolution of multidrug resistant (MDR) microbes poses a worldwide threat to the health of human beings. The reason for the rapid development of antibiotic resistance in microbes lies in their unwise and excessive use in recent decades (Balderrama-González et al. 2021). The microbes responsible for nosocomial infections, such as *S. aureus*, *E.coli*, *A. baumannii*, *P. aeruginosa*, and *K. pneumoniae*, pose a critical threat to the patient's health. The attack of bacteria on the host's immune defense machinery is observed to be the major cause of death in patients admitted to the intensive care unit (ICU) (Al-Anazi and Al-Jasser 2014; Ubagai et al. 2015). The widespread use of broad-spectrum antibiotics resulted in the MDR isolates which further complicated the problem as bacteria spread epidemically among the patients (Hemeg 2017).

As a result, there has been a boom in the discovery of novel and more effective antimicrobial agents. The production of nanomedicines via combinatorial research of nano- and biotechnology appeared as an attractive strategy for bacteria theranostics (Park et al. 2017). Nanomaterials possess great ability for improving the pharmacokinetics and effectiveness of therapeutics because of their high accumulation at disease sites, long-term blood circulation, and great flexibility for functionalization due to larger surface area (Zheng et al. 2017). NPs with sizes ranging from 1 to 100 nm provided an attractive area in current biomedical applications. The reports suggested that antibiotic material in the form of NPs can enhance physicochemical and biological properties. Metals and metal oxide (MO) nanoparticles exhibited a crucial role with promising antimicrobial properties. The use of metals as antibacterial agents dates back thousands of years, with copper salt as the oldest mentioned astringent (Hemeg 2017). NPs such as copper, gold, and silver as well as certain metal oxides (MOs) found their application for diagnostic and therapeutic purposes over the years. They are found to be competent against MDR bacterial strains as nanobactericides and nanocarriers (Ren et al. 2020).

Hitherto, a diversity of metals and MO NPs, such as silver (Ag), gold (Au), copper (Cu), Ag_2O, CuO, MgO, TiO_2, and ZnO, were proved to exhibit selective toxicity on bacteria, while showing a minimal impact on mammalian cells (Fernández-Arias et al. 2019; Gold et al. 2018). The results revealed that metal and MO NPs could induce reactive oxygen species (ROS)-mediated cytotoxicity and DNA damage, membrane damage, and interfere with cellular functions that lead to damage to bacterial cells (Zhang et al. 2018).

The conventional physical (namely, laser ablation, mechanical milling, and sputtering) and chemical (such as chemical reduction, pyrolysis, vapor deposition, and atomic condensation) methods employed for the synthesis of NPs resulted in pure materials. However, these strategies involved the use of toxic and hazardous chemicals with extended reaction time which is against the green principles of synthetic chemistry (Eltaweil et al. 2022; Hu et al. 2021; Ahmed et al. 2021). In addition, NPs synthesized by chemical and physical methods tend to be unstable and agglomerate easily due to the lack of capping agents. Thus, researchers developed unique "green chemistry"-aided methodologies using biological systems such as fungi, bacteria, and plant extract in response to the escalating environmental concerns (Vanlalveni et al. 2021; Nawabjohn et al. 2022). The nanomaterials are developed by mixing biological substances with a metal salt solution (Gui et al. 2023).

Moreover, the use of microorganisms for NP synthesis is time-consuming with several preparative steps and is also expensive (Chugh et al. 2021). Accounting for the benefits of readily scalable, less expensive, and rapid procedure, green synthesis using plant extract was found to be the most promising method (Alharbi et al. 2022; Ahmad et al. 2019). Plant extract contains a wide range of biomolecules that act as reducing, capping, and stabilizing agents (Hamedi and Shojaosadati 2019; Taher et al. 2020; Ganguli et al. 2023). Various plant parts such as leaves, flowers, bark, roots, and even fruit extract have been employed for the synthetic procedure.

This chapter summarizes an account of the activity of extensively reported metals and MO NPs as antibacterial materials, namely, Ag, Cu, and Au NPs, synthesized utilizing plant extracts, over the last five years. In addition, their mode of action as antibacterial agents has also been discussed.

5.2 Nanoparticles: An Overview

5.2.1 Introduction to Nanoparticles

Since the introduction of nanotechnology, the field has witnessed remarkable and revolutionary developments. Owing to an exponential increase in the commercial applications of nonmaterial, the technology whisked the research area in the twenty-first century. Extremely small size with large surface areas modified their chemical and biological properties compared to the bulk material. The unique properties of NPs resulted in their role in human lives through industry, agriculture, food, clothing, medicine, and cosmetics. The NPs found their role in a broad category of applications such as in solar cells (Thirugnanasambandan et al. 2018), insulating substances (Contreras et al. 2017), the removal of dyes and pollutants (Zhao et al. 2018), high-sensitivity biosensors (Lu et al. 2010), high-energy batteries (Shcherbakov et al. 2014), high-power magnetic materials (Maksoud et al. 2019), and long-lasting biomedical materials (Yang et al. 2019; Marti et al. 2019). Also, their role in cancer treatment, antioxidant agents, and against multidrug resistant microbes has been explored (Valsalam et al. 2019; El-Batal et al. 2018; Mosallam et al. 2018).

5.2.2 Types of Nanoparticles

The NPs can be categorized based on their size, morphology, and physical and chemical properties (Fig. 5.1). Based on the dimensions NPs can be divided into four major classes, namely, a) Zero-dimensional NPs such as quantum dots and fullerenes, b) one-dimensional NPs: nanotubes, nanofibers, nanowires, and nanorods are included in this category, c) two-dimensional NPs: Examples are nanosheets,

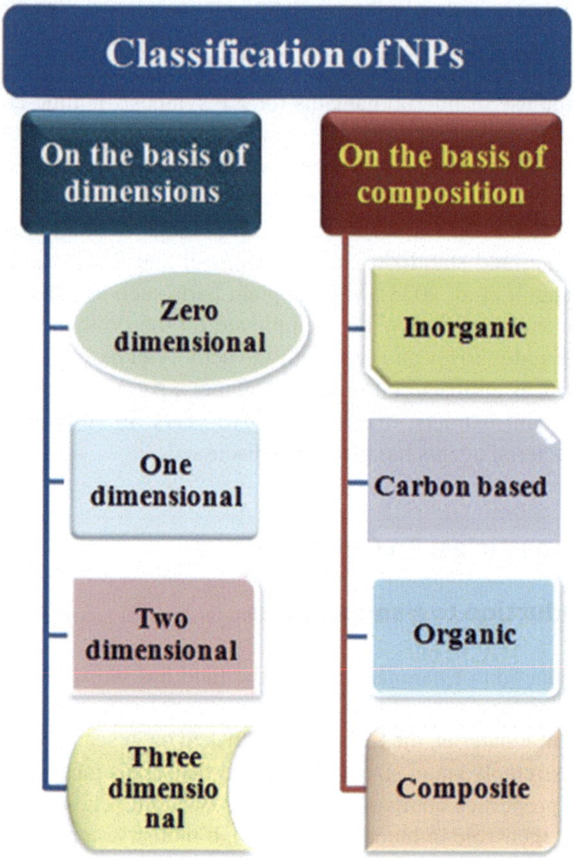

Fig. 5.1 Schematic representation for classification of NPs

nanofilms, and nanolayers, and d) three-dimensional NPs or bulk particles: This category comprises bulk powders, dispersions of NPs, arrays of nanowires, nanotubes, etc. Based on the composition/chemical or physical properties, NPs can be categorized into four major classes (Joudeh and Linke 2022; Majhi and Yadav 2021)

1. **Inorganic NPs**: This class includes NPs in the form of metals, metal oxides, mixed metal oxides, ceramics, and semiconductors. Examples are silver (Ag), aluminum (Al), gold (Au), copper (Cu), copper oxide (CuO), zinc oxide (ZnO), magnesium aluminum oxide ($MgAl_2O_4$), iron oxide (Fe_2O_3), titanium dioxide (TiO_2), silica (SiO_2), etc.
2. **Carbon-based NPs**: Materials such as graphene, fullerene, carbon nanotubes (CNTs: single-walled and multi-walled), activated carbon, and carbon black are considered under this class. Sometimes these particles are considered together as organic NPs.

3. **Organic NPs**: All the organic material excluding carbon-based NPs, for example, dendrimers, cyclodextrin, micelles, and polymeric and lipid-based NPs.
4. **Composite NPs**: This class includes the materials obtained as the combination of the above three classes, with complicated structures like a metal-organic framework.

5.2.3 Applications of Metallic Nanoparticles

The localized surface plasmon resonance provides unique optical and electrical properties to the metallic NPs. Along with this, the thermal, magnetic, and biological properties exhibited by metallic NPs made them the material of choice for researchers. The literature revealed that NPs enjoy a rich realm of chemical, electronic, biological, manufacture, environmental, and pharmaceutical applications (Khan et al. 2019). Some common roles of metallic NPs are as follows:

1. Application in biological fields such as drugs, drug carriers, and diagnostic agents.
2. Application in photocatalysis, optic, and electronic devices.
3. In the manufacture of fabrics, food packaging, and pharmaceutical industries.
4. Application in environment with their role in the removal of hazardous heavy metals water treatment, sensors for trace elements, and photodegradation purposes.
5. Application in energy generation and storage,
6. Applications in food industry for food production, preservation, and packaging.
7. Applications in agriculture as nano fertilizers and pesticides.
8. As catalysts in synthesis.

5.3 Green Synthesis of Metallic Nanoparticles

5.3.1 Principles of Green Synthesis

Green synthesis is designed to avoid the use and production of hazardous substances following the 12 principles of green chemistry (Anastas and Eghbali 2010). The researchers consider environmental safety and design or use the procedures that involve minimization or the elimination of the use of harmful chemicals, solvents, or reagents for synthesis. The methodologies employed for the synthesis of NPs include chemical, physical, and biological procedures. The whole synthesis of NPs is broadly divided into two categories, namely, top-down (destructive method from large molecules to NPs) and bottom-up approaches (constructive method from simple molecules to NPs). All other methodologies rest under the shade of these two approaches as depicted in Fig. 5.2 (Khan et al. 2019; Ghoshal and Singh 2022). The "bottom-up" approach is favored over the "top-down" approach as the method is convenient, easy and requires less time for synthesizing NPs (Singh et al. 2018).

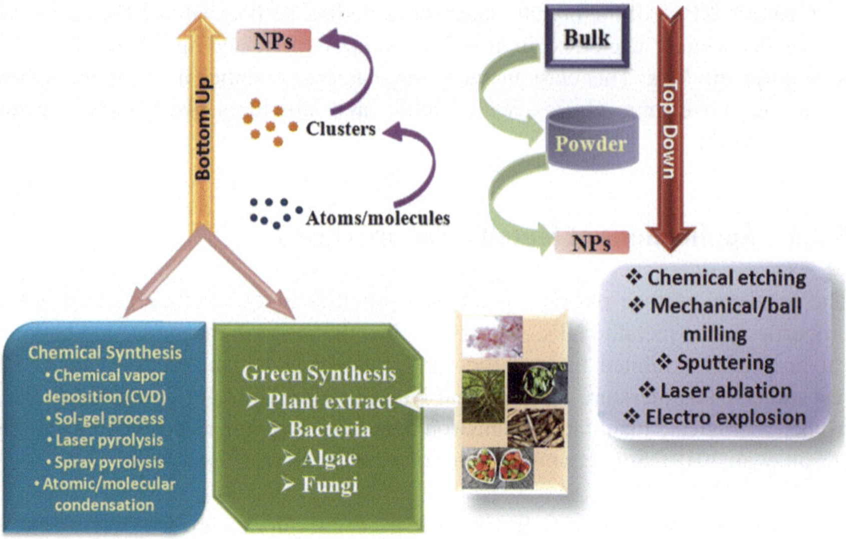

Fig. 5.2 Depiction of various synthetic procedures for NPs

Green synthetic technology involves the use of bacteria, algae, fungi, and plant extracts. This biological approach attracted the attention of the research community owing to its reliable, benign, and environmentally friendly characteristics (Hebbalalu et al. 2013; Velusamy et al. 2016).

5.3.2 Role and Advantages of Plant Extracts in Nanoparticle Synthesis

The plant extracts provide a medium that is nontoxic, economical, and fast as compared to the synthesis using microorganisms. Phytochemicals such as terpenes, flavanoids, polyphenols, alkaloids, and carbohydrates play an important role in the synthesis of NPs (Carrillo-López et al. 2016; Makarov et al. 2014). These chemicals together played the role of reducing, capping, and stabilizing agents for NP synthesis (Khan et al. 2019). The plant extracts are advantageous as they are widely available, safe to handle and do not produce any toxic waste or by-product after the reaction (Gavade et al. 2015). However, there are many factors that affect the reaction rate and shape of NPs, including the type of phytochemicals, extract concentration, pH, reaction time, and temperature (Bajpai and Kumari 2015). Various advantages of using plant extracts are shown in Fig. 5.3.

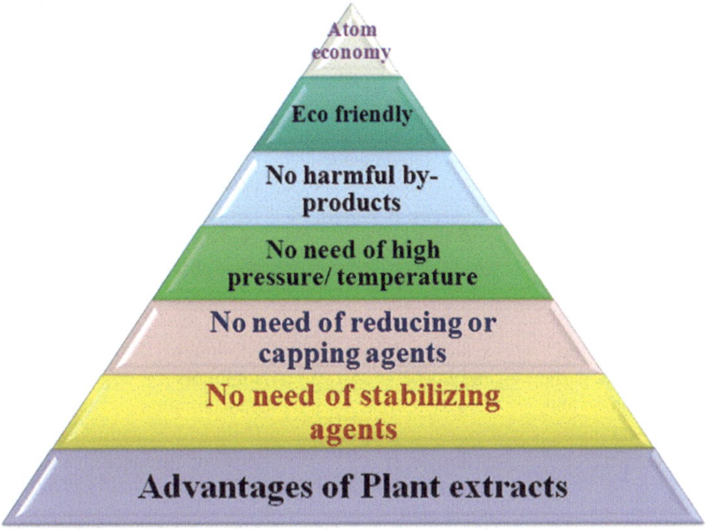

Fig. 5.3 Advantages of the plant extract

5.4 Case Studies: Plant-Mediated Metallic Nanoparticles and Antibacterial Activity

5.4.1 Silver Nanoparticles (AgNPs)

Silver NPs (AgNPs) owing to their lower toxicity against normal cells were mostly studied for their antimicrobial activity (Sampath et al. 2023). There have been many ways reported for their synthesis, including chemical, physical, and biological methods, with the biological process employing plant extract being considered the best. The literature consists of several reports on the phytochemical synthesis of these NPs and their role as antibacterial material. Table 5.1 represents the synopsis of research work reported in the literature and quoted in this chapter, consisting of plant extracts used for synthesizing AgNPs and the microbes against which they are active. The interesting part of this study is that all the AgNPs were synthesized using silver nitrate ($AgNO_3$) as the precursor. AgNPs were also reported to be used for coating biofilms, to increase the biocompatibility and antibacterial properties of these films (Gugala et al. 2017; Jaiswal and Mishra 2018; Schwass et al. 2018; Wang et al. 2018).

The leaf extract of Piper betel possess reducing and stabilizing characteristics that were used to synthesize the AgNPs. These biosynthesized AgNPs were used in pectin and chitosan-based biofilms for wound healing applications. Owing to the good bacteriostatic ability of composite film their wound healing property was found to increase. The effectiveness of the composite was examined against two Gram-positive (Gram+ve) (*S. aureus*, *B. cereus*) and Gram-negative (Gram-ve) (*P. aeruginosa*, *K. pneumonia*) bacteria. The enhanced antibacterial activity of

Table 5.1 Summary of the reported literature for antibacterial activity of AgNPs

S. no.	Plant extract	Bacteria under study	Method of determination	Reference
1.	Piper betel leaf	*S. aureus, Bacillus cereus, P. aeruginosa,* and *K. pneumoniae*	Disc diffusion method in terms of zone of inhibition (ZOI)	Nguyen et al. (2023)
2.	*Ophiorrhiza mungos, Ophiorrhiza harrisiana*, and *Ophiorrhiza rugosa* leaf extract	*S. aureus, B. cereus, E. coli,* and *Vibrio cholerae*	Disc diffusion method in terms of zone of inhibition (ZOI)	Ganguli et al. (2023)
3.	Potato, orange, red pepper, radish, garlic, and white onion and apple extract	*S. aureus, B. cereus,* and *E. coli*	Minimum inhibitory concentration (MIC)	Wasilewska et al. (2023)
4.	Carrot extract	*S. aureus, L. monocytogens, A. baumannii, B. spizizenni, E. coli, S. typhimurium, K. pneumoniae*, and *P. aeruginosa*	Zone of clearance	Fareed et al. (2023)
5.	Banana flour and agar blend	*L. monocytogens, E. coli*	Cell viability test	Orsuwan (2023)
6.	*Commelina forskaolii*	*E. fecalis, P. aeruginosa*	MIC	Vijay et al. (2023)
7.	*Syzygium aromaticum* (clove)	*S. typhimurium, S. sciuri*	MIC and ZOI	Makky et al. (2023)
8.	Strawberry seed extract	*E. coli, S. typhimurium, S. sonnei, H. halophile, S. aureus*, and *B. subtilis*	Disc diffusion method (ZOI)	Ali et al. (2022)
9.	Oxalis griffithii leaf extract	*B. subtilis* and *E. coli*	Well diffusion assay (ZOI)	Singla et al. (2022)
10.	Erythrina abyssinica	*E. coli* and *Salmonella*	Disc diffusion method (ZOI)	Muchanyereyi et al. (2022)
11.	Allium fistulosam, Tabernaemontana divaricate, Basella alba leaf extract	*S. aureus, E. fecalis, E. coli*, and *S. typhii*	Well diffusion assay (ZOI)	Vinodhini et al. (2022)
12.	Fenugreek leaves	*S. aureus, E. coli, P. aeruginosa*, and *V. cholera*	ZOI	Ghoshal and Singh (2022)
13.	Sumac leaf extract	*S. aureus* and *E. coli*	Reduction in bacterial growth	Stular et al. (2021)
14.	Acacia cyanophylla	*E. coli*	MIC (Microdilution method)	Jalab et al. (2021)

(continued)

Table 5.1 (continued)

S. no.	Plant extract	Bacteria under study	Method of determination	Reference
15.	Lignin	*S. aureus*, *S. epidermis*, *P. aeruginosa*, *K. pneumoniae*, and *A. baumannii*	MIC (broth microdilution method)	Slavin et al. (2021)
16.	Zizyphus mauritiana fruit extract	*B. subtilis*, *S. boydii*, and *E. coli*	ZOI	Kabir et al. (2020)
17.	Ocimum canum leaf extract	*E. coli*	ZOI	Tailor et al. (2020)
18.	Carya illinoinensis leaf extract	*E. coli*, *P. aeruginosa*, *S. aureus*, and *L. monocytogens*	Microdilution method (MIC) and agar well diffusion assay	Dalir et al. (2020)
19.	Hibiscus rosasinensis leaf extract	*E. coli* and *S. aureus*	Agar well diffusion assay (ZOI)	Singh et al. (2020)
20.	Sporotrichum thermophile	*B. subtilis*, *M. luteus* and *E. coli* DH5α	Agar well diffusion assay (ZOI)	Shankar et al. (2020)
21.	Chaetoceros sp., Skeletonema sp., Thalassiosira sp. of marine diatoms	*E. coli*, *B. subtilis*, *S. aureus*, *S. pneumonia*, *Aeromonas* sp.	Agar well diffusion assay (ZOI) and microdilution assay (MIC)	Mishra et al. (2020)
22.	Green tea extract (Camelia sinensis)	*S. mutans*	MIC	Rodrigues et al. (2020)
23.	Parkia speciosa leaf extract	*E. coli*, *S. aureus*, *P. aeruginosa*, and *B. subtilis*	ZOI	Ravichandran et al. (2019)
24.	Berberis vulgaris leaf and root extract	*E. coli* and *S. aureus*	Disc diffusion assay and MIC	Behravan et al. (2019)
25.	Oscilatoria sp. extract (blue green algae)	*E. coli*, *S. aureus*, *P. aeruginosa*, *Citrobactor* sp., *S. typhii*, and *B. cereus*	Agar well diffusion assay (ZOI)	Adebayo-Tayo et al. (2019)
26.	Persea Americana (avocado) seed extract	*E. coli*	Disc diffusion technique	Girón-Vázquez et al. (2019)
27.	Gracilaria birdiae (seaweed)	*E. coli* and *S. aureus*	MIC	Passos de Aragao et al. (2019)
28.	Justicia spicigera leaf extract	*B. cereus*, *K. pneumoniae* and *E. aerogenes*	Disc diffusion method (ZOI)	Mazariegos et al. (2019)

AgNPs owing to their interaction with cell walls resulted in the disruption of metabolic processes leading to bacterial death (Yin et al. 2020; Nguyen et al. 2023). The biosynthesis of AgNPs using aqueous leaf extracts of *Ophiorrhiza* species was reported by Ganguli et al. The antibacterial activity was examined against two

Gram+ve (*S. aureus* and *B. cereus*) and two Gram-ve (*E. coli* and *V. cholerae*) bacteria. The results revealed the size-dependent activity, with a larger zone of inhibition (ZOI) for smaller-sized AgNPs. Among all the bacteria examined, *V. cholera* was highly susceptible to AgNPs (Ganguli et al. 2023). The reports also depicted the use of fruit or vegetable extracts to synthesize AgNPs with good antibacterial activities. The results revealed that the size of AgNPs is dependent upon the type of extract used for their synthesis as depicted in the following order:

Potato/white onion (9 and 10 nm) < garlic/orange juice (10–30 nm) < apple juice (around 30 nm)

The NPs synthesized in the presence of onion, potato, garlic, and apple extracts were found to be effective for positive bacteria (*S. aureus* and *B. cereus*), and much lower activity against Gram-ve bacteria, e.g., *E. coli* (Wasilewska et al. 2023). Carrots contain various biomolecules that can play the role of stabilizing and capping agents for the synthesis of NPs. Thus, the carrot extract is employed for the synthesis of AgNPs, thus synthesized particles were found to be effective against seven strains of pathogenic bacteria and three macrolide-resistant clinical isolates. However, the higher concentration of the NPs was found to be toxic for biological application, but they can be efficiently used for cleaning surgical equipment and surfaces of intensive care units (Fareed et al. 2023). The antimicrobial activity of the AgNPs synthesized using banana/agar-based films was examined against the food-borne pathogenic *L. monocytogenes* and *E. coli* microbes. The AgNPs showed extended bacteriostatic activity against Gram+ve bacteria *L. monocytogenes* compared to Gram-ve *E. coli* (Orsuwan 2023). The uses of medicinal plants for the synthesis of AgNPs acquired much attention owing to their sustainable, nontoxic nature, and ease of availability. Rathi and Gopalakrishnan delineated the synthesis of AgNPs from the aqueous extract of *Commelina forskaolii* (Vijay et al. 2023). The NPs thus synthesized exhibited a potential antibacterial activity against *E. fecalis* and *P. aeruginosa* with MIC values of 62.5 and 125 μg/ml, respectively. The activity of these AgNPs was subjected to the lipo-solubility of the active components present in them, which degraded the bacterial cell membrane leading to cell death (Elemike et al. 2017). The synergistic effect of AgNPs and natural products improved the antimicrobial potential of NPs. In this regard, *Syzygium aromaticum* (clove) is used for the bioreduction of silver ions into AgNPs (Sa-AgNPs). The antibacterial activity was evaluated as minimum inhibitory concentration (MIC) and minimum bactericidal concentration (MBC). The reports suggested an outstanding activity against the Gram-ve *S. typhimurium* with the prevention of biofilm formation for Gram+ve, *S. sciuri*. Moreover, the combinatorial effect of AgNPs with bacteriophage (ZCSE6) presented the potential role of these two against the contaminating bacteria (Makky et al. 2023). Nazir et al. used the strawberry seed extract for the biosynthesis of AgNPs. The NPs were tested against *S. typhimurium, B. subtilis, E. coli, H. halophile, S. aureus*, and *S. sonnei* (Ali et al. 2022). The AgNPs were found to be active against all microbes compared with standard drug streptomycin. The antibacterial potential of NPs resulted from their ability to bind with cellular structures leading to dysfunction and the production of ROS in the prokaryotic cells causing bacterial death (Burduşel et al. 2018; Bibi et al. 2019; Muhammad et al. 2020; Anjum et al. 2020). *Oxalis griffithii*'s methanolic leaf extract was successfully applied for synthesizing AgNPs. The antimicrobial activity of AgNPs against *B. subtilis* and *E. coli* demonstrated their effectiveness against both the microbes. The larger ZOI was found for *E. coli which* could be due to Gram-ve bacteria's unique cell wall construction (Singla et al. 2022). The *Erythrina abyssinica* extract functioned as both reducing and capping agents for the green synthesis of AgNPs (Muchanyereyi et al. 2022). The antibacterial efficacy was examined on *Salmonella* and *E. coli* showing remarkable activity against *E. coli* at a lower dosage compared to that for *Salmonella.* The antimicrobial mechanism may involve the binding of NPs to the elements of fundamental biomolecules in the bacteria, or the cytoplasmic membrane disorganization following the leakage of important molecules resulting in microbial death (Salehi et al. 2016; Patil et al. 2012). AgNPs were also synthesized using *Tabernaemontana divaricate*, *Basella alba*, and *Allium fistulosum* plant extracts. The synthesized NPs were investigated against *S. aureus, E. faecalis, E. coli*, and *S. typhi.* The bactericidal action in terms of ZOI displayed the highest activity for *S. typhi* (Vinodhini et al. 2022). Fenugreek leaf extract acted as a reducing agent for the conversion of silver ions to AgNPs. The AgNPs were tested against *S. aureus, P. aeruginosa, E. coli*, and *V. cholera. P. aeruginosa* was found to be highly susceptible among all the tested strains. The efficacy of the AgNPs was found to be size-dependent with small-sized particles more active than large size. This might be attributed to the increase in surface area with the decrease in the size resulting in a good attachment with the bacterial surface (Ghoshal and Singh 2022). The in situ synthesis of Ag NPs on the cotton fibers was carried out using sumac leaf extract. These fibers were tested for antibacterial and ultraviolet (UV) protection properties. Antibacterial activity was evaluated on the functionalized samples toward the Gram+ve bacteria *S. aureus* and the Gram-ve bacteria *E. coli.* The results depicted the dual effect of Ag NPs and sumac leaf extract as antibacterial against *S. aureus* and *E. coli.* The excellent biocidal activity was achieved even at low concentrations of AgNPs. However, in the absence of AgNPs, the sumac leaf extract was active against *S. aureus* but is ineffective against *E. coli* (Stular et al. 2021). *Acacia cyanophylla* plant extract was used for AgNPs synthesis providing an eco-friendly and economical approach. The synthesized particles possess high MIC values against *E. coli* isolates ranging from 3.125 to 12.5 μg/mL (Jalab et al. 2021). It was reported that the high tendency of AgNPs toward sulfur and phosphorous resulted in their interaction with the proteins containing sulfur, thereby affecting the cell survival of microbes (Behravan et al. 2019). Lignin was employed as a reducing and capping agent for the green synthesis of Ag-ligninNPs (AgLNPs). The synthesized particles were able to inhibit the growth of MDR clinical isolates, including *S. aureus*, *K. pneumoniae*, *P. aeruginosa*, *S. epidermidis*, and *A. baumannii.* All the tested MDR bacterial strains are responsible for severe nosocomial infections that commonly include skin and urinary tract infections and pneumonia (Maragakis and Perl 2008; Nataro et al. 2000). AgLNPs were reported to be more effective against MDR strains than ATCC strains. However,

AgNPs were effective only against Gram+ve *S. aureus* and *S. epidermidis,* and Gram-ve *A. baumannii* whereas lignin alone was unable to produce any antibacterial effect on its own even when tested up to 50 μg/mL (Slavin et al. 2021). *Zizyphus mauritiana* fruit extract was employed for synthesizing Ag/AgCl-NPs. Its antibacterial potential was evaluated in terms of ZOI for *B. subtilis*, *S. boydii*, *S. enteritidis*, and *E. coli* which was obtained in the range of 9–14 mm, with higher activity for *B. subtilis*. The report suggested that small-sized NPs increased the permeability and binding with the functional groups of DNA and proteins leading to the destruction of the cells (Kabir et al. 2020; Sharma et al. 2009). Tailor et al. reported the synthesis of AgNPs using the leaf extract of *Ocimum canum,* a widely distributed medicinal herb. The AgNPs antibacterial examination showed the highest ZOI of 2.45 cm at 30 ppm against the pathogenic bacterium *E. coli* (Tailor et al. 2020). Djahaniani et al. used *Carya illinoinensis* leaf extract for the formation of AgNPs. The antibacterial activity was evaluated in terms of MIC which resulted in more inhibition against Gram-ve bacteria (*E. coli* and *P. aeruginosa*) than Gram+ve microbes (*S. aureus* and *L. monocytogenes*). The report suggested that a thicker peptidoglycan membrane prevented the entry of Ag ions to the cytoplasm; therefore, higher MIC was essential for the inhibition action (Malanovic and Lohner 2016). The antibacterial activity of the leaf extract was also evaluated but the AgNPs have considerably more antibacterial effect against the tested microbes (Dalir et al. 2020). Bio-reduction of $AgNO_3$ to AgNPs was performed using an aqueous leaf extract of *Hibiscus rosasinensis* (HRS). The method is environment-friendly, and cost-efficient with low waste production. The AgNPs exhibited better antibacterial activity against *S. aureus* than *E. coli* with ZOI of 19 and 12 mm, respectively (Singh et al. 2020). The green synthesis of AgNPs was achieved using *Sporotrichum thermophile* BJTLRMDU7. The biomolecules like enzymes/proteins, vitamins, polysaccharides, and amino acids present in the culture filtrate resulted in the reduction of Ag^+/Ag^0 (Ottoni et al. 2017; Elgorban et al. 2016; Saxena et al. 2016). The antibacterial study revealed higher inhibition against Gram-ve bacteria than Gram+ve bacteria. The cell wall of negative bacteria is made up of lipopolysaccharides at the exterior, but in the case of positive bacteria, a thick layer of peptidoglycan is present. Thus, it is harder for silver ions to penetrate due to which AgNPs are more toxic to *E. coli* than *B. subtilis* (Shankar et al. 2020). The work of Tiwari et al. demonstrated the green synthesis of AgNPs from *Skeletonema* sp., *Chaetoceros* sp., and *Thalassiosira* sp. The antibacterial efficiency of biosynthesized AgNPs examined against *Aeromonas* sp., *E. coli*, *S. aureus*, *B. subtilis*, and *S. pneumonia* exhibited a broad-spectrum antibacterial activity. The AgNPs synthesized from *Thalassiosira* sp. show strong antibacterial activity compared to those obtained from *Skeletonema* sp. and *Chaetoceros* sp. (Mishra et al. 2020). The green tea (GT) extract was employed for the synthesis of AgNPs, forming GT-AgNPs, whose surface is coated with silica that resulted in light-colored $Ag@SiO_2$ nanoparticles. These NPs found active against *S. mutans* prevented biofilm formation without any cytotoxicity on human dental cells; hence, Ag was found to be good in the treatment of dental caries (Rodrigues et al. 2020). Green synthesis of AgNPs was achieved by the bio-reduction of $AgNO_3$ using an aqueous leaf extract of *Parkia speciosa*. The observed antibacterial activity was maximum for *S. aureus* followed by *B. subtilis*, *E. coli*, and *P. aeruginosa*. However, the precursor and leaf extract alone were found to be antibacterially weak compared to AgNPs (Ravichandran et al. 2019). *Berberis vulgaris* leaf and root extract demonstrated the synthesis of AgNPs. The antibacterial effect was examined against *S. aureus* and *E. coli* where higher efficacy was found against *E. coli* than *S. aureus*. An increase in the concentration and a decrease in particle size increased the activity owing to their easier absorption through the cell membrane (Behravan et al. 2019). A green route synthesis for AgNPs was reported using the methanol extract of *Oscillatoria* species. The antibacterial activity of the biosynthesized particles was evaluated against seven strains of pathogenic bacteria. The ZOI depicted the highest activity against *P. aeruginosa,* while the lowest inhibition was against *Citrobacter* species. The AgNPs were found to reduce the biofilm formation against all the tested pathogens (Adebayo-Tayo et al. 2019). Luque et al. studied the green synthesis of AgNPs with different concentrations of *Persea americana* seed extract. The NPs were found to be efficiently active against *E. coli* at different concentrations of AgNPs, namely, 6.01, 334.11, and 823.34 μg/mL. The results explained that the size, concentration, and structural properties of AgNPs affected their antibacterial activity (Girón-Vázquez et al. 2019). AgNPs were synthesized via a polysaccharide isolated from red marine algae (*Gracilaria birdiae*). The polysaccharide acted as both reducing and stabilizing agent. The AgNPs synthesized via this method were found to be stable without any agglomeration for four months. They were tested for antimicrobial activity against *E. coli* and *S. aureus* with good activity against *E. coli* (Passos de Aragao et al. 2019). The aqueous extract from *Justicia spicigera* was used for the biosynthesis of AgNPs. The antimicrobial activity of biosynthesized AgNPs was tested against foodborne bacteria (*B. cereus*, *K. pneumoniae*, and *E. aerogenes*). The mechanism of action of AgNPs was not completely clear; however, recent studies mention that surface charge neutralization, oxidative stress induction, the increase in cell permeability, and metal ion release could be the possible factors for their activity against these bacteria (Mazariegos et al. 2019).

5.4.2 Mechanism of Action for the AgNPs

The currently accepted antibacterial mechanistic routes include:

1. Direct interaction with the bacteria, leading to the disruption of membrane potential and integrity.
2. Triggering of the host immune responses.
3. Inhibition of biofilm formation.
4. Generation of ROS and.
5. Inhibition of RNA and protein synthesis through the induction of intracellular effects (Lee et al. 2019).

The precise mechanism for antibacterial activity of AgNPs is still not fully understood. The four proposed antimicrobial pathways of AgNPs are as follows:

1. Adhesion toward the surface membrane of microbes.
2. Penetration of AgNPs into the cells and disruption of biomolecules leading to intracellular damage.
3. Generation of ROS which trigger the oxidative stress of cell and.
4. Disruption of the signal transduction pathways of the cells (Elkodous et al. 2019).

An overview of the action mechanism of AgNPs is depicted in Fig. 5.4. The reports revealed that AgNPs generated biologically active silver ions (Ag^+) that resulted in the binding with the bacterial cell walls via a "Trojan Horse" mechanism (Singh and Park 2015). The anchoring and penetration of AgNPs to the bacterial cell wall resulted in structural changes thereby varying the permeability of the cell wall. Studies have proposed the formation of pits with the destruction of cross-linking bonds in the cell wall. These pits resulted in the leakage of internal cell content causing cell disintegration (Franci et al. 2015; Agnihotri et al. 2014). In particular, AgNPs smaller than 10 nm are more toxic to bacteria as they may penetrate the cytoplasm, and damage the bacterial cell by interacting with biomolecules such as proteins, lipids, and DNA (Liao et al. 2019). The electrostatic attraction and affinity of silver ions toward sulfur proteins resulted in their adherence to the cell wall and cytoplasmic membrane. These adhered ions increased the cytoplasmic permeability which leads to the disruption of the bacterial envelope (Khorrami et al. 2018). Sadoon et al. (2020) have shown the role of silver ions in the dissociation of double strands of DNA into single strand by disrupting their H-bonds.

This activity together with the interaction between AgNPs and respiratory enzymes lead to the generation of ROS such as hydrogen peroxide, hydroxyl, and

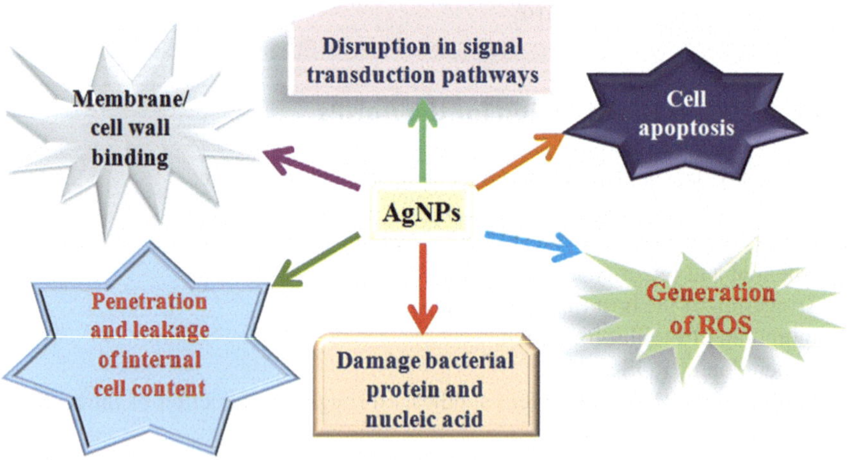

Fig. 5.4 Pathways for activity of AgNPs against different bacteria

superoxide radicals resulting in oxidative stress and damage to proteins and nucleic acids. It has been shown that ROS induced by AgNPs leads to cell apoptosis and can act as a principal agent in the provocation of cell membrane disruption and DNA modification (Salleh et al. 2020). The ability of silver ions to cause genetic disruption and enzymatic inhibition is responsible for interrupting the division and growth of bacterial cells (Reshma et al. 2017). Also, the AgNPs cause disruption of the signal transduction which leads to cell apoptosis and termination of cell multiplication leading to cell death in bacteria (Li et al. 2019).

5.4.3 Copper and Copper Oxide Nanoparticles (cu/CuONPs)

Among all the metallic NPs, Cu NPs have attracted the attention of researchers because of their antifungal, antibacterial properties, and more affordable synthesis compared to gold and silver (Abbas and Fairouz 2022). Various green synthetic methods for Cu/CuONPs reported were relatively affordable and environment-friendly and can be used for antimicrobial purposes. A brief insight into the plant extract-derived Cu/CuONPs is delineated in Table 5.2.

The aqueous extract of *Trianthema portulacastrum* demonstrated the synthesis of chitosan-based copper oxide (CH-CuO) NPs. The CH-CuONPs were evaluated for their antibacterial potential against MDR *E. coli*, *P. aeruginosa* (Gram-ve), *E. faecium*, and *S. aureus* (Gram+ve). Their activity was tested in terms of ZOI with a maximum value for *E. coli* (24 ± 1.99 mm), while *S. aureus* (17 ± 1.54 mm) was found to be least susceptible. An increase in NPs concentration depicted a significant reduction in the biofilm formation. The order of biofilm inhibition was as follows:

E. coli > *P. aeruginosa* > *E. faecium* > *S. aureus*.

The reports suggested that metal oxide (CuO) possesses the ability to produce ROS which can damage the cell membrane by leaking out the intracellular components, thus restricting cell growth (Letchumanan et al. 2021; Sarfraz et al. 2023). The leaf extracts of *Athrixia phylicoides* were employed for the synthesis of CuO-NPs and these were examined against Gram+ve and -ve bacterial strains. The lower MIC values were obtained for Gram+ve *B. cereus* and *S. aureus* suggesting them to be most susceptible to CuO-NPs, whereas the Gram-ve bacterial strains were resistant with high MIC values (Kaningini et al. 2023). Kaliyan et al. reported the use of the petal extract of *Hibiscus cannabinus* for synthesizing CuO nanospindle. The good antibacterial activity of these particles was noticed against *K. pneumonia* followed by *S. aureus* and *E. coli;* however, *B. cereus* was least affected by these particles. The study suggested that the morphology of CuO nanospindle was helpful in effective binding on the surface of bacteria through electrostatic interaction. This was followed by the release of copper ions and the leakage of intercellular constituents resulting in cell death (Kalaiyan et al. 2023). Abbas and Fairouz successfully explored the ginger extract to synthesize CuNPs within a short time. CuNPs have

Table 5.2 Summary of the reported literature for antibacterial activity of Cu/CuONPs

S. no.	Copper precursor	Plant extract	Bacterial strain	Method of analysis	Reference
1.	$CuSO_4.5H_2O$	*Trianthema portulacastrum* leaf extract	*E. coli*, *P. aeruginosa*, *E. faecium*, and *S. aureus*	ZOI	Sarfraz et al. (2023)
2.	$Cu(NO_3)_2.3H_2O$	*Athrixia phylicoides*	*B. cereus*, *S. aureus*, *E. coli*, and *S. typhimurium*	Broth microdilution (MIC)	Kaningini et al. (2023)
3.	$Cu(NO_3)_2.3H_2O$	*Hibiscus cannabinus* flower petal extract	*B. cereus*, *S. aureus*, *E. coli*, and *K. pneumoniae*	Well diffusion assay (ZOI)	Kalaiyan et al. (2023)
4.	$CuSO_4$	Ginger root extract	*S. aureus*, *E. coli*	ZOI	Abbas and Fairouz (2022)
5.	$CuSO_4.5H_2O$	*Ephedra alata*	*S. aureus*, *B. subtilis*	Disc diffusion method (ZOI)	Atri et al. (2023)
6.	$CuSO_4.5H_2O$	*Eupatorium adenophorum* leaf extract	*E. faecalis*, *S. aureus*, *E. coli*, and *S. typhi*	Agar well diffusion method (ZOI)	Maheo et al. (2022)
7.	$CuCl_2.2H_2O$	*Moringa oleifera* leaf extract	*S. aureus*, *B. subtilis*, and *E. coli*	Disc diffusion method (ZOI)	Preethi and Philominal (2022)
8.	$CuSO_4.5H_2O$	*Adhatoda vasica* leaf extract	*E. coli*, *P. aeruginosa*, *K. pneumoniae*, and *S. aureus*	Agar well diffusion (ZOI), broth dilution (MIC & minimum bactericidal concentration MBC)	Bhavyasree and Xavier (2020)
9.	$Cu(NO_3)_2$	L-ascorbic acid and chitosan	*E. coli* and *S. aureus*	Agar well diffusion method (ZOI)	Vázquez et al. (2019)
10.	$Cu(NO_3)_2$	Sugarcane juice	*E. coli*, *P. aeruginosa*, *B. subtilis*, and *S. aureus*	Agar well diffusion method (ZOI)	Mary et al. (2019)
11.	$CuSO_4.5H_2O$	*Piper retrofractum* vahl fruit extract	*E. coli* and *S. aureus*	Disc diffusion assay (ZOI)	Amaliyah et al. (2020)
12.	$CuSO_4.5H_2O$	*Angelica keiskei* (Miq.) Koidz. Leaf extract	*B. cereus*, *S. aureus*, *E. coli*, and *S. typhimurium*	Well diffusion assay(ZOI) and serial dilution method (MIC)	Krishnaraj et al. (2022)

(continued)

Table 5.2 (continued)

S. no.	Copper precursor	Plant extract	Bacterial strain	Method of analysis	Reference
13.	$CuSO_4$	*Persea americana* seed extract	*E. coli*, *Streptococcus* sp., *Klebsiella* sp., *Rhizobium*	Agar well diffusion (ZOI)	Rajeshkumar and Rinitha (2018)
14.	$CuSO_4.5H_2O$	Tilia extract	*P. aeruginosa*, *E. coli*, *B. subtilis*, and *S. aureus*	Well diffusion assay(ZOI)	Hassanien et al. (2018)
15.	Copper acetate	*Thespesia populnea*	*S. aureus*, *S. pyogenes*, and *P. aeruginosa*	Disc diffusion assay (ZOI)	Narayanan et al. (2022)
16.	Copper acetate	*Moringa oleifera* leaf extract	*K. pneumoniae*, *S. aureus* and *A. baumannii*	Agar well diffusion (ZOI)	Bai et al. (2022)
17.	$Cu(NO_3)_2.3H_2O$	*Mussaenda frondosa*	*S. aureus*, *B. subtilis*, *E. coli*, *P. aeruginosa*, and *P. vulgaris*	Disc diffusion (ZOI) and macrobroth dilution assay (MIC)	Manasa et al. (2021)

acquired a lot of attention because of their antifungal and antibacterial properties. The microbial tests performed on two pathogenic strains of Gram-ve and Gram+ve bacteria showed higher efficacy for *S. aureus* than *E. coli* due to a thick cell divider with several fatty acids (LTA) and mucosal peptides (Abbas and Fairouz 2022). The aqueous extract of *Ephedra alata* offered an eco-friendly, simple, and economical synthetic route for CuONPs. The extract served the role of a reducing and capping agent. Antibacterial activities were evaluated using the disk diffusion technique against *S. aureus* and *B. subtilis*. The observations concluded that CuONPs synthesized via the plant extract exhibited higher antibacterial activity with an average ZOI of 20.4 and 16 mm compared to chemically obtained CuONPs, which had inhibition zones of 16 and 11.6 mm for *S. aureus* and *B. subtilis*, respectively (Atri et al. 2023). Prasad T. et al used chitosan and aqueous leaf extract of *Eupatorium adenophorum* to synthesize the CuONPs. The chitosan used in this method played the dual role of being a controller of nucleation as well as a stabilizer. The antibacterial potential of NPs was tested against two Gram+ve bacteria (*E .faecalis* and *S. aureus*) and two Gram-ve bacteria (*E. coli* and *S. typhi*). The order of ZOI for these microbes was of the order *E. faecalis* (18 mm), *S. aureus* (12 mm), *E. coli* (11 mm), and *S. typhi* (12 mm). The plausible mechanism for the growth inhibition involved cell wall destruction, inhibition of protein, and inhibition of nucleic acid synthesis (Maheo et al. 2022). *Moringa oleifera* leaf extract was used for the synthesis of CuO and Ag-doped CuONPs, here the phytochemicals and biomolecules (proteins and amino acids) present in the leaves played the role of reducing agents (Venkatachalam and Jesuraj 2021; Iqbal et al. 2020). The concentration of 100 μg/mL for NPs was found to be optimal for the highest activity against all the examined strains. With an increase in the concentration of 20% Ag-doped CuONPs, maximum inhibition was observed against *S. aureus*, *B. subtilis*, and *E. coli,* whereas CuONPs showed maximum inhibition against *B. subtilis* and *E. coli.* The test results suggested that the release of Ag^+ and Cu^{2+} ions from NPs resulted in the damage of DNA and oxidative stress, which finally led to the leakage in the cytoplasm followed by cell death (Preethi and Philominal 2022). The greener methods were also effective in the synthesis of nanocomposites. Bhavyasree and Xavier employed the leaf extract of *Adhatoda vasica* for synthesizing copper oxide/carbon (CuO/C) nanocomposites under ambient conditions. The leaf extract served the multipurpose role of a capping agent, a reducing agent, and a source of carbon for the nanocomposites. The nanocomposites showed significant antibacterial activity against the pathogenic bacterial strains *E. coli*, *K. pneumoniae*, *P. aeruginosa*, and *S. aureus*. The ZOI covered a diameter of 11–14 mm showing good activity against all the tested bacteria (Bhavyasree and Xavier 2020). Silver and copper oxide NPs were prepared using L-ascorbic acid as the reducing agent. The AgNPs synthesized via this method were not effective, whereas CuONPs possess antibacterial activity against prominent pathogen bacteria, namely, *E. coli* and *S. aureus.* Their activity was almost similar to that of a conventional drug Ceftriaxone and could therefore potentially be used in medicinal applications (Vázquez et al. 2019). Interestingly, sugarcane juice (SCJ) was employed as an eco-friendly medium to synthesize CuONPs. Synthesis of well-dispersed CuONPs suggested the role of SCJ as a capping and stabilizing agent that prevented the agglomeration of NPs. The ZOI for the synthesized NPs against *E. coli*, *P. aeruginosa*, *S. aureus,* and *B. subtilis* was observed to be 5, 8, and 9 mm, respectively (Mary et al. 2019). Sabarudin et al. have used the medicinal fruit extract of *Piper retrofractum* Vahl as an eco-friendly reagent for synthesizing CuNPs using copper sulfate as a precursor in this synthesis. The synthesized CuNPs exhibited better antibacterial potential than the positive control (Gentamicin) against *E. coli* than *S. aureus* as indicated by larger inhibition areas (Amaliyah et al. 2020). The leafextract of *Angelica keiskei* (Miq.) Koidz was employed for the production of AuNPs and CuNPs. The antibacterial evaluation was carried out with the synthesized NPs and their precursors ($CuSO_4.5H_2O$) and $HAuCl_4.3H_2O$. The results delineated that NPs exhibit good ZOI against *E. coli*, *S. typhimurium*, *S. aureus*, and *B. cereus*. The plant extract was also examined for its antibacterial properties, which did not show any zone of inhibition against bacteria (Krishnaraj et al. 2022). The eco-friendly synthesis of CuNPs was studied by using *Persea americana* seed extract. The results showcased the maximum ZOI against Gram+ve bacteria *Streptococcus* with a diameter of 22.23 ± 0.15 mm and the lowest ZOI against *Rhizobacterium* with a zone diameter of 9.27 ± 0.15 mm. The activity against *Streptococcus* might be useful as this species is responsible for various infections such as wound and skin infection, sepsis, and endocarditis (Rajeshkumar and Rinitha 2018). Tilia extract was found to be a cost-effective method for producing CuNPs under optimum conditions. The tilia extract provided the bio capping of NPs thus making them stable. The CuNPs obtained via green synthesis were active against Gram-ve and Gram+ve bacteria over the concentration range of 25–200 mg/mL. The antibacterial potential was found to be concentration dependent as it increases with an increase in the concentration of NPs. Compared with the standard drug Fluconazole and Cefipime, the CuNPs exhibit efficient activity against *Bacillus*, *S. aureus*, *B. subtilis*, *E. coli*, and *P. aeruginosa* (Hassanien et al. 2018). The aqueous bark extract of *Thespesia populnea* was used for the conversion of copper acetate to CuONPs. They exhibited good antimicrobial activity against skin infection-causing bacteria such as *S. aureus*, *S. pyogenes*, and *P. aeruginosa* (Narayanan et al. 2022). The application of *Moringa oleifera* extract (a medicinal plant) was studied for the synthesis the CuONPs. The antibacterial activity of these NPs depicted the ZOI of 20, 14, and 18 mm for *K. pneumoniae*, *S. aureus*, and *A. baumannii*, respectively. Also, the CuONPs resulted in an excellent antibiofilm activity at 1000 μg/mL concentration (Bai et al. 2022). Different parts of the *Mussaenda frondosa* plant were applied for the synthesis of CuONPs. The antibacterial activity was evaluated by the disc diffusion technique and efficacy was studied against two Gram+ve bacteria (*S. aureus* and *B. subtilis*) and three Gram-ve bacteria (*E. coli*, *P. aeruginosa*, and *P. vulgaris*). Results of MIC values were in the order of *S. aureus* (96 μg/mL) followed by *B. subtilis* (112 μg/mL), *P. vulgaris* (138 μg/mL), *P. aeruginosa* (154 μg/mL), and *E. coli* (305 μg/mL) (Manasa et al. 2021).

5.4.4 Mechanism of Action for Copper Nanoparticles

Research works suggested that the surface of CuNPs played a crucial role in antibacterial activity. The interactions between the negative charges of microorganisms and the positive charge of NPs resulted in their attachment to the lipophilic cell membrane. The attachment further altered the cell permeability by affecting the nutrient flow across the membrane (Sarfraz et al. 2023; Vázquez et al. 2019; Azam et al. 2012). The antimicrobial potential of CuNPs is associated with various mechanisms that have been incorporated in Fig. 5.5, namely, damaging the microbial DNA, altering bacterial protein synthesis, and altering membrane integrity. To inhibit the formation of biofilm, the extracellular matrix is destroyed which leads to

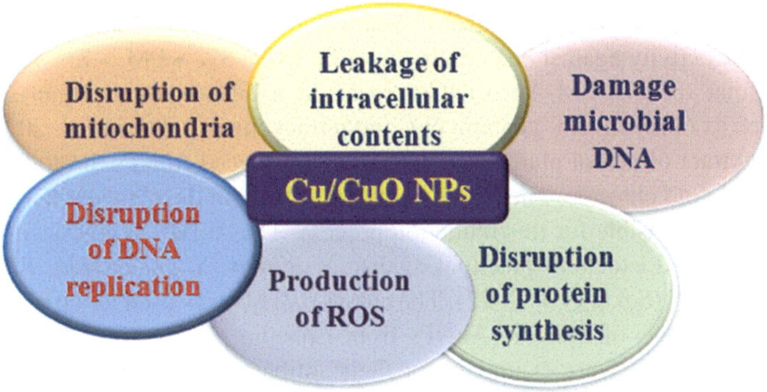

Fig. 5.5 Possible mode of action for CuNPs as antibacterial agents

the weakening of the biofilm allowing the entry of the drug into the inner layer of the biofilm (Rashki et al. 2023). The antimicrobial activity is also affected by the particle size, as smaller particles with large surface area can further enter the cell and react with the DNA and proteins to denature them.

Cu-NPs in metallic and oxide form are believed to alter the cell membrane permeability by producing ROS or releasing copper ions which damage the cell membrane, and via disruption in DNA replication (Maheo et al. 2022; Zhang et al. 2019; Wang et al. 2022). The liberated cupric ions also bind to the DNA molecule thus altering and interrupting the biochemical activities of the cell leading to cell death (Manasa et al. 2021). Also, the interaction between CuNPs and -SH (sulfhydryl) group of proteins and enzymes was supposed to destroy the bacterial cell (Krishnaraj et al. 2022). The morphology of CuO is supposed to affect the antibacterial activities as the nanospindles were reported to effectively bind on the surface of bacterial strains through electrostatic interaction thereby destroying the bacterial cell by leakage of intracellular constituents (Kalaiyan et al. 2023, 2021).

5.4.5 Gold Nanoparticles (AuNPs)

Metal-based nanoparticles, mainly Au and Ag owing to their exceptional photothermal, and biocompatible properties, found an essential role in commercial and biomedical applications. The nanosize of these materials provided with large surface area, and their integration with drugs to target specific sites for treatment and imaging made it useful for cancer treatments and other clinical applications (Das et al. 2021). Due to this, Au is a symbolic material that can be studied for its activity such as an antioxidant, anticancerous, and antibacterial agent (Vinay et al. 2021). Table 5.3 provides a brief overview of the antibacterial activity of AuNPs discussed in this chapter.

The unprecedented use of galangin (GA) from *Suaeda maritima* (SM) was reported for the synthesis of AuNPs, AgNPs, and silver-gold nanocomposites (Ag-AuNCs). Their antibacterial potential was evaluated in terms of ZOI and MIC against *S. aureus*, *K. pneumonia*, *P. mirabilis*, and *E. coli*. The maximum ZOI for nanocomposites was obtained for *E. coli* and *S. aureus*. The inhibition test results depicted the activity against bacteria in the order EC > SA > PM > KP. The results revealed that the GA-SM-Ag-AuNCs presented a higher activity than standard drugs such as clindamycin, gentamycin, and tetracycline (Ashtari et al. 2023). The aqueous extract of herbal plant *Physalis minima* produced AuNPs through the bioreduction of gold chloride. The phytochemicals present in the plant extract were the key factors responsible for this reduction process. The AuNPs synthesized via this method exhibited excellent potential against *S. aureus*, *S. pneumonia*, *P. aeruginosa*, and *E. coli* (Sekar et al. 2022). The $HAuCl_4.3H_2O$ was used as a precursor for the green synthesis of Ephedra-AuNPs from Ephedra plant extract. Furthermore, the Ephedra-AuNPs were examined for their antibacterial activity against a wide spectrum of bacteria as mentioned in Table 5.3. The ZOI for *E. coli* (26 ± 0.6) was observed to be much better than that of standards Ampicillin-Sulbactam (SAM) and

Table 5.3 Summary of reported literature for antibacterial activity of AuNPs

S. no.	Gold precursor	Plant extract	Microbes analyzed	Method of determination	Reference
1.	$HAuCl_4$	*Suaeda maritima*	*S. aureus*, *K. pneumoniae*, *P. mirabilis*, and *E. coli*	Agar disc diffusion (ZOI) and macrobroth dilution technique (MIC and MBC)	Ashtari et al. (2023)
2.	$AuCl_3$	*Physalis minima*	*S. aureus*, *P. aeruginosa*, *S. pneumoniae*, and *E. coli*	Agar well diffusion (ZOI)	Sekar et al. (2022)
3.	$HAuCl_4{:}3H_2O$	Ephedra extract	*S. aureus*, *L. moncytogenes*, *B. subtilis*, *E. coli*, *P. aeruginosa*, and *S. typhimurium*	Agar well diffusion (ZOI)	Al-Radadi (2023)
4.	$HAuCl_4{:}3H_2O$	*Pistacia vera* hull extract	*E. coli*, *S. aureus*, *P. mirabilis*, *P. aeruginosa*, *E. faecalis*, *K. pneumonia*, and *A. baumannii*	Microbroth dilution Assay (MIC)	Shirzadi-Ahodashti et al. (2023)
5.	$HAuCl_4{:}3H_2O$	*Andrographis paniculata* leaf extract	*E. coli*, *S. aureus*, and *E. faecalis*	Agar well diffusion method (ZOI)	Paramasivam et al. (2023)
6.	$HAuCl_4{:}3H_2O$	*Sarcophyton crassocaule* marine coral	*V. cholerae* and *S. aureus*	Agar well diffusion assay (ZOI)	Rokkarukala et al. (2023)
7.	$HAuCl_4$	*Dudleya brittonii* extract	*E. coli*, *S. derby*, *S. enteritidis*, *S. typhimurium*, *Y. enterocolitica*, *Y. pseudotuberculosis*, and *C. difficile*	Agar disc diffusion method (ZOI)	Kuppusamy et al. (2022)
8.	$AuCl_3$	*Jatropha integerrima* Jacq. flower extract	*B. subtilis*, *S. aureus*, *E. coli*, and *K. pneumoniae*	Agar well diffusion assay (ZOI) and two-fold dilution (MIC)	Suriyakala et al. (2022)
9.	$HAuCl_4{:}3H_2O$	*Olea europaea* leaf extract	*Klebsiella* and *S. aureus*	ZOI	Sellami et al. (2022)
10.	$HAuCl_4$	*Citrus limon* juice	*S. aureus* and *E. coli*	Kirby-Bauer disc diffusion method (ZOI)	Anwar et al. (2021)

(continued)

Table 5.3 (continued)

S. no.	Gold precursor	Plant extract	Microbes analyzed	Method of determination	Reference
11.	$HAuCl_4$	*Curcuma pseudomontana* isolated curcumin	*P. aeruginosa*, *S. aureus*, *B. subtilis*, and *E. coli*	Disc diffusion method (ZOI)	Muniyappan et al. (2021)
12.	$HAuCl_4{:}3H_2O$	Licorice root extract	*B. subtilis*, *S. aureus*, *E. coli*, *P. aeruginosa* and *S. typhi*	Well diffusion technique (ZOI)	Al-Radadi (2021)
13.	$HAuCl_4$	*Elaeocarpus ganitrus* seed extract	*P. desmolyticum* and *S. aureus*	ZOI	Vinay et al. (2021)
14.	$HAuCl_4{:}3H_2O$	*Sambucus wightiana*	*B. subtilis*, *S. flexneri*, *E. coli*, *P. aeruginosa*, *S. typhi*, *S. enteritidis*, *S. aureus*, *S. epidermidis*, and *K. pneumonia*	Agar well diffusion assay (ZOI)	Khuda et al. (2021)
15.	$HAuCl_4$	*Lignosus rhinocerotis* sclerotial and chitosan	*P. aeruginosa*, *E. coli*, *S. aureus*, and *Bacillus* sp.	Agar well diffusion assay (ZOI)	Katas et al. (2019)
16.	$HAuCl_4$	*Clerodendrum infortunatum* leaf extract	*S. aureus*, *B. cereus*, *K. pneumoniae*, *E. coli*, and *P. aeruginosa*	Broth dilution method (MIC)	Nagaraj et al. (2022)
17.	$HAuCl_4$	*Ribes nigrum* fruit extract	*S. aureus*, *P. aeruginosa*, and *E. coli*	MIC and MBC	Dobrucka et al. (2019)
18.	$HAuCl_4$	*Camellia japonica* leaf extract	*B. subtilis*, *S. aureus*, *S. faecalis*, *K. pneumoniae*, *P. aeruginosa*, and *E. coli*	Agar well diffusion assay (ZOI)	Sharma et al. (2019)
19.	$HAuCl_4$	*Anacardium occidentale* leaves extract	*E. coli* and *B. subtilis*	ZOI	Sunderam et al. (2019)
20.	$HAuCl_4$	*Acorus calamus* rhizome extract	*E. coli* and *S. aureus*	Percentage inhibition	Ganesan and Prabu (2019)

Ceftazidime (CAZ) with values of 22 ± 0.5 and 24 ± 0.5 mm, respectively (Al-Radadi 2023). The aqueous extract of *Pistacia vera* hull was utilized for the fabrication of gold nanoparticles (AuNPs@PV). The observations highlighted the higher activity of NPs for Gram+ve (owing to the absence of an outer membrane), including *E. faecalis* (0.5 μg/mL), followed by *S. aureus* (4.2 μg/mL), *P. aeruginosa* (4.29 μg/mL), *A. baumannii* and *P. mirabilis* (8.59 μg/mL). AuNPs were also explored for their potential against multidrug-resistant isolates (Shirzadi-Ahodashti et al. 2023) The synergistic application of plant science and nanotechnology resulted in the synthesis of AuNPs via *Andrographis paniculata* leaf extract. The antibacterial efficacy of synthesized AuNPs was explored for *S. aureus*, *E. faecalis*, and *E. coli* with good activity against all the tested strains (Paramasivam et al. 2023). For the very first time, Cherian and Mohan et al. explored the aqueous extract of marine soft coral *Sarcophyton crassocaule* (SCE) for the fabrication of AuNPs. The antibacterial effectiveness of NPs showed that *V. cholerae* was more susceptible than *S. aureus* which may be attributed to a much thinner layer of cell wall for Gram-ve cells compared to Gram+ve bacteria (Rokkarukala et al. 2023).

Dudleya brittonii (DB) extract was used to reduce $HAuCl_4$ and $AgNO_3$ into AuNPs and AgNPs, respectively. These were explored against many bacteria (Table 5.3) using the agar disc diffusion method. The Ag and AuNPs showed the highest inhibitory activity against *Y. pseudotuberculosis* (20 ± 3 mm). AgNPs exhibited moderate activity against all other microbes but AuNPs represent weak activity against *E. coli* and no bactericidal activity against *Salmonella* sp. and *Y. enterocolitica*. The study concluded that with an increase in the concentration of NPs, effectiveness against resistant bacterial strains can be increased (Kuppusamy et al. 2022). The AuNPs obtained by using *Jatropha integerrima* jacq. Flower extract were accessed for their antibacterial potential in terms of MIC against *B. subtilis*, *S. aureus*, *E. coli*, and *K. pneumoniae* with values of 5.0, 10, 2.5, and 2.5 μg/mL, respectively, making them a promising therapeutic candidate. The effective antibacterial activity against Gram-ve bacteria could be attributed to the fact that Gram+ve organisms have a rigid cell wall that prevents easy entry while the cell walls of Gram-ve bacteria can be destroyed easily (Suriyakala et al. 2022; Renuka et al. 2020). *Olea europaea* leaf extract demonstrated the fabrication of gold nanoparticles (OL-AuNPs). The antibacterial and antibiofilm activities of OL-AuNPs were investigated against two different bacterial strains which resulted in suppressed proliferation and biofilm inhibition in all tested bacteria. Moreover, molecular docking studies revealed that OL-AuNPs can bind and inhibit the proteins of *S. aureus* that are involved in the cell wall and fatty acid biosynthesis. It was observed that Gram-ve bacteria were more vulnerable to OL-Au NPs than Gram+ve bacteria (Sellami et al. 2022). An interesting approach for the synthesis of AuNPs was reported on cotton cloth. The hydroxyl (-OH) functional groups of the cellulose macromolecules were responsible for the slow reduction of Au ions to NPs. The synthesized AuNPs exhibited a clear ZOI against both *E. coli* (16 mm) and *S. aureus* (18 mm) species (Anwar et al. 2021). The AuNPs were synthesized and co-functionalized using *Curcuma pseudomontana* isolated curcumin (CUR). The antibacterial potential of CUR-AuNPs was assessed against *P. aeruginosa*, *S. aureus*, *B. subtilis*, and *E. coli*. The ZOI of CUR-AuNPs against all the tested microbes at a concentration of 300 μg/mL was found in the following order:

E. coli (28 mm) > *B. subtilis* (26 mm) > *S. aureus* (25 mm) > *P. aeruginosa* (23 mm).

Also at a minimum concentration of 100 μg/mL, *P. aeruginosa* showed a maximum ZOI of 13 mm, followed by *E. coli*, *B. subtilis*, and *S. aureus* (Muniyappan et al. 2021). Licorice root extract was employed as an environment-friendly material for the synthesis of AuNPs. The synthesized AuNPs were analyzed for their antibacterial activity against Gram+ve and Gram-ve bacteria and were found to exhibit good activity against Gram-ve strains (Al-Radadi 2021). The AuNPs synthesized from the *Elaeocarpus ganitrus* seeds extract via a hydrothermal process depicted a good level of inhibition for Gram-ve (*P. desmolyticum*) and Gram+ve (*S. aureus*) strains (Vinay et al. 2021). The aqueous extract of *Sambucus wightiana* presented a rapid and eco-friendly synthesis of AuNPs. The ZOI against *E. coli* (25 mm), *S. epidermis* (23 mm), and *S. enteritidis* (18 mm) proposed effective antibacterial properties of AuNPs. The report suggested that cellular uptake, aggregation, and the biosorption of AuNPs by the bacteria were responsible for membrane damage and toxicity (Khuda et al. 2021). The reducing and stabilizing nature of *Lignosus rhinocerotis* sclerotial extract (LRE) and chitosan (CS) was studied for the production of AuNPs from chloroauric acid ($HAuCl_4$). The results obtained in this research delineated the impact of the extract on the antibacterial potential of synthesized NPs. The antibacterial tests performed against *P. aeruginosa*, *E. coli*, *S. aureus*, and *Bacillus* sp. showed that CS-AuNPs possess good antibacterial potential against Gram-ve bacteria. The LRE obtained from cold and hot water extraction (CWE and HWE) did not show any antibacterial activity. Also, the CS-AuNPs synthesized using CLRE were more effective against bacteria than those prepared using HLRE. The antibacterial properties were attributed to the smaller particle size of the CS-AuNPs obtained using CLRE with increased contact with the bacterial surface thereby enhancing antimicrobial efficiency (Katas et al. 2019). Ethanolic leaf extract of *Clerodendrum infortunatum* was used for the synthesis of AuNPs under ambient conditions. The reported NPs showed higher inhibition toward the microorganisms (Table 5.3) at much lower concentrations (approx. 20 fold) compared to pure clerodin (Nagaraj et al. 2022). The AuNPs synthesized using fruit extract of *Ribes nigrum* were assayed against standard and clinical strains of bacteria (*S. aureus*, *P. aeruginosa*, and *E. coli*). The synthesized AuNPs showed a promising inhibitory activity with MIC values in the range of 13–26 μg/mL with no activity for fruit extract alone (Dobrucka et al. 2019). The synthesis of AuNPs was reported at room temperature using *Camellia japonica* leaf extract. And the antimicrobial activity was evaluated against seven microbial strains (Table 5.3). The maximum activity was observed against *B. subtilis* (15 ± 0.82 mm) and *P. aeruginosa* (15 ± 0.82 mm) (Sharma et al. 2019). Interestingly the aqueous extract of cashew leaves was reported for the fabrication of AuNPs. The good antimicrobial potential of AuNPs against *E. coli* and *B. subtilis* paved the way for future studies on the antibacterial application of these particles in food packing, wound dressing, and antihelmintic applications (Sunderam et al. 2019). The extract of *Acorus calamus* rhizome has been used as a reducing agent for chloroauric acid to synthesize AuNPs. The antibacterial activity of these microbes was examined on the cotton fabrics against *S. aureus* and *E. coli* bacterial strains. The outcome of this study depicted that the cotton fabric coated with extract having AuNPs had higher antibacterial activity than other test samples against *E. coli*. This might be attributed to the presence of impregnated AuNPs on the surface of cotton fabric (cellulosic matrix) assisted by hydroxyl groups present in the *A. calamus* extract (Ganesan and Prabu 2019).

5.4.6 Mechanism of Action for au NPs

Reports explained that bare AuNPs did not possess antibacterial properties; however, the plant extract-derived AuNPs are shown to possess antibacterial activities that arise from the plant extract components found on their surface (Muniyappan et al. 2021). Also, the surface functionalized AuNPs with drugs, vaccines, and antibiotics exhibited an enhanced antibacterial activity compared with the standard antibiotics or drugs (Zhang et al. 2020).

The antibacterial efficacy of AuNPs produced by various plant extracts was supposed to follow any of these possible ways (Fig. 5.6, (a) the inhibition of the metabolic process by altering the membrane potential and lowering adenosine triphosphate (ATP) synthase activity; (b) the denaturation of bacterial cell wall and associated constituents (Rokkarukala et al. 2023); (c) the interruption of respiratory function; (d) the destruction of membrane permeability that resulted in lysis and cell death (Ashtari et al. 2023); (e) not allow the binding of tRNA with ribosomal subunit, thus affecting its biological process; and (f) ROS production blocks the metabolic enzyme, damaging the DNA and cellular constituents, and finally leading to the death of bacterial cells (Sathiyaraj et al. 2021).

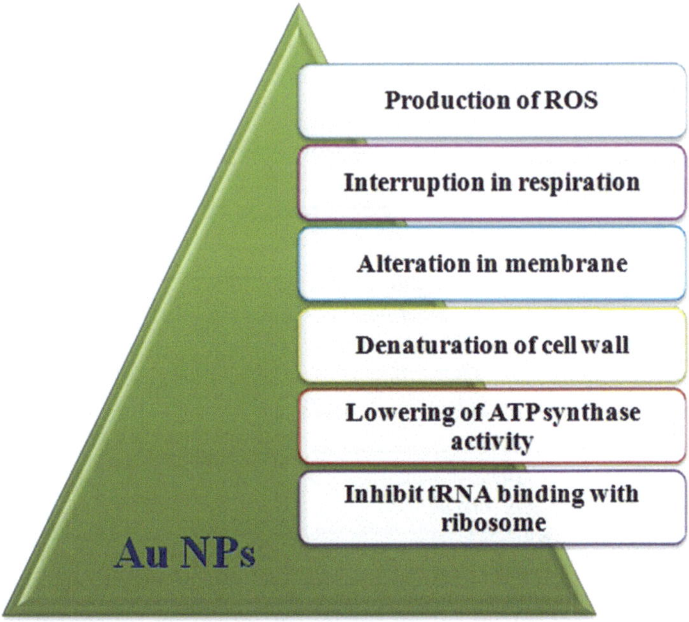

Fig. 5.6 Different antibacterial pathways for AuNPs

5.5 Conclusion and Future Perspectives

This chapter described how numerous plant part extracts were used in the green production of metallic NPs. The concepts of green chemistry were appreciated by the synthesis because they produced no side products or harmful waste material; therefore, being environmentally benign has piqued the interest of the research community. The metallic particles synthesized using these extracts were found to have good to exceptional activity against a wide range of bacterial species. Furthermore, the mode of action against the bacterial strains has been explored, and the activity has been highlighted in many ways. It is concluded that among various pathways, ROS production and membrane permeability alteration are exhibited by all of the reported metal NPs. This chapter provided insight into the biosynthesis of metallic NPs, the plant extract utilized for their synthesis, bacterial strains analyzed, and possible mode of action pathways reported in the literature. The investigations will allow readers and researchers to investigate the antibacterial capabilities of metallic NPs through various changes in order to combat the growing menace of multidrug-resistant pathogens.

Glossary

A. baumannii *Acinetobacter baumannii*
B. cereus *Bacillus cereus*
B. spizizenni *Bacillus spizizenni*
B. subtilis *Bacillus subtilis*
C. difficile *Clostridioides difficile*
E. aerogenes *Enterobacter aerogenes*
E. coli *Escherichia coli*
E. faecium *Enterococcus faecium*
E. fecalis *Enterococcus fecalis*
H. halophile *Halomonas halophile*
K. pneumonia *Klebsiella pneumonia*
L. monocytogenes *Listeria monocytogenes*
M. luteus *Micrococcus luteus*
P. aeruginosa *Pseudomonas aeruginosa*
P. desmolyticum *Pseudomonas desmolyticum*
P. mirabilis *Proteus mirabilis*
P. vulgaris *Proteus vulgaris*
S. aureus *Staphylococcus aureus*
S. boydii *Shigella boydii*
S. derby *Salmonella derby*
S. enteritidis *Salmonella enteritidis*
S. epidermidis *Staphylococcus epidermidis*
S. epidermis *Staphylococcus epidermis*

S. flexneri *Shigella flexneri*
S. mutans *Streptococcus mutans*
S. pyogenes *Streptococcus pyogenes*
S. sciuri *Staphylococcus sciuri*
S. sonnei *Shigella sonnei*
S. typhii *Salmonella typhii*
S. typhimurium *Salmonella typhimurium*
V. cholerae *Vibrio cholera*
Y. enterocolitica *Yersinia enterocolitica*
Y. pseudotuberculosis *Yersinia pseudotuberculosis*

References

Abbas AH, Fairouz NY (2022) Characterization, biosynthesis of copper nanoparticles using ginger roots extract and investigation of its antibacterial activity. Mater Today Proc 61:908. https://doi.org/10.1016/j.matpr.2021.09.551

Adebayo-Tayo B, Salaam A, Ajibade A (2019) Green synthesis of silver nanoparticle using Oscillatoria sp. extract, its antibacterial, antibiofilm potential and cytotoxicity activity. Heliyon 5(10):e02502. https://doi.org/10.1016/j.heliyon.2019.e02502

Agnihotri S, Mukherji S, Mukherji S (2014) Size-controlled silver nanoparticles synthesized over the range 5–100 nm using the same protocol and their antibacterial efficacy. RSC Adv 4(8):3974. https://doi.org/10.1039/c3ra44507k

Ahmad S, Munir S, Zeb N, Ullah A, Khan B, Ali J, Bilal M, Omer M, Alamzeb M, Salman SM, Ali S (2019) Green nanotechnology: a review on green synthesis of silver nanoparticles—an ecofriendly approach. Int J Nanomedicine 14:5087. https://doi.org/10.2147/IJN.S200254

Ahmed F, Ali I, Ali HS, Yasmeen S, Ullah S, Burki S, Adil M, Nisar J, Shah MR (2021) Synthesis and characterization of a plant growth regulator based silver nanoparticles for the ultrasensitive detection of environmentally toxic Hg^{2+} ions in tap water. New J Chem 45(38):18039. https://doi.org/10.1039/D1NJ03393J

Al-Anazi KA, Al-Jasser AM (2014) Infections caused by Acinetobacter baumanniiin recipients of hematopoietic stem cell transplantation. Front Oncol 4:186. https://doi.org/10.3389/fonc.2014.00186

Alharbi NS, Alsubhi NS, Felimban AI (2022) Green synthesis of silver nanoparticles using medicinal plants: characterization and application. J Radiat Res Appl Sci 15(3):109. https://doi.org/10.1016/J.JRRAS.2022.06.012

Ali F, Younas U, Nazir A, Hassan F, Iqbal M, Hamza B, Mukhtar S, Khalid A, Ishfaq A (2022) Biosynthesis and characterization of silver nanoparticles using strawberry seed extract and evaluation of their antibacterial and antioxidant activities. J Saudi Chem Soc 26(6):101558. https://doi.org/10.1016/j.jscs.2022.101558

Al-Radadi NS (2021) Facile one-step green synthesis of gold nanoparticles (AuNp) using licorice root extract: antimicrobial and anticancer study against HepG2 cell line. Arab J Chem 14(2):102956. https://doi.org/10.1016/j.arabjc.2020.102956

Al-Radadi NS (2023) Ephedra mediated green synthesis of gold nanoparticles (AuNPs) and evaluation of its antioxidant, antipyretic, anti-asthmatic, and antimicrobial properties. Arab J Chem 16(1):104353. https://doi.org/10.1016/j.arabjc.2022.104353

Amaliyah S, Pangesti DP, Masruri M, Sabarudin A, Sumitro SB (2020) Green synthesis and characterization of copper nanoparticles using Piper retrofractum Vahl extract as bioreductor and capping agent. Heliyon 6(8):e04636. https://doi.org/10.1016/j.heliyon.2020.e04636

Anastas P, Eghbali N (2010) Green chemistry: principles and practice. Chem Soc Rev 39:301. https://doi.org/10.1039/b918763b

Anjum F, Shahid M, Jilani MI, Oranab S, Farooq S, Nazir A, Naz S, Iqbal M (2020) Evaluation of antioxidant potential and cytotoxic behavior of different varieties of Allium sativum. Pol J Environ Stud 29(6):4447. https://doi.org/10.15244/pjoes/116106

Anwar Y, Ullah I, Ul-Islam M, Alghamdi KM, Khalil A, Kamal T (2021) Adopting a green method for the synthesis of gold nanoparticles on cotton cloth for antimicrobial and environmental applications. Arab J Chem 14(9):103327. https://doi.org/10.1016/j.arabjc.2021.103327

Ashtari A, Kazemi N, Ghasemi E, Chamkouri N, Koolivand Z, Dahdouh E, Golzarian F (2023) Gold nanoparticles, silver nanoparticles, and silver-gold nanocomposites using Suaeda maritima: phytochemical analyses, biosynthesis, characterization, and biological activity. Results Chem 5:100983. https://doi.org/10.1016/j.rechem.2023.100983

Atri A, Echabaane M, Bouzidi A, Harabi I, Soucase BM, Chaabane RB (2023) Green synthesis of copper oxide nanoparticles using Ephedra Alata plant extract and a study of their antifungal, antibacterial activity and photocatalytic performance under sunlight. Heliyon 9(2):e13484. https://doi.org/10.1016/j.heliyon.2023.e13484

Azam A, Ahmed AS, Oves M, Khan MS, Memic A (2012) Size-dependent antimicrobial properties of CuO nanoparticles against Gram-positive and -negative bacterial strains. Int J Nanomedicine 7:3527. https://doi.org/10.2147/IJN.S29020

Bai B, Saranya S, Dheepaasri V, Muniasamy S, Alharbi NS, Selvaraj B, Undal VS, Gnanamangai BM (2022) Biosynthesized copper oxide nanoparticles (CuO NPs) enhances the anti-biofilm efficacy against K. pneumoniae and S. aureus. J King Saud Univ Sci 34(6):102120. https://doi.org/10.1016/j.jksus.2022.102120

Bajpai K, Kumari M (2015) A green approach to prepare silver nanoparticles loaded gumacacia/poly(acrylate) hydrogels. Int J Biol Macromol 80:177. https://doi.org/10.1016/j.ijbiomac.2015.06.048

Balderrama-González AS, Pinón-Castillo HA, Ramírez-Valdespino CA, Landeros- Martínez LL, Orrantia-Borunda E, Esparza-Ponce HE (2021) Antimicrobial resistance and inorganic nanoparticles. Int J Mol Sci 22(23):12890. https://doi.org/10.3390/ijms222312890

Behravan M, Panahi AH, Naghizadeh A, Ziaee M, Mahdavi R, Mirzapour A (2019) Facile green synthesis of silver nanoparticles using Berberis vulgaris leaf and root aqueous extract and its antibacterial activity. Int J Biol Macromol 124:148. https://doi.org/10.1016/j.ijbiomac.2018.11.101

Bhavyasree PG, Xavier TS (2020) Green synthesis of copper oxide/carbon nanocomposites using the leaf extract of adhatodavasicanees, their characterization and antimicrobial activity. Heliyon 6(2):e03323. https://doi.org/10.1016/j.heliyon.2020.e03323

Bibi I, Nazar N, Ata S, Sultan M, Ali A, Abbas A, Jilani K, Kamal S, Sarim FM, Khan MI, Jalal F, Iqbal M (2019) Green synthesis of iron oxide nanoparticles using pomegranate seeds extract and photocatalytic activity evaluation for the degradation of textile dye. J Mater Res Technol 8(6):6115. https://doi.org/10.1016/j.jmrt.2019.10.006

Burdușel AC, Gherasim O, Grumezescu AM, Mogoantă L, Ficai A, Andronescu E (2018) Biomedical applications of silver nanoparticles: an up-to-date overview. Nano 8(9):681. https://doi.org/10.3390/nano8090681

Carrillo-López LM, Soto-Hernández RM, Zavaleta-Mancera HA, Vilchis-Néstor AR (2016) Study of the performance of the organic extracts of Chenopodium ambrosioides for Ag nanoparticle synthesis. J Nanomater 2016:1. https://doi.org/10.1155/2016/4714162

Chugh D, Viswamalya VS, Das B (2021) Green synthesis of silver nanoparticles with algae and the importance of capping agents in the process. J Genet Eng Biotechnol 191(19):1. https://doi.org/10.1186/S43141-021-00228-W

Contreras JE, Rodriguez EA, Taha-Tijerina J (2017) Nanotechnology applications for electrical transformers—a review. Electr Power Syst Res 143:573. https://doi.org/10.1016/j.epsr.2016.10.058

Dalir SJ, Djahaniani H, Nabati F, Hekmati M (2020) Characterization and the evaluation of antimicrobial activities of silver nanoparticles biosynthesized from Carya illinoinensis leaf extract. Heliyon 6(3):e03624. https://doi.org/10.1016/j.heliyon.2020.e03624

Das S, Langbang L, Haque M, Belwal VK, Aguan K, Singha Roy A (2021) Biocompatible silver nanoparticles: an investigation into their protein binding efficacies, anti-bacterial effects and cell cytotoxicity studies. J Pharm Anal 11(4):422. https://doi.org/10.1016/J.JPHA.2020.12.003

Dobrucka R, Dlugaszewska J, Kaczmarek M (2019) Antimicrobial and cytostatic activity of biosynthesized nanogold prepared using fruit extract of Ribes nigrum. Arab J Chem 12(8):3902. https://doi.org/10.1016/j.arabjc.2016.02.009

El-Batal AI, Al-Hazmi NE, Mosallam FM, El-Sayyad GS (2018) Biogenic synthesis of copper nanoparticles by natural polysaccharides and *Pleurotus ostreatus* fermented fenugreek using gamma rays with antioxidant and antimicrobial potential towards some wound pathogens. Microb Pathog 118:159. https://doi.org/10.1016/j.micpath.2018.03.013

Elemike EE, Fayemi OE, Ekennia AC, Onwudiwe DC, Ebenso EE (2017) Silver nanoparticles mediated by Costusafer leaf extract: synthesis, antibacterial, antioxidant and electrochemical properties. Molecules 22(5):701. https://doi.org/10.3390/molecules22050701

Elgorban AM, Al-Rahmah AN, Sayed SR, Hirad A, Mostafa AA, Ahkali AH (2016) Antimicrobial activity and green synthesis of silver nanoparticles using Trichoderma viride. Biotechnol Biotechnol Equip 30(2):299. https://doi.org/10.1080/13102818.2015.1133255

Elkodous MA, El-Sayyad GS, Abdelrahman IY, El-Bastawisy HS, Mohamed AE, Mosallam FM, Gobara HANM, Baraka A, Elsayed MA, El-Batal AI (2019) Therapeutic and diagnostic potential of nanomaterials for enhanced biomedical applications. Colloids Surf B: Biointerfaces 180:411. https://doi.org/10.1016/j.colsurfb.2019.05.008

Eltaweil AS, Abdelfatah AM, Hosny M, Fawzy M (2022) Novel biogenic synthesis of a Ag@ Biochar nanocomposite as an antimicrobial agent and photocatalyst for methylene blue degradation. ACS Omega 7(9):8046. https://doi.org/10.1021/ACSOMEGA.1C07209

Fareed N, Nisa S, Bibi Y, Fareed A, Ahmed W, Sabir M, Alam S, Sajjad A, Kumar S, Hussain M, Syed A, Bahkali AH, Elgorban AM, Qayyum A (2023) Green synthesized silver nanoparticles using carrot extract exhibited strong antibacterial activity against multidrug resistant bacteria. J King Saud Univ Sci 35(2):102477. https://doi.org/10.1016/j.jksus.2022.102477

Fernández-Arias M, Boutinguiza M, Del Val J, Covarrubias C, Bastias F, Gómez L, Maureira M, Arias-González F, Riveiro A, Pou J (2019) Copper nanoparticles obtained by laser ablation in liquids as bactericidal agent for dental applications. Appl Surf Sci 507:145032. https://doi.org/10.1016/j.apsusc.2019.145032

Franci G, Falanga A, Galdiero S, Palomba L, Rai M, Morelli G, Galdiero M (2015) Silver nanoparticles as potential antibacterial agents. Molecules 20(5):8856. https://doi.org/10.3390/molecules20058856

Ganesan RM, Prabu GH (2019) Synthesis of gold nanoparticles using herbal Acorus calamus rhizome extract and coating on cotton fabric for antibacterial and UV blocking applications. Arab J Chem 12(8):2166. https://doi.org/10.1016/j.arabjc.2014.12.017

Ganguli S, Howlader S, Atique Ullah AKM, Bhuiyan FR, Akhi AA, Hasan A, Dey K, Islam S, Ali F, Chakraborty AK, Bhattacharjee S, Dey BK (2023) Size controlled biosynthesis of silver nanoparticles using Ophiorrhiza mungos, Ophiorrhiza harrisiana and Ophiorrhiza rugosa aqueous leaf extract and their antimicrobial activity. Heliyon 9(5):e16072. https://doi.org/10.1016/j.heliyon.2023.e16072

Gavade SJM, Nikam GH, Dhabbe RS, Sabale SR, Tamhankar BV, Mulik GN (2015) Green synthesis of silver nanoparticles by using carambola fruit extract and their antibacterial activity. Adv Nat Sci Nanosci Nanotechnol 6:045015. https://doi.org/10.1088/2043-6262/6/4/045015

Ghoshal G, Singh M (2022) Characterization of silver nano-particles synthesized using fenugreek leave extract and its antibacterial activity. Mater Sci Energy Technol 5:22. https://doi.org/10.1016/j.mset.2021.10.001

Girón-Vázquez NG, Gómez-Gutiérrez CM, Soto-Robles CA, Nava O, Lugo-Medina E, Castrejón-Sánchez VH, Vilchis-Nestor AR, Luqu PA (2019) Study of the effect of Persea Americana seed in the green synthesis of silver nanoparticles and their antimicrobial properties. Results Phys 13:102142. https://doi.org/10.1016/j.rinp.2019.02.078

Gold K, Slay B, Knackstedt M, Gaharwar AK (2018) Antimicrobial activity of metal and metal-oxide based nanoparticles. Adv Ther 1(3):1700033. https://doi.org/10.1002/adtp.201700033

Gugala N, Lemire JA, Turner RJ (2017) The efficacy of different anti-microbial metals at preventing the formation of, and eradicating bacterial biofilms of pathogenic indicator strains. J Antibiot 70(6):775. https://doi.org/10.1038/ja.2017.10

Gui F, Mo W, Guo X, Cao F, Zhai T, Hong C, Guan X, Huang B, Pan X (2023) Biosynthesis of nanocrystalline silver chloride with high antibacterial activity using bacterial extracts. Advanced Agrochem 2(1):88. https://doi.org/10.1016/j.aac.2022.12.002

Hamedi S, Shojaosadati SA (2019) Rapid and green synthesis of silver nanoparticles using Diospyros lotus extract: evaluation of their biological and catalytic activities. Polyhedron 171:172. https://doi.org/10.1016/J.POLY.2019.07.010

Hassanien R, Husein DZ, Al-Hakkani MF (2018) Biosynthesis of copper nanoparticles using aqueous Tilia extract: antimicrobial and anticancer activities. Heliyon *4*(12):e01077. https://doi.org/10.1016/j.heliyon.2018.e01077

Hebbalalu D, Lalley J, Nadagouda MN, Varma RS (2013) Greener techniques for the synthesis of silver nanoparticles using plant extracts, enzymes, bacteria, biodegradable polymers, and microwaves. ACS Sustain Chem Eng 1(7):703. https://doi.org/10.1021/sc4000362

Hemeg HA (2017) Nanomaterials for alternative antibacterial therapy. Int J Nanomedicine 10(12):8211. https://doi.org/10.2147/ijn.s132163

Hu D, Yang X, Chen W, Feng Z, Hu C, Yan F, Chen X, Qu D, Chen Z (2021) Rhodiola rosea rhizome extract-mediated green synthesis of silver nanoparticles and evaluation of their potential antioxidant and catalytic reduction activities. ACS Omega 6(38):24450. https://doi.org/10.1021/ACSOMEGA.1C02843

Iqbal S, Javed M, Bahadur A, Qamar MA, Ahmad M, Shoaib M, Raheel M, Ahmad N, Akbar MB, Li H (2020) Controlled synthesis of Ag-doped CuO nanoparticles as a core with poly (acrylic acid) microgel shell for efficient removal of methylene blue under visible light. J Mater Sci Mater Electron 31(11):8423. https://doi.org/10.1007/s10854-020-03377-9

Jaiswal S, Mishra P (2018) Antimicrobial and antibiofilm activity of curcumin-silver nanoparticles with improved stability and selective toxicity to bacteria over mammalian cells. Med Microbiol Immunol 207(1):39. https://doi.org/10.1007/s00430-017-0525-y

Jalab J, Abdelwahed W, Kitaz A, Al-Kayali R (2021) Green synthesis of silver nanoparticles using aqueous extract of Acacia cyanophylla and its antibacterial activity. Heliyon *7*(9):e08033. https://doi.org/10.1016/j.heliyon.2021.e08033

Joudeh N, Linke D (2022) Nanoparticle classification, physicochemical properties, characterization, and applications: a comprehensive review for biologists. J Nanobiotechnol 20:262. https://doi.org/10.1186/s12951-022-01477-8

Kabir SR, Asaduzzaman AKM, Amin R, Tanbirul Haque ASM, Ghose R, Rahman MM, Islam J, Amin MB, Hasan I, Debnath T, Chun BS, Zhao XD, Khan MKR, Alam MT (2020) Zizyphus mauritiana fruit extract-mediated synthesized silver/silver chloride nanoparticles retain antimicrobial activity and induce apoptosis in mcf-7 cells through the Fas pathway. ACS Omega 5(32):20599. https://doi.org/10.1021/acsomega.0c02878

Kalaiyan G, Prabu KM, Suresh N, Suresh S (2023) Green synthesis of copper oxide spindle like nanostructure using Hibiscus cannabinus flower extract for antibacterial and anticancer activity applications. Results Chem 5:100840. https://doi.org/10.1016/j.rechem.2023.100840

Kalaiyan G, Suresh S, Prabu KM, Thambidurai S, Kandasamy M, Pugazhenthiran N, Kumar SK, Muneeswaran T (2021) Bactericidal activity of *Moringa oleifera* leaf extract assisted green synthesis of hierarchical copper oxide microspheres against pathogenic bacterial strains. J Environ Chem Eng 9(1):104847. https://doi.org/10.1016/j.jece.2020.104847

Kaningini AG, Motlhalamme T, More GK, Mohale KC, Maaza M (2023) Antimicrobial, antioxidant, and cytotoxic properties of biosynthesized copper oxide nanoparticles (CuO-NPs) using Athrixiaphylicoides DC. Heliyon *9*(4):e15265. https://doi.org/10.1016/j.heliyon.2023.e15265

Katas H, Lim CS, Azlan AYHN, Buang F, Busra MFM (2019) Antibacterial activity of biosynthesized gold nanoparticles using biomolecules from Lignosus rhinocerotis and chitosan. Saudi Pharm J 27(2):283. https://doi.org/10.1016/j.jsps.2018.11.010

Khan I, Saeed K, Khan I (2019) Nanoparticles: properties, applications and toxicities. Arab J Chem 12(7):908. https://doi.org/10.1016/j.arabjc.2017.05.011

Khorrami S, Zarrabi A, Khaleghi M, Danaei M, Mozafari M (2018) Selective cytotoxicity of green synthesized silver nanoparticles against the MCF-7 tumor cell line and their enhanced antioxidant and antimicrobial properties. Int J Nanomedicine 13:8013. https://doi.org/10.2147/ijn.s189295

Khuda F, Haq ZU, Ilahi IUR, Khan A, Fouad H, Khalil AAK, Ullah Z, Sahibzada MUK, Shah Y, Abbas M, Iftikhar T, Batiha GES (2021) Synthesis of gold nanoparticles using Sambucus wightiana extract and investigation of its antimicrobial, anti-inflammatory, antioxidant and analgesic activities. Arab J Chem 14(10):103343. https://doi.org/10.1016/j.arabjc.2021.103343

Krishnaraj C, Young GM, Yun S (2022) In vitro embryotoxicity and mode of antibacterial mechanistic study of gold and copper nanoparticles synthesized from Angelica keiskei (Miq.) Koidz. leaves extract. Saudi J Biol Sci 29(4):2552. https://doi.org/10.1016/j.sjbs.2021.12.039

Kuppusamy P, Kim S, Kim SJ, Song KD (2022) Antimicrobial and cytotoxicity properties of biosynthesized gold and silver nanoparticles using D. brittonii aqueous extract. Arab J Chem 15(11):104217. https://doi.org/10.1016/j.arabjc.2022.104217

Lee NY, Ko WC, Hsueh PR (2019) Nanoparticles in the treatment of infections caused by multidrug-resistant organisms. Front Pharmacol *10*:1153. https://doi.org/10.3389/fphar.2019.01153

Letchumanan D, Sok SP, Ibrahim S, Nagoor NH, Arshad NM (2021) Plant-based biosynthesis of copper/copper oxide nanoparticles: an update on their applications in biomedicine, mechanisms, and toxicity. Biomol Ther 11(4):564. https://doi.org/10.3390/biom11040564

Li L, Li L, Zhou X, Yu Y, Li Z, Zuo D, Wu Y (2019) Silver nanoparticles induce protective autophagy via Ca^{2+}/CaMKKβ/AMPK/mTOR pathway in SHSY5Y cells and rat brains. Nanotoxicology 13(3):369. https://doi.org/10.1080/17435390.2018.1550226

Liao C, Li Y, Tjong SC (2019) Bactericidal and cytotoxic properties of silver nanoparticles. Int J Mol Sci 20:449. https://doi.org/10.3390/ijms20020449

Lu YC, Xu Z, Gasteiger HA, Chen S, Hamad-Schifferli K, Shao-Horn Y (2010) Platinum-gold nanoparticles: a highly active bifunctional electrocatalyst for rechargeable lithium-air batteries. J Am Chem Soc 132:12170. https://doi.org/10.1021/ja1036572

Maheo AR, Vithiya BSM, Prasad TAA (2022) Biosynthesis and characterization of eupatorium adenophorumand chitosan mediated copper oxide nanoparticles and their antibacterial activity. Results Surf Interfac 6:100048. https://doi.org/10.1016/j.rsurfi.2022.100048

Majhi KC, Yadav M (2021) Synthesis of inorganic nanomaterials using carbohydrates. Green Sustainable Process for Chemical and Environmental Engineering and Science Green Inorganic Synthesis, 109–135. https://doi.org/10.1016/B978-0-12-821887-7.00003-3

Makarov VV, Love AJ, Sinitsyna OV, Makarova SS, Yaminsky IV, Taliansky ME, Kalinina NO (2014) Green nanotechnologies: synthesis of metal nanoparticles using plants. Acta Nat 6(1):35. https://doi.org/10.32607/20758251-2014-6-1-35-44

Makky S, Rezk N, Abdelsattar AS, Hussein AH, Eid A, Essam K, Kamel AG, Fayez MS, Azzam M, Agwa MM, El-Shibiny A (2023) Characterization of the biosynthesized Syzygiumaromaticum-mediated silver nanoparticles and its antibacterial and antibiofilm activity in combination with bacteriophage. Results Chem 5:100686. https://doi.org/10.1016/j.rechem.2022.100686

Maksoud MIAA, El-ghandour A, El-Sayyad GS, Awed AS, Fahim RA, Atta MM, Ashour AH, El-Batal AI, Gobara M, Abdel-Khalek EK, El-Okr MM (2019) Tunable structures of copper substituted cobalt nanoferrites with prospective electrical and magnetic applications. J Mater Sci Mater Electron 30:4908. https://doi.org/10.1007/s10854-019-00785-4

Malanovic N, Lohner K (2016) Gram-positive bacterial cell envelopes: the impact on the activity of antimicrobial peptides. Biochim Biophys Acta 1858(5):936. https://doi.org/10.1016/j.bbamem.2015.11.004

Manasa DJ, Chandrashekar KR, Kumar DJM, Niranjana M, Navada KM (2021) Mussaenda frondosa L. mediated facile green synthesis of copper oxide nanoparticles–characterization, photocatalytic and their biological investigations. Arab J Chem 14(6):103184. https://doi.org/10.1016/j.arabjc.2021.103184

Maragakis LL, Perl TM (2008) Acinetobacter baumannii: epidemiology, antimicrobial resistance, and treatment options. Clin Infect Dis 46(8):1254. https://doi.org/10.1086/529198

Marti M, Frigols B, Salesa B, Serrano-Aroca A (2019) Calcium alginate/graphene oxide films: reinforced composites able to prevent staphylococcus aureus and methicillinresistant Staphylococcus epidermidis infections with no cytotoxicity for human keratinocyte HaCaT cells. Eur Polym J 110:14. https://doi.org/10.1016/j.eurpolymj.2018.11.012

Mary APA, Ansari AT, Subramanian R (2019) Sugarcane juice mediated synthesis of copper oxide nanoparticles, characterization and their antibacterial activity. J King Saud Univ Sci 31(4):1103. https://doi.org/10.1016/j.jksus.2019.03.003

Mazariegos EB, Salas BV, Mendoza DG, Abdelmoteleb A, Camacho OT, Duran CC, Gutiérrez-Miceli F (2019) Silver nanoparticles from Justicia spicigera and their antimicrobial potentialities in the biocontrol of foodborne bacteria and phytopathogenic fungi. Rev Argent Microbiol 51(2):103. https://doi.org/10.1016/j.ram.2018.05.002

Mishra B, Saxena A, Tiwari A (2020) Biosynthesis of silver nanoparticles from marine diatoms Chaetoceros sp., Skeletonema sp., Thalassiosira sp., and their antibacterial study. Biotechnol Rep *28*:e00571. https://doi.org/10.1016/j.btre.2020.e00571

Mosallam FM, El-Sayyad GS, Fathy RM, El-Batal AI (2018) Biomolecules-mediated synthesis of selenium nanoparticles using *Aspergillus oryzae* fermented Lupin extract and gamma radiation for hindering the growth of some multidrug-resistant bacteria and pathogenic fungi. Microb Pathog 122:108. https://doi.org/10.1016/j.micpath.2018.03.013

Muchanyereyi NM, Gusha C, Mujuru M, Guyo U, Nyoni S (2022) Synthesis of silver nanoparticles using plant extracts from Erythrina abyssinica aerial parts and assessment of their anti-bacterial and anti-oxidant activities. Results Chem 4:100402. https://doi.org/10.1016/j.rechem.2022.100402

Muhammad I, Pandian S, Hopper W (2020) Antibacterial and antioxidant activity of p-quinone methide derivative synthesized from 2, 6-di-tert-butylphenol. Chem Int 6(4):260. https://ssrn.com/abstract=3558545

Muniyappan N, Pandeeswaran M, Amalraj A (2021) Green synthesis of gold nanoparticles using Curcuma pseudomontana isolated curcumin: its characterization, antimicrobial, antioxidant and anti- inflammatory activities. Environ Chem Ecotoxicol 3:117. https://doi.org/10.1016/j.enceco.2021.01.002

Nagaraj B, Musthafa SA, Muhammad S, Ramanujam GM, Chung WJ, Alodaini HA, Hatamleh AA, Al-Dosary MA, Ranganathan V (2022) Anti-microbial and anti-cancer activity of gold nanoparticles phytofabricated using clerodin enriched clerodendrum ethanolic leaf extract. J King Saud Univ Sci 34(4):101989. https://doi.org/10.1016/j.jksus.2022.101989

Narayanan M, Hussain FAJ, Srinivasan B, Sambantham MT, Al-Keridis AL, AL-mekhlafi FA (2022) Green synthesizes and characterization of copper-oxide nanoparticles by Thespesia populnea against skin-infection causing microbes. J King Saud Univ Sci 34(3):101885. https://doi.org/10.1016/j.jksus.2022.101885

Nataro JP, Blaser MJ, Cunningham-Rundles S (2000) Persistent bacterial infections. ASM Press, Washington, D.C. https://doi.org/10.1128/9781555818104

Nawabjohn MS, Sivaprakasam P, Anandasadagopan SK, Begum AA, Pandurangan AK (2022) Green synthesis and characterisation of silver nanoparticles using Cassia tora seed extract and investigation of antibacterial potential. Appl Biochem Biotechnol 194(1):464. https://doi.org/10.1007/S12010-021-03651-4/FIGURES/8

Nguyen TTT, Tran NTK, Le TQ, Nguyen TTA, Nguyen LTM, Tran TV (2023) Passion fruit peel pectin/chitosan based antibacterial films incorporated with biosynthesized silver nanoparticles for wound healing application. Alex Eng J 69:419. https://doi.org/10.1016/j.aej.2023.01.066

Orsuwan A (2023) Effect of cellulose nanocrystals and green synthesized silver nanoparticles on mechanical properties and antimicrobial activity of banana flour/agar composite films. Heliyon *9*(4):e15102. https://doi.org/10.1016/j.heliyon.2023.e15102

Ottoni CA, Simoe MF, Fernandes S, Santos JG, Silva ES, Souza RFB, Maiorano AE (2017) Screening of filamentous fungi for antimicrobial silver nanoparticles synthesis. AMB Express 7(1):31. https://doi.org/10.1186/s13568-017-0332-2

Paramasivam V, Paulpandian P, Venkatachalam K, Hussain S, Kangal A, Farraj DAA, Elshikh MS, Balaji P (2023) Cytotoxicity and antimicrobial efficiency of gold (Au) nanoparticles formulated by green approach using Andrographis paniculata leaf extract. J King Saud Univ Sci 35(5):102687. https://doi.org/10.1016/j.jksus.2023.102687

Park KS, Kim H, Kim S, Lee K, Park S, Song J, Min C, Khanam F, Rashu R, Bhuiyan TR, Ryan ET, Qadri F, Weissleder R, Cheon J, Charles RC, Lee H (2017) Nanomagnetic system for rapid diagnosis of acute infection. ACS Nano 11(11):11425. https://doi.org/10.1021/acsnano.7b06074

Passos de Aragao A, Maria de Oliveira T, Quelemes PV, Perfeito MLG, Araujo MC, Santiago JAS, Cardoso VS, Quaresma P, Leite JRSA, Alves da Silva D (2019) Green synthesis of silver nanoparticles using the seaweed Gracilariabirdiae and their antibacterial activity. Arab J Chem 12(8):4182. https://doi.org/10.1016/j.arabjc.2016.04.014

Patil SV, Borase HP, Patil CD, Salunke BK (2012) Biosynthesis of silver nanoparticles using latex from few euphorbian plants and their antimicrobial potential. Appl Biochem Biotechnol 167(4):776. https://doi.org/10.1007/s12010-012-9710-z

Preethi DRA, Philominal A (2022) Antimicrobial and antiurolithiatic activities of pure and silver doped copper oxide nanoparticles using *Moringa oleifera* leaf extract on struvite urinary stones. Appl Surf Sci Adv 12:100351. https://doi.org/10.1016/j.apsadv.2022.100351

Rajeshkumar S, Rinitha G (2018) Nanostructural characterization of antimicrobial and antioxidant copper nanoparticles synthesized using novel Persea Americana seeds. OpenNano 3:18. https://doi.org/10.1016/j.onano.2018.03.001

Rashki S, Ghanbari M, Khudhair ZH, Marzhoseyni Z, Bameri Z, Afsharikhah S, Abdulsahib WK, Nazari-Alam A, Salavati-Niasari M (2023) Evaluate the effect of graphitic carbon nitride nanosheets decorated with copper nanoparticles on biofilm and icaA gene expression in Staphylococcus aureus isolated from clinical samples. Arab J Chem 16(8):104882. https://doi.org/10.1016/j.arabjc.2023.104882

Ravichandran V, Vasanthi S, Shalini S, Shah SAA, Tripathy M, Paliwal N (2019) Green synthesis, characterization, antibacterial, antioxidant and photocatalytic activity of Parkia speciosa leaves extract mediated silver nanoparticles. Results in Physics 15:102565. https://doi.org/10.1016/j.rinp.2019.102565

Ren E, Zhang C, Li D, Pang X, Liu G (2020) Leveraging metal oxide nanoparticles for bacteria tracing and eradicating. *View 1*(3):20200052. https://doi.org/10.1002/VIW.20200052

Renuka R, Devi KR, Sivakami M, Thilagavathi T, Uthrakumar R, Kaviyarasu K (2020) Biosynthesis of silver nanoparticles using Phyllanthus emblica fruit extract for antimicrobial application. Biocatal Agric Biotechnol 24:101567. https://doi.org/10.1016/j.bcab.2020.101567

Reshma V, Syama S, Sruthi S, Reshma SC, Remya NS, Mohanan PV (2017) Engineered nanoparticles with antimicrobial property. Curr Drug Metab 18(11):1040. https://doi.org/10.2174/1389200218666170925122201

Rodrigues MC, Rolim WR, Viana MM, Souza TR, Gonçalves F, Tanaka CJ, Bueno-Silva B, Seabr AB (2020) Biogenic synthesis and antimicrobial activity of silica-coated silver nanoparticles for esthetic dental applications. J Dent 96:103327. https://doi.org/10.1016/j.jdent.2020.103327

Rokkarukala S, Cherian T, Ragavendran C, Mohanraju R, Kamaraj C, Almoshari Y, Albariqi A, Sultan MH, Alsalhi A, Mohan S (2023) One-pot green synthesis of gold nanoparticles using Sarcophytoncrassocaule, a marine soft coral: assessing biological potentialities of antibacterial, antioxidant, anti-diabetic and catalytic degradation of toxic organic pollutants. Heliyon 9(3):e14668. https://doi.org/10.1016/j.heliyon.2023.e14668

Sadoon AA, Khadka P, Freeland J, Gundampati RK, Manso RH, Ruiz M, Krishnamurthi VR, ThallapuranamSK CJ, Wang Y (2020) Silver Ions Caused Faster Diffusive Dynamics of Histone-Like Nucleoid-Structuring Proteins in Live Bacteria. Appl Environ Microbiol 86(6):02479. https://doi.org/10.1128/aem.02479-19

Salehi S, Shandiz S, Ghanbar F, Darvish M, Ardestani M, Mirzae A, Jafari M (2016) Phytosynthesis of silver nanoparticles using Artemisia marschalliana Sprengel aerial part extract and assessment of their antioxidant, anticancer and antibacterial properties. Int J Nanomedicine 11:1835. https://doi.org/10.2147/IJN.S99882

Salleh A, Naomi R, Utami ND, Mohammad AW, Mahmoudi E, Mustafa N, Fauzi MB (2020) The potential of silver nanoparticles for antiviral and antibacterial applications: a mechanism of action. Nano 10(8):1566. https://doi.org/10.3390/nano10081566

Sampath G, Govarthanan M, Rameshkumar N, Krishnan M, Alotaibi SH, Nagarajan K (2023) A comparative analysis of in vivo toxicity, larvicidal and catalytic activity of synthesized silver nanoparticles. Appl Nanosci 13(3):2379. https://doi.org/10.1007/s13204-021-02004-1

Sarfraz MH, Muzammil S, Hayat S, Khurshid M, Sayyid AH (2023) Fabrication of chitosan and Trianthema portulacastrum mediated copper oxide nanoparticles: antimicrobial potential against MDR bacteria and biological efficacy for antioxidant, antidiabetic and photocatalytic activities. Int J Biol Macromol 242:124954. https://doi.org/10.1016/j.ijbiomac.2023.124954

Sathiyaraj S, Suriyakala G, Gandhi AD, Babujanarthanam R, Almaary KS, Chend TW, Kaviyarasu K (2021) Biosynthesis, characterization, and antibacterial activity of gold nanoparticles. J Infect Public Health 14(12):1842. https://doi.org/10.1016/j.jiph.2021.10.007

Saxena J, Sharma PK, Sharma MM, Singh A (2016) Process optimization for green synthesis of silver nanoparticles by Sclerotinia sclerotiorum MTCC 8785 and evaluation of its antibacterial properties. Springer Plus 5(1):861. https://doi.org/10.1186/s40064-016-2558-x

Schwass D, Lyons KM, Love R, Tompkins GR, Meledandri CJ (2018) Antimicrobial activity of a colloidal AgNP suspension demonstrated in vitro against monoculture biofilms: toward a novel tooth disinfectant for treating dental caries. Adv Dent Res 29(1):117. https://doi.org/10.1177/0022034517736495

Sekar V, Al-Ansari MM, Narenkumar J, Al-Humaid L, Arunkumar P, Santhanam A (2022) Synthesis of gold nanoparticles (AuNPs) with improved anti-diabetic, antioxidant and antimicrobial activity from Physalis minima. J King Saud Univ Sci 34(6):102197. https://doi.org/10.1016/j.jksus.2022.102197

Sellami H, Khan SA, Amjad H, Hirad AH, Ansari SA, Egeh MA (2022) Olea europaea mediated bioengineered biocompatible gold nanoparticles for antimicrobial, cytotoxic applications, and molecular docking study. J King Saud Univ Sci 34(6):102133. https://doi.org/10.1016/j.jksus.2022.102133

Shankar A, Kumar V, Kaushik NK, Kumar A, Malik V, Singh D, Singh B (2020) Sporotrichum thermophile culture extract-mediated greener synthesis of silver nanoparticles: eco-friendly functional group transformation and anti-bacterial study. Curr Res Green Sustain Chem 3:100029. https://doi.org/10.1016/j.crgsc.2020.100029

Sharma TSK, Selvakumar K, Hwa KY, Sami P, Kumaresan M (2019) Biogenic fabrication of gold nanoparticles using Camellia japonica L. leaf extract and its biological evaluation. J Mater Res Technol 8(1):1412. https://doi.org/10.1016/j.jmrt.2018.10.006

Sharma VK, Yngard RA, Lin Y (2009) Silver nanoparticles: green synthesis and their antimicrobial activities. Adv Colloid Interf Sci 145(1):83. https://doi.org/10.1016/j.cis.2008.09.002

Shcherbakov MR, Neshev DN, Hopkins B, Shorokhov AS, Staude I, Melik- Gaykazyan EV, Decker M, Ezhov AA, Miroshnichenko AE, Brener I, Fedyanin AA, Kivshar YS (2014) Enhanced third-harmonic generation in silicon nanoparticles driven by magnetic response. Nano Lett 14(11):6488. https://doi.org/10.1021/nl503029j

Shirzadi-Ahodashti M, Mizwari ZM, Mohammadi-Aghdam S, Ahmadi S, Ebrahimzadeh MA, Mortazavi-Derazkola S (2023) Optimization and evaluation of anticancer, antifungal, catalytic, and antibacterial activities: biosynthesis of spherical-shaped gold nanoparticles using Pistacia vera hull extract (AuNPs@PV). Arab J Chem 16(1):104423. https://doi.org/10.1016/j.arabjc.2022.104423

Singh A, Gaud B, Jaybhaye S (2020) Optimization of synthesis parameters of silver nanoparticles and its antimicrobial activity. Materials Science for Energy Technologies 3:232. https://doi.org/10.1016/j.mset.2019.08.004

Singh J, Dutta T, Kim KH, Rawat M, Samddar P, Kumar P (2018) Green synthesis of metals and their oxide nanoparticles: applications for environmental remediation. J Nanbiotechnol 16(1):84. https://doi.org/10.1186/s12951-018-0408-4

Singh S, Park IS (2015) Comparative study on antimicrobial efficiency of $AgSiO_2$, ZnAg and Ag-Zeolite for the application of fishery plastic container. J Mater Sci 4(4):2169. https://doi.org/10.4172/2169-0022.1000180

Singla S, Jana A, Thakur R, Kumari C, Goyal S, Pradhan J (2022) Green synthesis of silver nanoparticles using oxalis griffithiiextract and assessing their antimicrobial activity. OpenNano 7:100047. https://doi.org/10.1016/j.onano.2022.100047

Slavin YN, Ivanova K, Hoyo J, Perelshtein I, Owen G, Haegert A, Lin YY, LeBihan S, Gedanken A, Häfeli UO, Tzanov T, Bach H (2021) Novel lignin-capped silver nanoparticles against multidrug-resistant bacteria. ACS Appl Mater Interfaces 13(19):22098. https://doi.org/10.1021/acsami.0c16921

Stular D, Savio E, Simonci B, Sobak M, Jerman I, Poljansek I, Ferri A, Tomsi B (2021) Multifunctional antibacterial and ultraviolet protective cotton cellulose developed by in situ biosynthesis of silver nanoparticles into a polysiloxane matrix mediated by sumac leaf extract. Appl Surf Sci 563:150361. https://doi.org/10.1016/j.apsusc.2021.150361

Sunderam V, Thiyagarajan D, Lawrence AV, Mohammed SSS, Selvaraj A (2019) In-vitro antimicrobial and anticancer properties of green synthesized gold nanoparticles using Anacardium occidentale leaves extract. Saudi J Biol Sci 26(3):455. https://doi.org/10.1016/j.sjbs.2018.12.001

Suriyakala G, Sathiyaraj S, Babujanarthanam R, Alarjani KM, Hussein DS, Rasheed RA, Kanimozhi K (2022) Green synthesis of gold nanoparticles using Jatropha integerrima Jacq. flower extract and their antibacterial activity. J King Saud Univ Sci 34(3):101830. https://doi.org/10.1016/j.jksus.2022.101830

Taher M, Shaari SS, Susanti D, Arbain D, Zakaria ZA (2020) Genus Ophiorrhiza: a review of its distribution, traditional uses, phytochemistry, biological activities and propagation. Molecules 25(11):2611. https://doi.org/10.3390/MOLECULES25112611

Tailor G, Yadav BL, Chaudhary J, Joshi M, Suvalka C (2020) Green synthesis of silver nanoparticles using Ocimum canumand their anti-bacterial activity. Biochem Biophys Rep 24:100848. https://doi.org/10.1016/j.bbrep.2020.100848

Thirugnanasambandan T, Pal K, Sidhu A, Elkodous MA, Prasath H, Kulasekarapandian K, Ayeshamariam A, Jeevanandam J (2018) Aggrandize efficiency of ultra-thin silicon solar cell via topical clustering of silver nanoparticles. Nanostructures Nano-objects 16:224. https://doi.org/10.1016/j.nanoso.2018.07.003

Ubagai T, Nakano R, Nakano A, Kamoshida G, Ono Y (2015) Gene expression analysis in human polymorphonuclear leukocytes stimulated by LPSs from nosocomial opportunistic pathogens. Innate Immun 21(8):802. https://doi.org/10.1177/1753425915605892

Valsalam S, Agastian P, Arasu MV, Al-Dhabi NA, Ghilan AKM, Kaviyarasu K, Ravindran B, Chang SW, Arokiyaraj S (2019) Rapid biosynthesis and characterization of silver nanoparticles from the leaf extract of tropaeolum majus T L. and its enhanced in-vitro antibacterial, antifungal, antioxidant and anticancer properties. J Photochem Photobiol B Biol 191:65. https://doi.org/10.1016/j.jphotobiol.2018.12.010

Vanlalveni C, Lallianrawna S, Biswas A, Selvaraj M, Changmai B, Rokhum SL (2021) Green synthesis of silver nanoparticles using plant extracts and their antimicrobial activities: a review of recent literature. RSC Adv 11(5):2804. https://doi.org/10.1039/D0RA09941D

Vázquez ME, López JR, Medina-Rodelo D, Jiménez-Edeza M, Castañeda-Ruelas GM, López AM, Herrera-Ramírez JM, Méndez PF (2019) Electrochemical study, structural characterization and antimicrobial activity of silver and copper oxide (CuO) nanoparticles synthesized by a green method using l-ascorbic acid and chitosan. Int J Electrochem Sci 14(7):6366. https://doi.org/10.20964/2019.07.04

Velusamy P, Kumar GV, Jeyanthi V, Das J, Pachaiappan R (2016) Bio-inspired green nanoparticles: synthesis, mechanism, and antibacterial application. Toxicol Res 32(2):95. https://doi.org/10.5487/tr.2016.32.2.095

Venkatachalam A, Jesuraj JP (2021) *Moringa oleifera* leaf extract-mediated green synthesis of nanostructured alkaline earth oxide (MgO) and its physicochemical properties. J Chem 2021:1. https://doi.org/10.1155/2021/4301504

Vijay R, Drisya VM, Selta DRF, Rathi MA, Gopalakrishnan VK, Alkhalifah DHM, Hozzein WN (2023) Synthesis and characterization of silver nanomaterial from aqueous extract of Commelina forskaolii and its potential antimicrobial activity against Gram negative pathogens. J King Saud Univ Sci 35(1):102373. https://doi.org/10.1016/j.jksus.2022.102373

Vinay SP, Udayabhanu, Sumedha HN, Shashank M, Nagaraju G, Chandrasekhar N (2021) In-vitro antibacterial, antioxidant and cytotoxic potential of gold nanoparticles synthesized using novel Elaeocarpus ganitrus seeds extract. J Sci Adv Mater Devices 6(1):127. https://doi.org/10.1016/j.jsamd.2020.09.008

Vinodhini S, Vithiya BSM, Prasad TAA (2022) Green synthesis of silver nanoparticles by employing the Allium fistulosum, Tabernaemontana divaricate and Basella alba leaf extracts for antimicrobial applications. J King Saud Univ Sci 34(4):101939. https://doi.org/10.1016/j.jksus.2022.101939

Wang X, Han Q, Yu N, Wang T, Wang C, Yang R (2018) GO-AgCl/Ag nanocomposites with enhanced visible light-driven catalytic properties for antibacterial and biofilm-disrupting applications. Colloids Surf B: Biointerfaces 162:296. https://doi.org/10.1016/j.colsurfb.2017.11.060

Wang Y, Li H, Yuan X, Jiang Y, Xiao Z, Li Z (2022) Review of copper and copper alloys as immune and antibacterial element. Trans Nonferrous Metals Soc China 32(10):3163. https://doi.org/10.1016/s1003-6326(22)66011-4

Wasilewska A, Klekotka U, Zambrzycka M, Zambrowski G, Swięcicka I, Kalska- Szostko B (2023) Physico-chemical properties and antimicrobial activity of silver nanoparticles fabricated by green synthesis. Food Chem 400:133960. https://doi.org/10.1016/j.foodchem.2022.133960

Yang Z, Hao X, Chen S, Ma Z, Wang W, Wang C, Yue L, Sun H, Shao Q, Murugadoss V, Guo Z (2019) Long-term antibacterial stable reduced graphene oxide nanocomposites loaded with cuprous oxide nanoparticles. J Colloid Interface Sci 533:13. https://doi.org/10.1016/j.jcis.2018.08.053

Yin IX, Zhang J, Zhao IS, Mei ML, Li Q, Chu CH (2020) The antibacterial mechanism of silver nanoparticles and its application in dentistry. Int J Nanomedicine 15:2555. https://doi.org/10.2147/IJN.S246764

Zhang J, Mou L, Jiang X (2020) Surface chemistry of gold nanoparticles for health-related applications. Chem Sci 11(4):923. https://doi.org/10.1039/c9sc06497d

Zhang S, Wang Y, Song H, Lu J, Yuan Z, Guo J (2019) Copper nanoparticles and copper ions promote horizontal transfer of plasmid-mediated multi-antibiotic resistance genes across bacterial genera. Environ Int 129:478. https://doi.org/10.1016/j.envint.2019.05.054

Zhang Y, Gu AZ, Xie S, Li X, Cen T, Li D, Chen J (2018) Nano-metal oxides induce antimicrobial resistance via radical-mediated mutagenesis. Environ Int 121(2):1162. https://doi.org/10.1016/j.envint.2018.10.030

Zhao L, Deng J, Sun P, Liu J, Ji Y, Nakada N, Qiao Z, Tanaka H, Yang Y (2018) Nanomaterials for treating emerging contaminants in water by adsorption and photocatalysis: systematic review and bibliometric analysis. Sci Total Environ 627:1253. https://doi.org/10.1016/j.scitotenv.2018.02.006

Zheng K, Setyawati MI, Leong DT, Xie J (2017) Antimicrobial gold nanoclusters. ACS Nano 11(7):6904. https://doi.org/10.1021/acsnano.7b02035

Functionalized Quantum Dots as Antimicrobial Agents

6

Namrata Singh, Rajeshree Shinde, Bhanushree Gupta, Kamil Kuca, Nigar Mehboob, and Archi Sawant

Abstract

It can be challenging to treat disorders linked to infectious immunodeficiency when bacteria develop antibiotic resistance. Nanomaterials with dedicated antibacterial properties have played a promising role in treating bacterial infections. In comparison with traditional antibiotics, quantum dots (QDs) exhibit not only exceptional structural stability but also photoluminescence characteristics for imaging and photodynamic therapy. Excessive use of QDs in biomedical research is related to their low toxicity. To boost their antimicrobial efficacy, QDs normally are functionalized, and this chapter will summarize various QD functionalization and their effective role against antimicrobial agents. Emphasis will be given to carbon QDs as they have proved to possess good bacteriostatic and bactericidal properties. Graphene quantum dots (GQDs) stand out among other

N. Singh (✉)
Department of Engineering Sciences, Ramrao Adik Institute of Technology, DY Patil University, Navi Mumbai, India

R. Shinde
Department of Science, Sir Vithaldas Thackersey College of Home Science, SNDTWU, Mumbai, India

B. Gupta
Department of Chemistry, Centre for Basic Sciences, Pt. Ravishankar Shukla University, Raipur, Chhattisgarh, India

K. Kuca
Department of Chemistry, Faculty of Science, University of Hradec Kralove, Hradec Kralove, Czech Republic

Aumed, a.s, Praha, Czech Republic

N. Mehboob · A. Sawant
School of Pharmacy, DY Patil deemed to be University, Navi Mumbai, Maharashtra, India

M. Y. Wani et al. (eds.), *Nanotechnology Based Strategies for Combating Antimicrobial Resistance*, https://doi.org/10.1007/978-981-97-2023-1_6

graphene derivatives because of the amazing optical, electrical, and electrochemical properties. QDs therefore provide greater potential for antibacterial application than nanoparticles, but challenges in QD application under physiological conditions are often ignored. Herein, we will discuss challenges associated with QD application in parallel to their effective antibacterial activity. A variety of functionalized QDs and their antimicrobial mechanism will be discussed.

Keywords

Quantum dots · Antibacterial · Antimicrobial · Functionalized GQDs · Nanomaterials

6.1 Introduction

A novel class of carbon nanomaterials with a size less than 10 nm, carbon dots (CDs), also known as graphene quantum dots (GQDs), carbon quantum dots (CQDs), carbon nanodots (CNDs), and carbonized polymer dots (CPDs), have steadily emerged as the rising star of new nanocarbon members (Rajendiran et al. 2019). The exceptional biological characteristics of CDs, such as their low cytotoxicity and high biocompatibility, allow for their use in a number of applications, including as biosensors, bioimaging, and the administration of drugs and molecules. QDs are a relatively new type of luminous nanomaterial. QDs exhibit distinct physical and chemical properties as compared to typical materials, such as great stability, an incredibly narrow emission range, and a high quantum yield (Wang et al. 2022). As a result, they are used in biosensors, real-time tracking, multi-color labeling, and imaging. QDs are currently attracting a lot of attention, indicating that QDs could be used in antibacterial research as an effective alternative to existing antibiotic medications (Jahangir et al. 2019). Antibiotic resistance in microorganisms is a complex issue, particularly when treating infectious immunodeficiency illnesses. Nanomaterials (NMs) with specialized antibacterial properties have played a novel role in treating bacterial infections in recent years, when compared to bulk materials. Among NMs, quantum dots (QDs), specifically carbon-containing QDs such as graphene oxide QD (GOQD), graphene QD (GQD), and carbon QD (CQD), have demonstrated bacteriostatic and bactericidal activities against antibiotic-resistant bacteria via photodynamic (PD) effects at a specific wavelength of light (Kováčová et al. 2022a; Owusu et al. 2019). Different kinds of functionalized quantum dots are depicted in Fig. 6.1.

As nanomedicine, a wide range of nanomaterials with unique intrinsic biological characteristics have been produced. Several nanoparticles have been found to have antibacterial effects. Metal and metal oxide nanoparticles, in particular, have been extensively researched for their possible use as a biocide (Jhonsi et al. 2018a). Despite the fact that these metal (Ag and Au) and metal oxide (CuO, Fe2O3, and ZnO) nanoparticles have antibacterial activity, the release of metal ions has created a need to replace them with safer nanomaterials. Carbon-based nanoparticles that

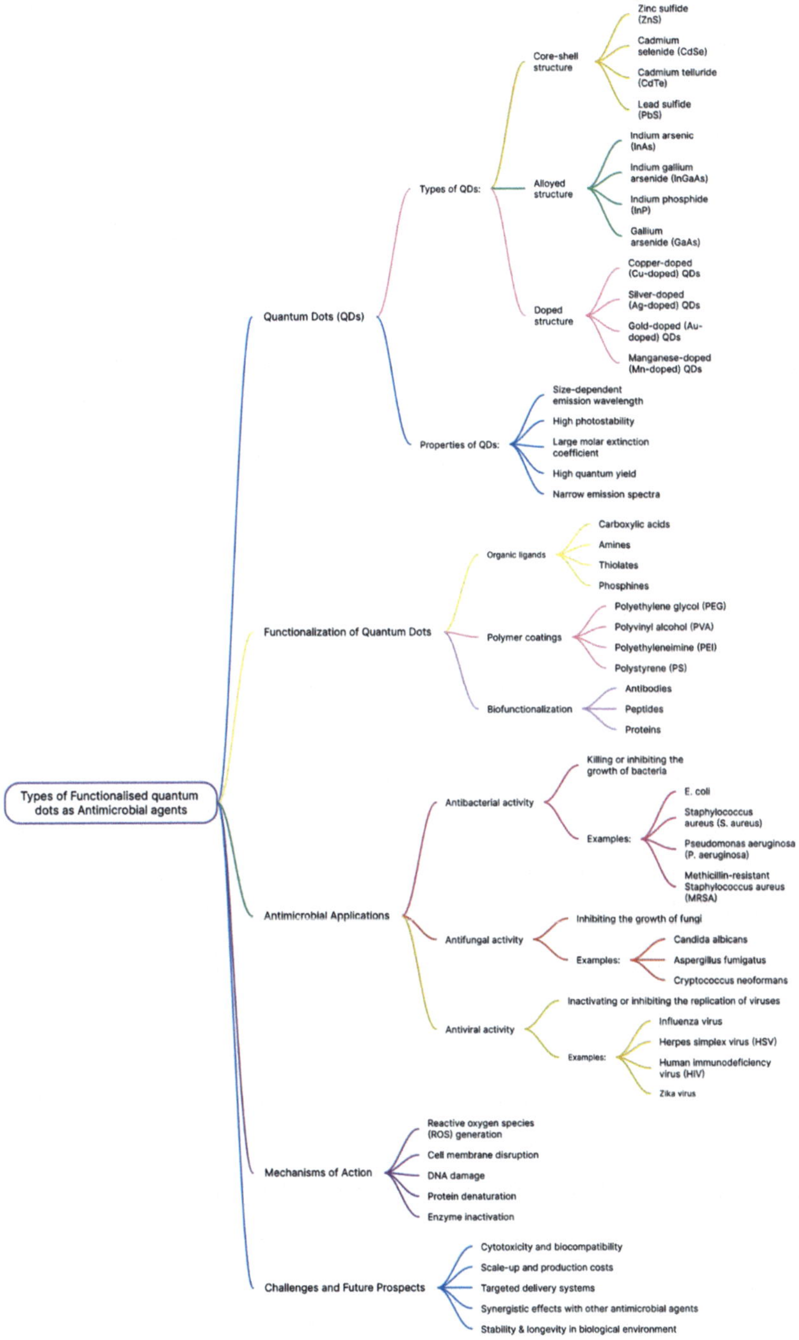

Fig. 6.1 Different kinds of functionalized quantum dots, along with their potential applications, obstacles, and microbial action

are less toxic have recently been found as promising antibacterial agents (Anand et al. 2019).

QD antibacterial action is primarily influenced by their size, surface charge, ability to withstand oxidative stress, and ability to cause hyperthermia. By damaging bacterial cell membranes and proteins, as well as by preventing bacterial gene expression, the majority of bacteria can be eliminated. QDs have received a considerable deal of attention in recent years for biomedical applications due to their ease of production, programmable fluorescence, low cytotoxicity, and high biocompatibility. Carbon QDs (sp2/sp3 carbon) and graphene QDs (sp2 carbon) can be functionalized with various chemicals and functional groups for biolabeling, sensing, tissue imaging, and antimicrobial applications. QDs can be synthesized using both bottom-up and top-down methods. Bottom-up synthesis is preferred because of the high synthetic diversity from a variety of precursors and the ease of functionalization with desired functional groups. The affinity, specificity, and antibacterial capabilities of QDs can be altered by adjusting the surface charge and functional properties. We recently reported on the synthesis and surface modification of a wide range of QDs derived from multiple precursors for a number of applications, including bacterial labeling and bacterial inactivation. The surface chemistry related to antibacterial activity and low cytotoxicity gives the chance in designing and developing better CQDs for biofilm infection as discussed (Li et al. 2020). Due to stronger electrostatic interactions and longer surface coronas, CQD-2.5w had much higher activity than CQD600 and CQD1w in eradicating mature Staphylococcus aureus biofilm. Furthermore, the cytotoxicity test shows that L929 cells may maintain high survival rates. Lin et al. (Fang et al. 2023). have discussed the various antibacterial methods that FQDs can employ, such as size effect, reactive oxygen species, covalent and non-covalent interactions, and photothermal therapy. Also, surface modification, doping, and combining with other nanomaterials are incorporated into the design of antibacterial FCDs. CD-delivered drugs have improved antibacterial effectiveness, higher stability, improved cellular absorption, and decreased adverse effects in comparison with their free versions. In this article, we examine CDs loaded with antimicrobial agents for a range of antiviral, antifungal, and antibacterial uses (Wang et al. 2022). Further in this article, we will discuss challenges associated with QD application in parallel to their effective antibacterial activity. A variety of functionalized QDs and their antimicrobial mechanism will also be discussed. Various functionalized quantum dots and their uses are depicted in Fig. 6.2.

6.2 Functionalized QDs and Antimicrobial Action

QDs are always functionalized with polymers and photosensitizers to raise their antibacterial efficacy and improve their adhesion to bacterial components. QDs are among the most promising nanomaterials due to their unique physical and chemical properties. The efficiency of the inhibitory activity varies depending on the ligand, size, shape, zeta potential, and charge transfer impact of functionalized QDs. QDs restrict bacterial growth or kill bacteria by various mechanisms such as ROS

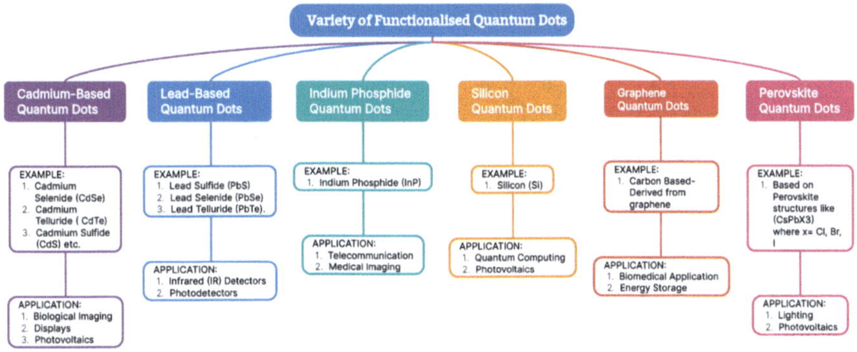

Fig. 6.2 Various functionalized quantum dots and their uses

generation, cell structure disintegration, and genomic DNA fragmentation and condensation, resulting in cytoplasm leakage. Functionalized QDs can be inserted into the cell wall/membrane, destroying the cell barrier and producing a huge amount of ROS, causing damage to the cell components. The antibacterial efficacy of C-dots is strongly related to surface charge and ROS production. The production of spermidine-functionalized carbon quantum dots (Spd-CQDs) with a high surface charge (zeta potential up to +60.6 mV) was recently reported (Jian et al. 2020). Antibacterial mechanism of QDs include (a) destruction of cell walls/cell membranes; (b) production of reactive oxygen species (ROS) to destroy the cells; and (c) binding with nucleic material (DNA/RNA) to inhibit cell proliferation.

Figure 6.3 shows the overall antimicrobial mechanism of quantum dots. The core antibacterial concept of QDs is their electron charge transfer property and the ease of availability of an electron conjugation system with microbe biomatrices. Furthermore, QDs are remarkable oxidizing agents, producing free radicals in microbiological cells.

6.2.1 ROS Production

Photodynamic Effect: Reactive oxygen species (ROS) are generated when quantum dots are exposed to light, and this is one of the main mechanisms. It is common to refer to this process as the "photodynamic effect." ROS can be categorized into various species, such as hydroxyl radicals (OH^−), superoxide radicals (O_2^−), and singlet oxygen (~1O_3). Microbial cells may endure damage from these highly oxidative reactive species.

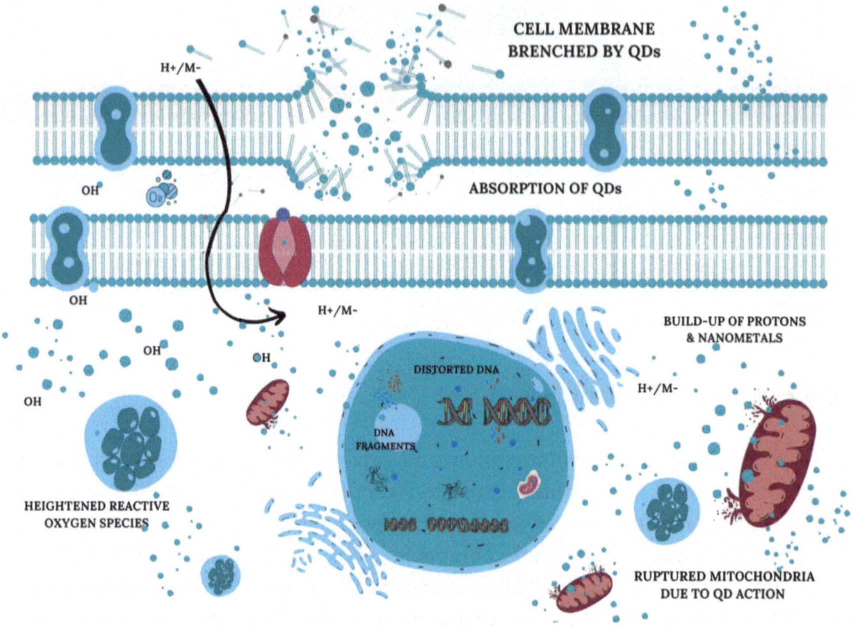

Fig. 6.3 Overall antimicrobial mechanism of quantum dots (QDs). Inducing free radicals, breaking down cell walls and membranes, attaching to genetic material, and preventing the synthesis of energy are all ways that QDs have antimicrobial effects

6.2.2 Microbial Membranes in Interaction

Disruption of Cell Membranes: The lipid bilayer of microbial cell membranes can interact with quantum dots. Membrane integrity may be compromised by the interaction and result in structural damage.

Loss of Permeability Control: Damage to the cell membrane may enhance permeability, which might enable ions and other components of the cell to escape.

6.2.3 Blocking of Intracellular Mechanisms

ROS emitted by quantum dots has a tendency to oxidatively damage microbial DNA. This damage impedes transcription and DNA replication, which in turn prevents the microbial cell from proliferating.

Protein Dysfunction: ROS can also cause denaturation and dysfunction of proteins, which can impact vital cellular functions in microbes.

6.2.4 Effects Dependent on Size

Effects of Quantum Size: The antibacterial activity of quantum dots is dependent on their size. Smaller quantum dots are more potent in causing harm because they can enter microbial cells more deeply.

Increased Surface Area: More active sites for interaction between quantum dots and microbial cells could result in stronger antimicrobial effects.

6.2.5 Functionalization of Surfaces

Improving Specificity: Quantum dots' surfaces can be changed to improve their affinity for particular microbiological targets. This could entail adhering to biomolecules or ligands that enable specific interactions.

6.2.6 Photothermal Consequences

Temperature Elevation: Quantum dots may occasionally display photothermal effects, which raise the temperature in the area around microbial cells. Microorganisms can become inactive due to elevated temperatures.

6.3 Polymer QDs

Since polymers improve GQD adhesion to bacterial cell membranes, polymer-modified GQDs have the potential to exhibit enhanced antibacterial action. A few common methods for polymer modification of CDs are as follows: (1) The polymer is used directly as a precursor for the synthesis of P-CDs; (2) the polymer reacts with a carbon source as a passivator to synthesize P-CDs; (3) the polymer is used to functionalize the surface of CDs; and (4) the polymer and CDs combine to form a composite hydrogel/membrane (Chen et al. 2016; Zhou et al. 2017). Tiwari et al. (Vibhute et al. 2022)synthesized FQDs by the one-pot pyrolysis method. The study proves the higher antibacterial efficiency of polymer QDs over non-functionalized carbon QDs. Polymer QD was successfully employed in the bioimaging of the bacterial cell through fluorescence microscopy.

6.3.1 Graphene QDs

Graphene quantum dots (GQDs) are a subclass of carbon quantum dots (CQDs) with a structure consisting of one or more graphene layers with a diameter of less than 10 nm and a higher crystallinity than CQDs. CQDs are significant in the field of medicine as well. CQDs are employed in the detection of intracellular ions, pathogens, vitamins, enzymes, proteins, nucleic acids, and toxins, as well as in the assessment of biological pH values. Although there are many biomedical uses for CQDs, we would want to concentrate on their antibacterial qualities and those of their polymer composites. CQDs' antimicrobial activity is derived from noninvasive photodynamic treatment (PDT). PDT can trigger a particular biological reaction at the cellular or subcellular level, including necrosis, a nonprogrammed pathway, or programmed death, such as apoptosis. One highly promising new class of antibacterial nanoparticles is CQDs (Kováčová et al. 2022b). CQDs are significant in the field of medicine as well. CQDs are employed in the detection of intracellular ions, toxins, pathogens, vitamins, enzymes, proteins, nucleic acids, and biological pH values. Zhang et al. developed (Zhang et al. 2020) efficient antibacterial composite, and silver nanoparticles (AgNPs) were coated with graphene quantum dots (GQDs) as a shell layer (GQDs@AgNPs). Because GQDs and AgNPs have a synergistic antibacterial effect, the antibacterial activity of the GQDs@AgNPs composite against E. coli and S. aureus was greater than that of single GQDs and single AgNPs. The authors investigated the effects of chemical reduction on the antimicrobial activity of graphene oxide quantum dots and the improvement of bactericidal performance through the formation of nanocomposites with metal oxide nanoparticles (Nichols and Chen 2020; Rojas-Andrade et al. 2020; Catanio et al. 2022). Functionalized GQDs were shown to be effective antibacterial drugs. Through electron transfer, the massive π-conjugated system of GQDs readily bonded to the bacterial cell wall. GQDs increased oxidative stress, which in turn caused damage to the cell walls and subsequent leakage of cell contents and cell death. Nevertheless, the antibacterial activity of unmodified GQDs is somewhat low. GQDs have a variety of structural dimensions, including spherical, linear, pentagonal, and hexagonal. By adding different functional groups (polymers and ligands), it is feasible to tune and construct GQDs with an edge effect and fine size. Modifications to the structural orientation of modified GQDs enable their attachment to the microbial biomolecular lattice. Additionally, photostable GQDs, which are colored blue and green, aid in the bioimaging of the bacteria. The alterations in surface functioning led to a significant reduction in cytotoxicity and phototoxicity. It was suggested that the elimination of oxidative phenanthrenequinone -like moieties, which function to obstruct the growth of the bacterial electron transport chain and produce reactive oxygen species, was the cause of the decrease in cytotoxicity. As evidenced by the marked decline in the activity of photocatalytic degradation of organic dyes, the complete loss of phototoxicity was found to result from the diminished formation of hydroxyl radicals as a result of the newly formed conjugation to hydroxyl groups that rendered homolytic cleavage unfavorable. When compared to other examined microorganisms, such as Staphylococcus aureus, Klebsiella pneumoniae, and Pseudomonas aeruginosa, CQDs more successfully limit the development of E. coli and C. albicans (Jhonsi et al. 2018b). Nie et al. showed that CQDs are efficient photosensitizers for in vitro PDT and that visible light illumination ($\lambda \geq 420$ nm, 65 ± 5 mW/cm2; 60 min) results in detection limit inactivation (99.9999 + %) of E. coli and S. aureus (Nie et al. 2020). The antibacterial effect of photoexcited GQDs using blue light (465–475 nm) significantly affected the viability of E. coli or

S. aureus. Sun et al. designed an antibacterial system combining GQDs with a low level of H2O2, which could inhibit the growth of E. coli and S. aureus (106 CFU/mL) bacteria and assessed the antibacterial efficacy of patches from GQDs in vivo using Kunming mice as a model (Ristic et al. 2014). When photoexcited for only 3 min of 670 nm laser exposure, Kuo et al. demonstrated that a nitrogen-doped graphene quantum dot (N-GQD), acting as a photosensitizer, was able to generate a greater amount of ROS than a nitrogen-free GQD in PDT, indicating greatly increased antimicrobial properties. After 3 min of exposure, the N-GQD (5.1%) effectively exhibited antibacterial activity, leading to 100% eradication (Kuo et al. 2017).

6.3.2 Cadmium QDs

Zhang et al. (Anon n.d.) reported AP-loaded CdSe NPs (AP-CdSe NPs) showed strong antibacterial activity against multidrug-resistant (MDR) *Escherichia coli* (*E. coli*) and *Staphylococcus aureus* (*S. aureus*) in vitro and in vivo. The quantitative analysis of apoptosis by flow cytometry analysis further confirmed that MDR *E. coli* and *S. aureus* treated with AP-CdSe NPs had death rates of 98.76% and 99.13%, respectively. Toxicity analysis indicated that AP-CdSe NPs had few side effects. Heavy metal ion hybrids in nano-quantum confinement make up CdTe QDs. Antimicrobial activity was demonstrated by CdTe QDs with sizes varying from 2.5 nm to 8 nm against microorganisms that were resistant to many drugs (Fang et al. 2012; Moradi Alvand et al. 2019). The lattice dimension shared by bacterial protein molecules and nucleic acids is thought to underlie the interaction of CdTe QDs. Additionally, CdTe QDs exhibit efficient electron transport within bandgaps, resulting in photoluminescence that is employed in light-activated therapeutics; one electron is excited from the valence band to the conducting band at a wavelength of 450–650 nm. Cadmium-based quantum dots (QDs) are semiconducting nanoparticles with a narrow-band gap that varies with size. They are particularly effective in the evident and some UV wavelengths. The amount of released Cd2+ inside the cell is determined by the chemical composition and structure of cadmium-based QDs, which can cause a series of stress responses.

6.3.3 Peptide- and Drug-Functionalized Fluorescent QDs

Sharma et al. (Li et al. 2021) reported a method for creating a vector derived from nanoconjugates that effectively inject antibiotics into bacterial cells. Utilizing an electrostatic method, the negatively charged CDTe@CdS quantum dots (QDs) in the nanoconjugate have their surface functionalized with the cationic BP-100 (KKLFKKILKYL-amide), a known cell-penetrating peptide (CPP). When compared to bare ampicillin, which has no cytotoxic impact, the QD-CPP-Amp conjugate's internalization and bacterial debilitation are two to four times more potent. Strong methods that tend to be used in biological and biochemical investigations utilizing spectroscopy and microscopy are fluorescence and photoluminescence. Regrettably, many fluorophores are susceptible to photobleaching, while certain ones have quantum yields, which makes them susceptible to ambient dielectric properties and/or quenching. Low-dimensional inorganic transition metal dichalcogenides (TMDCs) are a new and promising class of semiconducting materials that are highly fluorescent, optically stable, and safe for the environment. They comprise a series of triatomic building blocks (MX2, where M = Mo or W and X = Se or S). Two TMDC members, MoS2 and WS2, share crucial characteristics in common, including small size (1–6 nm), high fluorescence intensity (20–30 times higher than organic dyes), and emission that can be tailored based on size. Large Stokes shifts are conceivable due to the QDs' narrow emission and wide range of excitation wavelengths, which are beneficial for tissue imaging and higher signal-to-noise ratios for optimal performance. The past decade in particular has seen an enormous spike in multidrug resistance to conventional antibiotics, which has prompted interest in both AMPs and the use of different inorganic nanomaterials as antimicrobial agents. As an outcome, attempts have been made to harness AMPs with higher chemical and proteolytic stability, better potency, and improved selectivity. Building on these earlier investigations, we created two hybrid AMPs from the parent peptides. By using solution NMR spectroscopy to examine the conjugated peptides' conformational arrangement in detail, it was possible to connect the conjugates' interactions with bacterial and model lipid membranes. In addition to examining the antimicrobial effects against Pseudomonas aeruginosa (P. aeruginosa) and Candida albicans (C. albicans), additional research evaluated the biological ramifications of these interactions in terms of cell toxicity. Particularly in critical care units, surgical patients' rooms, and among immunocompromised people, these two opportunistic pathogens are the root cause of fatal infections. AMPs are regarded as a new class of super antibiotics with great promise for combating bacterial resistance because of their ability to quickly destroy both planktonic and biofilm forms of bacteria while having a minimally harmful effect on eukaryotic cells. Numerous nanoparticles and quantum dots have demonstrated their efficacy in combating an extensive range of infections and illnesses. Nanoparticles have the potential to be functionalized and conjugated with potentially active antimicrobial peptides so as to broaden their applications in drug discovery and delivery systems, such as imaging and diagnostics (Madrakian et al. 2017). QD-GS24 and QD-T7 probes were shown to be able to distinguish between B16 and HeLa cells, respectively, using fluorescence and confocal microscopy imaging. Comparable free transferrin may facilitate the uptake of QD-GS24 or QD-T7 into cells, according to quantitative flow cytometry analysis. The unique optical and spectroscopic properties of quantum dots (QDs), which are nanocrystalline semiconductor materials, include strong luminescence, long luminescent lifetimes, resistance to photobleaching, narrow and tunable emission, and broad absorption. Because of these extraordinary attributes, QDs are currently employed in countless biological applications as an alternative to protein and small-molecule fluorophores. A freshly emerging area of study is the interaction of QDs with the intricate chemical structure and biology of the peptide arena. In addition to acting as ligands for the production of water-soluble QDs, peptides engineered with the proper cysteines or histidines are additionally being utilized to tag protein ligands and biosensors to QD surfaces. The translocation of functionalized QDs into cells for intracellular imaging applications has been made possible by the incorporation of cell-penetrating peptides on QD surfaces. The creation of innovative protease tests has demonstrated the usefulness of QDs containing fluorescently labeled peptide substrates. Furthermore, peptides labeled with QDs that bind to cellular receptors offer an alternative to imaging at the whole-cell and single-molecule levels using antibodies when seeking to investigate the distribution and trafficking of receptors.

6.3.4 ZnO QDs

ZnO QDs have a high surface charge and range in size from 2 to 16 nm. Numerous investigations on ZnO QDs have confirmed their biocompatibility and biosafety in relation to human cells (Garcia et al. 2018; Wang et al. 2018). ZnO QDs are widely used in the treatment of bacterial and fungal infections due to their unique features. ZnO QDs' ability to limit the growth of microbiological organisms can be explained by three possible mechanisms: (1) The hydroxyl radical, superoxide anion, and perhydroxyl radical are among the ROS species that are induced by ZnO QD electron-hole pairs; (2) ZnO QDs directly rupture the phospholipid layer, resulting in cell content and cytoplasm leakage; and (3) Zn^{2+} accumulates in the cell, damaging the DNA and causing cell apoptosis (Premanathan et al. 2011). Pure ZnO QDs had to be obtained before hydrophilic QDs could be produced. 150 milliliters of pure ethanol was used to dissolve 5.0 g of zinc acetate dihydrate, and the resulting solution was refluxed at 80 °C for a period of 2 h to guarantee full solubilization. Following room temperature cooling of the prepared solution, a 20 mL potassium hydroxide-ethanol solution (1.75 mol/L) was progressively added to the solution previously mentioned. The amalgamated solution gradually turns milky white and becomes colorless and transparent, as can be noticed clearly. ZnO QDs were effectively created by silane surface modification using APTES, which gave rise to ultra-stable luminescence and good water solubility. When energized at 355 nm, the as-prepared APTES-modified ZnO QDs (M-ZnO QDs) demonstrated robust emission of yellow fluorescence that is specifically quenched by Cu^{2+}. In order to detect Cu2+ in an aqueous solution, we have developed a fluorescence sensor based on M-ZnO QDs, implementing into account this property. Significant progress in the synthesis of ZnO QDs has made it achievable to precisely manipulate their dimensions, form, and surface attributes. Achieving this degree of control, ZnO QDs can now have their properties tweaked to fit the needs of a variety of applications, notably the biomedical sciences and electronics. Due to quantum confinement, ZnO QDs have size-dependent bandgap behavior, thereby rendering their optical properties particularly intriguing. Light-emitting diodes (LEDs), solar cells, and sensors are just a few optoelectronic devices that have benefited from this special trait. ZnO QDs' biocompatibility has also sparked interest in prospective applications in drug delivery, theranostic uses, and biomedical imaging. In the field of nanoscience and technology, zinc oxide quantum dots, or ZnO QDs, have drawn a lot of attention as a fascinating class of nanomaterials with special and adjustable properties. All things accounted for, quantum dots are semiconductor nanocrystals that exhibit quantum confinement effects, which lead to size-dependent optical and electronic characteristics. Zinc oxide, or ZnO, is a wide-bandgap semiconductor with remarkable electrical, optical, and photochemical properties. ZnO QDs are distinguished by their zinc oxide content. ZnO QDs exhibit great potential for resolving issues and delivering opportunities spanning a range of fields as nanotechnology matures. Considering regard to ZnO QD synthesis, properties, surface modifications, and applications, this review attempts to offer a thorough examination of the most recent advancements. We hope to shed light on the various applications where ZnO QDs have demonstrated their potential and emphasize the obstacles that scientists are working to overcome in order to make the most of this nanomaterial by exploring its complexities. Zinc Oxide Quantum Dots are Revolutionizing Implant Coatings, Drug Delivery Systems, and Wound Healing in Medicine. Zinc oxide quantum dots (ZnO QDs) represent an avenue for therapy for wound healing. They are great additions to topical formulations and wound dressings due to their high surface area and antimicrobial activity, which actively fight infections. Zinc oxide nanoparticles (ZnO QDs) have the capacity to produce reactive oxygen species (ROS), which hasten tissue regeneration and improve wound healing. Studies indicate that they could help develop next-generation wound care products that are more efficient and have a quicker healing time. Due to their ability to stop bacteria from colonizing implant surfaces, zinc oxide quantum dots, or ZnO QDs, are essential in the fields of orthopedics and implantology. By preventing microbial adhesion, their antimicrobial qualities when integrated into implant coatings lower the chance of infections. ZnO QDs are released from coatings under controlled conditions, which boosts the stability and functionality of implanted devices over time by serving as a persistent antimicrobial barrier. This application has the potential to drastically boost the success rates and safety of various implants.

6.3.5 Amino-Functionalized QDs

Dey and Mondal (Mondal and De 2021) have studied the impact of MoS2 quantum dots modified with hydrophobic amino acids on their antibacterial properties. Depending on the amino acid residue, functionalization increases the antibacterial activity and selectivity. While MoS2 quantum dots with a phenylalanine-functionalized ligand exhibit nearly equal antibacterial efficacy against both MRSA and Pseudomonas aeruginosa (PA), MoS2 quantum dots with a leucine-functionalized ligand exhibit selective antibacterial activity against methicillin-resistant Staphylococcus aureus (MRSA). Amino-functionalized quantum dots (QDs) are a luminescent nanomaterial with antibacterial properties. QDs can be functionalized to improve their antibacterial activity and selective ability. Functionalized QDs could be a viable alternative to traditional antibiotics. A one-step hydrothermal synthesis was used to create amino functionalization. The amino group produced more reactive oxygen species, demonstrating their superior efficacy for two-photon PDT. Furthermore, the amino's intrinsic luminescence properties and high photostability show that it is an effective two-photon contrast agent for tracking bacteria during two-photon biomedical imaging. A simple one-pot hydrothermal approach was used to derive amino-functionalized carbon quantum dots (N-CQDs), a form of efficient fluorescent probes. N-methyl ethylenediamine and gluconic acid were incorporated as the carbon and nitrogen sources, respectively. Following that, the N-CQDs were used to calculate isotretinoin (INN). X-ray diffraction (XRD), transmission electron microscopy (TEM), dynamic light scattering (DLS), Fourier transform infrared spectroscopy (FT-IR), UV-vis absorption, and photoluminescence spectroscopy were used to characterize the synthesized N-CQDs. The as-prepared N-CQDs are used as a highly selective and sensitive probe for INN detection because INN effectively quenches the emission of N-CQDs. In the range of 0.08–70.0 μmol L^{-1} for INN determination, a linear response was seen under ideal circumstances. Quantum dots with functionalized amino groups (QDs) are a unique class of nanomaterials that leverage the size-dependent properties and adaptable surface chemistry of quantum dots. Amino groups are added to the surface of quantum dots, giving them a new level of functionality and a plethora of applications, particularly in the field of antimicrobial therapy (Zaid and Abdullah 2017). The addition of amino functionalization confers antimicrobial and biocompatible characteristics, thereby establishing these nanostructures as viable candidates for the fight against microbial infections. Amino-functionalized quantum dots are promising antimicrobial agents owing to their distinct properties. These include the ability to interact with microbial membranes, foster reactive oxygen species (ROS), and have the potential to release antimicrobial payloads to satisfy controlled conditions. Amino-functionalized QDs' diverse properties make them an adaptable platform for conquering problems brought on by different microbial strains. The increasing prevalence of antibiotic resistance and the restricted effectiveness of conventional antibiotics highlight the urgent requirement for novel therapeutic strategies. Because pathogens may end up resistant, it is necessary to investigate different methodologies to deal with this ongoing issue.

The excellent photostability and biocompatibility of amino-functionalized quantum dots (QDs) have led to their widespread use in bioimaging. This part examines the latest developments in using these QDs for fluorescence imaging, highlighting their usefulness for applications such as in vivo imaging, tracking, and cellular labeling. Amino-functionalized quantum dots (QDs) are essential for encapsulating and imparting therapeutic agents in drug delivery systems. The use of these QDs for targeted drug delivery has advanced recently, and this section scrutinizes these breakthroughs, highlighting how they can improve drug solubility, stability, and controlled release. Amino-functionalized quantum dots (QDs) offer novel approaches in fields such as drug delivery, imaging, and medicine that could profoundly impact a person's everyday life. Their special qualities promise better health outcomes by raising the efficacy of therapeutic interventions, medical treatments, and diagnostics. Amino-functionalized QDs are likely to lead to developments that augment and overhaul a number of facets of daily life as these applications acquire more thoroughly.

6.3.6 Nanocomposite Material QDs

Ag/In/S-ZnS QDs are hybrid nanomaterials featuring a shell-core structure and modified up to 300 ns (Mir et al. 2018) N-acetyl-l-cysteine-capped Ag/In/S QDs demonstrated outstanding bactericidal action (Gram-negative bacteria), much lower than non-capped Ag/In/S QDs (Soheyli et al. 2019). A unique nanocomposite made of cross-linked chitosan (cl-Ch-pMAc@ZnO/CdSQDs) embedded in ZnO/CDs QDs was reported, which showed outstanding antibacterial activity (Midya et al. 2019). The enhanced adsorption efficiency of the organic-inorganic combination polymeric nanocomposite materials has attracted a lot of attention. This can occasionally result in extra perks through reciprocal benefits in guest-host interaction. The incorporation of QDs inside a polymer matrix creates very interesting nanocomposites that could be used in future optoelectronic devices because they combine the properties of QDs with the technological feasibility of polymers. By creating ZnO/CdS quantum dots on anionically functionalized chitosan (poly (methacrylic acid) cross-linked chitosan (cl-Ch-pMAc) in the presence of diethylene glycol Di methacrylate (DEGDMA) cross-linker), a novel nanocomposite (cl-Ch-pMAc@ZnO/CdSQDs) was produced under microwave irradiation. It was found that the aforementioned nanocomposite effectively inhibited the growth of Gram-negative E. coli and B. subtilis (Gram-positive) bacteria. Nanocomposite material has been prepared and comprises amine-functionalized graphene oxide (NH_2-GO) incorporation with water-soluble CdS quantum dot nanoparticles to form a new composite material (NH_2-GO/QDs). This composite mixture shows highly homogenous without precipitation and has been characterized by using Raman, Fourier transform infrared spectroscopy (FTIR), and scan electron microscopy (SEM-EDX). Furthermore, GO, or reduced graphene oxide (rGO), in combination with other nanoparticles may improve performance in certain functional electrochemical applications, such as extending the shelf life of the selected biorecognition element. The application of quantum dots, in particular aqueous synthesis in the field of electrochemical biosensors, and quantum dot (QD) composite with graphene are still relatively recent discoveries. Amine-functionalized graphene oxide is easily dispersed in water-soluble CdS quantum dots thanks to a novel composite material based on the functionalization of graphene oxide with amines through covalent functionalization. The cyclic voltammetry (CV) technique was then employed in order to electrodeposit this bio-composite material on a screen-printed electrode (SPCE). These altered electrodes were instrumental as a platform for developing electrochemical DNA biosensors by immobilizing biomolecules. For use as an electrode modifier, a nanomaterial bio-composite based on amine-functionalized graphene oxide with CdS quantum dots has been effectively created. Raman characterization demonstrates that following the functionalization stage, ethylenediamine has effectively reduced the graphene oxide. Moreover, the FTIR analysis shows that covalent bonds are being formed between the CdS quantum dots and the amine-functionalized graphene oxide. After additional electrodeposition of composite material onto a screen-printed electrode, the modified electrode demonstrates a viable technique for creating a sensing material platform with good sensing capabilities. With methylene blue (MB) as a mediator, a faster hybridization kinetic was also discovered, enhancing the analytical capabilities of the electrochemical biosensors.

6.3.7 Lead-Based QDs

Lead-based quantum dots (QDs) can be tuned to emit in the biological tissue's transparent region (700–1100 nm), making them a promising candidate for optical bioimaging. However, in order to be used as biolabels, these QDs must retain their luminescence and colloidal stability in water, physiologically saline buffers, varying pH levels, and biological product media. To achieve this, four different surface modification strategies were tried: (1) silica encapsulating; (2) ligand substitution with polyvinylpyrrolidone; (3) polyethylene glycol-oleate (PEG-oleate) interaction into the oleate ligands on the surface of the QDs; and (4) intercalation of poly(maleicanhydride-alt-1-octadecene) (PMAO) into the oleate ligands on the surface of the QDs and further cross-linking of the PMAO. Lead chalcogenides, such as PbS or PbSe, are the primary constituents of lead-based quantum dots (QDs), which are produced via colloidal chemistry methods. Precursor compounds are dissolved in a solvent, and then, they are broken down under carefully regulated conditions to form nanocrystals. The growth parameters are then adjusted to provide exact control over the size and shape of the nanocrystals. Organic ligand surface passivation improves stability and inhibits unchecked growth. The size, shape, and optical characteristics of the resultant quantum dots are verified by characterization methods such as TEM and spectroscopy. Tailored QDs with potential applications in optoelectronics and antimicrobial interventions are ensured by this synthesis approach. Because of quantum confinement effects, lead-based quantum dots (QDs) have specific optical and electronic properties because of their wavelength-comparable dimensions. As a result, behavior at the nanoscale is influenced by quantum size effects. Quantum confinement allows for size-tunable emission and absorption spectra, which allow for the customization of energy levels during synthesis. With careful consideration of toxicity and biocompatibility, this property is beneficial in imaging for enhanced contrast, and sensing for analyte detection, and may hold promise for antimicrobial interventions. Lead-based QDs' tunability opens up a wide range of uses in biomedicine and nanotechnology. Due to their ability to produce reactive oxygen species upon light activation for microbial damage, lead-based quantum dots show promise for use in photodynamic therapy (PDT) and other antimicrobial applications. By utilizing special qualities for membrane interaction and cellular disruption, they could also be used in antibacterial coatings to prevent growth. Inducing oxidative stress, generating reactive oxygen species, and interacting with microbial membranes are the antimicrobial mechanisms. These characteristics demonstrate the potential of lead-based quantum dots in various approaches to the fight against microbial infections. Prospective investigations into the efficaciousness of lead-based quantum dots (QDs) in antimicrobial applications center on mitigating toxicity issues, guaranteeing biocompatibility, and examining their ability to combat microbial infections. The goal of optimizing synthesis techniques is to regulate surface characteristics such as size, shape, and properties for customized features in particular interventions. It is anticipated that refining surface modifications will improve microbe interaction, stability, and biocompatibility. The main objective is to strike a balance between safety and antimicrobial efficacy in order to fully utilize lead-based QDs in novel and secure biomedical applications.

6.3.8 Indium Phosphide QDs

Indium phosphide quantum dots (InP QDs) have emerged as a viable alternative to hazardous metal ion-based QDs. To reach their full potential in biological applications, InP QDs must be able to generate cationic surface charge while maintaining stability and biocompatibility. Since the development of effective antibiotics is not keeping up with the evolution of multidrug-resistant (MDR) bacteria, infections caused by these bacteria are becoming a growing imminent danger to human health. Multidrug-resistant (MDR) bacterial infections are becoming a serious threat to human health because these bacteria are evolving faster than the creation of effective antibiotics. High-efficiency bactericidal activity against an abundance of bacterial species, such as Staphylococcus aureus, Bacillus cereus, Escherichia coli, and Pseudomonas aeruginosa, originates from PDT in the presence of InP QDs. When InP QDs absorb light, they produce superoxide (O_2^-), which effectively and selectively kills MDR bacteria while sparing mammalian cells. InP QDs are bio- and blood-compatible over a broad therapeutic window, according to the cytotoxicity evaluation. The near-infrared wavelengths that our InP QDs are made of, which have a high transmission through skin and tissues, activate them. The QDs are non-toxic on a macroscale, as demonstrated by body weight and organ tissue histology. Serum indicators of oxidative stress and inflammation show that the InP QD treatment had no discernible impact on mouse health at concentrations that decreased the viability of drug-resistant bacteria in subcutaneous abscesses. In mouse subcutaneous abscesses, the InP QD treatment reduced bacterial viability by more than three orders of magnitude. When InP QDs are exposed to light, they produce reactive oxygen species (ROS), which demonstrate antimicrobial activity. Microbial cell membranes and DNA are harmed by this oxidative stress, which reduces the viability of the cells. Membrane permeability is compromised when InP QDs interact with microbial membranes, causing structural integrity to be disrupted. By hindering vital cellular processes, this disruption enhances the antimicrobial effect even more. Within microbial cells, InP QDs have the potential to disrupt intracellular functions. Their existence can cause cellular stress, interfere with enzyme function, and interfere with signaling pathways within cells, all of which can lead to microbial inhibition. InP QDs' antimicrobial mechanism is influenced by their size-dependent characteristics. It is possible that smaller QDs will cause more damage to cellular structures because they can enter microbial cells more easily. Comprehending the antimicrobial action mechanisms of InP QDs lays anatomical foundations for additional investigation. Subsequent investigations may concentrate on refining these processes, customizing InP QDs, and accomplishing higher effectiveness and selectivity in the fight against microbial infections. One of the main antimicrobial strategies involves the interaction of indium phosphide quantum dots (InP QDs) with microbial cell membranes. InP QDs cause the microbial cell membranes to become compromised in this mechanism. Cell death results from this disruption, which allows the contents of the cell to leak out. Effectively, InP QDs prevent microbial growth by disrupting the structural integrity of microbial membranes. As a result of the weakened cell membrane integrity, the microorganisms are unable to carry out their regular activities, which ultimately leads to their death. The mechanism of cell membrane disruption elucidates the antimicrobial agent potency of InP QDs and highlights their potential to combat microbial infections as well.

6.3.9 Silicon QDs

Antimicrobial properties of silicon quantum dots (Si QDs) have been discovered. They have the ability to degrade cell walls and membranes, generate reactive oxygen species, bind to gene material, and prevent energy production. Si QDs are additionally applied in biological applications. Si QDs, for example, have been used in the controlled dissolution of antibiotics. It has been found that silicon quantum dots, or Si QDs, have antimicrobial properties. They can break down cell walls and membranes, construct reactive oxygen species, cling themselves to genetic material, and stop the production of energy. Moreover, biological applications employ Si QDs. For instance, Si QDs have been employed in the regulated dissolution of antibiotics. At low concentrations, the Si-PEIPDs were able to completely suppress S. aureus and E. coli without external assistance, and the breakdown of membranes contributed to this result. Because silicon quantum dots (SiQDs) are easily functionalized and conjugated with biomolecules through a variety of chemical modifications, they are an excellent choice for in vivo labeling and detection. However, reflecting their surface single bonds, a large proportion of native SiQDs are not exceptionally soluble in water or compatible with hydrophilic solvents, which prohibits their deployment in biological and biomedical disciplines. SiQDs have one of the most widening spectral ranges with emission tunability and are composed of abundant, nontoxic material. Moreover, printing, roll-to-roll, stamping, spraying, painting, and other low-temperature deposition techniques are plausible for colloidal SiQDs on an assortment of substrates, ranging from glass, plastic, wood, and textiles. More importantly, SiQDs' nontoxicity and biodegradability create new avenues for application in areas where environmental and health ordinances are scrupulously adhered to. These include in vivo bioimaging, drug delivery, lab-on-a-chip sensing, optoelectronic toys, smart textiles, photocatalysis for food, waste, and water sterilization, phototherapeutics, and cosmetics, among other areas. For advantageous electron and charge transport, SiQDs with the proper functionalization would have been granted an active morphology. In order to increase native SiQDs' water dispersibility, photostability, and, if advisable, fluorescence quantum yield, hydrophilic ligands must be added. This work reports the synthesis of water-soluble allylamine-modified silicon quantum dots (NH2@SiQDs) using the microemulsion technique. They were also used to build a basic enzyme-free biosensor that detects glucose with great sensitivity. The freshly invented techniques for glucose sensing not only do not require the use of an enzyme but also have outstanding reliability, good reproducibility, and cost-effectiveness.

6.3.10 Perovskite QDs

A comparative analysis was conducted to examine the antibacterial properties of ZnO and CsPbBr3 perovskite nanoparticles against rod-shaped, Gram-negative Escherichia coli, the strain O157: bacteria cells. According to experimental findings, CsPbBr3 nanoparticles have superior antibacterial activity than ZnO nanoparticles. Because of their exceptional optoelectronic attributes, ultrahigh photoluminescence quantum yield (PLQY), optical tunability, narrow-band emission, and expansive color gamut, perovskite quantum dots have found extensive application in the field of optoelectronics. By altering the halogen species and ratio, lead-based halide perovskite may create a modulating effect on the photoluminescence spectrum. The resulting modulation effect can be used to predict optical properties that are relatively controllable and facilitate adaptation of the material's corresponding absorption and emission spectra. Perovskite quantum dots have numerous advantages such as improved stability, variable emission, and absorption spectra, and excessive PLQY, considering doping them is effortless. In the same way, perovskites' exceptional characteristics have prompted their use in the field of luminescence solar concentrators (LSC). The primary emission peak of CsPbCl3:Yb nanocrystals is located at 980 nm, whereas the primary absorption peak is located at 400 nm. This reveals a significant Stokes shift, which is extremely beneficial and has little impact on self-absorption throughout transport. The energy conversion efficiency of larger-size LSCs based on quantum cutting is predicted to be four times higher than that of traditional perovskite nanocrystals, based on theoretical calculations. The Mn2+/Yb3+ codoped CsPbCl3 nanocrystals demonstrate the requisite effects (Wu et al. 2020). Perovskite quantum dots on the surface have emerged rapidly and proven very popular, in accordance with numerous studies. However, there are still a lot of hurdles that need to be addressed before employing experimental perovskite quantum dots on luminescence solar concentrators (LSCs). The primary issue arising from the material is the stability of perovskite quantum dots, which is fundamental to future industrialization and will be contingent on its ability to withstand light, water, oxygen, heat, and other aspects. While there is currently progress in alleviating the corresponding stability concerns with some of the current encapsulation and passivation solutions, it still appears to be trivial in terms of the development of prospective practical applications. Everyone's second concern is the problem's toxicity; lead can enter the body directly through the skin and is extremely difficult to eradicate from the body, which will have an enormous detrimental impact on numerous bodily systems. Although lead-free perovskite quantum dots have garnered scrutiny, their optical and electrical characteristics are typically inferior to those of lead-based perovskite quantum dots. Theory and practice ought to be studied in tandem in order to solve this problem. In the case of environmentally friendly quantum dots, the issue of biological toxicity is comparatively unnecessary, but the material's toxicity problem severely restricts the possibilities for further applications. Enumeration of quantum dots (QDs) and the microbial mechanisms through which they interact with fungi and bacteria is presented in Table 6.1.

Table 6.1 Enumeration of quantum dots (QDs) and the microbial mechanisms through which they interact with fungi and bacteria

Types of quantum dots	Microorganisms (bacteria and fungi)	Mechanism of microbial action	Reference
Graphene QD	Escherichia coli Staphylococcus aureus	Generation of free radicals which destroy cell walls	Vibhute et al. (2022)
Cadmium telluride QD	Escherichia coli	Moreover, the release of heavy metal ions into the cells diminishes the expression of the superoxide dismutase gene. The QDs are inserted into the cell membrane to induce membrane stress	Jahangir et al. (2019)
Cadmium selenium QD	Pseudomonas aeruginosa	Cd^{2+} internalization culminates from genomic and cell toxicity	Wang et al. (2022)
Zinc oxide QD	Escherichia coli	After ZnO QDs are photoexcited, an electron hole pair emerges and the electron trapped in the oxygen then induces an excessive amount of reactive oxygen species	Wang et al. (2022)
Lead-based QD	Escherichia coli	Demonstrated by their ability to penetrate microbial cell membranes, induce membrane stress, and release lead ions that cause cellular disruption	Vibhute et al. (2022)
Indium phosphide QD	Escherichia coli Candida albicans	By penetrating microbial cell membranes and exerting membrane stress, indium phosphide QDs indicate antimicrobial effects cause a halt in microbial growth	Wang et al. (2022)
Silicon QD	Escherichia coli Saccharomyces cerevisiae	Their presence triggers cellular effects that may mess with the viability and function of microbes and illustrate microbial activity	Jahangir et al. (2019)
Perovskite QD	Pseudomonas strains	Permeate cell membrane and act as a microbe concomitantly, liberated heavy metal ions block expression of superoxide dismutase gene	Jahangir et al. (2019)

6.3.11 Cytotoxicity and Biocompatibility of QDs for Future Scope

An essential consideration that must be made before using CDs in the biomedical industry is their toxicity. Numerous studies have demonstrated the exceptional biocompatibility of CDs. Wei et al. used zebrafish eggs to assess the biological toxicity and created CDs utilizing a natural gynostemma. The results showed that CDs can be biologically imaged in zebrafish and had no influence on the embryo's growth [86]. Wu et al. described a nitrogen-rich CD (NCD)-mediated DNA nanostructure self-assembly technique for bioimaging and anticancer treatment (Midya et al. 2019). Several reports further manifest that the QDs have excellent biocompatibility and can be applied in biomedical applications (Wu et al. 2020; Cai et al. 2020). Hence, QDs have great scope for medicinal applications. In order to fully utilize quantum dots (QDs), which are becoming more and more significant in a variety of applications, it is critical to understand how they affect cellular systems. This talk explores QDs' potential for the future while highlighting two important factors: cytotoxicity and biocompatibility. Assessing the interactions of QDs with biological entities provides important information about their safety and directs their development for various applications. It is possible to optimize the responsible and efficient use of QDs in a variety of fields in the future by comprehending and addressing these factors. Quantum dots' (QDs') future potential is reliant upon proving that they are biocompatible, particularly in therapeutic and medical settings. To create formulations safe for use in biological systems, it is critically important that we examine the interactions between QDs and living organisms, tissues, and organs. In order to minimize immunogenic reactions, improve biocompatibility, and shrink side effects, research should focus on fine-tuning QD surface modifications. To further expand the uses of QDs in medicine, from targeted drug delivery systems to advanced imaging techniques, comprehensive biocompatibility profiles must satisfy the criteria established. The release of heavy metals has substantially hampered the biomedical applications of conventional semiconductor quantum dots, which usually involve the creation of elements from the II-VI or III-V groups (e.g., CdSe). Because they are more photostable than organic dyes and less toxic than semiconductor quantum dots, carbon luminescent nanomaterials have recently sparked tremendous research interest. The lateral dimensions of graphene quantum dots (GQDs), a type of zero-dimensional material, are less than 100 nm, contrary to sustaining the same single-atom layer as graphene. GQDs have the potential to be vectors for the delivery of protein or drug molecules to cells due to their high surface area and good biocompatibility. GQDs' exceptional luminescence makes them suitable for probes with fluorescent properties for bioimaging in tandem. Additionally, GQDs can be functionalized with different chemical groups to create multifunctional structures by covalently joining them with a variety of other materials, including drug molecules, proteins, and nanotubes. This will increase the range of applications for which they can be used in the biomedical field.

6.4 Conclusion

QDs are carbon nanostructures with promising antibacterial properties; yet, significant technological difficulties remain, restricting their effective design and practical application. Nanomaterial design techniques that remove unwanted biomolecule adsorption, preserve the basal plane from contamination, and boost antibacterial activity without compromising biocompatibility should be developed. Any structural flaws in the carbon network will have a substantial impact on the characteristics and effectiveness of the nanomaterial. As a result, one of the most difficult issues for large-scale manufacturing of graphene-based nanomaterials is to constantly maintain a high quality and uniform size for the synthesized nanomaterials while keeping the synthesis process low cost and reproducible. Because of the ease of synthesis and functionalization of QD precise control over charge density and functional group is now possible. This is a fantastic chance to solve bacterial MDR concerns by producing innovative QDs with antibacterial activity against specific bacteria. Due to the fact that different types of carbon nanomaterials have distinct properties based on lateral size, surface charge, and surface functionalization, their cytotoxicity and biocompatibility for different cell lines should be thoroughly investigated for safe biomedical applications. Acute and chronic cytotoxicity experiments employing animal models should also be carried out for these carbon nanoparticles.

Acknowledgments and Funding NS is grateful to the Institute Head and Head of the Department of Engineering Sciences, Ramrao Adik Institute of Technology, DY Patil University, Nerul, Navi Mumbai, for providing the RAIT research grant 2023-2024. This work was also supported by the Excellence project PrF UHK-2216/2023-2024 and by the MPO projects (Aumed as.). We also acknowledge the research support by Mr. Sudhir Krishnaraj Thackersey Research Grant-2021-23&#39, at Sir Vithaldas Thackersey College of Home Science (Autonomous), SNDTWU, Mumbai.

Conflict of Interest The authors declare that there is no conflict of interests.

References

Anand A, Unnikrishnan B, Wei SC, Chou CP, Zhang LZ, Huang CC (2019) Graphene oxide and carbon dots as broad-spectrum antimicrobial agents–a minireview. Nanoscale Horiz 4(1):117–137

Anon (n.d.). https://www.tandfonline.com/doi/full/10.2147/IJN.S295928

Cai T, Wang J, Li W, Hills-Kimball K, Yang H, Nagaoka Y, Yuan Y, Zia R, Chen O (2020) Mn2+/Yb3+ Codoped CsPbCl3 perovskite nanocrystals with triple-wavelength emission for luminescent solar concentrators. Adv Sci 7(18):2001317

Catanio AT, Bergmann EV, Kimura NM, Petrucci T, Freitas CF, Herculano LS, Malacarne LC, Astrath NG (2022) Spectroscopic and photothermal characterization of graphene quantum dots for antimicrobial applications. J Appl Phys 131:15

Chen W, Hu C, Yang Y, Cui J, Liu Y (2016) Rapid synthesis of carbon dots by hydrothermal treatment of lignin. Materials 9(3):184

Fang M, Lin L, Zheng M, Liu W, Lin R (2023) Antibacterial functionalized carbon dots and their application in bacterial infections and inflammation. J Mater Chem B 11(39):9386–9403
Fang TT, Li X, Wang QS, Zhang ZJ, Liu P, Zhang CC (2012) Toxicity evaluation of CdTe quantum dots with different size on Escherichia coli. Toxicol In Vitro 26(7):1233–1239
Garcia IM, Leitune VCB, Visioli F, Samuel SMW, Collares FM (2018) Influence of zinc oxide quantum dots in the antibacterial activity and cytotoxicity of an experimental adhesive resin. J Dent 73:57–60
Jahangir MA, Gilani SJ, Muheem A, Jafar M, Aslam M, Ansari MT (2019) Quantum dots: next generation of smart nano-systems. Pharm Nanotechnol 7:234–245
Jhonsi MA, Ananth DA, Nambirajan G, Sivasudha T, Yamini R, Bera S, Kathiravan A (2018a) Antimicrobial activity, cytotoxicity and DNA binding studies of carbon dots. Spectrochim Acta A Mol Biomol Spectrosc 196:295–302
Jhonsi MA, Ananth DA, Nambirajan G, Sivasudha T, Yamini R, Bera S et al (2018b) Antimicrobial activity, cytotoxicity and DNA binding studies of carbon dots. Spectrochim Acta A Mol Biomol Spectrosc 5(196):295–302
Jian HJ, Yu J, Li YJ, Unnikrishnan B, Huang YF, Luo LJ, Ma DHK, Harroun SG, Chang HT, Lin HJ, Lai JY (2020) Highly adhesive carbon quantum dots from biogenic amines for prevention of biofilm formation. Chem Eng J 386:123913
Kováčová M, Špitalská E, Špitálský Z (2022a) Light-activated polymer nanocomposites doped with a new type of carbon quantum dots for antibacterial applications. In: Soria F, Rako D, de Graaf P (eds) Urinary stents. Springer, Cham. https://doi.org/10.1007/978-3-031-04484-7_25
Kováčová M, Špitalská E, Špitálský Z (2022b) Light-activated polymer nanocomposites doped with a new type of carbon quantum dots for antibacterial applications. In: Urinary stents: current state and future perspectives. Springer International Publishing, Cham, pp 315–324
Kuo WS, Chen HH, Chen SY, Chang CY, Chen PC, Hou YI et al (2017) Graphene quantum dots with nitrogen-doped content dependence for highly efficient dual-modality photodynamic antimicrobial therapy and bioimaging. Biomaterials 1(120):185–194
Li P, Yang X, Zhang X, Pan J, Tang W, Cao W, Zhou J, Gong X, Xing X (2020) Surface chemistry-dependent antibacterial and antibiofilm activities of polyamine-functionalized carbon quantum dots. J Mater Sci 55:16744–16757
Li W, Song P, Xin Y, Kuang Z, Liu Q, Ge F, Zhu L, Zhang X, Tao Y, Zhang W (2021) The effects of luminescent CdSe quantum dot-functionalized antimicrobial peptides nanoparticles on antibacterial activity and molecular mechanism. Int J Nanomedicine:1849–1867
Madrakian T, Maleki S, Gilak S, Afkhami A (2017) Turn-off fluorescence of amino-functionalized carbon quantum dots as effective fluorescent probes for determination of isotretinoin. Sens Actuators B Chem 247:428–435
Midya L, Patra AS, Banerjee C, Panda AB, Pal S (2019) Novel nanocomposite derived from ZnO/CdS QDs embedded crosslinked chitosan: an efficient photocatalyst and effective antibacterial agent. J Hazard Mater 369:398–407
Mir IA, Radhakrishanan VS, Rawat K, Prasad T, Bohidar HB (2018) Bandgap tunable AgInS based quantum dots for high contrast cell imaging with enhanced photodynamic and antifungal applications. Sci Rep 8(1):9322
Mondal A, De M (2021) Amino acid-functionalized MoS2 quantum dots for selective antibacterial activity. ACS Appl Nano Mater 4(12):13947–13954
Moradi Alvand Z, Rajabi HR, Mirzaei A, Masoumiasl A, Sadatfaraji H (2019) Rapid and green synthesis of cadmium telluride quantum dots with low toxicity based on a plant-mediated approach after microwave and ultrasonic assisted extraction: synthesis, characterization, biological potentials and comparison study. Mater Sci Eng C Mater Biol Appl 98:535–544
Nichols F, Chen S (2020) Graphene oxide quantum dot-based functional nanomaterials for effective antimicrobial applications. Chem Rec 20(12):1505–1515
Nie X, Jiang C, Wu S, Chen W, Lv P, Wang Q et al (2020) Carbon quantum dots: a bright future as photosensitizers for in vitro antibacterial photodynamic inactivation. J Photochem Photobiol B Biol 206:111864

Owusu EGA, MacRobert AJ, Naasani I, Parkin IP, Allan E, Yaghini E (2019) Photoactivable polymers embedded with cadmium-free quantum dots and crystal violet: efficient bactericidal activity against clinical strains of antibiotic-resistant bacteria. ACS Appl Mater Interfaces 11:12367–12378

Premanathan M, Karthikeyan K, Jeyasubramanian K, Manivannan G (2011) Selective toxicity of ZnO nanoparticles toward gram-positive bacteria and cancer cells by apoptosis through lipid peroxidation. Nanomedicine 7(2):184–192

Rajendiran K, Zhao Z, Pei DS, Fu A (2019) Antimicrobial activity and mechanism of functionalized quantum dots. Polymers 11(10):1670

Ristic BZ, Milenkovic MM, Dakic IR, Todorovic-Markovic BM, Milosavljevic MS, Budimir MD et al (2014) Photodynamic antibacterial effect of graphene quantum dots. Biomaterials 35(15):4428–4435

Rojas-Andrade MD, Nguyen TA, Mistler WP, Armas J, Lu JE, Roseman G, Hollingsworth WR, Nichols F, Millhauser GL, Ayzner A, Saltikov C (2020) Antimicrobial activity of graphene oxide quantum dots: impacts of chemical reduction. Nanoscale Adv 2(3):1074–1083

Soheyli E, Azad D, Sahraei R, Hatamnia AA, Rostamzad A, Alinazari M (2019) Synthesis and optimization of emission characteristics of water-dispersible ag-in-s quantum dots and their bactericidal activity. Colloids Surf B: Biointerfaces 182:110389

Vibhute A, Nille O, Kolekar G, Rohiwal S, Patil S, Lee S, Tiwari AP (2022) Fluorescent carbon quantum dots functionalized by poly l-lysine: efficient material for antibacterial, bioimaging and antiangiogenesis applications. J Fluoresc 32(5):1789–1800

Wang Y, He L, Yu B, Chen Y, Shen Y, Cong H (2018) ZnO quantum dots modified by pH-activated charge-reversal polymer for tumor targeted drug delivery. Polymers 10(11):1272

Wang ZX, Wang Z, Wu FG (2022) Carbon dots as drug delivery vehicles for antimicrobial applications: a minireview. ChemMedChem 17(13):e202200003

Wu D, Li BL, Zhao Q, Liu Q, Wang D, He B, Wei Z, Leong DT, Wang G, Qian H (2020) Assembling defined DNA nanostructure with nitrogen-enriched carbon dots for theranostic cancer applications. Small 16(19):1906975

Zaid MHM, Abdullah J (2017) Preparation and characterization of amine functionalized graphene oxide with water soluble quantum dots for sensing material. In: AIP conference proceedings. (Vol. 1877, No. 1). AIP Publishing

Zhang L, Liu L, Wang J, Niu M, Zhang C, Yu S, Yang Y (2020) Functionalized silver nanoparticles with graphene quantum dots shell layer for effective antibacterial action. J Nanopart Res 22:1–12

Zhou Y, Sharma SK, Peng Z, Leblanc RM (2017) Polymers in carbon dots: a review. Polymers 9:67

7 Antimicrobial Peptide-Based Nanomaterials in Combating Multidrug-Resistant Bacteria

Anindita Thakur, Ramakrishnan Ganesan, and Jayati Ray Dutta

Abstract

The escalated increase in morbidity and mortality in the recent era has resulted due to the impact of antimicrobial resistance (AMR) among microorganisms. There could be more than the predicted deaths resulting due to cancer as reported by the United Nations Environment Program, which estimates ten million annual deaths from AMR by 2050. Thus, there is a dire need to develop new antimicrobial agents and strategies to address such a pressing issue. Naturally obtained small multifunctional supramolecular peptides and antimicrobial peptides (AMPs) are found in all living organisms as an innate immune system. Therefore, to combat AMR, a multimodal target is a prerequisite. Such AMP-based nanomaterials would not only have an antimicrobial activity but would also have the potential to reinvigorate old drugs. In this chapter, we will emphasize the overview of the importance and progress of AMPs, the mechanism of action of AMPs, the technique of designing AMP-based nanomaterials, and the application and classification of such AMPs. A brief discussion has also been provided on the challenges of developing AMP-based nanomaterials and strategies to overcome them. Finally, the chapter summarizes future prospects and major potentials in harnessing antimicrobial activities via AMP-based nanomaterials.

A. Thakur · J. Ray Dutta (✉)
Department of Biological Sciences, Birla Institute of Technology and Science (BITS), Pilani, Hyderabad Campus, Hyderabad, Telangana, India
e-mail: jayati@hyderabad.bits-pilani.ac.in

R. Ganesan
Department of Chemistry, Birla Institute of Technology and Science (BITS), Pilani, Hyderabad Campus, Hyderabad, Telangana, India

M. Y. Wani et al. (eds.), *Nanotechnology Based Strategies for Combating Antimicrobial Resistance*, https://doi.org/10.1007/978-981-97-2023-1_7

Keywords

Antimicrobial resistance (AMR) · Antimicrobial peptides (AMPs) · Nanomaterial · Host defense peptides (HDPs) · Nanocarrier

7.1 Introduction

Injudicious use of conventional antibiotics has made them precarious in treating infectious diseases resulting in the rapid propagation of antimicrobial resistance (AMR), thereby abundantly imparting to the mortality rate. There could be more than the predicted deaths resulting due to cancer as reported by the United Nations Environment Program, which estimates ten million annual deaths from AMR by 2050 (de Kraker et al. 2016).

Treatment of bacterial infections with antimicrobial agents gives them ample opportunities to develop resistance (Chinemerem Nwobodo et al. 2022; Varela et al. 2021; Belete 2019). One such commonly reported is the production of β-lactamase; in addition to these, other resistance development pathways are altering the drug binding site, using quorum sensing for biofilm formation, and preventing cell membrane penetration (Saha and Sarkar 2021; Peterson and Kaur 2018; Zhao et al. 2020; Pang et al. 2019). Sometimes resistance can also develop due to certain genes present in the chromosome of the bacterial family or in the extrachromosomal DNA and plasmids (Peterson and Kaur 2018; Nikaido 2009). However, certain reports also indicate the development of resistance through transmissible elements (Peterson and Kaur 2018). The red alert pathogens are identified to be the most common cause of such life-threatening infections, acronym as ESKAPE pathogens; these microbes include *Enterococcus faecium, Staphylococcus aureus, Klebsiella pneumoniae, Acinetobacter baumannii, Pseudomonas aeruginosa,* and *Enterobacter species* (Kalpana et al. 2023; Masoud et al. 2022; Chaudhary et al. 2023; Idris and Nadzir 2023; Mukhopadhyay et al. 2020). This escalated rise in AMR among microorganisms thus ends the golden era of antibiotics for treating infectious diseases and begins the new era of discovering an alternative to antibiotics. Thereby to combat AMR, a multimodal target is a prerequisite.

Discovered in the 1980s, host defense peptides (HDPs), also known as antimicrobial peptides (AMPs), are innate immune responses found across all life forms (Haney et al. 2019; Lei et al. 2019; Lopes et al. 2022; Wu et al. 2022; Nizet et al. 2001). AMPs serve as essential defense molecules, combating pathogens and maintaining health within diverse species. In higher organisms, AMPs also play an immunomodulatory role (Konovalova et al. 2018). With around 5–50 amino acid residues, these compact peptides exhibit remarkable efficacy against drug-resistant bacteria, capitalizing on unique inter-species distinctions. AMPs boast a broad spectrum of activity, targeting bacteria, viruses, fungi, yeast, and even cancer cells (Huan et al. 2020).

The most common AMPs are cationic ones consisting of +2 to +9 charge, though anionic AMPs also exist due to the presence of amino acids that are acidic in nature like glutamic acid and aspartic acid. Some AMPs are completely hydrophobic (Lei et al. 2019). The mode of action of AMPs depends on their interaction with the

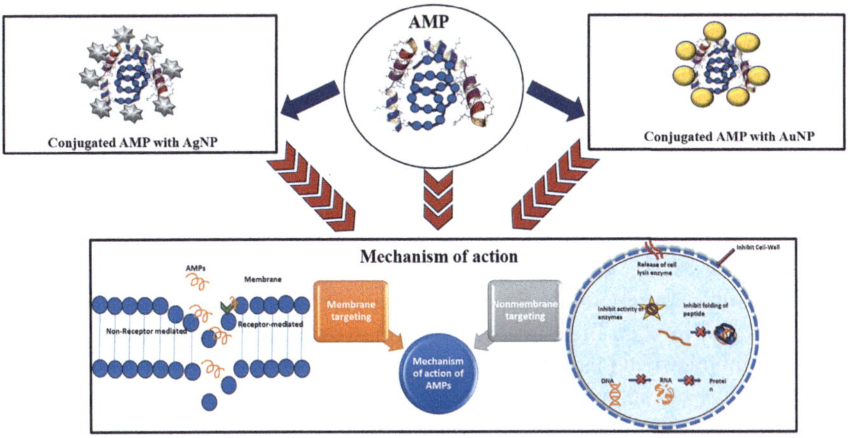

Fig. 7.1 Role of AMPs

bacterial membrane or cell wall. The cationic AMPs majorly lead to non-enzymatic membrane lysis, commonly called bacterial cell membrane or bacterial cell wall disruption (Benfield and Henriques 2020). This has been reported as the major mode of action. Other mechanisms involve targeting intercellular components along with interference with the vital intracellular processes (Le et al. 2017). Thus, AMPs can have unprecedented activity against MDR bacteria (Le et al. 2017).

As of May 17, 2023, the updated data repository of antimicrobial peptides (DRAMP) documents a total of 6105 AMPs, encompassing both natural and synthetic variants (Kang et al. 2019). While AMPs hold promise against multidrug-resistant (MDR) strains, unresolved challenges persist. These include scarcity of naturally occurring AMPs, vulnerability to protease degradation, and pH sensitivity (Lei et al. 2019; Kang et al. 2012). Notably, oral AMP administration exhibits notable toxicity (Lei et al. 2019; Kang et al. 2012). Hence, ongoing endeavors aim to curtail expenses, explore alternative production methods, and harness nano-based delivery systems. These innovations seek to enhance antimicrobial efficacy while ensuring safe AMP delivery and mitigating adverse effects. Figure 7.1 represents the role of AMPs.

7.2 The History of Antimicrobial Peptides

AMPs are molecules that are biologically conserved in organisms spanning from microorganisms to mammals. In the year 1922, Alexander Fleming successfully identified lysozyme, which was the initial AMP (Huan et al. 2020). Nevertheless, the finding of lysozyme was eclipsed by Fleming's discovery of penicillin in the year 1928 (Gaynes n.d.; Fleming 1929). Florey, Chain, and Fleming realized the medical potential of penicillin and were awarded the Nobel Prize in Medicine in 1945. In the 1940s, the onset of the golden age of antibiotics emerged with a decline in curiosity regarding the potential therapeutic value of natural antimicrobial peptides. As multidrug-resistant microorganisms become more prevalent, the fascination with AMPs as host defense molecules was rekindled in the 1960s (Cole and Nizet 2016). In the year 1981, alpha-helical AMPs termed cecropins derived from the silk moth *Cecropia* were studied, and in 1987, magainin was isolated from the African clawed amphibian *Xenopus laevis* (Fleming 1929; Zasloff 1987). During the course of the 1990s, over 300 peptides were reported in the field of antimicrobial peptides (Zasloff 1987; Sahoo et al. 2021). Since then, AMPs have been widely discovered and studied in nearly all organisms. A staggering 6105 AMPs have been submitted to the Antimicrobial Peptide Database at present. AMP was presumably first discovered in plant life, subsequently followed by the identification of bombinin in amphibians and lactoferrin in milk in the 1960s.

In 1981, Hans Boman reported in a seminal study that injecting microorganisms inside the *Cecropia* silk insect unleashed an efficient AMP. This finding of "cecropins" represented the primary significant report of alpha-helical AMPs. In 1987, Zasloff and colleagues isolated and characterized cationic AMPs, dubbed "magainins," from the African clawed amphibian *Xenopus laevis* (Kumar et al. 2018; McMillan and Coombs 2020). Hoffmann's group demonstrated in the mid-1990s that the genetic deletion of AMP synthesis rendered fruit flies susceptible to a vast fungal infection, thereby providing the first evidence of the crucial function of AMPs in insect host defense (Lemaitre et al. 1996). Accelerated clinical and scientific interest in 1994 along with the discovery of AMPs in the mammalian epidermis and the subsequent generation of mice deficient in one of these cathelicidins, which demonstrated the importance of AMPs to mammalian host defense (Nizet et al. 2001; Gallo and Hooper 2012).

Since then, AMPs have been exhaustively characterized and identified in virtually all multicellular organisms studied for this activity (Zasloff 2002).

7.3 Importance and Development of AMPs

AMR poses a grave hazard to animal husbandry and aquaculture in addition to medicine. The pressing concern arises from multi-drug resistant pathogens harboring a spectrum of resistance genes and plasmids. The use of both medical and non-medical antibiotics contributes to this predicament, with particular apprehension surrounding non-medical applications. The surge in antibiotic demand, driven by global population growth, finds application in agriculture, aquaculture, and animal husbandry for growth promotion and infection control. Well-known human antibiotics such as bacitracin, tetracycline, tylosin, and bambermycin are extensively utilized in poultry farming, compounding AMR escalation (Mehdi et al. 2018). A staggering 95% of ingested antibiotics are excreted, permeating soil and water and catalyzing AMRs' rapid ascent within the natural ecosystem (Cabello et al. 2016; Khmelevtsova et al. 2020). The aforementioned factors underscore a critical concern for both public health and economic well-being. Effective strategies must be devised to counteract the adaptability of bacteria, demanding innovative treatment approaches. AMPs emerge as versatile tools in combating AMR. They can stand alone or complement traditional antibiotics and antivirals, achieving synergistic outcomes. Despite the characterization of over 2500 AMPs, Food and Drug Administration (FDA) approval for human medicinal use remains elusive. A handful of AMPs, including daptomycin, gramicidin, vancomycin, dalbavancin, oritavancin, colistin, and telavancin, have gained FDA recognition (Walas et al. 2023; Talapko et al. 2022; Thakur et al. 2022; Mehta et al. 2022). Gramicidin D, authorized in 1955, is a neosporin ingredient, targeting bacterial conjunctivitis (Stevenson 2009; Hallett et al. 1956). Daptomycin, sanctioned in 2003, safeguards against infectious diseases (Carpenter and Chambers 2004). Colistin, approved in 1962, combats Gram-negative bacteria. Oritavancin and dalbavancin, endorsed in 2009, tackle Gram-positive bacterial infections. Telavancin, by inhibiting bacterial cell wall formation, proves effective against vancomycin-resistant strains. The viability of peptides as drugs hinges on their stability, gauged by their elimination half-life. For FDA-approved drugs, large peptides boast an average half-life of approximately 50 h, while small peptides exhibit around 37 h. These FDA-endorsed peptides demonstrate robust in vivo stability, while unstable AMPs lack this biological resilience (Hamamoto et al. 2002; Nguyen et al. 2010; Wang et al. 2015a, b). Some of the FDA-approved AMPs are listed in Table 7.1.

7.4 Classification of AMPs

The diversity of naturally obtained AMPs can be classified based on multiple factors described in Fig. 7.2.

1. **Classification on the Basis of Source:** The different sources from which AMPs are majorly obtained and are utilized to a large extent are mammals, microorganisms, amphibians, and insects. Among all the majority of AMPs, which act as defense peptides, are obtained from mammalian and microbial species.
   (a) **Mammalian Obtained Antimicrobial Peptides**: AMPs from mammals are mostly found in humans, sheep, cattle, and other vertebrates. Two majorly obtained AMPs are defensin and cathelicidins. The defensin peptide synthesis by humans is known as HDPs. HDPs are the ones that help in protecting humans from microbial infections. Cathelicidins are an example of one such AMP that shows different expressions at different stages of life. This is so because cathelicidins LL-37 have been identified in the skin of newborn infants; however, in the case of an elderly person, beta-defensin 2 has been detected (Gschwandtner et al. 2014; Field 2005). The eyes, ears, epidermis, mouth, lungs, intestine, respiratory system, and urethra all feature HDBs (Zhang et al. 2017). HDPs not only serve in antimicrobial activity but also have active participation in immune regulation, wound healing, and apoptosis (Wang 2014).
   (b) **Microorganisms Obtained Antimicrobial Peptides:** Biologically expressed AMPs from microorganisms such as bacteria and fungi have

Table 7.1 List of FDA-approved AMPs

AMP drugs	Company name	Application	Antimicrobial target	Mechanism of Action
Bacitracin	Baciim	Topical administration, infections of the eyes, wounds, and localized areas	Gram-positive bacteria	Inhibition of peptidoglycan synthesis and interference in the cell wall biosynthesis process
Daptomycin	Cubicin	Intravenous, skin infections caused by bacteria	Gram-positive bacteria	Inhibition of DNA, RNA, and protein synthesis and interference, in the cell membrane biosynthesis process
Dalbavancin	Dalvance, Xydalba	Intravenous, acute skin infections caused by bacteria	Gram-positive bacteria	Cell wall biosynthesis is hampered
Oritavancin	Orbactiv	Intravenous, skin infections caused by bacteria	Gram-positive bacteria	Gram-positive bacterial cell membrane gets damaged, and transglycosylation and transpeptidation are inhibited
Enfuvirtide	Fuzeon	Subcutaneous, infections caused by HIV-1	Virus	6HB structure is inhibited along with viral fusion
Teicoplanin	Targocid	Intravenous and intramuscular and bacterial infections	Gram-positive bacteria	Inhibit bacterial cell wall biosynthesis
Telavancin	Vibativ	Intravenous, skin infections caused by bacteria	Gram-positive bacteria	Inhibition of peptidoglycan synthesis and interference in the cell wall biosynthesis process
Telaprevir	Incivo, Incivek	Oral administration, against hepatitis C	Virus	NS3/4A protease inhibitor and inhibits the viral replication of hepatitis C virus
Vancomycin	Vancocin	Intravenous and oral and bacterial infections	Gram-positive bacteria	Inhibit bacterial cell wall biosynthesis
Boceprevir	Victrelis	Oral administration, against hepatitis C	Virus	NS3/4A protease inhibitor and inhibits the viral replication of hepatitis C virus
Natural interferon-α or multiferon	Intron, roferon-A	Subcutaneous, intramuscular, for the treatment of renal cell carcinoma at the metastatic stage	Virus	Induce immune response against virus-infected cells

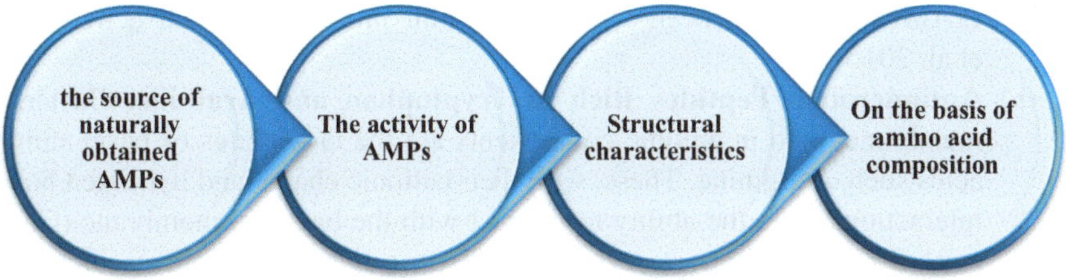

Fig. 7.2 Classification of AMPs

gained more importance than chemically synthesized ones due to their lower price. Microbes, which are mostly used for the expression system of such AMPs, are yeast species such as *Pichia pastoris* and *Saccharomyces cerevisiae* and bacterial species such as *Escherichia coli* and *Bacillus subtilis* (Cao et al. 2018; Parachin et al. 2012). In the case of *E. coli* system, although there is an ease of synthesis of AMPs with lower price, the toxicity, proteolytic degradation, and purification of the AMPs make it an incompatible expression system, hence making it necessary to use fusion tags (Yu et al. 2015).

2. **Classification on the Basis of Activity:** According to the ADP3 database, AMPs can be categorized into eight categories, and these categories are:
   (a) **Antibacterial Peptides**: They account for almost 60% of the AMPs. They exhibit a wide range of activity against the majority of microorganisms that are pathogenic, such as VRE, MRSA, *E. coli*, *Salmonella*, and *Vibrio parahaemolyticus* in aquatic products (Li et al. 2019).
   (b) **Antifungal Peptide:** This statistically accounts for 26% of the total AMPs. Antifungal peptides (AFPs) are the ones that are potent against enhanced drug-resistant fungal infections. They are employed against widespread pathogenic fungi, such as *Candida albicans* and *Aspergillus* (Madanchi et al. 2020). Yeast species including *Zygosaccharomyces rouxii* and *Zygosaccharomyces bailii* can be inhibited by two synthetically synthesized radish AMPs (Shwaiki et al. 2020).

   The other AMPs acted against viruses, parasites, cancer, and HIV accounting for almost 2–5% of the total AMPs.
3. **Classification on the Basis of Amino Acid-Rich Species.**
   (a) **Antimicrobial Peptides Rich in Proline (PrAMPs):** The nonpolar amino acid proline, upon entering the cytoplasm, targets ribosomes for degradation and inhibits the binding of aminoacyl-tRNA to peptidyl transferase or captures the decoding release factors in the ribosome during the termination of the translation (Seefeldt et al. 2015). Therefore, interfering with protein synthesis, pPR-AMP1, a proline-rich AMP found in crustaceans (*Scylla paramamosain*), exhibits antimicrobial activity in Gram-positive and Gram-negative bacteria (Imjongjirak et al. 2017). PrAMPs are primarily known for their antimicrobial activity against Gram-positive bacteria.

PrAMPs have also been shown to exhibit immunostimulant activity (Li et al. 2016).

(b) **Antimicrobial Peptides Rich in Tryptophan and Arginine:** Bacterial membranes rich in anionic components are the target sites of basic amino acids such as arginine. These, with their cationic charge and hydrogen bond interactions, gain the ability to interact with the bacterial membrane (Chan et al. 2006). In contrast, tryptophan (Trp), a non-polar amino acid, influences the interface region of the lipid bilayer. AMPs that are rich in Trp and Arg amino acids such as Octa 2 impede the proliferation of Gram-positive microorganisms, *S. aureus,* and Gram-negative bacteria *Pseudomonas aeruginosa* (Strøm et al. 2002; Bacalum et al. 1861).

(c) **Antimicrobial Peptides Rich in Histidine:** One of the common basic amino acids, His, has a potential membrane permeability ability, thereby causing cell membrane rupture followed by cell death. Certain histidine-rich AMPs also exhibit strong activity by inhibiting the production of tumor necrosis factor-alpha (TNF- α), this compound demonstrates anti-inflammatory properties (Dong et al. 2019).

(d) **Antimicrobial Peptides Rich in Glycine:** The amino acid glycine has a R group due to which it has been categorized as a non-polar amino acid. The commonly known AMPs rich in glycine existing in nature are attacins and diptericins (Lee et al. 2001; Kwon et al. 2008). One such example of AMPs rich in glycine derived from cathelicidins a host defense peptide of Salmonid undergoes microbicidal activity via phagocytosis (D'Este et al. 2016).

4. **Classification on the Basis of Structures of AMPs:** AMPs can be categorized into four major categories on the basis of the structures, and these include linear α- helical peptides, linear extension structures, β-sheets, and both α-helix and β-sheet peptides (Lei et al. 2019; Talapko et al. 2022). There is also the existence of other complex topological structures, which have recently been reported by researchers.

7.5 Mechanism of Action of AMP

The two major mechanisms of action of AMPs are (a) AMP targeting membrane (b) AMP target non-membranous structures, mainly inhibiting nucleic acid synthesis or certain functional proteins. AMPs employ both bacteriostatic and bactericidal effects and are preferred over conventional antibiotics as microbes have a lesser ability to develop resistance against them, and this is explained in Fig. 7.3.

1. **Membrane Targeting Mechanism**: Microbial cell membranes can be impaired either via (a) receptor-mediated or (b) non-receptor-mediated. The majority of the antimicrobial peptide kills microorganisms via a non-receptor-mediated mechanism.

(a) **Receptor-Mediated Mechanism of AMPs**: These groups of AMPs are majorly produced by bacteria. Therefore, their activity is mostly imparted to

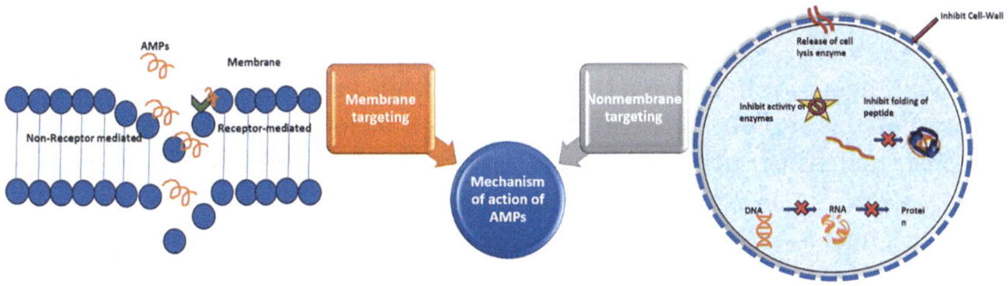

Fig. 7.3 Mechanism of action of AMPs

a specific targeted bacterium only. Such AMPs are composed of two domains. The first domain binds to a receptor, and the other domain forms the pore (Benfield and Henriques 2020; Breukink et al. 1999). The AMP, nisin Z is the first defined receptor-mediated AMP. This bacteriocin targets lipid II, a cell wall precursor. These remain anchored to the cell membrane as the receptor (Breukink et al. 1999; Shai 2002; Fleury et al. 1996). Nisin not only has the ability to form pores in the bacterial membrane resulting in membrane permeabilization at nanomolar concentration but also has a high affinity toward lipid II.

(b) **Non-Receptor-Mediated Mechanism of AMPs:** These groups of AMPs are most abundantly found, and members of these groups have a broad spectrum of amino acid composition, structure, and length. This is a reason why they do not have a specific receptor as their target; instead, the target was found to be the bacterial membrane in total. Both Gram-positive and Gram-negative bacteria possess a net negative charge on their outer surface as the outer surface is mainly composed of teichoic acid and lipopolysaccharide in Gram-positive and Gram-negative bacteria, respectively (Benfield and Henriques 2020; Brock et al. 2003). Along with it, the inner membrane of Gram-positive and Gram-negative bacteria is composed of negatively charged phospholipids. Therefore, AMPs with a net positive charge bond to negative-charged bacterial cells predominantly (Shai 1999; Tossi et al. 2000; Wu et al. 1999). Therefore, electrostatic attraction serves as the major driving force here, whereas such kind of electrostatic force between AMPs and mammalian cell membranes is relatively weak. This is because animal membranes are composed of Zwitter ionic phospholipids such as sphingomyelin and phosphatidylcholine (Verkleij et al. 1973). Additionally, the membrane stability is also enhanced by the cholesterol content in the mammalian cell membrane.

2. **Nonmembrane Active Mechanism:** Beyond membrane targeting, certain AMPs focus on vital cellular components and major metabolic pathways, inducing microbial demise. These peptides adeptly infiltrate cell membranes, ensuring minimal harm before halting processes such as protein and nucleic acid synthesis. They can also degrade essential proteins and enzymes, effectively dismantling key cellular functions (Brogden 2005). One such example is the PrAMPs.

Proline and arginine-enriched AMPs orchestrate bacterial demise by disrupting protein synthesis. A prime illustration is buforin II, derived from frog histones, which engages with *E. coli*'s DNA or RNA, while targeting intracellular chaperone proteins (Park et al. 1998). Similarly, non-membrane active AMPs such as copsin, defensin, and plectasin from *Coprinopsis cinerea* intervene in vital pathways (Essig et al. 2014; Schmitt et al. 2010; Tu et al. 2015). For instance, plectasin, a fungus-derived peptide, obstructs bacterial cell wall synthesis by binding to lipid II precursor (Hsu et al. 2004; Schneider et al. 2010). Further, certain AMPs, such as pyrrhocoricin, apidaecin, and drosocin from insects, thwart chaperone protein biosynthesis by interacting with GroEL and DnaK, key players in protein assembly and folding (Otvos et al. 2000).

7.6 AMP-Based Nanomaterials

While AMPs offer potent antimicrobial potential, they come with drawbacks: in vivo toxicity, neurotoxicity, uncertain pharmacokinetics, susceptibility to protease degradation, and costly bulk synthesis (Ageitos et al. 2017; Nizet 2006; Navon-Venezia et al. 2002). To enhance their bioavailability, AMPs are combined with nanomaterials (Patel et al. 2014). This synergy reduces cytotoxicity, shields against enzymatic degradation, and heightens efficacy at precise targets, making AMPs in nanostructures a promising solution. Nanoparticles emerge as AMP nanocarriers, exemplified by silver oxide nanoparticles (AgO NPs) aiding odorranain-A-OA1 to synergistically combat Gram-negative bacteria like *E. coli* (Ali et al. 2017). Another strategy involves combining AMPs with nano-delivery techniques to mitigate cytotoxicity and enzyme degradation, heightening targeted cell efficacy (Umerska et al. 2017). Encapsulation of GIBIM-P5S9K peptide with poly(lactic-co-glycolic acid) (PLGA) NPs demonstrated safe to use against *S. aureus*, *E. coli*, and *P. aeruginosa*, with no hemolysis within therapeutic doses (Cruz et al. 2017). The preponderance of nanostructures is used to deliver AMPs for the treatment of infections. This includes micelles, liposomes, liquid crystalline systems, polymeric NPs, hydrogels, microspheres, nanofibers, mesoporous silicon NPs, carbon nanotubes, and quantum dots.

AMP Delivery Through Nanostructures Nanocarrier-mediated AMP delivery offers multifaceted benefits. Systems such as liquid crystals, liposomes, and amphiphilic micelles adopt ring-like structures, safeguarding AMPs from degradation. Hydrogels and nanofibers optimize AMP pharmacokinetics, ensuring even distribution and controlled release. Certain nanocarriers, such as metal crystalline materials and mesoporous silica NPs, boost AMP loading capacity through enhanced surface area and porosity. Polymeric NPs, including dendrimers, counter enzyme degradation through polymer combinations or multifunctionality (Tang et al. 2021). Consequently, AMP-functionalized nanostructures hold immense promise as antimicrobial agents, boasting versatility, and expansive applications (Seena et al. 2023).

7.7 Advancements in AMP Delivery Via Nanomaterials

The frontier of AMP delivery has progressed with inorganic nanomaterials, notably AgNPs and gold nanoparticles (AuNPs). Employing polymeric NPs improves drug safety, bioavailability, enzymatic degradation prevention, and aggregation deterrence, amplifying drug pharmacokinetics. Chitosan NPs, a widely embraced biocompatible carrier, exhibit broad-spectrum antibacterial activity, synergizing effectively with co-functionalized antimicrobial agents (Teixeira et al. 2020).

AuNPs Conjugated with AMPs AuNPs exhibit remarkable bacterial efficacy coupled with facile synthesis, making them a promising player in nanomedicine. Their conjugation with biomolecules and in vivo stability bolsters their potential as antimicrobial agents. Recent research delves into combining Au nanostructures with AMPs for delivery, such as photoluminescent Au nanodots (AuNDs) fused with AMP surfactin (SFT) and 1-dodecanethiol (DT) (Chen et al. 2015). This ultra-small 2.5 nm structure demonstrated potent antimicrobial action, particularly against multidrug-resistant bacteria, achieved by disrupting bacterial cell membranes. Biocompatibility was affirmed through cytotoxic and in vitro studies, with SFT/DT-AuNDs showcasing accelerated wound healing and improved epithelialization, exemplified in MRSA-infected wounds. In another approach, a one-step method yielded AMP-conjugated AuNPs (*Cecropin melittin)* with cysteine (KWKLFKKIGAVLKVLC) (CM-SH) and without cysteine (KWKLFKKIGAVLKVL) (CM*)* approximately 14 nm in size, exhibiting low polydispersity for high AMP concentration (Rai et al. 2016) (Fig. 7.4). This conjugation displayed robust antimicrobial activity, circulation stability, protection against proteolytic enzymes, and reduced cytotoxicity compared to AMPs alone (Comune et al. 2017; Rai et al. 2022a; Comune et al. 2021; Rai et al. 2022b). These strides underscore the potential of AuNPs as effective and biocompatible vehicles for AMP delivery, igniting advancements in antimicrobial nanomedicine.

AgNPs Conjugated with AMPs Recent investigations into AgNPs have shed light on certain limitations when used in isolation. However, when amalgamated with AMPs, often with polymer augmentation, their antibacterial prowess intensifies while cytotoxicity diminishes. Notably, supra-molecular peptidyl ion amphiphiles, featuring aldehyde functional groups, aid in reducing metallic silver and nucleic nanoparticles, inducing the spontaneous creation of uniform silver ions at distribution-matched distances. This amalgamation exhibits robust antimicrobial traits and reduced cytotoxicity. In the context of combating *S. aureus*-induced wound infections, a study explored the synergy between plant peptide MBP1 and AgNPs (Salouti et al. 2016). Through micro- and macro-dilution methodologies, the combined antibacterial potency of AgNPs with MBP1 was evaluated, showcasing a synergistic effect. This collaborative approach proved highly effective in a mouse model of *S. aureus*-infected wounds, demonstrating accelerated healing compared to singular treatments.

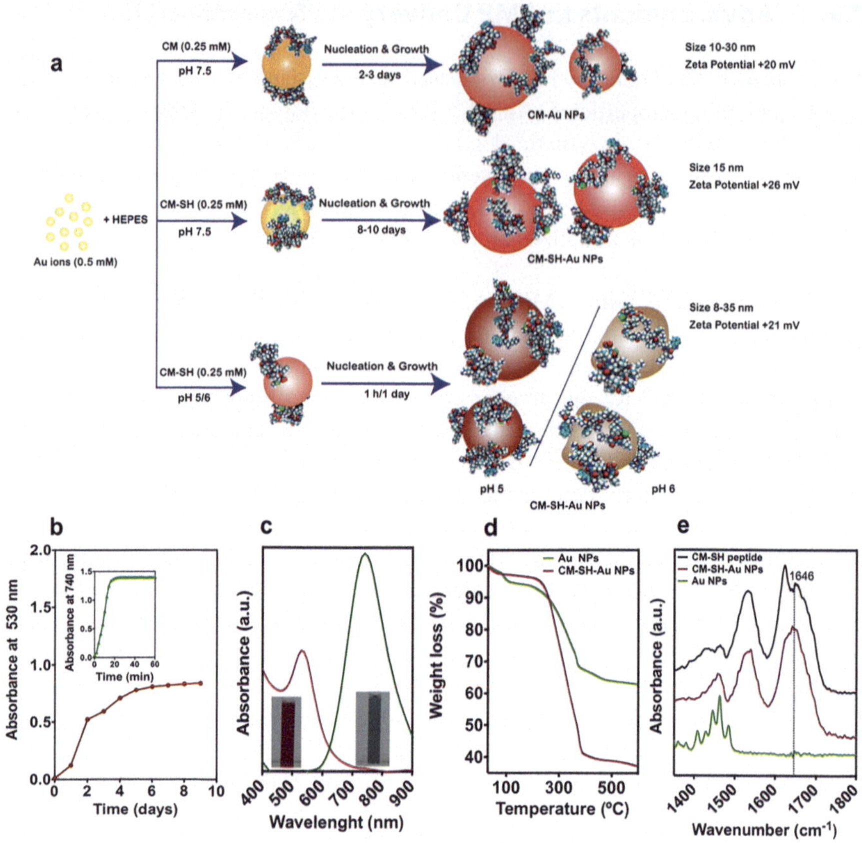

Fig. 7.4 (**a**) Schematic representation of AuNP synthesis in the presence of CM-SH and CM peptides, under different experimental conditions. The different peptide configurations do not represent the real configurations of conjugated peptides on the synthesized NPs. (**b**) Time-dependent absorbance for the synthesis of CM-SH-AuNPs (wine line) and AuNPs (inset, green line) in HEPES buffer in the presence and absence of CM-SH peptide, respectively. (**c**) UV-visible spectra of AuNPs synthesized in HEPES buffer in the presence (wine line, CM-SH-AuNPs) or absence (green line) of CM-SH peptides. Insets show the color of AuNP suspensions synthesized in both conditions. (**d**) TGA. (**e**) FTIR analyses of AuNPs synthesized in HEPES buffer in the presence (CM-SH-AuNPs; 9 days) and absence (AuNPs; 9 days) of CM-SH peptide (Rai et al. 2016)

7.8 Therapeutic Applications of AMPs

AMPs exhibit rapid and wide-ranging antimicrobial efficacy, and their reduced potential for microbial resistance positions them as promising contenders for treating infectious diseases (Mahlapuu et al. 2016). Beyond infections, AMPs hold potential for wound healing, anti-tumor activity, and immune enhancement (Rai et al. 2022b). Combined with traditional antibiotics, AMPs can augment their effectiveness. However, only a few selected AMPs have gained FDA approval for clinical use, including gramicidins, polymyxins, nisins, omiganan, daptomycin, telavancin, and dalbavancin. These are employed to address diverse infections, ranging from surface wounds to complicated skin infections. Further applications emerge, with AMPs such as LL-37 tackling sexually transmitted infections and serving as novel contraceptive microbicides (Madanchi and Sardari 2019). Synthetic peptide VRP hinders sperm motility for contraception without harming vaginal epithelial cells (Ghosh et al. 2018). In oncology, AMPs hold potential as anti-cancer peptides (ACPs), exemplified by cecropin B, which targets stomach, lung, and leukemia cancers (Seyfi et al. 2020). ACPs interact with tumor cell phosphatidylserines, disrupting their negatively charged plasma membrane (Ghosh et al. 2018). Notably, human lactoferricin (hLF1–11) AMP offers promise against bacterial and fungal infections in immunocompromised patients undergoing stem cell transplantation, proposed for intravenous administration (Mehta et al. 2022; van der Velden et al. 2009). This multifaceted potential underscores AMPs' growing significance across medical domains.

Application of Nano-Delivery System Loaded with AMPs Two prevalent nanotechnology-based drug delivery approaches encompass modified and unmodified nano-drug delivery systems (Biswaro et al. 2018). The unmodified system employs physical alterations, facilitating AMP adsorption onto nanocarriers for passive, non-directional drug delivery. This approach offers size and shape control for nanocarriers, enabling versatile loading of different AMP molecular sizes. Recent research demonstrated the inhibitory impact on MRSA through monolaurin-lipid nanocapsules loaded with AMPs AP114 and AP138 (Umerska et al. 2017). Nanostructure-AMP synergy reduced toxicity, enhanced pharmacokinetics, and curbed drug resistance probability (Umerska et al. 2017). Chemically modified nano-drug delivery systems excel in targeted drug delivery. Nanocarriers are endowed with ligands or fluorescent components to interact with specific human cell sites, facilitating site-specific treatment. Histatin 5 and sodium cholate molecules serve as such chemically modified ligands. Histatin 5 acts as both a targeting ligand and antifungal agent, binding to fungal cell walls through the SSAl/2p receptor (Park et al. 2017). Sodium cholate synergistically enhances AMP activity by electrostatically interacting with bacterial membranes, ensuring efficient AMP delivery to the lungs' aqueous phase (Souza et al. 2018). Another synergistic activity of nanoparticles conjugated with AMPs is acidic sophorolipid nanoparticles (ASLNP) with LL37 peptide in the presence of trypsin. This combinatory therapy helps in the eradication of both Gram-positive and Gram-negative bacteria present in the human serum. To efficiently kill a broad spectrum of bacteria in the human serum (5%) and in the presence of trypsin, the amount of ASL NPs required is 500 ug/ml, while the LL37 peptide required is 15–25 ug/ml. These formulations not only eradicate planktonic cells but are also useful for the eradication of biofilm formed by *S. aureus*. The AFM studies have also shown the damages in the outer and inner membranes of *E. coli* cells when subjected to a combinatory treatment of ASLNPs and LL37 peptide. This combination has also proved non-cytotoxic effect when in human cell line (Seena et al. 2023). Recent advances involve AuNPs coupled with DNA aptamers for intracellular antimicrobial peptide delivery in mammalian cells. This conjugated molecule not only exhibited intracellular antibacterial action but also remained stable in human serum. In HeLa cell lines infected with *S. typhimurium*, histidine-tagged DNA aptamer-AuNP, and C-terminal hexahistidine-tagged A3-APO conjugates successfully diminished bacterial infection and elevated cell viability (Yeom et al. 2016; Lee et al. 2017). These innovations underscore the potency of AMP-loaded nanocarriers in precise and effective drug delivery strategies.

7.9 Factors Affecting the Activity of AMPs

Therapy involving a combination of AMP along with certain adjuvants involves simple modulation of environmental factors, and such formulated compounds help in enhancing the antimicrobial activity of the AMPs. As the majority of AMPs are applied topically, such as in dressings, moisturizers, oral, and eye drops, they are susceptible to a variety of environmental conditions.

(a) **The Dependence of AMPs on Variation of Ionic Strength and Ions**: Considering the ionic nature of the very first phase of adhesion to the membranes of microorganisms, it has become widely accepted that excessive monovalent concentrations are detrimental of salts greater than 100 mM impede AMP activity and may have clinical significance (Wei et al. 2007; Fehlbaum et al. 1996). For instance, those suffering from cystic fibrosis frequently develop pulmonary infections caused by *P. aeruginosa*, which is associated with elevated levels of sodium concentration (about 120 mM) observed in these infected individuals (Goldman et al. 1997). Consequently, an assortment of research efforts has concentrated on enhancing AMP sodium tolerance. Numerous strategies, such as enhancing cationic and peptide stability and evaluating AMPs isolated from aquatic organisms, have been investigated. Beta-defensins and magainins are typically less efficient when the molar concentration of sodium chloride exceeds 100 mM, but AMPs that are resistant to salt have been reported to be effective at doses of sodium chloride as high as 300 mM and even 450 mM in the case of AMPs from certain aquatic organisms (Fedders et al. 2008). The determination of the strength of the ion as opposed to the salt abundance is preferable considering the buffering species ionic in nature has the ability to frequently donate considerably to the ion's potency, especially when polyvalent solutions that include phosphate are employed. Divalent ions frequently inhibit AMP activity at concentrations several orders of magnitude below anticipated based on the calculation of the strength of ions, and certain ion impacts belong to a particular group. At low mM concentrations, the divalent ions such as calcium and magnesium severely suppress the activity of AMPs. Several mammalian AMPs, for instance, were found to have drastically reduced activity, especially against Gram-negative microbes (Shin et al. 2002). Therefore, AMPs have to initially traverse the outer membrane of Gram-negative bacteria, lipopolysaccharide (LPS), in order to adhere to the plasma membrane. On binding LPS, the peptides are believed to replace divalent calcium and magnesium ions and trigger a self-promoted absorption process (Hancock and Lehrer 1998).

(b) **The Dependence of AMPs on pH:** Several peptides have investigated the effect of pH on the AMP function as an additional environmental variable to consider (Wu et al. 2008; Lee et al. 1997). Variations in pH may directly influence ionic interactions between AMPs and the cell membranes via altering protons that are attached to the functional groups that are embedded in the surface of the bacterium or on the peptide, whereas changes in ionic strength can indirectly affect activity. Regarding the pKas of functional categories, an alteration in the pH might impact peptide binding by altering the overall electron density of AMP or desired biomolecules on the bacterial cell membrane as well as the cell wall. Numerous AMPs are stable and continue to exhibit their antimicrobial activity over an extensive pH spectrum. AMPs exhibit higher activity at low pH due to their alkaline properties. This is due to the release of protons from histidine at an acidic pH, which encourages electrostatic bonding with anionic surfaces, such as LPS and the anionic groups of phospholipids, and consequently improves antimicrobial activity. Nevertheless, in accordance with the Henderson Hasselbalch equation, as the pH changes, so do the concentrations of the various species that are charged in the buffer. The variation in the strength of ions along with pH is capable of being substantial due to z2 term in the ionic strength equation, particularly with polyvalent ions, for example, phosphate. Consequently, it is necessary to consider ionic concentration when conducting a pH analysis. Moreover, electrostatic interaction constitutes one of the many crucial factors in the assembly of peptides. The electrostatic state of amino acid and constituent functional groups is affected by pH. Consequently, pH adjustment is an extremely prevalent approach for regulating peptide aggregation and disintegration (Walkenhorst et al. 2013). The MIC and MBC of the AMP C18G-His peptide at different pH are represented in Table 7.2. At pH 5, maximum antibacterial activity is exhibited and is reduced with the increase in pH (Hitchner et al. 2019).

(c) **The Dependence of AMPs on Protease:** These enzymes have an awful impact on AMPs. In particular, the protease chlamydial has protease-like activity factor (CPAF) released by *Chlamydia* preventing LL-37, which has the most potent inhibiting effect on chlamydial infection (Tang et al. 2015). To address this issue, research has focused on the engineering of AMP carriers. The existence of a chitosan silica strong foundation protects KR-12 peptide from being hydrolyzed by alpha trypsin, as well as the extent of protection is 38 percent greater than that of unbound KR-12 (Diosa et al. 2020). Nonetheless, several enzymes, including protease 65, esterase 66, and phosphatase 67, cleave the peptide's blocking group and initiate the peptide's self-assembly, which has a positive effect on AMPs (Tian et al. 2015).

Table 7.2 Antimicrobial activity of the AMP C18G-His peptide (Hitchner et al. 2019). Antimicrobial activity (μM)

	S. aureus		*P. aeruginosa*		*A. baumannii*		*E.coli*	
pH	MIC	MBC	MIC	MBC	MIC	MBC	MIC	MBC
5	0.47	0.94	3.75	3.75	0.23	0.94	0.47	0.94
6	15	15	7.5	7.5	1.88	3.75	1.88	1.88
7	15	>15	>15	>15	7.5	7.5	3.75	7.5
8	15	>15	>15	>15	>15	>15	15	15

7.10 Challenges and Advantages in AMP-Based Nanomaterials

Nanostructures integrated with antimicrobial peptides confront several challenges. While combined therapy shows promise against microbial infections and antimicrobial-resistant strains, the substantial cost presents a significant drawback. Recent US FDA-approved synthetic peptidyl drugs, estimated at USD 300 to 500 per gram, underscore the economic limitations and unpredictability of nanotechnology commercialization (Bray 2003; da Costa et al. 2015). Challenges in synthesis, including extreme temperatures affecting ion pairing between nanocarriers and peptides, can compromise conjugated peptide stability in intravenous systems. Safety and cytotoxicity concerns persist when applying nanocarrier-conjugated AMPs in human diseases. Although these conjugates offer potential cytotoxicity reduction compared to AMPs alone, a thorough evaluation is imperative due to uncertainties in the nano-drug delivery systems. Clinical AMP applications also grapple with low bioavailability and metabolic stability, favoring tropical formulations. Proteolytic cleavage of AMPs further hampers intravenous formulations, resulting in a short half-life within the body. These multifaceted challenges necessitate comprehensive investigation to unlock the full potential of AMP-loaded nanostructures. The conventional AMP delivery method presents multiple limitations, including cytotoxicity and vulnerability to enzymatic degradation upon in vivo administration. To address these challenges, a synergistic approach involving nanostructures and AMPs has emerged. This conjugated system enhances AMP sensitivity by combining them with nanocarriers in drug delivery, elevating their antimicrobial effectiveness. Notably, special nanocarriers such as AuNPs and AgNPs have been reported as effective carriers for antibiotics. AgNPs conjugated with odorranain-A-OA1 exhibited antibacterial activity against *E. coli*, highlighting the potential of NP-AMP synergy in boosting antimicrobial action. Nanostructures in AMP delivery provide a compelling solution to cytotoxicity. AMPs conjugated with nanocarriers effectively mitigate cytotoxicity, shield AMPs from proteolytic degradation, and enhance their efficacy against targeted cells. NP-based delivery of conjugated AMPs also regulates peptide dissociation rates, prolonging their in vivo effectiveness. Nano-delivery systems help AMPs overcome natural barriers, such as enzymes and digestive fluids in the gastrointestinal tract and intestinal mucosa, resulting in reduced initial impact and facilitated release of antimicrobial agents at the target site.

Challenges extend beyond AMP delivery to include addressing emerging concerns about resistant microbes in the environment, driving the quest for potent antimicrobial agents. AMPs have shown significant benefits over traditional antibiotics, with nanoparticle-enhanced AMPs offering promising additive therapy against drug-resistant microbes, effectively minimizing adverse effects while maintaining efficacy at lower doses. In this combinatory therapy, NPs not only enhance cell permeability but also increase AMP bioavailability. While laboratory and animal studies show promising outcomes, the clinical potential of AMP-nanoparticle conjugates requires further exploration, necessitating molecular-level understanding of their mechanisms. This pursuit of knowledge could yield innovative treatments, expanding AMPs' applications beyond bacterial infections and multi-drug-resistant bacteria to encompass anti-biofilm agents.

7.11 Conclusion and Future Prospects

The preceding discussion has illuminated a comprehensive understanding of AMPs and their potential in addressing the global challenge of antimicrobial resistance. In terms of the marketing of AMPs as therapeutics, their administration or delivery methods are also a significant concern. It is now widely acknowledged the fact that the therapeutic value of a medication is dependent upon its mode of administration. In this vein, it has been discovered that nanostructure-based drug delivery systems are particularly promising. Because of their elevated retention and permeability effect, nanostructured AMPs are more efficacious because they increase the bioavailability of the drug, facilitate sustained release, and enhance the retention of the AMPs in circulation. Due to the enhanced retention and penetration effect (EPR effect), reduced renal elimination of nanostructured AMPs results in total material utilization in the bloodstream. While significant challenges exist in incorporating AMPs into clinical practice, their multitarget action, synergistic interactions, and lower susceptibility to resistance development offer a hopeful path for combating this pressing health concern. Thus, this dialog offers valuable insights into considerations for selecting nano-delivery systems for AMPs. The choice of suitable nanocarriers hinges on factors such as cytotoxicity, localized delivery, degradation protection, and enhanced bioactivity. This elaboration underscores the advantages of AMPs over conventional antibiotics in terms of their mode of action and clinical utility. Different types of nanocarriers, including liposomes, polymeric NPs, microspheres, and micelles, provide distinct advantages in augmenting AMP efficacy. Nanocarriers confer benefits such as degradation resistance, targeted therapy, and improved pharmacokinetics. The unique mechanism of AMPs, involving interaction with bacterial cell membranes, demonstrates resistant potential due to their reduced vulnerability to mutations. Additionally, the broad-spectrum nature of AMP activity across cellular membranes poses challenges for microbial resistance development. The future trends in AMP design exhibit accuracy and forward-thinking. Hybrid AMPs, combining membrane-permeabilizing and intracellularly active peptides, hold promise for a dual mechanism of action. This innovation could overcome limitations that individual AMP might face in targeting specific microbial functions. Multivalent AMPs, featuring multiple AMP units, present an exciting avenue for enhanced interactions with microbial surfaces and improved antimicrobial properties. The immune-regulating capabilities of certain AMPs further enhance their potential as comprehensive antimicrobial agents. The integration of nanocarriers and AMPs not only facilitates AMP delivery but also synergizes with other therapeutic agents, bolstering infection-combating effects, especially against antibiotic-resistant bacteria. As AMP research continues to expand, the integration of nanocarriers and peptides promises to be a captivating realm in medicine. Furthermore, the natural metabolic degradation of AMPs into monomeric amino acids, unlike antibiotics, has the potential to address resistance challenges and advance new avenues for infection treatment. Thus, effective AMP delivery systems can play a pivotal role in propelling these peptides toward clinical trials and addressing challenges related to their application and therapeutic potential. From the above discussion, a deep understanding of AMPs and their potential in addressing AMR has been brought to light. The challenges mentioned regarding the incorporation of AMPs into clinical practice are indeed significant, but the promising aspects of their multitarget action, synergistic interactions, and a lower likelihood of resistance development make them an optimistic avenue for combating the global health issue. Therefore, the above discussion provides valuable insights into the various considerations for choosing nano-delivery systems for AMPs. Therefore, the future trends of AMP design are quite accurate and forward-thinking. The development of hybrid AMPs that combine a membrane-permeabilizing peptide with an intracellularly active peptide holds significant potential. This strategy allows for a dual mechanism of action, targeting both the bacterial cell membrane and intracellular components. By doing so, hybrid AMPs can potentially overcome some of the limitations that individual AMPs might face in terms of targeting specific microbial functions. Additionally, the concept of multivalent AMPs is another promising avenue. These multivalent structures involve the presentation of multiple AMP units, which can lead to enhanced interactions with microbial surfaces and improved overall antimicrobial properties.

The immune regulatory capabilities of certain AMPs further enhance their potential as comprehensive antimicrobial agents. The combination of nanocarriers and AMPs holds promise not only for delivering AMPs but also for loading other therapeutic agents, which can lead to synergistic effects in combating infections, including those caused by antibiotic-resistant bacteria. As the field of AMP research continues to grow and evolve, the integration of nanocarriers and peptides is likely to be an exciting area of exploration in medicine. Additionally, the metabolic degradation of AMPs results in the formation of monomeric amino acids, which can be utilized by essential biosynthetic pathways. This natural degradation process contrasts with antibiotics, which might undergo detoxification and face challenges in complete renal clearance. This approach has the potential to address challenges related to antibiotic resistance and open up new possibilities for treating infections. Therefore, the use of an effective AMP delivery system has the ability to play a crucial role in promoting these peptides toward clinical trials and eventually in implementing and addressing challenges related to their application and potential therapeutic benefits.

Furthermore, there is a growing recognition that AMPs possess multifaceted capabilities beyond merely safeguarding mammals against infectious agents. They actively shape the characteristics of microorganisms, playing a pivotal role in the maintenance of overall microbial equilibrium, which, in turn, profoundly impacts a range of well-being factors. The escalating prevalence of resistance to conventional pharmaceuticals, coupled with their overuse, has ignited a surge in research concerning the development of AMPs. These peptides are poised to emerge as a novel class of antibacterial agents, offering a more targeted approach to combating pathogens. As we navigate toward the future, their potential is magnified due to their selectivity and effectiveness. Exploring the regulation of endogenous AMP production represents a promising avenue in the quest to address a broad spectrum of animal and human ailments, even in the face of existing limitations. This approach holds the promise of a comprehensive solution, underscoring the intricate interplay between AMPs and health, while paving the way for innovative treatments.

References

Ageitos J, Sánchez-Pérez A, Calo-Mata P, Villa T (2017) Antimicrobial peptides (AMPs): ancient compounds that represent novel weapons in the fight against bacteria. Biochem Pharmacol 133:117–138

Ali SS, Morsy R, El-Zawawy NA, Fareed MF, Bedaiwy MY (2017) Synthesized zinc peroxide nanoparticles (ZnO2-NPs): a novel antimicrobial, anti-elastase, anti-keratinase, and anti-inflammatory approach toward polymicrobial burn wounds. Int J Nanomedicine 6059-73

Bacalum M, Janosi L, Zorila F, Tepes AM, Ionescu C, Bogdan E et al (1861) (2017) modulating short tryptophan- and arginine-rich peptides activity by substitution with histidine. Biochim Biophys Acta Gen Subj 7:1844–1854. https://doi.org/10.1016/j.bbagen.2017.03.024

Belete TM (2019) Novel targets to develop new antibacterial agents and novel alternatives to antibacterial agents. Human Microb J 11:100052. https://doi.org/10.1016/j.humic.2019.01.001

Benfield AH, Henriques ST (2020) Mode-of-action of antimicrobial peptides: membrane disruption vs intracellular mechanisms. Front Med Technol 2. https://doi.org/10.3389/fmedt.2020.610997

Biswaro LS, da Costa Sousa MG, Rezende TMB, Dias SC, Franco OL (2018) Antimicrobial peptides and nanotechnology, recent advances and challenges. Front Microbiol 9:855. https://doi.org/10.3389/fmicb.2018.00855

Bray BL (2003) Large-scale manufacture of peptide therapeutics by chemical synthesis. Nat Rev Drug Discov 2(7):587–593. https://doi.org/10.1038/nrd1133

Breukink E, Wiedemann I, Cv K, Kuipers O, Sahl H-G, De Kruijff B (1999) Use of the cell wall precursor lipid II by a pore-forming peptide antibiotic. Science 286(5448):2361–2364

Brock TD, Madigan MT, Martinko JM, Parker J (2003) Brock biology of microorganisms. Prentice-Hall, Upper Saddle River (NJ)

Brogden KA (2005) Antimicrobial peptides: pore formers or metabolic inhibitors in bacteria? Nat Rev Microbiol 3(3):238–250

Cabello FC, Godfrey HP, Buschmann AH, Dölz HJ (2016) Aquaculture as yet another environmental gateway to the development and globalisation of antimicrobial resistance. Lancet Infect Dis 16(7):e127–ee33. https://doi.org/10.1016/S1473-3099(16)00100-6

Cao J, de la Fuente-Nunez C, Ou RW, Torres MT, Pande SG, Sinskey AJ et al (2018) Yeast-based synthetic biology platform for antimicrobial peptide production. ACS Synth Biol 7(3):896–902. https://doi.org/10.1021/acssynbio.7b00396

Carpenter CF, Chambers HF (2004) Daptomycin: another novel agent for treating infections due to drug-resistant gram-positive pathogens. Clin Infect Dis 38(7):994–1000

Chan DI, Prenner EJ, Vogel HJ (2006) Tryptophan- and arginine-rich antimicrobial peptides: structures and mechanisms of action. Biochim Biophys Acta 1758(9):1184–1202. https://doi.org/10.1016/j.bbamem.2006.04.006

Chaudhary S, Ali Z, Tehseen M, Haney EF, Pantoja-Angles A, Alshehri S et al (2023) Efficient in planta production of amidated antimicrobial peptides that are active against drug-resistant ESKAPE pathogens. Nature. Communications 14:1. https://doi.org/10.1038/s41467-023-37003-z

Chen WY, Chang HY, Lu JK, Huang YC, Harroun SG, Tseng YT et al (2015) Self-assembly of antimicrobial peptides on gold nanodots: against multidrug-resistant bacteria and wound-healing application. Adv Funct Mater 25(46):7189–7199

Chinemerem Nwobodo D, Ugwu MC, Oliseloke Anie C, Al-Ouqaili MTS, Chinedu Ikem J, Victor Chigozie U et al (2022) Antibiotic resistance: the challenges and some emerging strategies for tackling a global menace. J Clin Lab Anal 36(9):e24655. https://doi.org/10.1002/jcla.24655

Cole JN, Nizet V (2016) Bacterial evasion of host antimicrobial peptide defenses. Microbiol Spectr 4(1). https://doi.org/10.1128/microbiolspec.VMBF-0006-2015

Comune M, Rai A, Chereddy KK, Pinto S, Aday S, Ferreira AF et al (2017) Antimicrobial peptide-gold nanoscale therapeutic formulation with high skin regenerative potential. J Control Release 262:58–71. https://doi.org/10.1016/j.jconrel.2017.07.007

Comune M, Rai A, Palma P, TondaTuro C, Ferreira L (2021) Antimicrobial and pro-angiogenic properties of soluble and nanoparticle-immobilized LL37 peptides. Biomater Sci 9(24):8153–8159. https://doi.org/10.1039/D1BM01034D

da Costa JP, Cova M, Ferreira R, Vitorino R (2015) Antimicrobial peptides: an alternative for innovative medicines? Appl Microbiol Biotechnol 99(5):2023–2040. https://doi.org/10.1007/s00253-015-6375-x

Cruz J, Flórez J, Torres R, Urquiza M, Gutiérrez J, Guzmán F et al (2017) Antimicrobial activity of a new synthetic peptide loaded in polylactic acid or poly (lactic-co-glycolic) acid nanoparticles against Pseudomonas aeruginosa, Escherichia coli O157: H7 and methicillin resistant Staphylococcus aureus (MRSA). Nanotechnology 28(13):135102

D'Este F, Benincasa M, Cannone G, Furlan M, Scarsini M, Volpatti D et al (2016) Antimicrobial and host cell-directed activities of Gly/Ser-rich peptides from salmonid cathelicidins. Fish Shellfish Immunol 59:456–468. https://doi.org/10.1016/j.fsi.2016.11.004

Diosa J, Guzman F, Bernal C, Mesa M (2020) Formation mechanisms of chitosan-silica hybrid materials and its performance as solid support for KR-12 peptide adsorption: impact on KR-12 antimicrobial activity and proteolytic stability. J Mater Res Technol 9(1):890–901

Dong N, Wang C, Zhang T, Zhang L, Xue C, Feng X et al (2019) Bioactivity and bactericidal mechanism of histidine-rich β-hairpin peptide against gram-negative bacteria. Int J Mol Sci 20:16. https://doi.org/10.3390/ijms20163954

Essig A, Hofmann D, Münch D, Gayathri S, Künzler M, Kallio PT et al (2014) Copsin, a novel peptide-based fungal antibiotic interfering with the peptidoglycan synthesis. J Biol Chem 289(50):34953–34964

Fedders H, Michalek M, Grötzinger J, Leippe M (2008) An exceptional salt-tolerant antimicrobial peptide derived from a novel gene family of haemocytes of the marine invertebrate Ciona intestinalis. Biochem J 416(1):65–75

Fehlbaum P, Bulet P, Chernysh S, Briand J-P, Roussel J-P, Letellier L et al (1996) Structure-activity analysis of thanatin, a 21-residue inducible insect defense peptide with sequence homology to frog skin antimicrobial peptides. Proc Natl Acad Sci 93(3):1221–1225

Field CJ (2005) The immunological components of human milk and their effect on immune development in infants. J Nutr 135(1):1–4. https://doi.org/10.1093/jn/135.1.1

Fleming A (1929) On the antibacterial action of cultures of a penicillium, with special reference to their use in the isolation of B. influenzae. Br J Exp Pathol 10(3):226

Fleury Y, Dayem MA, Montagne JJ, Chaboisseau E, Le Caer JP, Nicolas P et al (1996) Covalent structure, synthesis, and structure-function studies of mesentericin Y 10537, a defensive peptide from gram-positive bacteria Leuconostoc mesenteroides. J Biol Chem 271(24):14421–14429

Gallo RL, Hooper LV (2012) Epithelial antimicrobial defence of the skin and intestine. Nat Rev Immunol 12(7):503–516

Gaynes R (n.d.) The discovery of penicillin—new insights after more than 75 years of clinical use. (1080–6040 (Print))

Ghosh P, Bhoumik A, Saha S, Mukherjee S, Azmi S, Ghosh JK et al (2018) Spermicidal efficacy of VRP, a synthetic cationic antimicrobial peptide, inducing apoptosis and membrane disruption. J Cell Physiol 233(2):1041–1050. https://doi.org/10.1002/jcp.25958

Goldman MJ, Anderson GM, Stolzenberg ED, Kari UP, Zasloff M, Wilson JM (1997) Human β-defensin-1 is a salt-sensitive antibiotic in lung that is inactivated in cystic fibrosis. Cell 88(4):553–560

Gschwandtner M, Zhong S, Tschachler A, Mlitz V, Karner S, Elbe-Bürger A et al (2014) Fetal human keratinocytes produce large amounts of antimicrobial peptides: involvement of histone-methylation processes. J Invest Dermatol 134(8):2192–2201. https://doi.org/10.1038/jid.2014.165

Hallett JW, Wolkowicz MI, Leopold IH (1956) Ophthalmic use of neosporin. Am J Ophthalmol 41(5):850–853

Hamamoto K, Kida Y, Zhang Y, Shimizu T, Kuwano K (2002) Antimicrobial activity and stability to proteolysis of small linear cationic peptides with D-amino acid substitutions. Microbiol Immunol 46(11):741–749

Hancock RE, Lehrer R (1998) Cationic peptides: a new source of antibiotics. Trends Biotechnol 16(2):82–88

Haney EF, Straus SK, Hancock REW (2019) Reassessing the host defense peptide landscape. Front Chem 7. https://doi.org/10.3389/fchem.2019.00043

Hitchner MA, Santiago-Ortiz LE, Necelis MR, Shirley DJ, Palmer TJ, Tarnawsky KE et al (2019) Activity and characterization of a pH-sensitive antimicrobial peptide. Biochim Biophys Acta Biomembr 1861(10):182984. https://doi.org/10.1016/j.bbamem.2019.05.006

Hsu S-TD, Breukink E, Tischenko E, Lutters MA, De Kruijff B, Kaptein R et al (2004) The nisin–lipid II complex reveals a pyrophosphate cage that provides a blueprint for novel antibiotics. Nat Struct Mol Biol 11(10):963–967

Huan Y, Kong Q, Mou H, Yi H (2020) Antimicrobial peptides: classification, design, application and research Progress in multiple fields. Front Microbiol 11. https://doi.org/10.3389/fmicb.2020.582779

Idris FN, Nadzir MM (2023) Multi-drug resistant ESKAPE pathogens and the uses of plants as their antimicrobial agents. Arch Microbiol 205(4):115. https://doi.org/10.1007/s00203-023-03455-6

Imjongjirak C, Amphaiphan P, Charoensapsri W, Amparyup P (2017) Characterization and antimicrobial evaluation of SpPR-AMP1, a proline-rich antimicrobial peptide from the mud crab Scylla paramamosain. Dev Comp Immunol 74:209–216. https://doi.org/10.1016/j.dci.2017.05.003

Kalpana S, Lin W-Y, Wang Y-C, Fu Y, Lakshmi A, Wang H-Y (2023) Antibiotic resistance diagnosis in ESKAPE pathogens–a review on proteomic perspective. Diagnostics 13(6):1014

Kang S-J, Kim D-H, Mishig-Ochir T, Lee B-J (2012) Antimicrobial peptides: their physicochemical properties and therapeutic application. Arch Pharm Res 35(3):409–413. https://doi.org/10.1007/s12272-012-0302-9

Kang X, Dong F, Shi C, Liu S, Sun J, Chen J et al (2019) DRAMP 2.0, an updated data repository of antimicrobial peptides. Sci Data 6(1):148. https://doi.org/10.1038/s41597-019-0154-y

Khmelevtsova LE, Sazykin IS, Azhogina TN, Sazykina MA (2020) The dissemination of antibiotic resistance in various environmental objects (Russia). Environ Sci Pollut Res 27(35):43569–43581. https://doi.org/10.1007/s11356-020-10231-2

Konovalova MV, Zubareva AA, Lutsenko GV, Svirshchevskaya EV (2018) Antimicrobial peptides in health and disease (review). Appl Biochem Microbiol 54(3):238–244. https://doi.org/10.1134/S0003683818030079

de Kraker ME, Stewardson AJ, Harbarth S (2016) Will 10 million people die a year due to antimicrobial resistance by 2050? PLoS Med 13(11):e1002184. https://doi.org/10.1371/journal.pmed.1002184

Kumar P, Kizhakkedathu JN, Straus SK (2018) Antimicrobial peptides: diversity, mechanism of action and strategies to improve the activity and biocompatibility in vivo. Biomol Ther 8(1):4

Kwon YM, Kim HJ, Kim YI, Kang YJ, Lee IH, Jin BR et al (2008) Comparative analysis of two attacin genes from Hyphantria cunea. Comp Biochem Physiol B Biochem Mol Biol 151(2):213–220. https://doi.org/10.1016/j.cbpb.2008.07.002

Le CF, Fang CM, Sekaran SD (2017) Intracellular targeting mechanisms by antimicrobial peptides. Antimicrob Agents Chemother 61:4. https://doi.org/10.1128/aac.02340-16

Lee B, Park J, Ryu M, Kim S, Joo M, Yeom J-H et al (2017) Antimicrobial peptide-loaded gold nanoparticle-DNA aptamer conjugates as highly effective antibacterial therapeutics against Vibrio vulnificus. Sci Rep 7(1):13572

Lee IH, Cho Y, Lehrer RI (1997) Effects of pH and salinity on the antimicrobial properties of clavanins. Infect Immun 65(7):2898–2903

Lee JH, Cho KS, Lee J, Yoo J, Lee J, Chung J (2001) Diptericin-like protein: an immune response gene regulated by the anti-bacterial gene induction pathway in Drosophila. Gene 271(2):233–238. https://doi.org/10.1016/s0378-1119(01)00515-7

Lei J, Sun L, Huang S, Zhu C, Li P, He J et al (2019) The antimicrobial peptides and their potential clinical applications. Am J Transl Res 11(7):3919–3931

Lemaitre B, Nicolas E, Michaut L, Reichhart JM, Hoffmann JA (1996) The dorsoventral regulatory gene cassette spätzle/Toll/cactus controls the potent antifungal response in drosophila adults. Cell 86(6):973–983. https://doi.org/10.1016/s0092-8674(00)80172-5

Li C, Zhu C, Ren B, Yin X, Shim SH, Gao Y et al (2019) Two optimized antimicrobial peptides with therapeutic potential for clinical antibiotic-resistant Staphylococcus aureus. Eur J Med Chem 183:111686. https://doi.org/10.1016/j.ejmech.2019.111686

Li W, Sani MA, Jamasbi E, Otvos L Jr, Hossain MA, Wade JD et al (2016) Membrane interactions of proline-rich antimicrobial peptide, Chex1-Arg20, multimers. Biochim Biophys Acta 1858(6):1236–1243. https://doi.org/10.1016/j.bbamem.2016.02.035

Lopes BS, Hanafiah A, Nachimuthu R, Muthupandian S, Md Nesran ZN, Patil S (2022) The role of antimicrobial peptides as antimicrobial and antibiofilm agents in tackling the silent pandemic of antimicrobial resistance. Molecules. Basel, Switzerland 27:9. https://doi.org/10.3390/molecules27092995

Madanchi H, Sardari S (2019) The role of antimicrobial peptides in the prevention of sexually transmitted infection (STI). Womens Health 8(2):192–194

Madanchi H, Shoushtari M, Kashani HH, Sardari S (2020) Antimicrobial peptides of the vaginal innate immunity and their role in the fight against sexually transmitted diseases. New Microbes New Infect 34:100627. https://doi.org/10.1016/j.nmni.2019.100627

Mahlapuu M, Håkansson J, Ringstad L, Björn C (2016) Antimicrobial peptides: an emerging category of therapeutic agents. Front Cell Infect Microbiol 194

Masoud SS, Kovacevich A, Gangji R, Nyawale H, Nyange M, Ntukula A (2022) Extent and resistance patterns of ESKAPE pathogens isolated in pus swabs from hospitalized patients. Can J Infect Dis Med Microbiol 2022:3511306. https://doi.org/10.1155/2022/3511306

McMillan KAM, Coombs MRP (2020) Review: examining the natural role of amphibian antimicrobial peptide magainin. Molecules. (Basel, Switzerland) 25:22. https://doi.org/10.3390/molecules25225436

Mehdi Y, Létourneau-Montminy MP, Gaucher ML, Chorfi Y, Suresh G, Rouissi T et al (2018) Use of antibiotics in broiler production: global impacts and alternatives. Anim Nutr (Zhongguo xu mu shou yi xue hui) 4(2):170–178. https://doi.org/10.1016/j.aninu.2018.03.002

Mehta K, Sharma P, Mujawar S, Vyas A (2022) Role of antimicrobial peptides in treatment and prevention of mycobacterium tuberculosis: a review. Int J Pept Res Ther 28(5):132

Mukhopadhyay S, Bharath Prasad AS, Mehta CH, Nayak UY (2020) Antimicrobial peptide polymers: no escape to ESKAPE pathogens—a review. World J Microbiol Biotechnol 36(9):131. https://doi.org/10.1007/s11274-020-02907-1

Navon-Venezia S, Feder R, Gaidukov L, Carmeli Y, Mor A (2002) Antibacterial properties of dermaseptin S4 derivatives with in vivo activity. Antimicrob Agents Chemother 46(3):689–694

Nguyen LT, Chau JK, Perry NA, de Boer L, Zaat SA, Vogel HJ (2010) Serum stabilities of short tryptophan-and arginine-rich antimicrobial peptide analogs. PLoS One 5(9):e12684

Nikaido H (2009) Multidrug resistance in bacteria. Annu Rev Biochem 78:119–146. https://doi.org/10.1146/annurev.biochem.78.082907.145923

Nizet V (2006) Antimicrobial peptide resistance mechanisms of human bacterial pathogens. Curr Issues Mol Biol 8(1):11–26

Nizet V, Ohtake T, Lauth X, Trowbridge J, Rudisill J, Dorschner RA et al (2001) Innate antimicrobial peptide protects the skin from invasive bacterial infection. Nature 414(6862):454–457. https://doi.org/10.1038/35106587

Otvos LOI, Rogers ME, Consolvo PJ, Condie BA, Lovas S et al (2000) Interaction between heat shock proteins and antimicrobial peptides. Biochemistry 39(46):14150–14159

Pang Z, Raudonis R, Glick BR, Lin T-J, Cheng Z (2019) Antibiotic resistance in Pseudomonas aeruginosa: mechanisms and alternative therapeutic strategies. Biotechnol Adv 37(1):177–192. https://doi.org/10.1016/j.biotechadv.2018.11.013

Parachin NS, Mulder KC, Viana AAB, Dias SC, Franco OL (2012) Expression systems for heterologous production of antimicrobial peptides. Peptides 38(2):446–456. https://doi.org/10.1016/j.peptides.2012.09.020

Park CB, Kim HS, Kim SC (1998) Mechanism of action of the antimicrobial peptide buforin II: buforin II kills microorganisms by penetrating the cell membrane and inhibiting cellular functions. Biochem Biophys Res Commun 244(1):253–257

Park SC, Kim YM, Lee JK, Kim NH, Kim EJ, Heo H et al (2017) Targeting and synergistic action of an antifungal peptide in an antibiotic drug-delivery system. J Control Release 256:46–55. https://doi.org/10.1016/j.jconrel.2017.04.023

Patel A, Patel M, Yang X, Mitra KA (2014) Recent advances in protein and peptide drug delivery: a special emphasis on polymeric nanoparticles. Protein Pept Lett 21(11):1102–1120

Peterson E, Kaur P (2018) Antibiotic resistance mechanisms in bacteria: relationships between resistance determinants of antibiotic producers, environmental bacteria, and clinical pathogens. Front Microbiol 9. https://doi.org/10.3389/fmicb.2018.02928

Rai A, Ferrão R, Marta D, Vilaça A, Lino M, Rondão T et al (2022a) Antimicrobial peptide-tether dressing able to enhance wound healing by tissue contact. ACS Appl Mater Interfaces 14(21):24213–24228. https://doi.org/10.1021/acsami.2c06601

Rai A, Ferrão R, Palma P, Patricio T, Parreira P, Anes E et al (2022b) Antimicrobial peptide-based materials: opportunities and challenges. J Mater Chem B 10(14):2384–2429. https://doi.org/10.1039/D1TB02617H

Rai A, Pinto S, Velho TR, Ferreira AF, Moita C, Trivedi U et al (2016) One-step synthesis of high-density peptide-conjugated gold nanoparticles with antimicrobial efficacy in a systemic infection model. Biomaterials 85:99–110. https://doi.org/10.1016/j.biomaterials.2016.01.051

Saha M, Sarkar A (2021) Review on multiple facets of drug resistance: a rising challenge in the 21st century. J Xenobiot 11(4):197–214

Sahoo A, Swain SS, Behera A, Sahoo G, Mahapatra PK, Panda SK (2021) Antimicrobial peptides derived from insects offer a novel therapeutic option to combat biofilm: a review. Front Microbiol 12. https://doi.org/10.3389/fmicb.2021.661195

Salouti M, Mirzaei F, Shapouri R, Ahangari A (2016) Synergistic antibacterial activity of plant peptide MBP-1 and silver nanoparticles combination on healing of infected wound due to Staphylococcus aureus. Jundishapur J Microbiol 9(1):e27997

Schmitt P, Wilmes M, Pugnière M, Aumelas A, Bachère E, Sahl HG et al (2010) Insight into invertebrate defensin mechanism of action: oyster defensins inhibit peptidoglycan biosynthesis by binding to lipid II. J Biol Chem 285(38):29208–29216. https://doi.org/10.1074/jbc.M110.143388

Schneider T, Kruse T, Wimmer R, Wiedemann I, Sass V, Pag U et al (2010) Plectasin, a fungal defensin, targets the bacterial cell wall precursor lipid II. Science 328(5982):1168–1172

Seefeldt AC, Nguyen F, Antunes S, Pérébaskine N, Graf M, Arenz S et al (2015) The proline-rich antimicrobial peptide Onc112 inhibits translation by blocking and destabilizing the initiation complex. Nat Struct Mol Biol 22(6):470–475. https://doi.org/10.1038/nsmb.3034

Seena S, Ferrão R, Pala M, Roelants SLKW, Soetaert W, Stevens CV et al (2023) Acidic sophorolipid and antimicrobial peptide based formulation as antimicrobial and antibiofilm agents. Biomater Adv 146:213299. https://doi.org/10.1016/j.bioadv.2023.213299

Seyfi R, Kahaki FA, Ebrahimi T, Montazersaheb S, Eyvazi S, Babaeipour V et al (2020) Antimicrobial peptides (AMPs): roles, functions and mechanism of action. Int J Pept Res Ther 26(3):1451–1463. https://doi.org/10.1007/s10989-019-09946-9

Shai Y (1999) Mechanism of the binding, insertion and destabilization of phospholipid bilayer membranes by α-helical antimicrobial and cell non-selective membrane-lytic peptides. Biochim Biophys Acta 1462(1-2):55–70

Shai Y (2002) Mode of action of membrane active antimicrobial peptides. Pept Sci 66(4):236–248. https://doi.org/10.1002/bip.10260

Shin SY, Yang S-T, Park EJ, Eom SH, Song WK, Kim Y et al (2002) Salt resistance and synergistic effect with vancomycin of α-helical antimicrobial peptide p18. Biochem Biophys Res Commun 290(1):558–562

Shwaiki LN, Arendt EK, Lynch KM (2020) Anti-yeast activity and characterisation of synthetic radish peptides Rs-AFP1 and Rs-AFP2 against food spoilage yeast. Food Control 113:107178. https://doi.org/10.1016/j.foodcont.2020.107178

Souza LMP, Nascimento JB, Romeu AL, Estrada-López ED, Pimentel AS (2018) Penetration of antimicrobial peptides in a lung surfactant model. Colloids Surf B Biointerfaces 167:345–353. https://doi.org/10.1016/j.colsurfb.2018.04.030

Stevenson CL (2009) Advances in peptide pharmaceuticals. Curr Pharm Biotechnol 10(1):122–137

Strøm MB, Rekdal O, Svendsen JS (2002) Antimicrobial activity of short arginine- and tryptophan-rich peptides. J Pept Sci 8(8):431–437. https://doi.org/10.1002/psc.398

Talapko J, Meštrović T, Juzbašić M, Tomas M, Erić S, Horvat Aleksijević L et al (2022) Antimicrobial peptides-mechanisms of action, antimicrobial effects and clinical applications. Antibiotics (Basel, Switzerland) 11:10. https://doi.org/10.3390/antibiotics11101417

Tang L, Chen J, Zhou Z, Yu P, Yang Z, Zhong G (2015) Chlamydia-secreted protease CPAF degrades host antimicrobial peptides. Microbes Infect 17(6):402–408

Tang Z, Ma Q, Chen X, Chen T, Ying Y, Xi X et al (2021) Recent advances and challenges in Nanodelivery Systems for Antimicrobial Peptides (AMPs). Antibiotics 10(8):990

Teixeira MC, Carbone C, Sousa MC, Espina M, Garcia ML, Sanchez-Lopez E et al (2020) Nanomedicines for the delivery of antimicrobial peptides (AMPs). Nanomaterials (Basel, Switzerland) 10:3. https://doi.org/10.3390/nano10030560

Thakur A, Sharma A, Alajangi HK, Jaiswal PK, Lim Y-B, Singh G et al (2022) In pursuit of next-generation therapeutics: antimicrobial peptides against superbugs, their sources, mechanism of action, nanotechnology-based delivery, and clinical applications. Int J Biol Macromol 218:135

Tian X, Sun F, Zhou XR, Luo SZ, Chen L (2015) Role of peptide self-assembly in antimicrobial peptides. J Pept Sci 21(7):530–539

Tossi A, Sandri L, Giangaspero A (2000) Amphipathic, α-helical antimicrobial peptides. Pept Sci 55(1):4–30

Tu J, Li D, Li Q, Zhang L, Zhu Q, Gaur U et al (2015) Molecular evolutionary analysis of β-defensin peptides in vertebrates. Evol Bioinforma 11:EBO.S25580

Umerska A, Cassisa V, Bastiat G, Matougui N, Nehme H, Manero F et al (2017) Synergistic interactions between antimicrobial peptides derived from plectasin and lipid nanocapsules containing monolaurin as a cosurfactant against Staphylococcus aureus. Int J Nanomedicine 5687-99

Varela MF, Stephen J, Lekshmi M, Ojha M, Wenzel N, Sanford LM et al (2021) Bacterial resistance to antimicrobial agents. Antibiotics 10(5):593

van der Velden WJ, van Iersel TM, Blijlevens N, Donnelly JP (2009) Safety and tolerability of the antimicrobial peptide human lactoferrin 1-11 (hLF1-11). BMC Med 7(1):1–8

Verkleij A, Zwaal R, Roelofsen B, Comfurius P, Kastelijn D, Van Deenen L (1973) The asymmetric distribution of phospholipids in the human red cell membrane. A combined study using phospholipases and freeze-etch electron microscopy. Biochim Biophys Acta 323(2):178–193

Walas N, Slown S, Amato HK, Lloyd T, Bender M, Varghese V et al (2023) The role of plasmids in carbapenem resistant E. coli in Alameda County, California. BMC Microbiol 23(1):147. https://doi.org/10.1186/s12866-023-02900-2

Walkenhorst WF, Klein JW, Vo P, Wimley WC (2013) pH dependence of microbe sterilization by cationic antimicrobial peptides. Antimicrob Agents Chemother 57(7):3312–3320

Wang G (2014) Human antimicrobial peptides and proteins. Pharmaceuticals (Basel, Switzerland) 7(5):545–594. https://doi.org/10.3390/ph7050545

Wang J, Yadav V, Smart AL, Tajiri S, Basit AW (2015a) Toward oral delivery of biopharmaceuticals: an assessment of the gastrointestinal stability of 17 peptide drugs. Mol Pharm 12(3):966–973

Wang J, Yadav V, Smart AL, Tajiri S, Basit AW (2015b) Stability of peptide drugs in the colon. Eur J Pharm Sci 78:31–36

Wei G-X, Campagna AN, Bobek LA (2007) Factors affecting antimicrobial activity of MUC7 12-mer, a human salivary mucin-derived peptide. Ann Clin Microbiol Antimicrob 6(1):1–10

Wu D, Fu L, Wen W, Dong N (2022) The dual antimicrobial and immunomodulatory roles of host defense peptides and their applications in animal production. J Anim Sci Biotechnol 13(1):141. https://doi.org/10.1186/s40104-022-00796-y

Wu G, Ding J, Li H, Li L, Zhao R, Fan X, et al (2008) Effects of cations and pH on antimicrobial activity of thanatin and sthanatin against Escherichia coli ATCC25922 and B. subtilis ATCC 21332. Curr Microbiol 57(6):552–557

Wu M, Maier E, Benz R, Hancock RE (1999) Mechanism of interaction of different classes of cationic antimicrobial peptides with planar bilayers and with the cytoplasmic membrane of Escherichia coli. Biochemistry 38(22):7235–7242

Yeom J-H, Lee B, Kim D, Lee J-k, Kim S, Bae J et al (2016) Gold nanoparticle-DNA aptamer conjugate-assisted delivery of antimicrobial peptide effectively eliminates intracellular salmonella enterica serovar Typhimurium. Biomaterials 104:43–51

Yu H, Li H, Gao D, Gao C, Qi Q (2015) Secretory production of antimicrobial peptides in Escherichia coli using the catalytic domain of a cellulase as fusion partner. J Biotechnol 214:77–82. https://doi.org/10.1016/j.jbiotec.2015.09.012

Zasloff M (1987) Magainins, a class of antimicrobial peptides from Xenopus skin: isolation, characterization of two active forms, and partial cDNA sequence of a precursor. Proc Natl Acad Sci USA 84(15):5449–5453. https://doi.org/10.1073/pnas.84.15.5449

Zasloff M (2002) Antimicrobial peptides of multicellular organisms. Nature 415(6870):389–395

Zhang F, Cui X, Fu Y, Zhang J, Zhou Y, Sun Y et al (2017) Antimicrobial activity and mechanism of the human milk-sourced peptide Casein201. Biochem Biophys Res Commun 485(3):698–704. https://doi.org/10.1016/j.bbrc.2017.02.108

Zhao X, Yu Z, Ding T (2020) Quorum-sensing regulation of antimicrobial resistance in bacteria. Microorganisms 8:3. https://doi.org/10.3390/microorganisms8030425

Nanocomposites in Combating Antimicrobial Resistance

8

Meera Varghese, Aleena Ann Mathew, and Manoj Balachandran

Abstract

Extensive and improper usage of antibiotics has resulted in the outbreak of multidrug-resistant microorganisms and increasing antimicrobial resistance (AMR), which has become a significant threat to global health and health care. Resistant microorganisms adapt various resistance mechanisms like modifying the structure of antibiotics, altering the target, inhibiting the internalization of antibiotics, ejection of antibiotics from bacterial cells, etc. By lowering or completely disabling the efficacy of antibiotics, AMR may become a primary cause of mortality if left unattended. Developing effective antiresistance strategies to combat AMR is an urgent need of time. Nanomaterials have great potential to inactivate pathogens, and their mechanism of antimicrobial activity is different from antibiotics. With these unique mechanisms of antimicrobial action, nanomaterials are less prone to develop AMR. Developing nanocomposites can provide synergistic effects to improve the properties and strengthen the antimicrobial capability of individual nanomaterials. In this chapter, contemporary developments in the application of antimicrobial composites such as carbon nanocomposites, metallic nanocomposites, nonmetallic nanocomposites, metalloid nanocomposites, polymer nanocomposites, ceramic nanocomposites, and their hybrid forms to prevent the evolution of AMR will be discussed. The current research direction, prospects, and possible strategies to explore nanocomposites as potent antimicrobial agents to conquer AMR will be highlighted.

Keywords

Antibacterial · Nanocomposite · Synergistic effect · AMR

M. Varghese · A. A. Mathew · M. Balachandran (✉)
Department of Physics and Electronics, CHRIST (Deemed to be University), Bengaluru, Karnataka, India
e-mail: manoj.b@christuniversity.in

M. Y. Wani et al. (eds.), *Nanotechnology Based Strategies for Combating Antimicrobial Resistance*, https://doi.org/10.1007/978-981-97-2023-1_8

8.1 Introduction

The invention of antibiotics has brought health standards by serving as an efficient infection-fighting tool for several decades. However, the irrational prescription and handling of antibiotics caused the bacteria to acquire novel resistance mechanisms, which made them more virulent and resistant to antibiotics-based treatments (Haenni et al. 2022). Recent reports show that if necessary steps are not taken, AMR can become a significant cause of death by 2050. Unhygienic environments and poor-quality medications also cause the spread of AMR. Hence, safe and effective means should be identified to combat drug-resistant pathogens to ensure global health and protection from infectious diseases (Tang et al. 2023). Pathogens-free surface coatings, textiles, medical devices, etc., are also equally important in the journey of fighting against AMR (Pietsch et al. 2020; Dehari et al. 2023; Chu et al. 2022; Rajaramon et al. 2023). The battle against the progress of AMR necessitates not only the elimination of resistant microbes but also the development of new antimicrobials that can tackle the adverse effects of AMR. In this point of view, nanomaterials have great importance as they can serve as potential antimicrobials because the mechanisms of antibacterial activity of nanomaterials are different from conventional antibiotics; hence, they are less prone to AMR (Dennis et al. 2022). Designing nanocomposites by combining different materials can enhance the antibacterial properties through the synergistic effects of the individual components in the composites. Improved dispersibility, biocompatibility, and selectivity are some properties of interest in therapeutic applications (Li et al. 2021; Riley and Narayan 2021).

In many cases, composites have been utilized to enhance the desired properties of the final material. Certain studies shed light on the fact that nanocomposites can not only eradicate multidrug-resistant pathogens but are also capable of deteriorating the resistance attempts of bacterial species. Apart from the basic antibacterial studies, attempts were made to develop antibacterial coatings, wound coverings, packaging, textiles, etc. Herein, we discuss the inevitable role of different nanocomposites in combating AMR, the antibacterial mechanisms, potential applications, and prospects.

8.1.1 Nanocomposite Materials: Fundamentals and Types

Nanocomposites are materials that incorporate nanoparticles into a matrix material, creating a new material with enhanced properties. These nanoparticles are typically on the nanometer scale and can be composed of various materials, such as metals, polymers, ceramics, or carbon-based materials. Nanocomposites find applications in a wide range of areas owing to their unique combination of properties arising from the unique characteristics of the nanoparticles, such as their small size, high surface area, and quantum effects. It includes a matrix material and reinforced material at an aspect ratio, which influences the properties of the nanocomposite (Karak 2019; Fu et al. 2019).

8.1.2 Basics of Nanocomposites

In nanocomposites, the terms matrix and reinforcement refer to two key components that constitute the composite material. These components work together to impart specific properties to the nanocomposite. The matrix is the continuous phase or bulk material that surrounds and holds together the reinforcement phase in a composite material. The matrix provides structural integrity, defines the shape of the material, and protects the reinforcement. It also plays a remarkable role in determining specific bulk properties of the composite. The reinforcement phase comprises small particles, fibers, or structures dispersed within the matrix. In the context of nanocomposites, these reinforcements are nanoparticles with at least one dimension in the nanoscale range. The reinforcement imparts unique and enhanced properties to the composite. Due to their small size, high surface area, and unique properties, nanoparticles can significantly influence the mechanical, thermal, electrical, or other specific characteristics of the composite material (Omanović-Mikličanin et al. 2020; Veeman et al. 2021).

8.1.3 Types of Nanocomposite Materials

Nanocomposites are materials that contain a matrix or host material reinforced with nanoscale fillers or nanoparticles. These nanoscale reinforcements can be of various types, and the specific properties and applications of nanocomposites depend on the combination of the matrix and nanoparticles. The main types of nanocomposite materials are carbon nanocomposites, metal nanocomposites, metalloid nanocomposites, polymer nanocomposites, and ceramic nanocomposites (Fig. 8.1). A carbon nanocomposite refers to a material that combines carbon-based nanoparticles with a matrix material, resulting in a composite material with enhanced properties. Carbon nanocomposites are known for their unique electrical, thermal, and mechanical characteristics (Zahran and Marei 2019). A metal nanocomposite is a type of composite material that incorporates metal nanoparticles or nanoscale metallic structures into a matrix material. This combination results in a material with unique and enhanced properties compared to the pure matrix material (Pyrzynska 2020). Metalloid composites combine metalloid elements with other materials to make composite materials with unique properties. Metalloids are elements that exhibit features of both metals and nonmetals and are found in a specific region on the periodic table. Common metalloids include boron, silicon, germanium, arsenic, antimony, and tellurium (Shameem et al. 2021). Polymer nanocomposites are materials that consist of a polymer matrix reinforced with nanoscale fillers or nanoparticles. Adding nanoparticles to polymers enhances various properties, making polymer nanocomposites suitable for multiple applications (Kiranakumar et al. 2022). Ceramic nanocomposites are materials that combine ceramic matrices with nanoscale reinforcements, resulting in materials with enhanced properties compared to traditional ceramics. Incorporating nanoparticles or nanofillers into

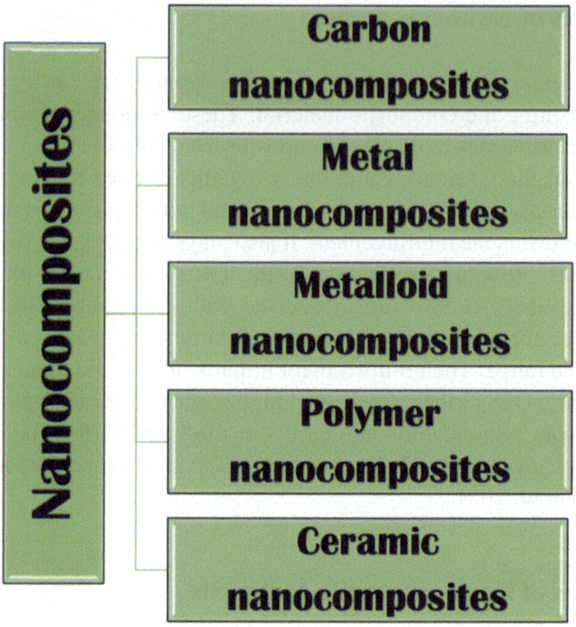

Fig. 8.1 Classification of nanocomposites

ceramics imparts unique characteristics, such as improved mechanical strength, toughness, and thermal properties (Li et al. 2020a).

8.2 Mechanisms of Antimicrobial Resistance

Since the inception of antibiotics, the healthcare sector has deemed its victory over infectious diseases. However, the inappropriate and indiscriminate use of antimicrobial agents has led to the emergence of a formidable challenge of resistance. The current surge in antimicrobial resistance prominently observed among bacterial pathogens that commonly occur, poses a formidable threat to the successful therapeutic outcomes of patients who are critically ill. It is of paramount importance that we take immediate and decisive action to address this issue (Chinemerem Nwobodo et al. 2022).

Developing effective antimicrobial agents or alternative tools that can combat public health challenges requires a thorough understanding of the resistance mechanisms of bacteria. Acquiring a profound comprehension of these mechanisms is crucial in anticipating latent or obscure resistance mechanisms that can be applied to emerging multidrug-resistant pathogens (Biondo 2023). Therefore, it is imperative to prioritize further research in this area to ensure that effective interventions are developed.

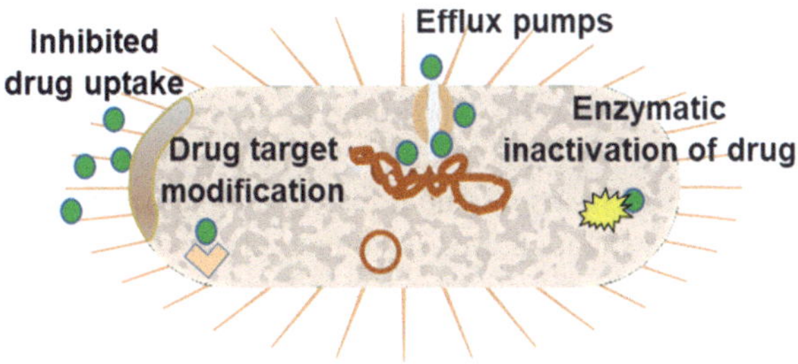

Fig. 8.2 Mechanisms of antibiotic resistance

8.2.1 How Microorganisms Develop Resistance?

Comprehending the underlying mechanisms by which bacteria develop antibiotic resistance assumes paramount importance in the context of antimicrobial resistance. Bacteria have, over time, evolved intricate biochemical pathways to gain resistance to specific categories of antimicrobial agents. These pathways can be broadly classified into four categories: drug uptake limitation, drug target modification, drug inactivation, and drug efflux (Fig. 8.2). Gram-positive bacteria are less likely to employ drug uptake limitation and drug efflux mechanisms since they lack lipopolysaccharide in the outer membrane. In contrast, Gram-negative bacteria exhibit all four mechanisms due to their unique structural characteristics (Biondo 2023; Mohanty et al. 2021; Uddin et al. 2021; Adeniji et al. 2022).

The outer membrane of Gram-negative bacteria acts as a barrier to permeability, rendering it challenging for certain antibiotics to enter. Additionally, bacterial efflux pumps actively transport many antibiotics out of the cell, which contributes to the inherent resistance of Gram-negative bacteria (Altınöz and Altuner 2019). The rise of antibiotic resistance is also being aided by bacterial enzymes. These enzymes are grouped based on their specific functions in different biochemical processes. They may alter antibiotic targets, convert antibiotics, modify intracellular targets through enzymatic means, or catalyze cellular metabolic reactions (Egorov et al. 2018; Schaenzer and Wright 2020).

8.2.2 Challenges Posed by Antimicrobial Resistance

Antimicrobial resistance represents a consequential global matter and poses a grave threat to the health and well-being of individuals across the planet. The widespread occurrence of antibiotic-resistant microorganisms can result in untreatable infections and create major challenges in routine medical procedures such as chemotherapy, surgeries, and organ transplants. This could lead to severe consequences such as prolonged hospital stays, increased mortality risks, and elevated healthcare costs (Dadgostar 2019). The development of antibiotic resistance among infectious microbial agents requires undivided attention, as initial antibiotics may prove ineffective, necessitating the use of second- or third-line antibiotics, which can lead to more significant hazards and increased costs. Additionally, the impact of AMR on livestock output is also significant. The surge in antimicrobial resistance poses a significant threat to animal treatment efficacy, leading to more severe infections. This, in turn, could result in a substantial loss in animal production, which could exacerbate the economic situation. Resistant microorganisms and their corresponding resistance genes have the potential to enter soil, air, water, and sediments through various modes of transmission. Environmental research has identified the presence of resistant organisms and genes, particularly within water bodies (Ahmad et al. 2021; Taneja and Sharma 2019; Larsson and Flach 2022). Hence, it is vital to take immediate and effective action to mitigate the spread of AMR to preserve public health and safety.

Nanocomposites have emerged as a promising solution to combat microbial drug resistance. These innovative materials incorporate various types of nanoparticles that work in tandem to fight off microbes (Kankala et al. 2020). By utilizing multiple mechanisms, nanocomposites make it difficult for microbes to develop resistance, as it would require multiple simultaneous gene mutations within the same microbial cell. Notably, the risk of resistance to the numerous antimicrobial agents within these nanocomposites is also low (Sivam et al. 2023). Moreover, many of the nanocomposites can precisely target antimicrobial agents to the site of infection, allowing for higher drug dosages to be administered and reducing resistance while minimizing negative effects on the patient (Hatae et al. 2023).

8.3 Carbon Nanocomposites

Carbon nanomaterials and composites are widely known for their inherent antimicrobial properties, biocompatibility, and nontoxicity (Mathew et al. 2023; Varghese et al. 2022; Varghese and Balachandran 2021). The large surface area of carbon nanomaterials is favorable for bacterial adhesion, followed by inactivation. Their combination meth metal nanoparticles can improve the dispersibility of metal nanoparticles and reduce the toxic effects. Several studies have been conducted to explore the antibacterial potential of carbon nanomaterials in combination with organic and inorganic materials (Joseph et al. 2023).

In a recently reported work, graphene nanoplatelets/CeO_2 nanotiles nanocomposites are more efficient than CeO_2 nanotiles alone in inhibiting *P. aeruginosa* and *S. aureus* bacteria. The tremendous enhancement is attributed to the combined action of the individual components in the composite. The authors presented a hypothetical mechanism to explain the bacterial toxicity of the composite. With their sharp blade kind of morphology, CeO_2 nanotiles can rupture the bacterial cell to enter it. Similarly, the sharp edges of nanoplatelets can damage the bacterial membrane by exerting stress on it. Another factor contributing to bacterial toxicity is the ability of graphene to act as an excellent electron acceptor, which induces ROS (Reactive Oxygen Species) generation via charge imbalance in the bacterial membrane. Besides, the CeO_2 nanotiles in the composite release Ce^{4+}, and their interaction with the thiol groups in bacteria can enlarge specific proteins that control nutrient transport. Consequently, bacteria lack sufficient nutrients, leading to their death (ur Rehman et al. 2022).

Sajjad et al. synthesized GO/Ag_2O through a wet chemical method. The antibacterial efficiency of the composite against *E. coli* and *S. aureus* was higher than that of pure GO and Ag_2O. The more excellent antibacterial activity is mainly because of the increased ROS production and Ag^+ ions release in the composite. With their positively charged characteristics, Ag + ions can effectively interact with the bacterial cell membrane as the presence of some lipids negatively charges it. The absorption of heat energy at the heterojunction of the composite helps the bonding of carbon, silver, and oxygen and can boost the bactericidal properties of the material. It also supports the electron transitions in Ag_2O. As an electron acceptor, GO suppresses the charge recombination and helps the diffusion of Ag^+ ions to the bacterial cell. ROS formation and oxidative stress disintegrate the bacterial cell by interfering with enzymes, protein synthesis, and DNA. The overall antibacterial capability of the final material is highly improved by the combined effects of GO and Ag_2O (Sajjad et al. 2019).

Munawar et al. designed a nanocomposite in which W, Ag, and Zn-doped SnO_2 nanocrystals are anchored on GO sheets through the Sol-gel approach followed by ultrasonication. The antibacterial test results showed that all samples could inhibit *S. aureus* and *K. pneumoniae* bacteria. Compared to doped and undoped composites, the GO composites exhibited superior antibacterial activity, higher charge transfer rate, and lower charge recombination. $SnZnO_2$/GO nanocomposites provided the highest inhibition zone among all the samples. The leaching of positive metal ions may explain the enhancement in the antibacterial action, increased bacterial interaction with uneven block-type morphology, and increased ROS production in the composite. The factors above lead to oxidative stress and cytotoxicity in the bacterial cell via DNA damage, inhibition of respiratory electron transport, and peroxidation of membrane lipid and enzyme protein (Munawar et al. 2022).

By the hydrothermal method, Kumar et al. could successfully synthesize rGO-CuO and rGO-Cu_2O of different shapes and phases, and the obtained materials were tested against *Klebsiella* and *S. aureus* bacteria. Higher antibacterial activity was observed with gram-negative *Klebsiella* bacteria. Compared to pristine CuO, the two composites offered better antibacterial results. The release of Cu^{2+} ions ruptured the cell membrane of bacteria. The oxidation states of CuO and Cu_2O NPs play an essential role in bacterial inactivation. Large surface area, presence of sharp edges, and high electrical conductivity of rGO promote biological and chemical activities such as charge transfer during bacterial inactivation through contact (Kumar et al. 2019). In another work, it has been reported that the synergistic action of cubic Ag NPs and rGO in the Ag/rGO nanocomposite offered excellent concentration-dependent antibacterial action toward *B. subtilis* and *E. coli, S. aureus* bacteria (Devi et al. 2021). The rGO-Ag nanocomposite prepared by Karthik et al. was examined against *Salmonella enterica (S. enterica), B. cereus, E. coli, P. aeruginosa,* and *Vibrio parahaemolyticus (V. parahaemolyticus).* The nanocomposite showed an enhanced bacterial inactivation through the symbiotic effect compared to the Ag NPs and rGO alone. The antibacterial mechanisms include free radical action and resultant oxidative stress that induce structural and functional disorders in bacteria (Karthik et al. 2020).

Using the hydrothermal method, Dhandapani et al. synthesized rGO-ZnO nanocomposite that shows bacterial toxicity against *E. coli* and *S. aureus.* Bacterial inactivation occurred in dark and light conditions. The composite showed higher bacterial toxicity compared to ZnO nanorods. rGO sheets with high surface area and free radical production capability inhibited bacterial proliferation (Dhandapani et al. 2021). Zhang et al. synthesized rGO functionalized with ZnO NPs by exfoliation of oxidized graphite in the presence of Zn^{2+}. The composite material could destroy gram-negative (*E. coli* and *Serratia marcescens (S. marcescens))* and gram-positive (*B. subtilis*) bacteria with better efficiency than GO and rGO. Experimental results showed that intracellular ROS plays a significant role in the Zn^{2+} action to induce antibacterial activity. In composite, there was a dramatic increase in the ROS that favors antibacterial activity. The composite material also highly suppresses the recombination of electrons and holes (Zhang et al. 2019).

Qiao et al. reported an effective treatment strategy for bone infections (osteomyelitis) with Fe_3O_4/CNT/gentamicin nanocapturers. Gentamicin (Gent) was used as a chemotherapy agent with the nanomaterials. The in vivo studies were conducted on a rabbit model. The magnetic properties of the material realized targeted therapy and MRSA-infected tibia osteomyelitis of the rabbit model was treated with the support of microwave irradiation. Figure 8.3a shows the macroscopic images of tibia and femur specimens of MRSA-infected rabbit model treated for 14 days. MV and MF denote treatment with microwave or magnetic field. Among all the femur and tibia specimens, those treated with Fe_3O_4/CNT/Gent + MV and Fe_3O_4/CNT/Gent + MV + MF appear infection-free without any pustules or swelling. Similarly, the specimens mentioned above showed a highly reduced count of MRSA and lymphocytes (Fig. 8.3b, c). The MRI images (Fig. 8.3d) and the WBC counts (Fig. 8.3e) also indicate the potential of Fe_3O_4/CNT/ gentamicin nanocapturers in treating osteomyelitis (Qiao et al. 2020).

Sadhukhan et al. synthesized rGO/CdO composite by coprecipitation and hydrothermal method and studied its antibacterial efficiency against various gram-positive and gram-negative bacteria with MIC values, 3.93 ± 0.28 μg/ml for *E. coli*, 4.011 ± 0.95 μg/ml for *B. subtilis*, 9.62 ± 2.95 μg/ml for *K. pneumoniae*, 4.959 ± 1.00 μg/ml for *S. aureus,* 4.426 ± 0.05 μg/ml for *Listeria monocytogenes (L. monocytogenes),* and 6.17 ± 1.13 μg/ml for *S. typhimurium* bacteria. Interestingly, the results were comparable to the performance of streptomycin, a commercially available antibiotic. The comparison between the bactericidal properties of the rGO/CdO composite and rGO alone revealed the superior bactericidal efficiency of the composite (Sadhukhan et al. 2019). Naseem et al. synthesized rGO/ZnO nanocomposite from Graphite flakes and $Zn(NO_3)_2{\cdot}6H_2O$ using modified Hummer's and chemical reduction methods. The bacterial toxicity was tested on *S. aureus, E. coli,* and *P. aeruginosa*. The bactericidal effect was more robust with rGO/ZnO nanocomposite compared to GO, which shows improved biocompatibility of the nanocomposite (Naseem et al. 2020). Alayande et al. fabricated rGO-CuO nanocomposite

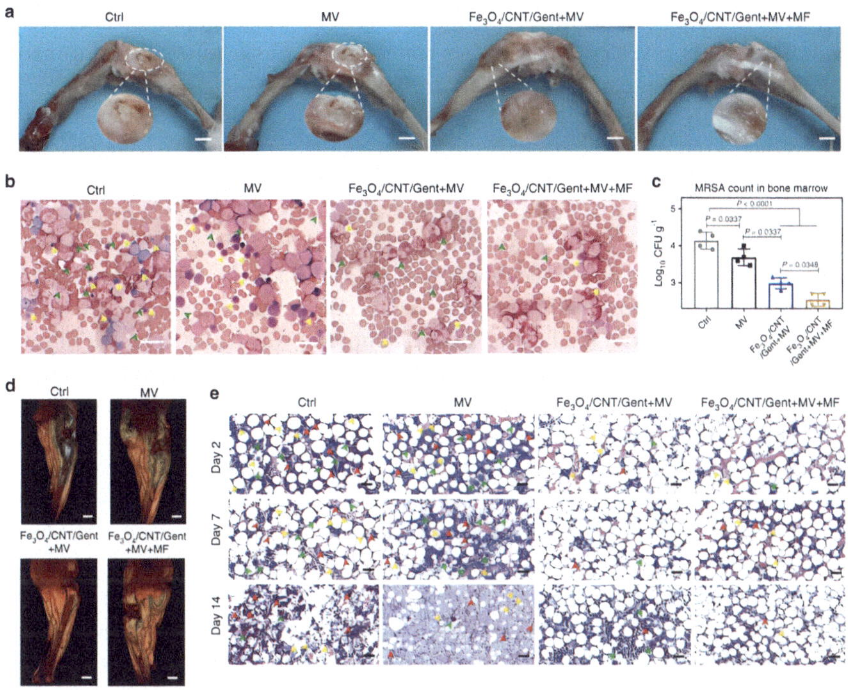

Fig. 8.3 (**a**) Macroscopic images of tibia and femur specimens of MRSA-infected rabbit model treated for 14 days. (**b**) MRI images, (**c**) MRSA counts, (**d**) WBC counts, and (**e**) H&E staining images of MRSA-infected bone marrow

film and tested it against *P. aeruginosa*. ROS-free bacterial inactivation was observed due to the electronic structure and electron transfer from the bacterial cell. Incorporating CuO into rGO improved the roughness and surface area of the film. The electron conductivity of CuO can bridge the nonconducting lipid layer of the bacterial cell. Due to the negative resting potential of the bacterial cells, they can quickly lose electrons, and with Cu, they can have a circuit effect that maintains electron transfer. At a higher pH level, nanorods of rGO-CuO were formed, and bacterial toxicity was enhanced due to the nano-knife impact, which can rupture the cell wall of bacteria (Alayande et al. 2020).

Zheng et al. prepared a nanocomposite of lanthanum hydroxide and graphene oxide that could eradicate bacterial strains through synergistic effects. Unlike the antibiotics and silver nanoparticles, continued exposure of *E. coli* to the material was confirmed for not inducing AMR. This can be ascribed to the unique antibacterial mechanism of the material involving lipid peroxidation, dephosphorylation, and peptidoglycan disruption (Zheng et al. 2019). Tan et al. synthesized rGO/Ag nanocomposite that can destroy regular and MDR bacteria. The rGO/Ag nanocomposite exhibited higher antibacterial efficacy than GO Ag, NPs, and rGO. The rGO/Ag composite induced a photothermal effect and offered a synergistic antibacterial performance over *E. coli* and MDR bacterium *K. pneumoniae.* E. coli and K.

pneumoniae could be destroyed with 15 and 30 μg/ml samples, respectively. Cytotoxic studies revealed that rGO/Ag nanocomposite is cytotoxic toward mammalian cells at higher concentrations (Tan et al. 2020).

The positively charged nitrogen-doped CQDs were tested against a set of gram-positive and gram-negative bacteria. Effective antibacterial results were observed with *S. epidermidis,* MRSA, *S. aureus* (ATCC6538), and *S. aureus* (ATCC43300). However, the material was inactive against *S. paratyphi-β, E. coli, E. faecalis,* and *P. aeruginosa* bacteria (Zhao et al. 2019). Fang et al. hydrothermally synthesized CD/Ag composite NPs and tested them against *E. coli* bacteria. The reduction of Ag ions resulted in the negatively charged CDs becoming positively charged in the composite. The bacterial eradication of the composite was better than that of pristine CDs (Fang et al. 2019). Al-Shawi et al. hydrothermally synthesized S, N-GQDs/NiO composite. They tested antibacterial effects against *E. coli, P. aeruginosa, S. aureus,* and *methicillin-resistant S. aureus* bacteria, which exhibited zones of inhibition 10,16, 19, and 14 mm, respectively. The antibacterial competence of the material was excellent compared to the NiO NPs alone. ROS formation is the principal mechanism behind the antibacterial activity of the composite (Al-Shawi et al. 2021).

Nitrogen-doped sponge-like carbon spheres synthesized by Xi et al. effectively inhibited *E. coli, S. aureus,* and multidrug-resistant bacteria. The laser exposure created more active sites on the material and improved photocatalytic activity and antibacterial efficiency (Xi et al. 2019). According to the report of Cao et al., the CNT/Cu composite showed excellent antibacterial response against *S. aureus* and *E. coli. In contrast, pure* CNTs did not exhibit any significant antibacterial properties. A potent antibacterial effect against S. aureus is ascribed to the lack of bacterial cell walls (Cao et al. 2021). The carboxylated CNTs/Biogenic ZnO Composite prepared by Hussain et al. exhibited excellent toxicity toward *B. cereus, P. aeruginosa,* and *E. coli* bacteria. Compensating the limitations of CNTs and ZnO NPs, the hybrid material has become highly antibacterial and less toxic to biological cells (Hussain et al. 2022).

In another work, the authors hydrothermally prepared different composites by combining Sulfur, GO, CNT, and SnO_2 and assessed them against *E. coli* and *Pseudomonas graminis (P. graminis).* The samples showed a dose-dependent antibacterial activity against the tested microorganisms. Results showed that sulfonated samples have a superior bactericidal effect compared to nonsulfonated samples (Pandiyan et al. 2019). Jiang et al. hydrothermally prepared carbon spheres immobilized Ag NPs that can eradicate *E. coli* and *S. aureus* bacteria. The composite is further anchored on cellulose nanofibers to gain a composite paper with antibacterial properties. The composite paper showed enhanced effects compared to the carbon spheres anchored paper (Jiang et al. 2020).

Yang et al. developed economic carbon black NMs and doped them with single metal atoms (Cu, Ni, Au) to assess the photothermal antibacterial properties of the final material. The electronic structure of carbon black was modified by doping such that the photothermal efficacy and resultant bacterial inactivation were highly improved. Under light irradiation (808 nm), extra electrons were excited, and the relaxation of electrons and holes in the carbon black matrix released thermal energy, which is utilized to inactivate tested microorganisms, *E. coli* and *B. subtilis*. An interesting fact was noted by the authors that there is no ROS or resultant oxidative stress involved in this antibacterial activity. A gradual decrease in the antibacterial performance was noticed with the increasing bandgap of the materials such that, from higher to lower in the order of Cu-doped sample, Ni-doped sample, Au-doped sample, and finally carbon black, which offered the weakest antibacterial efficacy (Yang et al. 2021). Du et al. designed a graphene/nitrocellulose membrane having antibacterial efficiency >98% and > 90% toward *E. coli* and *S. aureus*, respectively, and can be used for wound dressing. Experiments conducted on a rat model showed that the as-prepared membrane can recover skin tissue with 93.03 ± 1.08% wound closure. The photographs of wounded skin tissue treated by different concentrations of the composite material, percentage of wound closure, images of histological sections of the newly formed tissue are shown in Fig. 8.4 (Du et al. 2021).

Cao et al. developed a composite hydrogel composed of 3-aminobenzene boronic acid, graphene oxide, carboxymethyl chitosan, and cellulose nanocrystalline network, which could be successfully utilized for targeted photothermal eradication of *S. aureus* bacteria under near-infrared light with an efficiency of 99.99%. The sample also exhibited appreciable wound-healing properties on *S. aureus-infected* tissues (Cao et al. 2023). Benko et al. reported the discovery of antibacterial carbon

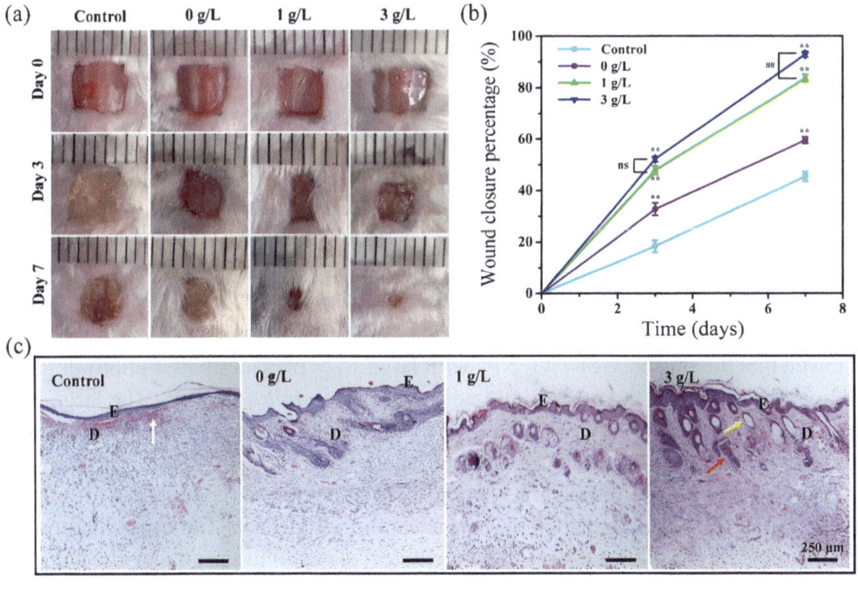

Fig. 8.4 (**a**) Photographs of wounded skin tissue treated by graphene/nitrocellulose membranes with different concentrations of the composite material. (**b**) Percentage of wound closure. (**c**) Hematoxylin and eosin-stained images of histological sections of the newly formed tissue after seven days. Epidermis and dermis layers are indicated by the letters D and E. The yellow and white arrows indicate the hair follicle and erythrocyte exudation (Du et al. 2021)

nanotubes showing more potent effects on methicillin-resistant *S. aureus* bacteria (Benko et al. 2021). Gao et al. designed lamellar-sized GO nanosheets that exhibited strong dose-dependent antibacterial potential toward methicillin-resistant *S. aureus* bacteria (Gao et al. 2021). In another work, ZnO NPs doped imidazolium were tested against *S. aureus*, *E. faecalis,* vancomycin-resistant *enterococci,* and methicillin-resistant *S. aureus* bacteria. Integrating imidazolium into ZnO nanoparticles improved bacterial inhibition through synergistic effects (Eskandari et al. 2023).

Rifai et al. fabricated a nanodiamond-coated 3D-printed Titanium inhibiting *S. aureus* colonization. The addition of nanodiamonds improved the wettability and promoted bacterial adhesion, favoring antibacterial activity (Rifai et al. 2019). Zhu et al. synthesized a nanocarbon composite (CDs/rGO/P25) by combining rGO and TiO2 with CDs and observed its excellent antibacterial efficiency on gram-negative *E. coli* and gram-positive *S. aureus* bacteria. The morphological changes of the bacteria with and without the treatment of the sample under light and dark conditions were studied with SEM images. The light irradiation of the bacteria treated with CDs/rGO/P25 resulted in shape distortion and severe damage to the bacterial membrane. The bacterial toxicity of the material is mainly attributed to the membrane damage induced on bacterial cells. The high photocatalytic property of the material is due to the synergistic effects of the material (Zhu et al. 2021).

8.4 Metallic Nanocomposites

Metallic nanocomposites have played a significant role in the history of antimicrobial treatments. In response to the global call to combat AMR, metallic nanocomposites have gained more interest owing to their strong antimicrobial potential. Several metals are indispensable for the smooth functioning of their biological processes in microorganisms. Hence, both starvation and excessiveness of metals can lead to bacterial death. Different forms of metals like metal ions, metal nanoparticles, or complexes can impose toxic effects on bacteria (Frei et al. 2023).

Pan et al. investigated the influence of relative sizes and alignment of carbon and TiO_2 NPs on the antibacterial efficiency of the composite material. The results showed that the presence of TiO_2 NPs with large sizes provides better photo-derived antibacterial activity than the small ones against *B. subtilis* bacteria. The large size of TiO_2 NPs prevents them from entirely enfolding carbon, allowing their active contribution to the photoexcited antibacterial activity (Pan et al. 2020). Through the co-pyrolysis of polystyrene/$AgNO_3$ fibers, Ji et al. synthesized a tubular structured Ag-Carbon-TiO_2 composite that inactivates *S.aureus* bacteria. The oxygen atoms in TiO_2 substitute the carbon atoms to create defect states in the band gap and help the electron-hole pairs to get excited by visible light exposure.

Moreover, surface plasmon resonance and interband transitions of silver NPs promote UV–vis light absorption of TiO_2. Doping with silver protects TiO_2 from phase transition. The electron-hole pair separation efficacy and electron transmission are highly enhanced in the composite. The holes produced by silver NPs can directly oxidize organic molecules. The tubular shape of the composite offers increased contact surface area with bacteria. Moreover, the abundant oxygen defects stimulate oxidative stress in the bacterial cell and cause damage to the cell wall, DNA, and other intracellular components of bacteria (Ji et al. 2020).

Eleftheriadou and coworkers tested polyol-coated ZnO, CuO, and CuZn nanoparticles against multidrug-resistant (MDR) *Pseudomonas aeruginosa*. The nanoparticles combined with the antibiotics ciprofloxacin and meropenem exhibited efflux pump inhibition potential, which enhanced the antibacterial effects on the tested bacterial strains. The schematic illustration of the efflux pump inhibition action of PEGylated ZnO, PEGylated CuO, and CuZn bimetallic nanoparticles in triethylene glycol is shown in Fig. 8.5.

Zafar et al. designed Ag/TiO2/Fe2O3/CS nanocomposite encapsulated by the antibiotic Ciprofloxacin to inactivate multidrug-resistant *E. coli* bacterial strains. The positively charged nanoparticles could easily adhere, damage the negatively charged bacterial cell wall, and induce ROS effects to destroy the bacteria (Zafar et al. 2022). The green synthesized Ag-ZnO nanocomposites prepared by Arya et al. showed bacterial toxicity toward multidrug-resistant Enteroaggregative *E. coli*. The antibacterial action of the material was observed with interruption of biofilm formation and dose-dependent antioxidant activity (Arya et al. 2023). Wang et al. designed and tested mercaptophenylboronic acid-activated Au nanoparticles against different bacterial strains and their MDR types isolated from urine or saliva samples. The material displayed selective antibacterial activity toward Gram-positive *Staphylococcus aureus*, MDR *Staphylococcus aureus*, *Staphylococcus epidermidis*, and MDR *Staphylococcus epidermidis* with minimum drug resistance and maximum biosafety (Wang et al. 2020). Mickymaray reported dose-dependent antibacterial potential of green synthesized Ag nanoparticles against multidrug-resistant Pseudomonas aeruginosa and Acinetobacter baumanii bacteria with a lowest concentration of 6.25 μg (Mickymaray 2019). According to a recent report, Ag and ZnO nanoparticles and their combination with antibiotics were effective against multidrug-resistant Acinetobacter baumanni and Enterococcus bacterial species (Adeniji et al. 2022).

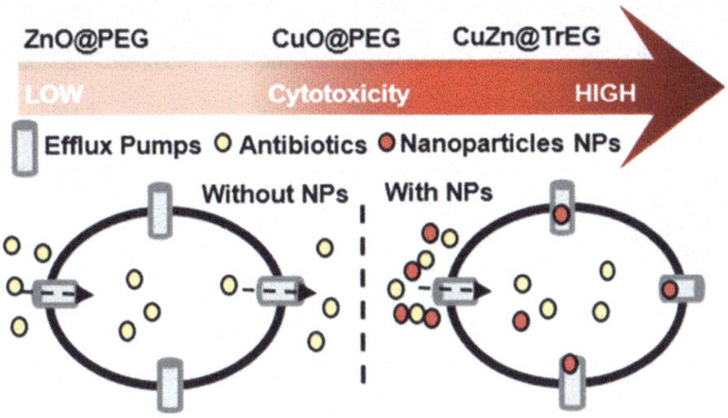

Fig. 8.5 Schematic illustration of the efflux pump inhibition action of PEGylated ZnO, PEGylated CuO, and CuZn bimetallic nanoparticles in trimethylene glycol

Wu et al. reported a nanocomposite of AgNPs inside ZnO/SiO2 nanocages (AgZn @ mesSiO2) capable of inhibiting MRSA. Through a controllable release and recapture of silver ions, the composite can reduce the biotoxicity of silver with enhanced sensitivity toward bacterial strains (Fig. 8.6) (Wu et al. 2020).

Vanitha et al. hydrothermally prepared Ce-doped ZnO/rGO without any surfactant. The PL studies showed that Ce doping caused the emission peaks to occur in the UV-Vis region with numerous active sublevels. The composite acted as an excellent antibacterial against *B. subtitles* and *Vibrio harveyi (V. harveyi)* bacteria. The antibacterial activity was compared between Ce-doped ZnO/rGO and other samples such as ZnO/rGO, Ce/ZnO, ZnO, and GO. The Ce-doped ZnO/rGO showed superior antibacterial activity over other samples. The electron-hole pairs formed on the surface of ZnO promote ROS formation, which expedites the death of bacteria through the inactivation of DNA, protein, and enzymes in the cell. The high surface area of rGO favors improved oxygen and water adsorption, and these molecules react with electron-hole pairs to form excess ROS. Additionally, the edges of rGO and ZnO can exert combined efforts to develop cavities on bacterial cell walls. Cerium can promote ROS formation by trapping the electrons and forming oxygen vacancies in the presence of its ions (Vanitha et al. 2018). $CuCo_2S_4$ nanozymes prepared by Li et al. exhibited incredible peroxidase-like properties and superior

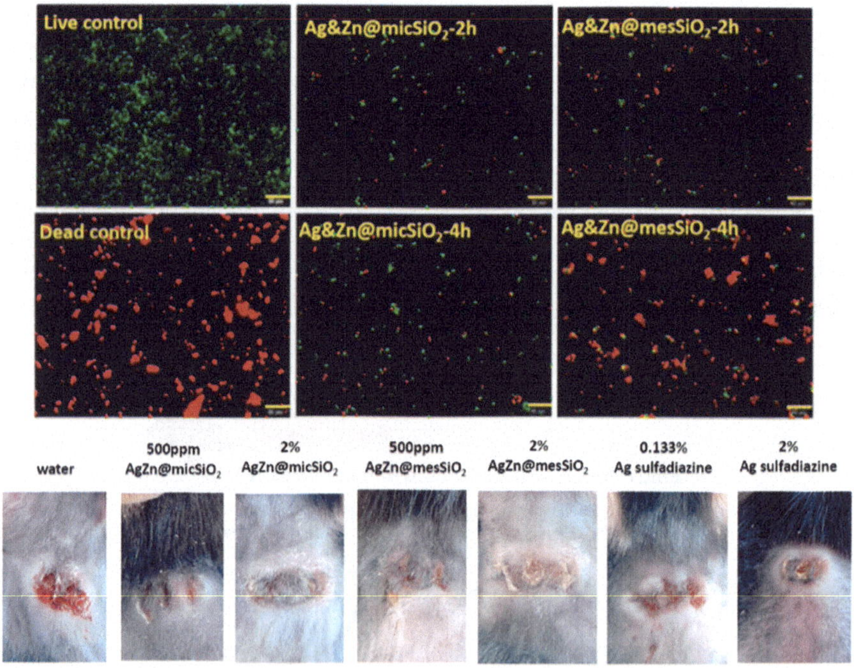

Fig. 8.6 The antibacterial efficiency of AgZn @ mesSiO_2 against MRSA. In vivo antibacterial investigation on wounds infected by MRSA and treated with different AgZn @ mesSiO_2 and other antimicrobial agents (Wu et al. 2020)

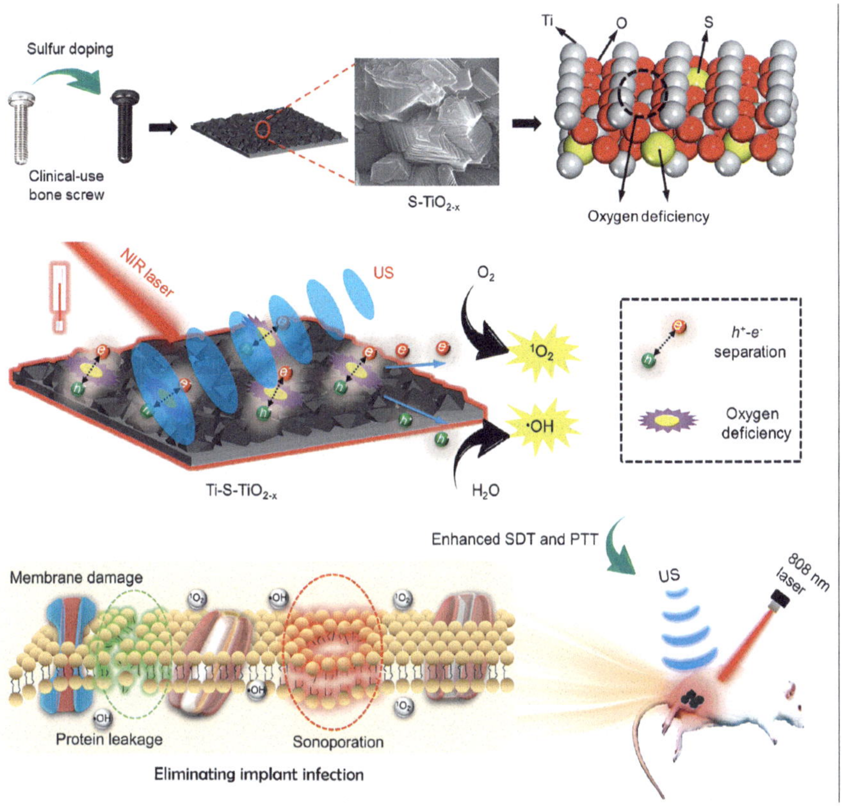

Fig. 8.7 Illustrates a clinically used bone screw with enhanced sonocatalytic–photothermal antibacterial activity for treating bone infections (Su et al. 2020)

antibacterial activity against MRSA compared to CoS and CuS nanoparticles. The $CuCo_2S_4$ nanozymes could also completely destroy the MRSA biofilms (Li et al. 2020b). Su et al. suggested a 15-minute photosonotherapy using a sulfur-doped titanium implant (Ti − S − TiO2 − x) with the aid of ultrasound and near-infrared light. The therapy was effective against *S. aureus* with 99.995% antibacterial efficiency. It was also confirmed successful in curing bone infection with better osseointegration. Clinical bone screws developed with the material above were coated with bacteria and implanted at the tibial plateau of rats. Through the sonocatalytic–photothermal therapy, the antibacterial efficiency of the bone screws was extended to 99.26%, which is schematically shown in Fig. 8.7 (Su et al. 2020).

8.5 Metalloid Nanocomposites

Metalloid nanocomposites are nanocomposites whose physical and chemical characteristics lie amid metal and nonmetal classes. The most widely recognized metalloids are tellurium, silicon, boron, antimony, arsenic, polonium, and germanium.

Metalloid nanoparticles are often effective for therapeutic purposes and have various impacts on both cells and tissues. Microorganisms use antimicrobial metalloid composites as a means of competitive advantage over other microbial community members. As a result, resistance is being evolved under intense selection pressure.

Mukheem et al. fabricated a boron-nitride-doped Polyhydroxyalkanoate/chitosan (PHA/Ch-hBN) nanocomposite via solvent casting technique. The time-kill approach was used to test the antibacterial action of nanocomposites against MDR microorganisms, methicillin-resistant *Staphylococcus aureus,* and *Escherichia coli.* Lactate dehydrogenase (LDH) assay in human keratinocytes (HaCaT) cell lines revealed that the synthesized nanocomposites showed substantial cell viability and antibacterial activity, attributed to efficient and stable scaffold for biomedical applications (Fig. 8.8) (Mukheem et al. 2019). Antibiofilm activities against urinary tract pathogens were investigated by El-Batal et al. using silver boron nanomaterials with gamma rays and PVP polymer. These nanomaterials are synthesized by an eco-friendly method and display an average diameter of 85.25 nm with an antibacterial

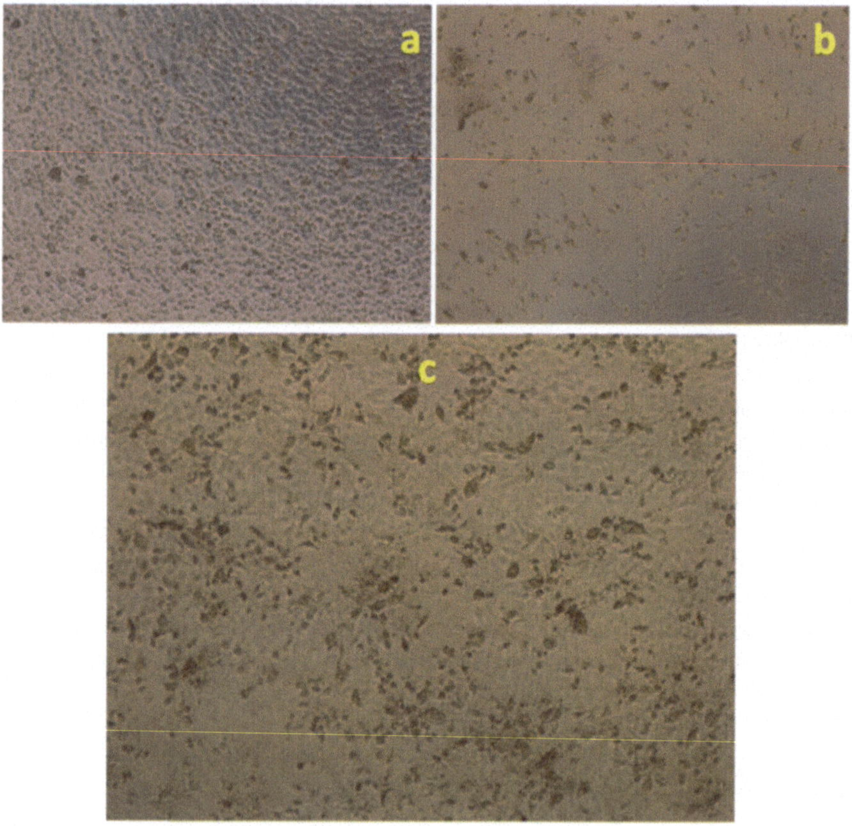

Fig. 8.8 LDH assay of PHA/Ch-hBN nanocomposite in HaCaT cell line (**a**—negative controls, **b**—positive controls, and **c**—nanocomposite) (Mukheem et al. 2019)

action against *Staphylococcus aureus*, *Candida albicans*, and *Escherichia coli* (El-Batal et al. 2019). Arslan et al. prepared biocompatible diamond-like carbons—Germanium nanocomposite with a mean size of 400 nm. The biocompatibility test was done on the human fibroblast cell line by MTT assay, and the antibacterial analysis was done against *Staphylococcus aureus* and *Pseudomonas aeruginosa*. They suggested a possible novel bionanocomposite film material for various biomedical applications with nonporous and dense characteristics (Arslan et al. 2022). A potential alternative for bone repair of contaminated bone deformities was synthesized by Yingqi Ji et al. with remarkable antibacterial features. They reported that the osteogenic and biocompatible bioactive material Germanium–silicocarnotite composite can successfully prevent the proliferation of *Staphylococcus aureus* and *Escherichia coli*, and no toxic effect is also demonstrated in rat bone marrow-derived mesenchymal stem cells (Ji et al. 2023).

Mohammed et al. reported an effective new antimicrobial agent against multidrug-resistant pathogens by tellurium oxide/graphene oxide and hydroxyapatite nanocomposite. The ternary nanocomposite exhibits a mean grain size of 20.8 nm and is biocompatible in human osteoblast cell lines. The antibacterial action of synthesized composite is inspected by *Staphylococcus aureus* and *Escherichia coli* (Al Mogbel et al. 2022). Singh et al. developed a chemotherapeutic alternative to treat lung infections by specific nasal delivery from Tellurium-doped Cerium dioxide (TeO_2–CeO_2) nano wools. Cubic fluorite-shaped nanomaterials were synthesized by thermo-mechanical route and mean diameter of 200 nm. The antibacterial performance of the synthesized materials was tested against *Bacillus subtilis* and *Klebsiella pneumoniae*. When the bactericidal capability of the antibiotic cefotaxime was assessed using the disc diffusion approach, nano wools were shown to be more effective. Toxicity studies on human embryonic kidney cell lines demonstrated that they were hardly harmful at an IC_{50} value of 0.27 mg/ml (Singh et al. 2023). Ikram et al. prepared tellurium (Te) and starch-doped strontium oxide (SrO) nanocomposite via a chemical coprecipitation route. The Te doping in SrO-starch composite enhanced the crystallinity and catalytic properties also investigated. Compared with ciprofloxacin, the nanocatalyst established significant antibacterial action against *Staphylococcus aureus*.

Furthermore, molecular docking investigations were carried out of produced nanocomposites against specific enzyme targets, namely β-lactamase and DNA Gyrase (Ikram et al. 2022). In another work, Mathai et al. fabricated antibacterial and anti-inflammatory tellurium dioxide-niobium pentoxide bioimplants through a low-cost, low-temperature thermal decomposition method. The produced bioimplant coating had good biomimetic features and a highly adhesive nature. The developed coatings have shown inhibition zones in the antibacterial assay against *Escherichia coli* and *Staphylococcus aureus* at a minimum inhibition concentration of 5 μg/ml (Mathai and Shaji 2023).

8.6 Polymer Nanocomposites

Polymer nanocomposites are one of the major categories of materials with diverse physical and chemical features with outstanding applications. Polymer nanocomposites with antifouling properties inhibit bacteria growth and are a novel alternative to antibiotics. Ghosal et al. investigated the antibacterial effectiveness of synthesized hydrophobic polymeric nanocomposite in gram-positive *Listeria monocytogenes, Staphylococcus aureus,* and gram-negative *Klebsiella pneumoniae* and *Escherichia coli.* Antibacterial photodynamic activity of carbon quantum dots and polycaprolactone nanocomposite demonstrated that they can create singlet oxygen at 470 nm and are effective for wound dressing (Ghosal et al. 2021). Sethy et al. investigated eco-friendly and less expensive polyacrylic acid/Graphene Oxide hybrid nanocomposites through a polymerization approach. Outstanding tensile properties, distinguished chemical resistance features, and thermally stable characteristics have been established for the synthesized nanocomposite and layered structure confirmed in HRTEM analysis. The antibacterial action of nanocomposite was tested against *Escherichia coli* and proved they can be used for packaging applications (Sethy et al. 2021).

Antimicrobial and degradable chitosan–polyaniline nanocomposite with remarkable electrical conductivity was studied by Mohammadi et al., and the polyaniline particle has a spherical shape at a size range from 45 to 100 nm. The antifungal and antibacterial activity of synthesized film was tested against *Aspergillus niger* fungi *Escherichia coli* bacteria and observed that the zone of inhibition increased with increasing polyaniline concentration and can be applicable in the food packaging industry (Fig. 8.9) (Mohammadi et al. 2019). Çimen et al. reported nanofiber from polycaprolactone (PCL) with polylactide acid (PLA) comprising silver nanostructures with *Thymus vulgaris L.* (thyme) extract as bacterial resistant and can be applied in the biomedical field. Hydrophobic biodegradable nanofiber structures at an average size of 280 nm exhibit high tensile strength properties and are bacterial resistant to *Escherichia coli* and *Staphylococcus aureus*. The toxic and anticancer characteristics were investigated using L929 and SK-MEL-30 cells. The utilization of *Thymus vulgaris L.* extract-encapsulated silver nanoparticle-doped PCL/PLA nanofibers generated by electrospinning offers a tool for cancer treatment in skin tumor cell lines (Güneş Çimen et al. 2022). Salahuddin et al. investigated nanocomposites, which can potentially be effectual carriers for the long-term delivery of hazardous anticancer drugs. An extended-release formulation of Norfloxacin antibiotic was produced, encapsulated, and loaded poly (lactic acid) nanocomposites-based Titanium dioxide (PLA/TiO2). The nanocomposites exhibit a high antibacterial capability against *Klebsiella pneumonia, Staphylococcus aureus, Escherichia coli, Pseudomonas aeruginosa,* and *Salmonella* (Salahuddin et al. 2020).

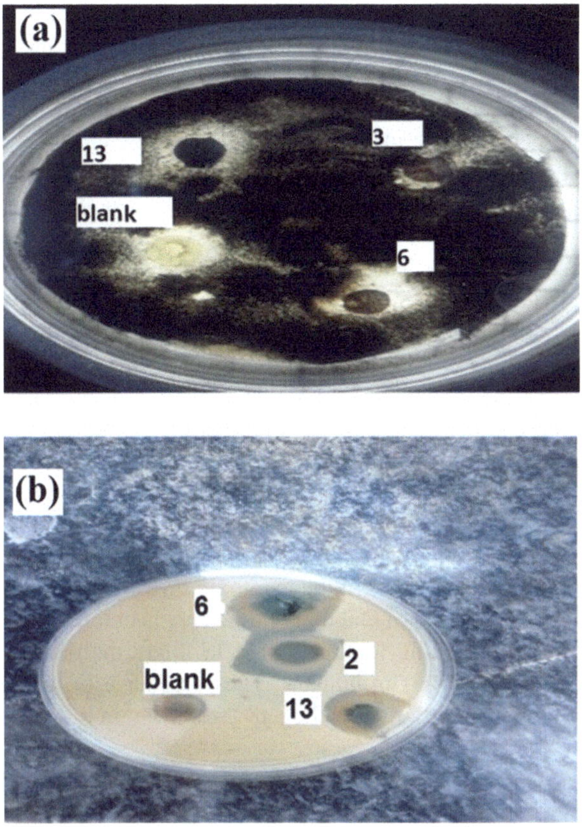

Fig. 8.9 Antifungal and antibacterial test against chitosan–polyaniline film on the development of (**a**) *Aspergillus niger* and (**b**) *Escherichia coli* (Mohammadi et al. 2019)

8.7 Ceramic Nanocomposites

Ceramic nanocomposites are ceramic fiber-reinforced materials and contain carbides, nitrides, and oxides in combination with metallic or nonmetallic elements, and they are utilized in the biomedical and therapeutic fields. Reda et al. synthesized calcium-doped zinc oxide ceramic nanostructures via a sol-gel approach, which are effective against the developing multidrug-resistant *Candida auris*. The ceramic nanomaterial calcined at 600 °C had particle sizes in the nanoscale range and exhibited negative charge in zeta potential. Furthermore, the nanoparticles were effective against gram-positive *Staphylococcus aureus* and gram-negative *E. coli* and can be applied in multidrug resistance (Reda and Fayed 2023). Kaliaraj et al. reported calcia stabilized zirconia (CaSZ) and silver-CaSZ nanocomposite and effectually inhibited the growth of *Pseudomonas aeruginosa*. The Alizarin red-based assay improved osteoblast-like cell adhesion by the synthesized nanocomposite and exhibited hemocompatible and biocompatible features. Hence, the silver-calcia stabilized zirconia nanocomposite synthesis becomes a potential bioimplant in the biomedical field. Zirconium dioxide (ZrO_2), zinc oxide (ZnO), and titanium oxide (TiO_2) nanocomposites on stainless steel via radio frequency sputtering method were prepared by Lee et al. The nanocomposite coating has shown remarkable corrosion shielding performance in body fluid and saliva by electrochemical impedance studies. The cell viability studies have been done in mouse fibroblasts, and the nanocomposite coating has shown better cellular attachment than the uncoated surface. The antibacterial action of synthesized samples was tried against *Staphylococcus aureus* and *Escherichia coli* bacteria (Lee et al. 2022). Hydroxyapatite/graphene oxide/zirconia oxide nanocomposite was investigated by Khalili et al. via electrophoretic deposition. The antibacterial test was done in nanocomposite by gram-positive *Staphylococcus aureus* and gram-negative *Escherichia coli,* and the test results verified that the topographical layer effectually reduces bacterial proliferation (Khalili et al. 2020). Ayanwale et al. reported the synthesis of ZrO_2-Ag_2O via the sol-gel method and assessed the antifungal effect against *Candida dubliniensis, Candida tropicalis, Candida albicans*, and *Candida glabrata* by disc diffusion technique. The average size ranges from 14 to 70 nm, having spherically shaped particles. The antifungal activity of ZrO_2-Ag_2O nanoparticles revealed that they were efficient against *Candida* species. In the cytotoxic investigation, there was no significant variation in viability as a function of time or concentration on human mononuclear cells. These intriguing findings could facilitate the development of new treatments for fungal illnesses in humans. The percentage of inhibition of ZrO_2, Ag_2O, and Z-A against different pathogens is shown in Fig. 8.10 (Ayanwale et al. 2021).

Rahmanpour et al. explored the structure for the first time and demonstrated its effectiveness in wound healing due to the necessity to treat bacterial-infected wound dressings. The Aluminum oxide (Al_2O_3)/ TiO_2 nanocomposites coated with chitosan prepared by fruit extract of oak and tested its wound-healing studies in mice compete with ointment mupirocin®. The spherical TiO_2 nanoparticles produced on the alumina sheet-like structure had a size of 4–6 nm. Biomedical experiments proved that the synthesized nanocomposite has beneficial effects on antibacterial death and, as a result, has a superior ability to treat bacterial infection compared with a commercial ointment (Rahmanpour et al. 2022). Almotiri et al. examined the antibacterial activity of synthesized alumina (Al_2O_3) and hydroxyapatite (HAP) and were entrenched through graphene oxide against *Staphylococcus aureus* and *Escherichia coli* bacteria. The cell viability assay was done on osteoblast cells and disclosed $98.6 \pm 1.3\%$ cell viability after treatment with the nanocomposite (Almotiri and Alkhamisi 2022). PMMA/Si3N4/TaC nanocomposites were formed via the casting technique potential tools for employment in diverse electrical and optical nanodevices by Mohammed et al. The nanocomposite displayed a zone of inhibition against *K. aerogenes* and *S. aureus* bacteria in a dose-dependent manner (Mohammed and Habeeb 2023).

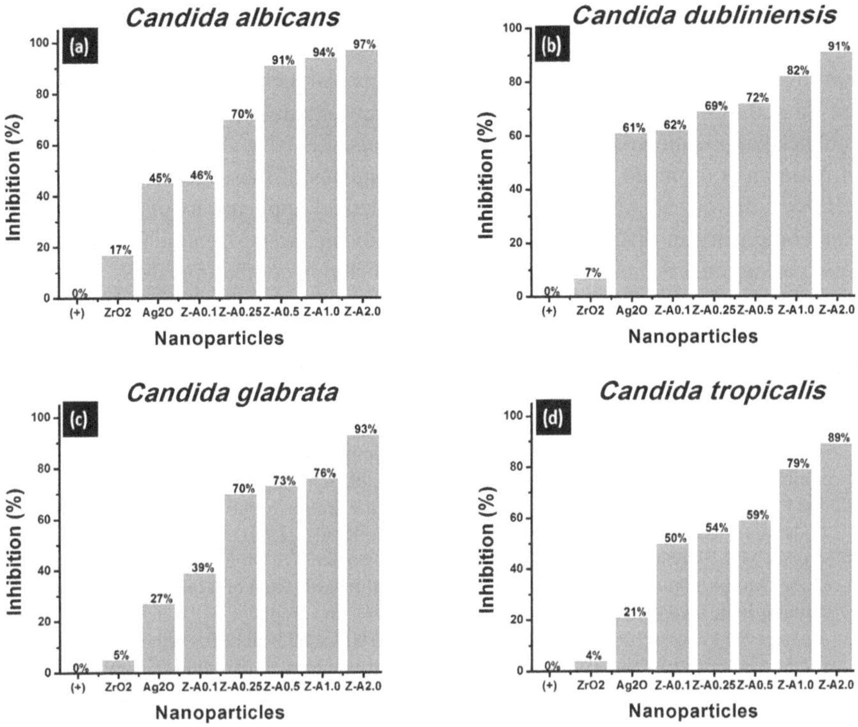

Fig. 8.10 The percentage of inhibition of ZrO_2, Ag_2O, and Z-A nanostructures against (**a**) *C. albicans*, (**b**) *C. dubliniensis*, (**c**) *C. glabrata*, and (**d**) *C. tropicalis* (Ayanwale et al. 2021)

8.8 Conclusion and Future Prospects

Challenging the public health sector by threatening infectious treatments and other routine medical interventions, AMR has become one of the most crucial global issues that demands immediate attention from the research community. To solve this problem, effective and economical strategies less prone to AMR must be developed and promoted. In this context, nanomaterials and their composites are hopeful materials with their potential and unique antibacterial properties, which are unlikely to develop AMR. Many nanocomposites have been reported with their antibacterial capability toward drug-resistant bacteria. Some of the materials have successfully been utilized to combat multidrug-resistant microbes. The synergistic effects in the composite materials enhanced bactericidal effects with improved properties desirable for medical applications. Specific applications of the nanocomposites as antibacterial coatings, medical utensils, textiles, etc., have been explored, which need to be investigated more for the extensive and fruitful application of these antibacterial agents. It is also relevant to ensure the long-term effects of the antibacterial nanocomposites along with the biosafety. The impacts of these nanocomposites on the environment and their probable side effects are also worth investigating. A potential path might be the creation of metallophores that firmly bind antibacterial nanocomposites but can be liberated in response to particular stimuli for the systematic administration of nanocomposite-based antimicrobial therapies. Most of the studies are limited to investigating limited types of microorganisms, which may affect the enhanced application of antimicrobial nanocomposites. One of the significant areas of research is to unveil the antibacterial mechanisms of these materials to provide reliable solutions to AMR. Integrating the practical applications of antimicrobial nanocomposites in diverse fields, such as healthcare, water treatment, agriculture, food packaging, etc., can be a rapid and radical step in combating AMR.

References

Adeniji OO, Ojemaye MO, Okoh AI (2022) Antibacterial activity of metallic nanoparticles against multidrug-resistant pathogens isolated from environmental samples: nanoparticles/antibiotic combination therapy and cytotoxicity study. ACS Appl Bio Mater 5(10):4814–4826

Ahmad I, Malak HA, Abulreesh HH (2021) Environmental antimicrobial resistance and its drivers: a potential threat to public health. J Glob Antimicrob Resist 27:101–111

Al Mogbel MS, Elabbasy MT, Abd El-Kader MFH, Mohamed RS, Moustapha ME, Menazea AA (2022) Morphological, mechanical, and antibacterial investigation of a ternary nanocomposite contains hydroxyapatite, tellurium(IV) oxide (Te2O4), and graphene oxide in vitro. Appl Phys A Mater Sci Process 128(4):301. https://doi.org/10.1007/s00339-022-05338-x

Alayande AB, Obaid M, Kim IS (2020) Antimicrobial mechanism of reduced graphene oxide-copper oxide (rGO-CuO) nanocomposite films: the case of Pseudomonas aeruginosa PAO1. Mater Sci Eng C 109:110596. https://doi.org/10.1016/J.MSEC.2019.110596

Almotiri RA, Alkhamisi MM (2022) Physico-chemical behavior and microstructural manipulation of nanocomposites containing hydroxyapatite, alumina, and graphene oxide. Appl Phys A Mater Sci Process 128(4):351. https://doi.org/10.1007/s00339-022-05479-z

Al-Shawi SG, Alekhina NA, Aravindhan S, Thangavelu L, Kartamyshev NV, Zakieva RR (2021) Synthesis of NiO nanoparticles and sulfur, and nitrogen co doped-graphene quantum dots/NiO nanocomposites for antibacterial application. J Nanostruct 11(1):181–188. https://doi.org/10.22052/JNS.2021.01.019

Altınöz E, Altuner EM (2019) Antibiotic resistance and efflux pumps. Int J Innov Res Rev 3(2):1–9. [Online]. Available: http://www.injirr.com/article/view/35

Arslan ME et al (2022) Structural, biocompatibility, and antibacterial properties of Ge–DLC nanocomposite for biomedical applications. J Biomed Mater Res Part B Appl Biomater 110(7):1667–1674

Arya PR et al (2023) Antioxidant, antibacterial and antibiofilm potential of green synthesized silver-zinc oxide nanocomposites from Curcuma longa extract against multi-drug-resistant enteroaggregative E. coli. Med Sci Forum, MDPI:23

Ayanwale AP, Estrada-Capetillo BL, Reyes-López SY (2021) Evaluation of antifungal activity by mixed oxide metallic nanocomposite against candida spp. PRO 9(5):773. https://doi.org/10.3390/pr9050773

Benko A et al (2021) Conductive all-carbon nanotube layers: results on attractive physicochemical, anti-bacterial, anticancer and biocompatibility properties. Mater Sci Eng C 120:111703

Biondo C (2023) Bacterial antibiotic resistance: the Most critical pathogens. Pathogens 12(1):1310. https://doi.org/10.3390/pathogens12010116

Cao Y, Javadhesari SM, Mohammadnejad S, Raise A, Akbarpour MR (2021) Microstructural characterization and antibacterial activity of carbon nanotube decorated with Cu nanoparticles synthesized by a novel solvothermal method. Ceram Int 47(18):25729–25737

Cao Y et al (2023) Composite hydrogel for the targeted capture and Photothermal killing of bacteria toward facilitating wound healing. Langmuir 39(18):6413–6424

Chinemerem Nwobodo D et al (2022) Antibiotic resistance: the challenges and some emerging strategies for tackling a global menace. J Clin Lab Anal 36(9):e24655. https://doi.org/10.1002/jcla.24655

Chu X, Yang F, Tang H (2022) Recent advance in polymer coatings combating bacterial adhesion and biofilm formation. Chin J Chem 40(24):2988–3000

Dadgostar P (2019) Antimicrobial resistance: implications and costs. Infect Drug Resist 12:3903–3910

Dehari D, Chaudhuri A, Kumar DN, Nath G, Agrawal AK (2023) Fiber and textile in drug delivery to combat multidrug resistance microbial infection. In: Fiber and textile engineering in drug delivery systems. Elsevier, pp 359–387

Dennis SA, Sudhamshu AV, Mathew AA, Abraham JE, Daniel EC, Manoj B (2022) Polymer-based nanomaterial as a bacteriostatic agent on gram-positive bacteria. In: International conference on nanotechnology: opportunities and challenges. Springer, pp 353–358

Devi AP, Padhi DK, Mishra PM, Behera AK (2021) Bio-surfactant assisted room temperature synthesis of cubic Ag/RGO nanocomposite for enhanced photoreduction of Cr (VI) and antibacterial activity. J Environ Chem Eng 9(2):104778. https://doi.org/10.1016/J.JECE.2020.104778

Dhandapani P et al (2021) Biological mediated synthesis of RGO-ZnO composites with enhanced photocatalytic and antibacterial activity. J Hazard Mater 409:124661. https://doi.org/10.1016/J.JHAZMAT.2020.124661

Du S et al (2021) A nanoporous graphene/nitrocellulose membrane beneficial to wound healing. ACS Appl Bio Mater 4(5):4522–4531

Egorov AM, Ulyashova MM, Rubtsova MY (2018) Bacterial enzymes and antibiotic resistance. Acta Nat 10(4):33–48. https://doi.org/10.32607/20758251-2018-10-4-33-48

El-Batal AI, El-Sayyad GS, Al-Hazmi NE, Gobara M (2019) Antibiofilm and antimicrobial activities of silver boron nanoparticles synthesized by PVP polymer and gamma rays against urinary tract pathogens. J Clust Sci 30(4):947–964. https://doi.org/10.1007/s10876-019-01553-4

Eskandari F, Mofidi H, Asheghi B, Mohammadi F, Gholami A (2023) Bringing resistance modulation to methicillin-resistant Staphylococcus aureus (MRSA) and vancomycin-resistant enterococci (VRE) strains using a quaternary ammonium compound coupled with zinc oxide nanoparticles. World J Microbiol Biotechnol 39(7):193

Fang HY, Huang WM, Chen DH (2019) One-step synthesis of positively charged bifunctional carbon dot/silver composite nanoparticles for killing and fluorescence imaging of gram-negative bacteria. Nanotechnology 30(36):365603. https://doi.org/10.1088/1361-6528/ab1fef

Frei A, Verderosa AD, Elliott AG, Zuegg J, Blaskovich MAT (2023) Metals to combat antimicrobial resistance. Nat Rev Chem 7(3):202–224

Fu S, Sun Z, Huang P, Li Y, Hu N (2019) Some basic aspects of polymer nanocomposites: a critical review. Nano Mater Sci 1(1):2–30

Gao Y et al (2021) Graphene oxide Nanosheets with efficient antibacterial activity against methicillin-resistant Staphylococcus aureus (MRSA). J Biomed Nanotechnol 17(8):1627–1634

Ghosal K, Kováčová M, Humpolíček P, Vajďák J, Bodík M, Špitalský Z (2021) Antibacterial photodynamic activity of hydrophobic carbon quantum dots and polycaprolactone based nanocomposite processed via both electrospinning and solvent casting method. Photodiagn Photodyn Ther 35:102455. https://doi.org/10.1016/j.pdpdt.2021.102455

Güneş Çimen C, Dündar MA, Demirel Kars M, Avcl A (2022) Enhancement of PCL/PLA electrospun nanocomposite fibers comprising silver nanoparticles encapsulated with thymus vulgaris L. molecules for antibacterial and anticancer activities. ACS Biomater Sci Eng 8(9):3717–3732. https://doi.org/10.1021/acsbiomaterials.2c00611

Haenni M et al (2022) Environmental contamination in a high-income country (France) by antibiotics, antibiotic-resistant bacteria, and antibiotic resistance genes: status and possible causes. Environ Int 159:107047

Hatae AC, Roque-Borda CA, Pavan FR (2023) Strategies for lipid-based nanocomposites with potential activity against Mycobacterium tuberculosis: Microbial resistance challenge and drug delivery trends. OpenNano 13:100171

Hussain S et al (2022) Green production and interaction of Carboxylated CNTs/biogenic ZnO composite for antibacterial activity. Bioengineering 9(9):437. https://doi.org/10.3390/bioengineering9090437

Ikram M et al (2022) Facile synthesis of starch and tellurium doped SrO nanocomposite for catalytic and antibacterial potential: in silico molecular docking studies. Int J Biol Macromol 221:496–507. https://doi.org/10.1016/j.ijbiomac.2022.09.034

Ji L, Qin X, Zheng J, Zhou S, Xu T, Shi G (2020) Synthesis of ag–carbon–TiO2 composite tubes and their antibacterial and organic degradation properties. J Sol-Gel Sci Technol 93(2):291–301. https://doi.org/10.1007/S10971-019-05197-8/FIGURES/12

Ji Y, Yang S, Sun J, Ning C (2023) Realizing both antibacterial activity and Cytocompatibility in Silicocarnotite bioceramic via germanium incorporation. J Funct Biomater 14(3):154. https://doi.org/10.3390/jfb14030154

Jiang Q, Luo B, Wu Z, Wang X (2020) Antibacterial composite paper with corn stalk-based carbon spheres immobilized AgNPs. Mater Sci Eng C 113:111012

Joseph N, Mathew AA, Daniel EC, Balachandran M (2023) Polymer-carbon nanocomposite: synthesis, optical and biocidal properties. Results Chem 5:100826

Kankala RK, Lin W-Z, Lee C-H (2020) Combating antibiotic resistance through the synergistic effects of mesoporous silica-based hierarchical nanocomposites. Nano 10(3):597

Karak N (2019) Fundamentals of nanomaterials and polymer nanocomposites. In: Nanomaterials and polymer nanocomposites: raw materials to applications. Elsevier, pp 1–45. https://doi.org/10.1016/B978-0-12-814615-6.00001-1

Karthik C, Swathi N, Pandi Prabha S (2020) Green synthesized rGO-AgNP hybrid nanocomposite—an effective antibacterial adsorbent for photocatalytic removal of DB-14 dye from aqueous solution. J Environ Chem Eng 8(1):103577. https://doi.org/10.1016/J.JECE.2019.103577

Khalili A, Naeimi F, Fakhrizadeh AA (2020) Electrodeposited hydroxyapatite/graphene oxide/zirconia oxide composite coatings: characterization and antibacterial activity. Adv Ceram Prog 6(4):8–14

Kiranakumar HV et al (2022) A review on electrical and gas-sensing properties of reduced graphene oxide-metal oxide nanocomposites. Biomass Convers Biorefinery:1–11. https://doi.org/10.1007/s13399-022-03258-7

Kumar S, Ojha AK, Bhorolua D, Das J, Kumar A, Hazarika A (2019) Facile synthesis of CuO nanowires and Cu2O nanospheres grown on rGO surface and exploiting its photocatalytic, antibacterial and supercapacitive properties. Phys B Condens Matter 558:74–81. https://doi.org/10.1016/J.PHYSB.2019.01.040

Larsson DGJ, Flach C-F (2022) Antibiotic resistance in the environment. Nat Rev Microbiol 20(5):257–269

Lee M et al (2022) ZrO2/ZnO/TiO2 nanocomposite coatings on stainless steel for improved corrosion resistance, biocompatibility, and antimicrobial activity. ACS Appl Mater Interfaces 14(11):13801–13811. https://doi.org/10.1021/acsami.1c19498

Li C, Sun W, Lu Z, Ao X, Li S (2020a) Ceramic nanocomposite membranes and membrane fouling: a review. Water Res 175:115674

Li D et al (2020b) Bimetallic CuCo2S4 nanozymes with enhanced peroxidase activity at neutral ph for combating burn infections. Chembiochem 21(18):2620–2627

Li J et al (2021) Near infrared photothermal conversion materials: mechanism, preparation, and photothermal cancer therapy applications. J Mater Chem B 9(38):7909–7926

Mathai S, Shaji PS (2023) Anti-inflammatory effects of tellurium dioxide-niobium pentoxide nanocomposite coatings on titanium implants. Mater Today Proc

Mathew AA, Varghese M, Balachandran M (2023) Biosafety and toxicity evaluation of carbon nanomaterials. In: Carbon nanostructures in biomedical applications. Springer, pp 363–398

Mickymaray S (2019) One-step synthesis of silver nanoparticles using Saudi Arabian desert seasonal plant Sisymbrium irio and antibacterial activity against multidrug-resistant bacterial strains. Biomol Ther 9(11):662

Mohammadi B, Pirsa S, Alizadeh M (2019) Preparing chitosan–polyaniline nanocomposite film and examining its mechanical, electrical, and antimicrobial properties. Polym Polym Compos 27(8):507–517. https://doi.org/10.1177/0967391119851439

Mohammed AA, Habeeb MA (2023) Modification and development of the structural, optical and antibacterial characteristics of PMMA/Si3N4/TaC nanostructures. SILICON 15:1–12. https://doi.org/10.1007/s12633-023-02426-2

Mohanty H, Pachpute S, Yadav RP (2021) Mechanism of drug resistance in bacteria: efflux pump modulation for designing of new antibiotic enhancers. Folia Microbiol (Praha) 66(5):727–739. https://doi.org/10.1007/s12223-021-00910-z

Mukheem A et al (2019) Boron nitride doped polyhydroxyalkanoate/chitosan nanocomposite for antibacterial and biological applications. Nano 9(4):645. https://doi.org/10.3390/nano9040645

Munawar T et al (2022) Transition metal-doped SnO2 and graphene oxide (GO) supported nanocomposites as efficient photocatalysts and antibacterial agents. Environ Sci Pollut Res:1–22

Naseem T et al (2020) Reduced graphene oxide/zinc oxide nanocomposite: from synthesis to its application for wastewater purification and antibacterial activity. J Inorg Organomet Polym Mater 30:10. https://doi.org/10.1007/s10904-020-01529-2

Omanović-Mikličanin E, Badnjević A, Kazlagić A, Hajlovac M (2020) Nanocomposites: a brief review. Health Technol (Berl) 10:51–59

Pan N, Okonjo PA, Wang P, Tang Y, Sun YP, Yang L (2020) Optical and photodynamic properties of carbon/TiO2 hybrid dots in different nanoscale configurations. Chem Phys Lett 743:137208. https://doi.org/10.1016/j.cplett.2020.137208

Pandiyan R, Mahalingam S, Ahn YH (2019) Antibacterial and photocatalytic activity of hydrothermally synthesized SnO 2 doped GO and CNT under visible light irradiation. J Photochem Photobiol B Biol 191:18–25. https://doi.org/10.1016/j.jphotobiol.2018.12.007

Pietsch F et al (2020) Selection of resistance by antimicrobial coatings in the healthcare setting. J Hosp Infect 106(1):115–125

Pyrzynska K (2020) Nanomaterials in speciation analysis of metals and metalloids. Talanta 212:120784

Qiao Y et al (2020) Treatment of MRSA-infected osteomyelitis using bacterial capturing, magnetically targeted composites with microwave-assisted bacterial killing. Nat Commun 11(1):4446

Rahmanpour A, Farahpour MR, Shapouri R, Jafarirad S, Rahimi P (2022) Synthesis and characterization of alumina-based nanocomposites of TiO2/Al2O3/chitosan with antibacterial properties accelarate healing of infected excision wounds. Colloids Surf A Physicochem Eng Asp 644:128839. https://doi.org/10.1016/j.colsurfa.2022.128839

Rajaramon S et al (2023) Multi-functional approach in the design of smart surfaces to mitigate bacterial infections: a review. Front Cell Infect Microbiol 13:587

Reda AE, Fayed B (2023) The synthesis of calcium doped zinc oxide ceramic nanoparticles via sol–gel effective against the emerging multidrug-resistant Candida auris. J Aust Ceram Soc 59:1–9

Rifai A et al (2019) Engineering the interface: nanodiamond coating on 3D-printed titanium promotes mammalian cell growth and inhibits Staphylococcus aureus colonization. ACS Appl Mater Interfaces 11(27):24588–24597

Riley PR, Narayan RJ (2021) Recent advances in carbon nanomaterials for biomedical applications: a review. Curr Opin Biomed Eng 17:100262

Sadhukhan S et al (2019) Green synthesis of cadmium oxide decorated reduced graphene oxide nanocomposites and its electrical and antibacterial properties. Mater Sci Eng C 99:696–709. https://doi.org/10.1016/J.MSEC.2019.01.128

Sajjad S, Arshad F, Uzair B, Leghari SAK, Noor S, Maaza M (2019) GO/Ag2O composite nanostructure as an effective antibacterial agent. ChemistrySelect 4(35):10365–10371

Salahuddin N, Abdelwahab M, Gaber M, Elneanaey S (2020) Synthesis and Design of Norfloxacin drug delivery system based on PLA/TiO2 nanocomposites: antibacterial and antitumor activities. Mater Sci Eng C 108:110337. https://doi.org/10.1016/j.msec.2019.110337

Schaenzer AJ, Wright GD (2020) Antibiotic resistance by enzymatic modification of antibiotic targets. Trends Mol Med 26(8):768–782. https://doi.org/10.1016/j.molmed.2020.05.001

Sethy PK, Mohapatra P, Patra S, Bharatiya D, Swain SK (2021) Antimicrobial and barrier properties of polyacrylic acid/GO hybrid nanocomposites for packaging application. Nano-Structures Nano-Objects 26:100747. https://doi.org/10.1016/j.nanoso.2021.100747

Shameem MM, Sasikanth SM, Annamalai R, Raman RG (2021) A brief review on polymer nanocomposites and its applications. Mater Today Proc 45:2536–2539

Singh A, Sharma K, Sharma M, Modi SK, Nenavathu BP (2023) Novel tellurium doped CeO2 nano wools as a next generation antibacterial therapeutic agent. Mater Chem Phys 307:128172. https://doi.org/10.1016/j.matchemphys.2023.128172

Sivam V, Rangasamy A, Dara PK (2023) Nanoparticle approach to control AMR. In: Handbook on antimicrobial resistance: current status, trends in detection and mitigation measures. Springer, pp 925–946

Su K et al (2020) Rapid photo-sonotherapy for clinical treatment of bacterial infected bone implants by creating oxygen deficiency using sulfur doping. ACS Nano 14(2):2077–2089

Tan S, Wu X, Xing Y, Lilak S, Wu M, Zhao JX (2020) Enhanced synergetic antibacterial activity by a reduce graphene oxide/Ag nanocomposite through the photothermal effect. Colloids Surf B Biointerfaces 185:110616. https://doi.org/10.1016/J.COLSURFB.2019.110616

Taneja N, Sharma M (2019) Antimicrobial resistance in the environment: the Indian scenario. Indian J Med Res 149(2):119

Tang KWK, Millar BC, Moore JE (2023) Antimicrobial Resistance (AMR). Br J Biomed Sci 80:11387

Uddin TM et al (2021) Antibiotic resistance in microbes: history, mechanisms, therapeutic strategies and future prospects. J Infect Public Health 14(12):1750–1766. https://doi.org/10.1016/j.jiph.2021.10.020

ur Rehman S, Niazi RK, Zulqurnain M, Mansoor Q, Iqbal J, Arshad A (2022) Graphene nanoplatelets/CeO2 nanotiles nanocomposites as effective antibacterial material for multiple drug-resistant bacteria. Appl Nanosci 12:1–12. https://doi.org/10.1007/s13204-022-02422-9

Vanitha M, Joni IM, Camellia P, Balasubramanian N (2018) Tailoring the properties of cerium doped zinc oxide/reduced graphene oxide composite: characterization, photoluminescence study, antibacterial activity. Ceram Int 44(16):19725–19734. https://doi.org/10.1016/J.CERAMINT.2018.07.226

Varghese M, Balachandran M (2021) Antibacterial efficiency of carbon dots against gram-positive and gram-negative bacteria: a review. J Environ Chem Eng 9(6):106821. https://doi.org/10.1016/J.JECE.2021.106821

Varghese M, Kootery KP, Sarojini S, Balachandran M (2022) Antibacterial effect of phosphorous-doped carbon nanomaterial derived from Alstonia Venenata. In: International conference on nanotechnology: opportunities and challenges. Springer, pp 327–331

Veeman D et al (2021) Sustainable development of carbon nanocomposites: synthesis and classification for environmental remediation. J Nanomater 2021:1–21

Wang L et al (2020) Mercaptophenylboronic acid-activated gold nanoparticles as nanoantibiotics against multidrug-resistant bacteria. ACS Appl Mater Interfaces 12(46):51148–51159

Wu H et al (2020) Self-detoxifying hollow zinc silica nanospheres with tunable Ag ion release-recapture capability: a nanoantibiotic for efficient MRSA inhibition. Compos Part B Eng 202:108415

Xi J et al (2019) Light-enhanced sponge-like carbon nanozyme used for synergetic antibacterial therapy. Biomater Sci 7(10):4131–4141

Yang Y et al (2021) Single-atom doping in carbon black nanomaterials for photothermal antibacterial applications. Cell Rep Phys Sci 2(8):100535

Zafar N et al (2022) Moringa concanensis-mediated synthesis and characterizations of ciprofloxacin encapsulated into ag/TiO2/Fe2O3/CS nanocomposite: a therapeutic solution against multidrug Resistant E. Coli strains of livestock infectious diseases. Pharmaceutics 14:8. https://doi.org/10.3390/pharmaceutics14081719

Zahran M, Marei AH (2019) Innovative natural polymer metal nanocomposites and their antimicrobial activity. Int J Biol Macromol 136:586–596

Zhang W et al (2019) One-step sonochemical synthesis of a reduced graphene oxide—ZnO nanocomposite with antibacterial and antibiofouling properties. Environ Sci Nano 6(10):3080–3090. https://doi.org/10.1039/C9EN00753A

Zhao C et al (2019) Nitrogen-doped carbon quantum dots as an antimicrobial agent against Staphylococcus for the treatment of infected wounds. Colloids Surf B Biointerfaces 179:17–27. https://doi.org/10.1016/j.colsurfb.2019.03.042

Zheng H et al (2019) Engineered graphene oxide nanocomposite capable of preventing the evolution of antimicrobial resistance. ACS Nano 13:10. https://doi.org/10.1021/acsnano.9b04970

Zhu W, Mi J, Fu Y, Cui D, Lü C (2021) Multiple-cores@ shell clustered carbon dots/P25/rGO nanocomposite as robust visible-light photocatalyst for organic pollutant degradation and water disinfection. Appl Surf Sci 538:148087

Carbon Nanomaterials as Antimicrobial Agents to Combat Multidrug Resistance

9

Iram Saba, Kaiser Wani, Khalid Mujasam Batoo, Suriya Rehman, and Saif Hameed

Abstract

Microorganisms have developed resistance to common antimicrobial medications because of incorrect and excessive usage, demanding innovative strategies to combat microbial infections. This chapter aims to explore the potential of carbon nanomaterials as promising antimicrobial agents in overcoming multidrug resistance. The discussion begins by introducing the complexities of multidrug resistance, emphasizing the urgency for novel therapeutic approaches. The subsequent sections provide an exhaustive examination of carbon nanomaterials, highlighting the distinctive characteristics and structural variety of carbon nanostructures such as fullerenes, carbon nanotubes, and graphene oxides. In this chapter, we thoroughly detailed the antibacterial activity and method of working with carbon nanoparticles. The integrity of microorganisms' cellular membranes, metabolic activity, and shape are drastically affected when they are close to car-

I. Saba
Research and Scientific Center, Sultan Bin Abdulaziz Humanitarian City, Riyadh, Saudi Arabia

Amity Institute of Biotechnology, Amity University Haryana, Gurugram (Manesar), India

K. Wani
Department of Biochemistry, College of Science, King Saud University, Riyadh, Saudi Arabia

University Institute of Biotechnology, Chandigarh University, Mohali, India

K. M. Batoo
King Abdullah Institute for Nanotechnology, King Saud University, Riyadh, Saudi Arabia

S. Rehman
Department of Epidemic Disease Research, Institute of Research and Medical Consultations (I.R.M.C.), Imam Abdulrahman Bin Faisal University, Dammam, Saudi Arabia

S. Hameed (✉)
Amity Institute of Biotechnology, Amity University Haryana, Gurugram (Manesar), India

M. Y. Wani et al. (eds.), *Nanotechnology Based Strategies for Combating Antimicrobial Resistance*, https://doi.org/10.1007/978-981-97-2023-1_9

bon nanostructures. Carbon-based nanostructures can be assessed for antibacterial activity because of their substantial surface-to-volume ratio, huge interior volume, and other distinctive chemical and physical features. These carbon nanostructures demonstrate a range of antibacterial activities through membrane disruption, oxidative stress induction, or other mechanisms. In conclusion, the chapter emphasizes the need to develop novel treatment strategies for multidrug-resistant bacterial infections, and in that context, breakthroughs in carbon nanomaterials herald a new era in the search for effective antimicrobial drugs.

Keywords

Carbon nanotubes · Graphene oxide · Nanomaterials · Multidrug resistance · Antibiotics

9.1 Introduction

More than 75 years have passed since antibacterial agents were first used in their modern form. Their continued ubiquity may be attested to by the abundance of new drugs produced and distributed throughout the 1960s and 1970s. The decline in the effectiveness of antimicrobials against bacteria and, by extension, their capacity to treat diseases caused by bacteria are a major cause for worry in today's medical practice. According to a statement that was released by the General Assembly of the United Nations in September 2016, it is conceivable to expect that by the year 2050, illnesses that cannot be treated and are caused by microorganisms that are resistant to several drugs will be the primary cause of death (O'Neill 2016). "Multidrug-resistant (MDR), extensively drug-resistant (XDR), and pan-drug-resistant (PDR) bacteria" have all emerged because of newly identified resistance mechanisms (O'Neill 2016). Pan-drug-resistant bacteria are hazardous since they resist almost all antibiotics (Basak et al. 2016). The three mechanisms of antibiotic resistance that are most common are the mutation of antibiotic binding sites, the alteration of enzymes that are responsible for antibiotic metabolism, and changes in membrane permeability (caused by efflux transporters and porins). The synthesis of enzymes by bacteria is an essential factor in the development of antibiotic resistance because enzymes irreversibly change antibiotics.

Lactamases, nucleotidyltransferases, and acetyltransferases are some examples of enzymes that fall within this category (Schaenzer and Wright 2020; Santajit and Indrawattana 2016). Bacteria may regulate the amount of medicine that accumulates inside their cells by changing their membranes in various ways, including producing porins and efflux pumps (Munita and Arias 2016). Porins are channels of protein present in Gram-negative bacteria membranes that enable hydrophilic molecules to flow through passively (Reygaert 2018). Porins may be identified by their characteristic porin shape. Changing their appearance may confer some resistance to certain drugs. Furthermore, since efflux pumps rapidly remove the antibiotic from the cell, the medicine concentration is never high enough to impact the bacteria. Finally, certain bacteria may acquire antibiotic resistance by preventing the medication from reaching its intended target (Li et al. 2015).

According to a report from the World Health Organization (ESKAPE) in 2016 (Santajit and Indrawattana 2016), the most harmful bacteria are "*Enterococcus faecium, Staphylococcus aureus, Klebsiella pneumoniae, Acinetobacter baumannii, Pseudomonas aeruginosa,* and other species of *Enterobacter*." Because these bacteria have evolved mechanisms that allow them to withstand several therapies and even transfer their resistance to other species, new pharmaceuticals must be produced to battle them as soon as possible. In addition, infections, particularly those caused by bacteria resistant to many medicines, make the development of new antibiotics essential (Worthington and Melander 2013). To make matters even more problematic, bacteria can change themselves and acquire antibiotic resistance almost immediately as they are brought into a clinical environment (Peterson and Kaur 2018).

Nanostructured materials are now being created to overcome these resistances and others. Nanostructured materials can transport antimicrobial agents, deliver novel drugs, or act as antimicrobial agents. Because they possess antibacterial capabilities, nanoparticles (whether metallic, organic, or made of carbon nanotubes) can overcome drug resistance mechanisms in bacteria and impede biofilm formation and other critical functions. Antimicrobials and nanoparticles derived from plants are also being investigated as potential means of toxicity reduction. To combat bacteria resistant to many drugs, nanoparticles and antimicrobials derived from natural sources (or other compounds that have been repurposed) might potentially impair "bacterial efflux pumps, biofilm development, quorum sensing, and perhaps plasmid curing" (Baptista et al. 2018).

Furthermore, carbon-based nanoparticles have been found to have significant antibacterial properties. Initial investigations revealed that "fullerenes, single-walled carbon nanotubes (SWCNTs), and graphene oxide (GO) nanoparticles have potent microbicidal properties" (Wang et al. 2014). These "allotropic forms of Carbon" were discovered over the past two decades and have since been used in various scientific disciplines (Wang et al. 2014).

Additionally, it has been demonstrated that carbon nanomaterials' size and surface area are essential characteristics that determine their antibacterial activity. Expanding the nanoparticles' total surface area by decreasing their size boosted their capacity for interacting with bacteria (Kang et al. 2008). Nanoparticles' antibacterial efficacy may be influenced by various factors, including their structure, surface modification, inherent qualities, and the species of microorganisms they are exposed to (Buzea et al. 2007). Carbon nanoparticles are thought to cause oxidative stress, which destroys bacterial membranes (Hajipour et al. 2012; Gurunathan et al. 2012; Shvedova et al. 2012; Vecitis et al. 2010). Recent investigations (Manke et al. 2013; Pacurari et al. 2012) show that the considerable antimicrobial activity displayed by carbon-based nanomaterials is due to physical contact between these nanoparticles and bacteria rather than oxidative stress. This contradicts the commonly held idea that the antibacterial capacity of nanomaterials made from carbon is caused by oxidative stress. The chemical reactions between microbial cells and carbon-based nanoparticles determine how carbon-based nanoparticles hinder bacterial growth. Researchers discovered a link between the agglomeration of cells from bacteria with carbon nanotubes and the immediate interaction among the living things and the nanomaterials, which may result in cell death (Kalaskar and Venkataramana 2012). This chapter provides a succinct overview of the antibacterial properties of carbon-based nanoparticles, in addition to a discussion of the fundamental mechanism that underpins this activity.

9.2 Multidrug Resistance

Multidrug resistance in microorganisms, including bacteria, fungi, parasites, and viruses, is a serious concern for healthcare workers (Saba et al. 2022; Fatima et al. 2023). This opposition renders treating sufferers more difficult, effectively raising morbidity and fatality rates. The lengthy drug research and clinical testing procedures contribute to the scarcity of new antimicrobial treatments. In recent years, just a few new agents have been authorized for usage. The rise and spread of germs resistant to various antimicrobial treatments, dubbed "superbugs," is a major source of worry. Pathogen mutations and the overuse of antimicrobials predominantly cause this (Purkait et al. 2021). Methicillin-resistant existence of vancomycin-resistant bacteria *Staphylococcus aureus* is a problem that demands attention. Pathogens Enterococci, Staphylococci, and Streptococci can cause notoriously difficult-to-treat infections. Several mechanisms can be used to classify how bacteria become resistant to antibiotics. These include modifying or deactivating the antibiotic, altering the antibiotic's target site (such as penicillin-binding proteins), modifying metabolic pathways to avoid the antibiotic's effects, decreasing drug accumulation by increasing clearance, and producing substances that inhibit antibiotics. Various investigations (Ramana et al. 2013a; Sarada et al. 2014; Ramana and Rao 2009) have documented these mechanisms.

Due to their extensive and diverse characteristics, nanoparticles have become an indispensable vector for drug delivery targeting disease-causing pathogens. Due to their nanoscale dimension and relatively large surface area, their functional profile in the field of drug delivery is unique. Nanoparticles provide a new avenue that can be explored for the development of innovative therapies that can combat infections caused by MDR strains. Several studies indicate that inefficient membrane transfer must be surmounted to enhance conventional antibiotics' efficacy. However, since endocytosis allows nanoparticle-loaded pharmaceuticals to cross the membrane, nanoparticles increase the efficacy of these therapies. Additionally, nanoparticles can assist in membrane penetration by employing alternative pathways. Due to their interactions with the lipids on the cell's surface, adding gold nanoparticles to protein-based drugs, for instance, can increase membrane penetration (Nidhin et al. 2019). Because nanoparticles provide physical protection against the mechanisms of bacterial resistance, their therapeutic potential has been significantly increased. In addition, the possibility of incorporating multiple medication combinations into nanoparticles provides an extraordinarily complex mode of action, thereby reducing the likelihood that microbes will develop resistance to the drugs (Ramana et al. 2013b; Kang et al. 2007).

Carbon nanomaterials or CNMTs, have been investigated as a possible new class of antibiotics throughout the last decade. Functionalized carbon nanoparticles and nanocomposites have been shown to have antibacterial action. Carbon nanocomposites, including nanotubes and fullerenes, and graphene, in addition to analogues from these materials, have shown substantial promise as antibacterial agents. The large surface area per volume of CNMTs contributes significantly to their antibacterial activity. This ratio promotes easy and effective adhesion to microorganisms, which results in antimicrobial action through various mechanisms, including damage to the physiological cell membrane, oxidative stress, disruption of bacterial enzyme function, and influence on bacterial respiration.

9.3 Carbon Nanomaterials

Carbon nanostructures (CNSs) and their analogs have attracted a significant amount of attention among the many different nanomaterials because of the unique characteristics that they possess as well as the likelihood that they could be used in an array of both current and future uses that are just getting started (Peng et al. 2012). This is because carbon nanostructures (CNSs) and their analogs have the potential to be used in a broad range of present and future applications, which are only getting started. Figure 9.1 shows how carbon may uniquely join to one another. This might lead to the development of very low-dimensional structures like "fullerenes (0D), nanotubes (1D), graphene sheets (2D), and diamond-like carbon (3D)" (Peng et al. 2012). Carbon nanostructures are well known for their exceptional electrical conductivity, superior mechanical strength, high thermal conductivity, enormous surface area, photoluminescent capabilities (Li et al. 2011), outstanding transparency, and structural stability (Cao and Rogers 2009). All these features distinguish carbon nanostructures from other materials. Furthermore, CNSs have a vast surface area. Carbon nanoarchitectures offer enormous promise for a wide range of applications due to their one-of-a-kind features. "A thin layer semiconductor (Choi et al. 2010), transparent conducting electrodes (Wu et al. 2013), photovoltaics (Hecht et al. 2011), supercapacitors (Wang et al. 2008), biosensors (Seo et al. 2013), drug delivery (Seo et al. 2015), tissue engineering (Stancu et al. 2014), and photothermal treatment (Goenka et al. 2014; Yang et al. 2010) are some of these applications." CNSs' bactericidal activity often relies on the involvement of both physical and chemical processes (Dizaj et al. 2015). CNSs can significantly destroy the membranes and cell walls of the bacterium with which they have come into contact. Carbon nanostructures, such as graphene sheets, may physiologically separate cells from the microenvironments, potentially resulting in cell death (Prasad et al. 2017). Chemical interactions between CNSs and microbe surfaces may damage compounds, including reactive oxygen compounds (ROS) and oxidative damage in the cell. "Interactions between CNSs and organisms may result in an electron transfer phenomenon in which electrons are progressively emptied from the outer surface of

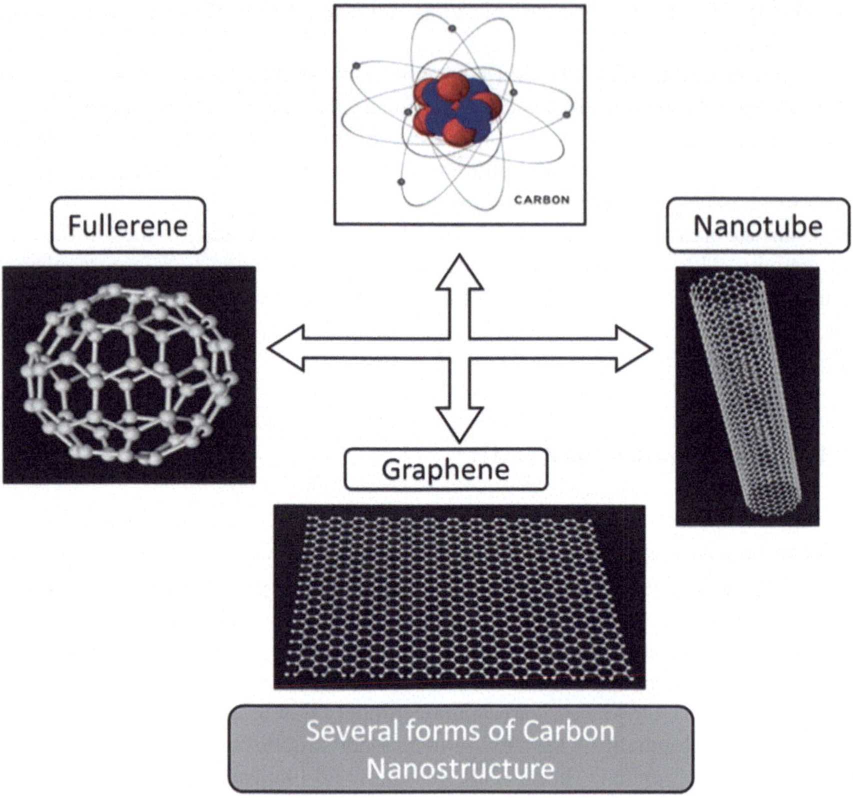

Fig. 9.1 Several forms of carbon nanostructures

microorganisms, causing ROS-independent oxidative stress and mortality" (Al-Jumaili et al. 2017). This may be a consequence of the process known as electron transfer.

Carbon nanostructures and their several forms are shown in Fig. 9.1.

9.3.1 Structure and Properties of Carbon Nanostructures

The structures of CNMs range from 0D "(Nanodiamonds, fullerenes) to 2D (graphene, graphene oxide, reduced graphene oxide)" and beyond (Bortolamiol et al. 2014). "Hydrophobicity/hydrophilicity, surface charge, electrical, optical, and usually mechanical properties of carbon nanomaterials" may be altered by grafting various surface functional groups onto them (Bortolamiol et al. 2014). The most prevalent pathway due to purifying operations is simple oxidation, which results in the attachment of oxygen-containing functional groups (Bortolamiol et al. 2014). Nanodiamonds are 0D nanostructures of sp3 hybridized carbon atoms at their core, usually surrounded by sp2 carbon atoms and functional groups (Shenderova and

Gruen 2012). Their "specific surface area is between 200 and 600 m^2/g" (Zeiger et al. 2015). They also have excellent mechanical and optical qualities. The sp3/sp2 ratio determines their qualities. Graphene is a two-dimensional substance having an atomic layer structure that is planar. It has the same basic structure as fullerene and carbon nanotubes (CNTs) (Xin et al. 2019), consisting of hexagonal arrays of sp2-hybridized carbon atoms. Several stacks graphene Few-layer Graphene (FLG) is made up of two to ten graphene layers, while graphite nanoplates are made up of 10 or more graphene sheets. Graphene oxide (GO) is a single layer of graphene with many oxygenation functionalities that may be reduced to produce reduced graphene oxide (rGO). With a theoretical specific surface area of 2600 m^2, graphene possesses outstanding physicochemical qualities and is very hydrophobic. It is insulating yet extremely hydrophilic substance that is simpler to spread than its nonfunctionalized version due to its g-1, strong electrical mobility, and mechanical characteristics (Shearer et al. 2016).

"Fullerenes are spherical 0D nanostructures composed mostly of sp2 hybridized carbon atoms (C60, C70, C76, C78, C84, and so on)" (Khandelwal et al. 2019). Their structure and diameter affect fullerenes' optical, mechanical, chemical, and electrical capabilities (Khandelwal et al. 2019). Finally, a carbon nanotube (also known as a CNT) may be considered a graphene sheet that has been coiled up and is generally closed by hemi fullerenes. "Single-Wall Carbon Nanotubes, or SWCNTs, have a diameter of a few nanometers or less" (Khandelwal et al. 2019). Based on the structure of their electronic molecules, their behavior may be either semiconductor or metal. The outer diameter of Multiwall Carbon Nanotubes (MWCNTs), made up of concentric arrays of SWCNTs, may reach 100 nm (Yang et al. 2019). "Intrinsic variables like a substance's diameter, length, number of walls, and functional groups, like extrinsic factors," play an important part in defining its qualities (Yang et al. 2019).

9.3.2 Carbon Nanomaterials and their Antimicrobial Mechanisms

As shown in Fig. 9.2, many ways have been proposed as plausible explanations for carbon nanoparticles' antibacterial activity. One of the most important chemical pathways is oxidative stress, which is caused by increased production of reactive oxygen species (ROS). It is also one of the most prevalent forms of stress. ROS are oxygen and free radical-based compounds that induce enzyme deactivation and oxidative damage to cell components such as lipids, proteins, and nucleic acids (Kehrer and Klotz 2015). Furthermore, carbon nanostructures that readily take electrons can interact with bacterial membranes through the electron transfer mechanism (Panda et al. 2018). There are physical processes, as has been stated several times. Using nanomaterials with piercing boundaries, such as graphene, to break down and penetrate the bacterial cell wall and induce the release of intracellular chemicals is known as "sharp edge insertion," and it is one of the techniques classified as "sharp edge insertion techniques." Another technique that might induce membrane rupture

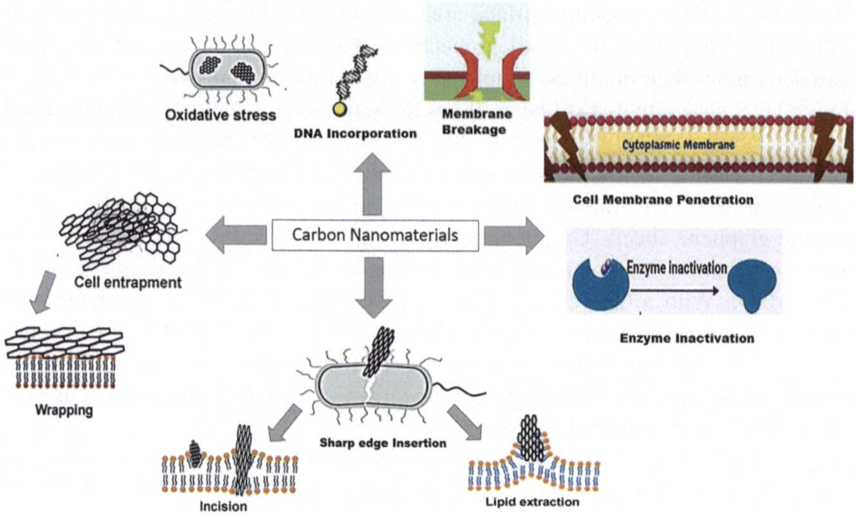

Fig. 9.2 Bactericidal action of carbon nanomaterials

is when an organism touches a bacterium's membrane (Li et al. 2014). Another technique is cell membrane penetration, followed by lipid extraction from the host cell.

As a result, the nanomaterial sheets may pass through the lipid bilayer and enter the bacterial cell. Furthermore, the nanomaterial may be causing dysfunction by interfering with membrane protein interactions (Luan et al. 2015). Damage to a cell's DNA, like membrane damage, may result in cell death. Furthermore, cell entrapment or wrapping, which happens when bacteria are contained inside the carbon nanomaterial and removed from their developing media (Akhavan and Ghaderi 2010), restricts bacterial growth. Furthermore, certain carbon nanoparticles have a deleterious influence on cell wall development. Some of them, for example, attach fresh components that might have been introduced to the peptidoglycan layer within the cell wall, preventing the layer from growing.

Figure 9.2 depicts the bactericidal action of carbon nanomaterials.

9.3.2.1 Antimicrobial Properties of Carbon Nanotubes

Carbon nanotubes (CNTs) are one of the most exciting novel nanostructures (Varghese et al. 2013), demonstrating a wide range of exceptional electrical, thermal, and mechanical capabilities despite their small size and high aspect ratio (>1000). CNTs may also permeate bacterial cell membranes and have significant antibiotic activity. Consequently, CNTs are regarded as a potential material for use in a variety of sectors, including nanoscale electronics (Ibrahim 2013), tissue engineering (Elsehly et al. 2016), polymer composites (Elsehly et al. 2017), and effluent treatment filters (Elsehly et al. 2018; Saba et al. 2021). Furthermore, one of the most important study topics in biomedical nanoparticle applications is the impact of CNTs in the biological environment (Taylor and Webster 2011). As a result, carbon nanotubes (CNTs) have sparked growing interest in the biomedical profession for their potential diagnostic and therapeutic uses. Carbon nanotubes (CNTs) can be diagnostic and therapeutic tools (Ji et al. 2010). Pathogenic organisms, including bacteria, are vulnerable to CNTs' antibacterial activity, varied mechanisms, and cytotoxic effects (Olivi et al. 2013). The "diameter, length, aggregation, concentration, surface functional groups, buffer solution, contact time, intensity, and cell type" have all been shown to have a major impact on influencing how CNTs respond in the presence of bacteria (Kang et al. 2009). Figure 9.3 depicts the mechanism of antimicrobial activity of carbon nanotubes.

C.NTs are to blame for damaging the microorganism's cell membrane and cell wall (Prasad et al. 2017). Furthermore, CNTs may cause oxidative stress in cells, eventually leading to the organism's extinction. Because of its excellent "electrical conductivity (Abdulhameed et al. 2021), high transparency, and structural stability (Aravind et al. 2011)," the material's antibacterial activity rises as it approaches the nanoscale range. This is due to their higher surface area-to-volume ratio. Only two of the various germs that may cause serious illnesses in humans are "*Escherichia coli* (gram-negative bacteria) and *Staphylococcus aureus*" (gram-positive bacteria) (Wu et al. 2014). Antibiotics and other chemotherapeutic agents are examples of antimicrobial medications often used to treat these illnesses.

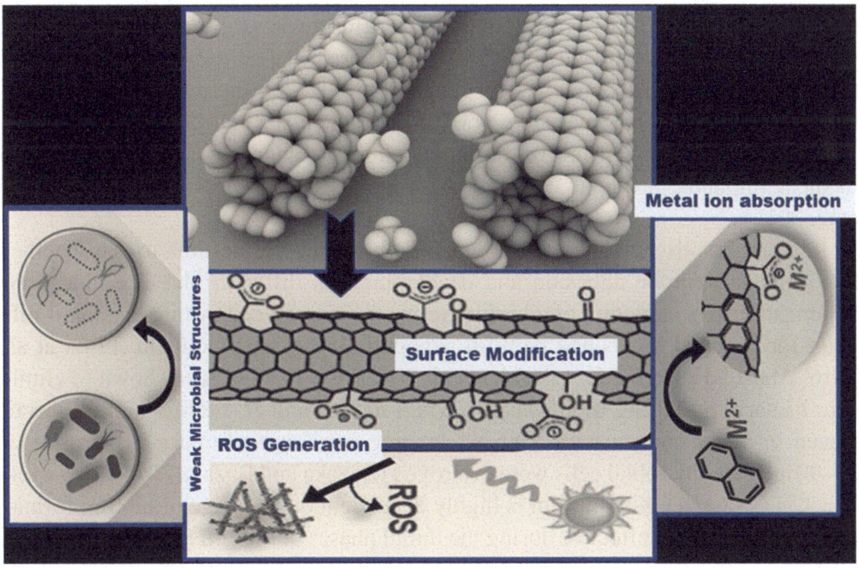

Fig. 9.3 Mechanism of antimicrobial activity of carbon nanotubes

In contrast, owing to incorrect and excessive antibiotic usage, microbial infections become resistant to standard antimicrobial treatments, increasing the chance of harmful health consequences. As a result, doing research and creating alternatives to current antimicrobial medicines is critical. Furthermore, these novel medicines must be safe for healthy cells and other living things. Commercial multiwall CNTs were chosen for this investigation over single-walled CNTs owing to their lower overall cost and simplicity of mass manufacture (Mukherjee et al. 2016).

The bactericidal activity of carbon nanotubes is achieved by combining physical and chemical processes (Ji et al. 2016). This is the core concept of the activity. When CNTs touch the microorganism's cell wall and membrane, they can inflict significant structural damage. Furthermore, they may physiologically isolate a cell from its environment, which creates hazardous substances such as reactive oxygen species and puts the cell under oxidative stress, eventually leading to the organism's death. As a result, determining the length of the nanotubes during contact with the cell membrane is critical. Studies show shorter tubes are more successful than longer tubes in eradicating germs (Ji et al. 2016). The interaction of CNTs and cells in a liquid media varies greatly from that of CNTs and cells on solid surfaces. Another critical aspect in the process of killing microorganisms is the diameter of a conduit. Smaller dimensions can produce cell membrane injury due to their contact with the cell surface (Corredor et al. 2013).

CNTs are well known in the scientific world for their outstanding "dielectric breakdown strength and field electron emission capabilities" (Smith and Rodrigues 2015). Using CNTs as a discharge electrode in a corona plasma system may considerably lower the plasma field's threshold voltage (Smotlacha and Pincak 2016). As a result, despite its "low operating voltage and ozone concentration, the CNT-corona discharge system" can eradicate bioaerosols (Smotlacha and Pincak 2016). The CNT-based corona discharge system efficiently inactivates 97% of *E. coli* bioaerosol cells at discharge voltages of 7.5 kV, much greater than the inactivation effectiveness when utilizing stainless steel electrodes (34%). Furthermore, by combining CNS with dielectric polymers like polydimethylsiloxane (PDMA) and polyvinylidene fluoride (PVDF) fibers, further applications ("nanoscale aerosol filters, biomedical implants, artificial muscles and robotics, sensors") are possible (Yang et al. 2011).

Researchers also investigated how CNTs break down biofilms to see whether they impede microbe adhesion and development at different phases of bacterial colonization. The structure of a biofilm protects bacterial cells while making it difficult for physical forces and harmful nanoparticles to separate them (Lolla et al. 2016; Abd-El-Haleem 2023; Bazaka et al. 2012). CNTs have been shown to eliminate bacteria by killing biofilm cells. When *E. coli* and *B. subtilis* biofilms were placed in direct contact with SWCNT coatings and studied under a microscope, 80–90% of the microbial cells were destroyed (Bazaka and Bazaka 2014). However, how CNTs interact with biofilm is highly dependent on the stage of biofilm formation. CNTs are most effective during the initial phases of biofilm formation (Ahmed et al. 2012). When the structure of a mature biofilm shields microorganisms, they are less susceptible to CNTs than bacteria in earlier biofilm stages.

When the biofilm matures, the soluble exopolymer compounds generated by microbial cells may be the key to making carbon tubes less toxic (Dong and Yang 2014). It is critical to note that the mobility of CNTs might have an anti-adhesive effect. This makes the base unstable and prevents bacteria from attaching properly. The length of the CNT is also thought to impact its capacity to inhibit biofilm development. This suggests that longer tubes are more flexible and may vibrate, preventing germs from settling (Rodrigues and Elimelech 2010). Furthermore, vertically aligned arrays of immaculate CNTs displayed great resistance to *B. subtilis*-induced biofilms (only 6.1% of the substrate was covered by biofilms), enabling only single microcolonies to develop on the surface. These groups are made up of tubes that are significantly smaller than a normal bacterial cell (2 m), making microorganisms unable to penetrate nanotubes (Malek et al. 2016). CNTs greatly influence biofilms in liquid systems, which is fascinating. CNTs inhibit microbial biofilms in a concentration-dependent manner. For example, 50 g mL1 SWCNTs decreased biofilm by 81.1%, whereas 200 g mL1 SWCNTs completely removed biofilm (Yick et al. 2015).

9.3.2.2 Antimicrobial Properties of Fullerenes

Fullerenes are a type of CNM whose structures can take on different forms, such as ellipses and hollow spheres. Buckminsterfullerene is a type of round fullerene that can also be called C60, buckyballs, or bucky onions (Alekseeva et al. 2019). This substance resembles a soccer ball because it comprises joined rings that look like a cage (Fig. 9.1). At each corner of this arrangement, there is a carbon atom with one bond and two bonds. This makes a universal vertex. This building is made up of 20 hexagons and 12 pentagons. Fullerene's structure is like graphite's, which makes sense when considering how graphite plates are grouped in hexagonal rings (Harris 2013). The biological processes of bacteria are slowed down by fullerenes and compounds made from them. Due to how fullerenes interact with the bacteria's cell wall, "Gram-positive bacteria are usually more susceptible to fullerenes than Gram-negative bacteria" (Zhang et al. 2019). Fullerenes work as antimicrobials against Gram-negative bacteria like *Pseudomonas putida* by lowering the amount of unsaturated fatty acids in the bacterial cell wall and increasing the amount of cyclopropane fatty acids simultaneously. Fullerenes, on the other hand, break down the cell walls of Gram-positive bacteria. This changes the structure of the phospholipids and makes the cell membrane more porous (Fang et al. 2007). The electric force depending on the shape of the fullerenes "(C60, C60-OH, C60-COOH, and C60-NH2)" is another critical factor in how *E. coli* and *Shewanella oneidensis* react to fullerene's antibiotic effects (Fang et al. 2007). At a minimum dose of 10 mg/L, C60-NH2 breaks down the structure of bacterial cells and makes it easier for Gram-positive and Gram-negative bacteria to find food (Tang et al. 2007). Researchers have found that electrostatic pull significantly affects the cytotoxic action of fullerene derivatives (Chen et al. 2016). This is similar to how oxidative stress has been found to have a negligible effect on the cytotoxic action of fullerene derivatives.

Fullerene particles can also change the bacterial membrane's breathing chain. Another way to stop germs from growing is to do this (Lahir 2017). It is possible that fullerene nanoparticles do not damage the bacterial barrier directly but instead mess with how cells use energy. For example, high levels of fullerene products could make it easier for cells to take in O_2, leading to more of it turning into H_2O_2, which breaks the respiratory chain (Lahir 2017). People have also used models to figure out how fullerenes get through the walls of microbes. Molecular dynamic models showed that how C60 moves depends significantly on how the target pathogen's lipids are made. C60 slightly tends to move into homogeneous bilayers of incomplete core lipopolysaccharides, but it is not thermodynamically more likely to move into complete core lipopolysaccharides. The same simulation showed that small changes in temperature, environmental ion concentration, lipopolysaccharide core sugar length, or the number of phospholipid flaws cause significant changes in how C60 interacts with surface membranes (Hsu et al. 2016). It is important to remember that the bio-reactivity of nanoparticles toward biological targets depends not only on the shape of the cell wall but also on the enzymes and metabolic processes of the microbe (Krishnamoorthy et al. 2012). Considering how bacteria's surroundings and cell activities can affect toxicology results will help explain why different studies on fullerene have shown different results.

Fullerenes are very hard to break down in water, but fullerene substitutes can make an even suspension in water (Zhou et al. 2008). Researchers have found that fullerene droplets, like groups of C60, may have different molecular effects on germs than bulk solid fullerene (Huh and Kwon 2011; Lyon and Alvarez 2008). People have said that in marine systems, fullerene particles do not have to break into microbe cells or make ROS. Instead, they can make particles poisonous by interacting with chemicals (Lyon and Alvarez 2008). So, it was shown that a fullerene water solution is an excellent photoinduced antibacterial agent that can stop the growth of gram-positive germs even at low doses (C60 = 1.8102 mM) (Heredia et al. 2022). Furthermore, it has been shown that the C70 solution is more photoactive than nC60 because, at wavelengths between 300 and 650 nm, it makes more 1O2 than nC60. This means there is always 1.690.05 times as much 1O2 as nC60 (Moor et al. 2015).

In the same way, a C60/pyrrolidinium solution is a potent, broad-spectrum photosensitizer that can kill more than 99.99% of bacterial and fungus cells in vitro (Tegos et al. 2005). Also, UV light (310–400 nm) on dissolved polyhydroxylated fullerene (fuller) made ROS, which made it much easier for bacteriophage MS2 to stop working (by up to 4 logs). However, Fullerol did not do much for living things without UV because it did not make many ROS (Badireddy et al. 2007). One problem with soluble functionalized fullerenes is that they tend to absorb in the blue and green bands of the visible spectrum instead of the red and far-red bands, which are better at getting into tissues (Yin et al. 2015). Also, fullerenes like C60 that have not been changed have a high hydrophobicity and a natural tendency to stick together, which makes them less photoactive (Sharma et al. 2011).

It is important to note that how fullerene solutions are made can also affect how well they kill germs. For instance, "liquid fullerene suspensions were made in four different ways: using tetrahydrofuran (THF) as a solvent (THF/nC60), sonicating C60 dissolved in toluene with water (son/nC60), stirring C60 powder in water (aq/nC60), and using a solubilizing agent (PVP/C60)" (Lyon et al. 2006). All four

Fig. 9.4 Mechanism of antimicrobial activity of fullerenes

fullerene products could kill *B. subtilis* bacteria, but THF/nC60 seemed to kill bacteria better than the others because of how it was removed (Lyon et al. 2006). The anti-microbial properties of fullerenes are depicted in Fig. 9.4.

9.3.2.3 Antimicrobial Properties of Graphene/Graphene Oxide

Graphene is the conventional term for a monolayer of carbon atoms packed into a two-dimensional lattice. Graphene is a kind of graphite. Graphene oxide (GO) nanosheets are produced when graphene undergoes chemical modification with suspended hydroxyl, epoxy, and carboxyl groups (Azimi et al. 2014). These nanosheets can be readily dispersed in water. According to the findings of several research projects, it has been hypothesized that both G and GO possess significant antibacterial capabilities. This is attributed to the extraordinary qualities that G and GO possess, such as their expansive surface areas and distinctive thermal, electrical, and physio-mechanical properties (Majidi et al. 2019). The effects of G and GO on microorganisms are shown in the production of reactive oxygen species and the activation of physical and chemical oxidation processes, which eventually lead to the extinction of the microbial population and a decrease in the resistance levels it has (Xie et al. 2020). The production of GO is simple, speedy, and cost-effective compared to other CNMs; when it is present in low quantities, GO demonstrates little cytotoxicity toward mammalian cells (Bolotin et al. 2008). According to Chen et al. (Chen et al. 2014), G and GO can dramatically reduce bacterial populations by 90% and macroconidia sprouting by 80%. These results were found in their study. When tested against "*E. coli, Salmonella typhimurium, B. subtilis,* and *Enterococcus faecalis,*" the results of Veerapandian et al. (Veerapandian et al. 2013) suggest that the minimum bactericidal concentration (MBC) of UV-irradiated graphene oxide (GO) is considerably lower than that of un-irradiated GO. This was determined by comparing the MBCs of the two types of graphene oxide.

Gram-negative bacteria, such as *E. coli*, have a higher resistance to the membrane damage that rGO may induce than Gram-positive bacteria, such as *S. aureus*, which lack an outer membrane (DuMont et al. 2011). G can also enter the lipid bilayer in various ways, depending on the size of the molecule. G molecules were assumed mainly a form parallel to the cell wall, while G molecules that were tiny assumed a horizontal shape (Dallavalle et al. 2015). In addition, the activity depends on the size of the flakes and the number of layers they include. Because smaller flakes are taken within the cells more, their influence on the functioning of the cells is magnified.

Additionally, research has shown that G with several layers has a less antibacterial effect than G with only one layer (Wang et al. 2013). The size of the surface and its roughness are further factors that influence how well an antibacterial agent functions. When there is a greater amount of surface area, there are more areas where germs may encounter it. The more abrasive the surface, the greater the contact the bacteria have with it. Additionally, the amount of carbon nanomaterial that is dispersed throughout the medium has a significant impact on the harmful effect. It should also be remembered that these nanomaterials have an antibiotic action that is selective for human pathogenic bacteria (Kumar et al. 2019), a crucial detail to keep in mind to prevent cell harm.

According to Zhang et al. (Zhang et al. 2016), Ag-GO composites have a remarkable antibacterial effect, dependent on "temperature, time, pH, and concentration." Thanks to this combination, the quantity of *E. coli and S. aureus* bacteria decreased by "99.99 and 99.96%," respectively (Zhang et al. 2016). The primary mechanisms were oxidative stress, production of reactive oxygen species (ROS), malfunction of lipopolysaccharides, cell membrane penetration, and replication of bacterial DNA (Tran et al. 2010). Ag-GO and ZnO-GO were put through tests against *E. coli*, *Enterococcus faecium*, *Staphylococcus aureus*, and *Klebsiella pneumonia* by Whitehead et al. (Whitehead et al. 2017). Ag-GO is superior to ZnO-GO in terms of its antibacterial activity. The minimum inhibitory concentrations (M.I.C.s) for *E. coli/E. faecium* and *S. aureus/K. pneumoniae* were 0.125 and 0.25 mg/mL, respectively. In addition, the presence of Ag in GO increased the antibacterial activity of Ag nanoparticles (NPs). Both GO/polyurethane and polyurethane/rGO/polyethyleneimine were investigated in Tang et al.'s (Tang et al. 2016) study. The high concentration of amine groups in PEI contributed to the increased bacteriostatic properties of the polyurethane/rGO/PEI composite. Graphene oxide, often known as GO, has contact-based antimicrobial activity, making it a potentially valuable material for creating antimicrobial surface coatings. However, it has not yet been determined how the physicochemical features of GO are connected to its antibacterial action. The more robust oxidative processes associated with the higher defect density of more miniature GO sheets are thought to be the cause of their more excellent antibacterial action compared to larger sheets. In contrast, GO interacted with bacteria in a cell entrapment mechanism when the tests were performed using suspensions. These results give helpful directions for the future development of graphene-based antimicrobial surface coatings, in which the antibacterial activity of the material may be increased by decreasing the sheet size of the graphene.

9.4 Conclusion

The advent and proliferation of multidrug-resistant bacteria have proven to be a grave public health concern that necessitates worldwide management. Nanomaterials present a unique and innovative approach to managing chronic multidrug-resistant planktonic bacterial and biofilm infections. The maximization of curative potential and minimization of host hazard necessitates the optimization of physical characteristics, specifically those about the size and surface charge. However, the present clinical application of nanoparticles is constrained by concerns regarding their enduring impact on individual and systemic well-being. Furthermore, the metabolic, toxicological, and stability concerns, as well as the elucidation of the underlying mechanisms at the genetic level, necessitate a more comprehensive investigation of the impact of NPs on gene expression in future research. Furthermore, it is imperative to elucidate nanoparticles' precise and intricate mode of action in relation to biological systems in forthcoming studies to fabricate nanomaterials possessing desirable physicochemical attributes that can effectively adapt to diverse biological environments for therapeutic benefits without any harmful consequences. In the foreseeable future, the realm of nano-antibiotics is poised to serve as a cutting-edge pharmaceutical solution in the fight against multidrug-resistant bacteria.

Several scientific studies have shown that carbon nanostructures have effective antibacterial capabilities. The size of these nanoparticles has a considerable impact on microbial neutralization. Fullerenes, SWCNTs, and GO nanoparticles, as well as their derivatives, are more effective antibacterial agents inside carbon nanostructures. The following mechanisms were proposed to explain their antibacterial efficacy: preventing the respiratory chain from impeding bacterial growth, impeding energy metabolism, engaging in physical interaction with the cell membrane, generating cell-CNTs/cell-GO aggregates, and instigating cell membrane disruption. Purification and functionalization of carbon nanostructures are required for biological and medical applications. Furthermore, their solubility in physiological fluids must be increased. Finally, employing carbon nanocomposites, made up of carbon nanostructures and metal nanoparticles, is a viable way to achieve disinfection goals. Nonetheless, further study is needed to understand the processes driving carbon nanostructures' antibacterial activity.

References

Abd-El-Haleem D (2023) Alpha-glucan: a novel bacterial polysaccharide and its application as a biosorbent for heavy metals. J Genet Eng Biotechnol 21:133. https://doi.org/10.1186/s43141-023-00609-3

Abdulhameed A, Wahab NZA, Mohtar MN, Hamidon MN, Shafie S, Halin IA (2021) Methods and applications of electrical conductivity enhancement of materials using carbon nanotubes. J Electron Mater 50:3207–3221

Ahmed F, Santos CM, Vergara RAMV, Tria MCR, Advincula R, Rodrigues DF (2012) Antimicrobial applications of electroactive PVK-SWNT nanocomposites. Environ Sci Technol 46:1804–1810

Akhavan O, Ghaderi E (2010) Toxicity of graphene and graphene oxide nanowalls against bacteria. ACS Nano 4:5731–5736

Alekseeva OV, Bagrovskaya NA, Noskov AV (2019) Physicochemical properties and biological activity of Polymethylmethacrylate/fullerene composites. Adv Nanostruct Compos 284:128–139

Al-Jumaili A, Alancherry S, Bazaka K, Jacob MV (2017) Review on the antimicrobial properties of carbon nanostructures. Materials 10:1066

Aravind SJ, Baskar P, Baby TT, Sabareesh RK, Das S, Ramaprabhu S (2011) Investigation of structural stability, dispersion, viscosity, and conductive heat transfer properties of functionalized carbon nanotube based nanofluids. J Phys Chem C 115:16737–16744

Azimi S, Behin J, Abiri R, Rajabi L, Derakhshan AA, Karimnezhad H (2014) Synthesis, characterization and antibacterial activity of chlorophyllin functionalized graphene oxide nanostructures. Sci Adv Mater 6:771–781

Badireddy AR, Hotze EM, Chellam S, Alvarez P, Wiesner MR (2007) Inactivation of bacteriophages via photosensitization of fullerol nanoparticles. Environ Sci Technol 41:6627–6632

Baptista PV, McCusker MP, Carvalho A et al (2018) Nano-strategies to fight multidrug resistant bacteria—"a Battle of the titans". Front Microbiol 9:1441

Basak S, Singh P, Rajurkar M (2016) Multidrug resistant and extensively drug resistant bacteria: a study. Journal of pathogens 2016:4065603

Bazaka K, Jacob MV, Crawford RJ, Ivanova EP (2012) Efficient surface modification of biomaterial to prevent biofilm formation and the attachment of microorganisms. Appl Microbiol Biotechnol 95:299–311

Bazaka O, Bazaka K (2014) Surface modification of biomaterials for biofilm control. Biomater Med Device Assoc Infect 6:103–132

Bolotin KI, Sikes K, Jiang Z et al (2008) Ultrahigh electron mobility in suspended graphene. Solid State Commun 146:351–355

Bortolamiol T, Lukanov P, Galibert A-M et al (2014) Double-walled carbon nanotubes: quantitative purification assessment, balance between purification and degradation and solution filling as an evidence of opening. Carbon 78:79–90

Buzea C, Pacheco II, Robbie K (2007) Nanomaterials and nanoparticles: sources and toxicity. Biointerphases 2:MR17–MR71

Cao Q, Rogers JA (2009) Ultrathin films of single-walled carbon nanotubes for electronics and sensors: a review of fundamental and applied aspects. Adv Mater 21:29–53

Chen J, Peng H, Wang X, Shao F, Yuan Z, Han H (2014) Graphene oxide exhibits broad-spectrum antimicrobial activity against bacterial phytopathogens and fungal conidia by intertwining and membrane perturbation. Nanoscale 6:1879–1889

Chen Q, Ma Z, Liu G, Wei H, Xie X (2016) Antibacterial activity of cationic cyclen-functionalized fullerene derivatives: membrane stress. Dig J Nanomater Biostruct 11:753–761

Choi W, Lahiri I, Seelaboyina R, Kang YS (2010) Synthesis of graphene and its applications: a review. Crit Rev Solid State Mater Sci 35:52–71

Corredor C, Hou W-C, Klein SA et al (2013) Disruption of model cell membranes by carbon nanotubes. Carbon 60:67–75

Dallavalle M, Calvaresi M, Bottoni A, Melle-Franco M, Zerbetto F (2015) Graphene can wreak havoc with cell membranes. ACS Appl Mater Interfaces 7:4406–4414

Dizaj SM, Mennati A, Jafari S, Khezri K, Adibkia K (2015) Antimicrobial activity of carbon-based nanoparticles. Adv Pharm Bull 5:19

Dong X, Yang L (2014) Inhibitory effects of single-walled carbon nanotubes on biofilm formation from Bacillus anthracis spores. Biofouling 30:1165–1174

DuMont AL, Nygaard TK, Watkins RL et al (2011) Characterization of a new cytotoxin that contributes to Staphylococcus aureus pathogenesis. Mol Microbiol 79:814–825
Elsehly E, Chechenin N, Bukunov K et al (2016) Removal of iron and manganese from aqueous solutions using carbon nanotube filters. Water Sci Technol Water Supply 16:347–353
Elsehly EM, Chechenin N, Makunin A, Motaweh H, Leksina E (2017) Functionalized carbon nanotubes based filters for chromium removal from aqueous solutions. Water Sci Technol 75:1564–1571
Elsehly EM, Chechenin N, Makunin A, Shemukhin A, Motaweh H (2018) Enhancement of CNT-based filters efficiency by ion beam irradiation. Radiat Phys Chem 146:19–25
Fang J, Lyon DY, Wiesner MR, Dong J, Alvarez (2007) Effect of a fullerene water suspension on bacterial phospholipids and membrane phase behavior. Environ Sci Technol 41:2636–2642
Fatima Z, Purkait D, Rehman S, Rai S, Hameed S (2023) Multidrug resistance: a threat to antibiotic era. Biological and Environmental Hazards, Risks, and Disasters Elsevier: 197–220
Goenka S, Sant V, Sant S (2014) Graphene-based nanomaterials for drug delivery and tissue engineering. J Control Release 173:75–88
Gurunathan S, Han JW, Dayem AA, Eppakayala V, Kim J-H (2012) Oxidative stress-mediated antibacterial activity of graphene oxide and reduced graphene oxide in Pseudomonas aeruginosa. Int J Nanomed:5901–5914
Hajipour MJ, Fromm KM, Ashkarran AA et al (2012) Antibacterial properties of nanoparticles. Trends Biotechnol 30:499–511
Harris PJ (2013) Fullerene-like models for microporous carbon. J Mater Sci 48:565–577
Hecht DS, Hu L, Irvin G (2011) Emerging transparent electrodes based on thin films of carbon nanotubes, graphene, and metallic nanostructures. Adv Mater 23:1482–1513
Heredia DA, Durantini AM, Durantini JE, Durantini EN (2022) Fullerene C60 derivatives as antimicrobial photodynamic agents. J Photochem Photobiol C: Photochem Rev 51:100471
Hsu P-C, Jefferies D, Khalid S (2016) Molecular dynamics simulations predict the pathways via which pristine fullerenes penetrate bacterial membranes. J Phys Chem B 120:11170–11179
Huh AJ, Kwon YJ (2011) "Nanoantibiotics": a new paradigm for treating infectious diseases using nanomaterials in the antibiotics resistant era. J Control Release 156:128–145
Ibrahim KS (2013) Carbon nanotubes? Properties and applications: a review. Carbon Lett 14:131–144
Ji H, Sun H, Qu X (2016) Antibacterial applications of graphene-based nanomaterials: recent achievements and challenges. Adv Drug Deliv Rev 105:176–189
Ji S-R, Liu C, Zhang B et al (2010) Carbon nanotubes in cancer diagnosis and therapy. Biochim Biophys Acta 1806:29–35
Kalaskar A, Venkataramana K (2012) Determination of antimicrobial resistance pattern and production of extended-spectrum B-lactamases amongst Escherichia coli and Klebsiella pneumoniae from clinical isolates. J Med Bacteriol 1:17–24
Kang S, Herzberg M, Rodrigues DF, Elimelech M (2008) Antibacterial effects of carbon nanotubes: size does matter! Langmuir 24:6409–6413
Kang S, Mauter MS, Elimelech M (2009) Microbial cytotoxicity of carbon-based nanomaterials: implications for river water and wastewater effluent. Environ Sci Technol 43:2648–2653
Kang S, Pinault M, Pfefferle LD, Elimelech M (2007) Single-walled carbon nanotubes exhibit strong antimicrobial activity. Langmuir 23:8670–8673
Kehrer JP, Klotz L-O (2015) Free radicals and related reactive species as mediators of tissue injury and disease: implications for health. Crit Rev Toxicol 45:765–798
Khandelwal G, Kumar R, Kumar V (2019) Antimicrobial activities of graphene–polymer nanocomposites. Nanostructured polymer composites for biomedical applications. Elsevier: 429–45
Krishnamoorthy K, Veerapandian M, Zhang L-H, Yun K, Kim SJ (2012) Antibacterial efficiency of graphene nanosheets against pathogenic bacteria via lipid peroxidation. J Phys Chem C 116:17280–17287
Kumar P, Huo P, Zhang R, Liu B (2019) Antibacterial properties of graphene-based nanomaterials. Nanomaterials 9:737

Lahir Y (2017) Impacts of fullerene on biological systems. Clin Immunol Endocr Metab Drugs (Discontinued) 4:47–58
Li H, He X, Liu Y et al (2011) One-step ultrasonic synthesis of water-soluble carbon nanoparticles with excellent photoluminescent properties. Carbon 49:605–609
Li J, Wang G, Zhu H et al (2014) Antibacterial activity of large-area monolayer graphene film manipulated by charge transfer. Sci Rep 4:4359
Li X-Z, Plésiat P, Nikaido H (2015) The challenge of efflux-mediated antibiotic resistance in gram-negative bacteria. Clin Microbiol Rev 28:337–418
Lolla D, Lolla M, Abutaleb A, Shin HU, Reneker DH, Chase GG (2016) Fabrication, polarization of electrospun polyvinylidene fluoride electret fibers and effect on capturing nanoscale solid aerosols. Materials 9:671
Luan B, Huynh T, Zhao L, Zhou R (2015) Potential toxicity of graphene to cell functions via disrupting protein–protein interactions. ACS Nano 9:663–669
Lyon DY, Adams LK, Falkner JC, Alvarez PJ (2006) Antibacterial activity of fullerene water suspensions: effects of preparation method and particle size. Environ Sci Technol 40:4360–4366
Lyon DY, Alvarez PJ (2008) Fullerene water suspension (nC60) exerts antibacterial effects via ROS-independent protein oxidation. Environ Sci Technol 42:8127–8132
Majidi HJ, Babaei A, Bafrani ZA, Shahrampour D, Zabihi E, Jafari SM (2019) Investigating the best strategy to diminish the toxicity and enhance the antibacterial activity of graphene oxide by chitosan addition. Carbohydr Polym 225:115220
Malek I, Schaber C, Heinlein T, Schneider J, Gorb S, Schmitz R (2016) Vertically aligned multi walled carbon nanotubes prevent biofilm formation of medically relevant bacteria. J Mater Chem B 4:5228–5235
Manke A, Wang L, Rojanasakul Y (2013) Mechanisms of nanoparticle-induced oxidative stress and toxicity. Biomed Res Int 2013:942916
Moor KJ, Snow SD, Kim J-H (2015) Differential photoactivity of aqueous [C60] and [C70] fullerene aggregates. Environ Sci Technol 49:5990–5998
Mukherjee A, Majumdar S, Servin AD, Pagano L, Dhankher OP, White JC (2016) Carbon nanomaterials in agriculture: a critical review. Front Plant Sci 7:172
Munita JM, Arias CA (2016) Mechanisms of antibiotic resistance. Microbiol Spectr 4(2):10.1128. https://doi.org/10.1128/microbiolspec.VMBF-0016-2015
Nidhin M, Saneha D, Hans S, Varghese A, Fatima Z, Hameed S (2019) Studies on the antifungal activity of biotemplated gold nanoparticles over Candida albicans. Mater Res Bull 119:110563
Olivi M, Zanni E, De Bellis G et al (2013) Inhibition of microbial growth by carbon nanotube networks. Nanoscale 5:9023–9029
O'Neill J (2016) Tackling drug-resistant infections globally: final report and recommendations. https://apo.org.au/node/63983, Accessed July 9, 2023
Pacurari M, Qian Y, Fu W et al (2012) Cell permeability, migration, and reactive oxygen species induced by multiwalled carbon nanotubes in human microvascular endothelial cells. J Toxic Environ Health A 75:112–128
Panda S, Rout TK, Prusty AD, Ajayan PM, Nayak S (2018) Electron transfer directed antibacterial properties of graphene oxide on metals. Adv Mater 30:1702149
Peng J, Gao W, Gupta BK et al (2012) Graphene quantum dots derived from carbon fibers. Nano Lett 12:844–849
Peterson E, Kaur P (2018) Antibiotic resistance mechanisms in bacteria: relationships between resistance determinants of antibiotic producers, environmental bacteria, and clinical pathogens. Front Microbiol 9:2928
Prasad K, Lekshmi GS, Ostrikov K et al (2017) Synergic bactericidal effects of reduced graphene oxide and silver nanoparticles against Gram-positive and Gram-negative bacteria. Sci Rep 7:1591. https://doi.org/10.1038/s41598-017-01669-5
Purkait D, Hameed S, Fatima Z (2021) Pathogen-omics: challenges and prospects in research and clinical settings. Integrated omics approaches to infectious diseases 521–42. https://doi.org/10.1007/978-981-16-0691-5_28

Ramana K, Kalaskar A, Rao M, Rao SD (2013a) Aetiology and antimicrobial susceptibility patterns of lower respiratory tract infections (LRTI's) in a rural tertiary care teaching hospital in Karimnagar, South India. Am J Infect Dis Microbiol 1:101–105

Ramana K, Rao R (2009) Significance of screening for colonization and vancomycin resistance in Staphylococcus aureus isolated from anterior nares of school going children. Online J Health Allied Sci 8

Ramana K, Rao R, Sharada CV, Kareem M, Reddy LR, Mani MR (2013b) Modified Hodge test: a useful and the low-cost phenotypic method for detection of carbapenemase producers in Enterobacteriaceae members. J Nat Sci Biol Med 4:346

Reygaert WC (2018) An overview of the antimicrobial resistance mechanisms of bacteria. AIMS microbiology 4:482

Rodrigues DF, Elimelech M (2010) Toxic effects of single-walled carbon nanotubes in the development of E. coli biofilm. Environ Sci Technol 44:4583–4589

Saba I, Wani K, Rehman S, Singh V (2022) Platform technologies based on virus-like particles (VLPs) for infectious diseases. In: Nanotechnology for Infectious Diseases. Springer, pp 541–553

Saba I, Wani K, Syed A, Rehman S (2021) Role of microbial nanotechnology in bioremediation of heavy metals. Microbial nanotechnology: green synthesis and applications 307–19

Santajit S, Indrawattana N (2016) Mechanisms of antimicrobial resistance in ESKAPE pathogens. Biomed Res Int 2016:2475067

Sarada V, Rao R, Mani R, Ramana K (2014) Colistin, polymyxin B and tigecycline suspectibility to metallo betalactamase producing Acinetobacter baumannii isolated from tertiary health care hospital. Am J Microbiol Res 2:60–62

Schaenzer AJ, Wright GD (2020) Antibiotic resistance by enzymatic modification of antibiotic targets. Trends Mol Med 26:768–782

Seo DH, Han ZJ, Kumar S, Ostrikov K (2013) Structure-controlled, vertical graphene-based, binder-free electrodes from plasma-reformed butter enhance supercapacitor performance. Adv Energy Mater 3:1316–1323

Seo DH, Pineda S, Yick S, Bell J, Han ZJ, Ostrikov KK (2015) Plasma-enabled sustainable elemental lifecycles: honeycomb-derived graphenes for next-generation biosensors and supercapacitors. Green Chem 17:2164–2171

Sharma SK, Chiang LY, Hamblin MR (2011) Photodynamic therapy with fullerenes in vivo: reality or a dream? Nanomedicine 6:1813–1825

Shearer CJ, Slattery AD, Stapleton AJ, Shapter JG, Gibson CT (2016) Accurate thickness measurement of graphene. Nanotechnology 27:125704

Shenderova OA, Gruen DM (2012) Ultrananocrystalline diamond: synthesis, properties and applications. William Andrew, pp 103–127

Shvedova AA, Pietroiusti A, Fadeel B, Kagan VE (2012) Mechanisms of carbon nanotube-induced toxicity: focus on oxidative stress. Toxicol Appl Pharmacol 261:121–133

Smith SC, Rodrigues DF (2015) Carbon-based nanomaterials for removal of chemical and biological contaminants from water: a review of mechanisms and applications. Carbon 91:122–143

Smotlacha J, Pincak R (2016) Electronic properties of carbon nanostructures. Graphene Research 31. https://doi.org/10.5772/63633

Stancu EC, Stanciuc A-M, Vizireanu S et al (2014) Plasma functionalization of carbon nanowalls and its effect on attachment of fibroblast-like cells. J Phys D Appl Phys 47:265203

Tang X-Z, Mu C, Zhu W, Yan X, Hu X, Yang J (2016) Flexible polyurethane composites prepared by incorporation of polyethylenimine-modified slightly reduced graphene oxide. Carbon 98:432–440

Tang YJ, Ashcroft JM, Chen D et al (2007) Charge-associated effects of fullerene derivatives on microbial structural integrity and central metabolism. Nano Lett 7:754–760

Taylor E, Webster TJ (2011) Reducing infections through nanotechnology and nanoparticles. Int J Nanomed:1463–1473

Tegos GP, Demidova TN, Arcila-Lopez D et al (2005) Cationic fullerenes are effective and selective antimicrobial photosensitizers. Chem Biol 12:1127–1135

Tran N, Mir A, Mallik D, Sinha A, Nayar S, Webster TJ (2010) Bactericidal effect of iron oxide nanoparticles on Staphylococcus aureus. Int J Nanomed:277–283

Varghese S, Kuriakose S, Jose S (2013) Antimicrobial activity of carbon nanoparticles isolated from natural sources against pathogenic Gram-negative and Gram-positive bacteria. Journal of Nanoscience 2013:1

Vecitis CD, Zodrow KR, Kang S, Elimelech M (2010) Electronic-structure-dependent bacterial cytotoxicity of single-walled carbon nanotubes. ACS Nano 4:5471–5479

Veerapandian M, Zhang L, Krishnamoorthy K, Yun K (2013) Surface activation of graphene oxide nanosheets by ultraviolet irradiation for highly efficient anti-bacterials. Nanotechnology 24:395706

Wang J, Wei Y, Shi X, Gao H (2013) Cellular entry of graphene nanosheets: the role of thickness, oxidation and surface adsorption. RSC Adv 3:15776–15782

Wang J-T, Chen C, Wang E, Kawazoe Y (2014) A new carbon allotrope with six-fold helical chains in all-sp 2 bonding networks. Sci Rep 4:4339

Wang X, Zhi L, Müllen K (2008) Transparent, conductive graphene electrodes for dye-sensitized solar cells. Nano Lett 8:323–327

Whitehead K, Vaidya M, Liauw C et al (2017) Antimicrobial activity of graphene oxide-metal hybrids. Int Biodeterior Biodegradation 123:182–190

Worthington RJ, Melander C (2013) Combination approaches to combat multidrug-resistant bacteria. Trends Biotechnol 31:177–184

Wu J-Y, Li C-W, Tsai C-H, Chou C-W, Chen D-R, Wang G-J (2014) Synthesis of antibacterial TiO2/PLGA composite biofilms. Nanomed Nanotechnol Biol Med 10:e1097–ee107

Wu Y, Lin X, Zhang M (2013) Carbon nanotubes for thin film transistor: fabrication, properties, and applications. J Nanomater 2013:64

Xie Y-Y, Hu X-H, Zhang Y-W et al (2020) Development and antibacterial activities of bacterial cellulose/graphene oxide-CuO nanocomposite films. Carbohydr Polym 229:115456

Xin Q, Shah H, Nawaz A et al (2019) Antibacterial carbon-based nanomaterials. Adv Mater 31:1804838

Yang K, Zhang S, Zhang G, Sun X, Lee S-T, Liu Z (2010) Graphene in mice: ultrahigh in vivo tumor uptake and efficient photothermal therapy. Nano Lett 10:3318–3323

Yang S, Huang YC, Luo CH et al (2011) Inactivation efficiency of bioaerosols using carbon nanotube plasma. Clean–Soil Air Water 39:201–205

Yang S, Zhang X, Zhang D (2019) Electrospun chitosan/poly (vinyl alcohol)/graphene oxide nanofibrous membrane with ciprofloxacin antibiotic drug for potential wound dressing application. Int J Mol Sci 20:4395

Yick S, Mai-Prochnow A, Levchenko I et al (2015) The effects of plasma treatment on bacterial biofilm formation on vertically-aligned carbon nanotube arrays. RSC Adv 5:5142–5148

Yin R, Agrawal T, Khan U et al (2015) Antimicrobial photodynamic inactivation in nanomedicine: small light strides against bad bugs. Nanomedicine 10:2379–2404

Zeiger M, Jäckel N, Aslan M, Weingarth D, Presser V (2015) Understanding structure and porosity of nanodiamond-derived carbon onions. Carbon 84:584–598

Zhang H-Z, Zhang C, Zeng G-M, Gong J-L, Ou X-M, Huan S-Y (2016) Easily separated silver nanoparticle-decorated magnetic graphene oxide: synthesis and high antibacterial activity. J Colloid Interface Sci 471:94–102

Zhang X, Cong H, Yu B, Chen Q (2019) Recent advances of water-soluble fullerene derivatives in biomedical applications. Mini Revi Org Chem 16:92–99

Zhou G, Li Y, Xiao W et al (2008) Synthesis, characterization, and antibacterial activities of a novel nanohydroxyapatite/zinc oxide complex. J Biomed Mater Res A 85:929–937

Polymer-Based Nanomaterials Against Multidrug-Resistant Bacteria

10

Ana R. M. Ribeiro, Marta O. Teixeira, Elina Marinho, A. Francisca G. Silva, Susana P. G. Costa, and Helena P. Felgueiras

Abstract

Currently, there are major developments in the fields of medicine and biomedical engineering to mitigate the spreading of resistant bacteria and the development of bacterial biofilm; yet, challenges remain and many polymer-based opportunities are still to uncover. This important and complex problem, inherent to our society, is derived from the misuse and/or excessive use of antibiotics, allowing for several human pathogens to become resistant to conventional therapeutics. Hence, leading to the outbreak of multiple drug-resistant bacteria (MDRB). In the last few years, innovative treatment alternatives based on polymeric nanomaterials have emerged from nanofibers, nanoparticles, nanocomposites, and nanohydrogels that can be made up of natural and/or synthetic polymers. Their elevated drug loading capacity, various types of stimuli responsiveness, and smart targeting can increase local drug concentration, enlarging the efficient delivery of antimicrobial agents and reducing bacterial resistance. The present chapter explores new and improved polymer-based nanotechnology strategies and approaches with potential to circumvent antimicrobial resistance of MDRB, highlighting their mechanisms of action and valuable applications.

Keywords

Antibiotic resistance · Drug delivery · Multidrug-resistant bacteria · Nanomaterials · Polymers · Nanotechnology

A. R. M. Ribeiro · M. O. Teixeira · E. Marinho · H. P. Felgueiras (✉)
Centre for Textile Science and Technology (2C2T), University of Minho, Guimarães, Portugal
e-mail: helena.felgueiras@2c2t.uminho.pt

A. F. G. Silva · S. P. G. Costa
Centre of Chemistry, University of Minho, Braga, Portugal

M. Y. Wani et al. (eds.), *Nanotechnology Based Strategies for Combating Antimicrobial Resistance*, https://doi.org/10.1007/978-981-97-2023-1_10

10.1 Introduction

Nanomaterials play a significant role in drug development as innovative tools for combating the increased rates of microbial resistance, which has emerged over the last few decades. The fact that they feature unique properties make nanomaterials valuable in a wide range of fields, including electronics, materials science, medicine, and environmental science (Makabenta et al. 2021a). Bacterial infections promoted by multidrug-resistant bacteria (MDRB) and associated with the diminished potency of current antibiotics remain a leading factor of morbidity and mortality, directly impacting public health. They target the severely ill, hospitalized, and immunocompromised patients, who are prone to infections. Furthermore, there is an escalating concern with respect to biofilm-associated infections that are refractory to the presently available antimicrobial strategies, leaving almost no therapeutic option available (Hemeg 2017). Developing antimicrobial treatments using nanomaterials is a potential solution. Nanomaterials have distinct properties that set them apart and enhance their effectiveness against MDRB infections. These materials can disrupt various aspects of bacterial cells, including their machinery, membranes, and internal components. The unique characteristics of nanomaterials, such as their size, shape, and surface chemistry, create challenges for bacteria in responding to therapeutics. They can infiltrate bacterial membranes while carrying specialized drug payloads, making them valuable in improving treatment efficacy by interacting with the bacterial cellular system (Munir and Ahmad 2022). Polymer-based nanomaterials have shown great potential for developing novel drug delivery systems, like target-based nanocarriers, because of their ability to protect and enhance bioavailability of the encapsulated drugs and to increase the concentration of the drug at the infected site, thus reducing the associated toxicity (Ren et al. 2022). With increased knowledge and understanding of these nanomaterials, it became possible to integrate attractive properties within the carriers, namely controlled release and target delivery, and to use new delivery pathways (e.g., subcutaneous patches, implantable scaffolds, etc.). These nanocarriers can be functionalized with bioactive molecules, ligands, antibodies or antibiotics, to improve their antimicrobial effectiveness against severe bacterial infections caused by MDRB (Han et al. 2018; Adeniji et al. 2022). Researchers continue to explore and exploit these properties for various applications, driving innovation and advancements in many industries.

10.2 Nanomaterials

Although conventional antibiotics remain, to date, the most effective antibacterial agents, bacterial resistance arising from poorly managed administration/dosage has become a serious issue, particularly with the emergence of multidrug-resistant bacteria. Nanomaterials appear as a promising alternative not only to deliver the antibacterial molecules (conventional antibiotics, antimicrobial peptides, natural extracts, etc.) locally and in a controlled fashion, significantly decreasing systemic toxicity, but also by exploring carriers with intrinsic antibacterial features (e.g., silver nanoparticles). For these reasons, nanomaterials are the focus of investigation for many new strategies to fight MDRB (Li et al. 2018).

Nanomaterials present dimensions up to 100 nm (Saleh 2020). They have an almost unlimited range of morphologies and structures and can have several classifications based on shape, composition, origin, and size. When made of multiple components, the final nanomaterial's overall properties are expected to be enhanced over their individual parts (Barhoum et al. 2022).

In the biomedical field, polymers represent about 45% of all biomaterials in use. They can be processed in the form of fibers, particles, hydrogels, and films and can be combined to generate more complex systems like composites (Teixeira et al. 2021). Many polymers (natural and synthetic), used in nanomaterial strategies, possess intrinsic antibacterial activity, or can be modified to incorporate antibacterial components. It is also common to reinforce intrinsic antibacterial nanomaterials with other antibiotic molecules to strengthen their antimicrobial profiles, sometimes referred to as combinatorial strategies (León-Buitimea et al. 2022).

10.2.1 Polymer-Based Nanomaterials

Polymer-based nanomaterials are made of natural and/or synthetic polymers. Natural polymers are derived from living organisms. They are biodegradable, non-toxic, biocompatible, and usually less expensive than the synthetic. On its turn, the later exhibit excellent reproducibility and mechanical properties and their chemical structure can be tuned for reaching the requirements of specific sites of acting (e.g., manipulation of their degradation profile) (Teixeira et al. 2021). Additionally, some of the challenges of the natural (e.g., weak mechanical properties) and synthetic polymers (e.g., reduced cell affinity) use can be underlined by the combination of both types in a composite or hybrid formulation (Keshvardoostchokami et al. 2021).

Here, polymers were categorized for their use in nanomaterial strategies against MDRB in (1) polymers with no reported/significant antibacterial activity and (2) polymers with antibacterial activity against MDRB. In the first case, alginate is one of the most popular naturally derived options (Taghiloo et al. 2023; Baek et al. 2019). It is a water-soluble polymer, made of blocks of 1-4 linked L-guluronic and D-mannuronic acid residues. Alginate is obtained from brown algae in different salt forms of alginic acid (e.g., sodium, calcium, and magnesium salts). It has important gel-forming capacities by means of hydrogen bonding or ionic interaction with divalent cations. Even though this natural polymer is valuable and highly sought out as a polymer matrix for MDRB nanomaterial-based therapies, synthetic polymers exhibit superior mechanical properties, being more appealing for such uses [e.g., polylactic acid (PLA), poly(vinyl alcohol) (PVA), polycaprolactone (PCL), and poly (lactic-co-glycolic acid) (PLGA)] (Sebe et al. 2016; Thamer et al. 2021; Xi et al. 2018; Ghomi et al. 2023; Salim et al. 2023; Abdelghany et al. 2019; Masood et al. 2016). PLA is an aliphatic polymer obtained from natural raw matter such as potatoes and corn. Its biodegradability is slow due to the polymer's molecular weight and crystallinity. Because of that, PLA is commonly used in the production of surgical sutures and skin grafts (Peixoto et al. 2022; Rezić et al. 2021). PVA is a semi-crystalline synthetic polymer, with a high degree of swelling, that can be synthesized from various poly(vinyl esters) and poly(vinyl ethers) by hydrolysis. Due to its hygroscopic nature, PVA's elongation at break, tensile strength, and barrier properties may vary. Still, PVA is frequently used for its capacity for originating elastic and flexible hydrogels (Mok et al. 2020). PCL is a semicrystalline linear polyester prepared by ε-caprolactone ring-opening polymerization (generally derived from fossil carbon) and is recognized for its excellent mechanical properties. In physiological conditions, it is degraded by hydrolysis on its ester linkages. However, because of its hydrophobic nature and weak interaction with biological fluids, PCL is frequently used together with natural polymers (Peixoto et al. 2022; El Fawal et al. 2021). Lastly, PLGA is a synthetic copolymer of lactic acid and glycolic acid, approved by the Food and Drug Administration (FDA) for uses in humans. It is known for its encapsulation efficiency of biologically active compounds, being a popular choice for loading antibiotic molecules (Abdelghany et al. 2019; Masood et al. 2016).

Regarding the polymers with antibacterial activity against MDRB, the naturally derived ε-polylysine (ε-PL) and the chemically modified versions of chitosan stand out. ε-PL is a water-soluble cationic polymer composed of L-lysine units linked by ε-amino and α-carboxyl groups (Alfei and Schito 2020). Its antimicrobial activity is usually linked to the amino groups present in its structure. Studies indicated that this polymer possesses antibacterial activity against various resistant bacteria such as methicillin-resistant *Staphylococcus aureus* (MRSA), MDRB *Pseudomonas aeruginosa* (*P. aeruginosa*), colistin-resistant *Escherichia coli* (*E. coli*), and vancomycin-resistant *Enterococcus faecium* (*E. faecium*) (Mayandi et al. 2020). Chitosan is a polysaccharide derived from chitin's deacetylation. Some of its most limitative characteristics are the poor water solubility, high viscosity, and the high dosage required for inducing antibiofilm activity (Zhang et al. 2018). Nevertheless, by chemically modifying its composition via carboxymethylation, acylation, quaternization, or sulfhydrylation of amino groups in glucopyranose C2 position, chitosan's biological features can be improved, including its antimicrobial profile. Some cationic chitosan derivatives, *N*-(2-hydroxypropyl)-3-trimethylammonium chitosan chlorides (HTCC), reported significant antibacterial results against MRSA, with no evidence of cytotoxic effects. These HTCC were inspired by the strategy of preparing synthetic mimics of antimicrobial peptides (SMAMPs) (Hoque et al. 2016). Antimicrobial peptides (AMPs) are known for their ability to fight multidrug-resistant microbes, even though limitations can be recognized, namely low in vivo stability and high cost of manufacture. SMAMPs main purpose is to overcome those limitations. Among SMAMPs, cationic antimicrobial polymers are demonstrating promising antibacterial results against MDRB bacteria. For instance, the cationic chitosan derivatives mechanistic studies suggested that the polymers act by disrupting the bacteria membrane integrity. Furthermore, negligible tendency for bacteria to develop resistance was demonstrated (Hoque et al. 2016). Also, SMAMPs are being investigated in dual-antimicrobial therapies that consist in loading the

SMAMPs with commercial antibiotics. Namivandi-Zangeneh et al. investigated the efficacy of SMAMPs in combination with doxycycline and colistin, with synergistic killing effects against planktonic and bacterial biofilms being observed, particularly with doxycycline. Still, the greatest synergistic behavior was observed against the MDRB *P. aeruginosa* through coadministration of all compounds (Namivandi-Zangeneh et al. 2019). A similar strategy is also based on peptide mimetics, the peptoids, which differ from normal peptides at their side chains since they are attached to the backbone amide nitrogen instead of being attached to the α-carbon, this way improving their stability toward proteases. Likewise, peptoid polymers (polymers based on a *N*-substituted glycine backbone), that can be synthesized by automated solid-phase synthesis or classical polymerization, are presenting promising antibacterial results in MRSA infections combat. Xie et al. examined the poly-*N*-aminoethylglycine peptoid mechanisms of action against MRSA and concluded that the peptoid penetrated the bacterial cell through a transmembrane channel generated by the destabilization of the membrane bilayer, followed by DNA binding and the generation of high levels of reactive oxygen species (ROS), all hindering bacteria from acquiring resistance (Xie et al. 2021).

10.2.1.1 Nanoparticles

Polymeric nanoparticles (NPs) vary between 10 and 1000 nm in size (Saleh 2020). NPs are excellent carriers to control/sustain drug delivery. Polymeric NPs can function as matrix systems in which the drug is dispersed, also known as nanospheres, or work as protective layers confining/surrounding a drug-containing core, named nanocapsules (Antunes et al. 2021). The drug can be encapsulated, attached, entrapped, or dissolved in the NPs matrix/shell, with the protection conferred helping preventing toxicity. NPs are versatile (different shapes, sizes, and materials) and can be modified at the surface with targeting molecules for specific recognition or be engineered to respond to environmental stimuli (e.g., pH, temperature) (Bhatia 2016). Their antimicrobial activity is based on the release of antimicrobial agents or is induced through direct contact by means of cationic surfaces (e.g., quaternary phosphoniums, alkyl pyridiniums or quaternary ammonium compounds) (Elfadil et al. 2022)—positively charged polymers bind to the negative charged bacteria cell surface. Cross-linked quaternary ammonium poly-ethylenimine NPs presumed mechanism of action is related to the disruption of the Gram-negative bacteria outer membrane through the displacement of divalent cations in the negative lipopolysaccharide network. It is possible that afterwards, the cationic groups penetrate into the inner membrane to cause leakage, promoting the release of intracellular components (Beyth et al. 2008).

NPs have been applied as drug delivery vehicles in several MDRB combat strategies (Abdelghany et al. 2019; Arif et al. 2022; Jamil et al. 2015). Alginate-modified PLGA NPs loaded with amikacin and moxifloxacin drugs were used as a platform for intramacrophage drug delivery to treat multidrug-resistant tuberculosis (MDR-TB) (Abdelghany et al. 2019; Arif et al. 2022; Jamil et al. 2015). Likewise, to combat the MDRB *Helicobacter pylori* (*H. pylori*) associated with gastric infections, mannose-modified chitosan/poly(malic acid) NPs were investigated. The NPs generated holes in the bacterial cells' outer membrane, inducing its disruption. They were also found antiadherent toward biofilms and capable of disrupting their formation. The NPs-enhanced adherence toward the bacteria is due to the interaction between mannose and a specific subunit protein FimH on the bacteria surface (Abdelghany et al. 2019; Arif et al. 2022; Jamil et al. 2015).

10.2.1.2 Nanofibers

As with other nanomaterials, polymeric nanofibers' characteristics (morphology, diameter, orientation, organization, etc.) are highly dependent on the processing system and the polymers' intrinsic properties. Nanofibrous films can help combat bacteria by preventing their penetration (functioning as a barrier) and by enhancing drug loading and subsequent delivery via their large surface area to volume ratio. Hamdan *et al.* consider the best design for antimicrobial nanofibers to comprehend small diameters, positively charged, rough and hydrophobic surfaces, and large interconnected pores. These features should contribute to enhance surface area and improve the overall interaction with the bacteria cell wall (Hamdan et al. 2021). A variety of complex fiber organizations/architectures (hollow, co-axial, tri-axial, multilayer, side-by-side, porous and helical) can also allow for drug-controlled release systems to be generated (Miranda et al. 2022). The controlled release can be tuned by considering the polymer/drug polarity and solubility.

Nanofibers can load a variety of antibacterial cargo, including metal NPs, carbon nanofibers or nanotubes, AMPs, natural extracts, antibiotics, etc. Also, the nanofibers' surface can be functionalized/modified to establish specific interactions with bacteria. The addiction of electron-withdrawing groups (e.g., amines, carboxylic acids, aldehydes) to the nanofibers surface can improve their net charge and roughness, hence enhancing their antimicrobial properties (Hamdan et al. 2021). Nanofiber-based matrixes have been explored against MDRB with promising results. Curcumin-loaded hyaluronic acid-based nanofibers were studied against the ESKAPE pathogens [*E. faecium*, *Staphylococcus aureus* (*S. aureus*), *Klebsiella pneumoniae* (*K. pneumoniae*), *Acinetobacter baumannii* (*A. baumannii*), *P. aeruginosa,* and *Enterobacter species*]. All nanofibers presented antibacterial activity, with the strongest outcomes being reported against *S. aureus* and *Enterococcus faecalis* (*E. faecalis*) (Hamdan et al. 2021). In another study, curcumin-loaded poly(2-hydroxyethyl methacrylate) nanofibrous scaffold was produced to examine its effectiveness in treating wound infections caused by MRSA and extended spectrum β-lactamases (enzymes produced by certain bacteria, such as *E. coli*, that are effective against β-lactam antibiotics). The natural extract was delivered locally and in a controlled manner from the nanofiber-based matrix, revealing great antimicrobial potential (Ramalingam et al. 2015; Snetkov et al. 2022; Alshabanah et al. 2021). Electrospun gelatin nanofiber dressings containing ε-PL, cross-linked with polydopamine, for the treatment of second-degree burns were also found efficient against various bacteria species, including MRSA and vancomycin-resistant *E. faecium*. ε-PL selectively targeted the bacteria cytoplasmic membrane, while the dressing prompts wound healing (Mayandi et al. 2020).

10.2.1.3 Nanohydrogels

Nanohydrogels (or nanogels) are three-dimensional nanoscale materials, composed of hydrophilic polymers that can significantly absorb and retain fluids. Compared to hydrogels, the nanoscale size of the nanogels enables them to establish specific cell interactions. In addition, their highly absorbent structure separates them from solid polymer NPs (Keskin et al. 2021). Nanohydrogels made of natural and/or synthetic polymers may require physical and/or chemical cross-linking for improving their long-term stability, particularly in physiological environments. Chemically cross-linked nanohydrogels result from covalent interactions (imply covalent bonding between polymer chains or cross-linking precursors), while physically cross-linked ones result from reversible/noncovalent interactions (weak interactions between polymer chains, such as hydrogen bonding, hydrophobic or ionic interactions) (Soni et al. 2016).

Nanohydrogels can present various shapes and sizes, with their 3D structure being capable of encapsulating drugs, thus providing protection against the surroundings. Nanohydrogels can also be tailored to target deliver (Yin et al. 2020). Additionally, trigger-based drug release is frequently used to control drug liberation and it is related to the nanohydrogel hydration/dehydration capacity in response to environmental stimuli, like pH, temperature, presence of enzymes, etc. (Makhathini et al. 2022).

For their antibacterial profile, nanohydrogels can be subdivided in those that possess inherent antibacterial properties (from antibacterial polymers/peptides), those containing an antibacterial agent (traditional antibiotic molecule/bioactive agent), and in those containing more than one antibacterial compound with synergistic effects. These nanomaterials have been successfully applied as antibiotic drug carriers in complex strategies for combating MDRB. For instance, a poly(methacrylic acid) nanogel was prepared as a carrier to decrease the side effects of isoniazid and rifampin drugs used for intestinal tuberculosis treatment, caused by the MDRB *Mycobacterium tuberculosis* (*M. tuberculosis*). The results indicated a prolonged antibacterial activity being promoted by the nanogel long-term drug release and a decrease in the cytotoxic effects compared to the drugs alone (Chen et al. 2016; Weldrick et al. 2019). Furthermore, Weldrick et al. developed polyacrylic nanogels functionalized with a biocompatible cationic polyelectrolyte as carriers for the antibiotics tetracycline hydrochloride and lincomycin hydrochloride. The aim of the work was to engineer a drug carrier that could prevent antibiotic resistance. Data demonstrated an improved antimicrobial activity, including against strains resistant to those antibiotics in free form. The authors confirmed that the antimicrobial effectiveness was due to the localized delivery of a larger dose of antibiotic and the ability of the cationic carrier to induce specific electrostatic adhesion to the anionic bacteria cell wall (Chen et al. 2016; Weldrick et al. 2019).

10.2.1.4 Nanocomposites

Composites result from the combination of two or more components that differ in composition and/or morphology. Composites are made of a matrix (primary/continuous phase) and a reinforcing phase (load-carrying component) and are expected to exhibit enhanced properties over their individual parts. Nanocomposites contain nanoscaled components as the reinforcing phase, known as nanofillers. Polymers (natural and synthetic) are usually used as matrixes because of their flexibility, mechanical resilience, facility of processing, etc. (Shameem et al. 2021). Accordingly, polymer nanocomposites are characterized by their structural diversity/design, easy processing, and relatively low cost (compared to metal and ceramic composites) (Zagho et al. 2018). The reinforcement classification is made in accordance to the nanofiller dimension: two-dimensional layered (e.g., layered silicate, graphene in the form of sheets), one-dimensional fibrous (e.g., carbon nanofibers or nanotubes), and zero-dimensional spherical (e.g., silica NPs, quantum dots). The reinforcement structure and nature play an important role in the overall composite properties. Fu et al. calculated the interfacial area considering spherical nanofillers and concluded that the ultra-large interfacial area between the polymer and the nanofiller endows the composites with superior physical/mechanical properties, distinguishing them from other conventional composites. Hence, as a result of the nanoscale filler dimension, there is a dominance of interfacial regions and, consequently, the mechanical and physical properties are enhanced (Fu et al. 2019).

Apart from the materials' physical properties, the nanofiller presence can also be determinant for the nanocomposite's antibacterial activity. The polymer matrix can ensure this activity, but, generally, it is the nanofiller the major responsible for the nanocomposite antimicrobial performance. For instance, chitosan has been used as base material in a sandwich-like nanocomposite film; however, it was the immobilized silver (Ag) NPs (AgNPs) dispersed on graphene oxide that instigated the observed antibacterial activity (Gu et al. 2021). Antimicrobial nanofillers, such as metal NPs and polymeric nanofibers, are a frequent choice in nanocomposite production and they can also be combined with other bioactive agents. Yang et al. prepared gold NPs coated with 6-aminopenicillanic acid, naming the agent an "antibacterial intermediate" as it was part of the main structural core of β-lactam antibiotics. Here, PCL/gelatin nanofibers coated with the engineered NPs were found to reduce local infection in a dorsal wound model of rat exposed to the MDRB *E. coli* and *P. aeruginosa* (Yang et al. 2017). El-Aassar et al. developed a polymeric wound dressing membrane of chitosan, gelatin, and polyvinyl pyrrolidone (PVP) loaded with AgNPs. As the concentration of AgNPs increased in the nanocomposite, the antibacterial action against various MDRB [e.g., *P. aeruginosa*, *S. aureus*, *E. coli*, *Salmonella typhi*, *Bacillus subtilis* (*B. subtilis*)] was enhanced. This was possible by targeting the microorganisms' drug resistance genes, such as β-lactamase. Here, chitosan facilitated drug loading (e.g., penicillin and macrolide), which together with gelatin improved wound healing, while PVP was used to ensure the dressing elasticity and consequent adaptability to the injured site (El-Aassar et al. 2021). In another study, nanocomposite hydrogels of PVA loaded with cuprous oxide NPs grafted with sepiolite were evaluated against MRSA, MDRB *P. aeruginosa* and MDRB *K. pneumoniae*. The nanofillers were seen to improve many of the nanocomposite properties, namely swelling ratio, moisture retention capacity, and porosity, while fighting bacteria. The nanocomposites were most effective against

Gram-positive bacteria, with the NPs activity being related to the production of ROS (Bhowmick and Koul 2016; Noor ul Ain et al. 2022).

10.3 Multidrug-Resistant Bacteria

The discovery of antibiotics is considered one of the greatest and best achievements of the scientific world and the pharmaceutical industry; however, its use is not benign (Ahmad and Khan 2019; Provenzani et al. 2020). Antibiotics have saved millions of lives in recent decades, but simultaneous they have led to a significant raise of bacteria resistance, causing serious damage to the human health. The effects of antibiotics on the host via the gut microbiome (which play a crucial role in regulating host health) are immense and can affect multiple functions, including immune regulation and metabolic activities, and may also be associated with the appearance of health afflictions like asthma, diabetes, and obesity. *Clostridioides difficile* (*C. difficile*) infection is an example of a disease caused directly by the disruption of the gut microbiota by antibiotics, leading to events of diarrhea that can cause severe colon damage and, ultimately, death. Other serious effects can include nausea, severe and life-threatening allergic reactions, skin irritation, fungal infections, etc. (Patangia et al. 2022). Figure 10.1 shows the brief history of antimicrobial drug development.

Nowadays, MDRB are one of the most important public health problems and are among the ten leading causes of mortality globally, according to the World Health Organization (WHO) (Abdullahi et al. 2023; Yang et al. 2021). They are also associated with high medical costs and have a significant influence on the effectiveness of antimicrobial drugs (Fig. 10.2). Frequently, MDRB occur by the accumulation of

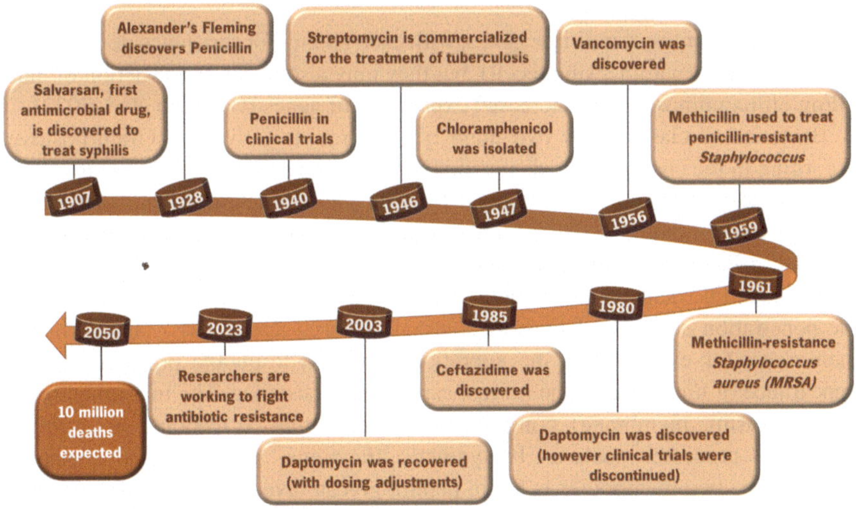

Fig. 10.1 History of antimicrobial drug discovery. (Adapted from da Cunha et al. (2019))

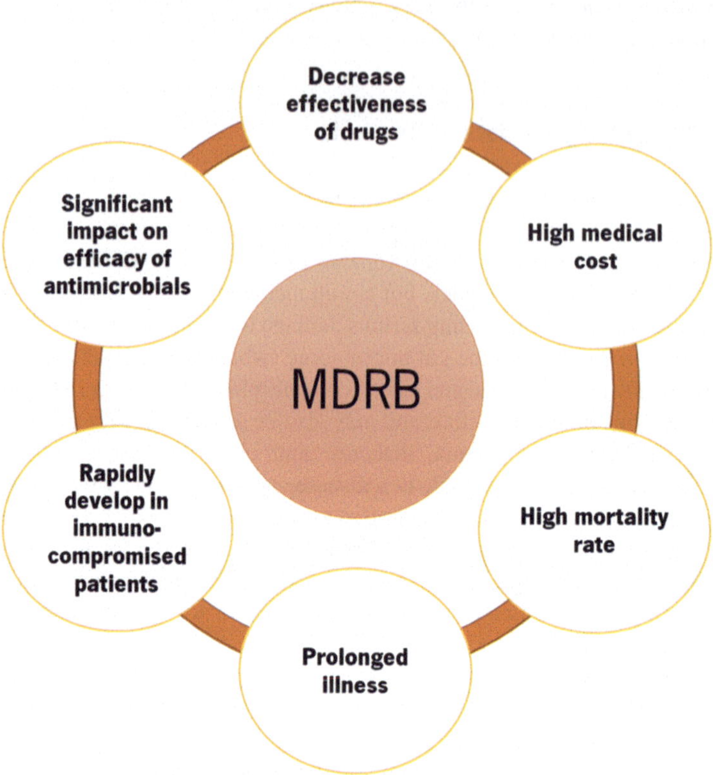

Fig. 10.2 Drug effectiveness and health implications of MDRB. (Adapted from Hetta et al. (2023) with CC BY 4.0 permission)

resistant R-plasmids of multiple genes, with each coding for resistance development to a single drug (within a single cell), but they may also occur by the action of multidrug efflux pumps, due to increased expression of genes (Catalano et al. 2022).

Bacteria are classified in two main types: Gram-negative and Gram-positive, according to the composition, disposition, and structure of the cell wall (Fig. 10.3). Gram-negative bacteria are enveloped by an asymmetric lipid bilayer membrane with the outer membrane being highly permeable, containing mainly lipopolysaccharides (LPS) and proteins, such as porins that allow the passage of small molecules like amino acids or saccharides. The function of the lipid bilayer is to reduce the penetration capacity of hydrophobic antibacterial agents. Any change in this membrane, for example via the alteration of the hydrophobic properties or the appearance of mutations in the porins, can trigger resistance. The inner membrane (or cytoplasmic membrane), composed of phospholipids, is responsible for multifunctional processes such as biosynthetic functions, structure, and transport. Between the two membranes is a thin peptidoglycan layer, the main component in the bacterial cell wall, which consists of polymeric glycan strands of two alternating β-(1→4)-linked sugars: *N*-acetylglucosamine and *N*-acetylmuramic acid,

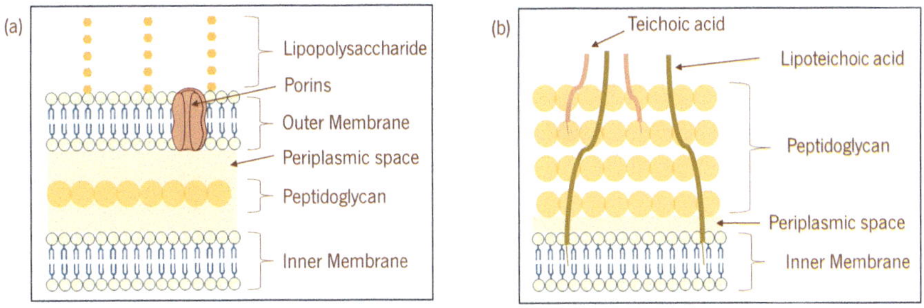

Fig. 10.3 Cell wall structure of (**a**) gram-negative and (**b**) gram-positive bacteria. (Adapted from Riu et al. (2022) with CC BY 4.0 permission)

surrounded by a periplasmic space. On the other hand, the cell wall of Gram-positive bacteria is made of peptidoglycan layers, thicker than in the Gram-negatives, ornamented with long anionic glycopolymers called teichoic acids and lipoteichoic acids. Teichoic acids are covalently linked to the peptidoglycan, whereas lipoteichoic acids are anchored to the inner membrane by a glycolipid anchor. These glycopolymers perform several functions, including controlling autolytic enzymes, regulating divalent cations, binding surface proteins, and protecting against antibacterial molecules. As the outer membrane is not present, this makes Gram-positive bacteria less resistant to antibiotics than Gram-negative (Tavares et al. 2020; Jubeh et al. 2020; Riu et al. 2022).

M. tuberculosis has been classified as an MDRB. MDR-TB is caused by strains of *M. tuberculosis* that do not respond to the two most powerful antituberculosis drugs, the isoniazid and the rifampicin (Espinosa-Pereiro et al. 2022). The cell wall of *M. tuberculosis*, compared to other Gram-positive bacteria, is thicker and more hydrophobic due to the presence of a highly branched heteropolysaccharide, the arabinogalactan, which covalently links the peptidoglycan to the outer layer of the mycolic acid via a phosphodiester bond (Italia et al. 2023; Jacobo-Delgado et al. 2023).

Antimicrobial resistance (AMR) is the resistance of a microorganism to an antimicrobial drug against which it was originally sensitive. It is a serious and growing global (public health) problem, which is expected to cause ten million deaths worldwide, annually by 2050. AMR is associated with the misuse and abuse of antibiotics, not only in humans but also in agriculture and ecosystems, animal farming, and industry, and affects all countries; although some (developed countries) are feeling the impact more than others (Cui et al. 2023). AMR occurs when microorganisms, such as bacteria, fungi, protozoa, and viruses, develop resistance to drugs (antimicrobial) that inhibit or kill them (such as antibiotics, antifungals, anthelminthics, antimalarials, antiprotozoals, and antivirals), spreading that resistance to future generations (Grenni et al. 2018; Pfab et al. 2021). Overuse of antibiotics is a significant driver of resistance, also leading to increased healthcare costs. Infection by antibiotic-resistant bacteria is behind many severe diseases, increased death rates, and an increased risk of complications and hospitalization. The drugs developed are

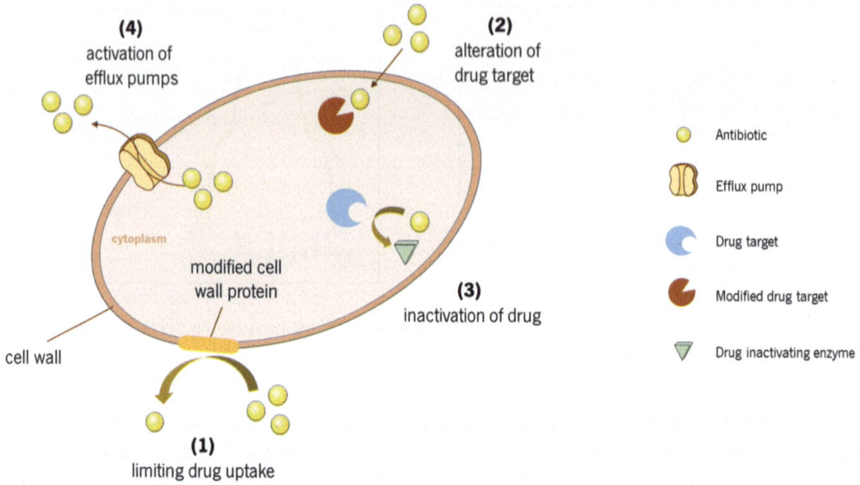

Fig. 10.4 Schematic representation of the bacteria resistance mechanisms. (Adapted from Rahman et al. (2022) with CC BY 4.0 permission)

ineffective and the infections persist in the body, increasing the risk of contagion to other people, being the search for alternative treatments a growing demand (Llor and Bjerrum 2014).

There are four main mechanisms described for the occurrence of antimicrobial resistance in bacterial cells (Fig. 10.4): (1) limiting drug uptake, (2) drug targeting modification, (3) inactivation or modification of the drug, and (4) activation of efflux pumps (Rahman et al. 2022). Due to the differences in the structure, Gram-negative bacteria use all four of the above mechanisms, while Gram-positive bacteria are less likely to limit the uptake of a drug (absence of LPS in the outer membrane) and do not possess the capacity for certain types of drug efflux mechanisms (Parmanik et al. 2022).

10.3.1 Limiting Drug Uptake

Gram-negative bacteria are resistant and less permeable to antimicrobials due to the presence of an outer cell wall and LPS layer. Porins facilitate the transport of hydrophilic molecules across the outer membrane, while hydrophobic molecules can diffuse across the lipid bilayer. This mechanism is not seen in Gram-positive bacteria due to the absence of an outer membrane to restrict drug entry, but this does not mean that resistance is not found (Nainu et al. 2021; Zhang and Cheng 2022).

10.3.2 Drug Targeting Modification

Modification of the antibiotic target is a common mechanism by which bacteria become resistant to antibiotics, and there are several components that can be targeted by these agents: point mutations in the gene that encode the target site (DNA gyrase and topoisomerase IV make them resistant to fluoroquinolones that act by inhibiting DNA transcription and replication), enzymatic alteration of the binding site, and bypassing the original site (Chawla et al. 2022).

10.3.3 Inactivation of the Drug

Inactivation of the drug by bacteria can be achieved by two ways: by transferring a chemical group (such as acetyl, adenyl, and phosphoryl groups to the compound) that prevents the antibiotic from binding to its target, or by degrading the drug by the action of β-Lactamases. These molecules are the most common example of enzymes that degrade the drug molecule by hydrolysis of the amide bond of the β-lactam ring, which renders it ineffective (Chawla et al. 2022; Darby et al. 2023).

10.3.4 Efflux Pumps

Efflux pumps pump the antibiotic out of the bacterial cells, reducing their intracellular concentration. Different types of efflux pumps, involved in antibiotic resistance, have been identified in both Gram-negative and Gram-positive bacteria, with distinct organizations (Chawla et al. 2022; Mohanty et al. 2021). There are six families of efflux pumps distinguished based on characteristics like structure and energy source: (1) ATP-binding cassette (ABC) superfamily, (2) the multidrug and toxic-compound extrusion (MATE) family, (3) the major facilitator superfamily (MFS), (4) the resistance-nodulation-division (RND) family, (5) the small multidrug resistance (SMR) family, and (6) the proteobacterial antimicrobial compound efflux (PACE) family (Baran et al. 2023). Among the six families, RND is considered the main efflux pump, only found in Gram-negative bacteria and is involved in the bacterial pathogenesis, virulence, and stress response. They also play an important role in resistance to many toxic substances, such as biocides and heavy metals (Auda et al. 2020). Efflux pumps in Gram-positive bacteria are usually encoded by genes based on chromosomes or plasmids. The MFS superfamily is the most widely studied in Gram-positive bacteria (Schindler and Kaatz 2016).

Various classes of antibiotics have been identified as ineffective toward some bacteria due to their resistance: β-Lactams (Carbapenems, Cephalosporins, Penicillins); Aminoglycosides (Amikacin, Gentamicin, Spectinomycin, Streptomycin, Tobramycin); Cationic peptides (Colistin, Polymyxin-B); Fluoroquinolones/Quinolones (Ciprofloxacin, Levofloxacin, Nalidixic acid, Norfloxacin, Ofloxacin); Glycopeptides (Teicoplanin, Vancomycin); Lincosamides (Clindamycin); Lipopeptides (Daptomycin); Macrolides (Azithromycin, Clarithromycin, Erythromycin); Oxazolidinones (Linezolid); Phenicols (Azidamphenicol, Chloramphenicol, Thiamphenicol); Rifamycins (Rifampin); Tetracyclines (Doxycycline, Minocycline, Tigecycline). Examples of the bacteria resistance mechanisms are described in Table 10.1.

Table 10.1 Examples of bacterial resistance mechanism against specific classes of antibiotics

Bacterium type	Name of bacterium	Typical diseases	Resistance to class of antibiotics	Resistance mechanism	Ref.
Gram-negative	*A. baumannii*	Bacteremia Gastrointestinal Meningitis Nosocomial infections Pneumonia Skin/wound infections Urinary tract infections (UTIs)	β-Lactams Aminoglycosides Cationic peptides Macrolides Fluoroquinolones/quinolones Rifamycins Tetracyclines	Drug inactivation Efflux pumps Membrane permeability Target modification	Vázquez-López et al. (2020), Vrancianu et al. (2020), Kyriakidis et al. (2021) and Saha and Sarkar 2021)
	E. coli	Blood stream infections UTIs	β-Lactams Aminoglycosides Cationic peptides Fluoroquinolones/quinolones Phenicols Rifamycins Tetracyclines	Drug inactivation Efflux pumps Target modification	Saha and Sarkar (2021), Tanwar et al. (2014) and Alarcon and Omenaca (2004)
	H. pylori	Chronic gastritis Duodenal and stomach ulcers Gastric cancer	β-Lactams Fluoroquinolones/quinolones Macrolides Rifamycins Tetracyclines	Alteration in penicillin-binding proteins (PBPs) Efflux pumps Membrane permeability rRNA mutations (Gerrits et al. 2006)	Gerrits et al. (2006), Lin et al. (2023) and Tshibangu-Kabamba and Yamaoka (2021)
	K. pneumoniae	Blood stream infections Pneumonia UTIs	β-Lactams Aminoglycosides Fluoroquinolones Rifamycins Tetracyclines	Biofilm formation Drug inactivation Efflux pumps Membrane permeability Target modification	Saha and Sarkar (2021), Tanwar et al. (2014), Galani et al. (2021), Ballén et al. (2021) and Li et al. (2022)
	Neisseria gonorrhoeae (*N. gonorrhoeae*)	Cervicitis Conjuctivitis Gonorrhea Mucosal tissues Pharyngitis Systemic disease Urethritis	β-Lactams Aminoglycosides Fluoroquinolones/quinolones Macrolides Tetracyclines	Drug inactivation Efflux pumps Membrane permeability Target modification	Saha and Sarkar (2021), Tanwar et al. (2014), Rubin et al. (2020) and Mlynarczyk-Bonikowska et al. (2022)
	P. aeruginosa	Bacteremia Burn and wound injuries Otitis externa Systemic disease UTIs	β-Lactams Aminoglycosides Cationic peptides Fluoroquinolones/quinolones	Drug inactivation Efflux pumps Membrane permeability	Saha and Sarkar (2021), Botelho et al. (2019), Karruli et al. (2023) and Mancuso et al. (2023)
	Shigella species	Diarrhea	β-Lactams Fluoroquinolones/quinolones Macrolides Phenicols Tetracyclines	Class beta-lactamases Chloramphenicol acetyltransferase Efflux pumps Drug inactivation Membrane permeability Plasmid-borne resistance	Tanwar et al. (2014) and Ranjbar and Farahani (2019)

Gram-positive	*C. difficile*	Diarrhea Sepsis	β-Lactams Fluoroquinolones/quinolones Glycopeptides Oxazolidinones Phenicols Tetracyclines	Biofilm formation Chloramphenicol acetyltransferase Drug inactivation Efflux pumps Ribosomal protection Target modification	Saha and Sarkar (2021), Spigaglia et al. (2018) and Wickramage et al. (2021)
	S. aureus	Bacteremia Blood stream infections Nosocomial infections Osteomyelitis Pneumonia Sepsis Wound infections	β-Lactams Aminoglycosides Fluoroquinolones/quinolones Glycopeptides Lincosamides Lipopeptides Macrolides Oxazolidinones Phenicols Tetracyclines	Drug inactivation Efflux pumps Limiting drug uptake Target modification	Saha and Sarkar (2021), Tanwar et al. (2014), Reygaert (2018) and Nandhini et al. (2022)
	Streptococcus pneumoniae (*S. pneumoniae*)	Bacteremia Meningitis Otitis media Pneumonia Septicemia	β-Lactams Fluoroquinolones/quinolones Macrolide Phenicols Tetracyclines	Alteration in PBPs Efflux pumps Ribosomal protection	Saha and Sarkar (2021), Tanwar et al. (2014) and Li et al. (2023)
Other	*M. tuberculosis*	Tuberculosis	Aminoglycosides Ethambutol Fluoroquinolones/quinolones Isoniazid Oxazolidinones Pyrazinamide Rifamycins	Drug inactivation Efflux pumps Prodrug activation Target modification	Saha and Sarkar (2021), Shantanam and Mueller (2018) and Gygli et al. (2017)

PBPs penicillin-binding proteins, *UTIs* urinary tract infections

10.3.5 Resistance Mechanisms Through Biofilm Formation

Biofilms are aggregates adhered to the surface of colonies of bacterial cells combined in a single location, forming a matrix of extracellular polymeric substance (EPS) that are adherent to each other (Shamim et al. 2023). Adherence to surfaces, differences between interfaces, decreased antimicrobial susceptibility, host defense mechanisms, and three-dimensional organizations are some important characteristics that biofilms need to survive (Khan et al. 2021). The presence of biological materials such as blood, interstitial fluid, respiratory secretions, saliva, tears, urine, and wound cultures influence the adhesion of bacteria to their surface and promote the formation of biofilm. Nutrients, pH, salinity, and temperature also can effect on biofilm formation (Roy et al. 2022; Liu et al. 2023).

Biofilm formation can be described in four main stages (Fig. 10.5): (1) attachment (adhesion of planktonic cells to the surface), (2) proliferation (colony formation and bacterial growth), (3) maturation (accumulation of bacteria cells, diatoms, fungi, protozoa, producing colonies that are surrounded by an EPS matrix composed of exogenous DNA, exopolysaccharides, lipids, and proteins), and (4) dispersion (biofilm maturation actively or passively breaks down to disperse the microorganisms into the environment as planktonic cells initiate a new cycle of biofilm formation. These microbial communities produce enzymes, which degrade the stability of the polysaccharides, causing the release of surface bacterial cells that colonize a new surface. For example, *P. aeruginosa* generates alginate lyase and *E. coli* produces *N*-acetyl-heparosan lyase for biofilm matrix degradation) (Shamim et al. 2023; Asma et al. 2022; de Brito et al. 2022).

Biofilms formed by MDRB are more serious and difficult to combat. A classic case of biofilm involvement in chronic infections is the cystic fibrosis, in which biofilms are derived from *P. aeruginosa*. This chronic infection is usually incurable and can lead to death. Additionally, biofilms are responsible for medical device related infections, meningitis, osteomyelitis, pneumonia, and chronic wounds (Liu et al. 2023; Asma et al. 2022; Scandorieiro et al. 2023). In biofilm formation,

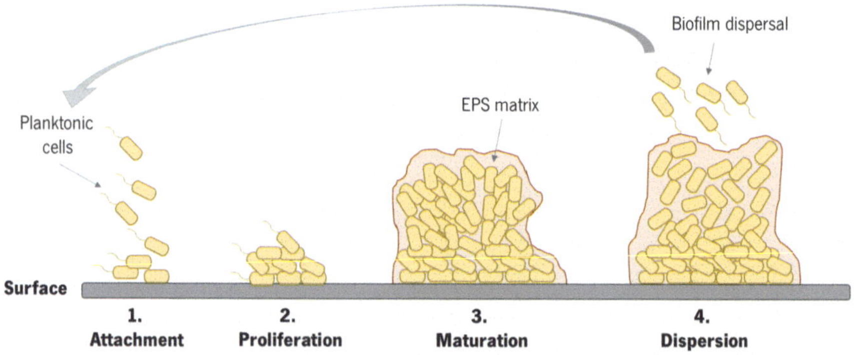

Fig. 10.5 The four steps in biofilm formation. (Adapted from de Brito et al. (2022) with CC BY 4.0 permission)

quorum sensing molecules play a significant regulatory role and can result in dispersion. Quorum sensing is a crucial communication channel between cells for biofilm development and varies in chemical nature between bacteria. Gram-negative bacteria produce *N*-acyl-homoserine lactones as a signaling molecule, while Gram-positive produce autoinducing peptides (Shamim et al. 2023).

10.4 Polymer-Based Nanomaterials Against Multidrug-Resistant Bacteria

In recent decades, nanomaterials have emerged as promising tools to overcome the MDRB problematic, complementing the action of conventional agents, facilitating their delivery (Makabenta et al. 2021a; Hemeg 2017; Munir and Ahmad 2022) and reducing their side effects (Yeh et al. 2020; Safhi et al. 2016). Nanoscale materials penetrate more easily through tissue barriers, ensuring drug protection until cellular absorption, and can facilitate targeted release (Spirescu et al. 2021). In general, the interactions established between nanomaterials and bacteria are based on electrostatic attraction, hydrophobic bonding, receptor-ligand interactions, and Van der Waals forces (Munir and Ahmad 2022; Nandhini and Rajan 2022).

Many polymer-based nanomaterials display a broad spectrum of antibacterial activity (Wang and Sun 2021); yet, that is dependent on the type of nanomaterial, its composition, and the type of targeted bacteria, since different antibacterial mechanism of action may need to be implemented (Díaz-Puertas et al. 2023; Chandrasekaran et al. 2020). In the case of polymeric NPs, properties such as chemistry, particle size and shape, surface-to-volume ratio, and zeta potential can affect their antibacterial action (Makhathini et al. 2022; Spirescu et al. 2021). Regarding polymeric nanofiber films, factors such as surface area to volume ratio, nano size, high porosity, and interconnected pores are the main effectors on their antibacterial activity (Hamdan et al. 2021). The smaller the diameter of the nanofibers, the greater the probability for them to alter the bacteria's conformation and, thus, increase their susceptibility (Abrigo et al. 2015). Another important parameter to take into consideration in the production of the nanomaterial are the hydrophobic domains of the polymers. The more hydrophobic the polymer that composes the nanomaterial, the more likely it is for the hydrophobic interactions between lipid and protein membranes to occur and, consequently, the greater the accumulation of polymers and the generation of pores in the membranes of bacteria (Pearce and O'Reilly 2021). However, modifying polymer matrices with NPs must be carefully analyzed in order to regulate their toxicity to mammalian cells (Gupta et al. 2018). In addition, it is important to note that the length and charge of the polymer directly influence its antibacterial activity (Ma et al. 2022). Polymers with long chains have a greater bacteriostatic activity against Gram-negative bacteria (Namivandi-Zangeneh et al. 2018). Thus, the nanomaterial can interfere with the different components of the bacteria and, in this way, condition the growth of the bacteria, trigger its death or prevent the formation of biofilms (Lam et al. 2018;

Costa et al. 2017; Chindera et al. 2016). It should be noted that the study of the antibacterial potential of polymer-based nanomaterials against MDRBs has focused mainly on research aimed at the use of polymer-based nanomaterials as drug delivery systems (Malzahn et al. 2014; Wang et al. 2020; Ejaz et al. 2020; Alhusein et al. 2016).

10.4.1 Mechanisms of Action

Polymer-based nanomaterials may exert a bactericidal effect by interfering with the structural stability of bacteria and/or by interacting with their intracellular components (Kamaruzzaman et al. 2018; Divya et al. 2017). Their antibacterial activity is associated with the intrinsic antibacterial properties of the nanomaterial that are directly related to the physicochemical properties of the polymer (Divya et al. 2017; MubarakAli et al. 2018). On the other hand, the antibacterial action of polymer-based nanomaterials can be associated with the properties of the incorporated drugs. Thus, functioning as delivery vehicles for antimicrobial agents such as antibiotics, natural extracts, AMPs, and photosensitizers (Wang and Sun 2021; Sedghi and Shaabani 2016; Salem et al. 2019; Khosravimelal et al. 2021; Usacheva et al. 2016). In addition, polymeric nanomaterials may present antibacterial adhesion action, preventing structures such as biofilms from forming and proliferating (Gupta et al. 2018).

The nanomaterials mode of action against MDRBs can be divided in three main types: (1) Membrane Targeting Mechanism; (2) Nonmembrane Targeting Mechanism; and (3) Mechanism of Action against Biofilms (Fig. 10.6).

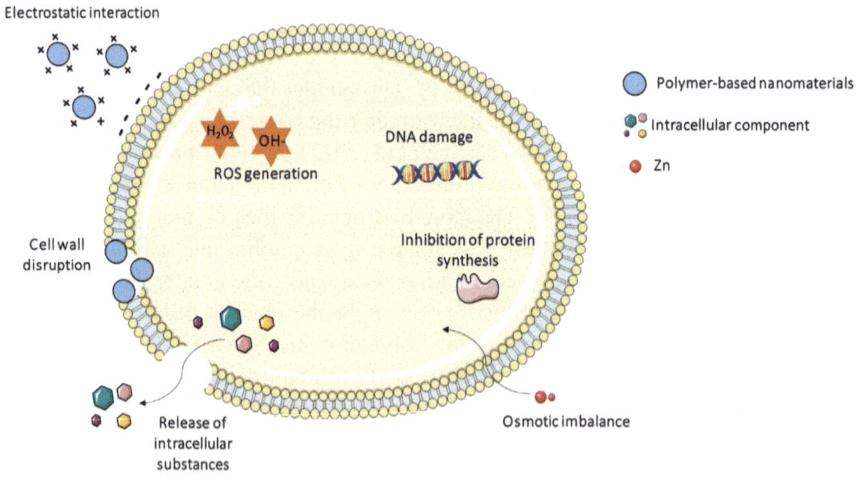

Fig. 10.6 Membrane and nonmembrane targeting antibacterial mechanisms associated with polymer-based nanomaterials. (Adapted from Rofeal et al. (2022) with CC BY 4.0 permission)

10.4.1.1 Membrane Targeting Mechanisms

Membrane and cell wall are key components for bacterial survival and successful infection/invasion of host cells (Pizarro-Cerdá and Cossart 2006; Nikolic and Mudgil 2023). In addition, cell membranes are an important target for many antibacterial formulations. As mentioned, Gram-positive bacteria have a phospholipid membrane and a thick wall rich in peptidoglycan, while Gram-negative bacteria have a higher phospholipid content present in both cell membranes. Given the different compositions of the bacterial surface, the action of antibacterial agents can be more or less effective (Chandrasekaran et al. 2020). Regardless, many polymer-based nanomaterials have shown excellent ability to disturb the bacterial surface, thus being effective against both Gram-negative and Gram-positive bacteria (Lam et al. 2018; Morena and Tzanov 2022; Ding et al. 2019).

Specifically, nanomaterials based on cationic polymers (e.g., chitosan) have been described as materials with excellent antibacterial capacity, namely against MDRBs by destabilizing the bacterial surface (Kamaruzzaman et al. 2018; Zhao et al. 2020; Hou et al. 2017; Nederberg et al. 2011; Thappeta et al. 2020). It is known that positively charged polymers [e.g., chitosan, polypyrrole, polyaniline, dextran, and polyethyleneimine (PEI)] can easily destabilize bacteria because of their chemical groups (e.g., amino groups of glucosamine) that have a high affinity toward the negatively charged components present along the bacterial surface (e.g., anionic lipids of cardiolipin, phosphatidylglycerol, and phosphatidylserine) (Hoque et al. 2016; Antunes et al. 2021; Nandhini and Rajan 2022; Chandrasekaran et al. 2020; Samal et al. 2012). These polymers can disrupt the bacterial surface (bacterial cell membrane or cell wall structure) and/or change the permeability of the membrane. These modifications occur due to the significant loss of bacterial membrane potential, electrostatic interactions, and the insertion of hydrophobic groups into the phospholipid bilayer. Insertion of hydrophobic groups causes translocation of anionic lipids, resulting in perforation and disruption of membranes. Associated with these strong interactions, osmotic imbalance and the leakage/efflux of intracellular substances (e.g., intracellular ions) are also observed, which trigger bacterial cells death (Pan et al. 2019; Nasri et al. 2021; Li et al. 2015).

It is important to highlight that the polycationic positive charge density is a crucial factor for an effective antibacterial action (Ma et al. 2022). The higher the charge density of the NPs, the greater the interaction with bacterial cells. In addition to the polymer charge, the surface area of the nanomaterial is also an important factor for a good antibacterial action (Silva et al. 2015). Of all the polymer-based nanomaterials studied against MDRB, chitosan NPs have been the most explored. The antibacterial activity of chitosan NPs is attributed in part to their high positive zeta potentials that enable interactions with the bacterial surface and to their large surface area that facilitate bacterial penetration and consequent membrane disruption (Costa et al. 2018). For instance, chitosan-based NPs loaded with oligolysine antimicrobial peptide (CSM5-K5) have been described as able to group the lipids of the anionic membrane, disrupting its structure and, thus, enabling a rapid bactericidal activity against MRSA, MDRB *E. coli*, and vancomycin-resistant *E. faecalis* (VRE)

(Thappeta et al. 2020). Based on the mechanism of action centered on the destabilization of the bacterial surface, many cationic nanostructures have been considered promising in fighting drug-resistant bacteria, namely *E. coli*, *P. aeruginosa*, *E. faecalis,* and *B. subtilis* (Zhao et al. 2020).

Polymer-based nanomaterials can also be modified with bioactive molecules or other nanomaterials to increase their antibacterial potential (Wang and Sun 2021; Jin et al. 2014). Poly (lactic-co-glycolic acid)–polyethyleneimine (PLGA–PEI) NPs loaded with clindamycin (antibiotic) interact through electrostatic interactions with the cell wall of MRSA bacteria, modifying their physicochemical properties. In this way, the entry and controlled release of the antibiotic into the bacteria is facilitated. This action allows a sufficient concentration of drug to be trapped in a specific location, thereby inhibiting bacterial growth and protein synthesis and causing bacteria death (Hasan et al. 2019a). Chitosan NPs when modified with carbohydrate groups (e.g., mannose) acquire improved action against several MDRBs. Mannose polysaccharide improves the interaction of NPs with some MDRBs (e.g., *P. aeruginosa*, and *S. aureus*), disrupting the bacterial cell membrane. Chitosan nanofibers loaded with AgNPs destabilize the cell membrane of MRSA bacteria by means of the affinity of AgNPs toward the negatively charge surface (Jin et al. 2014). The same potential against MRSA is observed in PVA/chitosan and in poly(2-hydroxy ethyl methacrylate) nanofibers functionalized with curcumin. The incorporation of this natural extract gives these structures excellent antibacterial potential. Even though curcumin's mechanisms of action require further analysis (Sedghi and Shaabani 2016; Ramalingam et al. 2014); it is known that curcumin treatment is associated with damage to the bacterial cell wall, disruption of the cell membrane, and consequent release of cytoplasmic contents, ultimately leading to cell lysis (Mun et al. 2014).

Some studies also show the antibacterial potential of polymeric nanohydrogels against MDRBs (Fasiku et al. 2022). In the case of nisin-loaded chondroitin sulfate nanogels, it is known that these have an antibacterial action against MRSA bacteria possibly due to nisin's ability to bind to the bacterial membrane (through physical interactions, such as hydrophobic or electrostatic bonds) and cause disruption of their membrane functions (Tayeferad et al. 2021).

10.4.1.2 Nonmembrane Targeting Mechanisms

Polymeric nanomaterials are also known for their antibacterial action based on a mechanism involving the production of ROS and/or interaction with intracellular elements (Chindera et al. 2016; Malzahn et al. 2014; Wang et al. 2022).

10.4.1.2.1 Production of Reactive Oxygen Species

Some polymeric nanomaterials can induce the production of elevated amounts of ROS, which consequently lead to bacterial death due to their ability to damage various bacterial cell components (Wang et al. 2022). ROS cause irreparable damage to the cell membrane and oxidation of polyunsaturated fatty acids and amino acids, which inhibit enzymatic activity, interfere with DNA synthesis, interrupt energy transduction (ATP), disturb cellular metabolism, and finally trigger cell lysis

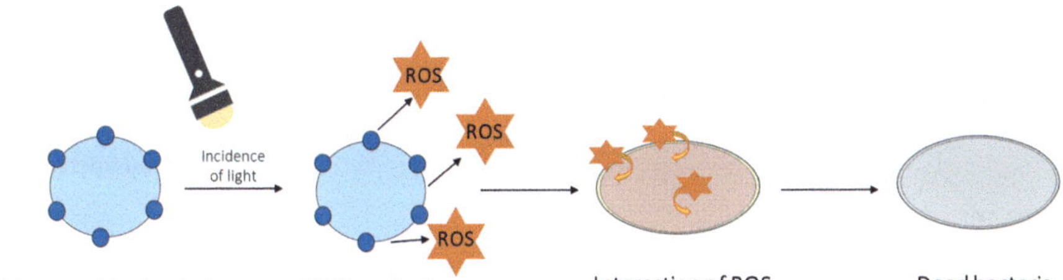

Fig. 10.7 Schematic representation of the antibacterial mechanism of polymeric NPs modified with photosensitizers for the production of ROS. (Adapted from Yuan et al. (2022) and Ma et al. (2016))

(Hamdan et al. 2021; Nandhini and Rajan 2022). Most polymeric nanomaterials (without photosensitizer properties) work as delivery vehicles for photosensitizers (small molecules linked to polymers by covalent or noncovalent interactions capable of absorbing light energy) so that bacterial inactivation is possible from the production of ROS (Fig. 10.7) (Henke et al. 2020; Guo et al. 2020). With these molecular modifications (incorporation of photosensitizers) and with the use of an energy source (e.g., light emitting diodes; at a specific wavelength), it is possible to generate different types of ROS (e.g., H_2O_2, O^{2-}, and OH). Thus, it is also possible to impar the bacteria functions, putting at risk their viability (Usacheva et al. 2016; Polat and Kang 2021). It should be noted that the photodynamic antibacterial inactivation technique is a noninvasive therapeutic and is considered quite effective against MDRB species (Yin et al. 2015). Therefore, associating this technique with polymer-based nanomaterials is an excellent alternative for fighting MDRBs.

10.4.1.2.2 Binding to and Damaging Intracellular Components

Bacteria are composed of several intracellular components that, in an interconnected way, allow the bacteria to function properly. Specifically, any DNA damage will adversely affect bacterial synthesis and replication. It is described that some cationic polymers (e.g., polyhexamethylene biguanide; PHMB) can inhibit cell division and selectively bind and condense bacterial chromosomes by directly interacting with DNA. After bonding with the DNA by strong electrostatic interactions (between the polymer's cationic charges and the negatively charged phosphate backbone of the DNA), the PHMB condenses the DNA and NPs are formed, triggering bacteria death (Fig. 10.8) (Chindera et al. 2016). In this case, the polymer can be used in its free form or in the form of NPs to contact with the DNA, with the last option being more appropriate for reducing polymer toxicity. PHMB NPs can also be conjugated with different antibiotics (e.g., nadifloxacin) to increase the antibacterial potential of this nanomaterial (by interrupting cleavage-relegation during the DNA replication process). In this way, PHMB-based nanomaterials kill and prevent the regrowth of bacteria, namely MRSA (Kamaruzzaman et al. 2018).

Chitosan NPs can be loaded with cells (e.g., mesenchymal stem cells) and the joint action of both elements (chitosan and proteins secreted by cells) make the wall

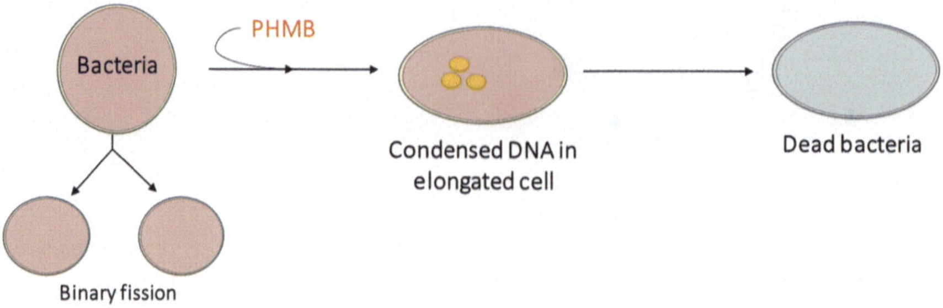

Fig. 10.8 Proposed antibacterial mode of action of polyhexamethylene biguanide (PHMB) against MRSA. (Adapted from Chindera et al. (2016) with CC BY 4.0 permission)

of bacterial cells more disruptive, like the multidrug-resistant *Vibrio cholerae*. This alteration facilitates the entry of chitosan into the bacterial cytoplasm, allowing the polymer to penetrate the bacterial nuclei and inhibiting mRNA and protein synthesis (Saberpour et al. 2020).

In addition to the described mechanisms, inhibition of the bacterial fatty acid biosynthesis pathway using PLGA NPs loaded with platensimycin (inhibitor of bacterial fatty acid synthases) has also been reported as an effective alternative to kill MRSA (Liu et al. 2020). This is possible because these biosynthetic pathways are extremely important for the production of lipids that will play a vital role in the structure and function of bacteria (Cronan and Thomas 2009).

A dysregulation in intracellular levels of metal ions (e.g., Zn) can disrupt several metabolic pathways and be lethal to bacteria. High concentrations of Zn, for example, can trigger the denaturation and malfunction of several proteins and lead to enzymatic inactivation, putting at risk the survival of the bacteria (Godoy-Gallardo et al. 2021). In this sense, increasing the concentration of zinc in these environments can be a good antibacterial strategy. Specifically, zinc nitrate can be incorporated into dextran nanogels to serve as an antimicrobial agent. This nanostructure works as a zinc-controlled release platform allowing excellent antibacterial action against MRSA (Malzahn et al. 2014).

10.4.1.3 Mechanisms of Action Against Biofilms

Biofilms are considered communities of bacteria that are surrounded/protected by a matrix (extracellular polymeric substances). This association and community organization makes these bacteria show greater resistance to various drugs and, hence, gain a greater capacity to resist to the host's immune system (Makabenta et al. 2021b). As previously mentioned, for the formation of a biofilm, several steps must take place, starting with bacterial adhesion. Bacterial adhesion and interaction with the nanomaterial surface are influenced by surface chemistry, surface charge, roughness, porosity, topography, and wettability of the nanostructure (James et al. 2019; Damiati et al. 2018). Although the mechanism of action is still not entirely clear, it is known that many polymeric nanomaterials have a major impact on preventing biofilm formation. They may interfere with the initial adhesion of bacteria to surfaces or penetrate and break down established biofilms. In addition, they facilitate the action of antibacterial agents and allow them not to be easily degraded/inactivated by the extracellular substances of the biofilm matrix and to be delivered directly to the site of infection (Fig. 10.9) (Lee et al. 2022; Ramalingam et al. 2021; Birk et al. 2021).

Many polymeric NPs (e.g., oxanorbornene polymer derivatives) with a highly cationic and hydrophobic nature play an important role against MRSA and in destabilizing *P. aeruginosa* and MRSA biofilms. This happens because these NPs have the ability to penetrate and disperse in biofilms, causing disruption of the bacterial membrane (Gupta et al. 2018). Poly(oxanorbornene imide)-based polymeric NPs (PONI; antimicrobial polymer) are described as able to penetrate and eradicate multispecies-based bacterial biofilms of *E. coli* and MRSA, without affecting the viability of fibroblasts (Makabenta et al. 2021b). PCL/GN nanofibers functionalized with an antibiotic (minocycline) and plant extract (*Gymnema sylvestre*) by allowing a sustained release of these bioactive components, potentiate their antibacterial action against MRSA and prevent the formation of MRSA/*P. aeruginosa* biofilms (Ramalingam et al. 2021). PCL/polyethylene oxide (PEO) nanofibers loaded with the antibiotic vancomycin appeared to have the ability to inhibit the proliferation and formation of MRSA biofilm on the surface of ossicular prostheses and otitis media (Jang et al. 2015). On the other hand, PLGA–PEI NPs (positively charged) with incorporated nitric oxide (NO) bind effectively to the surface of the biofilm matrix (negatively charged components), causing a strong dispersion of the MRSA biofilm and reducing its biomass. Interestingly, wounds infected by this biofilm and treated with this formulation had almost complete healing in 12 days, unlike the control (there was no closure of the wound within the same time frame) (Hasan et al. 2019b).

Chitosan NPs play an active role in combating the development of MDRB-derived biofilms, varying their mechanism of action and effectiveness depending on the bacterial strain (Costa et al. 2017; Darabpour et al. 2016). Thus these NPs can be effective in reducing MRSA biofilm formation and in preventing adhesion of MRSE (Costa et al. 2017). In addition, chitosan NPs increase the effectiveness of antimicrobial photodynamic inactivation (methylene blue used as a photosensitizer)

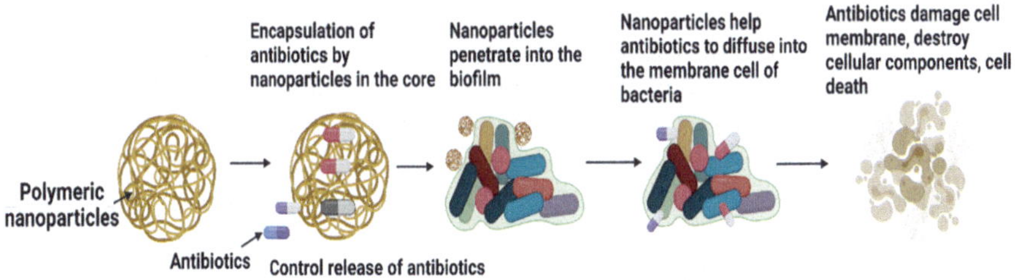

Fig. 10.9 Schematic representation of the mechanism of action of polymeric NPs functionalized with antibiotics toward the destruction of bacterial biofilms. (Adapted from Thambirajoo et al. (2021) with CC BY 4.0 permission)

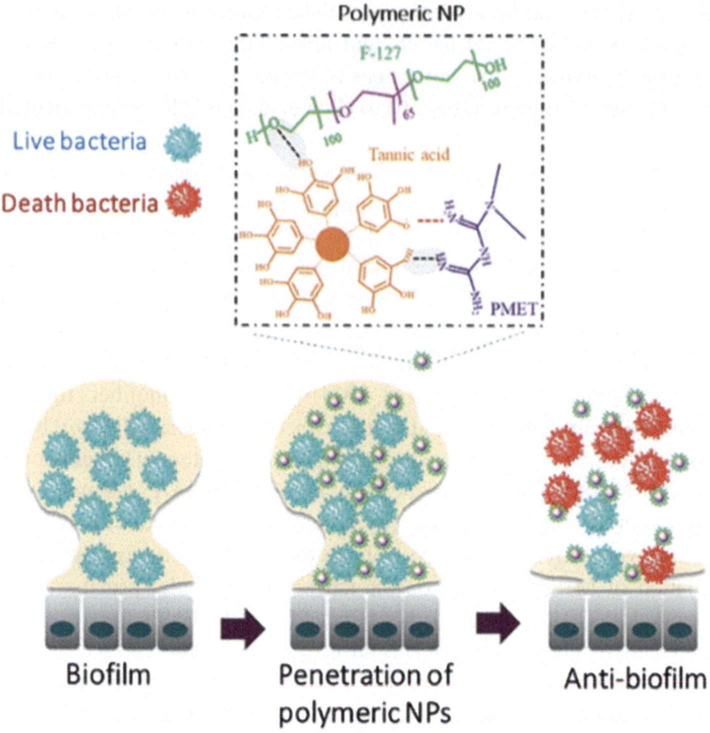

Fig. 10.10 Anti-MRSA biofilm action by polymeric NPs made of biocompatible surfactant F-127, tannic acid, and biguanide-based polymetformin (PMET). (Reprinted (adapted) with permission from Zhao et al. (2023). Copyright 2023 American Chemical Society)

against MDR *P. aeruginosa* biofilms, since they cause the disruption of the biofilm structure and subsequently allow deeper and greater penetration of the photosensitizer into the biofilms (Darabpour et al. 2016). Chitosan NPs can also be modified with mannose sugar (mannose functionalized chitosan nanosystem) to increase their antibiofilm potential and antiadherence activity. This modification makes the nanosystem to increase its affinity with some bacterial membrane proteins called adhesins, more specifically with lectins (strong affinity with mannose). Lectins are used by bacteria to adhere and develop a biofilm. In this way, the interactions of this polymeric nanosystem with the bacteria prevent the lectins from carrying out some of their important functions related to adhesion and motility, which will condition invasion, bacterial infection, and biofilm formation of resistant bacteria (e.g., *E. coli* and *Listeria monocytogenes*) and MDRBs (*P. aeruginosa* e *S. aureus*) (Ejaz et al. 2020). Also, the biodegradable polymer-coated AgNPs nanocomposite resultant from the combination of *N*,*N*-dimethyl-*N*-hexadecyl ammonium chitin tosylate polymer and silver (antimicrobial agent) have an excellent ability to eradicate biofilm of MRSA bacteria, and inactivate various drug-resistant bacteria such as vancomycin-resistant *E. faecium*, and β-lactam-resistant *K. pneumoniae* (Hoque

Table 10.2 Examples of polymer-based nanomaterials against multidrug-resistant bacteria for biomedical applications

Nanomaterial	Composition	Bacterial target	Mechanism of action	Application	Ref.
Nanocomposites	Black phosphorus (BP) nanosheets + hemoglobin self-assembled layer-wise onto electrospun poly-L-lactide (PLLA) nanofibers with charged quaternized chitosan (QCS) and hyaluronic acid	MRSA and *E. coli*	Synergistic effect of all components (positively charged QCS targets the negatively charged bacterial membrane increasing membrane permeability, which leads to the destruction of the cell structure); sharp edge of BP during interactions with bacteria causes physical damage to the bacterial membrane and induces RNA leakage, increases the intracellular ROS levels and bacterial sensitivity positively inhibiting bacterial growth	Antibacterial activity in diabetic wound healing	Zhao et al. (2023) and Jang et al. (2023)
	Graphene/chitosan coating	MDR *P. aeruginosa* and *K. pneumoniae*	Intracellular and cell surface damage	Antibiofilm formation	Muthuchamy et al. (2020)
	Lipase-sensitive methoxy poly (ethylene glycol)-block-poly(ε-caprolactone) (mPEG-PCL micelles loaded with photosensitizer hypocrellin A (HA)	MRSA	Release of HA (bacteria inhibitor) via light irradiation	New antibacterial treatment through photodynamic therapy	Guo et al. (2020)
	Chitosan/mannose (polysaccharide) coating	Resistant *E. coli*, Listeria monocytogenes, and MDR *P. aeruginosa* and *S. aureus*	Membrane interaction	Alternative approach to fight emerging resistant infections	Ejaz et al. (2020) and Rofeal et al. (2022)
	PCL nanofiber with curcumin-loaded chitosan nanoencapsulated particles	MRSA and extended spectrum β-lactamase	Not mentioned	Wound healing	Fahimirad et al. (2021)

(continued)

Table 1 (continued)

Nanomaterial	Composition	Bacterial target	Mechanism of action	Application	Ref.
	Cationic triblock polycarbonates micelle	MRSA	Cationic micelles interact with the negatively charged cell wall by means of an electrostatic bond, inhibiting cell wall synthesis and/or damage the cell wall, resulting in cell lysis Micelles may easily permeate the cytoplasmic membrane due to their relatively large volume, destabilizing the membrane as a result of electroporation, leading to cell death	Multiresistant antimicrobial drug	Nederberg et al. (2011)
Nanofibers	Gelatin-poly(citrate-siloxane)-ε-poly(L-lysine) nanofibrous scaffold	*E. coli* and MRSA	Direct killing	Antibacterial activity in bone tissue regeneration	Xi et al. (2020)
	Coaxial electrospun core of poly(vinylpyrrolidone) (PVP) encapsulating emodin at the core and cellulose acetate at the sheath	MRSA	Release of emodin that promotes pyruvate pathway imbalance induction, protein synthesis inhibition, and DNA synthesis suppression	Wound healing	Ye et al. (2020)
	PVA-loaded AMP proline-rich host defense peptide	*A. baumannii*	Not mentioned	Wound healing	Sebe et al. (2016)
	PCL/gum tragacanth loaded with curcumin	MRSA and extended spectrum β-lactamase	Not mentioned	Antimicrobial activity promoting wound healing in diabetic wounds	Mohammadi et al. (2016)
Nanohydrogels	Poly(citrate-glycol-siloxane) (PCGS) hybrid polymer/F127-ε-poly-L-lysine	MRSA	Not mentioned	Treating MDRB-infected wound and skin repair	Cheng et al. (2021)
	Hyaluronic acid-cholesterol with entrapped LVF (fluoroquinolone antibiotic)	MRSA	LVF-NHs allow the cellular uptake of the drug by facilitating contact between the drug and bacteria and by reducing intracellular inactivation of the drug and its efflux	Intracellular infection treatments	Montanari et al. (2014)
	Poly(methacrylic acid) (PMAA)/isoniazid and rifampin (antibiotics)	*M. tuberculosis*	PMMA nanogel loaded with INH and RMP showed pH sensitivity on gastric fluid, releasing the antibiotic content that entered the bacterial membrane through diffusion	Treatment for intestinal tuberculosis	Chen et al. (2016)

Nanomaterial	Composition	Bacterial target	Mechanism of action	Application	Ref.
Nanoparticles	Alginate loaded with toluidine blue	*P. aeruginosa* and MRSA	ROS production through photosensitizers	Photodynamic therapy for the management of chronic infections	Usacheva et al. (2016)
	Alginate/honey/calcium chloride (1:1:1) loaded with rifampicin (antibiotic)	*M. tuberculosis*	Intracellular environment causes the release of rifampicin inside the cell once the NP penetrates the bacterial cell wall	Tuberculosis treatment	Griffiths et al. (2010)
	Chitosan/bacteriophage	*P. aeruginosa*, Salmonella typhimurium (*S. typhimurium*) and Staphylococcus sciuri (*S. sciuri*)	Binding to negatively charged bacterial cell walls, which disrupts and changes the permeability of the membrane causing malformations (holes and ruptures) and killing the cell by triggering intracellular component leakage	Alternative approach to fight antimicrobial resistance	Abdelsattar et al. (2023)
	Polyacrylate/penicillin	MRSA	Direct interaction of NP with the bacterial cell at membrane level	Free-penicillin substitute (nanobiotic)	Turos et al. (2007a)
	Polyacrylate/*N*-thiolated β-lactam antibiotic	MRSA	Increased bioavailability (membrane permeability or enhanced local concentration) of the antibiotic agent as a consequence of the nanoparticle serving as a delivery vehicle	Antibiotic substitute (nanobiotic)	Turos et al. (2007b)

et al. 2019). With regard to hyaluronic acid-based nanohydrogels capable of codelivering NO and a novel AMP with sequence RKKKKLLRKKC (antibiofilm agent), it is known that they have demonstrated the ability to eradicate MRSA biofilms. This effect is due to the inhibition of the accessory regulator *Staphylococcal A* (SarA), a DNA-binding protein that regulates the expression of the virulence and adhesion factor of the MRSA biofilm. Thus, inhibition of SarA prevents adherence of the MRSA biofilm to the surface (Fasiku et al. 2022; Morrison et al. 2012).

Hydrophilic polymers [e.g., poly(ethylene glycol); PEG], due to their antifouling properties, have the potential to be used against the formation/development of biofilms (Wu et al. 2019; Mitra et al. 2021). Biguanide-derived polymeric NPs [nanosystem made from biocompatible F-127 surfactant, tannic acid, and biguanide-based polymetformin (PMET); FTP NPs] can effectively kill MRSA biofilm bacteria. This nanosystem proved to be effective in dispersing and removing the biofilm. Furthermore, it can eradicate preformed biofilm by penetrating its matrix and then killing the bacteria. It may also cause depolarization of the cytoplasmic membrane, permeabilization, and the formation of MRSA clusters. It should be noted that the incorporation of the hydrophilic PEG chain of F-127 on the surfaces of FTP NPs gives this structure an improved effectiveness in inhibiting the formation of biofilms, since this antifouling chain can facilitate their penetration into the biofilm (Fig. 10.10) (Li et al. 2020). Table 10.2 lists other examples of polymer-based nanomaterials with good antibacterial action and the respective mechanisms of action against MDRBs.

10.5 Conclusion and Future Perspectives

MDR pathogens are becoming an emerging public health crisis with economic impact, making many healthcare-related diseases difficult to treat using conventional strategies. Research and development of novel nanomaterials may assist in controlling diseases caused by these pathogens. Given the mechanical and biological properties, modification capacity and availability of various types of polymers (both natural and synthetic) make these materials excellent for developing new treatment methodologies against MDRB. The variety of polymeric nanomaterials enables their use in various contexts, demonstrating antibacterial properties that can target the most diverse types of MRB in order to decrease or eradicate the evolution of MDRB. The effectiveness of any polymeric nanomaterial on fighting MDR relies on (1) effectiveness in controlling infection, (2) low toxicity toward the host, and (3) long-term, sustainable antimicrobial activity. At the same time, improved knowledge of the mechanisms of action (for both bacteria resistance and bacteria fighting) to control MDR should facilitate the development of these novel therapies and, hopefully, provide a deeper understanding of the pathobiology of resistant microbial organisms.

Acknowledgments This research was funded by the Portuguese Foundation for Science and Technology (FCT) via grant UIDP/00264/2020 of 2C2T Strategic Project 2020–2023. A.R.M.R.,

M.O.T, A.F.G.S acknowledge FCT for PhD scholarships 2021.08627.BD, 2021.06906.BD, 2022.11171.BD, respectively. H.P.F. acknowledges FCT for Auxiliary Researcher contract 2021.02720.CEECIND.

References

Abdelghany S, Parumasivam T, Pang A et al (2019) Alginate modified-PLGA nanoparticles entrapping amikacin and moxifloxacin as a novel host-directed therapy for multidrug-resistant tuberculosis. J Drug Deliv Sci Technol 52:642–651. https://doi.org/10.1016/j.jddst.2019.05.025

Abdelsattar AS, Yakoup AY, Khaled Y et al (2023) The synergistic effect of using bacteriophages and chitosan nanoparticles against pathogenic bacteria as a novel therapeutic approach. Int J Biol Macromol 228:374–384. https://doi.org/10.1016/j.ijbiomac.2022.12.246

Abdullahi ZU, Musa SS, Abu-Odah H et al (2023) Bactericidal effects of snake venom phospholipases A2: a systematic review and analysis of minimum inhibitory concentration. Physiologia 3:30–42. https://doi.org/10.3390/physiologia3010003

Abrigo M, Kingshott P, McArthur SL (2015) Electrospun polystyrene fiber diameter influencing bacterial attachment, proliferation, and growth. ACS Appl Mater Interfaces 7:7644–7652. https://doi.org/10.1021/acsami.5b00453

Adeniji OO, Nontongana N, Okoh JC, Okoh AI (2022) The potential of antibiotics and nanomaterial combinations as therapeutic strategies in the management of multidrug-resistant infections: a review. Int J Mol Sci 23:15038. https://doi.org/10.3390/ijms232315038

Ahmad M, Khan AU (2019) Global economic impact of antibiotic resistance: a review. J Glob Antimicrob Resist 19:313–316. https://doi.org/10.1016/j.jgar.2019.05.024

Alarcon A, Omenaca F (2004) Antimicrobial resistance in Escherichia coli sepsis. Pediatr Infect Dis J 23:979–980. https://doi.org/10.1097/01.inf.0000141751.73083.7a

Alfei S, Schito AM (2020) Positively charged polymers as promising devices against multidrug resistant gram-negative bacteria: a review. Polymers (Basel) 12:1195. https://doi.org/10.3390/POLYM12051195

Alhusein N, Blagbrough IS, Beeton ML et al (2016) Electrospun Zein/PCL fibrous matrices release tetracycline in a controlled manner, killing *Staphylococcus aureus* both in biofilms and ex vivo on pig skin, and are compatible with human skin cells. Pharm Res 33:237–246. https://doi.org/10.1007/s11095-015-1782-3

Alshabanah LA, Hagar M, Al-Mutabagani LA et al (2021) Biodegradable nanofibrous membranes for medical and personal protection applications: manufacturing, anti-COVID 19 and anti-multidrug resistant bacteria evaluation. Materials (Basel) 14:3892. https://doi.org/10.3390/ma14143862

Antunes JC, Domingues JM, Miranda CS et al (2021) Bioactivity of chitosan-based particles loaded with plant-derived extracts for biomedical applications: emphasis on antimicrobial fiber-based systems. Mar Drugs 19:359. https://doi.org/10.3390/md19070359

Arif M, Ahmad R, Sharaf M et al (2022) Antibacterial and antibiofilm activity of mannose-modified chitosan/PMLA nanoparticles against multidrug-resistant *Helicobacter pylori*. Int J Biol Macromol 223:418–432. https://doi.org/10.1016/j.ijbiomac.2022.10.265

Asma ST, Imre K, Morar A et al (2022) Natural strategies as potential weapons against bacterial biofilms. Life 12:1618. https://doi.org/10.3390/life12101618

Auda IG, Ali Salman IM, Odah JG (2020) Efflux pumps of Gram-negative bacteria in brief. Gene Rep 20:100666. https://doi.org/10.1016/j.genrep.2020.100666

Baek S, Joo SH, Toborek M (2019) Treatment of antibiotic-resistant bacteria by encapsulation of ZnO nanoparticles in an alginate biopolymer: insights into treatment mechanisms. J Hazard Mater 373:122–130. https://doi.org/10.1016/j.jhazmat.2019.03.072

Ballén V, Gabasa Y, Ratia C et al (2021) Antibiotic resistance and virulence profiles of Klebsiella pneumoniae strains isolated from different clinical sources. Front Cell Infect Microbiol 11:1–11. https://doi.org/10.3389/fcimb.2021.738223

Baran A, Kwiatkowska A, Potocki L (2023) Antibiotics and bacterial resistance—a short story of an endless arms race. Int J Mol Sci 24:5777. https://doi.org/10.3390/ijms24065777

Barhoum A, García-Betancourt ML, Jeevanandam J et al (2022) Review on natural, incidental, bioinspired, and engineered nanomaterials: history, definitions, classifications, synthesis, properties, market, toxicities, risks, and regulations. Nanomaterials 12:177. https://doi.org/10.3390/nano12020177

Beyth N, Houri-Haddad Y, Baraness-Hadar L et al (2008) Surface antimicrobial activity and biocompatibility of incorporated polyethylenimine nanoparticles. Biomaterials 29:4157–4163. https://doi.org/10.1016/j.biomaterials.2008.07.003

Bhatia S (2016) Natural polymer drug delivery systems: nanoparticles, plants, and algae. Springer, Berlin

Bhowmick S, Koul V (2016) Assessment of PVA/silver nanocomposite hydrogel patch as antimicrobial dressing scaffold: synthesis, characterization and biological evaluation. Mater Sci Eng C 59:109–119. https://doi.org/10.1016/j.msec.2015.10.003

Birk SE, Boisen A, Nielsen LH (2021) Polymeric nano- and microparticulate drug delivery systems for treatment of biofilms. Adv Drug Deliv Rev 174:30–52. https://doi.org/10.1016/j.addr.2021.04.005

Botelho J, Grosso F, Peixe L (2019) Antibiotic resistance in *Pseudomonas aeruginosa*—mechanisms, epidemiology and evolution. Drug Resist Updat 44:100640. https://doi.org/10.1016/j.drup.2019.07.002

Catalano A, Iacopetta D, Ceramella J et al (2022) Multidrug resistance (MDR): a widespread phenomenon in pharmacological therapies. Molecules 27:1–18. https://doi.org/10.3390/molecules27030616

Chandrasekaran M, Kim KD, Chun SC (2020) Antibacterial activity of chitosan nanoparticles: a review. Processes 8:1–21. https://doi.org/10.3390/PR8091173

Chawla M, Verma J, Gupta R, Das B (2022) Antibiotic potentiators against multidrug-resistant bacteria: discovery, development, and clinical relevance. Front Microbiol 13:1–19. https://doi.org/10.3389/fmicb.2022.887251

Chen T, Li Q, Guo L et al (2016) Lower cytotoxicity, high stability, and long-term antibacterial activity of a poly(methacrylic acid)/isoniazid/rifampin nanogel against multidrug-resistant intestinal Mycobacterium tuberculosis. Mater Sci Eng C 58:659–665. https://doi.org/10.1016/j.msec.2015.08.055

Cheng W, Wang M, Chen M et al (2021) Injectable antibacterial antiinflammatory molecular hybrid hydrogel dressing for rapid MDRB-infected wound repair and therapy. Chem Eng J 409:128140. https://doi.org/10.1016/j.cej.2020.128140

Chindera K, Mahato M, Kumar Sharma A et al (2016) The antimicrobial polymer PHMB enters cells and selectively condenses bacterial chromosomes. Sci Rep 6:1–13. https://doi.org/10.1038/srep23121

Costa EM, Silva S, Vicente S et al (2017) Chitosan nanoparticles as alternative anti-staphylococci agents: bactericidal, antibiofilm and antiadhesive effects. Mater Sci Eng C 79:221–226. https://doi.org/10.1016/j.msec.2017.05.047

Costa EM, Silva S, Veiga M et al (2018) Exploring chitosan nanoparticles as effective inhibitors of antibiotic resistant skin microorganisms—from in vitro to ex vitro testing. Carbohydr Polym 201:340–346. https://doi.org/10.1016/j.carbpol.2018.08.083

Cronan JE, Thomas J (2009) Chapter 17. Bacterial fatty acid synthesis and its relationships with polyketide synthetic pathways, 1st edn. Elsevier, Amsterdam

Cui ZH, He HL, Zheng ZJ et al (2023) Phentolamine significantly enhances macrolide antibiotic antibacterial activity against MDR gram-negative bacteria. Antibiotics 12:760. https://doi.org/10.3390/antibiotics12040760

da Cunha BR, Fonseca LP, Calado CRC (2019) Antibiotic discovery: where have we come from, where do we go? Antibiotics 8:45. https://doi.org/10.3390/antibiotics8020045

Damiati L, Eales MG, Nobbs AH et al (2018) Impact of surface topography and coating on osteogenesis and bacterial attachment on titanium implants. J Tissue Eng 9:1–16. https://doi.org/10.1177/2041731418790694

Darabpour E, Kashef N, Mashayekhan S (2016) Chitosan nanoparticles enhance the efficiency of methylene blue-mediated antimicrobial photodynamic inactivation of bacterial biofilms: an in vitro study. Photodiagnosis Photodyn Ther 14:211–217. https://doi.org/10.1016/j.pdpdt.2016.04.009

Darby EM, Trampari E, Siasat P et al (2023) Molecular mechanisms of antibiotic resistance revisited. Nat Rev Microbiol 21:280–295. https://doi.org/10.1038/s41579-022-00820-y

de Brito FAE, de Freitas APP, Nascimento MS (2022) Multidrug-resistant biofilms (MDR): main mechanisms of tolerance and resistance in the food supply chain. Pathogens 11:1416. https://doi.org/10.3390/pathogens11121416

Díaz-Puertas R, Álvarez-Martínez FJ, Falco A et al (2023) Phytochemical-based nanomaterials against antibiotic-resistant bacteria: an updated review. Polymers (Basel) 15:1392. https://doi.org/10.3390/polym15061392

Ding X, Wang A, Tong W, Xu FJ (2019) Biodegradable antibacterial polymeric nanosystems: a new hope to cope with multidrug-resistant bacteria. Small 15:1–29. https://doi.org/10.1002/smll.201900999

Divya K, Vijayan S, George TK, Jisha MS (2017) Antimicrobial properties of chitosan nanoparticles: mode of action and factors affecting activity. Fibers Polym 18:221–230. https://doi.org/10.1007/s12221-017-6690-1

Ejaz S, Ihsan A, Noor T et al (2020) Mannose functionalized chitosan nanosystems for enhanced antimicrobial activity against multidrug resistant pathogens. Polym Test 91:106814. https://doi.org/10.1016/j.polymertesting.2020.106814

El Fawal G, Hong H, Mo X, Wang H (2021) Fabrication of scaffold based on gelatin and polycaprolactone (PCL) for wound dressing application. J Drug Deliv Sci Technol 63:102501. https://doi.org/10.1016/j.jddst.2021.102501

El-Aassar MR, Ibrahim OM, Fouda MMG et al (2021) Wound dressing of chitosan-based-crosslinked gelatin/polyvinyl pyrrolidone embedded silver nanoparticles, for targeting multidrug resistance microbes. Carbohydr Polym 255:117484. https://doi.org/10.1016/j.carbpol.2020.117484

Elfadil D, Elkhatib WF, El-Sayyad GS (2022) Promising advances in nanobiotic-based formulations for drug specific targeting against multidrug-resistant microbes and biofilm-associated infections. Microb Pathog 170:105721. https://doi.org/10.1016/j.micpath.2022.105721

Espinosa-Pereiro J, Sánchez-Montalvá A, Aznar ML, Espiau M (2022) MDR tuberculosis treatment. Medicina 58:1–34. https://doi.org/10.3390/medicina58020188

Fahimirad S, Abtahi H, Satei P et al (2021) Wound healing performance of PCL/chitosan based electrospun nanofiber electrosprayed with curcumin loaded chitosan nanoparticles. Carbohydr Polym 259:117640. https://doi.org/10.1016/j.carbpol.2021.117640

Fasiku VO, Omolo CA, Kiruri LW et al (2022) A hyaluronic acid-based nanogel for the co-delivery of nitric oxide (NO) and a novel antimicrobial peptide (AMP) against bacterial biofilms. Int J Biol Macromol 206:381–397

Fu S, Sun Z, Huang P et al (2019) Some basic aspects of polymer nanocomposites: a critical review. Nano Mater Sci 1:2–30. https://doi.org/10.1016/j.nanoms.2019.02.006

Galani I, Karaiskos I, Giamarellou H (2021) Multidrug-resistant Klebsiella pneumoniae: mechanisms of resistance including updated data for novel β-lactam-β-lactamase inhibitor combinations. Expert Rev Anti Infect Ther 19:1457–1468. https://doi.org/10.1080/14787210.2021.1924674

Gerrits MM, van Vliet AH, Kuipers EJ, Kusters JG (2006) Helicobacter pylori and antimicrobial resistance: molecular mechanisms and clinical implications. Lancet Infect Dis 6:699–709. https://doi.org/10.1016/S1473-3099(06)70627-2

Ghomi ER, Lakshminarayanan R, Chellappan V et al (2023) Electrospun aligned PCL/gelatin scaffolds mimicking the skin ECM for effective antimicrobial wound dressings. Adv Fiber Mater 5:235–251. https://doi.org/10.1007/s42765-022-00216-w

Godoy-Gallardo M, Eckhard U, Delgado LM et al (2021) Antibacterial approaches in tissue engineering using metal ions and nanoparticles: from mechanisms to applications. Bioact Mater 6:4470–4490. https://doi.org/10.1016/j.bioactmat.2021.04.033

Grenni P, Ancona V, Barra Caracciolo A (2018) Ecological effects of antibiotics on natural ecosystems: a review. Microchem J 136:25–39. https://doi.org/10.1016/j.microc.2017.02.006
Griffiths G, Nyström B, Sable SB, Khuller GK (2010) Nanobead-based interventions for the treatment and prevention of tuberculosis. Nat Rev Microbiol 8:827–834. https://doi.org/10.1038/nrmicro2437
Gu B, Jiang Q, Luo B et al (2021) A sandwich-like chitosan-based antibacterial nanocomposite film with reduced graphene oxide immobilized silver nanoparticles. Carbohydr Polym 260:117835. https://doi.org/10.1016/j.carbpol.2021.117835
Guo LY, Yan SZ, Tao X et al (2020) Evaluation of hypocrellin A-loaded lipase sensitive polymer micelles for intervening methicillin-resistant *Staphylococcus aureus* antibiotic-resistant bacterial infection. Mater Sci Eng C 106:110230. https://doi.org/10.1016/j.msec.2019.110230
Gupta A, Landis RF, Li CH et al (2018) Engineered polymer nanoparticles with unprecedented antimicrobial efficacy and therapeutic indices against multidrug-resistant bacteria and biofilms. J Am Chem Soc 140:12137–12143. https://doi.org/10.1021/jacs.8b06961
Gygli SM, Borrell S, Trauner A, Gagneux S (2017) Antimicrobial resistance in Mycobacterium tuberculosis: mechanistic and evolutionary perspectives. FEMS Microbiol Rev 41:354–373. https://doi.org/10.1093/femsre/fux011
Hamdan N, Yamin A, Hamid SA et al (2021) Functionalized antimicrobial nanofibers: design criteria and recent advances. J Funct Biomater 12:59. https://doi.org/10.3390/jfb12040059
Han J, Zhao D, Li D et al (2018) Polymer-based nanomaterials and applications for vaccines and drugs. Polymers (Basel) 10:1–14. https://doi.org/10.3390/polym10010031
Hasan N, Cao J, Lee J et al (2019a) Bacteria-targeted clindamycin loaded polymeric nanoparticles: effect of surface charge on nanoparticle adhesion to MRSA, antibacterial activity, and wound healing. Pharmaceuticals 11:236
Hasan N, Cao J, Lee J et al (2019b) PEI/NONOates-doped PLGA nanoparticles for eradicating methicillin-resistant *Staphylococcus aureus* biofilm in diabetic wounds via binding to the biofilm matrix. Mater Sci Eng C 103:109741. https://doi.org/10.1016/j.msec.2019.109741
Hemeg HA (2017) Nanomaterials for alternative antibacterial therapy. Int J Nanomedicine 12:8211–8225. https://doi.org/10.2147/IJN.S132163
Henke P, Dolanský J, Kubát P, Mosinger J (2020) Multifunctional photosensitizing and biotinylated polystyrene nanofiber membranes/composites for binding of biologically active compounds. ACS Appl Mater Interfaces 12:18792–18802. https://doi.org/10.1021/acsami.9b23104
Hetta HF, Ramadan YN, Al-Harbi AI et al (2023) Nanotechnology as a promising approach to combat multidrug resistant bacteria: a comprehensive review and future perspectives. Biomedicine 11:413. https://doi.org/10.3390/biomedicines11020413
Hoque J, Adhikary U, Yadav V et al (2016) Chitosan derivatives active against multidrug-resistant bacteria and pathogenic fungi: in vivo evaluation as topical antimicrobials. Mol Pharm 13:3578–3589. https://doi.org/10.1021/acs.molpharmaceut.6b00764
Hoque J, Yadav V, Prakash RG et al (2019) Dual-function polymer-silver nanocomposites for rapid killing of microbes and inhibiting biofilms. ACS Biomater Sci Eng 5:81–91. https://doi.org/10.1021/acsbiomaterials.8b00239
Hou Z, Shankar YV, Liu Y et al (2017) Nanoparticles of short cationic peptidopolysaccharide self-assembled by hydrogen bonding with antibacterial effect against multidrug-resistant bacteria. ACS Appl Mater Interfaces 9:38288–38303. https://doi.org/10.1021/acsami.7b12120
Italia A, Shaik MM, Peri F (2023) Emerging extracellular molecular targets for innovative pharmacological approaches to resistant Mtb infection. Biomolecules 13:999
Jacobo-Delgado YM, Rodríguez-Carlos A, Serrano CJ, Rivas-Santiago B (2023) Mycobacterium tuberculosis cell-wall and antimicrobial peptides: a mission impossible? Front Immunol 14:1194923. https://doi.org/10.3389/fimmu.2023.1194923
James GA, Boegli L, Hancock J et al (2019) Bacterial adhesion and biofilm formation on textured breast implant shell materials. Aesthetic Plast Surg 43:490–497. https://doi.org/10.1007/s00266-018-1234-7
Jamil B, Habib H, Abbasi S et al (2015) Cefazolin loaded chitosan nanoparticles to cure multi drug resistant Gram-negative pathogens. Carbohydr Polym 136:682–691. https://doi.org/10.1016/j.carbpol.2015.09.078

Jang CH, Cho YB, Jang YS et al (2015) Antibacterial effect of electrospun polycaprolactone/polyethylene oxide/vancomycin nanofiber mat for prevention of periprosthetic infection and biofilm formation. Int J Pediatr Otorhinolaryngol 79:1299–1305. https://doi.org/10.1016/j.ijporl.2015.05.037

Jang EJ, Patel R, Patel M (2023) Electrospinning nanofibers as a dressing to treat diabetic wounds. Pharmaceutics 15:1144. https://doi.org/10.3390/pharmaceutics15041144

Jin S, Nyoung D, Moon J et al (2014) Electrospun chitosan nanofibers with controlled levels of silver nanoparticles. Preparation, characterization and antibacterial activity. Carbohydr Polym 111:530–537. https://doi.org/10.1016/j.carbpol.2014.04.026

Jubeh B, Breijyeh Z, Karaman R (2020) Resistance of gram-positive bacteria to current antibacterial agents and overcoming approaches. Molecules 25:1340. https://doi.org/10.3390/molecules25122888

Kamaruzzaman NF, de Pina M, Chivu A, Good L (2018) Polyhexamethylene biguanide and nadifloxacin self-assembled nanoparticles: antimicrobial effects against intracellular methicillin-resistant *Staphylococcus aureus*. Polymers (Basel) 10:521. https://doi.org/10.3390/polym10050521

Karruli A, Catalini C, D'Amore C et al (2023) Evidence-based treatment of *Pseudomonas aeruginosa* infections: a critical reappraisal. Antibiotics 12:1–19. https://doi.org/10.3390/antibiotics12020399

Keshvardoostchokami M, Majidi SS, Huo P et al (2021) Electrospun nanofibers of natural and synthetic polymers as artificial extracellular matrix for tissue engineering. Nanomaterials 11:1–23. https://doi.org/10.3390/nano11010021

Keskin D, Zu G, Forson AM et al (2021) Nanogels: a novel approach in antimicrobial delivery systems and antimicrobial coatings. Bioact Mater 6:3634–3657. https://doi.org/10.1016/j.bioactmat.2021.03.004

Khan J, Tarar SM, Gul I et al (2021) Challenges of antibiotic resistance biofilms and potential combating strategies: a review. 3 Biotech 11:1–15. https://doi.org/10.1007/s13205-021-02707-w

Khosravimelal S, Chizari M, Farhadihosseinabadi B et al (2021) Fabrication and characterization of an antibacterial chitosan/silk fibroin electrospun nanofiber loaded with a cationic peptide for wound-dressing application. J Mater Sci Mater Med 32:114. https://doi.org/10.1007/s10856-021-06542-6

Kyriakidis I, Vasileiou E, Pana ZD, Tragiannidis A (2021) *Acinetobacter baumannii* antibiotic resistance mechanisms. Pathogens 10:1–31. https://doi.org/10.3390/pathogens10030373

Lam SJ, Wong EHH, Boyer C, Qiao GG (2018) Antimicrobial polymeric nanoparticles. Prog Polym Sci 76:40–64. https://doi.org/10.1016/j.progpolymsci.2017.07.007

Lee HW, Kharel S, Loo SCJ (2022) Lipid-coated hybrid nanoparticles for enhanced bacterial biofilm penetration and antibiofilm efficacy. ACS Omega 7:35814–35824. https://doi.org/10.1021/acsomega.2c04008

León-Buitimea A, Garza-Cárdenas CR, Román-García MF et al (2022) Nanomaterials-based combinatorial therapy as a strategy to combat antibiotic resistance. Antibiotics 11:1–34. https://doi.org/10.3390/antibiotics11060794

Li Z, Yang F, Yang R (2015) Synthesis and characterization of chitosan derivatives with dual-antibacterial functional groups. Int J Biol Macromol 75:378–387. https://doi.org/10.1016/j.ijbiomac.2015.01.056

Li S, Dong S, Xu W et al (2018) Antibacterial hydrogels. Adv Sci 5:1700527. https://doi.org/10.1002/advs.201700527

Li J, Zhong W, Zhang K et al (2020) Biguanide-derived polymeric nanoparticles kill MRSA biofilm and suppress infection in vivo. ACS Appl Mater Interfaces 12:21231–21241. https://doi.org/10.1021/acsami.9b17747

Li Y, Kumar S, Zhang L, Wu H (2022) Klebsiella pneumonia and its antibiotic resistance: a bibliometric analysis. Biomed Res Int 2022:10. https://doi.org/10.1155/2022/1668789

Li L, Ma J, Yu Z et al (2023) Epidemiological characteristics and antibiotic resistance mechanisms of Streptococcus pneumoniae: an updated review. Microbiol Res 266:127221. https://doi.org/10.1016/j.micres.2022.127221

Lin Y, Shao Y, Yan J, Ye G (2023) Antibiotic resistance in Helicobacter pylori: from potential biomolecular mechanisms to clinical practice. J Clin Lab Anal 37:e24885. https://doi.org/10.1002/jcla.24885

Liu X, Wang Z, Feng X et al (2020) Platensimycin-encapsulated poly(lactic-co-glycolic acid) and poly(amidoamine) dendrimers nanoparticles with enhanced anti-staphylococcal activity in vivo. Bioconjug Chem 31:1425–1437. https://doi.org/10.1021/acs.bioconjchem.0c00121

Liu X, Yao H, Zhao X, Ge C (2023) Biofilm formation and control of foodborne pathogenic bacteria. Molecules 28:2432. https://doi.org/10.3390/molecules28062432

Llor C, Bjerrum L (2014) Antimicrobial resistance: risk associated with antibiotic overuse and initiatives to reduce the problem. Ther Adv Drug Saf 5:229–241. https://doi.org/10.1177/2042098614554919

Ma BC, Ghasimi S, Landfester K, Zhang KAI (2016) Enhanced visible light promoted antibacterial efficiency of conjugated microporous polymer nanoparticles via molecular doping. J Mater Chem B 4:5112–5118. https://doi.org/10.1039/c6tb00943c

Ma J, Li K, Gu S (2022) Selective strategies for antibacterial regulation of nanomaterials. RSC Adv 12:4852–4864. https://doi.org/10.1039/d1ra08996j

Makabenta JMV, Nabawy A, Li CH et al (2021a) Nanomaterial-based therapeutics for antibiotic-resistant bacterial infections. Nat Rev Microbiol 19:23–36. https://doi.org/10.1038/s41579-020-0420-1

Makabenta JMV, Park J, Li CH et al (2021b) Polymeric nanoparticles active against dual-species bacterial biofilms. Molecules 26:1–11. https://doi.org/10.3390/molecules26164958

Makhathini SS, Mdanda S, Kondiah PJ et al (2022) Biomedicine innovations and its nanohydrogel classifications. Pharmaceutics 14:1–31. https://doi.org/10.3390/pharmaceutics14122839

Malzahn K, Jamieson WD, Dröge M et al (2014) Advanced dextran based nanogels for fighting *Staphylococcus aureus* infections by sustained zinc release. J Mater Chem B 2:2175–2183. https://doi.org/10.1039/c3tb21335h

Mancuso G, Midiri A, Gerace E et al (2023) Urinary tract infections: the current scenario and future prospects. Pathogens 12:623. https://doi.org/10.3390/pathogens12040623

Masood F, Yasin T, Bukhari H, Mujahid M (2016) Characterization and application of roxithromycin loaded cyclodextrin based nanoparticles for treatment of multidrug resistant bacteria. Mater Sci Eng C 61:1–7. https://doi.org/10.1016/j.msec.2015.11.076

Mayandi V, Wen Choong AC, Dhand C et al (2020) Multifunctional antimicrobial nanofiber dressings containing ϵ-polylysine for the eradication of bacterial bioburden and promotion of wound healing in critically colonized wounds. ACS Appl Mater Interfaces 12:15989–16005. https://doi.org/10.1021/acsami.9b21683

Miranda CS, Silva AFG, Pereira-Lima SMMA et al (2022) Tunable spun fiber constructs in biomedicine: influence of processing parameters in the fibers' architecture. Pharmaceutics 14:164. https://doi.org/10.3390/pharmaceutics14010164

Mitra D, Kang ET, Neoh KG (2021) Polymer-based coatings with integrated antifouling and bactericidal properties for targeted biomedical applications. ACS Appl Polym Mater 3:2233–2263. https://doi.org/10.1021/acsapm.1c00125

Mlynarczyk-Bonikowska B, Kowalewski C, Krolak-Ulinska A, Marusza W (2022) Molecular mechanisms of drug resistance and epidemiology of multidrug-resistant variants of Neisseria gonorrhoeae. Int J Mol Sci 23:10499. https://doi.org/10.3390/ijms231810499

Mohammadi MR, Rabbani S, Bahrami SH et al (2016) Antibacterial performance and in vivo diabetic wound healing of curcumin loaded gum tragacanth/poly(ε-caprolactone) electrospun nanofibers. Mater Sci Eng C 69:1183–1191. https://doi.org/10.1016/j.msec.2016.08.032

Mohanty H, Pachpute S, Yadav RP (2021) Mechanism of drug resistance in bacteria: efflux pump modulation for designing of new antibiotic enhancers. Folia Microbiol (Praha) 66:727–739. https://doi.org/10.1007/s12223-021-00910-z

Mok CF, Ching YC, Muhamad F et al (2020) Adsorption of dyes using poly(vinyl alcohol) (PVA) and PVA-based polymer composite adsorbents: a review. J Polym Environ 28:775–793. https://doi.org/10.1007/s10924-020-01656-4

Montanari E, D'Arrigo G, Di Meo C et al (2014) Chasing bacteria within the cells using levofloxacin-loaded hyaluronic acid nanohydrogels. Eur J Pharm Biopharm 87:518–523. https://doi.org/10.1016/j.ejpb.2014.03.003

Morena AG, Tzanov T (2022) Antibacterial lignin-based nanoparticles and their use in composite materials. Nanoscale Adv 4:4447–4469. https://doi.org/10.1039/d2na00423b

Morrison JM, Anderson KL, Beenken KE et al (2012) The staphylococcal accessory regulator, SarA, is an RNA-binding protein that modulates the mRNA turnover properties of late-exponential and stationary phase *Staphylococcus aureus* cells. Front Cell Infect Microbiol 2:26. https://doi.org/10.3389/fcimb.2012.00026

MubarakAli D, LewisOscar F, Gopinath V et al (2018) An inhibitory action of chitosan nanoparticles against pathogenic bacteria and fungi and their potential applications as biocompatible antioxidants. Microb Pathog 114:323–327. https://doi.org/10.1016/j.micpath.2017.11.043

Mun S, Kim S, Kong R et al (2014) Curcumin reverse methicillin resistance in *Staphylococcus aureus*. Molecules 19:18283–18295. https://doi.org/10.3390/molecules191118283

Munir MU, Ahmad MM (2022) Nanomaterials aiming to tackle antibiotic-resistant bacteria. Pharmaceutics 14:1–17. https://doi.org/10.3390/pharmaceutics14030582

Muthuchamy M, Govindan R, Shine K et al (2020) Anti-biofilm investigation of graphene/chitosan nanocomposites against biofilm producing *P. aeruginosa* and *K. pneumoniae*. Carbohydr Polym 230:115646. https://doi.org/10.1016/j.carbpol.2019.115646

Nainu F, Permana AD, Djide NJN et al (2021) Pharmaceutical approaches to antimicrobial resistance: prospects and challenges. Antibiotics 10:981. https://doi.org/10.3390/antibiotics10080981

Namivandi-Zangeneh R, Kwan RJ, Nguyen TK et al (2018) The effects of polymer topology and chain length on the antimicrobial activity and hemocompatibility of amphiphilic ternary copolymers. Polym Chem 9:1735–1744. https://doi.org/10.1039/c7py01069a

Namivandi-Zangeneh R, Sadrearhami Z, Dutta D et al (2019) Synergy between synthetic antimicrobial polymer and antibiotics: a promising platform to combat multidrug-resistant bacteria. ACS Infect Dis 5:1357–1365. https://doi.org/10.1021/acsinfecdis.9b00049

Nandhini P, Rajan M (2022) Antimicrobial activities of conducting polymers and their derivatives. In: Conducting polymers. Chemistries, properties and biomedical applications, pp 237–255. https://doi.org/10.1201/9781003205418-17

Nandhini P, Kumar P, Mickymaray S et al (2022) Recent developments in methicillin-resistant *Staphylococcus aureus* (MRSA) treatment: a review. Antibiotics 11:1–21. https://doi.org/10.3390/antibiotics11050606

Nasri N, Rusli A, Teramoto N et al (2021) Past and current progress in the development of antiviral/ antimicrobial polymer coating towards covid-19 prevention: a review. Polymers (Basel) 13:4234. https://doi.org/10.3390/polym13234234

Nederberg F, Zhang Y, Tan JPK et al (2011) Biodegradable nanostructures with selective lysis of microbial membranes. Nat Chem 3:409–414. https://doi.org/10.1038/nchem.1012

Nikolic P, Mudgil P (2023) The cell wall, cell membrane and virulence factors of *Staphylococcus aureus* and their role in antibiotic resistance. Microorganisms 11:259. https://doi.org/10.3390/microorganisms11020259

Noor ul Ain, Masood F, Noor M, Farooq M (2022) Fabrication and evaluation of antibacterial properties of Cu2O/sepiolite-PVA nanocomposite hydrogels against multidrug-resistant bacteria. Appl Clay Sci 229:106663. https://doi.org/10.1016/j.clay.2022.106663

Pan Y, Xia Q, Xiao H (2019) Cationic polymers with tailored structures for rendering polysaccharide-based materials antimicrobial: an overview. Polymers (Basel) 11:1283. https://doi.org/10.3390/polym11081283

Parmanik A, Das S, Kar B et al (2022) Current treatment strategies against multidrug-resistant bacteria: a review. Curr Microbiol 79:1–15. https://doi.org/10.1007/s00284-022-03061-7

Patangia DV, Anthony Ryan C, Dempsey E et al (2022) Impact of antibiotics on the human microbiome and consequences for host health. Microbiology 11:1–23. https://doi.org/10.1002/mbo3.1260

Pearce AK, O'Reilly RK (2021) Polymers for biomedical applications: the importance of hydrophobicity in directing biological interactions and application efficacy. Biomacromolecules 22:4459–4469. https://doi.org/10.1021/acs.biomac.1c00434

Peixoto T, Nunes J, Lopes MA et al (2022) Poly(lactic acid) composites with few layer graphene produced by noncovalent chemistry. Polym Compos 43:8409–8425. https://doi.org/10.1002/pc.27012

Pfab C, Schnobrich L, Eldnasoury S et al (2021) Repurposing of antimicrobial agents for cancer therapy: what do we know? Cancers (Basel) 13:3193. https://doi.org/10.3390/cancers13133193

Pizarro-Cerdá J, Cossart P (2006) Bacterial adhesion and entry into host cells. Cell 124:715–727. https://doi.org/10.1016/j.cell.2006.02.012

Polat E, Kang K (2021) Natural photosensitizers in antimicrobial photodynamic therapy. Biomedicine 9:1–30. https://doi.org/10.3390/biomedicines9060584

Provenzani A, Hospodar AR, Meyer AL et al (2020) Multidrug-resistant gram-negative organisms: a review of recently approved antibiotics and novel pipeline agents. Int J Clin Pharmacol 42:1016–1025. https://doi.org/10.1007/s11096-020-01089-y

Rahman MM, Alam Tumpa MA, Zehravi M et al (2022) An overview of antimicrobial stewardship optimization: the use of antibiotics in humans and animals to prevent resistance. Antibiotics 11:1–31. https://doi.org/10.3390/antibiotics11050667

Ramalingam N, Natarajan TS, Rajiv S (2014) Preparation and characterization of electrospun curcumin loaded poly (2-hydroxyethyl methacrylate) nanofiber—a biomaterial for multidrug resistant organisms. J Biomed Mater Res A 103:16–24. https://doi.org/10.1002/jbm.a.35138

Ramalingam N, Natarajan TS, Rajiv S (2015) Preparation and characterization of electrospun curcumin loaded poly(2-hydroxyethyl methacrylate) nanofiber-A biomaterial for multidrug resistant organisms. J Biomed Mater Res Pt A 103:16–24. https://doi.org/10.1002/jbm.a.35138

Ramalingam R, Dhand C, Mayandi V et al (2021) Core-shell structured antimicrobial nanofiber dressings containing herbal extract and antibiotics combination for the prevention of biofilms and promotion of cutaneous wound healing. ACS Appl Mater Interfaces 13:24356–24369. https://doi.org/10.1021/acsami.0c20642

Ranjbar R, Farahani A (2019) Shigella: antibiotic-resistance mechanisms and new horizons for treatment. Infect Drug Resist 12:3137–3167. https://doi.org/10.2147/IDR.S219755

Ren R, Lim C, Li S et al (2022) Recent advances in the development of lipid-, metal-, carbon-, and polymer-based nanomaterials for antibacterial applications. Nanomaterials 12:3855. https://doi.org/10.3390/nano12213855

Reygaert CW (2018) An overview of the antimicrobial resistance mechanisms of bacteria. AIMS Microbiol 4:482–501. https://doi.org/10.3934/microbiol.2018.3.482

Rezić I, Majdak M, Bilić VL et al (2021) Development of antibacterial protective coatings active against MSSA and MRSA on biodegradable polymers. Polymers (Basel) 13:1–17. https://doi.org/10.3390/POLYM13040659

Riu F, Ruda A, Ibba R et al (2022) Antibiotics and carbohydrate-containing drugs targeting bacterial cell envelopes: an overview. Pharmaceuticals 15:1–38. https://doi.org/10.3390/ph15080942

Rofeal M, Abdelmalek F, Steinbüchel A (2022) Naturally-sourced antibacterial polymeric nanomaterials with special reference to modified polymer variants. Int J Mol Sci 23:4101. https://doi.org/10.3390/ijms23084101

Roy S, Chowdhury G, Mukhopadhyay AK et al (2022) Convergence of biofilm formation and antibiotic resistance in *Acinetobacter baumannii* infection. Front Med 9:793615. https://doi.org/10.3389/fmed.2022.793615

Rubin DHF, Ross JDC, Grad YH (2020) The frontiers of addressing antibiotic resistance in Neisseria gonorrhoeae. Transl Res 220:122–137. https://doi.org/10.1016/j.trsl.2020.02.002

Saberpour M, Bakhshi B, Najar-Peerayeh S (2020) Evaluation of the antimicrobial and antibiofilm effect of chitosan nanoparticles as carrier for supernatant of mesenchymal stem cells on multidrug-resistant vibrio cholerae. Infect Drug Resist 13:2251–2260. https://doi.org/10.2147/IDR.S244990

Safhi MM, Sivakumar SM, Jabeen A et al (2016) Therapeutic potential of chitosan nanoparticles as antibiotic delivery system: challenges to treat multiple drug resistance. Asian J Pharm 10:S61–S66

Saha M, Sarkar A (2021) Review on multiple facets of drug resistance: a rising challenge in the 21st century. J Xenobiotics 11:197–214. https://doi.org/10.3390/jox11040013

Saleh TA (2020) Nanomaterials: classification, properties, and environmental toxicities. Environ Technol Innov 20:101067. https://doi.org/10.1016/j.eti.2020.101067

Salem DMSA, Sallam MAE, Youssef TNMA (2019) Synthesis of compounds having antimicrobial activity from alginate. Bioorg Chem 87:103–111. https://doi.org/10.1016/j.bioorg.2019.03.013

Salim SA, Badawi NM, El-Moslamy SH et al (2023) Novel long-acting brimonidine tartrate loaded-PCL/PVP nanofibers for versatile biomedical applications: fabrication, characterization and antimicrobial evaluation. RSC Adv 13:14943–14957. https://doi.org/10.1039/D3RA02244G

Samal SK, Dash M, Van VS et al (2012) Cationic polymers and their therapeutic potential. Chem Soc Rev 41:7147–7194. https://doi.org/10.1039/c2cs35094g

Scandorieiro S, Teixeira FMMB, Nogueira MCL et al (2023) Antibiofilm effect of biogenic silver nanoparticles combined with oregano derivatives against carbapenem-resistant Klebsiella pneumoniae. Antibiotics 12:1–32. https://doi.org/10.3390/antibiotics12040756

Schindler BD, Kaatz GW (2016) Multidrug efflux pumps of Gram-positive bacteria. Drug Resist Updat 27:1–13. https://doi.org/10.1016/j.drup.2016.04.003

Sebe I, Ostorhazi E, Fekete A et al (2016) Polyvinyl alcohol nanofiber formulation of the designer antimicrobial peptide APO sterilizes *Acinetobacter baumannii*-infected skin wounds in mice. Amino Acids 48:203–211. https://doi.org/10.1007/s00726-015-2080-4

Sedghi R, Shaabani A (2016) Electrospun biocompatible core/shell polymer-free core structure nanofibers with superior antimicrobial potency against multi drug resistance organisms. Polymer (Guildf) 101:151–157. https://doi.org/10.1016/j.polymer.2016.08.060

Shameem MM, Sasikanth SM, Annamalai R, Raman RG (2021) A brief review on polymer nanocomposites and its applications. Mater Today Proc 45:2536–2539. https://doi.org/10.1016/j.matpr.2020.11.254

Shamim A, Ali A, Iqbal Z et al (2023) Natural medicine a promising candidate in combating microbial biofilm. Antibiotics 12:299. https://doi.org/10.3390/antibiotics12020299

Shantanam S, Mueller (2018) Antibiotic resistance mechanisms in M. tuberculosis. Physiol Behav 176:139–148. https://doi.org/10.1007/s00204-016-1727-6.Antibiotic

Silva NC, Silva S, Sarmento B, Pintado M (2015) Chitosan nanoparticles for daptomycin delivery in ocular treatment of bacterial endophthalmitis. Drug Deliv 22:885–893. https://doi.org/10.3109/10717544.2013.858195

Snetkov P, Rogacheva E, Kremleva A et al (2022) In-vitro antibacterial activity of curcumin-loaded nanofibers based on hyaluronic acid against multidrug-resistant ESKAPE pathogens. Pharmaceutics 14:1186. https://doi.org/10.3390/pharmaceutics14061186

Soni KS, Desale SS, Bronich TK (2016) Nanogels: an overview of properties, biomedical applications and obstacles to clinical translation. J Control Release 240:109–126. https://doi.org/10.1016/j.jconrel.2015.11.009

Spigaglia P, Mastrantonio P, Barbanti F (2018) Antibiotic resistances of Clostridium difficile. Adv Exp Med Biol 1050:137–159. https://doi.org/10.1007/978-3-319-72799-8_9

Spirescu VA, Chircov C, Alexandru Mihai Grumezescu EA (2021) Polymeric nanoparticles for antimicrobial therapies: an up-to-date overview. Polymers (Basel) 13:724

Taghiloo S, Ghajari G, Zand Z, Kabiri-samani S (2023) Designing alginate/chitosan nanoparticles containing Echinacea angustifolia: a novel candidate for combating multidrug-resistant *Staphylococcus aureus*. Chem Biodivers 20:e202201008. https://doi.org/10.1002/cbdv.202201008

Tanwar J, Das S, Fatima Z, Hameed S (2014) Multidrug resistance: an emerging crisis. Interdiscip Perspect Infect Dis 2014:7. https://doi.org/10.1155/2014/541340

Tavares TD, Antunes JC, Padrão J et al (2020) Activity of specialized biomolecules against gram-positive and gram-negative bacteria. Antibiotics 9:1–16. https://doi.org/10.3390/antibiotics9060314

Tayeferad M, Boddohi S, Bakhshi B (2021) Dual-responsive nisin loaded chondroitin sulfate nanogel for treatment of bacterial infection in soft tissues. Int J Biol Macromol 193:166–172. https://doi.org/10.1016/j.ijbiomac.2021.10.116

Teixeira MO, Antunes JC, Felgueiras HP (2021) Recent advances in fiber–hydrogel composites for wound healing and drug delivery systems. Antibiotics 10:1–34. https://doi.org/10.3390/antibiotics10030248

Thambirajoo M, Maarof M, Lokanathan Y et al (2021) Potential of nanoparticles integrated with antibacterial properties in preventing biofilm and antibiotic resistance. Antibiotics 10:1338. https://doi.org/10.3390/antibiotics10111338

Thamer BM, Esmail GA, Al-Dhabi NA et al (2021) Fabrication of biohybrid electrospun nanofibers for the eradication of wound infection and drug-resistant pathogens. Colloids Surf A Physicochem Eng Asp 609:125691. https://doi.org/10.1016/j.colsurfa.2020.125691

Thappeta KRV, Vikhe YS, Yong AMH et al (2020) Combined efficacy of an antimicrobial cationic peptide polymer with conventional antibiotics to combat multidrug-resistant pathogens. ACS Infect Dis 6:1228–1237. https://doi.org/10.1021/acsinfecdis.0c00016

Tshibangu-Kabamba E, Yamaoka Y (2021) Helicobacter pylori infection and antibiotic resistance—from biology to clinical implications. Nat Rev Gastroenterol Hepatol 18:613–629. https://doi.org/10.1038/s41575-021-00449-x

Turos E, Reddy GSK, Greenhalgh K et al (2007a) Penicillin-bound polyacrylate nanoparticles: restoring the activity of β-lactam antibiotics against MRSA. Bioorg Med Chem Lett 17:3468–3472. https://doi.org/10.1016/j.bmcl.2007.03.077

Turos E, Shim JY, Wang Y et al (2007b) Antibiotic-conjugated polyacrylate nanoparticles: new opportunities for development of anti-MRSA agents. Bioorg Med Chem Lett 17:53–56. https://doi.org/10.1016/j.bmcl.2006.09.098

Usacheva M, Layek B, Rahman Nirzhor SS, Prabha S (2016) Nanoparticle-mediated photodynamic therapy for mixed biofilms. J Nanomater 2016:4752894. https://doi.org/10.1155/2016/4752894

Vázquez-López R, Solano-Gálvez SG, Vignon-Whaley JJJ et al (2020) *Acinetobacter baumannii* resistance: a real challenge for clinicians. Antibiotics 9:1–22. https://doi.org/10.3390/antibiotics9040205

Vrancianu CO, Gheorghe I, Czobor IB, Chifiriuc MC (2020) Antibiotic resistance profiles, molecular mechanisms and innovative treatment strategies of *Acinetobacter baumannii*. Microorganisms 8:1–40. https://doi.org/10.3390/microorganisms8060935

Wang Y, Sun H (2021) Polymeric nanomaterials for efficient delivery of antimicrobial agents. Pharmaceutics 13:2108. https://doi.org/10.3390/pharmaceutics13122108

Wang Z, Liu X, Peng Y et al (2020) Platensimycin-encapsulated liposomes or micelles as biosafe nanoantibiotics exhibited strong antibacterial activities against methicillin-resistant *Staphylococcus aureus* infection in mice. Mol Pharm 17:2451–2462. https://doi.org/10.1021/acs.molpharmaceut.0c00194

Wang Q, Shi Q, Li Y et al (2022) Visible light-regulated cationic polymer coupled with photodynamic inactivation as an effective tool for pathogen and biofilm elimination. J Nanobiotechnol 20:1–20. https://doi.org/10.1186/s12951-022-01702-4

Weldrick PJ, Iveson S, Hardman MJ, Paunov VN (2019) Breathing new life into old antibiotics: overcoming antibacterial resistance by antibiotic-loaded nanogel carriers with cationic surface functionality. Nanoscale 11:10472–10485. https://doi.org/10.1039/c8nr10022e

Wickramage I, Spigaglia P, Sun X (2021) Mechanisms of antibiotic resistance of Clostridioides difficile. J Antimicrob Chemother 76:3077–3090. https://doi.org/10.1093/jac/dkab231

Wu C, Zhou Y, Wang H, Hu J (2019) P4VP modified zwitterionic polymer for the preparation of antifouling functionalized surfaces. Nanomaterials 9:706. https://doi.org/10.3390/nano9050706

Xi Y, Ge J, Guo Y et al (2018) Biomimetic elastomeric polypeptide-based nanofibrous matrix for overcoming multidrug-resistant bacteria and enhancing full-thickness wound healing/skin regeneration. ACS Nano 12:10772–10784. https://doi.org/10.1021/acsnano.8b01152

Xi Y, Guo Y, Wang M et al (2020) Biomimetic bioactive multifunctional poly(citrate-siloxane)-based nanofibrous scaffolds enable efficient multidrug-resistant bacterial treatment/non-invasive tracking in vitro/in vivo. Chem Eng J 383:123078. https://doi.org/10.1016/j.cej.2019.123078

Xie J, Zhou M, Qian Y et al (2021) Addressing MRSA infection and antibacterial resistance with peptoid polymers. Nat Commun 12:5898. https://doi.org/10.1038/s41467-021-26221-y

Yang X, Yang J, Wang L et al (2017) Pharmaceutical intermediate-modified gold nanoparticles: against multidrug-resistant bacteria and wound-healing application via an electrospun scaffold. ACS Nano 11:5737–5745. https://doi.org/10.1021/acsnano.7b01240

Yang X, Ye W, Qi Y et al (2021) Overcoming multidrug resistance in bacteria through antibiotics delivery in surface-engineered nano-cargos: recent developments for future nano-antibiotics. Front Bioeng Biotechnol 9:696514. https://doi.org/10.3389/fbioe.2021.696514

Ye P, Wei S, Luo C et al (2020) Long-term effect against methicillin-resistant *Staphylococcus aureus* of emodin released from coaxial electrospinning nanofiber membranes with a biphasic profile. Biomolecules 10:1–16. https://doi.org/10.3390/biom10030362

Yeh YC, Huang TH, Yang SC et al (2020) Nano-based drug delivery or targeting to eradicate bacteria for infection mitigation: a review of recent advances. Front Chem 8:1–22. https://doi.org/10.3389/fchem.2020.00286

Yin R, Agrawal T, Khan U et al (2015) Antimicrobial photodynamic inactivation in nanomedicine: small light strides against bad bugs. Nanomedicine 10:2379–2404. https://doi.org/10.2217/nnm.15.67

Yin Y, Hu B, Yuan X et al (2020) Nanogel: a versatile nano-delivery system for biomedical applications. Pharmaceutics 12:290. https://doi.org/10.3390/pharmaceutics12030290

Yuan H, Li Z, Wang X, Qi R (2022) Photodynamic antimicrobial therapy based on conjugated polymers. Polymers (Basel) 14:3657. https://doi.org/10.3390/polym14173657

Zagho MM, Hussein EA, Elzatahry AA (2018) Recent overviews in functional polymer composites for biomedical applications. Polymers (Basel) 10:739. https://doi.org/10.3390/polym10070739

Zhang F, Cheng W (2022) The mechanism of bacterial resistance and potential bacteriostatic strategies. Antibiotics 11:1215. https://doi.org/10.3390/antibiotics11091215

Zhang G, Liu J, Li R et al (2018) Conjugation of inulin improves anti-biofilm activity of chitosan. Mar Drugs 16:151. https://doi.org/10.3390/md16050151

Zhao MH, Bin ZJ, Yang KW et al (2020) Quaternized polymer-based nanostructures confer antimicrobial efficacy against multidrug-resistant bacteria. New J Chem 44:3874–3881. https://doi.org/10.1039/c9nj06173h

Zhao Y, Tian C, Liu Y et al (2023) All-in-one bioactive properties of photothermal nanofibers for accelerating diabetic wound healing. Biomaterials 295:122029. https://doi.org/10.1016/j.biomaterials.2023.122029

11 Development of Nanoemulsion-Based Drug Carrier Molecules in the Fight Against Multi-Drug Resistance

Musa Marimani

Abstract

Microbial agents develop selection pressure against various therapies due to long-term monotherapy, misdiagnosis, and over-prescription of powerful drugs reversed for urgent and critical clinical conditions. Eventually, this leads to the emergence and dissemination of drug-resistant pathogens that give rise to a wide array of clinical manifestations that are unresponsive to conventional antimicrobial drugs. Devastatingly, this has far-reaching implications in healthcare facilities across the globe due to rapid proliferation and transmission of the drug-resistant mutant strains and associated medical conditions. In the recent past, combination therapy has been applied as a strategy to treat diseases caused by drug-resistant microbes. This method relies on the use of therapeutic agents that target different essential biomarkers that promote microbial virulence, replication, and transmission. Although this method is largely successful, the major limitation accompanying administration of combination therapy is increased off-target effects and toxicity. Consequently, drug-carrier molecules such as nanoemulsions, metal, cationic, and polymer-based nanoparticles have been designed to facilitate efficient delivery of therapeutic agents into target cells. Advantages of using these carrier molecules include protection of the therapeutic payload against enzymatic degradation, increased circulation of drugs in target cells, and prolonged therapeutic activity. These desirable properties advance their use in vitro and in vivo owing to their diversity and versatility. Notably, the present book chapter will specifically evaluate the use of nanoemulsions in combating multi-drug-resistant microorganisms.

M. Marimani (✉)
Anatomical Pathology, School of Pathology, Health Sciences, University of the Witwatersrand, Johannesburg, South Africa
e-mail: Musa.Marimani@wits.ac.za

M. Y. Wani et al. (eds.), *Nanotechnology Based Strategies for Combating Antimicrobial Resistance*, https://doi.org/10.1007/978-981-97-2023-1_11

Keywords

Drug resistance · Combination therapy · Nanoemulsions · Nanoparticles

11.1 Introduction

11.1.1 Nanoemulsions

Nanoemulsions are emulsion droplets with a size ranging from 20 to 500 nm in diameter (Fig. 11.1). Generally, emulsion particles have a mean radii that range between 100 nm and 100 μm (McClements and Rao 2011). Both conventional emulsions and nanoemulsions are metastable systems, implying that they tend to break down over time due to various destabilisation mechanisms such as coalescence, Ostwald ripening, flocculation, and gravitational separation (Komaiko and McClements 2016). Nevertheless, emulsions may be designed to have enhanced stability. This may be achieved by the addition of stabilisers and co-adjuvant molecules which significantly influence the physical characteristics and stability profile of nanoemulsions. As such, molecular interactions determine the structural stability and rheological behaviour of these particles. Nanoemulsions are advantageous as compared to traditional emulsions as they:

(a) Possess a small diameter.
(b) Have decreased droplet aggregation.
(c) Have improved optical clarity and stability.
(d) Possess enhanced biodistribution and stability of encapsulated compounds.

These properties are very useful in drug delivery and beverage production as the improved transparency of nanoemulsions produces commercial products with decreased light scattering. Notably, the structural and physico-chemical characteristics of the nanoemulsions need to be regulated depending on the required

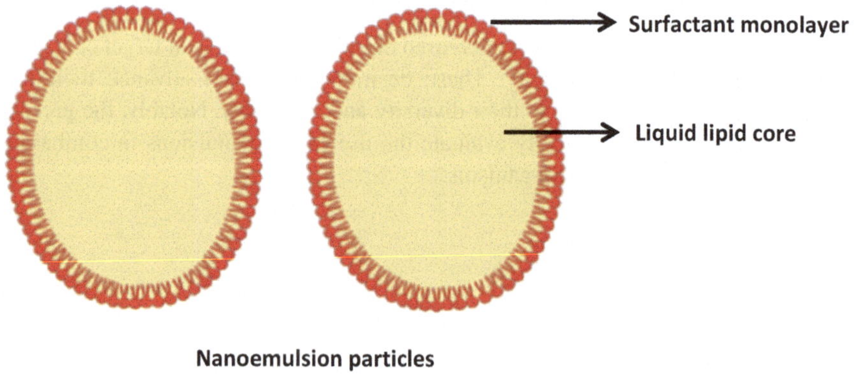

Fig. 11.1 Basic structure of nanoemulsions

product and application (Esmaeili and Gholami 2015; Mehmood 2015). This is particularly important for production of drug molecules, vitamins, and other commercial and pharmaceutical products. Interestingly, nanoemulsions stabilised with food proteins have been used as delivery vehicles owing to their:

(a) Enhanced binding to hydrophobic compounds.
(b) Impressive gelation features.
(c) Superior emulsifying qualities.

The proteins most commonly applied in emulsion stabilisation are milk-based and include whey and casein proteins. On the other hand, plant-derived proteins such as lentil, canola, pea, and soy have not been extensively studied, but are currently not regularly employed as emulsifiers. For this reason, there has been renewed interest to study and use plant-based proteins as emulsifiers (Yerramilli and Ghosh 2017). Previous studies have revealed that nanoemulsions have excellent properties relative to other non-nano-formulated delivery methods (Mehrnia et al. 2016; Luo et al. 2017; Pallathadka et al. 2023; Jo et al. 2015; Zhang et al. 2017; Ozogul et al. 2017; Moraes-Lovison et al. 2017). Nonetheless, disadvantages associated with the use of nanoemulsions include, among others, the following:

(a) Reduced instability due to Ostwald ripening.
(b) The need for high-quality surfactants and co-surfactants.
(c) Tendency to aggregate.

Typically, Ostwald ripening induces particle instability by initiating droplet growth following repeated particle aggregation events. However, this challenge can be overcome by utilising self-nanoemulsifying drug delivery methods that contain surfactants and co-surfactants. These systems lead to the formation of nanoemulsions by frequent stirring of the nanoemulsion constituents. This is followed by injection of the mixed nanoemulsion constituents into the aqueous phase. Strategic regulation of Ostwald ripening is pivotal for synthesis of nanoemulsion formulations with the desired size and stability.

Nanoemulsions are comprised of a surfactant monolayer and a liquid lipid core. The small size and structural simplicity of nanoemulsions allow incorporation of diverse chemical groups and therapeutic compounds. These properties promote stability and binding specificity and facilitate delivery of nanoemulsion-drug formulations into the desired cellular targets.

11.1.2 Nanoemulsions for Drug Delivery

The delivery of drugs using nanoemulsions is an attractive alternative strategy. This is because this method may be applied to enhance the biodistribution of hydrophobic therapeutical agents and bioactive food constituents. The majority of therapeutic compounds and bioactive components are hydrophobic, which decreases their solubility, absorption, and biodistribution upon administration (Karthik et al. 2017; Mu et al. 2013; Qian and McClements 2011). This inevitably leads to low drug activity (Karthik et al. 2017; Chatterjee et al. 2016; Dokania and Joshi 2015) and poor nutritional outcomes. To circumvent this, lipid-based systems are employed to facilitate delivery of therapeutic compounds and bioactive food constituents that have poor oral solubility, biodistribution, and absorption (Feeney et al. 2016). Several investigations have confirmed that the use of these delivery systems exceptionally enhanced the solubility, absorption, and biodistribution of bioactive foods and hydrophobic drugs (Belhaj et al. 2010; Neslihan Gursoy and Benita 2004; Mohsin et al. 2009; Yuan et al. 2008). Bioactive food components such as flavonoids and carotenoids have been prepared with improved qualities and produced desirable outcomes following co-administration with nanoemulsions (Liu et al. 2019). These ideal formulations may be conveniently utilised as they:

(a) Contain large interfacial regions,
(b) Display improved stability, and
(c) Shield compounds from enzymatic degradation (Madene et al. 2006).

In particular, these systems can be employed to facilitate efficient drug delivery via the transdermal and transmucosal methods (Rehman et al. 2017). For this purpose, the physical attributes of nanoemulsions such as the overall surface charge, size, polydispersity index, and pH are of critical importance in ensuring optimum delivery of the payload into the precise cellular location to achieve the desired therapeutic effect.

11.1.3 Properties of Nanoemulsions

There are two main types of self-emulsifying formulations, namely self-emulsifying drug delivery systems (SEDDS) and self-nanoemulsifying drug delivery systems (SNEDDS). The SEDDS produce a course emulsion, while SNEDDS generate particles at a nano scale. Both the SEDDS and SNEDDS are isotropic and consist of surfactants, co-surfactants, and oil. Successive dilution of the aqueous phase in SEDDS results in the formation of emulsions. Conversely, in vivo dilution of the aqueous phase in SNEDDS leads to the production of clear and fine nanoemulsions. Since the nanoemulsion or emulsion is produced after in vivo dilution of the aqueous phase, SEDDS and SNEDDS are typically referred to as emulsion or nanoemulsion pre-concentrates (Pouton 2000; Pouton and Porter 2008).

Generally, the nanoemulsion system is comprised of water, surfactants, co-surfactants, lipids, and oils. For production of nanoemulsion formulations, the oil phase may be comprised of triglycerides which include monoacylglycerols, diacylglycerols, or triacylglycerols, free fatty acids, mineral oils, and vegetable oils (Gonçalves et al. 2018). The selection of particular oil is dependent on the degree of drug solubility. Within a formulation, the phase with the highest drug delivering capacity is used for production of nanoemulsions (Qadir et al. 2016). The surfactants typically used for preparation of bioactive foods and therapeutic agents in nanoemulsion formulations include polyoxyethylene derivatives of sorbitan fatty acid esters, cremophor EL, lauroyl macrogolglycerides, polysaccharides, amphiphilic proteins, and phospholipids (Komaiko and McClements 2016; Singh et al. 2017; Kumar et al. 2019). With the inclusion of surfactants, a significantly lower interfacial tension is needed to produce nanoemulsions. This necessitates the use of co-solvents or co-surfactants such as propylene glycol, polyethylene glycol, ethylene glycol, diethylene glycol, glycerol, ethanol, and propanol (Singh et al. 2017; Khan et al. 2012). There are various methods used for preparation of nanoemulsion formulations and include:

1. Ultrasonication,
2. Microfluidisation, and
3. High pressure homogenisation.

Low energy techniques:

1. Phase inversion emulsification method:
   (a) Transitional phase inversion (TPI):
       - Phase inversion composition (PIC) and
       - Phase inversion temperature (PIT)
   (b) Catastrophic phase inversion (CPI):
       - Emulsion inversion point (EIP)
2. Self-nanoemulsification method.

These methods are categorised based on the nature of phase inversion, energy requirements, and self-emulsification.

11.1.4 Preparation of Nanoemulsions

Selection of appropriate ingredients is important for efficient preparation of nanoemulsions. Natural or tested ingredients may be exploited for assembling nanoemulsions. These include molecules such as polysaccharides, proteins, and surfactants (Sridhar et al. 2021). Non-toxic coatings and films are classified based on the nature of the structural component and may consist of lipids, proteins, polysaccharides, or composites. The stability of nanoemulsion formulations may be increased by:

(a) Optimising the chemical properties of the interphase (i.e. thickness, charge, and chemical reactivity),
(b) Regulating atmospheric factors (i.e. oxygen concentration, pH, light, and temperature), and
(c) The use of chelating agents or antioxidants (Liu et al. 2019).

Different components constitute a nanoemulsion system and adequate preparation, formulation, concentration, and stability are necessary for preparation of particles with desirable qualities.

11.1.4.1 Lipids and Oils

The use of oils and lipids in nanoemulsion formulations requires the application of waxes, acetoglycerides, organic fatty acids, neutral acids, resins, rice bran wax, and paraffin (Umaraw and Verma 2017). This application is mainly based on their structural and functional characteristics. Characteristics such as moisture content and hydrophobicity that shields the cargo (i.e. drugs or bioactive foods) from physiological enzymatic degradation are critical for optimal functionality of nanoemulsions. In particular, nanoemulsion formulations containing lipids also display an exceptionally high degree of solubilisation and absorption. Generally, nanoemulsions contain lipid/oil content ranging from 5 to 20% (Singh et al. 2017). However, the drawbacks associated with the use of lipid-based nanoemulsions are that they:

(a) Tend to have poor mechanical and optical properties,
(b) Are generally opaque and brittle,
(c) Consist of a thick layer, and
(d) Have poor binding affinity to foods containing hydrophilic surfaces.

11.1.4.2 Proteins

Protein molecules are amphiphilic as they possess both hydrophobic and hydrophilic amino acid groups. Consequently, in nanoemulsion formulations, most edible proteins have excellent emulsifying properties as they are capable of creating a protective layer surrounding the oil droplets. This layer protects and stabilises the oil droplets via electrostatic repulsion and steric hindrance (Costa et al. 2019). Nonetheless, it is essential to prepare nanoemulsions based on the intended application. This is because protein activity is extremely reliant on temperature, ionic strength, and pH. These parameters have an enormous influence on the electrical characteristics of protein molecules. For example, most proteins aggregate under a high salt concentration environment or near the isoelectric point. As a result, these conditions lead to the production of protein molecules with a weak electrostatic repulsion (Ozogul et al. 2022). Naturally available sources employed for production of edible protein coatings include collagen, casein, gelatin, mung bean protein, keratin, pea protein, rice bran protein, cotton seed, peanut protein, and wheat gluten (Flores-López et al. 2016). Protein molecules may form hydrogen bonds, ionic bonds, or hydrophobic interactions in an aqueous solution. This is because the hydrophilicity, hydrophilicity, or amphiphilicity properties of proteins in water depend on the nature of functional groups in the polypeptide chain. Although many edible protein films and coatings have a low moisture barrier, they display enhanced permeability.

11.1.4.3 Polysaccharides

Edible coatings derived from polysaccharides are mainly used for preparation of vegetables and fruit as they have improved adaptability and flexibility. This feature allows polysaccharide-based nanoemulsions to display improved stability and also have a high binding affinity. Polysaccharides with amphiphilic properties such as modified starch, modified alginate, modified cellulose, gum, gum arabic, carrageenan, and pectin have a superior emulsifying capacity (Versino et al. 2016). Advantages associated with the use of polysaccharides are that they are abundant in nature, are cost-effective and can be modified by incorporating various chemical groups to increase absorption, stability, and efficacy.

In addition, polysaccharides form strong hydrogen bonds in water, produce edible coatings with efficient mechanical properties, and have a high water retention capacity owing to their hydrophilicity. The utilisation of polysaccharides in formulations does not generally compromise the taste of food or drug activity as they rarely induce non-specific effects such as allergies, a feature induced by some protein molecules. The setbacks of employing polysaccharides in nanoemulsions are that they have a high water solubility and permeability profile. Moreover, polysaccharides are not recommended for use in some liquid formulations owing to their high permeability and low moisture barrier attributes (Kong et al. 2022).

11.1.4.4 Surfactants

Surfactants are chemical compounds that reduce the interfacial tension between two liquid solutions, a liquid or solid, or a liquid and gas. The role of surfactants is diverse and they may serve as foaming agents, emulsifiers, detergents, wetting agents, and dispersants. The presence of electrostatic repulsion enables ionic surfactants to mitigate droplet accumulation. There are various types of surfactants namely:

(a) Ionic surfactants,
(b) Non-ionic surfactants, and
(c) Zwitterionic surfactants.

The ionic surfactants exist as cationic or anionic. Cationic surfactants contain a positively charged functional group and include lauric arginate and lactoglobulin. By contrast, anionic surfactants possess a negatively charged functional group and include sodium lauryl sulfate, diacetyl tartaric acid esters, and citric acid. Edible anionic surfactants provide more advantageous qualities in the delivery of drugs and bioactive foods. These useful qualities include appropriate charge distribution, provision of adequate surface tension, and improved stability of nanoemulsion formulations. The downside of using ionic surfactants is that they may become irritating under high concentrations. This prevents widespread application in drug and food preparations that specifically require a high surfactant concentration (Qian and McClements 2011). Non-ionic surfactants are composed of an oxygen atom. This oxygen atom forms a covalent bond with hydrophilic residues, which are bonded to hydrophobic groups. These surfactants have high solubility and produce foam at low quantities. Polyoxyethylene ether, sorbitan monooleate, polyglycerol esters of fatty acids, ethoxylated sorbitan esters, and sucrose monopalmitate (Mushtaq et al. 2023) are examples of non-ionic surfactants and are an essential component in the production of edible nanoemulsion formulations. The advantage of using non-ionic surfactants is that they produce less toxic and more stable nanoemulsions without inducing irritation. Moreover, these surfactants also reduce aggregation, steric hindrance, and thermal fluctuations (Salvia-Trujillo et al. 2015).

The zwitterionic surfactants are made up of both the negatively and positively charged functional groups on the same molecule. Accordingly, they may have an overall negative, positive, or neutral charge depending on the pH value of the preparations. Phospholipids are the major zwitterionic surfactants applied in food and drug production (Qian and McClements 2011; Mushtaq et al. 2023). The utility of zwitterionic surfactants is that they:

(a) Have less toxicity and irritability,
(b) Have high foam stability,
(c) Possess a high surface activity,
(d) Can be degraded by biological systems,
(e) Can readily dissolve in hydrophobic and hydrophilic solutions,
(f) Have a reduced critical micelle concentration (CMC), and
(g) Can be modified to have potent therapeutic activity by adjusting the pH of the nanoemulsion formulations.

Unfortunately, the cost of producing zwitterionic surfactants is significantly higher relative to other surfactants (Sarkar et al. 2021). For this reason, they have only limited to specific applications that necessitate their widespread production. Nevertheless, zwitterionic surfactants are critical components that facilitate the stability of nanoemulsion particles.

11.1.4.5 Emulsifiers

Emulsifiers are additives used to blend substances that normally do not mix in a given formulation. For example, an emulsifying agent can be added to mix oil and water in a solution. This is because the additives have a polar (i.e. hydrophilic) and a non-polar (i.e. hydrophobic) group. During the preparation stage, the active components adhere to the oil droplets during homogenisation. Adherence promotes successive oil droplet dissolution by reducing the interfacial tension. This results in a decreased droplet size during homogenisation. For food and drug preparation, emulsifiers are mainly utilised to mitigate the disintegration of oil droplets by applying the top-down methods. These methods are applied to avert aggregation by restricting droplet accumulation. This is essential for maintaining long-term stability of nanoemulsions (Dasgupta et al. 2017; Jain et al. 2018). Without the incorporation of emulsifiers, the nanoemulsions would have low stability leading to rapid disintegration due to a high surface area. The selection of a specific emulsifying agent is contingent on the purpose, structure, and application of the nanoemulsion (Donsì and Ferrari 2016). The concentration of the emulsifying agent determines the interfacial tension of the oil droplet. This regulates the rate of droplet disintegration, coalescence, and adsorption (McClements and Jafari 2018). There are two classes of emulsifiers:

(a) Small amphiphilic molecules called surfactants and
(b) Polymers such as proteins (Espitia et al. 2019).

11.1.4.6 Ripening Inhibitors

Molecules with high hydrophobicity are incorporated into the nanoemulsions to prevent droplet expansion caused by Ostwald ripening. Ostwald ripening is defined as the change of a non-homogeneous structure over a period of time. This includes the dissolving of crystals in a liquid solution followed by expansion into larger crystal structures. Typically, the mechanism of action of ripening inhibitors is similar to that of essential oils (McClements and Jafari 2018) in that their adverse hydrophobicity inhibits ripening in nanoemulsion formulations. For nanoemulsion preparation, Ostwald ripening is mostly responsible for the growth of smaller oil droplets into larger droplets. This change in droplet size occurs during the continuous phase. Thus, the growth renders the nanoemulsion unstable and necessitates the use of ripening inhibitors to impede oil droplet expansion. The concentration of ripening inhibitors in the small oil droplets increases as the hydrophilic oil transitions from small to large droplets. This produces a concentration gradient that promotes the mobility of small hydrophilic oil molecules, thus, negating Ostwald ripening. Selected examples of ripening inhibitors are canola oil, corn oil, sesame oil, and sunflower oil (Majeed et al. 2016).

11.1.4.7 Plasticizers

Plasticizers are non-volatile, low molecular weight compounds that are incorporated into the polymer to reduce the brittleness and cohesiveness, improve thermoplastic properties and toughness, as well as to add flexibility to the polymer. Plasticizers are placed in the middle of the polymer structures. Depending on the structural integration, plasticizers are mainly divided into two groups, namely external and internal plasticizers. The internal plasticizers bind to polymers by reacting with the polymer functional groups or by co-polymerisation with the polymer structure. As opposed to their internal counterparts, external plasticizers are widely used because they are cheaper and their formulations range from highly flexible to semisolid depending on the concentration of plasticizer. Commonly used external plasticizers are esters formed by chemical reactions between acids and acid anhydrides or alcohol. The external plasticizers can be used as both primary and secondary materials. The primary plasticizers are added in large amounts to improve the flexibility, softness, and elongation of the polymer. Depending on the processing temperature, secondary plasticizers with lower compatibility can cause polymer degradation (Chaudhary et al. 2020). Films embedded with plasticizers produce large structures because they create sufficient space for movement of the polymer molecules. This causes a reduction in glass transition temperature and a lower elastic modulus and leads to the production of softer and more flexible polymers.

Plasticizers commonly used in film systems are oligosaccharides, monosaccharides, lipids, and polyols (Sharma et al. 2019).

Collectively, all these components are critical for preparation of nanoemulsions. Thus, appropriate concentration, dose regulation, size modulation, pH, temperature, and the production quality are essential for efficient synthesis of nanoemulsions. These factors also determine the functioning of nanoemulsions in the pharmaceutical, food, environmental, agricultural, and other sectors. With regard to synthesis of nanoemulsions to facilitate efficient drug transportation to target cells, it is imperative to avoid stimulating the innate immune response. This is because payload recognition by the innate immunity triggers a cascade of cellular pathways that lead to the production of proteins that bind and degrade the nanoemulsions before reaching the majority of target cells and tissues. Ultimately, this results in decreased concentration, reduced half-life of the particles in circulation, and poor therapeutic activity. As a result, strategic optimisation of the synthesis process, incorporation of critical

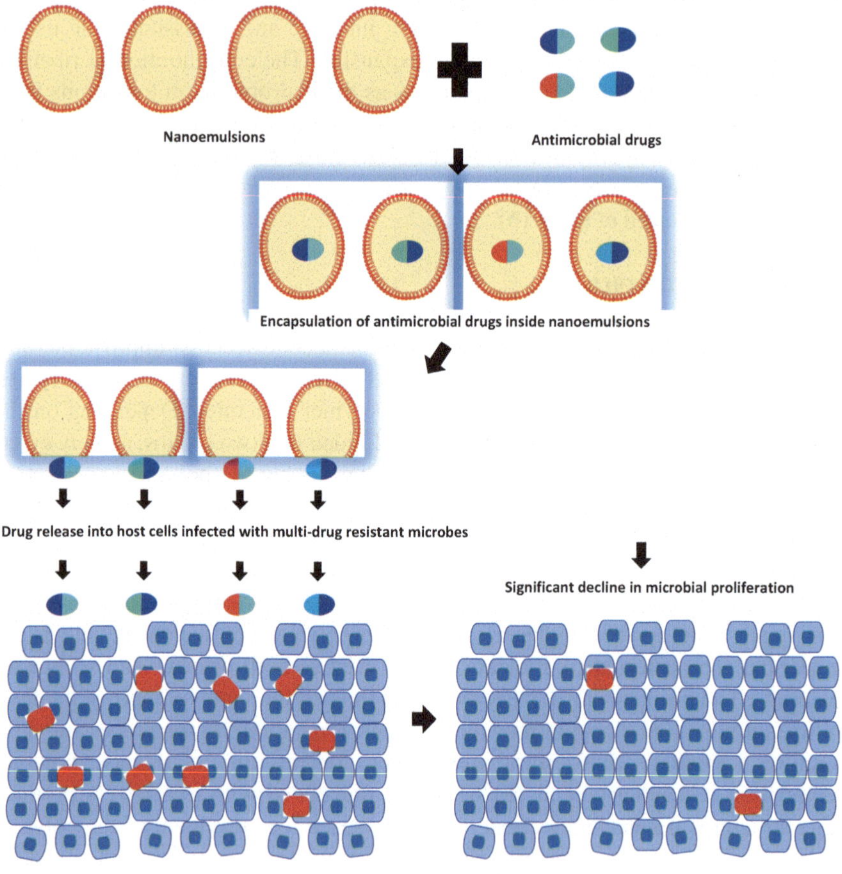

Fig. 11.2 Mode of action of nanoemulsion-mediated drug delivery against multi-drug-resistant microbes

components required for maintaining drug stability and activity, and detailed knowledge of the physiological conditions such as temperature, pH, and disease pathology are of paramount importance in abrogating microbial replication, transmission, and clinical manifestations. Desirable qualities that favour the use of nanoemulsions include among others:

(a) The relatively safe and regulated synthesis process,
(b) The cheaper production process, particularly when using natural plant materials,
(c) The ease of dose regulation,
(d) The small particle of the formulations and decreased toxicity, as well as,
(e) The adaptability to incorporate various components that bolster drug efficacy.

Owing to these ideal and diverse features, a number of studies have examined the use of nanoemulsions in dispatching drug molecules to wild-type and drug-resistant mutant pathogens with the aim of inhibiting microbial proliferation. In some of these research studies, a substantial decline in markers of virulence in cultured cells and animals was noted after the antimicrobial compounds were transported to target cells using nanoemulsion particles in both drug-naïve and -resistant strains. These encouraging findings give hope for treatment of diseases caused by drug-resistant pathogens. This places nanoemulsions (Fig. 11.2) in a good position to provide the much needed arsenal to mitigate the emergence, persistence, and transmission of mutant microbial strains. These promising research studies may serve as a benchmark for utilisation of nanoemulsions to mitigate the devastating clinical effects caused by multi-drug-resistant mutant strains. The basic nanoemulsion-mediated delivery mechanism of antimicrobial drugs is depicted in Fig. 11.2. This drug delivery strategy provides protection from enzymatic degradation and promotes durable therapeutic efficacy. These properties are ideally suited for abrogating the replication of drug-resistant strains.

The conjugation of nanoemulsions and antimicrobial drugs produces nanoemulsion-drug conjugates. The encapsulated drugs are delivered to the desired cellular location to attenuate the growth and spread of multi-drug resistant pathogens. An astronomical decrease in microbial replication and dissemination is achieved following efficient nanoemulsion-mediated transportation of therapeutic compounds.

11.2 The Application of Nanoemulsions Against Multi-Drug-Resistant Microorganisms

In the past decade, extensive genetic alterations have led to the emergence of multi-drug-resistant microorganisms and this constitutes a global public health crisis. To eradicate the shortcomings of conventional drugs in curbing the scourge of untreatable medical conditions associated with multi-drug-resistant strains, combination therapy has been utilised to repress microbial proliferation and resolve clinical manifestations. Discouragingly, long-term use of this treatment method generally leads to the development of unintended adverse effects. For example, although application of combination therapy robustly decreases the microbial load in patients co-infected with both Human Immunodeficiency Virus (HIV) and tuberculosis (TB), non-specific toxic effects such as heart and kidney complications, stomach cramps, hepatotoxicity, and hypersensitivity usually accompany their continual administration.

Historically, drugs were typically extracted from natural plants. In a research setting, the extracted plant materials are characterised. Structural characterisation is followed by identification of the active chemical moieties possessing antimicrobial efficacy and synthesis of desired compounds. In general, the conventional drugs manufactured from plants or via chemical synthesis have similar therapeutic targets and common mechanism of action. The majority of these drugs bind to critical biological molecules required for microbial replication and persistence. These biological molecules include DNA, RNA, and virulent proteins. Previously, this method was applied to treat a wide array of diseases caused by bacteria, fungi, viruses, and parasites. Nevertheless, this treatment plan has major flaws when the pathogens exert selection pressure on the conventional drugs and develop enormous genetic mutations that render the monotherapy ineffective against drug-resistant microbes. Consequently, the poor treatment outcomes of monotherapy in treating diseases caused by multi-drug-resistant strains and the devastating clinical complications that typically accompany combination therapy necessitate the use of alternative treatment approaches to resolve diseases caused by multi-drug-resistant pathogens. Accordingly, delivery vesicles such as cationic liposomes, polymeric nanoparticles, metallic nanoparticles, carbon nanodots, nanocapsules, nanosponges, and nanoemulsions have been designed and applied to deliver the therapeutic payload to target tissues and cells leading to disruption of microbial replication and alleviation of disease symptoms induced by multi-drug-resistant strains.

Ramalingam et al. (2018) used nanoemulsion drug particles to block the proliferation of *Acinetobacter baumannii* (*A. baumannii*) (Ramalingam and Lee 2018). The bacterial agent *A. baumannii* has developed mutations that confer drug resistance and has caused diseases that are of public health concern. The activity of the nanoemulsion formulations was tested against four strains of *A. baumannii* strains and antibacterial activity was assessed by minimum bactericidal concentration assay, biofilm and adherence assays, as well as scanning electron microscopy. Relative to untreated controls, administration of nanoemulsions led to an incredible decline in cell adherence, number of live cells, and biofilm formation (Ramalingam and Lee 2018). Also, a remarkable increase in cell surface disintegration and morphological changes was observed following application of nanoemulsion particles (Ramalingam and Lee 2018). These observations imply that nanoemulsions may be utilised to abrogate bacterial replication and may also be employed for generation of alternative antimicrobial compounds (Ramalingam and Lee 2018). A marked decline in different biomarkers of microbial replication supports the administration of nanoemulsion formulations to counter the growth of multi-drug-resistant microorganisms.

Biofilms consist of a group of bacterial cells attached to an extracellular matrix. The microbial diversity within a biofilm promotes emergence of drug-resistant strains due to increased gene transfer among various bacterial species. For this reason, infections caused by multi-drug-resistant bacteria existing within a biofilm pose a major public health challenge. Previously, natural essential oils were incorporated into cross-linked polymer nanoemulsion particles (Oz et al. 2021). These formulations also contained poly(lactic acid) and the antimicrobial compound carvacrol. This formulation was intended to inhibit the growth of multi-drug-resistant bacterial living in a biofilm community. Results from this undertaking indicated that the nanoemulsions infused with carvacrol drastically suppressed markers of bacterial replication in vitro (Oz et al. 2021). To be specific, noticeable decreases in the growth of methicillin-resistant *Staphylococcus aureus* (*S. aureus*), multi-drug-resistant *Escherichia coli* (*E. coli*), and *A. baumannii* were reported. Impressively, this antibacterial activity was specific as the efficacy was only observed in bacterial biofilms, and not in fibroblast cells (Oz et al. 2021). The sharp decrease in bacterial proliferation and absence of cytotoxicity support the use of nanoemulsions as drug carrier molecules to mitigate diseases caused by multi-drug-resistant microorganisms, particularly those found in complex environmental systems such as biofilm communities.

The plant-derived drug quercetin is a flavonoid that is utilised to counter the spread of multi-drug-resistant human leukemia cells (Marques et al. 2021). However, one of the limiting factors of employing this molecule is the poor biodistribution profile. Thus, quercetin was applied to both multi-drug and non-multi-drug-resistant cells using nanoemulsions as drug transportation systems (Marques et al. 2021). Compared to control cells, co-application of quercetin and nanoemulsions dramatically repressed markers associated with cancer spread including *ABCB1* gene silencing. Additionally, the combination of quercetin and vincristine, a chemotherapy drug, produced anticancer activity similar to that achieved by verapamil, an *ABCB1* gene inhibitory compound (Marques et al. 2021). Promising data from this study showed that quercetin can be utilised independently or in combination with other cancer drugs to achieve *ABCB1* gene knockdown and stimulate the inhibition of other biological indicators of malignancy.

The activity of nanoemulsions has been explored in persistent wound infections caused by vancomycin-resistant enterococci (VRE) and methicillin-resistant *S. aureus* (Chen et al. 2022). In this investigation, *S. aureus*, VRE, or a combination of both pathogens were introduced to induce porcine split skin wound infections. Following infection, the nanoemulsion formulation (X-1735) was applied to attenuate bacterial replication in wounds infected with methicillin-resistant *S. aureus* or VRE or co-infected with both drug-resistant strains (Chen et al. 2022). Data from this study highlighted that wounds treated with X-1735 displayed a substantial decrease in bacterial viability, bacterial load, and disease pathology at day 3 and 14 following administration. Notably, X-1735 also triggered a drastic decline in multistage inflammatory responses in treated wounds (Chen et al. 2022). The data suggest that this approach has potential clinical application for treating open wounds infected with methicillin-resistant *S. aureus*, VRE, or co-infected with both VRE and *S. aureus* (Chen et al. 2022).

In a similar study, nanoemulsions were utilised against multi-drug-resistant bacteria responsible for wound biofilm infections (Nabawy et al. 2022). The antimicrobial activity of these biodegradable cationic nanoemulsions containing antibacterial compounds triclosan and eugenol was tested in vitro against multi-drug-resistant *A. baumannii*, *Klebsiella pneumoniae* (*K. pneumoniae*), *Pseudomonas aeruginosa* (*P. aeruginosa*), and methicillin-resistant *S. aureus*. As compared to untreated wounds, administration of cationic nanoemulsions caused appreciable decreases in markers of microbial virulence including bacterial proliferation, bacterial load, and ability to form extensive biofilm networks in vitro (Nabawy et al. 2022). This antibacterial activity was also replicated in vivo after administration of cationic nanoemulsions to a multi-drug-resistant mouse model. In this animal model, the nanoemulsions exerted antibacterial efficacy of about 99% without triggering appreciable increases in cytotoxicity (Nabawy et al. 2022). These observations uncovered the utility of cationic nanoemulsions in treating wound infections caused by multi-drug-resistant bacteria and underscored their potential to serve as effective drug carrier particles (Nabawy et al. 2022).

The antimicrobial properties of garlic nanoemulsions have been interrogated against multi-drug-resistant *P. aeruginosa* strains collected from hatcheries and broiler farms (El-Oksh et al. 2022). The isolated strains conferred drug resistance ranging from 66 to 100%. These strains displayed a lower drug resistance level towards the antibiotics gentamicin and penicillin, while the highest resistance was exerted against sulfamethoxazole trimethoprim and tetracycline (El-Oksh et al. 2022). By contrast, the bacterial isolates were more sensitive to the drugs colistin and norfloxacin ciprofloxacin. The genes responsible for conferring drug resistance were *arr*, *oprL*, *tetA*, *mexR*, and *blaTEM*. A considerable decrease in bacterial growth, replication, viability, and cellular activity was noted after administration of garlic nanoemulsions (El-Oksh et al. 2022). Moreover, the lower minimum inhibitory concentration of the garlic nanoemulsions and dynamic light scattering analysis validated the beneficial therapeutic characteristics of the formulations. These outcomes revealed that the application of garlic nanoemulsions may be an excellent alternative treatment method to antibiotics as it dramatically reduced the virulence of various multi-drug-resistant *P. aeruginosa* strains (El-Oksh et al. 2022).

In a recent undertaking, the nanoemulsion nanoTTO derived from tea essential oil was co-administered with antibiotics to regulate the spread of multi-drug-resistant *E. coli* (Wei et al. 2022). Encouragingly, simultaneous application of this nanoemulsion and antibiotics led to a robust decline in bacterial growth and spread. Impressively, this formulation induced a sustained antimicrobial efficacy that persisted against 30 different bacterial strains (Wei et al. 2022). Detailed studies uncovered that the mechanism of action of nanoTTO was that it enhanced drug accumulation and concentration in multi-drug-resistant *E. coli*. This induced robust disruption of the outer and inner membranes and markedly repressed the AcrAB-TolC efflux pump associated with membrane permeability and drug resistance (Wei et al. 2022). The therapeutic hallmarks associated with co-administration of antibiotics and nanoTTO were also documented in the Galleria mellonella mouse model. In this animal model, the biomarkers associated with microbial virulence and persistence were remarkably reduced (Wei et al. 2022). The in vitro and in vivo therapeutic activities against multi-drug-resistant *E. coli* bolster the use of antibiotics in conjunction with nanoTTO formulations to achieve potent and sustained antimicrobial effects (Wei et al. 2022).

11.3 Conclusion and Future Perspectives

The ability of microorganisms to apply selection pressure, adapt, survive, and flourish in the presence of antimicrobial drugs has placed enormous pressure on various public health sectors across the globe. The rise in microbial drug resistance has led to the manifestation of clinical conditions that are untreatable using licensed therapeutics. This necessitates longer hospital stay, extended drug administration, and enrolment into a more intense treatment plan which usually involves combination therapy. Overall, this results in drug overuse which may lead to further drug tolerance and resistance. Unfortunately, this may also discourage the patients from adhering to the treatment plan due to extensive drug use and the cost-effectiveness of the treatment scheme. For this reason, other alternative methods have been examined to counter the shortcomings associated with long-term monotherapy and combination therapy.

To this end, drug carrier vesicles such as nanoemulsions, cationic nanoparticles, polymeric nanoparticles, metallic nanoparticles, carbon nanodots, and other molecules have been designed, developed, and tested for their suitability to efficiently transport antimicrobial compounds to the desired cellular location. Ideally, these drug carrier particles must be able to transport the therapeutic cargo and induce sustained antimicrobial activity with minimal non-specific cytotoxic effects. The nanoemulsions fall into the category of drug carrier molecules owing to their small size and ability to be amenable to chemical modification with the aim of improving their therapeutic qualities. The versatility of nanoemulsions allows them to transport plant-derived antimicrobial compounds and/or artificially designed drugs produced using a wide range of chemical synthesis procedures. Desirable qualities that support the use of nanoemulsion for drug transportation include:

(a) Small size of particles,
(b) Relatively cheaper production cost,
(c) The ease of dose regulation,
(d) The ability to incorporate and accommodate different chemical groups, and
(e) Flexibility in modulating the stability, pH, and temperature of the formulations.

These desirable and versatile drug qualities strengthen the application of nanoemulsions for drug transportation. Indeed, these ameliorated properties have led to their administration to impede the growth, survival, and persistence of both wild-type and drug-resistant mutant microbes. Ramalingam's research team successfully employed nanoemulsions to mitigate the growth and spread of drug-resistant *A. baumannii* strains (Ramalingam and Lee 2018). This resulted in noticeable attenuation in bacterial growth kinetics and deficiencies in morphological and microscopic characteristics (Ramalingam and Lee 2018). Natural essential oils and carvacrol have been incorporated to nanoemulsions for transportation into target bacterial cells. Upon deposition, the bacterial activities, replication, communication, and biofilm formation were markedly reduced in microbial communities existing within biofilm networks (Oz et al. 2021). Reassuringly, the reduction in microbial activities also affected pathogens of clinical relevance such as multi-drug-resistant *E. coli*, *A. baumannii*, and methicillin-resistant *S. aureus* (Oz et al. 2021). The adaptability of nanoemulsions was displayed after transporting the drug quercetin to impair the dissemination of multi-drug-resistant leukemia cells in vitro (Marques et al. 2021). The delivery of quercetin or the combination of quercetin and vincristine affected silencing of the *ABCB1* gene associated with the pathophysiology of cancer (Marques et al. 2021).

Other investigations exploited nanoemulsions to treat relentless wound infections emanating from infection with multi-drug-resistant bacteria (Chen et al. 2022). The beneficial clinical effects observed after nanoemulsion-mediated drug delivery included a drastic decrease in microbial load, microbial viability, and disease progression (Chen et al. 2022). These encouraging and durable therapeutic properties were observed in bacteria of global public health importance such as methicillin-resistant *S. aureus* and VRE (Chen et al. 2022). Likewise, the advantage of applying nanoemulsion formulations was probed in unresponsive wound infections resulting from contact with multi-drug-resistant bacteria found in biofilm systems (Nabawy et al. 2022). Delivery of eugenol and triclosan using cationic nanoemulsions led to an exceptional reduction in biofilm formation, bacterial load, replication, and virulence. The activities of multi-drug-resistant *K. pneumoniae*, *A. baumannii*, and *P. aeruginosa* were impeded and this antibacterial activity was achieved without a significant increase in cellular toxicity in cultured mammalian cells and treated mice (Nabawy et al. 2022). Intriguingly, garlic encapsulated with nanoemulsions has been deployed to weaken the virulence of *P. aeruginosa* isolates conferring multi-drug resistance (El-Oksh et al. 2022). This impeded bacterial growth and culminated in deteriorated pathogenicity associated with silencing of the *arr*, *oprL*, *tetA*, *mexR*, and *blaTEM* virulence genes (El-Oksh et al. 2022). In another interesting interrogation, nanoTTO extracted from tea essential oil was co-applied with antimicrobial agents resulting in diminished replication, virulence, and dissemination of multi-drug-resistant *E. coli* in cultured cells and mice (Wei et al. 2022). Astonishingly, the hallmarks associated with antibacterial activity were long-lasting and were also replicated in several bacterial strains (Wei et al. 2022). Notably, nanoTTO-mediated delivery of antibiotics dramatically inhibited the AcrAB-TolC efflux pump responsible for bacterial drug resistance (Wei et al. 2022).

Accordingly, comprehensive and rigorous testing of nanoemulsion-drug complexes in cell culture, mice, and non-human primates is essential to uncover the precise mechanism of action, pharmacokinetics, pharmacodynamics, cytotoxicity, and detailed clinical effects over an extended period of time. This detailed interrogation process will ensure that clinical trials involving human participants may commence swiftly with minimal hurdles. The urgency of conducting human clinical trials cannot be overstated as it may greatly benefit patients currently infected with multi-drug resistance pathogens that do not respond to traditional monotherapy. Thus, the current increase in global microbial resistance that gave rise to drug-resistant, multi-drug-resistant, and extensively drug-resistant strains underscores the immerse pressure exerted on both the private and public healthcare facilities. As a result, this necessitates the development of novel therapeutic approaches to mitigate and prevent emergence of mutant microbes. Taken together, the successful application of nanoemulsions to counter multi-drug resistance has produced exciting and fascinating research opportunities for the innovation, design, development, and testing of therapeutic compounds targeting known virulence factors or novel molecular biomarkers. Nonetheless, practical implementation of nanoemulsion as effective drug transport systems will inevitably require responsible drug use by the patients, medical professionals, and pharmaceutical companies. This will promote long-term antimicrobial activity and avert the emergence of drug resistance strains harbouring extensive genetic mutations.

In conclusion, the current global microbial drug resistance landscape requires undivided and decisive action to eradicate resistant pathogens and prevent the development of additional microbial strains containing genetic mutations associated with drug resistance. The co-application of nanoemulsion and therapeutic compounds has demonstrated the capacity to counter the replication, survival, and dissemination of multi-drug-resistant strains in cultured human cells and various animal models. The nanoemulsion formulations are diverse, scalable, and adaptable for delivering natural and chemically synthesised drug molecules to the intended cellular target to produce sustained therapeutic effects with minimal toxicity. In the same way, these drug-carrier particles are well-adapted to induce durable antimicrobial activity in multi-resistant microorganisms existing in independent communities as well as in complex biofilm systems.

Acknowledgements The authors are grateful for research funding offered by the National Research Foundation (NRF) of South Africa Thuthuka grant (reference TTK210331591452, grant number 138302) and the Carnegie DTA research grant (Grant code: RKSMM21).

Competing Interests The authors have no competing interests to declare.

References

Belhaj N, Arab-Tehrany E, Linder M (2010) Oxidative kinetics of salmon oil in bulk and in nanoemulsion stabilized by marine lecithin. Process Biochem 45:187–195. https://doi.org/10.1016/j.procbio.2009.09.005

Chatterjee B, Hamed Almurisi S, Ahmed Mahdi Dukhan A, Mandal UK, Sengupta P (2016) Controversies with self-emulsifying drug delivery system from pharmacokinetic point of view. Drug Deliv 23:3639–3652. https://doi.org/10.1080/10717544.2016.1214990

Chaudhary P, Fatima F, Kumar A (2020) Relevance of nanomaterials in food packaging and its advanced future prospects. J Inorg Organomet Polym Mater 30:5180–5192. https://doi.org/10.1007/s10904-020-01674-8

Chen J, Cao Z, Cannon J, Fan Y, Baker JRJ, Wang SH (2022) Effective treatment of skin wounds co-infected with multidrug-resistant bacteria with a novel nanoemulsion. Microbiol Spectr 10:e0250621. https://doi.org/10.1128/spectrum.02506-21

Costa C, Medronho B, Filipe A, Mira I, Lindman B, Edlund H et al (2019) Emulsion formation and stabilization by biomolecules: the leading role of cellulose. Polymers 11:1–18. https://doi.org/10.3390/polym11101570

Dasgupta N, Ranjan S, Ramalingam C (2017) Applications of nanotechnology in agriculture and water quality management. Environ Chem Lett 15:591–605. https://doi.org/10.1007/s10311-017-0648-9

Dokania S, Joshi AK (2015) Self-microemulsifying drug delivery system (SMEDDS)—challenges and road ahead. Drug Deliv 22:675–690. https://doi.org/10.3109/10717544.2014.896058

Donsì F, Ferrari G (2016) Essential oil nanoemulsions as antimicrobial agents in food. J Biotechnol 233:106–120. https://doi.org/10.1016/j.jbiotec.2016.07.005

El-Oksh AS, Elmasry DM, Ibrahim GA (2022) Effect of garlic oil nanoemulsion against multidrug resistant *Pseudomonas aeruginosa* isolated from broiler. Iraq J Vet Sci 36:877–888. https://doi.org/10.33899/ijvs.2022.132430.2094

Esmaeili A, Gholami M (2015) Optimization and preparation of nanocapsules for food applications using two methodologies. Food Chem 179:26–34. https://doi.org/10.1016/j.foodchem.2015.01.115

Espitia PJP, Fuenmayor CA, Otoni CG (2019) Nanoemulsions: synthesis, characterization, and application in bio-based active food packaging. Compr Rev Food Sci Food Saf 18:264–285. https://doi.org/10.1111/1541-4337.12405

Feeney OM, Crum MF, McEvoy CL, Trevaskis NL, Williams HD, Pouton CW et al (2016) 50 years of oral lipid-based formulations: provenance, progress and future perspectives. Adv Drug Deliv Rev 101:167–194. https://doi.org/10.1016/j.addr.2016.04.007

Flores-López ML, Cerqueira MA, de Rodríguez DJ, Vicente AA (2016) Perspectives on utilization of edible coatings and nano-laminate coatings for extension of postharvest storage of fruits and vegetables. Food Eng Rev 8:292–305. https://doi.org/10.1007/s12393-015-9135-x

Gonçalves A, Nikmaram N, Roohinejad S, Estevinho BN, Rocha F, Greiner R et al (2018) Production, properties, and applications of solid self-emulsifying delivery systems (S-SEDS) in the food and pharmaceutical industries. Colloids Surf A Physicochem Eng Asp 538:108–126. https://doi.org/10.1016/j.colsurfa.2017.10.076

Jain A, Ranjan S, Dasgupta N, Ramalingam C (2018) Nanomaterials in food and agriculture: an overview on their safety concerns and regulatory issues. Crit Rev Food Sci Nutr 58:297–317. https://doi.org/10.1080/10408398.2016.1160363

Jo Y-J, Chun J-Y, Kwon Y-J, Min S-G, Hong G-P, Choi M-J (2015) Physical and antimicrobial properties of trans-cinnamaldehyde nanoemulsions in water melon juice. LWT Food Sci Technol 60:444–451. https://doi.org/10.1016/j.lwt.2014.09.041

Karthik P, Ezhilarasi PN, Anandharamakrishnan C (2017) Challenges associated in stability of food grade nanoemulsions. Crit Rev Food Sci Nutr 57:1435–1450. https://doi.org/10.1080/10408398.2015.1006767

Khan AW, Kotta S, Ansari SH, Sharma RK, Ali J (2012) Potentials and challenges in self-nanoemulsifying drug delivery systems. Expert Opin Drug Deliv 9:1305–1317. https://doi.org/10.1517/17425247.2012.719870

Komaiko JS, McClements DJ (2016) Formation of food-grade nanoemulsions using low-energy preparation methods: a review of available methods. Compr Rev Food Sci Food Saf 15:331–352. https://doi.org/10.1111/1541-4337.12189

Kong I, Degraeve P, Pui LP (2022) Polysaccharide-based edible films incorporated with essential oil nanoemulsions: physico-chemical, mechanical properties and its application in food preservation—a review. Food Secur 11:11. https://doi.org/10.3390/foods11040555

Kumar M, Bishnoi RS, Shukla AK, Jain CP (2019) Techniques for formulation of nanoemulsion drug delivery system: a review. Prev Nutr Food Sci 24:225–234. https://doi.org/10.3746/pnf.2019.24.3.225

Liu Q, Huang H, Chen H, Lin J, Wang Q (2019) Food-grade nanoemulsions: preparation, stability and application in encapsulation of bioactive compounds. Molecules 24:4242. https://doi.org/10.3390/molecules24234242

Luo X, Zhou Y, Bai L, Liu F, Deng Y, McClements DJ (2017) Fabrication of β-carotene nanoemulsion-based delivery systems using dual-channel microfluidization: physical and chemical stability. J Colloid Interface Sci 490:328–335. https://doi.org/10.1016/j.jcis.2016.11.057

Madene A, Jacquot M, Scher J, Desobry S (2006) Flavour encapsulation and controlled release—a review. Int J Food Sci Technol 41:1–21. https://doi.org/10.1111/j.1365-2621.2005.00980.x

Majeed H, Antoniou J, Hategekimana J, Sharif HR, Haider J, Liu F et al (2016) Influence of carrier oil type, particle size on *in vitro* lipid digestion and eugenol release in emulsion and nanoemulsions. Food Hydrocoll 52:415–422. https://doi.org/10.1016/j.foodhyd.2015.07.009

Marques MB, Machado AP, Santos PA, Carrett-Dias M, Araújo GS, da Silva AB et al (2021) Anti-MDR effects of quercetin and its nanoemulsion in multidrug-resistant human leukemia cells. Anticancer Agents Med Chem 21(14):1911–1920. https://doi.org/10.2174/1871520621999210104200722

McClements DJ, Jafari SM (2018) Chapter 1—General aspects of nanoemulsions and their formulation. In: Nanoemulsions—formulation, applications and characterization. Academic Press, Boston, MA, pp 3–20. https://doi.org/10.1016/B978-0-12-811838-2.00001-1

McClements DJ, Rao J (2011) Food-grade nanoemulsions: formulation, fabrication, properties, performance, biological fate, and potential toxicity. Crit Rev Food Sci Nutr 51:285–330. https://doi.org/10.1080/10408398.2011.559558

Mehmood T (2015) Optimization of the canola oil based vitamin E nanoemulsions stabilized by food grade mixed surfactants using response surface methodology. Food Chem 183:1–7. https://doi.org/10.1016/j.foodchem.2015.03.021

Mehrnia M-A, Jafari S-M, Makhmal-Zadeh BS, Maghsoudlou Y (2016) Crocin loaded nanoemulsions: factors affecting emulsion properties in spontaneous emulsification. Int J Biol Macromol 84:261–267. https://doi.org/10.1016/j.ijbiomac.2015.12.029

Mohsin K, Long MA, Pouton CW (2009) Design of lipid-based formulations for oral administration of poorly water-soluble drugs: precipitation of drug after dispersion of formulations in aqueous solution. J Pharm Sci 98:3582–3595. https://doi.org/10.1002/jps.21659

Moraes-Lovison M, Marostegan LFP, Peres MS, Menezes IF, Ghiraldi M, Rodrigues RAF, Fernandes AM, Pinho SC (2017) Nanoemulsions encapsulating oregano essential oil: production, stability, antibacterial activity and incorporation in chicken pâté. LWT Food Sci Technol 77:233–240. https://doi.org/10.1016/j.lwt.2016.11.061

Mu H, Holm R, Müllertz A (2013) Lipid-based formulations for oral administration of poorly water-soluble drugs. Int J Pharm 453:215–224. https://doi.org/10.1016/j.ijpharm.2013.03.054

Mushtaq A, Mohd Wani S, Malik AR, Gull A, Ramniwas S, Ahmad Nayik G et al (2023) Recent insights into nanoemulsions: their preparation, properties and applications. Food Chem X 18:100684. https://doi.org/10.1016/j.fochx.2023.100684

Nabawy A, Makabenta JM, Schmidt-Malan S, Park J, Li C-H, Huang R et al (2022) Dual antimicrobial-loaded biodegradable nanoemulsions for synergistic treatment of wound biofilms. J Control Release 347:379–388. https://doi.org/10.1016/j.jconrel.2022.05.013

Neslihan Gursoy R, Benita S (2004) Self-emulsifying drug delivery systems (SEDDS) for improved oral delivery of lipophilic drugs. Biomed Pharmacother 58:173–182. https://doi.org/10.1016/j.biopha.2004.02.001

Oz Y, Nabawy A, Fedeli S, Gupta A, Huang R, Sanyal A et al (2021) Biodegradable poly(lactic acid) stabilized nanoemulsions for the treatment of multidrug-resistant bacterial biofilms. ACS Appl Mater Interfaces 13:40325–40331. https://doi.org/10.1021/acsami.1c11265

Ozogul Y, Yuvka İ, Ucar Y, Durmus M, Kösker AR, Öz M et al (2017) Evaluation of effects of nanoemulsion based on herb essential oils (rosemary, laurel, thyme and sage) on sensory, chemical and microbiological quality of rainbow trout (*Oncorhynchus mykiss*) fillets during ice storage. LWT Food Sci Technol 75:677–684. https://doi.org/10.1016/j.lwt.2016.10.009

Ozogul Y, Karsli GT, Durmuş M, Yazgan H, Oztop HM, McClements DJ et al (2022) Recent developments in industrial applications of nanoemulsions. Adv Colloid Interface Sci 304:102685. https://doi.org/10.1016/j.cis.2022.102685

Pallathadka H, Mustafa M, Sanchez DT, Sekhar Sajja G, Gour S, Naved M (2023) Impact of machine learning on management, healthcare and agriculture. Mater Today Proc 80:2803–2806. https://doi.org/10.1016/j.matpr.2021.07.042

Pouton CW (2000) Lipid formulations for oral administration of drugs: non-emulsifying, self-emulsifying and 'self-microemulsifying' drug delivery systems. Eur J Pharm Sci 11:S93–S98. https://doi.org/10.1016/S0928-0987(00)00167-6

Pouton CW, Porter CJH (2008) Formulation of lipid-based delivery systems for oral administration: materials, methods and strategies. Adv Drug Deliv Rev 60:625–637. https://doi.org/10.1016/j.addr.2007.10.010

Qadir A, Faiyazuddin MD, Talib Hussain MD, Alshammari TM, Shakeel F (2016) Critical steps and energetics involved in a successful development of a stable nanoemulsion. J Mol Liq 214:7–18. https://doi.org/10.1016/j.molliq.2015.11.050

Qian C, McClements DJ (2011) Formation of nanoemulsions stabilized by model food-grade emulsifiers using high-pressure homogenization: factors affecting particle size. Food Hydrocoll 25:1000–1008. https://doi.org/10.1016/j.foodhyd.2010.09.017

Ramalingam K, Lee VA (2018) Antibiofilm activity of an EDTA-containing nanoemulsion on multidrug-resistant *Acinetobacter baumannii*. Artif Cells Nanomed Biotechnol 46:737–743. https://doi.org/10.1080/21691401.2018.1468771

Rehman FU, Shah KU, Shah SU, Khan IU, Khan GM, Khan A (2017) From nanoemulsions to self-nanoemulsions, with recent advances in self-nanoemulsifying drug delivery systems (SNEDDS). Expert Opin Drug Deliv 14:1325–1340. https://doi.org/10.1080/17425247.2016.1218462

Salvia-Trujillo L, Rojas-Graü MA, Soliva-Fortuny R, Martín-Belloso O (2015) Use of antimicrobial nanoemulsions as edible coatings: impact on safety and quality attributes of fresh-cut Fuji apples. Postharvest Biol Technol 105:8–16. https://doi.org/10.1016/j.postharvbio.2015.03.009

Sarkar R, Pal A, Rakshit A, Saha B (2021) Properties and applications of amphoteric surfactant: a concise review. J Surfactant Deterg 24:709–730. https://doi.org/10.1002/jsde.12542

Sharma P, Shehin VP, Kaur N, Vyas P (2019) Application of edible coatings on fresh and minimally processed vegetables: a review. Int J Veg Sci 25:295–314. https://doi.org/10.1080/19315260.2018.1510863

Singh Y, Meher JG, Raval K, Khan FA, Chaurasia M, Jain NK et al (2017) Nanoemulsion: concepts, development and applications in drug delivery. J Control Release 252:28–49. https://doi.org/10.1016/j.jconrel.2017.03.008

Sridhar A, Ponnuchamy M, Kumar PS, Kapoor A (2021) Food preservation techniques and nanotechnology for increased shelf life of fruits, vegetables, beverages and spices: a review. Environ Chem Lett 19:1715–1735. https://doi.org/10.1007/s10311-020-01126-2

Umaraw P, Verma AK (2017) Comprehensive review on application of edible film on meat and meat products: an eco-friendly approach. Crit Rev Food Sci Nutr 57:1270–1279. https://doi.org/10.1080/10408398.2014.986563

Versino F, Lopez OV, Garcia MA, Zaritzky NE (2016) Starch-based films and food coatings: an overview. Starch/Stärke 68:1026–1037. https://doi.org/10.1002/star.201600095

Wei S, Tian Q, Zhao X, Liu X, Husien HM, Liu M et al (2022) Tea tree oil nanoemulsion potentiates antibiotics against multidrug-resistant *Escherichia coli*. ACS Infect Dis 8:1618–1626. https://doi.org/10.1021/acsinfecdis.2c00223

Yerramilli M, Ghosh S (2017) Long-term stability of sodium caseinate-stabilized nanoemulsions. J Food Sci Technol 54:82–92. https://doi.org/10.1007/s13197-016-2438-y

Yuan Y, Gao Y, Zhao J, Mao L (2008) Characterization and stability evaluation of β-carotene nanoemulsions prepared by high pressure homogenization under various emulsifying conditions. Food Res Int 41:61–68. https://doi.org/10.1016/j.foodres.2007.09.006

Zhang S, Zhang M, Fang Z, Liu Y (2017) Preparation and characterization of blended cloves/cinnamon essential oil nanoemulsions. LWT Food Sci Technol 75:316–322. https://doi.org/10.1016/j.lwt.2016.08.046

Nanoemulsions as Drug Carriers to Combat AMR

12

Partha Pratim Sarma and Pranjal K. Baruah

Abstract

Antimicrobial resistance (AMR) is an important health issue around the world. The substantial increase in AMR has severely restricted the existing conventional therapeutic options for multiple-drug-resistance pathogens. However, with the recent scientific revolution, novel antibiotics have been introduced to effectively combat pathogenic germs. Among these, nanoemulsions have attracted researchers due to their high stability, ease of synthesis, and enhanced adsorption rate. Nanoemulsions are thermodynamically stable colloidal dispersion framework composed of two relatively immiscible liquids mixed with emulsifying agents to achieve a single phase with average droplet size between 50 and 1000 nm. Because of the small size, it can easily permeate the biological membrane which enhances the therapeutic efficiency of drug molecules. Reduction in size of the droplets (10 to 100 nm) of the nanoemulsions significantly enhances their activity. The utilization of natural products including essential oil derived from plants ignited the interest of the pharmaceutical and food industries, owing to the possibility of minimal toxicities and the existence of a wide range of medicinal characteristics. In this article, we will provide a brief introduction to nanoemulsions along with their activity as antibacterial drug carriers. We will further discuss different types, advantages, disadvantages, and recent developments of nanoemulsions.

Keywords

Antimicrobial resistance · Drug delivery · Antibiotics · Nanoemulsion · Pathogens

P. P. Sarma · P. K. Baruah (✉)
Department of Applied Sciences, GUIST, Gauhati University, Guwahati, Assam, India
e-mail: pranjal@gauhati.ac.in

M. Y. Wani et al. (eds.), *Nanotechnology Based Strategies for Combating Antimicrobial Resistance*, https://doi.org/10.1007/978-981-97-2023-1_12

12.1 Introduction

The invention of antibiotics, which began with the introduction of penicillin over 70 years ago and improved to the advanced compounds in subsequent decades, modernized infectious disease treatments and drastically reduced the associated morbidity and death (Kalhapure et al. 2015). Antibiotics are regarded as crucial in almost every critical therapeutic field like major surgeries; particularly, organ transplantation, premature baby treatment, and treatment with chemotherapy in cancer patients couldn't be accomplished without properly preventing and treating bacterial infections (Kalhapure et al. 2015; Cars et al. 2011). However, several limitations are associated with modern as well as traditional antibiotic medication therapy that include limited bioavailability, poor penetration power, the emergence of antibiotics resistance bacteria, insufficient drugs concentrations at specific infection sites, significant side effects, higher frequency of administration as well as poor patient compliance (Kalhapure et al. 2015; Yeh et al. 2020; Huh and Kwon 2011; Sharma et al. 2012). These limitations along with the abuse and widespread use of antibiotics have ended up with the most critical limitation of antibacterial resistance of microorganisms (Kalhapure et al. 2015). Although there are various antimicrobial therapeutic alternatives, their effectiveness is limited due to the occurrence of antibiotic-resistant bacteria. Many traditional antibiotics have not been able to significantly improve the survival rate of people with infectious diseases (Yeh et al. 2020). In spite of numerous advancements in antibacterial medicine, the effectiveness of many of these medications is undoubtedly reduced by newly emerging novel strains of bacteria and the development of mutation-mediated resistance to drugs (Franklyne et al. 2021). Further, bacteria can also develop resistance by acquisition of resistance plasmids from the environment. Plasmids serve as the platforms for gene array assembly and recombination. Plasmids contain a broad range of genes, including the ones that contribute to antibiotic resistance. A resistance plasmid denotes any plasmid with one or more antimicrobial-resistant genes (Bennett 2008). Some examples of bacteria that have developed resistance against antibiotics are methicillin-resistant *Staphylococcus aureus* (MRSA) and Carbapenem-resistant *Enterobacteriaceae*; *Klebsiella pneumoniae* has developed resistance against carbapenem and tigecycline, which is the most efficient as well as a last-resort antibiotic for treating *K. pneumoniae (*Kalhapure et al. 2015*)*. Therefore, finding novel and efficient methods to improve drug therapy using the current antibiotics has been recognized as a key area of research priority worldwide (Kalhapure et al. 2015).

The innovation of novel antimicrobial substances to combat drug-resistant bacteria is very challenging but, in the meantime, it is very crucial and essential (Krishnamoorthy et al. 2018). Recent advancements in chemistry, biology, and medical research introduced new antibiotics and modernized medicine that are able to combat infectious bacteria efficiently. However, widespread antibiotic resistance has now reached a critical point, invalidating the most potent antimicrobial

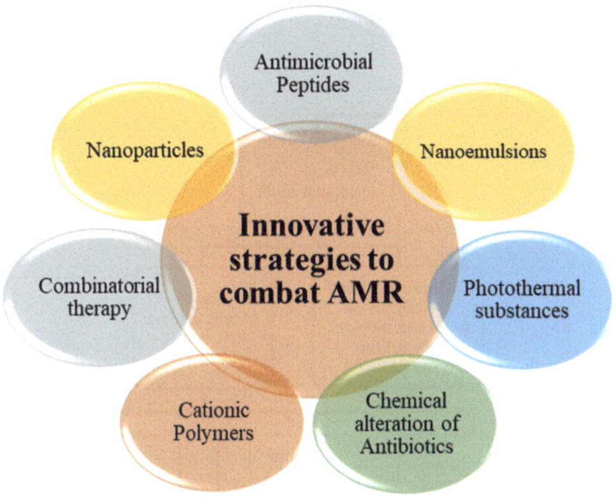

Fig. 12.1 Diverse innovative strategies to combat AMR

medications typically prescribed in clinics (Hakemi-Vala et al. 2017). Antibiotic-resistant bacteria have become a threat to human health nowadays, due to the excessive use of antibiotics. Among infectious microbes, bacteria persist as the leading factor of death in elders, children, and immunocompromised individuals (Yeh et al. 2020).

Diverse strategies like chemical alteration of antibiotics, photothermal substances, combinatorial therapy, antimicrobial peptides, nanoparticles, nanoemulsions, and cationic polymers have been revealed recently to be effective for combating Antimicrobial Resistance (AMR) and are represented in Fig. 12.1 (Yeh et al. 2020; Gebreyohannes et al. 2019).

The use of nanotechnology in the field of antibiotic drug delivery can provide a great opportunity to enhance the effectiveness of antibacterial treatment. Compared to conventional medications, nano systems for antibiotic administration as well as infectious site-targeting render several benefits including improved tissue targeting, extended antibiotic half-life, better solubility, improved stability, enhanced epithelial permeability, and minimal side effects (Yeh et al. 2020; Pizzolato-Cezar et al. 2019). Essential oils (EOs) along with other plant extracts have also drawn significant attention as antimicrobials nowadays (Tayeb et al. 2022). In this chapter, we will discuss the AMR developed by different classes of bacteria along with its adverse effect on human health. However, different strategies developed to combat AMR will also be discussed. Special emphasis will be made on the nanoemulsion-based drug carrier system to combat AMR. A schematic representation of the overall aim of the chapter is depicted on Fig. 12.2.

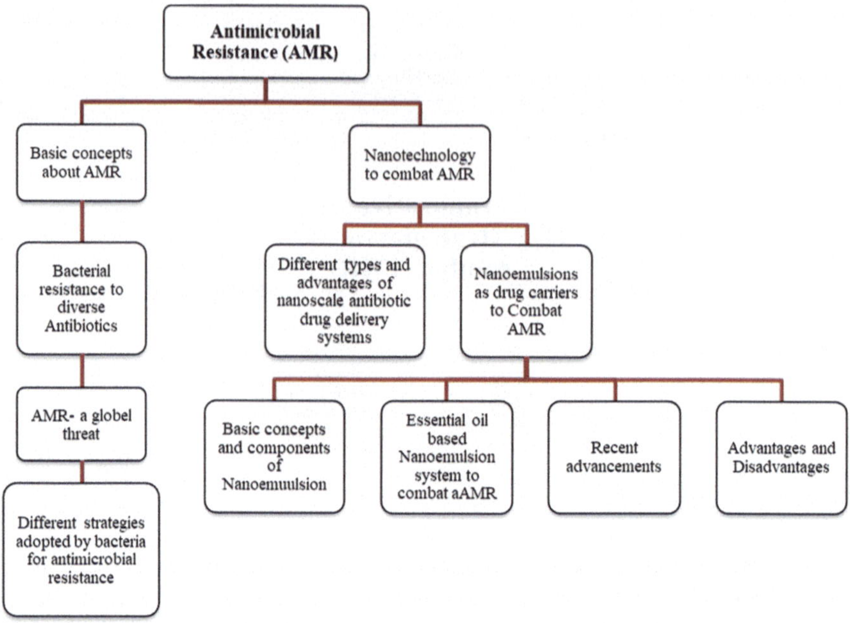

Fig. 12.2 Schematic representation of the overall aim of the chapter

12.2 Antimicrobial Resistance (AMR)

Antibiotics were widely utilized in human health and animal husbandry prior to the 1950s because of their low cost and few adverse effects. However, the widespread usage of antibiotics over a number of decades has given rise to the severe issue of drug resistance (Yeh et al. 2020). Antibacterial resistance is the capability of a particular microorganism to evade the pharmacological mechanism of action associated with antibacterial drugs and subsequently survive (Zong et al. 2022; Makabenta et al. 2020). Bacteria acquiring resistance genes that can inhibit the activity of drugs either by hindering drug uptake, altering the drug target, deactivating the medication, or promoting active efflux of drug molecules contribute to antibiotic resistance which results in an increased mortality rate (Tayeb et al. 2022; Larsson and Flach 2021). Antibacterial resistance leads to an alarming threat to curing bacterial infections as it causes significant economic loss for both patients and nations as well as severely increases mortality and morbidity rates (Krishnamoorthy et al. 2018). The uncontrolled usage of broad-spectrum antibiotics has been significantly associated with the prevalence of certain drug-resistant bacteria. Furthermore, certain Gram-negative as well as Gram-positive bacterial strains have acquired resistance not only to frequently employed antibiotics, but also to entire classes of numerous antibiotics. Extended-spectrum beta-lactamase (ESBL) producers and MRSA, in particular, have developed resistance to

Table 12.1 Examples of some antibiotic resistance bacteria along with the diseases caused by them

Name of the bacteria	Diseases caused	Resistance to antibiotics	Ref.
Klebsiella pneumoniae	Several types of hospital-acquired illnesses and epidemics, urinary tract infections, and pneumonia	Carbapenem and Tigecycline	Tayeb et al. (2022) and Paczosa and Mecsas (2016)
Acinetobacter baumannii	Hospital-acquired infections and epidemics	Carbapenem	Kohler et al. (2019) and Correa et al. (2018)
Pseudomonas aeruginosa	Responsible for pneumonia in critically ill or immunodeficient individuals	Carbapenem	Kohler et al. (2019)
Enterococcus faecium	Intra-abdominal infections, surgical wound infections, endocarditis, urinary tract infections, neonatal sepsis	Clindamycin, Aminoglycosides, Penicillin, Chloramphenicol, Vancomycin	Kohler et al. (2019) and Arias et al. (2010)
Helicobacter pylori	Gastritis, peptic ulcers, vitamin B12 deficiency, gastric adenocarcinoma	Clarithromycin	Kohler et al. (2019) and Shmuely et al. (2016)
Salmonella spp.	Bacteremia, gastroenteritis, paratyphoid, and typhoid	Fluoroquinolone	Kohler et al. (2019)
Enterococcus	Urinary tract infections, bacteremia, intra-abdominal infections, meningitis	Vancomycin	Kalhapure et al. (2015)
Streptococcus pneumonia	Pneumonia, bloodstream infections, sinusitis	Penicillin Tetracyclines	Kalhapure et al. (2015)

cephalosporin and methicillin, as well as to aminoglycosides, tetracycline, chloramphenicol, and macrolides (Krishnamoorthy et al. 2018; Nikaido 2009). *Helicobacter pylori* which are Gram-negative bacteria through the emerging mutation developed resistance against clarithromycin (Yeh et al. 2020; Alba et al. 2017). Some other examples of different bacterial species along with the diseases caused by them and the antibiotics to which they have become resistant are represented in Table 12.1.

12.3 Bacterial Resistance Mechanism Towards Traditional Antibiotics

To access the target site, the majority of antibiotics must, in principle, penetrate the microbial cell membrane. Once it has entered the bacterial cell, antibiotics bind with the target site and interfere with cellular biochemical processes, thereby impeding the cell growth or even causing cell death (Zong et al. 2022; Kohanski et al. 2010). Many pathogens, on the contrary, have significantly developed diverse

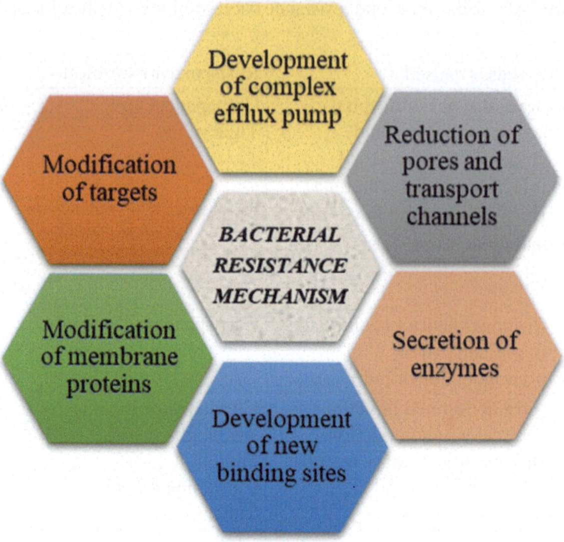

Fig. 12.3 Different resistance mechanisms of bacteria against antibiotics

strategies to prevent the action of antibiotics which include the protection as well as modifications of the target sites, diminution of transport channels and pores, modification of the structure and activity of the membrane proteins for preventing the admission of the antibiotics, production of enzymes in cytoplasm or periplasm to alter or disrupt the molecular structure of the antibiotics, creation of new binding sites to evade the susceptible therapeutic sites from binding with the drugs, modification of the target's function and structure, and developing or improving an intricate efflux pump for rapidly eliminating the drug from the bacterial cell (Zong et al. 2022; Blair et al. 2015; Munita and Arias 2016). Figure 12.3 represents the major strategies developed by bacteria to fight against traditional antibiotics.

12.4 Antimicrobial Resistance: A Global Threat

The World Health Organisation (WHO) recently identified AMR to be one of the major threats to human health and development worldwide (Zong et al. 2022). One of the primary factors responsible for antibacterial resistance is the improper application of broad-spectrum antibiotics, particularly in hospitals (Tayeb et al. 2022). The challenges faced in curing bacterial infections are exacerbated by the rapid growth and spread of antibiotic-resistant organisms and inefficient antimicrobial medications, leading to significant financial costs and increased mortality rates (Tayeb et al. 2022; Larsson and Flach 2021). Bacteria can cause fatal human diseases, and specifically superbug infection every year kills more than 700,000 people globally (Zong et al. 2022). Over the past few decades, AMR has emerged as an imminent threat to the successful treatment of the massively rising bacterial infections caused by fungi, parasites, viruses, and bacteria around the world. The number of antibiotic-resistant bacterial strain is terribly increasing and in recent decades the number has quadrupled (Krishnamoorthy et al. 2018). The WHO has warned that if proper measures are not taken worldwide to eradicate it at present, it will eventually turn into a fatal disease in the future. AMR causes an erosion in the therapeutic efficacy of antiparasitic, antibacterial, antifungal, and antiviral medications causing the treatment challenging, expensive, or perhaps impossible for the patient (Krishnamoorthy et al. 2018; Basak et al. 2016). The effect is perceived most acutely by vulnerable patients as it can lead to long-term morbidity and greater mortality. The scope of this problem is worldwide and its impact on animal and human health, as well as its prevalence in the wider community, is still largely unknown.

The rise of superbugs and the rapid expansion of antibiotic resistance to various environmental niches have made efficient prevention strategies even more challenging. For the prevention and control of antibiotic resistance, national, international, and local measures have been recommended (Soni and Sharma 2021). Proper understanding of the resistance mechanism of different drug-resistant bacterial strains, innovations in novel medications and vaccines along with a collaborative, multidisciplinary, and regulatory approach are strictly required to Combat AMR (Kalhapure et al. 2015). Antibacterial resistance is a very complicated issue with a wide range of underlying factors. To combat the enormously growing widespread antibiotic resistance, collaborative efforts from the patients, doctors, as well as individuals to worldwide policymakers and regulators are required (Soni and Sharma 2021). Drug resistance demands high-dose treatment with antibiotics that impose severe toxicity as well as the development of novel antibiotics, which leads to extensive research, time, and economic investments (Hakemi-Vala et al. 2017). Research into novel antimicrobials has been motivated by AMR, specifically the restricted antimicrobial effects of antibiotics and natural substances (Tayeb et al. 2022).

12.5 Nanotechnology to Combat AMR

Although antibiotics are essential to cure a number of diseases caused by bacterial infection, there are several side effects associated with the commonly prescribed antibiotics or antimicrobial drug therapy. The adverse effects caused by such antibiotics includes gastrointestinal effects like abdominal pain, nausea, loss of appetite, diarrhea and vomiting, allergic reactions, gonorrhoea, and neurologic disorders (Llor and Bjerrum 2014; Mohsen et al. 2020). Beside these, some other adverse effects include Clostridium difficile infections associated with clindamycin and amoxicillin, hepatotoxicity associated with clavulanate and amoxicillin, Linezolid causing hypertension, convulsions, and myelosuppression as well as Clostridium difficile infections (Llor and Bjerrum 2014; Mohsen et al. 2020). Hence, the development of an effective alternative to overcome the constraints associated with existing antibiotic drug delivery systems is of the utmost importance nowadays. In this regard, nanotechnology which deals with the design, fabrication as well as application of nanoscale materials is currently recognized as an innovative and fascinating approach to enhance the effectiveness of the treatment of infectious diseases (Kalhapure et al. 2015; Sarma et al. 2023). A wide range of nanoscale antibiotic drug delivery systems have already been introduced to fight against AMR which includes polymeric nanoparticles, metal nanoparticles, lipid polymer hybrid nanoparticles, carbon nanotubes, carbon nanorods, carbon nanosheets, nanoemulsions, nanohybrids, and solid lipid nanoparticles (Kalhapure et al. 2015; Sarma et al. 2023; Couvreur 2013; Barman et al. 2020a). The numerous advantages of introducing nanotechnology for the treatment of several chronic diseases like hypertension, AIDS, cancer, and inflammation through enhancing the bioavailability, specificity, efficacy, and solubility of drug molecules have been extensively reported (Kalhapure et al. 2015; Parhi et al. 2012; Gerson et al. 2014; Moretton et al. 2014). Therefore, nano-drug delivery systems provide a superior and revolutionary approach to eradicating numerous disadvantages related to the use of antimicrobial drug therapy, particularly the alarming global threat of resistance to antibiotics. Nanoemulsions have the potential to serve as an efficient antimicrobial agent themselves or as the carrier of antimicrobial drugs in order to boost the bioavailability as well as the activity of the drug (Yeh et al. 2020). The application of nano drug delivery systems precisely for encapsulating and conveying antimicrobial drugs is still in its infancy when compared to diseases like cancer and heart disease (Kalhapure et al. 2015).

12.6 Nanoemulsions: A Basic Introduction

Nanoemulsion can be defined as the fine, transparent water-in-oil or oil-in-water dispersion systems stabilized through an interfacial layer of surfactant particles with droplet sizes ranging from 20 to 500 nm (Franklyne et al. 2021). Nanoemulsions, also known as miniemulsion or ultrafine emulsion or submicron emulsions, are isotropic, heterogeneous systems with promising thermodynamic stability where a pair of immiscible liquids like oil and water is mixed to give a single phase employing the proper surfactants (Soni and Sharma 2021; Choradiya and Patil 2021; Tayeb and Sainsbury 2018). Amphiphilic surfactant molecules are incorporated at the liquid-liquid junction of the nanoemulsion to enhance stability and introduce surface properties (Tayeb et al. 2022). Prolonged stability, extended shelf-life, spontaneous emulsification, high degree of drug solubilization ability, large surface to volume ratio, and photostability, inherent antimicrobial properties, ability to target biofilms, potentials for organ as well as cellular targeting enable the nanoemulsions to serve as an efficient antibacterial agent as well as a drug delivery platform (Franklyne et al. 2021, 2016). Nanoemulsions also offer great potential in the field of biotechnologies, drug therapies, diagnostics, and cosmetics. The biological activities of the nanoemulsions are largely influenced by the surface parameters of the nanoemulsion droplets. Nanoemulsions can be obtained in a variety of dosage forms, including creams, liquids, aerosols, gels, sprays, and foams, and can be administered by a diverse route including oral, topical, pulmonary, intranasal, intravenous, and ophthalmic routes (Singh et al. 2017; Basha et al. 2020).

Three different forms of nanoemulsion are categorized according to the composition of the water and oil components: (Franklyne et al. 2021; Ali et al. 2017; Halnor et al. 2018)

1. Oil in water nanoemulsion (O/W): Here oil is dispersed in the aqueous phase. Water is regarded as the dispersion medium or continuous or external phase, while oil is considered as the dispersed phase or discontinuous or internal phase.
2. Water in oil nanoemulsion (W/O): Here droplet of water is dispersed in the oil phase. Oil is regarded as the dispersion medium or continuous or external phase, while water is considered as the dispersed phase or discontinuous or internal phase.
3. Bi-continuous nanoemulsion: Here microdomains of water and oil are interdispersed across the system.

12.7 Different Components of a Nanoemulsion System

Surfactant, a cosurfactant, oil, and water are the basic ingredients essential to formulate a nanoemulsion that yields an isotropic, thermodynamically stable, clear, transparent, or translucent solution having mean droplet sizes between 20 and 500 nm (Franklyne et al. 2016; Ali et al. 2017). In addition, the most crucial factor in choosing all of the components for nanoemulsion formulation is that each excipient must be pharmaceutically approved for topical application or oral treatment (Chime et al. 2014). Major components of a nanoemulsion system along with their different classes and examples are shown in Fig. 12.4.

IMPORTANT COMPONENTS OF NANOEMULSION WITH THEIR TYPES AND EXAMPLES	SURFACTANTS	*Nonionic Surfactants*	Polyoxyl 40, hydrogenated Castor Oil, Tween 20, 40, 60 etc.
		Anionic Surfactants	Carboxylate groups, Divalent ions, Soaps, Sulfonates etc.
		Cationic Surfactants	Quaternary ammonium compounds and Amines for eg Chlorhexidine.
		Zwitterionic Surfactants	Dodecyl betaine, Phospholipid, Amino acids like Leucine etc.
	COSURFACTANTS	*Short chain glycols*	Propylene glycol
		Medium chain alcohols	Acids or Amines
		Short chain alcohols	Ethanol and Butanol
	OILS	*Unsaturated fatty acids*	Linolenic acid, Oleic acid, Linoleic acid etc.
		Saturated fatty acids	Lauric, Myristic, Capric acid etc.
		Fatty acid esters	Methyl or Ethyl esters of lauric, Myristic

Fig. 12.4 Different components of a nanoemulsion system with some examples

12.7.1 Surfactants

A surfactant is a surface-active agent that serves as a junction between the aqueous phase and the hydrophobic portion or the hydrophilic and aqueous phases throughout the process of emulsification (Franklyne et al. 2021). Better solubility of oils in the surfactant is very essential for the better efficacy of the nanoemulsion system (Chime et al. 2014). Each surfactant molecule consists of two elements, one is a hydrophilic head and another one is the hydrophobic tail. Based on the charges present on the hydrophilic head, surfactants are divided into three different classes (Sutradhar and Amin 2014). Surfactants having no charged groups on their head are termed as nonionic and if it is positively charged, surfactant is termed as cationic. An anionic surfactant exhibits a net negative charge in its hydrophilic head (Sutradhar and Amin 2014). The properties of the nanoemulsions are largely influenced by the nature and type of the surfactant; therefore, the correct choice of surfactant is very essential (Chime et al. 2014). Oral and topical administration of nanoemulsion systems possessing excessive surfactant concentration may cause skin irritation and gastrointestinal disorders. Hence, it is also essential to maintain the proper concentration of the surfactant during the preparation of nanoemulsions (Chime et al. 2014; Kawakami et al. 2002). Nonionic surfactants are predominantly used because of their biocompatibility, less toxicity, less sensitivity to pH, and variations in ionic strength. However, use of the ionic surfactants is less frequent because of their toxicological concerns. Nonionic surfactants comparatively bear lesser toxicity than ionic ones and often possess lowered critical micelle concentration (Chime et al. 2014). Surfactants, in addition to emulsification, also demonstrated a wide range of pharmacological activities like anticancer, antimicrobial, antiaging, anti-inflammatory, etc. (Franklyne et al. 2021). A surfactant that interferes with the virion directly or hinders its penetration to the host is suitable for the formulation of a new antibacterial nanoemulsion or nano-delivery system. Tween 80, Tween 80, Tween 20, Poloxamer 124, Gelucire 44/14, Cremophor RH 40, Labrafil CS, Lauroglycol 90, Polyethylene glycol (MW > 4000), Labral M 2125 CS, Capryol 90, Softigen 701, Cremophor EL, and Labrasol are the several types of surfactants used in the nanoemulsion formulation (Chime et al. 2014).

The effectiveness of a combination of nonionic hydrophilic and lipophilic surfactants within the oil and water phase is determined by the hydrophilic-lipophilic balance (HLB) value (Hakemi-Vala et al. 2017). Surfactants having higher values of HLB are considered to be more hydrophilic. However, surfactants with lower HLB values are lipophilic. For nonionic surfactants, the HLB value varies from 0 to 20, whereas for ionic ones, it can exceed 40 (Hakemi-Vala et al. 2017). For the development of nanoemulsions with higher stability and small droplet size, optimization of the HLB value acts as the driving force (Hakemi-Vala et al. 2017).

12.7.2 Cosurfactants

Cosurfactants are generally short-chain alcohols, salts, and organic acids that are employed to enhance the emulsifying process of a particular surfactant. As cosurfactant molecules are weakly amphiphilic, they do not undergo self-aggregation but substantially concentrate and promote the aggregation of primary surfactant that takes place on the surfactant layer (Franklyne et al. 2021). Cosurfactants are employed in the formulation of nanoemulsion systems in order to minimize the concentration of the surfactant. To further minimize interfacial tension as well as to enhance the fluidity of the interface, medium to short chain length alcohols, generally possessing three to eight carbon atoms in the chain, are frequently used as cosurfactants (Chime et al. 2014; Tenjarla 1999). Additionally, they make the hydrocarbon tail more mobile which facilitates better penetration of the oil phase inside this region (Chime et al. 2014). Due to the partitioning among hydrophilic and hydrophobic phases, alcohols can also improve the miscibility of these two phases. Glycerin, ethanol, isopropyl alcohol, n-butanol, propanol, Carbitol, PEG 400, Transcutol P, propylene glycol, and ethylene glycol are the several types of cosurfactants employed in the preparation of nanoemulsions (Chime et al. 2014).

12.7.3 Oils

The drug's solubility within the oil phase is the key factor for assigning suitable oils for the preparation of nanoemulsions. In particular, this is crucial for developing oral formulations, since the solubility of the particular medication in the oil phase has a significant influence on the potential of the nanoemulsion to preserve the drug in the dissolved state. While oil in water nanoemulsions preferably solubilizes the lipophilic drugs, hydrophilic drugs are better soluble in water in oil nanoemulsion (Chime et al. 2014). However, because of the inability to solubilize the lipophilic drugs in large amounts, edible oils are generally not regarded as useful for the preparation of nanoemulsions. Moreover, the preparation of nanoemulsion based on oils bearing poor drug solubility requires an additional amount of oil to integrate the intended drug dose which in turn demands a higher concentration of surfactants to accomplish oil solubilisation, thereby introducing toxicity in the system. Captex 8000, Captex 355, Myritol 318, Witepsol, Sefsol-218, Isopropyl myristate, triacetin, Capryol 90, olive oil, and castor oil are the several types of oils employed in the preparation of nanoemulsions. Nowadays, several novel amphiphilic compounds possessing surfactant characteristics are gradually as well as efficiently omitting traditional medium-chained triglyceride oils (Chime et al. 2014; Chakravarty and Vora 2021).

12.8 Essential Oil Nanoemulsions as Antibiotics and Drug Carriers to Combat AMR

EO-based nanoemulsions have imminence potential for targeting microbes, particularly bacteria. They serve as a prominent vehicle for drug delivery as well as vaccines (Franklyne et al. 2021). Literature study reveals that essential EOs exhibit a natural broad spectrum of efficacy contrary to bacterial, fungal, and viral infections

(Tayeb et al. 2022; Gundel et al. 2018). The therapeutic value of the active components in the EOs has been greatly enhanced by the emulsification procedure (Franklyne et al. 2021; Gundel et al. 2018). Tayeb et al. developed a novel clove essential oil- (CEO) based oil in water nanoemulsion system and a water-in-oil nanoemulsion through encapsulation of CEO and meropenem for the investigation of antimicrobial activity as well as to deliver hydrophilic and hydrophobic compounds against the carbapenem-resistant *Klebsiella pneumonia* pathogen. The prepared nanoemulsions were negatively charged, monodispersed, and physically stable. The minimum bactericidal concentration (MBC) and minimum inhibitory concentration (MIC) of the nanoemulsion were calculated. For the development of CEO-based nanoemulsion, Imwitor 375 and tween 80 surfactants were used. The average droplet size was reported to be 138 nm. They have reported the MIC for CEO-based nanoemulsion as 0.16% CEO. For the preparation of CEO-meropenem nanoemulsion system, Imwitor 375, PGPR, and tween 80 were used. They have found the average droplet size as 183 nm and the MIC as 1 μg/mL meropenem with 0.08% CEO. The prepared nanoemulsions significantly increase the antibacterial activity of the meropenem and CEO and possess the great potential to serve as an antibiotics drug delivery system (Tayeb et al. 2022).

Contrary to commercial drugs, the majority of EOs are nontoxic and listed by FDA as widely recognized as safe. Numerous studies are now exploring the impact and effectiveness of EOs against several viral, acute, and chronic disorders (Franklyne et al. 2021). Biocidal activity against numerous bacterial strains is widespread in the nanoemulsions based on EOs derived from the herbal plants which represent a novel class of antibiotics for the prevention and treatment of bacterial infection caused by both Gram-positive and Gram-negative bacteria as well.

Encapsulation of EOs boosts solubility and stability, enhances therapeutic efficiency, and prevents the nanoemulsion system from deterioration processes (Gundel et al. 2018). The pharmaceutical as well as food industries have taken a keen interest in using goods of natural origin, especially because of the prospect of minimal toxicity of these compounds and the availability of a wide range of therapeutic attributes. EOs are generally intricate, volatile natural substances characterized by an intense odor (Gundel et al. 2018). EOs are chemically extracted from the Phenolic compounds, Phenylpropanoids, and terpenes along with their oxygenated derivatives (Gundel et al. 2018; Behera et al. 2020). Each of these compounds accounts for the various attributes of the EOs including antifungal, antibacterial, antiviral, anti-inflammatory, antiparasitic, antioxidant, and insecticidal properties (Gundel et al. 2018; Prabuseenivasan et al. 2006).

12.9 Some Recent Developments on Essential Oil Nanoemulsion System to Combat AMR

EO-based nanoemulsions exhibit immense potential for combating microbes, particularly viruses and bacteria (Franklyne et al. 2021). They serve as a delivery system for vaccinations and antiviral medications. EOs possess an inherent activity against viral, bacterial, and fungal pathogens. The emulsifying process significantly enhanced the potency of the active compounds in the EOs (Franklyne et al. 2021). Nowadays, numerous researches are going on in the development of plant-based nanoemulsion systems particularly using green methods for the treatment of bacterial infection as well as to use in antibacterial drug delivery.

Krishnamoorthy et al. in their study developed a nanoemulsion system based on *Cleome viscosa* seeds EO using the nonionic surfactant, Tween 80, and water. The droplet size was found to be 7 nm. They have reported that a surfactant:oil ratio (v/v) of 3:1 was efficient to combat the resistance mechanisms of drug resistance pathogens through the inhibition of the drug diffusion mechanism of drug resistance bacterial strains. Prepared nanoemulsion was reported to be effective against a wide range of Gram-positive and Gram-negative bacteria including drug-resistant *Streptococcus pyogenes*, methicillin-resistant *Staphylococcus aureus* (MRSA), and drug-resistant extended-spectrum beta-lactamase (ESBL)-producing *Klebsiella pneumoniae*, *Pseudomonas aeruginosa*, and *Escherichia coli* (Krishnamoorthy et al. 2018).

Carum carvi EO-based nanoemulsion gels were prepared by Alquarni et al. Using the nanoemulsion gel, they have investigated the activity of the *Carum carvi* EO for antimicrobial treatment. The average particle size of the nanoemulsion was reported as 137 nm with an encapsulation efficacy of 92%. They have blended the EO nanoemulsion with the carbopol gel which possesses in vitro cell viability as well as antibacterial activity against *Staphylococcus aureus* and *Escherichia coli*. They have further evaluated the efficacy of the prepared nano gel in transdermal drug delivery. The optimum pH for drug delivery was reported as 6.5 with the ability to rapidly release drug. Drug release ability was found to be higher in the case of the EO-based nano gel system as compared to the pure EO system, indicating the greater efficacy of nanoemulsion in drug delivery. Thus, the *Carum carvi* EO-loaded nano gel can act as a substitute for synthetic antibiotics in the therapy of antibacterial infections (Alqarni et al. 2023).

Gundel et al. derived a nanoemulsion containing *Ocimum basilicum* EO and evaluated the cytotoxicity in the human cell, stability, antioxidant, and antimicrobial properties. The nanoemulsion remained stable for up to 90 days. They have reported that the formulated nanoemulsion comprising basil oil possesses excellent physicochemical stability, efficacy, and safety and can be regarded as a potential alternative to the traditional antimicrobials. It also protects healthy human cells from the reduction of their cell viability. The prepared nanoemulsion was reported to be effective against Gram-positive *Staphylococcus aureus*, *Candida albicans* and *Candida tropicalis* as well as Gram-negative *Escherichia coli* and *Proteus mirabilis* bacteria. The mean particle size of the nanoemulsion was reported to be 119 nm (Gundel et al. 2018).

Fopase and his coworkers developed Lipopeptide, eucalyptus EO, and doxorubicin, an anticancer drug-loaded nanoemulsion system for controlled drug delivery. Lipopeptides generally possess antimicrobial and anticancer properties. However, EO derived from eucalyptus offers anti-inflammatory and antimicrobial activities. The particles of the prepared nanoemulsion were reported to be spherical with a mean size of 175 nm. The prepared nanoemulsion was found to be effective against *Enterococcus hirae* and *Escherichia coli*. They have also reported that the formulated nanoemulsion demonstrated broad-spectrum antimicrobial activity with an IC_{90} value comparable to that of Kanamycin (Fopase et al. 2020).

Silva et al. evaluated the antimicrobial and antibiofilm activities of EO extracted from the *tetragastris catuaba* leaves both in free as well as nanoemulsion encapsulated form against disease-causing pathogens that involve *Enterococcus faecalis* and *Listeria monocytogenes*. From the GC-MS analysis, they have reported that the free EO contains 54 different compounds among which β-caryophyllene is the major compound which is present in 25.6%. They have developed two different nanoemulsion systems, one is based on the EO and the other on β-caryophyllene using tween 80 and span 80 as surfactants. The average droplet size of the EO-based nanoemulsion system was reported around 40 nm and 41 nm for the β-caryophyllene-based nanoemulsion system. Their study reveals that the EO-based nanoemulsion system was found to be superior in controlling the growth as well as biofilm formation of *E. faecalis* and *L. monocytogenes*. However, nanoemulsion systems based on β-caryophyllene have not exhibited good activities against Listeria monocytogenes. The EO-based nanoemulsion was reported as highly stable and soluble in water, making the nanoemulsion system an efficient drug delivery system (Silva et al. 2022).

Peppermint oil nanoemulsion-loaded polycaprolactone nanofiber with enhanced antibacterial activity through an electrospinning technique has been recently reported by Naggar et al. For the preparation of the nanoemulsion, tween 40 was employed as a surfactant. The diameter of the nanoemulsion droplet was found ranging from 150 to 200 nm. The prepared nanocomposite film exhibits enhanced antifungal and antimicrobial activities along with significant potential to inhibit biofilm formation. The antimicrobial activity was evaluated against different human pathogenic bacteria including *Acinetobacter baumannii*, *Staphylococcus aureus*, *Pseudomonas aeruginosa*, and *Listeria monocytogenes* as well as fungi including *Candida albicans* and *Aspergillus niger*. They have reported that the fabricated nanocomposite loaded with peppermint oil nanoemulsion possesses excellent potential to act as antimicrobial mats that are useful for the development of dermal patches for treating skin diseases like acne and wound healing (Naggar et al. 2023).

Bamisaye et al. explored the antibacterial activity of sesame oil-based nanoemulsion system loaded with Ciprofloxacin drug as well as a nondrug-loaded nanoemulsion system against *Klebsiella pneumoniae* and *Bacillus subtilis*. The average droplet size was reported to be around 124 and 244 nm for nondrug-loaded and the drug-loaded nanoemulsion systems, respectively. When compared with the drug alone, ciprofloxacin-loaded nanoemulsion exhibits greater antibacterial properties with increased zones of inhibition and lower MIC and MBC values for both the two microorganisms. They have also investigated the toxicity of both the nanoemulsion systems in Wistar rates and found no hepatotoxic, hematotoxicity, and cardiotoxic effects as well as free from skin-irritating effects (Bamisaye et al. 2023).

12.10 Advantages and Disadvantages of Nanoemulsion in Drug Delivery

Nanoemulsions have the great potential to solubilize significant quantities of hydrophobic medicines when the surfactants and cosurfactants are properly adopted which enables substantial drug loading within the oil phase of the nanoemulsion system, leading it to serve as a reservoir of drug (Yukuyama et al. 2017; Hormann and Zimmer 2016). It is simple to customize the physicochemical attributes of a nanoemulsion system by tailoring different factors like oil-surfactants ratio, synthesis procedures, as well as surface modifications for particular drug delivery to a specific target through different targeting mechanisms (Yukuyama et al. 2017; Lallemand et al. 2012). Nanoemulsions, because of their small particle size and high surface area, can easily as well as readily interact with the target, providing enhanced drug release and carrier ability (Tayeb et al. 2022; Yukuyama et al. 2017). Because of the significantly small particle size, nanoemulsion does not suffer from the issue of flocculation, precipitations, creaming, and aggregation. However, small particle size imparts better penetration to the target site which leads to enhanced bioavailability of the drug (Khaleel et al. 2020; Bhatt and Madhav 2011). Apart from drug delivery, the application of nanoemulsions in food products has also gained significant importance nowadays (Barman et al. 2020b).

Nanoemulsions could be an ideal approach to combat the instability of weakly water-soluble and water-insoluble pharmaceuticals, relying upon compositions and synthetic procedures (Tayeb et al. 2022; Yukuyama et al. 2017). However, many drugs are chemically labile and generally very prone to oxidation, hydrolysis, and enzymatic degradations. These drugs can be protected from such degradation processes through encapsulation within the oil core rendering nanoemulsions as an excellent vehicle for a drug delivery system (Hakemi-Vala et al. 2017; Hormann and Zimmer 2016). Since it has no effect on the undesired parts of the body, nanoemulsion prolongs the drug retention time within the target area, resulting in minimal toxicities and side effects along with reduced drug loss (Sutradhar and Amin 2014; Khaleel et al. 2020). It can be prepared in multiple forms, including foams, creams, foams, sprays, and liquids (Hakemi-Vala et al. 2017; Bhatt and Madhav 2011). Nanoemulsions possess the ability to deliver peptides that are susceptible to hydrolysis by enzymes in the gastrointestinal tract and are suitable for both human and veterinary applications because they have no adverse effects on healthy human as well as animal cells (Khaleel et al. 2020). The application of nanoemulsions can result in longer-term efficacy along with dose reduction, which can lead to a reduction in typical side effects (Hakemi-Vala et al. 2017; Tayeb et al. 2022; Ali et al. 2017; Halnor et al. 2018). However, diverse functionalities of nanoemulsion provide a wide range of potential applications in the field of drug delivery including parenteral, ophthalmic, oral delivery, transdermal as well as cancer-targeted delivery of drugs (Ali et al. 2017; Yukuyama et al. 2017; Ganta et al. 2014). The antibacterial nanoemulsions with positive charges exhibit broad-spectrum efficacy against numerous pathogens. The electrostatic attraction between the positive charge of the nanoemulsion and the negatively charged microbes enables the nanoparticles to thermodynamically interact with lipid-containing microbes causing cell lysis or death (Khaleel et al. 2020).

However, for adjustment, an extensive compilation of surfactants and cosurfactants must be accomplished, and their stability is largely influenced by pH and temperature (Hakemi-Vala et al. 2017; Ganta et al. 2014). Ostwald ripening can lead to an increase in particle size of the nanoemulsion droplets which decreases the stability of the system (Chime et al. 2014). The concentration of surfactants and cosurfactants must be limited. Excessive surfactant and cosurfactant concentrations can impart toxicity to the nanoemulsion system (Hakemi-Vala et al. 2017; Halnor et al. 2018). Long-term storage of the nanoemulsion system is a major issue associated with nanoemulsions (Basha et al. 2020; Ali et al. 2017). Commercial production of the nanoemulsion system is very costly because of the requirement of sophisticated instruments (Chime et al. 2014). Figure 12.5 represents the major advantages and the limitations associated with the nanoemulsion system in the field of drug delivery.

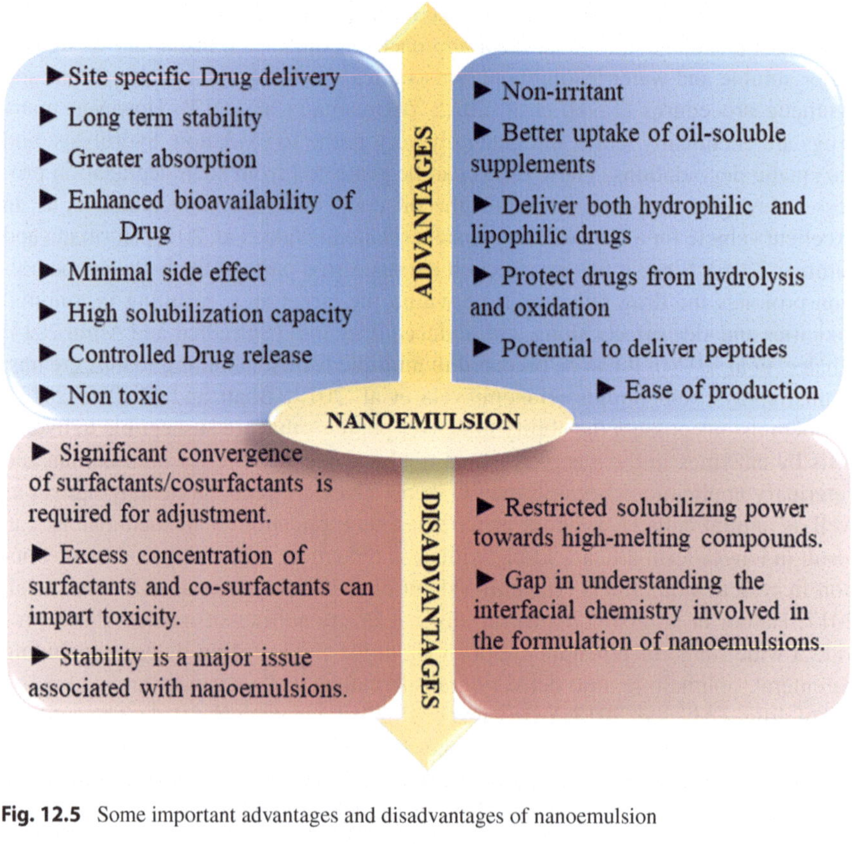

Fig. 12.5 Some important advantages and disadvantages of nanoemulsion

12.11 Mechanism of Nanoemulsion to Combat Bacterial Resistance

Nanoemulsions comprising EOs or antimicrobial agents or drugs can reduce viable bacteria cells more rapidly as compared to corresponding EOs or drugs alone. Due to the very small droplet size of nanoemulsion, it can easily transport the active antimicrobial components to the surface of the cell membrane. However, pure oil, which has limited water solubility, was unable to readily penetrate and interfere with the cell membrane (Hakemi-Vala et al. 2017). Nanoemulsion carries out the EOs containing the compounds possessing antimicrobial activity into close vicinity to the bacterial cell membrane. This allows the hydrophobic molecules in the EO or the active components in a nanoemulsion system to interfere with the cell membrane probably by affecting the integrity of the phospholipid bilayer or by interacting with the active transport proteins within the phospholipid bilayer (Hakemi-Vala et al. 2017; Franklyne et al. 2016). Alteration in the permeability of the destroyed cell membranes causes the leakage of proteins, potassium, and nucleic acids from the interior of the cells, resulting in cell death (Hakemi-Vala et al. 2017; Franklyne et al. 2016).

The cell membrane of the bacteria is composed of charged lipids possessing strong binding ability with high electrostatic attraction potential (Krishnamoorthy et al. 2018). However, nanoemulsions composed of antimicrobial nanodroplets also possess charged lipids within their outermost layer which can interact with the bacterial cell membrane and a core region comprised of bioactive health-promoting substances with the ability to damage the bacterial cells (Krishnamoorthy et al. 2018). Nanoemulsion particles are thermodynamically favored also to interact with lipid-containing microorganisms. This interaction is driven by the electrostatic interactions among the charges in the nanoemulsion system as well as negative charges on the bacterial cell membrane (Parhi et al. 2012). When the interaction is strong, it destabilizes the bacterial cell membrane leading to cell lysis or even cell death. A schematic representation of mechanism of nanoemulsion to combat bacterial resistance is shown in Fig. 12.6.

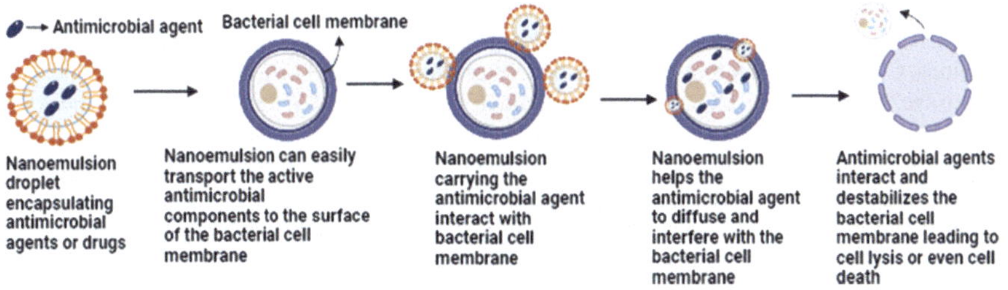

Fig. 12.6 A schematic representation of mechanism of nanoemulsion to combat bacterial resistance

12.12 Conclusion

Drug solubility in water is still a prevalent and persistent issue that complicates administration and bioavailability. Furthermore, individuals with chronic conditions, particularly those in the advanced phases of AIDS, find it difficult to swallow solid medicine formulations (Franklyne et al. 2021). Therefore, in transforming water-insoluble medicines into soluble and accessible one, nanoemulsion is of utmost interest. Nanoemulsions offer several intriguing advantages for the administration of active substances, and they are also receiving significant interest for improving the efficiency as well as delivery of antibacterial drugs. Significant research on nano-delivery systems has been conducted in the previous decades, leading to significant advancements in the field of drug delivery for different disease conditions. Nanoemulsions are of great importance to successfully combat AMR which has now become a global threat, but for further developments, multidisciplinary collaboration among academic research as well as pharmaceutical industries is highly recommended. However, there still exists a limited awareness concerning the development, production, fabrication, and modification of nanoemulsions, which can be notably due to the fact that the traditional aspects of stabilization and formulation of emulsions are only barely implemented in nanoemulsion systems. More potent nanoemulsion-based drug delivery systems as well as medicines were anticipated to become clinically available with advanced features which will serve as a promising tool for eradicating bacterial resistance towards antibiotics with reduced public hazards posed by intractable infectious illness and biofilm-related infections.

References

Alba C, Blanco A, Alarcon T (2017) Antibiotic resistance in *Helicobacter pylori*. Curr Opin Infect Dis 30(5):489–497. https://doi.org/10.1097/QCO.0000000000000396

Ali A, Ansari VA, Ahmad U, Akhtar J, Jahan A (2017) Nanoemulsion: an advanced vehicle for efficient drug delivery. Drug Res (Stuttg) 67(11):617–631. https://doi.org/10.1055/s-0043-115124

Alqarni MH, Foudah AI, Aodah AH, Alkholif FK, Salkini MA, Alam A (2023) Caraway nanoemulsion gel: a potential antibacterial treatment against *Escherichia coli* and *Staphylococcus aureus*. Gels 9(3):193. https://doi.org/10.3390/gels9030193

Arias CA, Contreras GA, Murray BE (2010) Management of multidrug-resistant enterococcal infections. Clin Microbiol Infect 16(6):555–562. https://doi.org/10.1111/j.1469-0691.2010.03214.x

Bamisaye A, Eromosele CO, Dare EO, Akinloye QA, Idowu MA, Ighodaro OM, Manicka S (2023) Fabrication, characterization, antimicrobial, toxicity and potential drug-delivery studies of PEGylated *Sesamum indicum* oil based nanoemulsion system. Beni-Suef Univ J Basic Appl Sci 12:33. https://doi.org/10.1186/s43088-023-00371-y

Barman K, Chowdhury D, Baruah PK (2020a) Bio-synthesized silver nanoparticles using *Zingiber officinale* rhizome extract as efficient catalyst for the degradation of environmental pollutants. Inorg Nano Met Chem 50(2):57–65. https://doi.org/10.1080/24701556.2019.1661468

Barman K, Chowdhury D, Baruah PK (2020b) Development of β-carotene loaded nanoemulsion using the industrial waste of orange (*Citrus reticulate*) peel to improve in vitro bioaccessibility of carotenoids and use as natural food colorant. J Food Process Preserv 44(5):e14429. https://doi.org/10.1111/jfpp.14429

Basak S, Singh P, Rajurkar M (2016) Multidrug resistant and extensively drug resistant bacteria: a study. J Pathog 2016:4065603. https://doi.org/10.1155/2016/4065603

Basha SK, Muzammil MS, Dhandayuthabani R, Kumari VS, Kaviyarasu K (2020) Nanoemulsion as oral drug delivery—a review. Curr Drug Res Rev 12(1):4–15. https://doi.org/10.2174/2589977511666191024173508

Behera A, Mittu B, Padhi S, Singh A (2020) Antimicrobial efficacy of essential oil nanoemulsions. In: Nanotechnological approaches in food microbiology. CRC, Boca Raton, FL, pp 293–309. https://doi.org/10.1201/9780429342776

Bennett PM (2008) Plasmid encoded antibiotic resistance: acquisition and transfer of antibiotic resistance genes in bacteria. Br J Pharmacol 153(S1):S347–S357. https://doi.org/10.1038/sj.bjp.0707607

Bhatt P, Madhav S (2011) A detailed review on nanoemulsion drug delivery system. Int J Pharm Sci Res 2(10):2482–2489. https://doi.org/10.13040/IJPSR.0975-8232.2(10).2482-89

Blair JM, Webber MA, Baylay AJ, Ogbolu DO, Piddock LV (2015) Molecular mechanisms of antibiotic resistance. Nat Rev Microbiol 13(1):42–51. https://doi.org/10.1038/nrmicro3380

Cars O, Hedin A, Heddini A (2011) The global need for effective antibiotics—moving towards concerted action. Drug Resist Updat 14(2):68–69. https://doi.org/10.1016/j.drup.2011.02.006

Chakravarty M, Vora A (2021) Nanotechnology-based antiviral therapeutics. Drug Deliv Transl Res 11:748–787. https://doi.org/10.1007/s13346-020-00818-0

Chime SA, Kenechukwu FC, Attama AA (2014) Nanoemulsions—Advances in formulation, characterization and applications in drug delivery. In: Application of nanotechnology in drug delivery. IntechOpen, London. https://doi.org/10.5772/58673

Choradiya BR, Patil SB (2021) A comprehensive review on nanoemulsion as an ophthalmic drug delivery system. J Mol Liq 339:116751. https://doi.org/10.1016/j.molliq.2021.116751

Correa A, del Campo R, Escandon-Vargas K, Perenguez M, Banos MR, Gomez CH, Pallares C, Perez F, Atias CA, Canton R, Villegas V (2018) Distinct genetic diversity of Carbapenem-resistant *Acinetobacter baumanniifrom* Colombian hospitals. Microb Drug Resist 24(1):48–54. https://doi.org/10.1089/mdr.2016.0190

Couvreur P (2013) Nanoparticles in drug delivery: past, present and future. Adv Drug Deliv Rev 65(1):21–23. https://doi.org/10.1016/j.addr.2012.04.010

Fopase R, Pathode SR, Sharma S, Dutta P, Pandey LM (2020) Lipopeptide and essential oil based nanoemulsion for controlled drug delivery. Polym Plast Technol Mater 59(18):2076–2086. https://doi.org/10.1080/25740881.2020.1784222

Franklyne JS, Mukherjee A, Chandrasekaran N (2016) Essential oil micro- and nanoemulsions: promising roles in antimicrobial therapy targeting human pathogens. Lett Appl Microbiol 63(5):322–334. https://doi.org/10.1111/lam.12631

Franklyne JS, Gopinath PM, Mukherjee A, Chandrasekaran M (2021) Nanoemulsions: the rising star of antiviral therapeutics and nanodelivery system—current status and prospects. Curr Opin Colloid Interface Sci 54:101458. https://doi.org/10.1016/j.cocis.2021.101458

Ganta S, Talekar M, Singh A, Coleman TP, Amiji MM (2014) Nanoemulsions in translational research-opportunities and challenges in targeted cancer therapy. AAPS Pharm Sci Tech 15(3):694–670. https://doi.org/10.1208/s12249-014-0088-9

Gebreyohannes G, Nyerere A, Bii C, Sbhatu DB (2019) Challenges of intervention, treatment, and antibiotic resistance of biofilm-forming microorganisms. Heliyon 5(8):e02192. https://doi.org/10.1016/j.heliyon.2019.e02192

Gerson T, Makarov E, Senanayake TH, Gorantla S, Poluektova LY, Vinogradov SV (2014) Nano-NRTIs demonstrate low neurotoxicity and high antiviral activity against HIV infection in the brain. Nanomed Nanotechnol Biol Med 10(1):177–185. https://doi.org/10.1016/j.nano.2013.06.012

Gundel SS, Velho MC, Diefenthaler MK, Favarin RF, Copetti PM, Fogaca AO, Klein B, Wagner R, Gundel A, Sagrillo MR, Ourique AF (2018) Basil oil-nanoemulsions: development, cytotoxicity and evaluation of antioxidant and antimicrobial potential. J Drug Deliv Sci Technol 46:378–383. https://doi.org/10.1016/j.jddst.2018.05.038

Hakemi-Vala M, Rafati H, Aliahmadi A, Ardalan A (2017) Nanoemulsions: a novel antimicrobial delivery system. In: Nano- and microscale drug delivery systems, pp 245–266. https://doi.org/10.1016/B978-0-323-52727-9.00013-3

Halnor VV, Pande VV, Borawake DD, Nagare HS (2018) Nanoemulsion: a novel platform for drug delivery system. J Mater Sci Nanotechnol 6(1):104

Hormann K, Zimmer A (2016) Drug delivery and drug targeting with parenteral lipid nanoemulsions—a review. J Control Release 223:85–98. https://doi.org/10.1016/j.jconrel.2015.12.016

Huh AJ, Kwon YJ (2011) "Nanoantibiotics": a new paradigm for treating infectious diseases using nanomaterials in the antibiotics resistant era. J Control Release 156(2):128–145. https://doi.org/10.1016/j.jconrel.2011.07.002

Kalhapure RS, Suleman N, Mocktar C, Seedat N, Govender T (2015) Nanoengineered drug delivery systems for enhancing antibiotic therapy. J Pharm Sci 104(3):872–905. https://doi.org/10.1002/jps.24298

Kawakami K, Yoshikawa T, Hayashi T, Nishihara Y, Masuda K (2002) Microemulsion formulation for enhanced absorption of poorly soluble drugs. II. In vivo study. J Control Release 81:75–82. https://doi.org/10.1016/S0168-3659(02)00050-0

Khaleel BS, Muzammil MS, Dhandayuthabani R, Kumari VS, Kaviyarasu K (2020) Nanoemulsion as oral drug delivery—a review. Curr Drug Res Rev 12(1):4–15. https://doi.org/10.2174/2589977511666191024173508

Kohanski MA, Dwyer DJ, Collins JJ (2010) How antibiotics kill bacteria: from targets to networks. Nat Rev Microbiol 8(6):423–435. https://doi.org/10.1038/nrmicro2333

Kohler V, Vaishampayan A, Grohmann E (2019) Problematic groups of multidrug-resistant bacteria and their resistance mechanisms. In: Antibacterial drug discovery to Combat MDR, pp 25–69. https://doi.org/10.1007/978-981-13-9871-1_2

Krishnamoorthy R, Athinarayanan J, Periasamy VS, Adisa AR, Al-Shuniaber MA, Gassem MA, Alshatwi AA (2018) Antimicrobial activity of nanoemulsion on drug-resistant bacterial pathogens. Microb Pathog 120:85–96. https://doi.org/10.1016/j.micpath.2018.04.035

Lallemand F, Daull P, Benita S, Buggage R, Garrigue JS (2012) Successfully improving ocular drug delivery using the cationic nanoemulsion, Novasorb. J Drug Deliv 2012:604204. https://doi.org/10.1155/2012/604204

Larsson DG, Flach C-F (2021) Antibiotic resistance in the environment. Nat Rev Microbiol 20(5):257–269. https://doi.org/10.1038/s41579-021-00649-x

Llor C, Bjerrum L (2014) Antimicrobial resistance: risk associated with antibiotic overuse and initiatives to reduce the problem. Ther Adv Drug Saf 5(6):229–241. https://doi.org/10.1177/2042098614554919

Makabenta JM, Nabawy A, Li C-H, Malan SS, Patel R, Rotello VM (2020) Nanomaterial-based therapeutics for antibiotic-resistant bacterial infections. Nat Rev Microbiol 19(1):23–36. https://doi.org/10.1038/s41579-020-0420-1

Mohsen S, Dickinson JA, Somayaji R (2020) Update on the adverse effects of antimicrobial therapies in community practice. Can Fam Physician 66(9):651–659

Moretton MA, Taira C, Flor S, Bernabeu E, Lucangioli S, Hocht C, Chiappetta DA (2014) Novel nelfinavir mesylate loaded D-α-tocopheryl polyethylene glycol 1000 succinate micelles for enhanced pediatric anti HIV therapy: *in vitro* characterization and *in vivo* evaluation. Colloids Surf B Biointerfaces 123(1):302–310. https://doi.org/10.1016/j.colsurfb.2014.09.031

Munita JM, Arias CA (2016) Mechanisms of antibiotic resistance. In: Kudva IT, Cornick NA, Plummer PJ, Zhang Q, Nicholson TL, Bannantine JP, Bellaire BH (eds) Virulence mechanisms of bacterial pathogens. ASM Press, Washington, DC, pp 481–511. https://doi.org/10.1128/9781555819286.ch17

Naggar MEE, Abdelgawad AM, Abdel-Sattar R, Gibriel AA, Hemdan BA (2023) Potential antimicrobial and antibiofilm efficacy of essential oil nanoemulsion loaded polycaprolactone nanofibrous dermal patches. Eur Polym J 184:111782. https://doi.org/10.1016/j.eurpolymj.2022.111782

Nikaido H (2009) Multidrug resistance in bacteria. Annu Rev Biochem 78(1):119–146. https://doi.org/10.1146/annurev.biochem.78.082907.145923

Paczosa MK, Mecsas J (2016) *Klebsiella pneumoniae*: going on the offense with a strong defense. Microbiol Mol Biol Rev 80(3):629–661. https://doi.org/10.1128/mmbr.00078-15

Parhi P, Mohanty C, Sahoo SK (2012) Nanotechnology-based combinational drug delivery: an emerging approach for cancer therapy. Drug Discov Today 17:1044–1052. https://doi.org/10.1016/j.drudis.2012.05.010

Pizzolato-Cezar LR, Okuda-Shinagawa NM, Machini MT (2019) Combinatory therapy antimicrobial peptide-antibiotic to minimize the ongoing rise of resistance. Front Microbiol 10:1703. https://doi.org/10.3389/fmicb.2019.01703

Prabuseenivasan S, Jayakumar M, Ignacimuthu S (2006) *In vitro* antibacterial activity of some plant essential oils. BMC Complement Altern Med 6(39):46. https://doi.org/10.1186/1472-6882-6-39

Sarma PP, Barman K, Baruah PK (2023) Green synthesis of silver nanoparticles using *Murraya koenigii* leaf extract with efficient catalytic, antimicrobial, and sensing properties towards heavy metal ions. Inorg Chem Commun 152:110676. https://doi.org/10.1016/j.inoche.2023.110676

Sharma A, Kumar Arya D, Dua M, Chhatwal GS, Johri AK (2012) Nano-technology for targeted drug delivery to combat antibiotic resistance. Expert Opin Drug Deliv 9(11):1325–1332. https://doi.org/10.1517/17425247.2012.717927

Shmuely H, Domniz N, Yahav J (2016) Non-pharmacological treatment of *Helicobacter pylori*. World J Gastrointest Pharmacol Ther 7(2):171–178. https://doi.org/10.4292/wjgpt.v7.i2.171

Silva RCS, Arruda IRS, Malafaia CB, Moraes MM, Beck TS, Camara CAG, Silva NH, Silva MV, Correia MTS, Frizzo CP, Machado G (2022) Synthesis, characterization and antibiofilm/antimicrobial activity of nanoemulsions containing *Tetragastris catuaba* (Burseraceae) essential oil against disease-causing pathogens. J Drug Deliv Sci Technol 67:102795. https://doi.org/10.1016/j.jddst.2021.102795

Singh Y, Meher JG, Raval K, Khan FA, Chaurasia M, Jain NK, Chourasia MK (2017) Nanoemulsion: concepts, development and applications in drug delivery. J Control Release 252:28–49. https://doi.org/10.1016/j.jconrel.2017.03.008

Soni H, Sharma S (2021) Current update on nanoemulsion: a review. Sch Int J Anat Physiol 4(1):6–13. https://doi.org/10.36348/sijap.2021.v04i01.002

Sutradhar KB, Amin ML (2014) Nanoemulsions: increasing possibilities in drug delivery. Eur J Nanomed 5(2):97–110. https://doi.org/10.1515/ejnm-2013-0001

Tayeb HH, Sainsbury F (2018) Nanoemulsions in drug delivery: formulation to medical application. Nanomedicine 13(19):2507–2525. https://doi.org/10.2217/nnm-2018-0088

Tayeb HH, Moqaddam SA, Hasaballah NH, Felimban RI (2022) Development of nanoemulsions for the delivery of hydrophobic and hydrophilic compounds against carbapenem-resistant *Klebsiella pneumoniae*. RSC Adv 12(40):26455–26462. https://doi.org/10.1039/D2RA03925G

Tenjarla S (1999) Microemulsions: an overview and pharmaceutical applications. Crit Rev Ther Drug Carrier Syst 16(5):461–521. https://doi.org/10.1615/CritRevTherDrugCarrierSyst.v16.i5.20

Yeh Y-C, Huang T-H, Yang S-C, Chen C-C, Fang J-Y (2020) Nano-based drug delivery or targeting to eradicate bacteria for infection mitigation: a review of recent advances. Front Chem 8:286. https://doi.org/10.3389/fchem.2020.00286

Yukuyama MN, Kato ET, Lobenberg R, Bou-Chacra NA (2017) Challenges and future prospects of nanoemulsion as a drug delivery system. Curr Pharm Des 23(3):495–508. https://doi.org/10.2174/1381612822666161027111957

Zong T-X, Silveira AP, Morais JA, Sampaio MC, Muehlmann LA, Zhang J, Jiang CS, Liu SK (2022) Recent advances in antimicrobial nano-drug delivery systems. Nanomaterials 12(11):1855. https://doi.org/10.3390/nano12111855

Liposomes as Efficient Drug Delivery Vehicles to Combat Antimicrobial Resistance

13

Ira Singh and Santosh Kumar

Abstract

Antimicrobial resistance (AMR) poses a growing global health threat, placing a significant economic burden on healthcare systems and jeopardizing public health. The rise of multidrug-resistant bacteria has compromised the efficacy of traditional antibiotics, emphasizing the critical need for innovative approaches to combat AMR. Liposomal delivery systems have emerged as promising solutions, capitalizing on their capacity to encapsulate a diverse range of drugs, biocompatibility, and biodegradability. These phospholipid bilayer-based nanoscale vesicles enable the encapsulation of both hydrophilic and hydrophobic medications, offering versatility in drug delivery. Liposomes, with modifiable surfaces for enhanced targeting and cellular uptake, prove ideal for delivering antibiotics directly to infection sites, minimizing adverse effects, and mitigating AMR development. Recent studies highlight the superior effectiveness of liposomal antibiotic formulations against drug-resistant bacteria compared to traditional antibiotics. Moreover, liposomes contribute to reduced antibiotic administration frequency, improved patient compliance, and enhanced medication stability, leading to cost savings in healthcare. Despite these advantages, challenges like rapid drug leakage from liposomes due to poor stability hinder their broader biological application. To address this, researchers are exploring surface modifications, such as fusing liposomes with various substances, with a particular focus on chitosan-liposome-based systems. This chapter delves into the development of diverse chitosan-liposome-based systems and their potential biological applications, emphasizing the need for further research to optimize liposome formulations and unlock their full therapeutic potential in combating AMR.

I. Singh · S. Kumar (✉)
Department of Chemistry, School of Basic and Applied Sciences, Harcourt Butler Technical University, Kanpur, Uttar Pradesh, India
e-mail: santoshk@hbtu.ac.in

M. Y. Wani et al. (eds.), *Nanotechnology Based Strategies for Combating Antimicrobial Resistance*, https://doi.org/10.1007/978-981-97-2023-1_13

Keywords

Antimicrobial resistance (AMR) · Liposomal delivery systems · Drug encapsulation · Multidrug-resistant bacteria · Targeted drug delivery

13.1 Introduction

Antimicrobial medications must be used to treat bacterial infections. However, the overuse of antibiotics in clinical settings and animal and agricultural production has long been established, placing significant pressure on the development of bacterial species. Bacterial strains that are resistant to antibiotics have resulted as a result (Rodríguez-Rojas et al. 2013; Luo et al. 2019). A possible clinical collapse must be prevented by immediately controlling and eliminating antibiotic resistance. A variety of action plans have been released by the World Health Organisation (WHO) and national health organizations to address this issue, with a focus on improving sanitation and drug disposal procedures as well as limiting the use of antibiotics (Metz and Shlaes 2014). Significant funds are also being allocated to the study and creation of fresh, efficient tactics to fight microorganisms that are resistant to drugs (MDR). This encompasses investigating biological strategies, adjuvant treatments, phages, and tiny compounds possessing antitoxin, anti-quorum, or antibacterial virulence characteristics. These cutting-edge methods are broadening the spectrum of viable therapeutic approaches that can be employed to address antibiotic resistance (Wang et al. 2020a). Unfortunately, the rates of medication resistance are developing faster than new antibiotics, biologicals, or adjuvant therapy can be developed. These treatments require a long time to develop. Thus, there is an urgent need for authorized fast-track methods to address this issue (Aslam et al. 2018). Considering the challenges in developing new medications, "nanobiotics" are mixtures of nanotechnology and currently licensed antibiotics. Drug delivery systems with enhanced pharmacokinetic and antibacterial properties have been created recently in this field (Zhang et al. 2010; Huh and Kwon 2011; Liu et al. 2016). The application of biomedical nanotechnology systems is greatly enhancing the therapeutic effects of conventional drugs. Additionally, these technologies may alter the way that already prescribed antibiotics function (Gupta et al. 2019). Of the several nanoplatforms currently available, research on liposomes—one of the most promising antibiotic delivery systems—is still ongoing (Pinto-Alphandary et al. 2000; Drulis-Kawa and Dorotkiewicz-Jach 2010). Since lipid-based nanosystems were first proposed as drug carriers in the 1970s, when significant advancements in liposome technology were made, interest in their potential use as efficient antibacterial drug delivery systems has developed (Drulis-Kawa and Dorotkiewicz-Jach 2010; Gonzalez Gomez and Hosseinidoust 2020; Lima et al. 2019). Ever since Bangham and Horne discovered liposomes in 1964 (Bangham and Horne 1964), a great deal of research has been done on the possibility of employing liposomes to transport drugs. The primary objective of this book is to demonstrate the benefits of using liposomes as carriers of antimicrobial agents, particularly their ability to fight antibiotic resistance in addition to eliminating infections (Brooks and Brooks 2014).

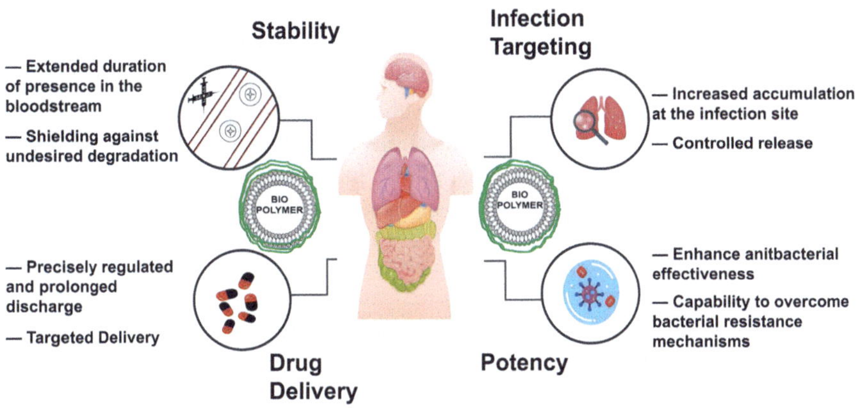

Fig. 13.1 This diagram illustrates the key advantages of employing liposomes as carriers for antibiotics. It underscores the benefits of utilizing liposomes as delivery systems, such as potential enhancements in drug solubility, increased bioavailability, extended release, precise targeting, and diminished toxicity, among various other advantages

Some of the most important characteristics of using liposomes as carriers of antibiotics are shown below (Fig. 13.1). These characteristics include, but are not limited to, increased accumulation at infected sites; precise, controlled, and prolonged discharge with extended duration in the bloodstream; low degradation rates; tendency to overcome resistance mechanisms by targeted bacterial infections; and increased biocompatibility.

13.2 Liposomes: Classifications, Advantages, and Limitations

The idea of liposomes has been around for a while; in the middle of the 1960s, scientists began to take an interest in it. The term "liposome" describes a structure where an internal aqueous phase is surrounded by lipids. The Greek terms "lipos" (meaning "fat") and "soma" (meaning "body" or "structure") are the source of this word (Bangham 1961, 1968). Liposomes are categorized based on their surface characteristics, membrane composition, and distribution of particle sizes (Euliss et al. 2006). Liposomes can be categorized into several categories based on their morphology, such as giant oligolamellar vesicles (GOVs), giant unilamellar vesicles (GUVs), big oligolamellar vesicles (SUVs), large unilamellar vesicles (LUVs), and multilamellar vesicles (MLVs) (Sharma and Sharma 1997; Pattni et al. 2015). Liposomes are a useful tool for biomedical applications due to their many properties. Their structure is similar to that of cellular membranes, and they can encapsulate both hydrophilic and hydrophobic drugs. They are also biocompatible, nontoxic, and able to change in size in response to various stimuli (Akbarzadeh et al. 2013; Fakhravar et al. 2016). As drug delivery vehicles, classic liposomes are greatly limited by their rapid clearance by macrophages in the reticuloendothelial system and from the bloodstream. Moreover, liposomes are poorly stable, poorly soluble, and have significant production costs. Liposomes have been demonstrated to effectively target specific cells or tissues in a range of medical therapies; nevertheless, they are readily complexed by positively or negatively charged substances. Liposomes are also quite successful at delivering medications to intestinal epithelial cells when compared to endocytosis; nevertheless, gastric juice can severely compromise the integrity of liposomes, which can lead to integrated drugs leaking out quickly (Allen and Cullis 2013).

13.3 Liposomes as Drug Delivery Vehicles Have Distinct Drawbacks

The use of liposomes as a vehicle for delivering bioactive compounds has garnered significant scientific interest; however, several obstacles have limited their application in industry and medicine. These drawbacks include the existence of organic residues, instability, and a large loss of cargo that is enclosed during storage (Banerjee et al. 2004; Fauvel and Yasri 2014). Many factors influence the stability of liposomes, especially the lipid bilayer (Babai et al. 1999; Voskuhl and Ravoo 2009). A liposome is composed of two different lipid layers, one of which is hydrophilic and the other is hydrophobic. These layers have an important role in the chemical and physical stability of liposomes as well as their susceptibility to different processes such as phospholipid oxidation, aggregation, degradation, and fusion (Senior et al. 1991; Zhao et al. 2011; Kim and Auh 1999).

13.4 Chitosan Is the Only Polycationic Polysaccharide

Chitosan (CS) possesses remarkable physical, chemical, and biological properties, such as its mucoadhesiveness, biocompatibility, biodegradability, low toxicity, and antibacterial capability (Panya et al. 2010; Kumar et al. 2014, 2015, 2016; Kumar and Koh 2012). CS is regarded as a very promising material for a variety of applications in the environmental and biological domains because of its exceptional qualities. These applications include the development of artificial organs and orthopedic devices, the creation of scaffolds to facilitate tissue regeneration, the production of wound dressings, and the use of CS to produce drug and gene-delivery vehicles (Kumar et al. 2017, 2018a, b; Garg et al. 2013; Sudheesh Kumar et al. 2012; Peter et al. 2010). CS is a hydrophilic, biocompatible, biodegradable polymer that possesses mucoadhesive properties that facilitate the easier passage of macromolecules through nasal and intestinal barriers. As a result of their structural similarity to cellular membranes, liposomes, on the other hand, may efficiently permeate cells with the encapsulated active components, assisting in overcoming biological barriers to cellular and tissue uptake (Alavi et al. 2017; Matta et al. 2001). Researchers in the pharmaceutical and biomedical industries have shown a great deal of interest in liposomes because of their remarkable properties, which include low toxicity, good biocompatibility, the ability to incorporate hydrophilic and hydrophobic drugs, proteins, peptides, antisense oligonucleotides, plasmid DNA, and targeted delivery of therapeutically active ingredients to the site of action (Ulrich 2002). CS is unique among polysaccharides such as starch, glycogen, and cellulose since it is the only polycationic polysaccharide that is readily available. Because of this specific feature, CS is a useful material to use in conjunction with other compounds of interest as it can enhance their properties.

13.4.1 The Rationale Behind Mixing Liposomes and Chitosan

The fact that CS is polycationic and may, through intermolecular electrostatic interaction, form polyelectrolyte complexes with negatively charged macromolecules also makes it a promising coating material for affordably boosting liposome stability. By doing this, a significant systemic flaw is removed. Numerous studies have revealed that applying a thin layer of CS to a surface lengthens liposome retention (Shin et al. 2013; Hao et al. 2017; Liu et al. 2015; Li et al. 2009; Park et al. 2014). Illustrated below (Fig. 13.2) is the combination of liposomes with CS to enhance properties for more effective delivery methods. When liposomes are coated with CS, electrostatic interactions have been shown to improve the membrane's integrity and stiffness, ultimately leading to an increase in physicochemical stability (Mertins and Dimova 2011). The covering of CS has two distinct effects on the liposomal membrane's fluidity: initially, it decreases, and then it increases (Tan et al. 2013). This phenomenon can be explained by the concentration-dependent conformation of polymer chains in solution. The stiffness and stability of the vesicle grow at initially rising concentrations, achieving a maximum effect at saturation concentration, as a result of the effective absorption of CS linear chains into the membrane surfaces. However, breaking through this concentration barrier causes asymmetric coils and self-aggregation of polymeric chains, which can improve membrane fluidity and ultimately cause liposome destabilization (Tan et al. 2013; Laye et al. 2008; Zhuang et al. 2010). Liposomal carriers are less likely to assemble when the CS layer creates repulsive interactions between them. This is achieved by surrounding the particles with conformational clouds, which act as significant steric obstacles

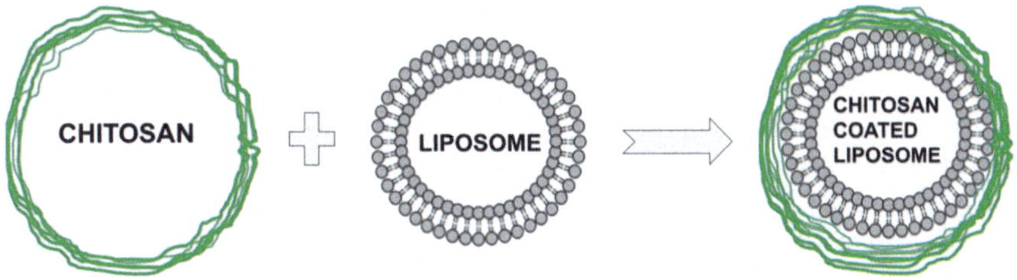

Fig. 13.2 A chitosan-coated liposome, demonstrating the presence of chitosan on the surface of the liposome

(Laye et al. 2008; Mady and Darwish 2010). The formation of a hydration shell on the surface of CS-coated liposomes (CS-Lip) may decrease the reticuloendothelial system's ability to identify and absorb carriers, extending the duration of circulation (Zhuang et al. 2010). Liposomes have drawn a lot of interest as a promising and therapeutically acceptable delivery mechanism for medications, proteins, and enzymes for the treatment of a range of illnesses (Torchilin 2005). Apart from boosting stability, CS-Lip demonstrates a remarkable ability to mucoadhere (Karn et al. 2011), which further improves paracellular cargo permeability (Lemmer and Hamman 2013). This leads to an improvement in the bioavailability of medications that are entrapped. Moreover, the CS backbone's functional groups enable a variety of physiologically active ligands to bind covalently to carriers' surfaces. This allows for targeted drug delivery, where medications can be precisely delivered to the desired location (Yang et al. 2013). These key characteristics thus demonstrate the potential of CS-Lip as an extremely efficient drug delivery system for targeted, prolonged, and controlled release of therapeutic compounds. A further benefit of combining CS and liposomes to create innovative biomedical systems is that CS can loosen the tight connections between epithelial cells, enabling the paracellular passage of materials across the epithelial barrier.

13.5 Traditional Problem and Statement

According to Van Leeuwenhoek, one of the major challenges in treating illness is the ability of antimicrobials to penetrate, aggregate, and kill bacteria over the whole thickness of the biofilm (Leewenhoeck 2017). This issue also entails avoiding enzymatic inactivation of the antimicrobials during transport via the bloodstream or inside the biofilm, as well as preventing the washout of antimicrobials in the dynamic milieu of the human body (Albayaty et al. 2018). Usually from inside the bloodstream, the antibiotic must first reach the pathogenic biofilm for efficient penetration and accumulation to take place. It is undesirable, nevertheless, because large concentrations of antimicrobial drugs circulating throughout the body might cause collateral tissue damage. Consequently, self-targeting carriers are being developed with the goal of reaching the biofilm at low blood concentrations and building up enough for a successful course of treatment (Forier et al. 2014). These elements make it more difficult to eradicate bacteria, which leads to infection recurrence. Liposomal nanocarriers offer unique advantages such as stealth qualities, protection against enzymatic degradation of encapsulated antimicrobials, and the capacity to penetrate bacteria and tissue cells, despite certain limits in resolving all the issues. Furthermore, liposomes have the potential to enter mammalian cells, something that certain antimicrobials cannot. This is particularly important because bacteria can hide inside macrophages, making treatment even more difficult. The primary issue with drug carriers' efficacy is resistance to antibiotics, medications, and antimicrobials. Treating pathogenic biofilms in the human body presents a number of challenges because some of them have lasted for generations.

This study attempts to give a thorough overview and summary of the difficulties faced when treating pathogenic biofilms with antimicrobials. Much research has investigated the possibility of liposomal antimicrobial-loaded nanocarriers for infection control. Next, we explore the possible applications of liposomal antimicrobial-loaded nanocarriers in addressing some of these issues, as well as the prerequisites for a successful clinical translation (Wang et al. 2020b). Over the past 25 years, the importance of lipids in cell physiology has gained widespread recognition, despite their limited role in the past as structural components. Lipids have demonstrated their powerful ability to work within cells and to regulate signals. By altering the pharmacodynamics and pharmacokinetics of the antibiotics, liposomes can be employed to modify their characteristics and increase their therapeutic index (Nisini et al. 2018).

Bacterial membranes and liposomes are similar in composition and structure. They can therefore combine with bacteria to provide larger concentrations of antimicrobial drugs straight to the site of action. Furthermore, by breaking the membranes of bacteria and facilitating the release of antimicrobial drugs, liposomes with a positively charged surface can interact with them to enhance their antibacterial effects. Compared to unformulated antimicrobials, this strategy is expected to increase the therapeutic index and increase the susceptibility of bacteria to antimicrobials delivered by delivery methods. These results show that there are good opportunities to improve the effectiveness and selectivity of antimicrobial therapies (Wang et al. 2020b). N-succinyl-chitosan liposomes and chitooligosaccharide coated on a four layered liposome using the layer-by-layer (LBL) method was reported (Seong et al. 2018). In prior work, quercetin was released from multilayered liposomes containing chitosan and hyaluronic acid more quickly at pH 5.5 than at pH 7.4, although there were no appreciable variations between the pH values (Jeon et al. 2015). Layered liposomes were used as transdermal drug delivery vehicles to enhance stability against surfactants and regulate drug release rate.

In comparison to uncoated liposomes, it exhibited enhanced stability against surfactants and comparatively higher drug release and skin penetration at pH 5.5. Folate-targeted liposome as a vehicle for nucleic acid therapies for tumor targeting, encapsulating low molecular weight CS/oligonucleotide polyplex FLS(CO) (Kang et al. 2014), was reported. Verification of FLS(CO) ability to deliver cargo polycations and nucleic acid polyplexes into the targeted tumor cell cytoplasm.

The purpose of this study was to look at how the behavior of dyskinesia rats was affected by levodopa nanoliposomes coated with CS and the underlying molecular mechanism (Cao et al. 2016). According to the study, levodopa liposomes coated with CS may be a more effective treatment than regular levodopa tablets for reducing Parkinson's disease-related dyskinesias. Co-delivering erlotinib and oxygen via liposomal complexes modified with aptamers is being investigated as a possible tactic to combat medication resistance brought on by hypoxia in lung cancer (Li et al. 2018; Zhao et al. 2018; Huang et al. 2017). The efficiency of cationic liposomes containing cholesterol, dimethyldioctadecylammonium bromide (DDAB), and dimyristoylphosphatidylcholine was tested against a range of oral bacteria biofilms using the extrusion technique. With regard to adsorption, Streptococcus salivarius showed the greatest amount. It is noteworthy that different species and strains of oral bacteria exhibited different levels of adsorption specificity, which was most noticeable when the liposomes contained more than 15 mol% DDAB.

One way to target infection sites is to develop drug delivery techniques that use organ or bacterial identifying features or to directly administer antimicrobial medicines to infected areas. Because the pulmonary route is easier to administer and encourages patient compliance, it is particularly appealing for some types of illnesses, like respiratory infections. Actually, because of its superior antibacterial action, aerosol antimicrobial therapy is becoming more and more popular in clinical settings. This is especially true for antibiotics like beta-lactams, colistin, aminoglycosides, and glycopeptides that have low lung penetration. Moreover, as described by Zarogoulidis and colleagues, the immunomodulatory impact linked to lower systemic toxicity has resulted in the approval of various aerosol antibiotics in recent years (Zarogoulidis et al. 2013).

13.6 Chitosan-Based Liposome

Although liposomes are easily produced and have been utilized for medication distribution for many years, a significant issue is the rapid release of active chemicals from the liposome's inner core due to the sudden disintegration of the outer lipid bilayer. Because of this decrease in their effectiveness, researchers are looking into surface modification techniques using both natural and synthetic polymers, with CS being a primary emphasis. While liposomes' surfaces have been modified with a variety of substances, including gelatin, hyaluronic acid, fibrin, pectin, alginate, and dextran, their effectiveness in ensuring the system's total stability throughout a range of physiological settings is limited. Chitosan can be easily added to liposome formulations to modify and diversify the liposomes' surface properties through electrostatic interactions, which enhances their physical stability. This section focuses on the development of different CS-liposome-based systems, with some of these systems being illustrated below.

13.6.1 Grape-Seed Extract CS-Coated Liposomes System

Because of its ability to efficiently inhibit the growth of gram-positive bacteria, grapeseed extract (GSE) is commonly used as a food preservative. Moreover, there are several other beneficial properties of GSE, including its capacity to prevent blood clots, lessen inflammation, treat ulcers, and even act as an anticarcinogen. In this case, Gibis et al. developed a technique (Gibis et al. 2016). The initial step in creating GSE-CS-coated liposomes as a drug delivery vehicle is dissolving lecithin in an acetate buffer to create a liposomal solution. After that, the solution is agitated for an entire night at room temperature. After dissolving lecithin in acetate buffer and stirring it overnight at room temperature, two different liposomal solutions were created. GSE was present in one of the solutions but not in the other. A 1% (w/w) solution of CS was then created by dissolving CS in acetate buffer and vigorously stirring with a magnetic stirrer for the duration of the night at room temperature. In order to make both GSE- and non-GSE-coated liposomes, the CS solution and the liposomal solutions were combined in a test tube, and then the mixture was well mixed by vortexing.

13.6.2 The Production and Examination of N-Succinyl-Chitosan

N-succinyl-chitosan (NSC), a pH-sensitive polymer electrolyte, was explored as a potential pH-sensitive drug release formulation for liposomes under skin pH conditions. NSC is a widely used polymer that adds a succinyl group to the amine group of CS to increase its solubility in water. By varying its succinylation degree, this material may be made to be pH-sensitive and is both safe and biocompatible for human use (Bashir et al. 2015).

13.6.3 Synthesis of N-Succinyl-Chitosan

For the production of NSC, 1 g of CS was dissolved in 20 mL of DMSO dimethyl sulfoxide. One gram of succinic anhydride was added to CS, and the mixture was continually stirred at 500 rpm and 60 °C for 4 h to generate the N-succinyl-chitosan NSC. After adding a 5% (w/v) NaOH solution to the mixture to bring its pH down to 5, the precipitate was recovered and combined with 50 mL of water. After that, the pH was increased to 10–12 by adding 5% (w/v) NaOH solution, which turned the solution yellow. After three acetone washes, the yellow solution was recrystallized from acetone to get N-succinyl-chitosan NSC, which was lyophilized to yield 1.26 g of light yellow NSC powder (Aiping et al. 2006).

13.6.4 The Process of Producing Multilayered Liposomes

A method called thin-film hydration was employed to create the core liposome. It was necessary to dissolve 500 M quercetin and 0.2 g of Egg PC in a 2:1 chloroform to methanol ratio. The inside surface of a round-bottom flask was coated with a lipid coating by using a rotary evaporator, which evaporated the solvent. After hydrating the film with 20 mL of distilled water using a probe sonicator, multi-lamellar vesicles were produced. The solution was centrifuged for 5 min, and then, in order to remove any contaminants, it was run through a 1.20-m (26 mm) filter. A homogeneous liposome with a size of no more than 100 nm was the end product. Based on testing, the average pH of the resulting core liposomes was 4. To create stock solutions, 10 mg/mL of NSC and COS were dissolved in distilled water and then filtered through a 5.00-m filter (26 mm; Minisart) to remove any remaining contaminants. In order to create COS concentrations of 0.3, 0.6, and 1.0 mg/mL for later studies, the resulting clear solutions were further diluted with distilled water. Similarly,

NSC concentrations of 1.0 and 1.5 mg/mL were created by further diluting the stock solution with distilled water (Seong et al. 2018).

13.7 Relevance/Impact

The study covered in the applications offered emphasizes the value and important effects of using liposomes to deliver antibiotics for the treatment of bacterial illnesses, particularly those brought on by bacteria that are drug-resistant. The study underlines the potential of liposomal antibiotic formulations to boost antibacterial activity, target drug delivery, and fight bacterial illnesses that are resistant to antibiotics. These discoveries have important ramifications for enhancing the effectiveness of antibiotic treatments and tackling the rising problem of antibiotic resistance in the medical field. The following portion provides a brief summary of their relevance and impact.

13.7.1 Antimicrobial Activity

To increase the antibacterial properties of liposomes, antibiotics can be added to their core. Since liposomes can only contain so much antibiotic material, chitosan can be used to alter them and improve their retention properties. Using chitosan-coated liposomes, Yang et al. (2015) demonstrated how to administer the antibiotic vancomycin hydrochloride intravenously. Treating gram-positive bacterial infections such as arthritis, osteomyelitis, pneumonia, and endocarditis was the aim of this approach. The researchers also produced a nanosuspension of vancomycin liposomes coated with CS using electrostatic deposition. A successful release of amoxicillin from CS-coated egg-derived liposomes was reported by Menikarachchi et al. (2016), suggesting that liposomes with CS or its derivatives added can have better antibacterial effects than liposomes without modification.

13.7.2 Increased Effectiveness Against Bacteria

The primary characteristic of antibiotic-loaded liposomes is their increased antibacterial activity in comparison to the corresponding antibiotic in its unbound form. Liposomal versions of antibiotics with enhanced potency, even against resistant bacteria, have been reported in numerous studies. For example, compared to the free drug, liposomal ciprofloxacin and gentamicin have been shown in multiple trials to have lower minimum inhibitory concentrations (MIC) against common resistant bacteria such as *P. aeruginosa*, *K. pneumoniae*, and *E. coli*. This suggests that using liposomal antibiotic formulations could be a useful tactic against bacterial illnesses that do not respond well to conventional antibiotic treatments (Mugabe et al. 2006; Gubernator et al. 2007; Bakker-Woudenberg et al. 2001). The authors of these investigations postulated that the wide and effective contact between the liposomes and the bacterial cell membrane is what gives liposomal formulations their increased antibacterial efficacy. Liposomes have a special ability to interact with bacteria because of their structural and compositional similarities to bacterial membranes. Antibiotics can be delivered to bacteria at high concentrations thanks to liposomes' ability to promote fusion with bacterial cell membranes. Liposomes have the potential to overcome antibiotic resistance processes due to their ability to interact with bacteria, which makes them a promising option for treating drug-resistant infections (Drulis-Kawa and Dorotkiewicz-Jach 2010).

13.8 Conclusion

One major challenge facing modern medicine in treating bacterial infections is the emergence of antibiotic-resistant organisms. Much research is being done to investigate novel methods of administering antibiotics that can increase their effectiveness against bacteria in response to this issue. Liposomes are among the most promising delivery systems now under investigation. These are minuscule, nanoscale particles that have the ability to encapsulate and carry medicines to their intended spot. Because liposomes can have different lipid compositions and structures, different liposomal formulations with better pharmacokinetic and pharmacodynamic qualities can be created. This is why liposomes are so versatile. These modifications facilitate the liposomes' efficient movement and metabolism throughout the body, as well as their effective interaction with bacteria to produce therapeutic benefits. Furthermore, direct interactions between the lipid bilayers of liposomes and the bacterial cell walls may result in the fusion of the two structures. By increasing the concentration of antibiotics within the bacterium, this fusion can ultimately improve the antibiotic's therapeutic efficacy. Numerous investigations have demonstrated that liposome-encapsulated antibiotics are capable of successfully circumventing a range of resistance mechanisms used by microbes, including enzymatic degradation, efflux mechanisms, and impenetrable outer membranes. These results demonstrate the great potential of liposomes as an effective drug delivery system, providing a way to fight bacterial illnesses that are currently incurable by making use of their special physicochemical characteristics and benefits as antibiotic carriers.

13.9 Future Perspective

The potential release of many more liposomal medications, such as liposomal vincristine, cisplatin, paclitaxel, annamycin, and hydrophobic derivatives of cisplatin, which are presently being developed as novel anticancer medications, has been made possible by the clinical approval of a number of liposomal formulations. It is imperative to assess the suitability of newly developed antitumor medicines for liposome encapsulation. Liposomal compositions hold great promise for medications that break down rapidly, like peptides, proteins, and cytokines. Therefore, there is a great deal of interest in investigating the application of liposomal formulations in this field. Antibacterial medications may be added to liposomal formulations to improve their delivery to areas of severe infection or to increase the medication's ability to enter intracellular organisms, among other advantages. Liposomal encapsulation may also make it possible to reevaluate some medications that were previously considered unsuitable because of issues with toxicity, pharmacokinetics, and poor solubility.

Even while liposomal drug formulations have shown success in producing passive targeting to regions with increased capillary permeability and prolonged drug release, there may be times when the formulation needs to be changed in order to improve its selectivity for the target cell. Researchers are investigating the possibility of using ligand-targeted liposomes, which entail attaching ligands or antibodies to the liposome's surface, to improve medication delivery to certain target cells. By using this method, the liposome's propensity to connect with particular receptors or target cell surface epitopes can be increased. The capacity to control the rate and amount of medication release at the intended site of action is a critical component of using liposomes for drug delivery. This issue is still highly important and urgent in the field of liposomal medication delivery. An important consideration that may impact both the drug's potential side effects and efficacy is choosing the best liposome release technique. As a result, creating methods to control medication release from liposomes has become a crucial topic of focus for scientists.

References

Aiping Z, Tian C, Lanhua Y, Hao W, Ping L (2006) Synthesis and characterization of N-succinyl-chitosan and its self-assembly of nanospheres. Carbohydr Polym 66(2):274–279

Akbarzadeh A, Rezaei-Sadabady R, Davaran S, Joo SW, Zarghami N, Hanifehpour Y, Samiei M, Kouhi M, Nejati-Koshki K (2013) Liposome: classification, preparation, and applications. Nanoscale Res Lett 8:1–9

Alavi S, Haeri A, Dadashzadeh S (2017) Utilization of chitosan-caged liposomes to push the boundaries of therapeutic delivery. Carbohydr Polym 157:991–1012

Albayaty YN, Thomas N, Hasan S, Prestidge CA (2018) Penetration of topically used antimicrobials through *Staphylococcus aureus* biofilms: a comparative study using different models. J Drug Deliv Sci Technol 48:429–436

Allen TM, Cullis PR (2013) Liposomal drug delivery systems: from concept to clinical applications. Adv Drug Deliv Rev 65(1):36–48

Aslam B, Wang W, Arshad MI, Khurshid M, Muzammil S, Rasool MH, Nisar MA, Alvi RF, Aslam MA, Qamar MU, Salamat MK (2018) Antibiotic resistance: a rundown of a global crisis. Infect Drug Resist 11:1645–1658

Babai I, Samira S, Barenholz Y, Zakay-Rones Z, Kedar E (1999) A novel influenza subunit vaccine composed of liposome-encapsulated haemagglutinin/neuraminidase and IL-2 or GM-CSF. I. Vaccine characterization and efficacy studies in mice. Vaccine 17(9–10):1223–1238

Bakker-Woudenberg IA, ten Kate MT, Guo L, Working P, Mouton JW (2001) Improved efficacy of ciprofloxacin administered in polyethylene glycol-coated liposomes for treatment of Klebsiella pneumoniae pneumonia in rats. Antimicrob Agents Chemother 45(5):1487–1492

Banerjee R, Tyagi P, Li S, Huang L (2004) Anisamide-targeted stealth liposomes: a potent carrier for targeting doxorubicin to human prostate cancer cells. Int J Cancer 112(4):693–700

Bangham AD (1961) A correlation between surface charge and coagulant action of phospholipids. Nature 192(4808):1197–1198

Bangham AD (1968) Membrane models with phospholipids. Prog Biophys Mol Biol 18:29–95
Bangham AD, Horne RW (1964) Negative staining of phospholipids and their structural modification by surface-active agents as observed in the electron microscope. J Mol Biol 8(5):660
Bashir S, Teo YY, Ramesh S, Ramesh K, Khan AA (2015) N-succinyl chitosan preparation, characterization, properties and biomedical applications: a state of the art review. Rev Chem Eng 31(6):563–597
Brooks BD, Brooks AE (2014) Therapeutic strategies to combat antibiotic resistance. Adv Drug Deliv Rev 78:14–27
Cao X, Hou D, Wang L, Li S, Sun S, Ping Q, Xu Y (2016) Effects and molecular mechanism of chitosan-coated levodopa nanoliposomes on behavior of dyskinesia rats. Biol Res 49:1–9
Drulis-Kawa Z, Dorotkiewicz-Jach A (2010) Liposomes as delivery systems for antibiotics. Int J Pharm 387(1–2):187–198
Euliss LE, DuPont JA, Gratton S, DeSimone J (2006) Imparting size, shape, and composition control of materials for nanomedicine. Chem Soc Rev 35(11):1095–1104
Fakhravar Z, Ebrahimnejad P, Daraee H, Akbarzadeh A (2016) Nanoliposomes: synthesis methods and applications in cosmetics. J Cosmet Laser Ther 18(3):174–181
Fauvel B, Yasri A (2014) Antibodies directed against receptor tyrosine kinases: current and future strategies to fight cancer. MAbs 6(4):838–851
Forier K, Raemdonck K, De Smedt SC, Demeester J, Coenye T, Braeckmans K (2014) Lipid and polymer nanoparticles for drug delivery to bacterial biofilms. J Control Release 190:607–623
Garg P, Kumar S, Pandey S, Seonwoo H, Choung PH, Koh J, Chung JH (2013) Triphenylamine coupled chitosan with high buffering capacity and low viscosity for enhanced transfection in mammalian cells, in vitro and in vivo. J Mater Chem B 1(44):6053–6065
Gibis M, Ruedt C, Weiss J (2016) In vitro release of grape-seed polyphenols encapsulated from uncoated and chitosan-coated liposomes. Food Res Int 88:105–113
Gonzalez Gomez A, Hosseinidoust Z (2020) Liposomes for antibiotic encapsulation and delivery. ACS Infect Dis 6(5):896–908
Gubernator J, Drulis-Kawa Z, Dorotkiewicz-Jach A, Doroszkiewicz W, Kozubek A (2007) In vitro antimicrobial activity of liposomes containing ciprofloxacin, meropenem and gentamicin against gram-negative clinical bacterial strains. Lett Drug Design Discov 4(4):297–304
Gupta N, Rai DB, Jangid AK, Kulhari H (2019) Use of nanotechnology in antimicrobial therapy. In: Methods in microbiology, vol 46. Academic Press, Cambridge, MA, pp 143–172
Hao J, Guo B, Yu S, Zhang W, Zhang D, Wang J, Wang Y (2017) Encapsulation of the flavonoid quercetin with chitosan-coated nano-liposomes. LWT- Food Sci Technol 85:37–44
Huang F, Gao Y, Zhang Y, Cheng T, Ou H, Yang L, Liu J, Shi L, Liu J (2017) Silver-decorated polymeric micelles combined with curcumin for enhanced antibacterial activity. ACS Appl Mater Interfaces 9(20):16880–16889
Huh AJ, Kwon YJ (2011) "Nanoantibiotics": a new paradigm for treating infectious diseases using nanomaterials in the antibiotics resistant era. J Control Release 156(2):128–145
Jeon S, Yoo CY, Park SN (2015) Improved stability and skin permeability of sodium hyaluronate-chitosan multilayered liposomes by layer-by-layer electrostatic deposition for quercetin delivery. Colloids Surf B Biointerfaces 129:7–14
Kang JH, Battogtokh G, Ko YT (2014) Folate-targeted liposome encapsulating chitosan/oligonucleotide polyplexes for tumor targeting. AAPS PharmSciTech 15:1087–1092
Karn PR, Vanić Z, Pepić I, Škalko-Basnet N (2011) Mucoadhesive liposomal delivery systems: the choice of coating material. Drug Dev Ind Pharm 37(4):482–488
Kim JC, Auh KH (1999) Computer simulation on particle agglomeration during the synthesis of titania powders. Model Simul Mater Sci Eng 7(3):447
Kumar S, Koh J (2012) Physiochemical, optical and biological activity of chitosan-chromone derivative for biomedical applications. Int J Mol Sci 13(5):6102–6116
Kumar S, Kumari M, Dutta PK, Koh J (2014) Chitosan biopolymer Schiff base: preparation, characterization, optical, and antibacterial activity. Int J Polym Mater Polym Biomater 63(4):173–177

Kumar S, Garg P, Pandey S, Kumari M, Hoon S, Jang KJ, Kapavarapu R, Choung PH, Sobral AJ, Chung JH (2015) Enhanced chitosan–DNA interaction by 2-acrylamido-2-methylpropane coupling for an efficient transfection in cancer cells. J Mater Chem B 3(17):3465–3475

Kumar S, Deepak V, Kumari M, Dutta PK (2016) Antibacterial activity of diisocyanate-modified chitosan for biomedical applications. Int J Biol Macromol 84:349–353

Kumar S, Silva JD, Wani MY, Gil JM, Sobral AJ (2017) Carbon dioxide capture and conversion by an environmentally friendly chitosan based meso-tetrakis (4-sulfonatophenyl) porphyrin. Carbohydr Polym 175:575–583

Kumar S, Prasad K, Gil JM, Sobral AJ, Koh J (2018a) Mesoporous zeolite-chitosan composite for enhanced capture and catalytic activity in chemical fixation of CO2. Carbohydr Polym 198:401–406

Kumar S, Wani MY, Koh J, Gil JM, Sobral AJ (2018b) Carbon dioxide adsorption and cycloaddition reaction of epoxides using chitosan–graphene oxide nanocomposite as a catalyst. J Environ Sci 69:77–84

Laye C, McClements DJ, Weiss J (2008) Formation of biopolymer-coated liposomes by electrostatic deposition of chitosan. J Food Sci 73(5):N7–N15

Leewenhoeck A (2017) An abstract of a letter from Mr. Anthony Leevvenhoeck at Delft, dated Sep. 17. 1683. Containing some microscopical observations, about animals in the scurf of the teeth, the substance call'd worms in the nose, the cuticula consisting of scales. Philos Trans R Soc Lond A 14(159):568–574

Lemmer HJ, Hamman JH (2013) Paracellular drug absorption enhancement through tight junction modulation. Expert Opin Drug Deliv 10(1):103–114

Li N, Zhuang C, Wang M, Sun X, Nie S, Pan W (2009) Liposome coated with low molecular weight chitosan and its potential use in ocular drug delivery. Int J Pharm 379(1):131–138

Li H, Li Y, Ao H, Bi D, Han M, Guo Y, Wang X (2018) Folate-targeting annonaceous acetogenins nanosuspensions: significantly enhanced antitumor efficacy in HeLa tumor-bearing mice. Drug Deliv 25(1):880–887

Lima R, Del Fiol FS, Balcão VM (2019) Prospects for the use of new technologies to combat multidrug-resistant bacteria. Front Pharmacol 10:692

Liu Y, Liu D, Zhu L, Gan Q, Le X (2015) Temperature-dependent structure stability and in vitro release of chitosan-coated curcumin liposome. Food Res Int 74:97–105

Liu D, Yang F, Xiong F, Gu N (2016) The smart drug delivery system and its clinical potential. Theranostics 6(9):1306

Luo W, Chen D, Wu M, Li Z, Tao Y, Liu Q, Pan Y, Qu W, Yuan Z, Xie S (2019) Pharmacokinetics/pharmacodynamics models of veterinary antimicrobial agents. J Vet Sci 20(5):e40

Mady MM, Darwish MM (2010) Effect of chitosan coating on the characteristics of DPPC liposomes. J Adv Res 1(3):187–191

Matta JL, Morales M, Armstrong RA, D'Antoni H (2001) Mechanisms of UV induced free radical oxidative damage in human skin fibroblasts. Free Radic Biol Med 31:S13

Menikarachchi MA, Katuwavila KA, Ekanayake EW, Thevanesam V, Karunaratne V, Karunaratne DN (2016) Release behaviour of amoxicillin from chitosan coated liposomes derived from eggs. J Natl Sci Found 44:167

Mertins O, Dimova R (2011) Binding of chitosan to phospholipid vesicles studied with isothermal titration calorimetry. Langmuir 27(9):5506–5515

Metz M, Shlaes DM (2014) Eight more ways to deal with antibiotic resistance. Antimicrob Agents Chemother 58(8):4253–4256

Mugabe C, Halwani M, Azghani AO, Lafrenie RM, Omri A (2006) Mechanism of enhanced activity of liposome-entrapped aminoglycosides against resistant strains of *Pseudomonas aeruginosa*. Antimicrob Agents Chemother 50(6):2016–2022

Nisini R, Poerio N, Mariotti S, De Santis F, Fraziano M (2018) The multirole of liposomes in therapy and prevention of infectious diseases. Front Immunol 9:155

Panya A, Laguerre M, Lecomte J, Villeneuve P, Weiss J, McClements DJ, Decker EA (2010) Effects of chitosan and rosmarinate esters on the physical and oxidative stability of liposomes. J Agric Food Chem 58(9):5679–5684

Park SN, Jo NR, Jeon SH (2014) Chitosan-coated liposomes for enhanced skin permeation of resveratrol. J Ind Eng Chem 20(4):1481–1485

Pattni BS, Chupin VV, Torchilin VP (2015) New developments in liposomal drug delivery. Chem Rev 115(19):10938–10966

Peter M, Binulal NS, Nair SV, Selvamurugan N, Tamura H, Jayakumar R (2010) Novel biodegradable chitosan–gelatin/nano-bioactive glass ceramic composite scaffolds for alveolar bone tissue engineering. Chem Eng J 158(2):353–361

Pinto-Alphandary H, Andremont A, Couvreur P (2000) Targeted delivery of antibiotics using liposomes and nanoparticles: research and applications. Int J Antimicrob Agents 13(3):155–168

Rodríguez-Rojas A, Rodríguez-Beltrán J, Couce A, Blázquez J (2013) Antibiotics and antibiotic resistance: a bitter fight against evolution. Int J Med Microbiol 303(6–7):293–297

Senior JH, Trimble KR, Maskiewicz R (1991) Interaction of positively-charged liposomes with blood: implications for their application in vivo. Biochim Biophys Acta Biomembr 1070(1):173–179

Seong JS, Yun ME, Park SN (2018) Surfactant-stable and pH-sensitive liposomes coated with N-succinyl-chitosan and chitooligosaccharide for delivery of quercetin. Carbohydr Polym 181:659–667

Sharma A, Sharma US (1997) Liposomes in drug delivery: progress and limitations. Int J Pharm 154(2):123–140

Shin GH, Chung SK, Kim JT, Joung HJ, Park HJ (2013) Preparation of chitosan-coated nanoliposomes for improving the mucoadhesive property of curcumin using the ethanol injection method. J Agric Food Chem 61(46):11119–11126

Sudheesh Kumar PT, Lakshmanan VK, Anilkumar TV, Ramya C, Reshmi P, Unnikrishnan AG, Nair SV, Jayakumar R (2012) Flexible and microporous chitosan hydrogel/nano ZnO composite bandages for wound dressing: in vitro and in vivo evaluation. ACS Appl Mater Interfaces 4(5):2618–2629

Tan C, Xue J, Eric K, Feng B, Zhang X, Xia S (2013) Dual effects of chitosan decoration on the liposomal membrane physicochemical properties as affected by chitosan concentration and molecular conformation. J Agric Food Chem 61(28):6901–6910

Torchilin VP (2005) Recent advances with liposomes as pharmaceutical carriers. Nat Rev Drug Discov 4(2):145–160

Ulrich AS (2002) Biophysical aspects of using liposomes as delivery vehicles. Biosci Rep 22(2):129–150

Voskuhl J, Ravoo BJ (2009) Molecular recognition of bilayer vesicles. Chem Soc Rev 38(2):495–505

Wang CH, Hsieh YH, Powers ZM, Kao CY (2020a) Defeating antibiotic-resistant bacteria: exploring alternative therapies for a post-antibiotic era. Int J Mol Sci 21(3):1061

Wang DY, Van der Mei HC, Ren Y, Busscher HJ, Shi L (2020b) Lipid-based antimicrobial delivery-systems for the treatment of bacterial infections. Front Chem 7:872

Yang KK, Kong M, Wei YN, Liu Y, Cheng XJ, Li J, Park HJ, Chen XG (2013) Folate-modified–chitosan-coated liposomes for tumor-targeted drug delivery. J Mater Sci 48:1717–1728

Yang Z, Liu J, Gao J, Chen S, Huang G (2015) Chitosan coated vancomycin hydrochloride liposomes: characterizations and evaluation. Int J Pharm 495(1):508–515

Zarogoulidis P, Kioumis I, Porpodis K, Spyratos D, Tsakiridis K, Huang H, Li Q, Turner JF, Browning R, Hohenforst-Schmidt W, Zarogoulidis K (2013) Clinical experimentation with aerosol antibiotics: current and future methods of administration. Drug Des Devel Ther 7:1115–1134

Zhang L, Pornpattananangkul D, Hu CM, Huang CM (2010) Development of nanoparticles for antimicrobial drug delivery. Curr Med Chem 17(6):585–594

Zhao W, Zhuang S, Qi XR (2011) Comparative study of the in vitro and in vivo characteristics of cationic and neutral liposomes. Int J Nanomedicine 6:3087–3098
Zhao D, Yu S, Sun B, Gao S, Guo S, Zhao K (2018) Biomedical applications of chitosan and its derivative nanoparticles. Polymers 10(4):462
Zhuang J, Ping Q, Song Y, Qi J, Cui Z (2010) Effects of chitosan coating on physical properties and pharmacokinetic behavior of mitoxantrone liposomes. Int J Nanomedicine 5:407–416

14 Photodynamic and Light-Response Nanomaterials Against Multidrug-Resistant Bacteria

Smritilekha Bera

Abstract

Antimicrobial resistance (AMR) is a life-threatening concern worldwide. To combat this issue, ongoing research is focused on advanced nanoparticle studies. The nanoparticles, with their unique quantum-size effects, shape characteristics, and tunable behavior, hold the potential to mitigate AMR. Light-responsive nanomaterials are particularly promising in addressing life-threatening antibacterial resistance, offering precise control at the atomic and molecular levels through light-induced reactive oxygen species. Furthermore, noninvasive light-responsive nanocomponents in photodynamic therapy can enhance drug effectiveness, shelf life, biocompatibility, and more. This chapter explores recent research on light-sensitive nanomaterials in the context of photodynamic therapy for antimicrobials to address the global health threat of AMR.

Keywords

Photodynamic · Light response nanomaterials · Multidrug resistance · Bacteria

14.1 Introduction

Antimicrobial resistance (AMR) in recent times is a very well-known glitch in the medicinal community (Prestinaci et al. 2015; Salam et al. 2023). To combat that, continuous research is in progress throughout the scientific community. With the growth of nanoparticle (NPs) research, there is an increase in hope that the ultra-small and controllable quantum-size effects, shape characteristics, surface-to-volume ratio, and tunable physicochemical behavior of nanomaterials might assist

S. Bera (✉)
School of Chemical Sciences, Central University of Gujarat, Gandhinagar, India

M. Y. Wani et al. (eds.), *Nanotechnology Based Strategies for Combating Antimicrobial Resistance*, https://doi.org/10.1007/978-981-97-2023-1_14

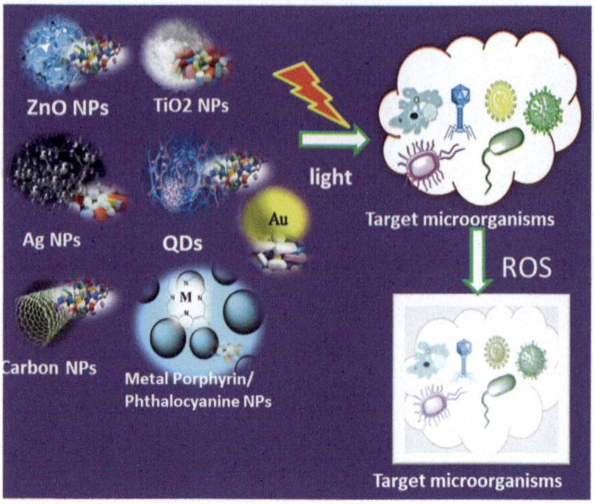

Fig. 14.1 A graphical representation of light-response antimicrobial nanomaterials

in easing the antimicrobial resistance of the developed drug molecules (Makabenta et al. 2021; Mba and Nweze 2021; Thambirajoo et al. 2021). Stimuli-responsive nanomaterials outperform conventional materials wherever they are applied. This superiority arises from their ability to precisely control the distribution of nanoparticles at the atomic and molecular levels in both space and time, primarily through a reactive oxygen species (ROS)-mediated mechanism facilitated by light (Pham et al. 2020; Shah et al. 2020; Blum et al. 2015; Shakiba M et al. 2015). Consequently, these responsive nanomaterials contribute to enhancing drug shelf life, promoting site-specific drug activity, increasing water solubility, improving biocompatibility, enhancing toxicological profiles, sustaining drug release, and enhancing the bioavailability of antibiotics at their intended target sites over an extended period. A variety of nanoparticles are capable of delivering photoactive molecules, and these photoactive nanoparticles have the potential to enhance the antimicrobial effects of photodynamic therapy through distinct mechanisms of action, thereby reducing AMR (Huang et al. 2012, 2014; Ganoth et al. 2015; Pellosi et al. 2017). In this chapter, the recent research on the development of light-sensitive nanomaterials for antimicrobial (Fig. 14.1) applications aimed at combating AMR will be discussed.

14.2 A Brief Description of Antibiotics

Antibiotics are essential life-saving drugs that treat bacterial infections and prevent potentially lethal complications, playing a crucial role in modern medicine. Their effectiveness in combating infections has saved innumerable lives since their discovery. Bacteria are ubiquitous in our environment, and while many are harmless or even beneficial, some can cause diseases when they enter the body and multiply uncontrollably (Doron and Gorbach 2008). An innate immune system plays a crucial role in defending against harmful bacteria. The white blood cells or leukocytes of the primary immune system recognize/eliminate pathogens and even control infections without the need for external intervention. However, there are cases when the immune system struggles to contain a bacterial infection, and that's when antibiotics come into play. These drugs target specific bacteria or interfere with their growth and reproduction, helping the immune system in its fight against the infection (Kohanski et al. 2010). The discovery of penicillin by Alexander Fleming in 1941 marked a turning point in medicine. After a short while, in 1945, Sir Alexander Fleming, Ernst Boris Chain, and Sir Howard Walter Florey were jointly given the Nobel Prize in physiology or medicine for their discovery of penicillin and its ability to treat numerous infectious diseases (Fleming 1929). Penicillin became the first widely used antibiotic and significantly reduced mortality rates from bacterial infections (Chain et al. 1940; Gaynes R. 2017). Following this success, there was a rapid development of various other antibiotics with different mechanisms of action to tackle a broader spectrum of bacterial diseases (Aminov 2010). Generally, the period 1945–1965 was called "Golden Era of Antibiotic Discovery" as during this period, many important antibiotic classes were indeed discovered, including those isolated from soil *Actinomycetes* such as Streptomyces. Notable antibiotics from this era include penicillin, streptomycin, chloramphenicol, tetracycline, erythromycin, and others. After the post-Golden Era of Antibiotic Discovery, there was a reduced antibiotic discovery activity though it is important to note that several antibiotics and antimicrobial agents were developed and introduced during this time also. Besides Linezolid, antibiotics like quinolones (ciprofloxacin), carbapenems (imipenem), and glycopeptides (vancomycin) were introduced in the years following the golden era.

Depending on the activities, antibiotics are classified as narrow-spectrum antibiotics that only work on one type of bacteria and are preferred since their impact on non-disease producing bacteria is minimal, for example, isoniazid. Unluckily, broad-spectrum antibiotics, which have a wide antimicrobial scope, are frequently utilized to recognize the infectious agent due to a lack of knowledge and/or diagnostic instruments. Generally, doctors prescribe broad-spectrum antibiotics such as tetracycline, aminoglycoside, chloramphenicol, etc. Antibiotics typically do not impact human and mammalian cells due to the absence of physically distinguishable targets in these cells. Figure 14.1 describes the development of different antibiotics and their progress since the "Golden era of antibiotics" (Fig. 14.2) (Hutchings et al. 2019; Rahman et al. 2022; Bera et al. 2016).

14.3 Mechanism of Action

Antibiotics interrupt critical bacterial activities or structures, and depending on these effects, antibiotics are classified as bactericidal or bacteriostatic (Kohanski et al. 2010). Based on their mechanisms of action and bacterial targets, antibiotics are categorized into a wide range of classes. Antibiotics often follow one of three types of bacterial mechanism: the membranes or cell wall of the bacterial cell; the

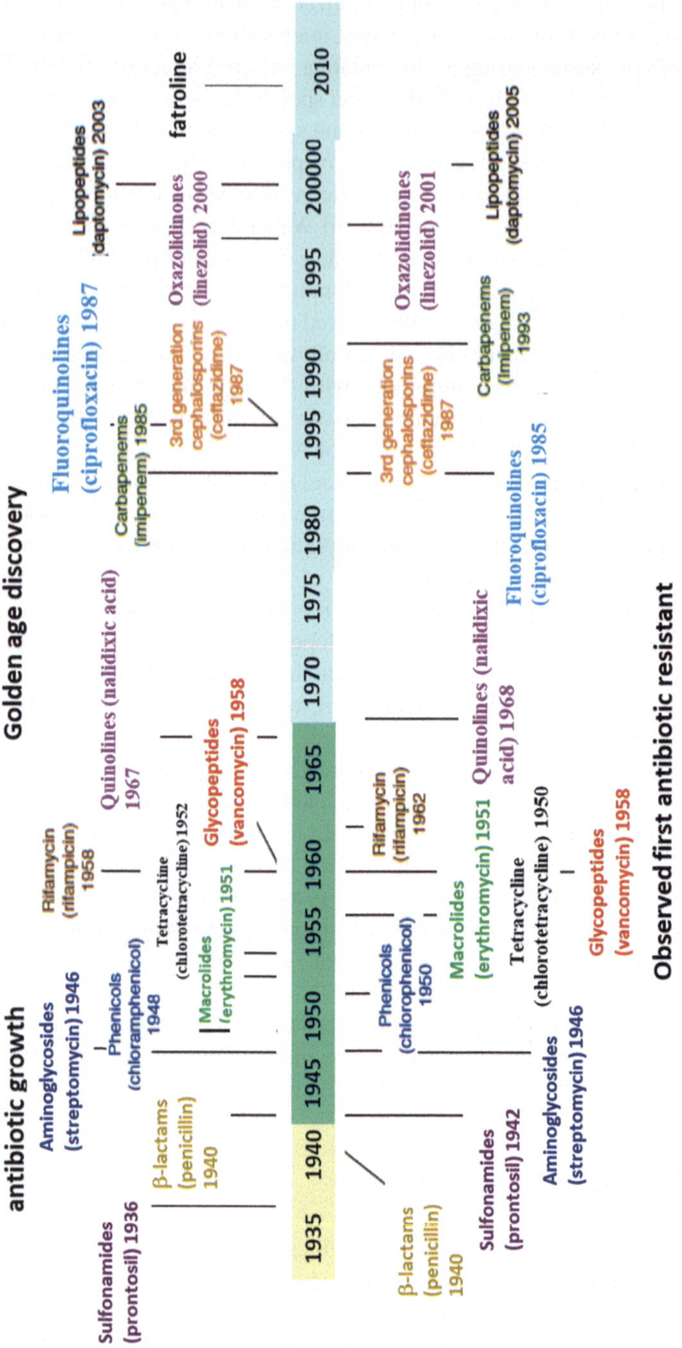

Fig. 14.2 Antibiotic development and its resistance

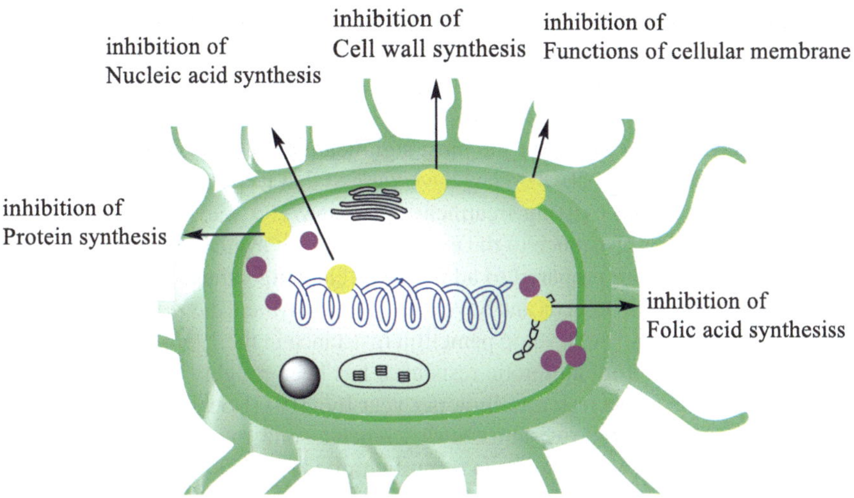

Fig. 14.3 Antimicrobial mechanisms of antibiotics against bacteria

mechanism that generates nucleic acids like DNA and RNA; and the mechanism that generates proteins like the ribosome (Fig. 14.3) (Etebu and Arikekpar 2016; Kırmusaoğlu et al. 2019; Kapoor et al. 2017). The probable mechanisms of action of antimicrobial agents are: the inhibition of

1. Cell wall synthesis: bacitracin, cephalosporins, penicillins, vancomycin, etc.
2. Functions of cellular membrane: polymyxins, etc.
3. Protein synthesis: macrolides, tetracyclines, clindamycin, chloramphenicol, aminoglycosides, etc.
4. Nucleic acid synthesis: rifampin, quinolones, etc.
5. Folic acid synthesis: sulfonamides, trimethoprim, etc.

14.4 Antibiotic Resistance

Antibiotics have had remarkable success, though antibiotic resistance (Fair and Tor 2014; Davies J. 2010) has muted the euphoria surrounding them, even more dangerous than their unwanted side effects. The utility of these crucial medications is currently in jeopardy because of an issue that emerged shortly after penicillin was first introduced. The "golden" pipeline of antibiotic discovery from the 1960s and 1970s quickly dried up as it became harder to find fresh chemicals. One of the most significant public health issues of the twenty-first century is antibiotic resistance, which results in high healthcare costs and rising morbidity and mortality. At first, insufficient dosage—rather than unchecked consumption and incorrect usage—was the risk factor for "antibacterial resistance" among patients. Antibacterial resistance, however, is now treated as if it were a form of "global terrorism" as a result of a group of dangerous and contagious bacterial strains gaining access to drug molecules through various pathways, indiscriminate use in agriculture and animal husbandry farms, irregular consumption, irrational antibiotic prescription, and poor adherence (Lamy et al. 2020). The World Health Organization (WHO) has emphasized that AMR stands as one of the most significant threats to our society (WHO Antimicrobial Resistance 2018). They have warned that if bacteria persist in developing resistance to the existing treatment options, WHO warned that if bacteria persist in developing resistance to the existing treatment available, and no successful new compounds are prepared to address this challenge, then we can enter into the "post-antibiotic era" (Zucca and Savoia 2010). *Staphylococcus aureus* (*S. aureus*) was the first "superbug" to develop penicillin resistance in the 1950s, and since then the global death toll from antibiotic resistance has been roughly 700,000 per year, with a potential increase to ten million per year by 2050 (Roundtree and Freeman 1956). Rising antibacterial resistance is very much a concern in the health sector as the development of new antibiotics is severely limited (Nikaido 2009). Even before the widespread use of penicillin, several observations revealed that bacteria may destroy it by enzymatic breakdown (Abraham and Chain 1940). In recent years, the enormous rise of multidrug-resistant (MDR) bacteria in hospitals and the population has resulted in a growing problem in which an increasing number of antibiotics are no longer therapeutically active (Taubes 2008; Prestinaci et al. 2015). For each antibiotic created during the Golden Age of Antibiotic Research, resistant bacterial strains were found 8 years after they were approved for use on the market. These resistances occur when harmful microorganisms come into repeated contact with an antibiotic, their susceptibility to it decreases or even disappears, which is called antibiotic resistance. The bacteria become resistant to some other antibiotics when they become resistant to one antibiotic which is known as cross-resistance. Although there is a continuous development of clinical antibiotics (Bera S et al. 2016) and approval is high, there is also a withdrawal rate for pharmaceutical therapeutic classes even after getting clinical approval (Outterson et al. 2013; Stephens et al. 2020).

14.5 Mechanism of Resistance

The multiple mechanisms of antibiotic resistance development have sparked interest in new treatments with unique modes of action(s), capable of preventing or delaying the establishment of resistance (Fig. 14.4). Today's microbiologists warn of a "medical disaster" that could return society to the days before penicillin when even seemingly minor diseases could be fatal. Drug resistance is thus caused by (a) altering the drug's target site. Methicillin resistance among *S. aureus* is one of the most significant antibiotic resistances in the world. (b) The antibiotic is inactivated by an enzyme: penicillins and cephalosporins are examples of beta-lactam antibiotics that can be inactivated by b-lactamases. (c) Active transport of the antibiotic out of the bacterial cell: tetracyclines, macrolides, and quinolones, among others, are actively transported out of the bacterial cell using efflux pumps. (d) Decreased

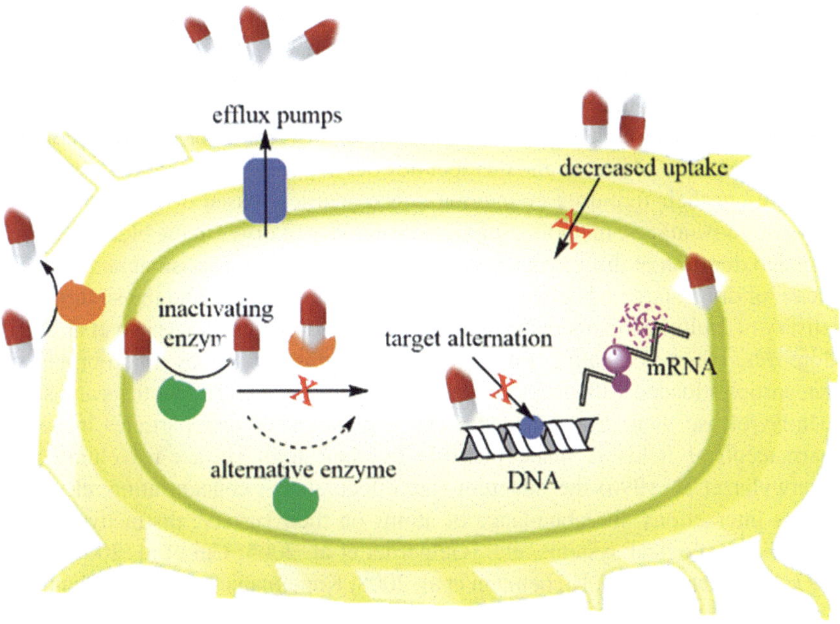

Fig. 14.4 Mechanisms of antibiotic resistance in bacteria

bacterial cell wall permeability to the antibiotic: modification in the porin proteins that form channels in the cell membrane, as in the case of *Pseudomonas aeruginosa*'s resistance to several penicillins and cephalosporins (Munita and Arias 2016; Uddin et al. 2021).

14.6 Nanoparticles in Drug Research

Although there is a continuous process of new antibacterial drug development or the modification of the old drug to overcome the resistant and unwanted side effects, the development of modern nanomedicine (Salata OV 2004, Chen et al. 2008) or nanotechnology emerges to enhance the antimicrobial drug development to combat microorganisms like bacteria, departing from conventional methods (Schairer DO et al. 2013; Seil JT, Webster TJ 2012). In 1959, physicist Richard Feynman introduced the concept of nanotechnology in his speech titled "There's Plenty of Room at the Bottom," which catalyzed significant progress in the field of nanotechnology (Feynman 1991). There is an ongoing endeavor to create novel antibiotics that are less harmful and more resistant to developing resistance. In addition to conventional antibacterial drug development, nanotechnology, which focuses on understanding and controlling phenomena that occur within the range of 1–100 nm, is revolutionizing various aspects of pharmaceuticals, including drug manufacturing, drug delivery (Desai et al. 2020; Desai et al. 2021), and medical diagnostics (Alonso 2004;

Burda et al. 2005; Nikalje 2015; Perrault and Chan 2009; Coleman et al. 2011; Afzal et al. 2022; Wang S et al 2013). It also introduces innovative techniques for drug synthesis, drug carriers, targeting methods, toxicity (Huang et al. 2024; Maurya et al. 2019) reduction, and precise cell targeting. When pharmaceuticals are loaded in nanoparticles through physical encapsulation, adsorption, or chemical conjugation, their pharmacokinetics and therapeutic effectiveness can be markedly enhanced compared to their free-drug counterparts. Nanoparticle-based drug delivery offers numerous advantages, including improved solubility of drugs in the bloodstream, an extended duration of drug circulation throughout the body, consistent and regulated release of drugs, targeted delivery to specific tissues and cells, and the simultaneous delivery of multiple therapeutic agents for combination therapy (Ansari et al. 2010). Moreover, when combating intracellular infections caused by microorganisms, nanoparticles loaded with drugs can enter host cells through endocytosis and subsequently release their drug payloads. The application of nanoparticles in various pharmaceutical fields is made possible by their high surface-to-volume ratio, (Leutwyler et al. 1996) the quantum size effect, shape, concentration, electrodynamics interactions, the abundance of atoms on their surface, distinctive interactions with biological systems, etc. (Gelperina et al. 2005; Cui et al. 2001; Fubini 1997; Zhang et al. 2008; Riehemann et al. 2008; Kopelman et al. 2005). Changes in these physical properties at the nanoscale directly impact the material's overall properties. As the size of nanoparticles decreases, the proportion of atoms on their surface relative to the total number of atoms in the material increases, intensifying their physical and chemical activity. This increased population of nanoparticles on the surface enhances antibacterial activity, with minimal adverse effects, in contrast to small molecule-based antimicrobial treatments. Nanotechnology's use in the field of medicine has undergone thorough investigation across various medical domains, particularly in the realm of drug delivery (Bera S and Mondal D 2019, 2022), over the past few decades. This exploration encompasses a wide array of applications, such as gene therapy, fluorescent bio-imaging agents, biological markers, enhancers for magnetic resonance imaging, DNA-templated probes, biosensors, hyperthermia-assisted tumor elimination, tissue engineering, as well as the development of antibacterial and antiviral agents (Hilliard et al. 2002; De et al. 2008).

14.7 Brief Description of Photodynamic Therapy and Light Stimuli

Drugs and prodrugs must overcome numerous biological barriers, such as the intestinal barrier and metabolic reactions, to have the desired pharmacological effect. To achieve high systemic bioavailability, oral medications need to have the best possible stability in the GI tract and be able to quickly pass intestinal epithelium (Alonso 2004). Similarly, effective drug distribution across the epithelium is necessary for drug delivery via the cutaneous, nasal, and pulmonary routes. While the majority of low molecular weight drugs (900 Da) are administered orally, intravenous (IV) administration is still the most direct and effective method for delivering medications that contain peptides, proteins, big molecules, and polynucleotides (Bruno et al. 2013). The blood-brain barrier (BBB) is another barrier that prevents big or hydrophilic molecules from diffusing into the cerebrospinal fluid. It is a significant barrier to the treatment of the majority of CNS and brain illnesses (Daneman and Prat 2015; Kadry et al. 2020). To transfer medications across the BBB, a number of nanoparticle-based techniques are being developed, including liposome, nanosphere, and cationic albumin nanoparticles (Silva 2008; Desai 2012). Many drugs, nevertheless, have the potential to be toxic to normal cells even after successful administration, leading to cell death or a dramatic alteration in cell function.

Light-activated compounds have been recognized for a significant duration as agents that can be activated by light to treat a wide spectrum of diseases (Carpenter BL et al. 2011). Notably, this approach has been historically applied to cutaneous infections, where sunlight exposure with photoactivated compounds was employed in the hope of healing the infected area. The success of this method in treating infections served as a proof of concept. In this context, Niels Finsen was awarded (Finsen 1901) the Nobel Prize in 1903 for his pioneering work in using light therapy to control skin manifestations of tuberculosis. This achievement underscored the substantial potential and practicality of light-based approaches for disease treatment. Recent advancements in light-controlled therapies, including photodynamic and photothermal therapies, offer promising strategies to prevent and combat bacterial infections, a significant contributor to illness and mortality (Hamblin and Hasan 2004).

Antibacterial photodynamic therapy (aPDT) has garnered increasing attention within the scientific community due to its potential to eliminate MDR pathogenic bacteria while having a low propensity to induce drug resistance (Zhi et al. 2020; Youf et al. 1995). In the realm of emerging antimicrobial therapeutic options, light-based methods show significant potential (Bera and Mondal 2018) (Wainwright et al. 2017). Interestingly, the use of light for therapeutic purposes has ancient roots, dating back to ancient Greece, Egypt, and India, where it was used to treat various skin ailments (Daniell and Hill 1991). The twentieth century marked a pivotal moment with Oscar Raab discovering the phototoxicity of acridine red against *Paramecium caudatum* and Tappenier and Jesionek documenting the photodynamic effects of eosin as a treatment for skin conditions. This laid the foundation for photodynamic therapy (PDT), which involves administering a nontoxic photosensitizer (PS) (Shekhar S et al. 2022) followed by targeted exposure to light of an appropriate wavelength. While PDT has been a clinical reality in anticancer treatments for over two decades (Shi et al. 2019), its application as an antimicrobial therapy emerged in the early 1990s, particularly in the healthcare sector, where it was first demonstrated as an effective approach against drug-resistant infections. This marked the beginning of a "photo-antimicrobial renaissance era." Notably, antimicrobial photodynamic therapy has shown efficacy against a range of MDR bacteria, regardless of their drug resistance profiles (Sabino et al. 2020; Plotino et al. 2019).

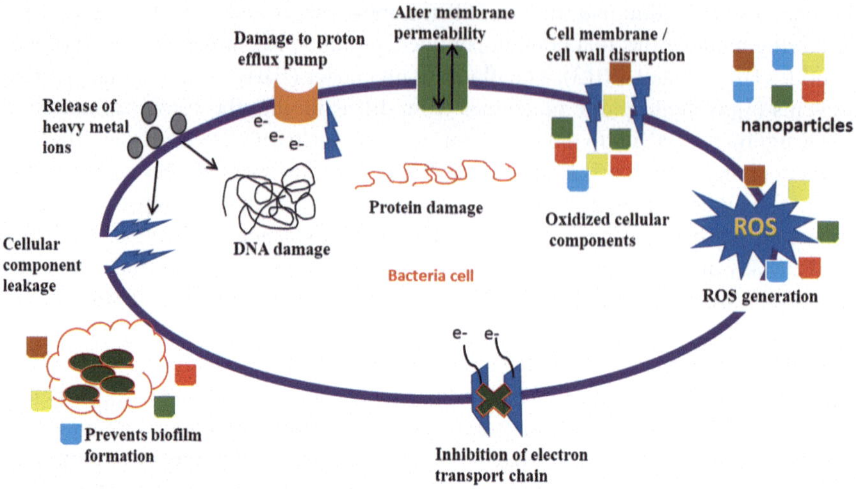

Fig. 14.5 Probable mechanistic antimicrobial action of nanoparticles

14.8 Mechanism of Action of Photodynamic Antimicrobial Nanoparticles

Understanding the mechanistic action of nanoparticles poses a challenge due to their variability depending on the specific material and the type of diseases being targeted (Fig. 14.5). Inorganic nanoparticles derived from metals or metal oxides, for instance, can engage in electrostatic interactions with the cell membranes of bacteria. This interaction results in damage to the bacterial membrane and the initiation of toxic oxidative stress within the bacteria. This stress is brought by the generation of cellular radicals and non-radical oxygen species, achieved through the partial reduction of oxygen. These species include the hydroxyl radical (•OH), superoxide anion (O_2^{2-}), and hydrogen peroxide (H_2O_2). These reactive oxygen species proceed to inflict damage on the bacterial DNA and other macromolecules, including the polyunsaturated phospholipids that make up the bacterial membrane, thereby disrupting the functioning of the bacterial cell. In this context, light emerges as a promising external stimulus capable of sensitizing a specific class of compounds. This sensitization leads to the conversion of triplet oxygen ($^3O_2$) found in and around tissues into reactive oxygen species (ROS) close to the diseased cells. This, in turn, results in cell death, oxidation of organic compounds, and the conversion of *Escherichia coli* (*E. coli*) cells into carbon dioxide. The oxidation of bacterial cell components, such as lipids and DNA, resulted in subsequent cell death (Maness et al. 2013; Nathan and Cunningham-Bussel 2013). Reactive nitrogen species (RNS), a different class of antibacterial nanomaterials, work in concert with ROS to damage cell membranes. By oxidizing ferrous heme into its corresponding ferric forms, RNS, which is produced from nitric oxide (NO) (Carpenter & Schoenfisch 2013; Han et al. 2013; Schairer DO et al. 2013) and superoxide anion

(O_2^{2-}), is involved in several mechanisms of action to directly damage the proteins that contain transition metal centers, such as haemoglobin, myoglobin, and cytochrome C (Kutner and Friedman 2013).

The mechanism of aPDT, specifically investigative photosensitizer (PS) and their application in combating both Gram-positive and Gram-negative bacteria is essential. In addition to conventional metal and metal oxide agents, a distinct category of organic compounds, including phthalocyanines (PC), phenothiazines, and porphyrin dyes referred to as photosensitizers, exhibits promise in photodynamic therapy (PDT) (Lan M et al. 2019). These photosensitizers, when exposed to light in the presence of oxygen, initiate a crucial process, generating two distinct types of reactive species collectively known as ROS/RNS (Mondal and Bera 2014). Consequently, light serves as an external catalyst, activating these specialized chemical compounds to convert the surrounding tissue oxygen ($^3O_2$) into ROS near infected cells, ultimately leading to cell demise.

PDT involves the administration of PSs locally or systemically, followed by exposure to a certain wavelength of light. PSs absorb light energy in their ground state and transform into the extremely energetic singlet state (1s*) as a result. The singlet-state PSs then release their stored energy and transition to a triplet state (3s*) (Sarbadhikary et al. 2021; Hu et al. 2021; Li et al. 2020). The triplet-state PSs interact with oxygen molecules during this transition, producing ROS (Fig. 14.6). Conventional drug-based PSs, however, have a number of inherent drawbacks that prevent their use in healthcare facilities, including poor water solubility, insufficient control of drug release, and low selectivity. Additionally, several drug-based PSs have the potential to harm healthy tissues and have unintended side effects (Kwiatkowski et al. 2018). Considering the growing challenge of bacterial resistance (Laxminarayan R and Chaudhury RR, 2016) to conventional therapies, current research explores the integration of PDT with existing bacterial infection treatments to enhance their effectiveness. PSs, light, and oxygen are the three essential components needed for PDT. Energy transfer, proton transfer, rearrangement, fluorescence, internal conversion, and intersystem crossing are just a few of the many stages that make up the whole process, which results in the production of biologically active ROS (Yu et al. 2022). Additionally, because of their extremely small size, high surface-to-mass ratio, high reactivity, and distinctive interactions with biological systems, nanoparticles have established extra benefits. The pharmacokinetics and therapeutic index of the photosensitizer in nano form can change due to physical encapsulation, adsorption, or chemical conjugation. When combined with silver, quantum dots (Gao et al. 2004; Kan S et al., 2003), iron oxide (Agarwala M et al, 2009), and gold-based hybrid nanoparticles (Nikoobakht B et al. 2003), these PSs demonstrated their potential to have enhanced antibacterial properties when exposed to the right wavelength of light. Another intriguing material for PDT is the utilization of nanoparticles as potential PS delivery systems. Controlled-release capabilities, great water solubility, favorable biocompatibility, and the capacity to produce high amounts of ROS are just a few of the notable qualities of nanomaterial-based PDT.

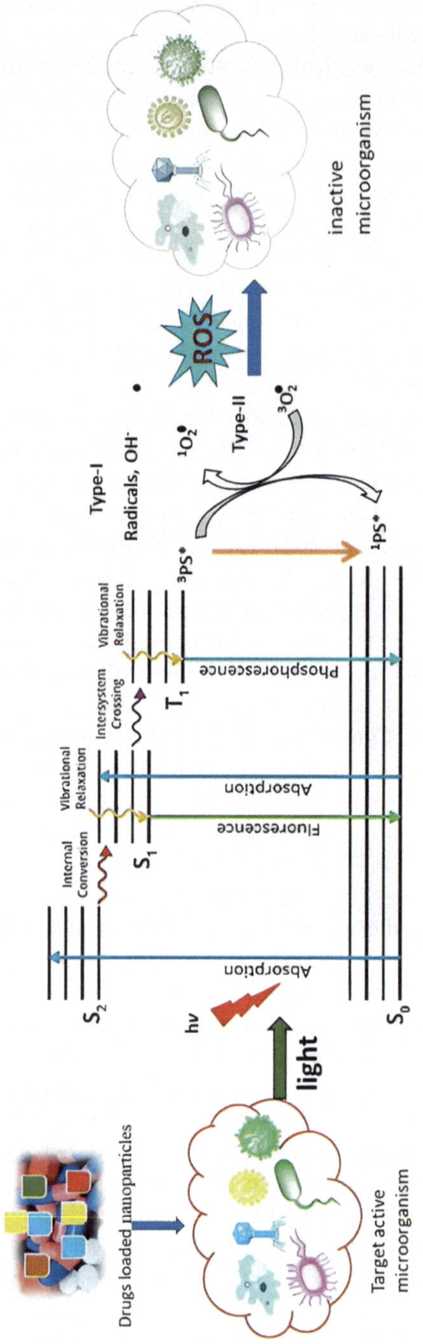

Fig. 14.6 Mechanism of PDT for bacterial inactivation

14.9 Strategies Employed for Nanoparticle-Based Photo-Controlled Delivery

Nanoparticle-based photo-controlled delivery systems have gained significant attention for their potential in targeted drug delivery and release (Roy et al. 2020). These systems utilize light-responsive nanoparticles to achieve controlled and spatiotemporal drug release upon exposure to specific light wavelengths. Some strategies for nanoparticle-based photo-controlled delivery are as follows:

Photoactivatable Nanoparticles Nanoparticles can be designed with light-responsive materials or coatings that undergo structural changes or release cargo upon light irradiation. These nanoparticles can be activated by specific wavelengths of light, such as ultraviolet (UV), visible, or near-infrared (NIR) light, triggering drug release or inducing changes in particle properties that enhance drug delivery (Bansal and Zhang 2014; Kim and Darafshe 2020).

Photochemical Internalization In this strategy, PSs are encapsulated within nanoparticles along with the therapeutic cargo. Upon light irradiation, the PSs generate ROS that disrupt the endosomal membrane, allowing for the cytosolic release of the encapsulated cargo. This technique improves the delivery of therapeutic agents that are typically sequestered within endosomes/lysosomes (Berg et al. 2010).

Light-Triggered Particle Disassembly Nanoparticles can be designed to disassemble or undergo structural changes upon light irradiation. Light-responsive linkers or crosslinkers can be incorporated within the nanoparticle structure, which, upon exposure to light, break or undergo conformational changes, leading to cargo release. This approach allows for controlled release at the target site (Long et al. 2021).

Optogenetics-Enabled Systems Optogenetics involves the use of light-sensitive proteins to control cellular activity. Nanoparticles can be engineered to carry light-sensitive proteins, such as channelrhodopsins or halorhodopsins, which respond to specific wavelengths of light. Upon light stimulation, these proteins can trigger specific cellular responses, such as ion flux or membrane depolarization, which can be utilized for drug release or cellular modulation (Ye and Fussenegger 2019; Huang et al. 2016; McCue and Kuhlman 2022).

NIR Light-Responsive Nanoparticles NIR light can penetrate tissues deeply, allowing for noninvasive light-based therapies. Nanoparticles can be engineered to respond to NIR light by incorporating NIR-absorbing materials, such as gold nanomaterials (Giljohann et al. 2010), or upconversion nanoparticles. Upon NIR light exposure, these nanoparticles generate localized heat or undergo energy transfer processes, leading to cargo release or enhanced drug delivery (Tang and Wang 2021).

Spatial Control with Light Patterning Sophisticated techniques like spatially controlled light patterning, such as photolithography or laser-induced techniques (Yang L et al. 2019), can be employed to precisely illuminate specific regions of the target site. By designing nanoparticles with spatially addressable photoresponsive components, drug release can be controlled with high spatial resolution, enabling localized and site-specific therapy (Barad et al. 2021).

External Light Activation In some cases, external light sources, such as lasers or light-emitting diodes (LEDs), are used to provide the required light stimulus for triggering drug release. These light sources can be externally applied to the target site, allowing for noninvasive and remotely controlled drug delivery (Tao et al. 2020; Zhang et al. 2023).

14.10 Antimicrobial Nanoparticles

Indeed, the use of antimicrobial nanoparticles has gained significant attention in recent years due to their potential to combat various types of infections and microbial growth. The literature often categorizes these nanoparticles into three main groups: inorganic, organic, and hybrid antimicrobial agents (Loomba and Scarabelli 2013). Here's an overview of each category:

Inorganic Antimicrobial Nanoparticles Inorganic nanoparticles, typically composed of metals or metal oxides, have shown antimicrobial properties due to their ability to release metal ions or exhibit surface interactions that inhibit microbial growth. Some commonly studied inorganic antimicrobial nanoparticles include silver (Varisco M et al. 2008), copper, titanium, zinc, etc. (Balderrama-González et al. 2021).

Organic Antimicrobial Nanoparticles Organic antimicrobial nanoparticles are usually derived from biopolymers, organic molecules, or polymers that exhibit antimicrobial properties. These nanoparticles often function by disrupting microbial membranes, inhibiting cell division, or interfering with essential cellular processes. Examples include carbohydrate nanoparticles, lipid-based nanoparticles, and polymeric nanoparticles (Palza H 2013; Li et al. 2021; Desai et al. 2017).

Hybrid Antimicrobial Nanoparticles Hybrid nanoparticles combine properties of both inorganic and organic materials, often to enhance their antimicrobial efficacy or to provide additional functionalities. These hybrids can exhibit synergistic effects and improved stability. Examples include: Metal-Organic Frameworks (MOFs) are hybrid materials composed of metal ions coordinated with organic ligands (Wang et al. 2017).

Silica-Coated Nanoparticles Inorganic nanoparticles can be coated with a layer of silica, which provides stability and allows for functionalization with organic molecules, including antimicrobial agents (Selvarajan et al. 2020).

Carbon-Based Nanomaterials Carbon-based nanomaterials like carbon nanotubes (CNTs), graphene, and carbon dots (Dong et al. 2020a); Dong XL et al 2020b) show promise in combating antimicrobial resistance. They can disrupt bacterial cell membranes, combat biofilms, and serve as carriers for antimicrobial agents. Functionalized CNTs and graphene can deliver antibiotics precisely, improving efficacy and reducing antibiotic resistance. These materials possess excellent photothermal properties, using NIR light to selectively kill drug-resistant bacteria while sparing surrounding tissue. They can also be utilized in photodynamic therapy, generating ROS when exposed to specific light wavelengths to damage bacterial cells and overcome resistance (Maleki et al. 2015; Jaiswal et al. 2023).

To the best of our knowledge, though, metals and metal oxides of zinc, silver, TiO_2, and copper in their bulk form are active antimicrobial agents, but their corresponding nanoparticles exhibit potent antibacterial effects against bacteria (Kemp et al. 2009). One of the metal nanoparticles widely used as an effective antimicrobial agent against bacteria, fungi, and viruses is silver nanoparticles, which show superior antimicrobial effects against various pathogens in vitro and in vivo models, to those equivalents of larger diameter bulk (Sirelkhatim et al. 2015). Titanium dioxide (TiO_2) nanomaterials (Tsai TM et al. 2010) are also effective against most resistant microorganisms, including *Bacillus* (de López et al. 2021). Zinc oxide (ZnO) nanoparticles with a broad-spectrum bactericidal effect demonstrate extensive antimicrobial activity against a variety of Gram-negative and Gram-positive microorganisms (Anghel et al. 2014). It's worth noting that while bulk forms of Fe_3O_4 and Au are typically inert and lack antimicrobial properties when reduced to the nanoscale, Fe_3O_4 exhibits anti-adherent characteristics that significantly reduce bacterial colonization by both Gram-negative and Gram-positive bacteria (Norman et al. 2008). Similarly, nanoparticles and nanorods derived from Au have shown bactericidal properties when appropriately functionalized for photothermally enhanced photodynamic therapy. Au NPs have been observed to bind effectively to various antibiotics, including ampicillin, vancomycin, the antibacterial enzyme lysozyme, and penicillin (Chamundeeswari et al. 2008; Chen et al. 2010). Furthermore, platinum-based nanomaterials (Gopal et al. 2010; Tseng YH et al 2013), particularly those in the size range of 1–3 nm, have demonstrated bactericidal activity against *P. aeruginosa* bacteria. Conversely, nanomaterials within the size range of 4–21 nm have displayed bacterio-compatible properties, similar to the antimicrobial activity observed with materials like copper oxide (CuO) (Ren G et al. 2009), ZnO (Vijayalakshmi K and Sivaraj D 2015), magnesium oxide (MgO), and magnesium fluoride (MgF_2) (Zhao et al. 2010).

14.11 Effect of Photodynamic and Light-Response Nanomaterials

Metal and metal oxide nanoparticles, especially silver nanoparticles, exhibit enhanced antibacterial activity due to their unique properties and interactions with bacteria. Exposure to UV or visible light boosts their antibacterial effects by increasing the release of silver ions and generating ROS. Light also excites electrons in silver nanoparticles, leading to ROS production and localized heating, which damages bacterial cells. Surface plasmon resonance (SPR) in silver nanoparticles enhances light interactions, further intensifying antibacterial effects (Le Ouay and Stellacci 2015). Combining silver nanoparticles with light helps penetrate biofilms by weakening their structure and facilitating bacterial cell disruption.

When subjected to UV light, the Ag-TiO_2/Ag/a-TiO_2 nanocomposite thin film exhibited an antibacterial efficacy against *E. coli* bacteria that was 5.1 times more potent than that of amorphous titanium oxide (a-TiO_2) in the absence of light. The nanocomposite film's ability to combat bacterial growth under solar light exposure was also significantly enhanced, surpassing the activity of Ag/a-TiO_2 by 1.35 times and that of a-TiO_2 by an impressive 6.90 times. This enhancement was attributed to the discharge of Ag + ions. The primary mechanism responsible for the release of Ag^+ ions involved the interdiffusion of water and silver nanoparticles through the pores within the TiO_2 layer. Furthermore, the nanocomposite film's (Akhavan 2009) durability outperformed that of the Ag/a-TiO_2 film by a factor of at least 11.

The effectiveness of silver-impregnated TiO_2/nylon-6 nanocomposite mats was assessed for their photocatalytic and antibacterial attributes against *E. coli*, as explored by Pant et al. Notably, the outcomes revealed promising photocatalytic and antimicrobial properties that were discernible even without the presence of UV light. Furthermore, the nanocomposite mats demonstrated exceptional reusability and robustness, positioning them as potential candidates for applications in environmental remediation. The underlying mechanism revolves around photoexcited electrons transitioning from the Ag nanoparticle surface to the conduction band of TiO_2. Subsequently, these electrons become ensnared by oxygen molecules and other species such as O_2•2, •OOH, and •OH, leading to the oxidation of organic bacterial compounds (Pant et al. 2011).

A light-triggered antimicrobial material was crafted by incorporating porphyrin and silver nanoparticles into polymethylmethacrylate. Antimicrobial coatings were created using polymethylmethacrylate doped with interacting porphyrin and silver nanoparticles. These coatings exhibit effectiveness against both Gram-positive and Gram-negative bacteria. Their antimicrobial activity is activated by blue light, with microbial eradication occurring upon illumination. The antimicrobial properties of coating stem from controlled silver release and the generation of active oxygen, facilitated by porphyrin in the presence of silver nanoparticles. The interaction with silver nanoparticles enhances porphyrin stability and positively influences the kinetics of silver release. Silver accounts for the antimicrobial impact against Gram-negative bacteria, while reactive oxygen targets Gram-positive bacteria. Moreover, the combination of photodynamic therapy with silver nanoparticles amplifies bacterial inactivation efficiency. The material's improved antimicrobial response is activated by local heating, porphyrin-silver energy transfer, and the combined effects of reactive oxygen and silver (Lyutakov et al. 2014).

Antimicrobial attributes were achieved through the controlled release of silver and the generation of active oxygen originating from porphyrin. The antimicrobial effects against *P. aeruginosa* and *S. aureus* using diverse coatings were examined in solution under varying treatment durations, both with and without illumination. It was discovered that thin films featuring silver nanostructure-coated TiO_2, carbon-doped TiO_2 [$TiO_2(C)$], and nitrogen-doped TiO_2 [$TiO_2(N)$] exhibited potent antibacterial properties in response to visible light, as reported by Wong et al. (Wong et al. 2010) In a synergistic antibacterial assessment, a reduction of approximately 5 log units in surviving bacteria was observed for *E. coli*, *S. pyogenes*, *S. aureus*, and *A. baumannii*. Following photocatalysis (1.2×10^3 lux, 10 min), the population of surviving *E. coli* (CFU) on corresponding substrates showed a statistically significant decrease compared to the Ag/TiO_2 and $Ag/TiO_2(C)$ groups, with a *p*-value less than 0.01.

Rapulenyane and colleagues developed a hybrid comprising glutathione-capped silver nanoparticles (GSH-AgNPs) and zinc phthalocyanine, resulting in a nanoconjugate. This nanoconjugate exhibited antimicrobial properties against *E. coli*, functioning even without light exposure. Furthermore, these nanoparticles demonstrated promise as bactericidal agents, effectively mitigating light toxicity within the 5–10 mM range (Rapulenyane et al. 2013). Silica-coated silver NPs ($Ag/mSiO_2$) were coupled with photosensitizer hematoporphyrin IX (HPIX) to get nanoconjugates. These nanoparticles were tested against the fungi *T. rubrum*. The $Ag/mSiO_2$ nanoparticles coupled with HPIX demonstrated significantly greater PDT efficacy (up to three orders of magnitude) compared to HPIX or $Ag/mSiO_2$ nanoparticles acting separately (Wijesiri et al. 2018).

Au NPs are a favored choice in drug delivery and nanomedicine (Ventola CL 2012; Zhu X et al. 2014) due to their biocompatibility, tunable size, and versatile properties. They can be synthesized in various shapes and sizes, facilitating customization for specific applications. Additionally, their surfaces are easily functionalized with molecules like drugs or antibodies for targeted drug delivery. They enhance drug solubility and can bear targeting ligands for precise cell targeting. Moreover, Au NPs serve as imaging agents in techniques like CT, photoacoustic imaging, and SERS, enabling real-time monitoring. They can also convert NIR light into heat, enabling localized heating for drug release or cancer cell ablation when combined with suitable lasers (Yeh et al. 2012).

Recent microscopic investigations of different bacterial strains have revealed that Au NPs do not penetrate the bacterial cells; instead, they adhere to the cell walls. These nanoparticles, owing to their resistance to degradation in biological settings, remain attached to the cell walls. Remarkably, they perform a dual role: they release the PS and enable the diffusion of singlet oxygen from the nanoparticles when they are exposed to light (Pajerski et al. 2019; Rosa MCD and Crutchley RJ 1997).

The covalently coupled toluidine blue O (TBO)-tiopronin Au conjugate exhibited a minimum bactericidal concentration (MBC) for *S. aureus* that was at least four times slower than that of free TBO. Notably, the Au conjugates demonstrated significantly enhanced activity compared to TBO under both 632.8 nm laser and white light exposure. For instance, with a concentration of 1 mM and under white light for 30 min, the conjugate displayed a substantial reduction in colony-forming units (CFU) by 4.5 log10, in contrast to 0.5 log10 for TBO at the same concentration (Gil-Tomas et al. 2007). Irrespective of light conditions, the MBC for the conjugate was 0.5 mM, while for TBO, it was 2 mM. Perni et al. introduced light-activated antimicrobial polysiloxane nanoparticles generated through a straightforward method involving swell-encapsulation-shrink processes, utilizing methylene blue (MB) and nanogold. These nanoparticles exhibited antimicrobial effectiveness against *Methicillin-resistant Staphylococcus aureus* (MRSA) (Dai et al. 2010) and *E. coli*, resulting in a reduction of up to 3.5 log10 in viable cell count upon 5 min of laser light exposure. This light irradiation employed a low-power 660 nm laser with short pulses (10 s) at the catheter entry point. The antimicrobial effect is attributed to the production of ROS and other ROS initiated by light through MB. Interestingly, the presence of 2 nM Au NPs notably bolstered the MB's capacity to exterminate bacteria (Perni et al. 2009a).

Pagonis et al. conducted experiments involving PLGA nanoparticles loaded with the PS MB for in vitro assessments against *Enterococcus faecalis*. The study included incubation of colloidal Au NPs with the PLGA complex for varying durations of 2.5, 5, and 10 min, followed by treatment with red light at 665 nm (60 J/ cm^2). The combination of MB-based nanoparticles and light exhibited synergistic effects, resulting in a reduction of colony-forming units (CFUs) in both the planktonic phase and root canals. These effects were notably more significant compared to controls and MB-loaded nanoparticles without light (Pagonis et al. 2010).

Silver-coated AuNPs with silica encapsulation are nanoparticles consisting of Au cores coated with Ag and encapsulated in silica. These silver-coated AuNPs (Au/AgNP/SiO_2) are functionalized by conjugating them with two substances—naphthalocyanine dihydroxide (Nc) and vancomycin. The functionalized nanoparticles (Au/AgNP/SiO_2/Nc-Van) have been used in targeted photodynamic therapy against Vancomycin-resistant Enterococci (VRE). PDT involves using light and a photosensitizing agent to kill targeted cells, in this case, the antibiotic-resistant bacteria VRE (Zhou et al. 2018). Cytotoxicity and phototoxicity tests were conducted to evaluate the safety and effectiveness of the functionalized nanoparticles. The results indicated that these nanoparticles were bactericidal (effective against bacteria) while not causing significant harm to human keratinocytes (skin cells).

TiO_2 nanoparticles exhibit antibacterial properties primarily due to their photocatalytic activity, which involves the generation of ROS when exposed to light, especially ultraviolet (UV) light. These ROS are highly reactive molecules that can damage bacterial cells in various ways, leading to cell death. The generated ROS assist the antibacterial activities as almost similar to silver ions. Here are some references related to photoactivated antibacterial TiO_2 nanoparticles:

The titania photocatalyst loaded with platinum (TiO_2/Pt) exhibits remarkable antimicrobial efficacy against soil-borne pathogenic microorganisms. Tseng et al. reported the antibacterial potential of TiO_2-Pt against nosocomially infected pathogens like *S. aureus*, *A. baumannii*, and exotoxin-producing *S. pyogenes*. Under visible light irradiation, these nanoparticles displayed significantly enhanced bactericidal capabilities against human pathogens, leading to a noteworthy 2-log reduction. This outperformed various other available photocatalysts such as ST01, P-25, C200, and BA-PW25, which achieved only a modest 0–1 log reduction under visible light exposure (Ammendolia and De Berardis 2022). Anehosur et al. found that incorporating 3 wt% of TiO_2 nanoparticles into a polymethyl methacrylate (PMMA) denture-based resin imparts antimicrobial properties against *S. aureus* when exposed to visible light. This combination of TiO_2 with PMMA leads to significantly enhanced denture hygiene compared to relying solely on solar energy or light (Anehosur et al. 2012).

The optimal visible light-induced inactivation of *E. coli* bacteria using carbon nanotube (CNT)-doped TiO_2 thin films was observed at a CNT content of 20 wt%. The bactericidal effect under visible light increased sequentially from MWNTs to TiO_2, to TiO_2/MWNTs annealed at 650 °C, where bacteria even exhibited slight growth on the MWNTs. Furthermore, TiO_2/MWNTs annealed at 400 °C displayed significantly enhanced antibacterial activity compared to the 100 °C annealing due to the formation of carbonaceous bonds at the heterojunction arrays at 400 °C, which were absent at 100 °C (Akhavan et al. 2010).

The core-shell-structured In_2O_3@$CaIn_2O_4$ substrate demonstrates enhanced visible-light-induced bactericidal capabilities compared to various other visible light-responsive photocatalysts, although its performance falls short of TiO_2. This heightened performance is attributed to more efficient photoexcited electron transfer between the interfaces of In_2O_3 and $CaIn_2O_4$, resulting in reduced electron-hole recombination during photocatalysis. Despite this, when compared to TiO_2-based photocatalysts, treatments based on In_2O_3@$CaIn_2O_4$ exhibit significantly lower levels of cell death and tissue damage, thus showcasing superior biocompatibility. This higher biocompatibility is demonstrated by the minimal fluctuations in platelet and white blood cell (WBC) counts observed during the treatment of mice with In_2O_3@$CaIn_2O_4$ (Chang et al. 2012).

TiO_2 photocatalysts have bactericidal effects on both antibiotic-resistant and antibiotic-sensitive bacteria when exposed to ultraviolet A (UV-A) radiation. MSSA and MRSA were similarly susceptible, and MDRAB's susceptibility was twice that of MDSAB ($P = 0.05$) in tests with *S. aureus* (MRSA/MSSA), *A. baumannii* (MDRAB/MDSAB), *E. faecalis* (VRE/VSE), *E. coli*, and the bacteriophage MS2. VSE was 2.4 times more susceptible than VRE ($P = 0.05$). TiO_2 successfully lowers the proportion of antibiotic-resistant bacteria in a suspension of CFU by 1–3 logs in the presence of UV-A. Under low-intensity UV light and in the dark, the effectiveness of the hybrid microspheres (PMS and TiO_2-HMS) against *E. coli* was studied. *E. coli* is killed by PMS to a greater or lesser extent in the dark and when exposed to UV light, ranging from 5.7% to 10.5%, respectively. Surprisingly, the hybrid nanoparticle destroyed more than 95% of the bacteria under UV light after only 2 h, and it also killed more than 50% of them in complete darkness (Jiang and Zeng 2010). TiO2/poly[2-(tert-butylamino)ethyl methacrylate/ethylene glycol dimethacrylate core-shell nanoparticles were synthesized for the surface-initiated photopolymerization utilizing titania as an initiator. Due to the combined antibacterial properties of a biocidal polymer shell and a light-driven anti-infective TiO_2 core, this nanocomposite's photocatalytic antibacterial activity is reported to be higher than that of pristine TiO_2 nanoparticles. These nanoparticles demonstrated great antibacterial efficacy (B95%) against Gram-positive *S. aureus* even in the dark. In comparison to TiO_2 nanoparticles, the TiO_2/biocidal polymer demonstrated better inhibition of bacterial growth against Gram-negative *E. coli* and Gram-positive *S. aureus* under UV irradiation (Kong et al. 2010). Liu and Yang observed that subjecting Gram-positive *Listeria* and Gram-negative *E. coli* to photoactivated ZnO nanoparticles, along with a 10-min incubation period, an illumination dose of 173 kJ/m^2, and a concentration of 1×10^{23} M, resulted in a reduction of bacterial counts. Specifically, the treatment led to a decrease of 3 log10 for *Listeria* and 4.7 log10 for *E. coli*. Notably, it was found that Listeria exhibited more resistance when exposed to light compared to when it was kept in the dark. Furthermore, the researchers noted that Gram-positive *B. subtilis* displayed a significantly higher susceptibility to the effects of ZnO nanoparticles when exposed to visible light, as compared to *E. coli* (Liu and Yang 2003).

14.12 Photoactivated Antibacterial Quantum Dots Nanoparticles

Quantum dots (QDs) are tiny nanocrystals composed of various elements like silicon, cadmium selenide, and graphene (Abdellatif AAH et al. 2022). They possess distinctive optical and emission characteristics that vary according to their size, typically ranging from 1 to 10 nm. When exposed to light, QDs of larger sizes emit light in the lower-energy red part of the spectrum, whereas smaller QDs emit light in the blue spectrum (Mohamed et al. 1926; Gidwani et al. 2021; Abdellatif et al. 1951; Murray et al. 1993; Efros and Brus 2021; Ekimov et al. 1993).

Additionally, QDs can serve as energy donors, transferring energy to triplet oxygen. This energy transfer results in the generation of ROS. However, it's important to note that the yield of singlet oxygen ($^1O_2$) produced in this process is relatively low.

Sdots, a novel type of nonmetallic quantum dots from the nanosulfur family, feature a polymeric sulfur core that was developed by Shivalkar et al. These Sdots display impressive antibacterial properties by generating ROS when exposed to sunlight or visible light. In just 6 h of sunlight exposure, they effectively inhibited over 90% of bacterial growth. Additionally, low-toxicity Sdots were used to create antibacterial fabric for combating bacterial infections efficiently. Notably, after 12 h of incubation in the presence of visible light, the treated fabric exhibited a substantial zone of inhibition, measuring up to 9 mm, against *E. coli* (Shivalkar et al. 2022).

Budimir et al. prepared hydrophobic carbon quantum dots/polyurethane nanocomposite (hCQD-PU) wherein the PU polymer matrix was combined with hydrophobic quantum dots (hCQDs), which can produce ROS when exposed to low-power blue light (470 nm). The hCQD-PU nanocomposite with a dose of 200 kGy radiation showed the best antibacterial action, eliminating Gram-positive *S. aureus* and Gram-negative *E. coli* bacteria after 15 min of exposure to the blue lamp (Budimir et al. 2021).

EDA-CQDs were produced by surface functionalizing the nanoparticles with 2,2′-(ethylenedioxy) bis(ethylamine) (EDA) in an amidation reaction. The results of this study reveal that carbon dots were very successful at killing bacteria when illuminated with visible light. The carbon dots were tested for their photoinduced bactericidal capabilities. The inhibitory effect could be seen even with basic ambient room lighting (Meziani et al. 2016).

Dong et al. assessed the antimicrobial effects of CDots (carbon dots) paired with MB or TB PSa at various concentrations, both with and without 1-h visible light exposure, against *E. coli* cells. They confirmed synergistic interaction between 5 g/mL CDots and 1 g/mL MB, resulting in complete inhibition of bacterial growth and a 6.2 log reduction in cell numbers. CDots alone at 5 g/mL had no antimicrobial effect, and 1 g/mL MB alone reduced cell numbers by only 1.86 log. Similar synergistic effects were observed with CDots/TB, validated using the checkerboard method and isobologram analysis. Increased intracellular ROS production was linked to these synergistic interactions (Dong et al. 2018).

Dong and group also created photoactive C-dots, which were then functionalized on the surface with 2,2′-(ethylenedioxy)bis(ethylamine) (EDA-CDots) to prevent the growth of *B. subtilis* biofilms and to render *B. subtilis* cells already present in biofilms inactive (Dong et al. 2021a). In order to correlate the results from the two studies on the inactivation of bacterial pathogens, Lagos et al. combined spectroscopic probing of these processes and attributes with an assessment of the photoinduced bactericidal action of CDots (Lagos et al. 2023). Awak et al. examined EDA-CDots, whose fluorescence quantum yields (FF) ranged from 7.5% to 27%, for their photoactivated antibacterial activity. The outcomes showed a correlation between the observed FF values and the efficiency of the photo-activated bactericidal action of CDots. The antimicrobial effects of EDA-CDots on the model bacterial species *E. coli* and *B. subtilis*, which belong to the Gram-positive and Gram-negative groups, respectively, were further assessed under settings with various Cdot concentrations and treatment periods. The potential for combining the chosen FF, concentration, and treatment duration to enhance the bactericidal efficacy of EDA-CDots was investigated, and the results' mechanistic ramifications were examined (Al Awak et al. 2017). When exposed to EDA-CDots at a concentration of 15.8 mg/mL for 3 h under laboratory light, a remarkable reduction of approximately 4.7 logs in viable B. subtilis cells was achieved, with an FF of 27%. Similarly, when using EDA-CDots with an FF of 17% at the same concentration

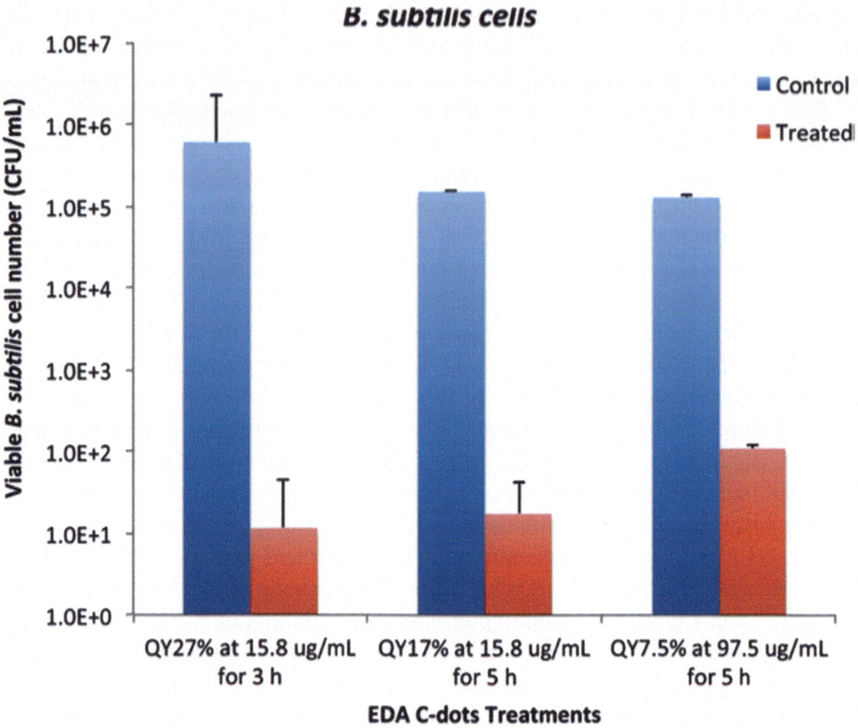

Fig. 14.7 Adjustable bactericidal effect of EDA-CDots by a combination of CDots' FF, concentration, and treatment time to achieve 3–5 log reduction in viable cell number of *B. subtilis* cells. (Image from Al Awak et al. 2017)

(15.8 mg/mL), but with a longer treatment duration of 5 h, a substantial reduction of 3.9 logs in viable cell numbers was observed. Furthermore, a 5-h treatment with EDA-CDots having the lowest FF (7.5%) at a higher concentration of 97.5 mg/mL also yielded a noteworthy viable cell reduction of 3.1 logs in *B. subtilis* cells (Fig. 14.7).

Cdots functionalized with oligomeric polyethylenimine were found to be highly responsive to visible light and demonstrated effective inactivation of MDR bacterial strains. These findings highlight CDots as a promising new class of agents for combating MDR bacteria, particularly noteworthy for their efficacy against highly ranked MDR strains such as *E. faecium* and *E. faecalis* (Abu Rabe et al. 2019). This work showed how a newly discovered class of CDots could be used to photodynamically inactivate foodborne bacteria. The outcomes showed that *L. monocytogenes* in suspensions and on stainless steel surfaces could be effectively photoinactivated using CDots. For Salmonella cells, it was far less effective; even so, Salmonella cells could still be inactivated by treatments with higher CDot concentrations and longer dwell durations (Dong et al. 2021b).

Sah et al. conducted an experiment utilizing single-walled CNTs that were linked with monoamino porphyrin to treat infections caused by *S. aureus*. After subjecting these nanotubes to 60 min of visible light exposure, the researchers observed a complete elimination of bacterial growth (Anju et al. 2019). In another related study, researchers employed multi-walled CNTs that were conjugated with TO. This approach effectively eradicated both planktonic cells and biofilms of *P. aeruginosa* and *S. aureus* (Sah et al. 2018). In both of these studies, the primary mechanism responsible for bacterial cell death was identified as damage to the cellular membrane.

The primary research focus has been directed towards designing nanoconjugates of quantum dots (QDs) and photosensitizers that can facilitate improved energy transfer, thereby augmenting the production of singlet oxygen ($^1O_2$). One study explored the use of graphene QDs conjugated with AgNPs, which exhibited enhanced antimicrobial activity against both *E. coli* and *S. aureus* when subjected to 450 nm irradiation. This effect was achieved through a synergistic combination of photodynamic and photothermal mechanisms (Viana et al. 2015). Conversely, in a different study, the conjugation of cadmium telluride QDs with zinc porphyrins hindered their uptake by cells. Consequently, no significant antimicrobial photodynamic effect was observed when targeting *C. albicans*. It is evident that further research is warranted to assess the efficacy of QD-photosensitizer nanoconjugates activated by either UV or NIR light for aPDT (Anju et al. 2019). PS molecules capture energy from activated QD surfaces, leading to the generation of ROS. To date, there is only one reported study in which X-ray-activated graphene QDs were tested against *E. coli* (Yu et al. 2020).

Graphene quantum dots (QDs) were coated with lipopolysaccharide, a significant component of the outer membrane of *E. coli*, and Protein A, a surface protein present in the cell wall of MRSA (Viana et al. 2015). This coating was aimed at enhancing the adhesion of the QDs to bacterial cells. Upon exposure to 800 nm light for just 15 s, both *E. coli* and MRSA cells were eliminated. Remarkably, the bacterial surfaces began to deteriorate after only 10 s of irradiation due to the generation of ROS. To visualize and study these changes in bacterial surfaces at different depths (ranging from 25 to 75 μm), researchers employed two-photon excitation microscopy, which is a three-dimensional fluorescence imaging technique.

Carbon-based nanoparticles (NPs) have become increasingly popular for PDT because of their excellent stability and capacity to generate ROS. Fullerenes such as C60, C70, and C84 have shown remarkable effectiveness as photosensitizers in combating bacterial infections (Tegos GP et al 2005). This efficacy is achieved through the functionalization of fullerenes with organic ligands, which enhances their water solubility and makes them suitable for PDT applications (Pan et al. 2017; Kuo et al. 2016; Sharma et al. 2011). Grinholc et al. demonstrated that C60 fullerene nanoparticles (Fig. 14.8), which were functionalized with a methylpyrrolidinium group, exhibited a substantial in vitro antibacterial effect against *E. coli*. In addition, these nanoparticles achieved a remarkable 2-log reduction in *S. aureus* levels in mouse infection models (Grinholc et al. 2015).

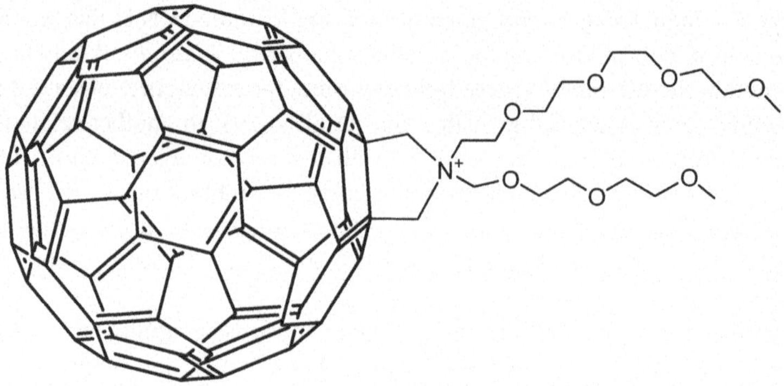

Fig. 14.8 Structure of C60-Fulleropyrrolidine derivative for C60 fullerene nanoparticles

The sharp corners and protruding edges of graphene oxide facilitate easy penetration of the bacterial membrane, resulting in membrane damage and the release of ribonucleic acids (Zhang et al. 2020). Recent research has shown that graphene oxide and nanographene oxide nanosheets can stimulate the generation of ROS and elevate local temperatures to 60 °C (Romero et al. 2020). This dual effect, which combines photothermal and photodynamic mechanisms, resulted in the complete elimination of both *S. aureus* and *E. coli* bacteria in vitro. Furthermore, the photodynamic properties of graphene oxide have been further improved by incorporating AgNPs onto its surface. These AgNPs themselves possess antibacterial properties, enhancing the overall antibacterial effectiveness (Xie et al. 2017). When subjected to 660 nm irradiation, the graphene oxide/AgNP/collagen nanohybrid displayed notable inhibition of *E. coli* (96.3%) and *S. aureus* (99.4%). Importantly, when used in vivo model, it did not lead to tissue necrosis. Additionally, black phosphorous nanosheets can absorb visible red and NIR wavelengths, resulting in the production of ROS (Luo et al. 2019). A nanofilm comprising black phosphorous nanosheets, modified with poly(4-pyridone-methyl styrene) endoperoxide (PPMS-EPO), can store ROS generated by black phosphorous. This nanofilm effectively deactivated both *E. coli* and *S. aureus* when exposed to 660 nm light irradiation (Tan et al. 2018). The antibacterial effectiveness of black phosphorous nanosheets was significantly improved by incorporating AgNPs/BP. When subjected to 808 nm NIR activation, a synergistic combination of photodynamic and photothermal effects led to the swift rupture of the membrane of MRSA (Fig. 14.9) (Ouyang et al. 2018).

14.13 Antibacterial Porphyrin Nanoparticles

Antibacterial porphyrin nanoparticles are nanoscale particles that contain porphyrin molecules with inherent antibacterial properties. Porphyrins are organic compounds known for their ability to generate ROS upon exposure to light, particularly in the presence of suitable catalysts like metal nanoparticles or complexes. Porphyrin-functionalized nanoparticles have gained significant attention in the field of

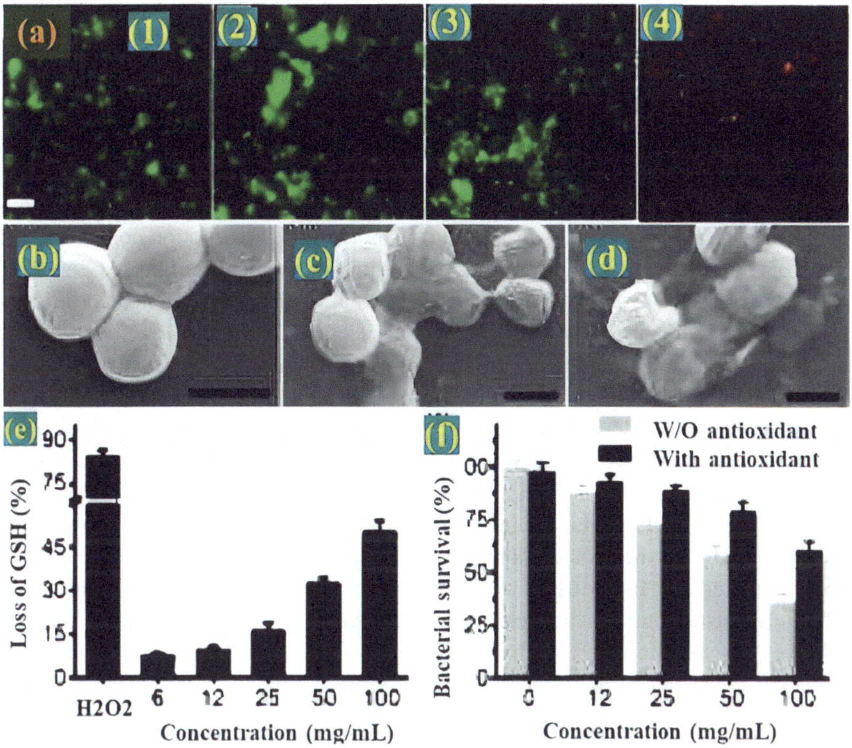

Fig. 14.9 (**a**) Bacterial membrane integrity after different treatments were examined through calcein-AM/PI staining: (1) blank control, (2) NIR, (3) Ag@BP nanohybrids, (4) Ag@BP nanohybrids + NIR (scale bar = 20 μm). SEM images of MRSA after different treatments: (**b**) blank control, (**c**) BP nanosheets + NIR, (**d**) Ag@BP nanohybrids + NIR. Scale bars equal 500 nm. The power density of NIR light was 0.8 W/cm^2, and the irradiation time was 5 min. (**e**) GSH depletion by Ag@BP nanohybrids at various concentrations. H_2O_2 (1 mM) was used as the positive control. (**f**) Antibacterial activity of Ag@BP nanohybrids in the dark. NAcetyl cysteine as an antioxidant was added to neutralize the oxidative stress induced by Ag@BP nanohybrids. (Image is adapted from Ouyang et al. 2018)

aPDT. Porphyrin molecules can be prone to degradation or aggregation, which can affect their photodynamic activity. By encapsulating or conjugating porphyrins onto nanoparticles, their stability can be improved, leading to prolonged activity and enhanced efficiency. Porphyrins can be loaded onto or encapsulated within nanoparticles, allowing for a higher payload of the photosensitizer. This increased loading capacity improves the availability of porphyrins for ROS generation, thereby enhancing the antibacterial efficacy of the therapy. Nanoparticles like Ag, Au, or other metal oxides are often used to decorate or encapsulate the porphyrin molecules, amplifying their antibacterial activity.

Nanoparticles exhibit synergistic interactions with pharmacodynamic and pharmacokinetic properties when combined with light, greatly enhancing their antibacterial effectiveness. While the photoactive behavior of inorganic compounds is well-documented, organic molecules also display photoactive characteristics. For instance, molecules such as porphyrins and phthalocyanines serve as PSs in PDT and generate ROS, notably singlet oxygen ($^1O_2$), when exposed to light energy in the presence of triplet oxygen. This phenomenon is attributed to their extended conjugated electronic structures, substantial molar absorption extinction constants, and favorable fluorescent quantum yields, rendering them ideal candidates for PDT (Yang et al. 2013; Yavuz et al. 2009; Carvalho et al. 2009). The augmented efficacy of PDT using nanoparticles stems from a dual mechanism: improved delivery of the photosensitizer to the target microorganisms and enhanced photosensitizer activity.

Ghiladi and their research group have successfully engineered rod-shaped nanocrystals composed of covalently linked porphyrin and cellulose, with an average length ranging from 100 to 400 nm. This innovative development initially demonstrated promising results by enhancing the lifetime and antibiotic activities of the covalently linked materials. Paradoxically, it also had the dual effect of reducing both the release of the biocidal agent into the environment and the aggregation of the nanomaterials. In terms of photocatalytic efficacy, these nanocrystals have exhibited outstanding performance in inactivating *M. smegmatis* and *S. aureus*. However, it's worth noting that their effectiveness against *E. coli* appears to be comparatively lower in contrast to the other tested microorganisms (Feese et al. 2011).

Mondal and their research team successfully synthesized nanocomposite films (Banerjee I, et al. 2012) using a porphyrin-CNT conjugate known as MWNT-PPIX. These films were designed for immobilizing antimicrobial agents and demonstrated remarkable efficiency, particularly against Gram-positive *S. aureus*. The nanocomposite films were fabricated from acid-functionalized MWNTs and protoporphyrin IX (PPIX), serving as the active antimicrobial agent. What's noteworthy is that this synthesis process was accomplished in a short period and with minimal quantities of materials. They also extended the application of the MWNT-PPIX conjugate by exploring its effectiveness as a photosensitizer for antiviral activity against Influenza A virus-infected mammalian cells. The results were highly promising, as MWNT-PPIX exhibited excellent efficacy in combating influenza viruses. Moreover, there was a limited risk of the virus developing resistance to this treatment, making it a potentially valuable approach for antiviral therapy (Banerjee et al. 2010).

In the fight against microbial contamination of aqueous solutions, Carvalho and coworkers devised a cationic nanomagnet-porphyrin hybrid, which proved to be an effective and stable nanomaterial. When exposed to white light at specific energy levels, this hybrid nanomaterial achieved complete photo-inactivation of various microorganisms (Fig. 14.10). Specifically, it caused total inactivation of *E. coli*, T4-like phage, and *E. faecalis* when subjected to white light with energy doses of 21.6, 43.2, and 14.4 J/cm^2, respectively, at a concentration of 20 μM. These findings suggest that these hybrid materials have significant potential for applications as novel photosensitizers in the disinfection of water or wastewater, offering a

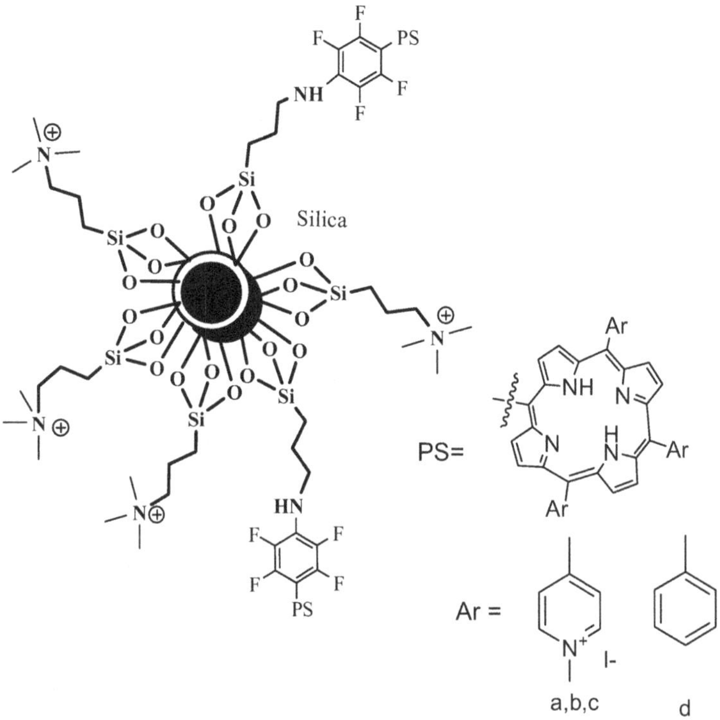

Fig. 14.10 Synthesis of the nanomagnet-porphyrin hybrids a–d

promising solution for combating microbial contamination in aqueous environments (Carvalho et al. 2010).

Nykong and colleagues conducted research on the potential of phthalocyanine (PC) as a candidate for PDT due to its efficient absorption coefficient ($\mu = 10^5$/M/cm) within the 600–800 nm range. In their study, they synthesized Zn-phthalocyanine (ZnPC)-ε-polylysine (ε-PL) conjugates to act as photodynamic sensitizers for the inactivation of *S. aureus* through a defense mechanism. The minimum inhibitory concentration (MIC) of these conjugates against *S. aureus* revealed that the presence of Ag NPs led to a reduction in the expression of the microorganism for both porphyrin conjugates (a) and (b). Additionally, their research indicated that the AgNP-conjugated (b) was a more effective photosensitizer compared to (a) in the context of PDT against *S. aureus* when exposed to light at an intensity of 39.6 mW/cm^2 (Nombona et al. 2012).

In the field of photodynamic antimicrobial chemotherapy, researchers have explored the use of various nanoparticles for drug delivery and photosensitization. Kuruppuarachchi et al. developed catpolyacrylamide nanoparticles with a size of 95–610 nm to serve as delivery materials in PDT for cancer and photodynamic antimicrobial chemotherapy. These nanoparticles were designed to deliver polylysine-bound tetrasulfonato-aluminum phthalocyanine-entrapped

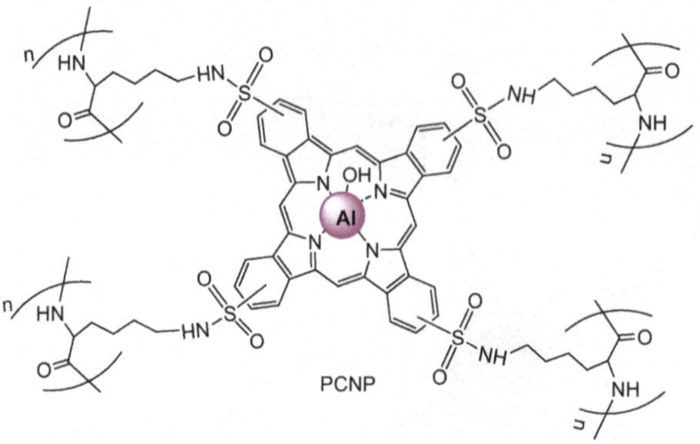

PCNP

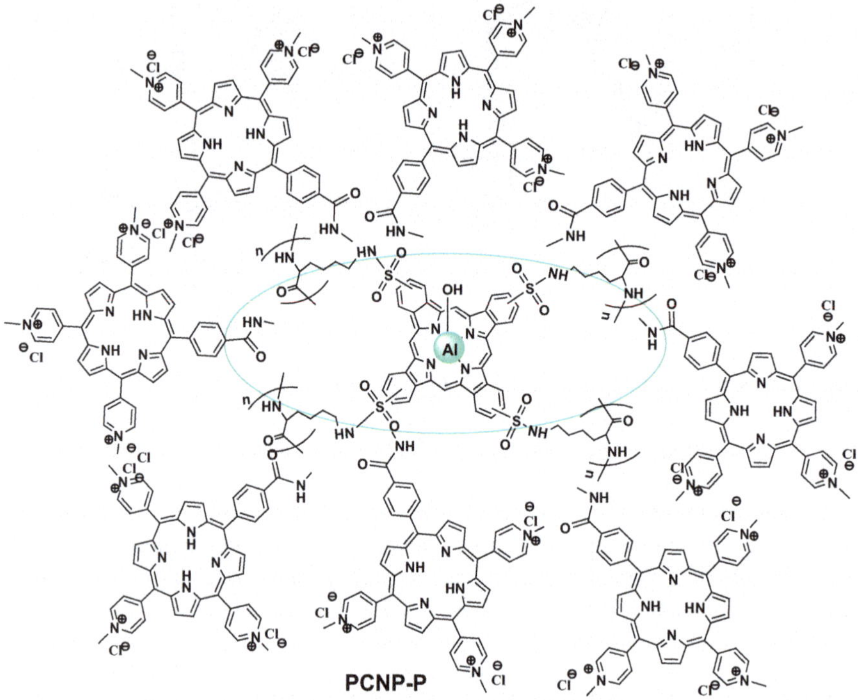

PCNP-P

Fig. 14.11 Confocal images of PCNP internalized in HT29 cells (after incubating for 20 h): (**a**) HT29 cells only, (**b**) PC emission, and (**c**) merged image and for PCNP-P (red, excited at 633 nm; green, excited at 405 nm): (**d**) PC emission, (**e**) porphyrin emission, (**f**) HT29 cells only, and (**g**) merged image. (Image is adapted from Kuruppuarachchi et al. 2011) nanoparticles (PCNP) and polylysine-bound tetrasulfonato-aluminum phthalocyanine-entrapped nanoparticles coated with a second, porphyrin-based, photosensitizer (PCNP-P). These conjugates exhibited absorbance signals at 424 nm (Soret band) and 680 nm (Q-band), corresponding to the porphyrin and PC components, respectively (Kuruppuarachchi et al. 2011). Flow cytometry and confocal microscopy were employed to assess the internalization of nanoparticles by HT29 human Caucasian colon adenocarcinoma cells. Cell viability assays involving PCNP-P and PCNP (Fig. 14.11) were conducted to investigate the impact of light activation on these photodynamic nanoparticles. The results indicated that these cancer cells could be harmed by light activation, both when the nanoparticles were in the external medium and after they had been internalized. This effect was observed for uptake times as short as less than 5 min and as long as 25 h (Fig. 14.12).

In 2013, Creanga et al. introduced ovoid, disk-shaped, and spherical Ag-porphyrin colloidal nanoparticles with sizes ranging from 65 to 120 nm and 30 to 49 nm. These nanoparticles were designed to function as antimicrobial agents and

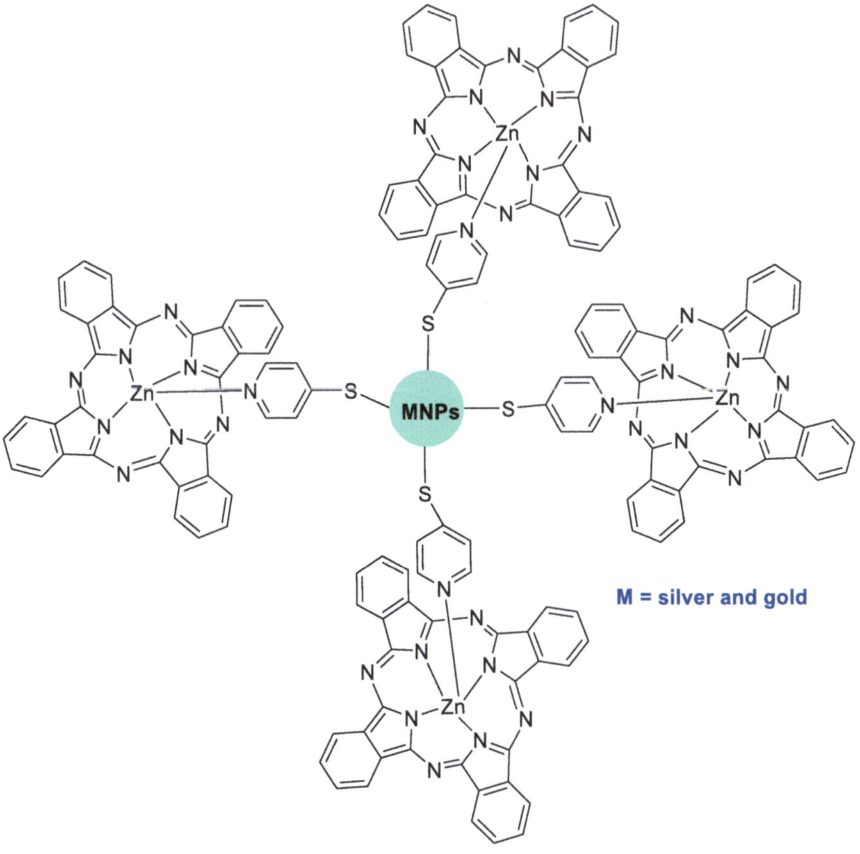

Fig. 14.12 Zn-phthalocyanine by conjugating them with Ag and Au NPs

photosensitizers in PDT. Additionally, they explored their potential application in photovoltaic cells (Creanga et al. 2013).

Ferro et al. have developed a novel approach using cationic amphiphilic cyclodextrins as nanoaggregates in PDT and as antimicrobial agents targeting MRSA and *E. coli*. Their research findings indicate that when the porphyrin unit is conjugated with these nanoaggregates, it leads to an enhancement in the photobactericidal activity. This suggests that the combination of cationic amphiphilic cyclodextrins with porphyrins can be a promising strategy to improve the efficacy of PDT and combat drug-resistant bacteria like MRSA and *E. coli* (Ferro et al. 2009).

Masilela et al. revealed significant advancements in improving the photophysicochemical and antimicrobial properties of bis-(1,6-hexanedithiol) silicon phthalocyanine and Zn-phthalocyanine by conjugating them with Ag NPs and Au NPs (Fig. 14.12). Notably, the Si-phthalocyanine derivative exhibited superior bioactivity compared to the zinc phthalocyanine counterpart when exposed to light against various pathogens. This suggests that the silicon-phthalocyanine complex is more effective in terms of antimicrobial action under light conditions. Additionally, the zinc phthalocyanine conjugate demonstrated the highest singlet and triplet oxygen quantum yield when combined with Au NPs. This finding highlights the potential of these conjugates to be used as PSs in PDT, where the generation of singlet and triplet oxygen species is essential for the photodynamic inactivation of pathogens (Masilela et al. 2013a).

The same research group has made recent advancements by investigating metal phthalocyanines (MPCs) that incorporate core metal atoms such as Sn, Cu, Zn, and Ti within the PC ring structure (Fig. 14.13). They synthesized monocysteinyl

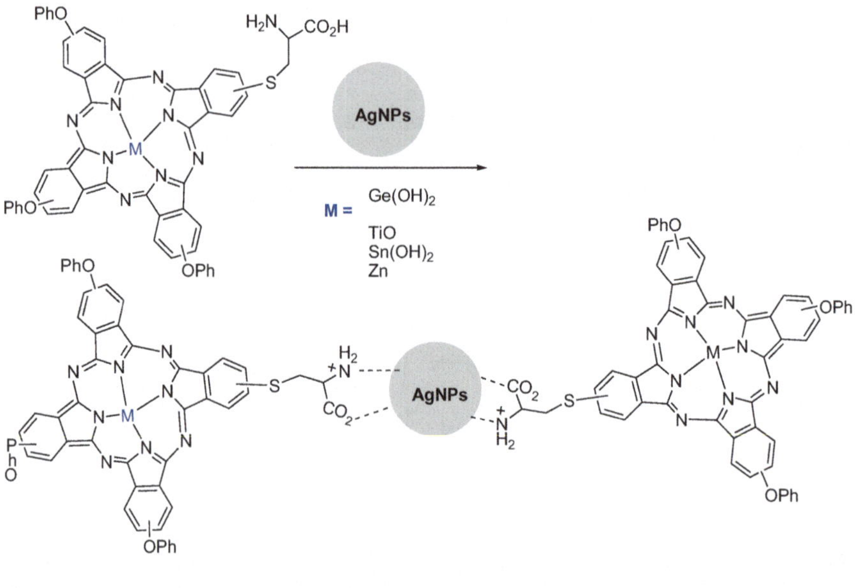

Fig. 14.13 Preparation of metal phthalocyanine nanomagnets

phthalocyanines (MCsPc) with variations including $(OH)_2$GeMCsPc, SnMCsPc, ZnMCsPc, and OTiMCPc (monocarboxy phthalocyanine). These MCsPc compounds were designed with both carboxylic acid and amino groups to facilitate coordination with Ag and Au NPs. In their research, it was observed that the antibacterial activity of $(OH)_2$GeMCsPC was at its maximum in the presence and absence of Ag NPs. Interestingly, the antibacterial properties of Ag NPs were observed in increasing order, irrespective of the presence of MPCs, with cubic, triangular, and spherical Ag NPs all showing antibacterial effects. Furthermore, the antibacterial activities of the various MPCs, including SnMCsPC, OTiMCsPC, ZnMCsPC, and $(OH)_2$GeMCsPC, were demonstrated in both the presence and absence of AgNPs (Masilela et al. 2013b).

14.14 Miscellaneous Photoactivated Antibacterial Nanoparticle

Chong et al. have reported the development of water-soluble red-emissive-conjugated polymer nanoparticles with a size of approximately 100 nm. These nanoparticles are composed of cationic fluorinated hydroxytelechelic polybutadiene (PBF) with fluorene and boron-dipyrromethene repeat units incorporated into their backbones. These nanoparticles were created in the presence of the negatively charged disodium salt of 3,3′-dithiodipropionic acid. One of the remarkable properties of these PBF nanoparticles is their ability to sensitize oxygen molecules, leading to the convenient formation of ROS when exposed to white light in the range of 400–800 nm for photoexcitation. This ROS generation capability makes these nanoparticles effective in rapidly killing nearby bacteria and cancer cells. The study demonstrated that when exposed to light at light intensities of 45 mW/cm^2 (equivalent to 10^8 J/cm^2) and 90 mW/cm^2 (equivalent to 216 J/cm^2), 60% cell death and a 90% reduction in bacterial count were achieved, respectively (Chong et al. 2012).

In a separate investigation, researchers led by Chen synthesized nanoparticles composed of zirconium (Zr) porphyrinoidmetal–organic materials. In this study, they introduced a partial substitution of zirconium clusters with Ti to enhance the ROS. Upon exposure to visible light, these nanoparticles exhibited a substantial reduction in the growth of various drug-resistant bacterial strains, including MDR strains of *E. coli* and *A. baumannii*, as well as methicillin-resistant strains of *S. aureus* and *S. epidermidis*, along with their respective nonresistant counterparts. Importantly, when subjected to biocompatibility testing in rats, no indications of inflammation were observed, and human umbilical vein endothelial cells maintained their viability (Chen et al. 2020).

5-aminolevulenic acid (ALA) formed a porphyrin-PDT (ALA-PDT) conjugated used for the antibiotic therapy of infected skin abscesses resulting in a complete cure in *M. fortuitum* treatment in vivo (Gong et al. 2016). In contrast, nanomicelles containing cobalt (II) porphyrin derivatives (Fig. 14.14) only produced a 67% cell reduction. An anti-TB prodrug is bound by a self-assembled nanoparticle with a

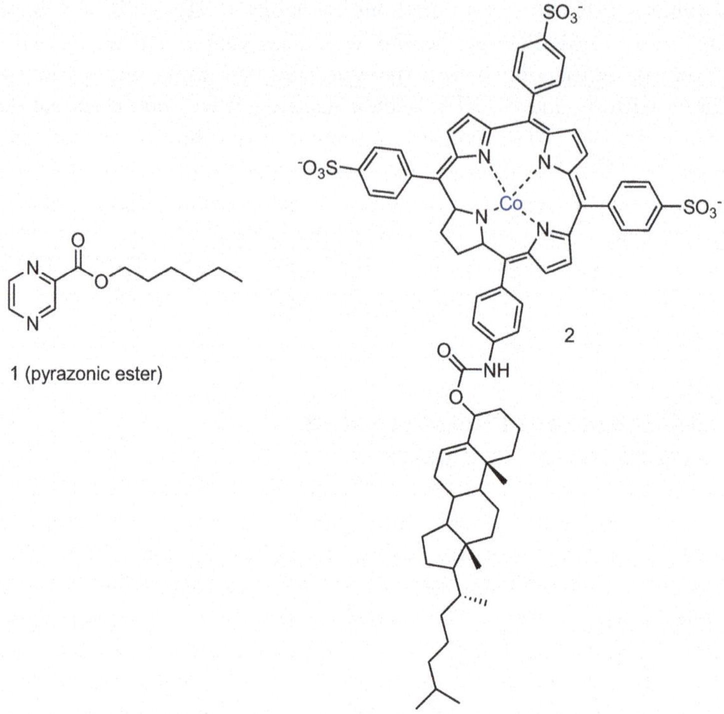

Fig. 14.14 Cobalt (II) porphyrin derivatives with Trojan horse moiety (cholesterol)

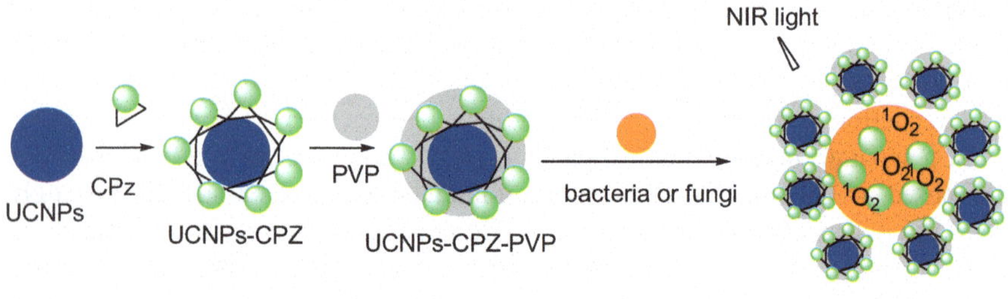

Fig. 14.15 $LiYF_4$:Yb/Er upconversion nanoparticles (UCNPs) for antimicrobial activity

photosensitizer and a Trojan horse moiety (cholesterol), which multiplies its action against mycobacteria by a factor of 1000 (Bhatti et al. 2014).

PDT using fullerene NPs also reduced the growth of *E. coli*, *P. aeruginosa,* and *A. baumannii* by over 90%, in comparison to a 50% reduction with water-soluble phthalocyanine (Shleeva et al. 2021). Zhang et al. employed $LiYF_4$:Yb/Er upconversion nanoparticles (UCNPs) along with the photosensitizer zinc β-carboxyphthalocyanine, which was coated with polyvinylpyrrolidone to prevent nanoparticle agglomeration. This approach exhibited potent antimicrobial activity against bacteria, including MRSA, MDR *E. coli*, and *C. albicans*. In vivo PDT using a murine model infected with multidrug-resistant MRSA demonstrated the NPs' effectiveness in promoting rapid healing of deep excisional wounds (Fig. 14.15) (Zhang et al. 2018).

The UCNPs-CPZ-PVP system, activated by near-infrared (NIR) light, effectively prevented CPZ aggregation and exhibited potent anti-infectious activity against drug-resistant bacteria, reducing methicillin-resistant Staphylococcus aureus levels by 4.7 log10 and MDR *E. coli* by 2.1 log10 following aPDT treatment (utilizing 50 μg/mL UCNPs-CPZ-PVP and 0.5 W/cm^2 980 nm light). Additionally, UCNPs-CPZ-PVP demonstrated strong antifungal properties against *Candida albicans* (Fig. 14.16).

Li et al. used β-$NaYF_4$:Yb, Tm UCNPs coupled with a fluorescent-conjugated polymer as a photosensitizer to overcome UV-vis light limitations. Activation of the photosensitizer by UCNPs via fluorescence resonance energy transfer resulted in a 90% eradication of *E. coli*. These UCNPs showed low toxicity to human alveolar basal epithelial cells. Additionally, surface modification with cations enhanced penetration depth and PDT efficacy by improving adherence to microbial cell walls (Li et al. 2016). Gu and colleagues synthesized UCNPs using $NaYF_4$:Yb, Er, loaded with the photosensitizer zinc phthalocyanine and introduced surface modifications with cationic N-octyl chitosan. When these nanoparticles were irradiated at 980 nm, they demonstrated greater efficacy compared to 660 nm visible light in reducing both superficial and deep skin tissue abscess volumes in mouse models infected with MRSA (Li et al. 2017).

According to Tawfik et al.'s findings, the MB-gold nanoparticles conjugate, when photoactivated, exhibited the most significant inhibitory impact on *S. aureus* activity. The gold nanoparticles demonstrated their effectiveness as a drug delivery system, amplifying the photodynamic antibacterial potential of methylene blue. A slight and statistically insignificant reduction in Staphylococcal bacteria activity

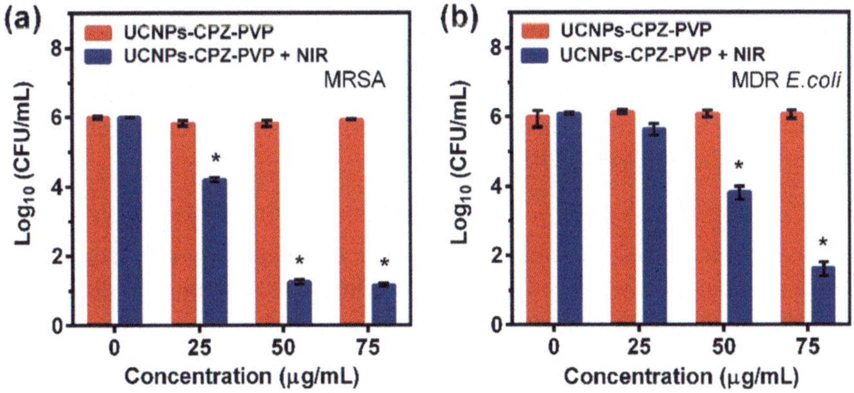

Fig. 14.16 In vitro activities of aPDT with UCNPs-CPZ-PVP nanocomposites. The antibacterial activities of the UCNPs-CPZ-PVP nanocomposites (at final concentrations of 0, 25, 50, and 75 μg/mL, respectively) against MDR bacteria MRSA (**a**) and MDR *E. coli* (**b**) upon 980 nm laser irradiation at a power density of 0.5 W/cm^2 for 10 min. (Image is adapted from Zhang et al. 2018)

was observed when subjected to laser treatment, resulting in a 25% survival rate within this group (Tawfik et al. 2015).

14.15 Mechanisms of Resistance with Photodynamic and Light-Response Nanomaterials

PDT and light-responsive nanomaterials also hold the potential for antibacterial applications. However, the emergence of bacterial resistance can pose challenges to their effectiveness. Here are some mechanisms of antibacterial resistance associated with photodynamic therapy and light-responsive nanomaterials:

Efflux Pump Systems Bacteria can develop efflux pump systems that actively pump out the photosensitizing agents or nanoparticles, reducing their intracellular concentration and limiting their antimicrobial activity (Sharma et al. 2019; Gupta et al. 2017; Varela et al. 2023; Sharma A et al. 2019).

ROS Detoxification Similar to cancer cells, bacteria can employ mechanisms to detoxify ROS generated during PDT or by light-responsive nanomaterials. They may produce enzymes, such as catalases and superoxide dismutases, which scavenge and neutralize ROS, reducing the bactericidal effect (Memar et al. 2018).

Biofilm Formation Bacterial biofilms are structured communities of microorganisms that are highly resistant to antimicrobial treatments. Biofilms can act as physical barriers, preventing the penetration of photosensitizing agents or nanomaterials and shielding bacteria from their antimicrobial effects (Høiby et al. 2010).

DNA Repair Systems Bacteria can activate DNA repair systems to counteract the DNA damage caused by PDT or ROS generated by light-responsive nanomaterials. Enhanced DNA repair mechanisms can lead to the survival of bacteria, reducing the therapeutic efficacy (Wigley 2013).

Reduced Uptake Some bacteria may exhibit decreased uptake or internalization of PSs or nanomaterials, which lowers their intracellular concentration and diminishes their antimicrobial activity (Ran et al. 2021).

Adaptive Responses Bacteria can activate various adaptive responses, such as the upregulation of stress response genes, which promote their survival under stressful conditions, including exposure to PDT or light-responsive nanomaterials (Foster 2005).

Genetic Mutations Genetic mutations can occur in bacteria, leading to alterations in the target sites of photosensitizers or nanomaterials. These mutations may render the bacteria less susceptible to the antimicrobial effects of the treatment (Ghorbani et al. 2018).

Heterogeneity Bacterial populations can exhibit heterogeneity, with subsets of bacteria displaying varying levels of resistance. This heterogeneity can contribute to treatment failure, as the resistant subpopulations may survive and repopulate the bacterial community (Liselot et al. 2019).

14.16 Conclusions

The development of new PSs and nanomaterials with improved antibacterial properties, combination therapies to target multiple bacterial pathways, and the use of light-responsive materials that can circumvent resistance mechanisms are a few of the strategies being investigated to address antibacterial resistance. Light-responsive drug delivery using nanotechnology can enhance drug effectiveness, especially for drugs needing precise release timing near tissues or for highly hydrolysable or toxic drugs. This approach offers advantages like controlled release and combating antibiotic resistance. Light-triggered nanocarriers improve antibiotic shelf life, site-specific activity, and bioavailability while reducing toxicity. These smart carriers can be applied to various drugs and may revolutionize treatments for conditions like cancer, diabetes, and metabolic diseases. It is significant to mention that the field of light-responsive nanomaterials and antibacterial photodynamic therapy is still developing. Ongoing research aims to comprehend and circumvent resistance mechanisms to improve the effectiveness of these treatments against bacterial infections.

Acknowledgments SB acknowledged the Central University of Gujarat for infrastructure facilities.

References

Abdellatif AAH, Younis MA, Alsharidah M, Rugaie OA, Tawfeek HM (1951) Int J Nanomedicine 2022:17

Abraham E, Chain E (1940) Nature 146:837

Abu Rabe DI, Al Awak MM, Yang F, Okonjo PA, Dong X, Teisl LR, Wang P, Tang Y, Pan N, Sun YP, Yang L (2019) Int J Nanomedicine 14:2655

Abu Rabe DI, Mohammed OO, Dong XL, Patel AK, Overton CM, Tang YA, Kathariou S, Sun YP, Yang LJ (2020) Mater Adv 1:321

Afzal O, Altamimi ASA, Nadeem MS, Alzarea SI, Almalki WH, Tariq A, Mubeen B, Murtaza BN, Iftikhar S, Riaz N, Kazmi I (2022) Nanomaterials (Basel) 12(24):4494

Agarwala M, Choudhury B, Yadav RNS (2009) Indian J Microbiol 54:365

Akhavan J (2009) Colloid Interface Sci 336:117

Akhavan O, Azimirad R, Safa S, Larijani MM (2010) J Mater Chem 20:7386
Al Awak MM, Wang P, Wang S, Tang Y, Sun YP, Yang L (2017) RSC Adv 7:30177
Alonso MJ (2004) Biomed Pharmacother 58:168
Aminov RI (2010) Front Microbiol 1:134
Ammendolia MG, De Berardis B (2022) Nanomaterials (Basel) 12(20):3616
Anehosur GV, Kulkarni RD, Naik MG, Nadiger RK (2012) J Gerontol Geriatric Res 1:1
Anghel A, Grumezescu M et al (2014) Molecules 19:8981
Anju VT, Paramanantham P, Sruthil Lal SB, Sharan A, Syed A, Bahkali NA, Alsaedi MH, Kaviyarasu K, Busi S (2019) Photodiagnosis Photodyn Ther 27:305
Ansari AA, Khan MN, Alhoshan M, Aldwayyan AS, Alsalhi MS (2010) Nanostructured materials, classification, properties, fabrication, characterization and their applications in biomedical sciences. In: Kestell AE, DeLorey GT (eds) Nanoparticles, properties, classification, characterization, and fabrication. Nova Science, Hauppauge
Balderrama-González AS, Piñón-Castillo HA, Ramírez-Valdespino CA, Landeros-Martínez LL, Orrantia-Borunda E, Esparza-Ponce HE (2021) Int J Mol Sci 22:12890
Banerjee I, Mondal D, Martin J, Kane RS (2010) Langmuir 26:17369
Banerjee I, Douaisi MP, Mondal D, Kane RS (2012) Nanotechnology 23:105101
Bansal A, Zhang Y (2014) Acc Chem Res 47:3052
Barad H-N, Kwon H, Alarcón-Correa M, Fischer P (2021) ACS Nano 15(4):5861–5875
Bera S, Mondal D (2018) Chapter-7. Stimuli-sensitive nanomaterials for antimicrobial drug delivery. In: Grumezescu AM (ed) Drug targeting and stimuli sensitive drug delivery systems. Elsevier, Amsterdam, pp 271–302. ISBN 9780128136904
Bera S, Mondal D (2019) J Ultrasound 22:131
Bera S, Mondal D (2022) ACS Omega 7(6):4724
Bera S, Mondal D, Palit S, Schweizer F (2016) Med Chem Commun 7(8):1499
Berg K, Weyergang A, Prasmickaite L, Bonsted A, Høgset A, Strand MT, Wagner E, Selbo PK (2010) Methods Mol Biol 635:133
Bhatti M, McHugh TD, Milanesi L, Tomas S (2014) Chem Commun 50:7649
Biener J, Wittstock A, Baumann TF, Weissmüller J, Bäumer M, Hamza AV (2009) Materials 2:2404
Blum AP, Kammeyer JK, Rush AM, Callmann CE, Hahn ME, Gianneschi NC (2015) J Am Chem Soc 137:2140–2154
Bruno BJ, Miller GD, Lim CS (2013) Ther Deliv 4:1443
Budimir M, Marković Z, Vajdak J, Jovanović S, Kubat P, Humpoliček P, Mičušik M, Danko M, Barras A, Milivojević D, Špitalsky Z, Boukherroub R, Marković BT (2021) Radiat Phys Chem 185:109499
Burda C, Chen XB, Narayanan R, El-Sayed MA (2005) Chem Rev 105:1025
Carpenter AW, Schoenfisch MH (2013) Chem Soc Rev 41:3742
Carpenter BL, Feese E, Sadeghifar H, Argyropoulos DS, Ghiladi RA (2011) Photchem Photobiol 88:5217
Carvalho CMB, Tome JPC, Faustino MAF, Neves MGPMS, Tome AC, Cavaleiro JAS et al (2009) J Porphyrins Phthalocyanines 13:574
Carvalho CMB, Alves E, Costa L, Tome JPC, Faustino MAF et al (2010) ACS Nano 4:7133
Cassidy CM, Tunney MM, McCarron PA, Donnelly RF (2009) J Photochem Photobiol B 95:71
Chain E, Florey HW, Gardner NG, Heatley NG, Jennings MA, Orr-Ewing J et al (1940) Lancet 236:226
Chamundeeswari M, Sobhana SSL, Jacob JP et al (2008) Biotechnol Appl Biochem 55:29
Chang WK, Sun DS, Chan H, Huang PT, Wu WS, Lin CH et al (2012) Nanomedicine 8:609
Chen G, Roy I, Yang C, Prasad PN (2008) Chem Rev 116:2826
Chen WY, Lin JY, Chen WJ, Luo L, Diau EW-G, Chen YC (2010) Nanomedicine 5:755
Chen M, Long Z, Dong R, Wang L, Zhang J, Li S, Zhao X, Hou X, Shao H, Jiang X (2020) Small 16:1906240
Chong H, Nie C, Zhu C, Yang Q, Liu L, Lv F et al (2012) Langmuir 28:2091
Coleman JN, Lotya M, O'Neill A, Bergin SD, King PJ, Khan U, Young K, Gaucher A, De S, Smith RJ et al (2011) Science 331:568

Creanga I, Fagadar-Cosma G, Palade A, Lascu A, Enache C, Birdeanu M et al (2013) Digest J Nanomater Bios 8:561
Cui Y, Wei Q, Park H, Lieber CM (2001) Science 293:1289
Dai T, Tegos GP, Zhiyentayev T, Mylonakis E, Hamblin MR (2010) Lasers Surg Med 42(1):38
Daneman R, Prat A (2015) Cold Spring Harb Perspect Biol 7:a020412
Daniell MD, Hill JS (1991) Aust N Z J Surg 61:340
Davies J (2010) Microbiol Mol Biol Rev 74:417
de López DC, Guerrero CM, Martínez F, Streitt C, José GM (2021) Antimicrobial effect of titanium dioxide nanoparticles. In: Antimicrobial resistance—a one health perspective. Intech Open, London. https://doi.org/10.5772/intechopen.90891
De M, Ghosh PS, Rotello VM (2008) Adv Mater 20:4225
Desai N (2012) AAPS J 14(2):282
Desai SK, Bera S, Singh M, Mondal D (2017) J Appl Polym Sci 134:7. https://doi.org/10.1002/app.44463
Desai SK, Mondal D, Bera S (2020) Carbohydr Res 495:108070
Desai SK, Mondal D, Bera S (2021) Sci Rep 11(1):8331
Dhandapani P, Maruthamuthu S, Rajagopal G (2012) J Photochem Photobiol B Biol 110:43
Dong X, Bond AE, Pan N, Coleman M, Tang Y, Sun YP, Yang L (2018) Int J Nanomedicine 13:8025
Dong XL, Ge L, Abu Rabe DI, Mohammed OO, Wang P, Tang YG, Kathariou S, Yang LJ, Sun YP (2020a) Carbon 170:137
Dong XL, Liang WX, Meziani MJ, Sun YP, Yang LJ (2020b) Theranostics 10:671
Dong X, Overton CM, Tang Y, Darby JP, Sun Y-P, Yang L (2021a) Front Bioeng Biotechnol 9:786077
Dong X, Wang P, Darby JP, Tang Y, Overton CM, Kathariou S, Sun YP, Yang L (2021b) Appl Environ Microbiol 87:e0104221
Doron S, Gorbach SL (2008) International Encyclopedia of public health, p 273
Efros AL, Brus LE (2021) ACS Nano 15:6192
Ekimov FH, Schanne-Klein MC, Ricard D, Flytzanis C, Kudryavtsev IA, Yazeva TV, Rodina AV, Efros Al L (1993) J Opt Soc Am B 10:100
Etebu E, Arikekpar I (2016) Int J Appl Microbiol Biotechnol Res 4:90–101
Fair RJ, Tor Y (2014) Perspect Med Chem 6:25
Feese E, Sadeghifar H, Gracz HS, Argyropoulos DS, Ghiladi RA (2011) Biomacromolecules 12:3528
Ferro S, Jori G, Sortino S, Stancanelli R, Nikolov P, Tognon G et al (2009) Biomacromolecules 10:2592
Feynman R (1991) Science 254:1300
Finsen NF (1901) Phototherapy. Arnold, London
Fleming A (1929) Br J Exp Pathol 10:226
Foster PL (2005) Mutat Res 569:3
Fubini B (1997) Environ Health Perspect 105:1013
Ganoth A, Merimi KC, Peer D (2015) Expert Opin Drug Deliv 12(2):223
Gao XH, Cui YY, Levenson RM, Chung LWK, Nie SM (2004) Nat Biotechnol 22:969
Gaynes R (2017) Emerg Infect Dis 23:849
Gelperina S, Kisich K, Iseman MD, Heifets L (2005) Am J Respir Crit Care 172:1487
Ghorbani J, Rahban D, Aghamiri S, Teymouri A, Bahador A (2018) Laser Ther 27:293
Gidwani B, Sahu V, Shukla SS, Pandey R, Joshi V, Jain VK, Vyas A (2021) J Drug Deliv Sci Technol 61:102308
Giljohann DA, Seferos DS, Daniel WL, Massich MD, Patel PC, Mirkin CA (2010) Angew Chem Int Ed 49:3280
Gil-Tomas J, Tubby S, Parkin IP, Narband N, Dekker L, Nair SP et al (2007) J Mater Chem 17:3739
Gong N, Tan Y, Li M, Lu W, Lei X (2016) Photodiagnosis Photodyn Ther 15:70
Gopal J, Hasan N, Manikandan M, Wu HF (2010) Sci Rep 3:1260
Grinholc M, Nakonieczna J, Fila G, Taraszkiewicz A, Kawiak A, Szewczyk G, Sarna T, Lilge L, Bielawski KP (2015) Appl Microbiol Biotechnol 99(9):4031

Gupta D, Singh A, Khan AU (2017) Nanoscale Res Lett 12(1):454
Hamblin MR, Hasan T (2004) Photochem Photobiol Sci 3(5):436
Han G, Martinez LR, Mihu MR, Friedman AJ, Friedman JM, Nosanchuk JD (2013) PLoS One 4:e7804
Harbarth S, Balkhy HH, Goossens H et al (2015) Antimicrob Resist Infect Control 4:1
Hilliard LR, Zhao X, Tan W (2002) Anal Chim Acta 470:51
Høiby N, Bjarnsholt T, Givskov M, Molin S, Ciofu O (2010) Int J Antimicrob Agents 35(4):322
Hu T, Wang Z, Shen W, Liang R, Yan D, Wei M (2021) Theranostics 11:3278
Huang YY, Sharma SK, Dai T et al (2012) Nanotechnol Rev 1(2):111
Huang YY, Sharma SK, Yin R, Agrawal T, Chiang LY, Hamblin MR (2014) J Biomed Nanotechnol 10(9):1918–1936
Huang Y, Guo X, Wu Y et al (2024) Nanotechnology's frontier in combatting infectious and inflammatory diseases: prevention and treatment. Sig Transduct Target Ther 9:34. https://doi.org/10.1038/s41392-024-01745-z
Huang K, Dou Q, Loh XJ (2016) RSC Adv 6:60896
Hutchings MI, Truman AW, Wilkinson B (2019) Curr Opin Microbiol 51:72–80
Jaiswal SK, Kadamannil NN, Jelinek R (2023) Curr Opin Colloid Interface Sci 66:101719
Jiang G, Zeng J (2010) J Appl Polym Sci 116:779
Kadry H, Noorani B, Cucullo L (2020) Fluids Barriers CNS 17:69
Kan S, Mokari T, Rothenberg E, Banin U (2003) Nat Mater 2:155
Kapoor G, Saigal S, Elongavan A (2017) J Anasthesiol Clin Pharmacol 33:300–305
Kemp MM, Kumar A, Clement D, Ajayan P, Mousa S, Linhardt RJ (2009) Nanomedicine (Lond) 4:421
Kim MM, Darafshe A (2020) Photochem Photobiol 96:280
Kırmusaoğlu S, Gareayaghi N, Kocazeybek BS (2019) Introductory Chapter: The action mechanisms of antibiotics and antibiotic resistance. https://doi.org/10.5772/intechopen.85211S
Kohanski MA, Dwyer DJ, Collins JJ (2010) Nat Rev Microbiol 8(6):423
Kong H, Song J, Jang A (2010) Environ Sci Technol 44:5672
Kopelman R, Koo Y-EL, Philbert M, Moffat BA, Reddy GR, McConville P et al (2005) J Magn Magn Mater 293:404
Kuo WS, Chang CY, Chen HH, Hsu CLL, Wang JY, Kao HF, Chou LCS, Chen YC, Chen SJ, Chang WT, Tseng SW, Wu PC, Pu YC (2016) ACS Appl Mater Interfaces 8(44):30467
Kuruppuarachchi M, Savoie H, Lowry A, Alonso C, Boyle RW (2011) Mol Pharm 8:920
Kutner AJ, Friedman A (2013) J Nanomed Nanobiotech 5:502
Kwiatkowski S, Knap B, Przystupski D, Saczko J, Kedzierska E, Knap-Czop K et al (2018) Biomed Pharmacother 106:1098–1107
Lagos KJ, García D, Cuadrado CF, de Souza LM, Mezzacappo NF, da Silva AP, Inada N, Bagnato V et al (2023) WIRES Nanomed Nanotechnol 15:e1887
Lamy B, Sundqvist M, Idelevich EA (2020) Clin Microbiol Infect 26(2):142–150
Lan M, Zhao S, Liu W, Lee CS, Zhang W, Wang P (2019) Adv Healthc Mater 8(13):e1900132
Laxminarayan R, Chaudhury RR (2016) PLoS Med 13(3):e1001974
Le Ouay B, Stellacci F (2015) Nano Today 10:339
Leutwyler WK, Bürgi SL, Burgl H (1996) Science 271:933–937
Li J, Zhao Q, Shi F, Liu C, Tang Y (2016) Adv Healthc Mater 5:2967
Li S, Cui S, Yin D, Zhu Q, Ma Y, Qian Z, Gu Y (2017) Nanoscale 9:3912
Li X, Lovell JF, Yoon J, Chen X (2020) Nat Rev Clin Oncol 17:657
Li X, Chen D, Xie S (2021) Adv Colloid Interface Sci 294:102475
Liselot D, Maarten F, Jan M (2019) Mol Cell 76:255
Liu H-L, Yang TC-K (2003) Process Biochem 39:475
Long K, Han H, Kang W et al (2021) J Nanobiotechnol 19:357
Loomba L, Scarabelli T (2013) Therapeut Deliv 4:859
Luo M, Fan T, Zhou Y, Zhang H, Mei L (2019) Adv Funct Mater 29(13):1808306
Lyutakov O, Hejna O, Solovyev A, Kalachyova Y, Svorcik V (2014) RSC Adv 4:50624

Makabenta JMV, Nabawy A, Li CH, Schmidt-Malan S, Patel R, Rotello VM (2021) Nat Rev Microbiol 19(1):23
Maleki DS, Mennati A, Jafari S, Khezri K, Adibkia K (2015) Adv Pharm Bull 5(1):19
Maness PC, Smolinski S, Blake DM, Huang Z, Wolfrum EJ, Jacoby WA (2013) Appl Environ Microbiol 65:4094
Masilela N, Antunes E, Nyokong T (2013a) J Photochem Photobiol 255:1
Masilela N, Antunes E, Nyokong T (2013b) J Porphyrins Phthalocyanines 17:417
Maurya A, Singh AK, Mishra G, Kumari K, Rai A, Sharma B, Kulkarni GT, Awasthi R (2019) Strategic use of nanotechnology in drug targeting and its consequences on human health: a focused review. Interv Med Appl Sci 11(1):38–54. https://doi.org/10.1556/1646.11.2019.04. PMID: 32148902; PMCID: PMC7044564
Mba IE, Nweze EI (2021) World J Microbiol Biotechnol 37(6):108
McCue AC, Kuhlman B (2022) Curr Opin Struct Biol 74:102377
Memar MY, Ghotaslou R, Samiei M, Adibkia K (2018) Infect Drug Resist 11:567
Meziani MJ, Dong X, Zhu L et al (2016) ACS Appl Mater Interfaces 8(17):10761
Mohamed WAA, Abd El-Gawad H, Mekkey S, Galal H, Hala H, Mousa H, Labib A (1926) Nanotechnol Rev 2021:10
Mondal D, Bera S (2014) Adv Nat Sci Nanosci Nanotechnol 5(3):033002
Munita JM, Arias CA (2016) Microbiol Spectr 4(2). https://doi.org/10.1128/microbiolspec.VMBF-0016-2015
Murray CB, Norris DJ, Bawendi MG (1993) J Am Chem Soc 115:8706
Nathan C, Cunningham-Bussel A (2013) Nat Rev Immunol 13:349
Nikaido H (2009) Annu Rev Biochem 78:119
Nikalje AP (2015) MedChem 5:81
Nikoobakht B, El-Sayed MA (1957) Chem Mater 2003:15
Nombona N, Antunes E, Chidawanyika W, Kleyi P, Tshentu Z, Nyokong T (2012) J Photochem Photobiol C 233:24
Norman RS, Stone JW, Gole A, Murphy CJ, Sabo-Attwood TL (2008) Nano Lett 8:302
Outterson K, Powers JH, Seoane-Vazquez E, Rodriguez-Monguio R, Kesselheim AS (2013) J Law Med Ethics 41(3):688
Ouyang J, Liu R-Y, Chen W, Liu Z, Xu Q, Zeng K, Deng L, Shen L, Liu Y-N (2018) J Mater Chem B 6(39):6302
Pagonis TC, Chen J, Fontana CR, Devalapally H, Ruggiero K, Song X et al (2010) J Endod 36:322
Pajerski W, Ochonska D, Brzychczy-Wloch M, Indyka P, Jarosz M, Golda-Cepa M, Sojka Z, Kotarba A (2019) J Nanopart Res 21:186
Palza H (2013) Int J Mol Sci 16:2099
Pan CL, Chen MH, Tung FI, Liu TY (2017) Acta Biomater 47:159
Pant HR, Pandeya DR, Nam KT, Baek WI, Hong ST, Kim HY (2011) J Hazard Mater 189:465
Pellosi DS, De Jesus PCC, Tedesco AC (2017) Expert Opin Drug Deliv 14(12):1395
Perni S, Piccirillo C, Pratten J, Prokopovich P, Chrzanowski W, Parkin IP et al (2009a) Biomaterials 30:89
Perrault SD, Chan WCW (2009) J Am Chem Soc 131:17042
Pham SH, Choi Y, Choi J (2020) Pharmaceutics 12(7):630
Plotino G, Grande NM, Mercade M (2019) Int Endod J 52(6):760–774
Prestinaci F, Pezzotti P, Pantosti A (2015) Pathog Glob Health 109(7):309
Rahman MM, Alam Tumpa MA, Zehravi M, Sarker MT, Yamin M, Islam MR, Harun-Or-Rashid M, Ahmed M, Ramproshad S, Mondal B et al (2022) Antibiotics 11(5):667
Ran B, Wang Z, Cai W, Ran L, Xia W, Liu W, Peng X (2021) J Am Chem Soc 143(43):17891
Rapulenyane N, Antunes E, Nyokon T (2013) New J Chem 37:1216
Ren G, Hu D, Cheng EWC, Vargas-Reus MA, Reip P, Allaker RP (2009) Int J Antimicrob Agents 33:587
Riehemann K, Schneider SW, Luger TA, Godin B, Ferrari M, Fuchs H (2008) Angew Chem Int Ed 48:872

Romero MP, Marangoni VS, de Faria CG, Leite IS, de Silva CCC, Maroneze CM, Pereira-da-Silva MA, Bagnato VS, Inada NM (2020) Front Microbiol 10:2995
Rosa MCD, Crutchley RJ (1997) Coord Chem Rev 233:351
Roundtree P, Freeman V (1956) Med J Aust 42:157
Roy W, Tomáš S, Dnyaneshwar K, Petr K (2020) Chem Rev 120:13135
Sabino CP, Wainwright M, Ribeiro MS, Sellera FP, dos Anjos C, Baptista MS, Lincopan N (2020) J Photochem Photobiol B 208:111893
Sah U, Sharma K, Chaudhri N, Sankar M, Gopinath P (2018) Colloids Surf B Biointerfaces 162:108
Salam MA, Al-Amin MY, Salam MT, Pawar JS, Akhter N, Rabaan AA, Alqumber MAA (2023) Healthcare (Basel) 11(13):1946
Salata OV (2004) J Nanobiotechnol 2:3
Sarbadhikary P, George BP, Abrahamse H (2021) Theranostics 11:9054
Schairer DO, Chouake JS, Nosanchuk JD, Friedman A (2013) J Virulence 3:271
Seil JT, Webster TJ (2012) Int J Nanomedicine 7:2767
Selvarajan V, Obuobi S, Ee PLR (2020) Front Chem 8:602. https://doi.org/10.3389/fchem.2020.00602
Shah RA, Frazar EM, Hilt JZ (2020) Curr Opin Chem Eng 30:103
Shakiba M, Chen J, Zheng G (2015) Appl Nanosci Photomed 24:511
Sharma P, Brown S, Walter G, Santra S, Moudgil B (2006) Adv Colloid Interface Sci 471:123
Sharma SK, Chiang LY, Hamblin MR (2011) Nanomedicine 6(10):1813
Sharma A, Gupta VK, Pathania R (2019) Indian J Med Res 149:129
Shekhar S, Chauhan M, Hu L, Yadav BS, Dutt R et al (2022) Nanomedicine 17(18):1213–1216
Shi X, Zhang CY, Gao J, Wang Z (2019) Wiley Interdiscip Rev Nanomed Nanobiotechnol 11:e1560
Shivalkar S, Arshad F, Sahoo AK, Palashuddin SK (2022) ACS Omega 7(37):33358
Shleeva M, Savitsky A, Kaprelyants A (2021) Appl Microbiol Biotechnol 105:4099
Silva GA (2008) BMC Neurosci 9:S4
Sirelkhatim A, Mahmud S, Seeni A, Kaus NHM, Ann LC, Bakhori SKM et al (2015) Nano Micro Lett 7:219
Stephens LJ, Werrett MV, Sedgwick AC, Bull SD, Andrews PC (2020) Future Med Chem:2037–2067
Tan L, Li J, Liu X, Cui Z, Yang X, Yeung KWK, Pan H, Zheng Y, Wang X, Wu S (2018) Small 14(9):1703197
Tang Y, Wang G (2021) J Photochem Photobiol C Photochem Rev 47:100420
Tao Y, Chan HF, Shi B, Li M, Leong KW (2020) Adv Funct Mater 30:2005029
Taubes G (2008) Science 321:356
Tawfik AA, Alsharnoubi J, Morsy M (2015) Photodiagnosis Photodyn Ther 12:215
Tegos GP, Demidova TN, Arcila-Lopez D, Lee H, Wharton T, Gali H, Hamblin MR (2005) Chem Biol 12(10):1127
Thambirajoo M, Maarof M, Lokanathan Y, Katas H, Ghazalli NF, Tabata Y, Fauzi MB (2021) Antibiotics 10:1338
Tsai TM, Chang HH, Chang KC, Liu LY, Tseng CC (2010) J Chem Technol Biotechnol 85:1642
Tseng YH, Sun DS, Wu WS, Chan H, Syue MS, Ho HC et al (2013) Biochim Biophys Acta 1830:3787
Uddin TM, Chakraborty AJ, Khusro A, Zidan BMRM, Mitra S, Emran TB, Dhama K, Ripon MKH, Gajdács M, Sahibzada MUK, Md Hossain J, Koirala N (2021) J Infect Public Health 14:1750
Varela MF, Stephen J, Bharti D, Lekshmi M, Kumar S (2023) Biomedicine 11:1448
Varisco M, Khanna N, Brunetto PS, Fromm KM (2008) Chem Med Chem 9:1221
Ventola CL (2012) Pharmacy Therapeut 37:512
Viana O, Ribeiro M, Rodas A, Rebouças J, Fontes A, Santos B (2015) Molecules 20(5):8893
Vijayalakshmi K, Sivaraj D (2015) RSC Adv 5:68461
Wainwright M, Maisch T, Nonell S, Plaetzer K, Almeida A, Tegos GP, Hamblin MR (2017) Lancet Infect Dis 17:e49–e55
Wang S, Lawson R, Ray PC, Yu H (2013) Toxicol Ind Health 27:547
Wang L, Hu C, Shao L (2017) Int J Nanomedicine 12:1227

WHO Antimicrobial Resistance (2018). https://www.whoint/news-room/fact-sheets/detail/antibiotic-resistance
Wigley D (2013) Nat Rev Microbiol 11:9
Wijesiri N, Yu Z, Tang H, Zhang P (2018) Photodiagnosis Photodyn Ther 23:202
Wong MS, Sun DS, Chang HH (2010) PLoS One 5:e10394
Xie X, Mao C, Liu X, Zhang Y, Cui Z, Yang X, Yeung KWK, Pan H, Chu PK, Wu S (2017) ACS Appl Mater Interfaces 9:26417
Yang S, Li NJ, Chen DY, Qi XX, Xu YJ, Xu Y et al (2013) J Mater Chem B 1:4628
Yang L, Wei J, Ma Z, Song P, Ma J, Zhao Y, Huang Z, Zhang M, Yang F, Wang X (2019) Nanomaterials 9:1789
Yavuz MS, Cheng YY, Chen JY, Cobley CM, Zhang Q, Rycenga M et al (2009) Nat Mater 8:935
Ye H, Fussenegger M (2019) Cold Spring Harb Perspect Med 9:a034371
Yeh YC, Creran B, Rotello VM (2012) Nanoscale 4(6):1871
Youf R, Müller M, Balasini A, Thétiot F, Müller M, Hascoët A, Jonas U, Schönherr H, Lemercier G, Montier T, Le Gall T (1995) Pharmaceutics 2021:13
Yu Y, Mei L, Shi Y, Zhang X, Cheng K, Cao F, Zhang L, Xu J, Li X, Xu Z (2020) J Mater Chem B 8:1371
Yu XT, Sui SY, He YX, Yu CH, Peng Q (2022) Biomater Adv 135:212725
Zhang L, Gu FX, Chan JM, Wang AZ, Langer RS, Farokhzad OC (2008) Clin Pharmacol Ther 83:761
Zhang Y, Huang P, Wang D, Chen J, Liu W, Hu P, Huang M, Chen X, Chen Z (2018) Nanoscale 10:15485
Zhang C, Wang Y, Ma J, Zhang Q, Wang F, Liu X, Xia T (2020) Sci Total Environ 721:137740
Zhang Y, Liu F, Zhang Y, Rogers JA (2023) PNAS 120:e2217734120
Zhao Y, Ye C, Liu W, Chen R, Jiang X (2010) Angew Chem Int Ed 53:8127
Zhi D, Yang T, O'Hagan J, Zhang S, Donnelly RF (2020) Photothermal therapy. J Control Release 325:52–71
Zhou Z, Peng S, Sui M, Chen S, Huang L, Xu H, Jiang T (2018) Colloids Surf B Biointerfaces 161:394
Zhu X, Radovic-Moreno AF, Wu J, Langer R, Shi J (2014) Nano Today 9:478
Zucca M, Savoia D (2010) Int J Biomed Sci 6:77

Applications of Ionic Liquids as Antimicrobials and Their Impact on Nanotechnology

15

Himani Singh, Richu, and Ashwani Kumar

Abstract

Ionic liquids (ILs) consist of individual ions and ion pairs. Due to the peculiar physical and chemical properties, ILs have shown several practical uses in the field of synthesis and catalysis. The adaptability of ionic liquid-based materials to be modified in different sizes, morphologies, and surface charges accounts for their prospective applications as antibacterial systems. Moreover, ionic liquids are viewed as greener materials. Current studies focus on adjusting biological characteristics to create innovative ionic liquid-based antibacterial materials. The best materials for extending this are nano-based systems. ILs can enhance the properties of currently used nanomaterials and could serve as models for creating nanoparticles. They are frequently employed to boost conductivity and reactivity while maintaining greater thermal and chemical stability, expanding the range of applications. Ionic liquids can be used to investigate new strategies for improving the current nanoparticles as well as to synthesize an entirely new class of nanoparticles. Hence, to achieve synergistic effects, structures incorporating both ILs and nanoparticles (NPs) have been designed. This chapter will discuss the ionic liquids and ionic liquid-based materials with an emphasis on their antimicrobial properties.

Keywords

Nanoparticles · Ionic liquids · Antimicrobial · Greener materials

H. Singh · Richu · A. Kumar (✉)
Department of Chemistry, University of Jammu, Jammu, Jammu and Kashmir, India

M. Y. Wani et al. (eds.), *Nanotechnology Based Strategies for Combating Antimicrobial Resistance*, https://doi.org/10.1007/978-981-97-2023-1_15

15.1 Introduction

15.1.1 Background and Significance

Ionic liquids are fascinating chemical substances that are used in a wide range of contemporary scientific fields. In addition to its physical and chemical characteristics, biochemists and medical experts have been paying close attention to ionic liquids because of their high biological activity. Researchers have also become interested in ionic liquids due to their antibacterial properties. Furthermore, ionic liquid-based goods and operational applications are being created for many novel industrial applications as well as academic uses involving nanotechnology-related procedures, processes, and products, particularly in nanobiotechnology and nanomedicine. The biological activities of ionic liquids are the focus of this chapter, with an emphasis on their mechanism as antimicrobial agents. The advantages of ILs as antimicrobials have been represented in Fig. 15.1. The wider range of uses for ionic liquids, potential advancements in chemistry and various bioaspects, potential in the more recent categories of nanobiotechnology products, and new trends, as well as opportunities are also discussed. The major goal of this chapter is to attract chemists, biologists, and medical professionals towards the benefits of ionic liquids.

15.1.2 Objectives of the Chapter

The current chapter presents an outline of the applications of ionic liquids as antimicrobials and their impact on nanotechnology. The principal goal of this chapter is to recognize the obligatory principles proficient in reconciling the extensive

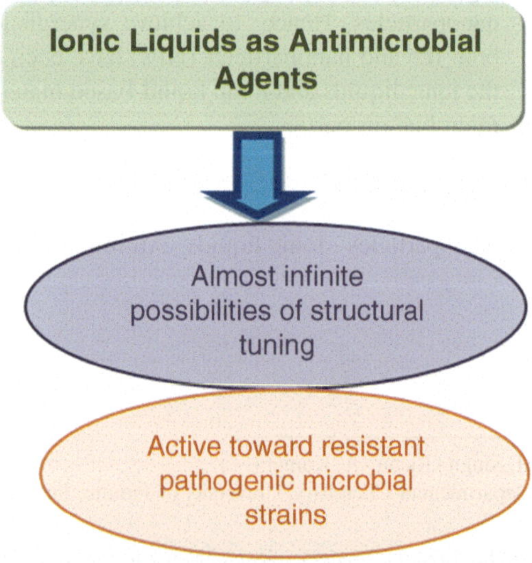

Fig. 15.1 Advantages of Ionic Liquids as antimicrobial agents

properties and phenoms unveiled by these systems. The various objectives of this chapter are as follows:

- To provide a detailed insight into the versatile properties and structural features of ionic liquids.
- To have a profound understanding of multidrug resistance, various problems allied to multidrug resistance, and the novel approaches required to combat multidrug resistance.
- To discuss the key antimicrobial applications of ILs and IL-based novel compounds.
- To analyse the mechanisms as well as biological effects of interaction between ILs and bacterial cells and the effects of ILs on microbial biofilms.
- To present the applications of ILs in nanotechnology, impact of nanotechnology on antimicrobial studies, synergy amid ILs and nanomaterials, and the novel nano technological advances implicated towards antimicrobial deliverance.
- To understand the role of ionic liquids in combating multidrug-resistant microorganisms in nanoscience and to provide the future prospects and challenges.

15.1.3 Overview of Ionic Liquids and Their Versatility

Ionic liquids (ILs) are an innovative class of solvents pleasing the provisions of biotechnology, as well as bioengineering, formerly owing to their atypical and tunable physicochemical prospects. Four generations of ILs have been decrypted reliant to their molecular organization and physicochemical features (Gomes et al. 2019). These have been grouped as presented below:

- The first-generation ILs came into public interest as a consequence of their peculiar physical properties. The majority of first-generation ILs include alkylpyridinium/dialkylimidazolium cations and halides anions. They were renowned as supreme solvents in electrochemistry. However, these were susceptible to moisture, air, and H_2O (MacFarlane et al. 2018).
- The second-generation ILs were stable towards air, H_2O, and moisture. Dialkylimidazolium/ammonium/phosphonium/alkylpyridinium were the widespread cations, whereas tetrafluoroborate/hexafluorophosphate were the typical anions comprising these ILs (Badgujar and Bhanage 2015). These ILs have established an extensive array of avenues in physicochemical fields such as reaction solvents, lubricants, etc. Nevertheless, the toxicity of ILs was not attended all through the first two generations (Egorova et al. 2017).
- The third-generation ILs incorporated natural as well as biodegradable ions or ions with familiar natal biological activity. This generation encompassed ILs with beleaguered and preferred biological properties, since the perception of task-specific ILs was pioneered during that period (Davis and Forrester 1999). These task-specific ILs were a breakthrough in developing antibacterial and antifungal compounds for their implication in biomedical and therapeutic avenues

(Giernoth 2010). These comprised covalently allied moieties which could be refrained for the scrupulous employment. Task-specific ILs obtained via didecyldimethylammonium bromide as well as sodium ibuprofen illustrated both anti-inflammatory and antibacterial activities (Hough et al. 2007).
- The fourth-generation ILs were established in 2018. These ILs were bio-companionable and infatuated exclusive and erratic features in mixtures or after amalgamation with different fluids.

15.2 Multidrug-Resistant Microbes: A Growing Challenge

15.2.1 Understanding Multidrug Resistance

Multidrug resistance may be described as antagonism or insensitivity of a microbe to the dispensed antimicrobial drugs (Singh 2013; Popęda et al. 2014). As per WHO, the invulnerable microbes (such as fungi, bacteria, parasites, and viruses) are competent to combat invasion by antimicrobial medicines, in turn are directed to inoperable cure ensuing in the diligence and dissemination of ailments. Although the expansion of multidrug resistance is an innate occurrence, there is an ample rise in varied immune-compromised diseases, such as HIV, diabetes, etc. Patients who have experienced organ transplantation, and excruciating burn, make the body an uncomplicated target for hospital-acquired virulent disorders, thereby leading to the additional stretch of MDR. Contemplations from WHO declaration have revealed significantly increased speeds of the antagonism (Table 15.1). Protracted medicine divulgence and persistent viral counterparts result in the origin of different unsusceptible strains and the persistence of ailments despite treatment. This has produced antiviral antagonism, a topic of apprehension in immune-compromised patients.

15.2.2 The Global Health Threat of Multidrug Resistance

In general, the antimicrobial resistance is allied to elevated death rates as well as high therapeutic overheads along with a remarkable impact on the efficacy of antimicrobial agents as depicted in Fig. 15.2. Multidrug resistance incites impediment in ailment management via escalating the risk of dispersion of defiant pathogens, consequently waning effectiveness of cure as well as, for this reason, ensuing in expanded time of illness in patient.

The outlay of treatment is moreover enlarged, attributable to multidrug resistance, as the microbial agents have become defiant to commercially accessible remedies that have prompted the employment of additional exclusive therapies. The success rate of contemporary therapeutic avenues such as melanoma chemotherapy as well as organ transplantation has abounded massively towards improvement of multidrug resistance. Disparities in the resistance silhouette of microbial agents along with the eminence of civic sanitation have a substantial blow on the usefulness of antimicrobial agents. Development of global trade and tourism escort to

Table 15.1 Various drug-resistant pathogens along with their peculiar diseases

Pathogen	Resistant to drug(s)	Diseases
Mycobacterium tuberculosis	Fluoroquinolone, isoniazid, and rifampicin	Tuberculosis
Escherichia coli	Fluoroquinolones and Cephalosporins	Blood stream infections and urinary tract infections
Streptococcus pneumonia	Penicillin	Otitis, pneumonia, and meningitis
Cryptococcus neoformans	Fluconazole (Rodero et al. 2003)	Cryptococcosis
Scopulariopsis spp.	Azoles, flucytosine, and amphotericin B (Cuenca-Estrella et al. 2003)	Onychomycosis
Aspergillus spp.	Azoles (Howard and Arendrup 2011)	Aspergillosis
Herpes simplex virus (HSV)	Valacyclovir, famciclovir, and acyclovir (Wutzler 1997)	Herpes simplex
Hepatitis B virus (HBV)	Lamivudine (Suppiah et al. 2014)	Hepatitis B
Varicella zoster virus	Valacyclovir and acyclovir	Chicken pox
Schistosomes	Oxamniquine and Praziquantel (Fallon and Doenhoff 1994; Qi and Cui 2013)	Schistosomiasis
Trichomonas vaginalis	Nitroimidazoles (Muzny and Schwebke 2013)	Trichomoniasis
Entamoeba	Metronidazole (Bansal et al. 2006)	Amoebiasis

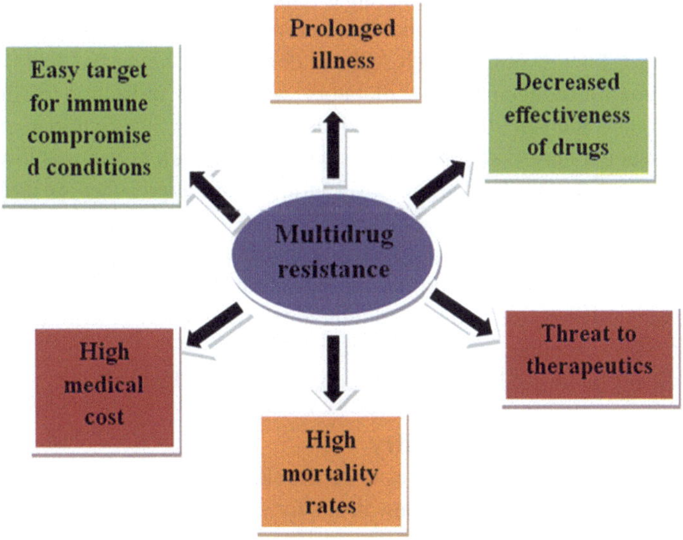

Fig. 15.2 Problems allied to multidrug resistance

augmented prospective of multidrug resistance to widen all over the globe and lessen in export and import of assorted goods upsetting the financial system of emergent nations (Fishbain and Peleg 2010).

15.2.3 Need for Novel Antimicrobial Approaches

Multidrug resistance in microorganisms has presently reached soaring and distressing levels. The sustained advancement in antibiotic innovation and expansion pragmatic throughout the preceding decades has led to a partial accessibility of novel substances for medical implication, thereby exacerbating the encumber of antibiotic resistance on morbidity as well as transience rates. In this picture, furthermore, numerous contemporary irrefutable practices are allied to an augmented menace of contagion in which antibiotics are crucial. In other words, organ and cell transplantations, anti-cancer medication, and implantation of prosthetic campaigns are fatally at threat of accomplishment (Hutchings et al. 2019). Therefore, it is comprehensible that there is an imperative necessity to ascertain and expand new antibiotics, in addition to widen modern antimicrobial advances, probably implication in synergy with typical chemotherapy. Although the attention in the field of antibiotic research and growth is recuperating, there are still numerous unmet medical requirements that have to be tackled in an apt way. Thus, the enlargement of antimicrobial resistance is an intricate matter that turned into a worldwide alarming apprehension. In order to lessen the increase and stretch of multidrug resistance, obliging approaches are mandatory because ailments which were curable previously are fetching sources of deaths in this period. Besides, emphasizing on areas which are liable to improper utilization of antimicrobials by means of execution of antibiotic stewardship is the necessity of the hour (Hamilton and Fishman 2014). Indeed, assorted antimicrobial stewardship programs are accomplished to optimize antimicrobial rehabilitation, lessen medication outlay, enliven medical effects and protection, and curtail or alleviate multidrug resistance (Owens 2008). Intercession through ASPs is done by confining the accessibility of chosen antimicrobial agents, by exploratory wide scale utilization of antibiotics (Moehring and Anderson 2012). For that reason, there is a vital requirement of synchronization at the universal, provincial, subregional, as well as national level to provide the potential growth. The various novel antimicrobial approaches are listed below:

15.2.3.1 Policymakers

(a) Group of awareness programmes for appropriate utilization of drugs.
(b) Enhance collaboration and information networking amid stakeholders.
(c) Emphasize on contagion impediment and control.
(d) Implementing law to bound sale of unprescribed medicines.

15.2.3.2 Scientists and Industry

(a) Innovative diagnostic procedures to be developed.
(b) Encouraging advancement of new drugs for efficient cure.

15.2.3.3 Health Workers

(a) Improving disease deterrence and control.
(b) Absurd treatment to be circumvent.
(c) Drugs to be prescribed merely while required.
(d) Accurate drug and dose must be given with reverence to ailment.

15.2.3.4 People

(a) Execution of treatment itinerary, though they feel healthier.
(b) Utilization of available drugs and allocation ought not to be done.
(c) Recurrent usage of drug must be prohibited.
(d) Self-medication should be circumvent.

15.3 Ionic Liquids: Properties and Characteristics

15.3.1 Definition and Structure of Ionic Liquids

A salt that has a melting point below some predetermined temperature, such as 100 °C (212 °F), is referred to as an ionic liquid (IL). Ionic liquids, unlike other materials, have intermolecular and intramolecular synergies that are considerably regulated by the conformation of cation/anion.

The fact that IL manufacturing is composed of cations and anions rather than molecules is its most distinctive feature. Most commonly used cations/anions are shown in Table 15.2.

Moreover, ILs solidify at lower temperatures because the charge on the ions is dispersed over wider surface area of the molecular moieties (Hayes et al. 2015). The choices of cations and anions in ILs have a significant impact on their features; for instance, cations influence stability while anions control functioning. The bulk of IL is influenced by dipole-dipole, van der Waals, and Coulombic interactions, among other things, due to the various amalgamations of inorganic anions and organic cations (Gebbie et al. 2017; Daschakraborty and Biswas 2014).

Table 15.2 Typical cations/anions comprising ionic liquids

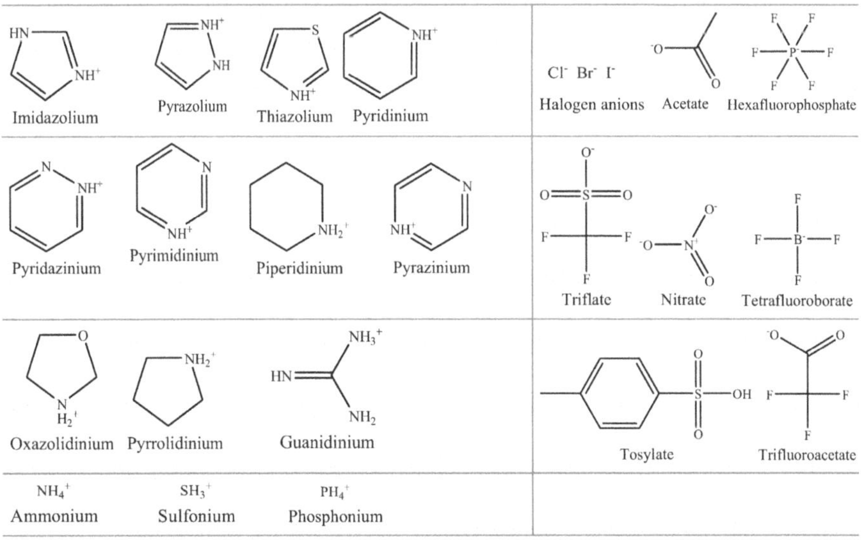

15.3.2 Key Properties Relevant to Antimicrobial Applications

Due to the numerous options created by combining and tuning the cation/anion structure, ionic liquids can satisfy the needs for various reaction types and setups, both technically and chemically in the lab and on an industrial scale (Frizzo et al. 2016).

The majority of ILs exhibit universal properties such as low melting points, little vapour pressure, remarkable solvation potential, and antibacterial activity. Additionally, by varying the cation/anion, their physicochemical and biological properties can be precisely "tuned." Ionic liquids often have broad activity spectra that can influence mycobacterium, fungus, Gram-negative, and Gram-positive bacteria. The molecular architecture of each ionic liquid determines the equilibrium between important variables that control antibacterial action, though.

Increasing the lipophilicity of ILs, which is frequently achieved by lengthening the linker alkyl chain, generally results in an increase in antibacterial activity. Pyridinium, imidazolium, and quaternary ammonium-based head groups are the antimicrobial ionic liquids' structural designs that have been reported the most frequently. Numerous head group structures with antibacterial action have been identified, including domiphen (Walkiewicz et al. 2010; Cybulski et al. 2011), morpholinium (Pernak et al. 2011), cholinium (Busetti et al. 2010), and quinolinium. Anions support the "chain-length effect" by having a secondary function in antibacterial action.

15.3.3 Classification of Ionic Liquids

ILs can be categorized into four different categories as given in Fig. 15.3. The solvent is referred to as "polar" if there are strong connections between substances, like H_2O and alcohols with short alkyl chains. In contrast, the solvent is referred to be "apolar" if these interactions are weak, such as with heptanes and other compounds. Due to their special physicochemical characteristics, like high thermal stability, non-flammability, a wide electrochemical window, as well as non-volatility, ILs have helped to stimulate awareness for avenues in electrochemistry, analytical/severance sciences, and biocatalytic transformations (Ao and Kim 2013; Zhao et al. 2005; Liu et al. 2005; Zafarani-Moattar et al. 2014). ILs are used in a variety of industrial processes. They are utilized in the chemical sector, medicines, gas treatment, cellulose processing, gas handling, nuclear fuel processing, batteries, solar thermal energy, waste recycling, food and by-products, etc. ILs have completely different structures than conventional solvents, whose properties depend on interactions between their constituent parts. As a model ionic liquid, derivatives of imidazolium have drawn a lot of attention and have undergone extensive research. Some scientists have looked at these ionic liquids in catalysis, synthesis reactions, and separation processes (Kumar and Katal 2018). For ILs possessing longer alkyl chains comprising imidazolium ion, their cytotoxicity may be a disadvantage when used as solubility promoters (de Almeida et al. 2015; Wang et al. 2010).

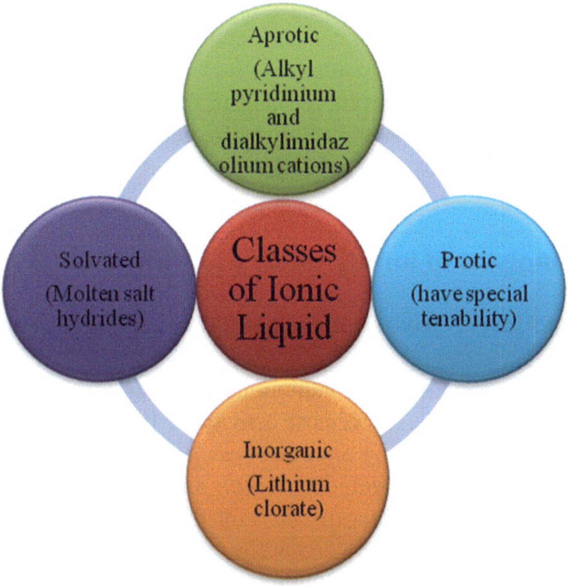

Fig. 15.3 Classes of Ionic liquids

Pyridinium-based ILs have also been studied, but there have been fewer of these investigations with these ILs. On the contrary with the imidazolium-reliant ILs, these exhibited meagre regioselectivity in the telomerization of butadiene with CH_3OH (Magna et al. 2003). Moreover, the rate of several Diels-Alder processes has been shown to be negatively impacted by these pyridinium salts (Biswas and Rao 2007). However, in several other eminent organic reactions, like Friedel-Crafts and Grignard reactions (Ghandi 2014; Busetti et al. 2010), the employment of these ILs is extremely effective. Additionally, pyridinium ions comprising ILs have demonstrated to be excellent catalysts for the production of a number of pharmaceutical compounds, including 3,5-bis(dodecyloxycarbonyl)-1,4-dihydropyridine derivatives, di-hydropyrimidinone, and 1,4-dihydropyridine (Obando et al. 2009; Pernak et al. 2003; Czekański et al. 2014).

The quaternary ammonium ILs are considered as "green" solvents for pharmaceuticals because they were found to have less harmful effects on species of varied organizational levels in comparison to pyridinium/imidazolium-based ILs. Attributable to their low melting temperatures and lesser viscosities, these ILs have also been utilized as electrolytes (Melo et al. 2013). They have also been used to increase both topical drug administration and antibacterial action (Zakrewsky et al. 2014).

Phosphonium ions comprising ILs are further modern and are appropriate for reactions that are suitably done at temperatures >100 °C, in view of the fact that they are thermally stable, in comparison to imidazolium/ammonium salts, (Dake et al. 2009) and have been utilized as catalysts as well as solvents in varied reactions (Cao and Alper 2010; Fan et al. 2012). Since the properties of the various classes of ILs greatly depend on the cation/anion constituting them, some of them could be used in oils, hydroalcoholic, or aqueous solutions. This opens up a wide range of options. These features are extremely relevant for industrial phenoms, and the thermophysical characteristics of ILs with varied additives are helpful for creating the industrial as well as chemical phenoms.

15.4 Mechanisms of Ionic Liquids as Antimicrobial Agents

15.4.1 Interactions Between Ionic Liquids and Microbes

ILs typically exhibit antibacterial commotion against an array of bacteria (mycobacteria), fungi, etc. Yet, the uppermost activity of ILs has been recognized for Gram-positive bacteria, including *Streptococcus* and *Staphylococcus,* and various Gram-positive bacteria, such as *Bacillus subtilis* and *Lactobacillus*. The molecular structure of each IL, which is primarily determined by the structure of the cation and in scrupulous by the length of the hydrocarbon chains, defines its biocidal capabilities (Pendleton and Gilmore 2015). Morpholinium, imidazolium, choline ions, pyridinium, quinolinium, and pyrrolidinium are among the most prevalent IL cationic head groups. When compared to ILs with heterocyclic nitrogen rings, nonaromatic ILs are typically less hazardous (Iwai et al. 2011; Fahri et al. 2020).

In general, adding substituents to the cationic head of these compounds can change their antibacterial activity and toxicity profile (Pernak et al. 2007). 1-alkyl quinolinium ILs are among the most effective antibacterial ILs, and quinolinium ILs are often more potent than imidazolium ILs (Cole et al. 2011). When a tail is present in the IL system, relevant consequences on activity have been discovered. The IL-cell surface affinity (Sekhon 2011) can be increased by adding more alkyl groups to the chain, which can also boost the antibacterial action. A surface-active compound's hydrophobicity determines the minimum concentration required to limit microbial growth (Łuczak et al. 2010). Alkyl chains of 10–16 carbon atoms have been found to have the strongest antibacterial properties in ILs (Garcia et al. 2014). The 1-alkylquinolinium ILs, which are quinoline-based ILs, demonstrated the best antimicrobial efficacy when the alkyl chain was 12- to 14-carbons long (Gilmore et al. 2013). The greatest antibacterial action against bacteria and fungi is shown by pyrrolidinium ILs with an alkyl chain of 14 carbon atoms. However, because the type of cation plays a crucial role in determining the necessary hydrophilic-lipophilic balance, the recital of ILs is highly reliant on the kind of cation. In view of the fact that they can easily permeate the cell wall (composed of peptidoglycan) of Gram-positive bacteria, ILs comprised of an alkyl chain of 8 carbons (trihexyl(alkyl)phosphonium ILs) are an efficient antibacterial (Cieniecka-Rosłonkiewicz et al. 2005). Although the cations and anions fundamental to ILs may both have innate antibacterial activity, changing the anion for a given cation does not seem to change how the ILs behave. Anions are normally chemical penetration enhancers; however, for ILs with tiny organic cations, the chaotropicity of the anion can be a significant component to their antimicrobial effect. Several phosphonium-based ILs experienced this behaviour (Sydow et al. 2018). The utilization of amino acids as anions, such as tryptophan, proline, phenylalanine, valine, and methionine, is an additional strategy that may be of interest. Together, Gram-positive/Gram-negative bacteria were effectively inhibited by ILs containing imidazolium cations and tryptophan anions. Furthermore, the choline-geranate ILs are becoming more and more popular within this context due to their remarkable antibacterial activity.

Moreover, the World Health Organization (WHO) has created a list of priority diseases for antibiotic resistance and is aiming to construct a worldwide programme to battle microbial resistance (Jamaledin et al. 2020). In an attempt to combat the menace of antimicrobial resistance, assorted systems were introduced as substitutes to antibiotics, together with the preface of antimicrobial and capable ILs (Forero et al. 2018) as their performance is on par with that of usual antibiotics. Usual antibiotics typically board bacteria together with their antibacterial potential (Lei et al. 2019) and their method of action can be alienated into the subsequent categories: (1) interfering with the synthesis of bacterial cell walls; (2) restraining the synthesis of proteins; (3) restraining the synthesis of nucleic acids; (4) disrupting the construction of the bacterial casing; and (5) inhibiting bacterial enzymes and metabolic pathways (Abushaheen et al. 2020). Ionic liquids, in contrast, quickly cause cell leakage and death by interfering with the bacterial membrane and producing pores (Hoque et al. 2019). The impact of ionic liquids on membrane characteristics can be used to identify these membrane changes that result in bacterial mortality. Ionic liquids have the potential to transform the prospective of the membrane, the viscoelasticity of the membrane, the alignment of the phospholipids, and ultimately the fluidity of the membrane. The latter influences protein diffusion and stability within the membrane, as well as the transport of molecules, their differentiation, migration, and adhesion. These alterations work together to cause bacterial cell death (Wang et al. 2015). Operationally, ionic liquids' ability to connect quickly to bacterial walls—which are primarily made of peptidoglycan—means that intricate cross-linking of bacterial cell walls has taken place. Prokaryotic cells, contrasting eukaryotic cells, don't have the capacity to ingest exogenous chemicals via pinocytosis or endocytosis; instead, absorption arises in the course of active/passive dispersion into the cell wall (Lu et al. 2020). Thus, IL-reliant materials can effortlessly serve as a requisite mediator to the cell wall in order to provide antibacterial characteristics.

Additionally, a molecule with a dual antibacterial action was reported using chlorin e6 (enclosing carboxyl moiety) as an anionic fragment and 1-vinyl-3-dodecyl imidazole as a cationic fragment. Further, an improved effect for eradicating the bacterial infection is demonstrated by the cationic component of the IL attaching to the wall composed of peptidoglycan and the anionic component being deprotonated in acidic phase (Wang et al. 2019). Besides the aforementioned factors, ionic liquids' structural characteristics, like chain length, can influence how effective they are against bacteria. Longer alkyl chains (over 10 carbon atoms) destabilize cell membranes by piercing the phospholipid bilayer and causing membrane commotion, which increases the antibacterial efficacy of ILs (Egorova et al.

2017). Shorter alkyl chains (under 10 carbon atoms) increase the hydrophobicity of the ionic liquid and reduce its antibacterial efficacy.

One review investigated into the antibacterial activities of alkoxyimidazolium/alkylimidazolium lactates with different alkyl residues, mainly 11–14 carbons. These findings support the effectiveness element (He et al. 2013). It suggested that an ionic liquid's threshold value (or ideal chain length) is reliant to its configuration and the proposal it will be used on (i.e. crosslinking density, hydrophobicity of the concluding made-up polymer, and compound). When such long-chain alkyl ILs are used, they tend to congregate in solutions, which leads to the creation of amphiphilic micelles (Prudêncio et al. 2020). ILs based on imidazolium have antimicrobial effectiveness and antibiofilm activity besides Gram-positive/Gram-negative bacteria, such as *Enterococcus faecalis, Staphylococcus aureus*, *Pseudomonas aeruginosa*, and *Escherichia coli*. The length of the hydrocarbon chain of ILs, along with the alteration of clusters, results in their increased lipophilicity. This increased lipophilicity allows the hydrocarbons to interact with and disrupt bacterial membranes. However, it lowers surface tension, increasing the likelihood that the molecule will connect with the cell membrane and then permeate the bacterial membrane to effectively suppress growth (Duman et al. 2019).

In addition to the conventional ionic liquids, erstwhile ILs, like those including the fumarate ion, are also soluble in aqueous and dipolar organic media, as is the case with [BMIM][MRF]. The antibacterial commotion in this case is associated with the characteristics of fumarate ions, the length of the hydrocarbon chain, and is dependent on the microorganisms under study. For example, an alkyl group comprising 4 carbons exhibits the greatest antibacterial commotion against *Escherichia coli*; further, the activity decreases as the length of chain increases. Alternatively, larger alkyl chain length exhibits stronger antibacterial activity in the case of *Bacillus subtilis*. These ionic liquids are more effective than traditionally available potassium sorbate as a reference for antibacterial activity when considering zone widths of growth inhibition. Fumarate anion ionic liquids can have both the necessary water solubility and significant antibioactivity by combining their hydrophobic and hydrophilic features. Whereas the hydrophilicity of the ILs is inferred via the carboxylate side, the hydrophobicity of such ILs could be transformed via changing the alkyl fragment of fumarate. Hence, the hydrophobic and hydrophilic components of the fumarate ion facilitate the interaction between cell walls of bacteria and ionic liquids. Further, Penicillin G and amoxicillin hydrolysates originated in ILs along with organic salts exhibit antibacterial potential in opposition to methicillin-resistant strains (*Escherichia coli/Staphylococcus aureus*). Also, their effortless production is realistic for their future built-up manufacture and could be a successful strategy to counter the developing microbial antibiotic resistance (Ferraz et al. 2020).

15.4.2 Disruption of Microbial Cell Membranes

Antimicrobial ILs and cationic biocides, like quaternary ammonium compounds, which disrupt membrane-bound proteins as their main mechanism of action, have

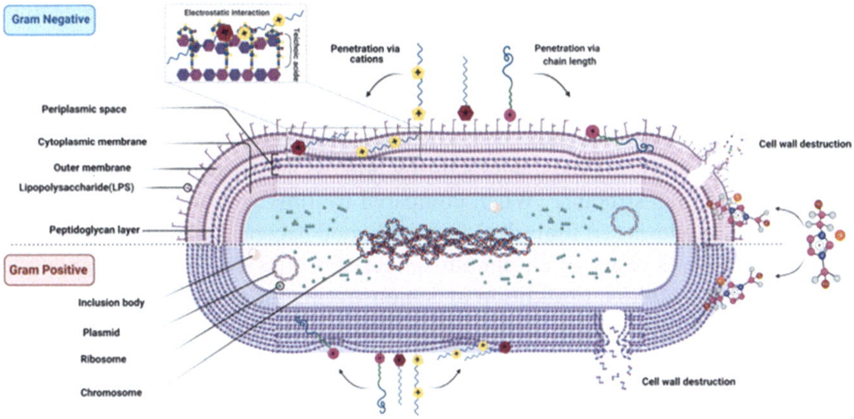

Fig. 15.4 Interaction of Ionic liquid with the bacterial cell wall (Fang et al. 2022)

structural and molecular similarities (Kumari et al. 2020). ILs primarily affect the cytoplasmic membrane of bacteria as well as the plasma lemma of yeast, according to mechanistic studies, which results in changes to the constitutional and dynamic characteristics of the outer coverings (Miskiewicz et al. 2018) and a commotion/reduction of membrane veracity (Samorì 2011). The majority of bacterial cells are negatively charged, which is frequently maintained via existence of divalent cations like Mg^{2+} and Ca^{2+}, whereas antimicrobial ILs which are mainly positively charged trigger the general mode of accomplishment. As a result, the cationic fragment of ILs interacts with the unfavourable constitutional proteins of the bacterial external coverings during the initial phase of antimicrobial action (Hodyna et al. 2016). Cations are therefore easily taken up by the cell layers, where they attach to the cytoplasmatic membrane and enter the cell's core (Benedetto and Ballone 2018). ILs also cause the cytoplasmic components to coagulate, which inhibits important enzymes (such as AMP deaminase, acetyl cholinesterase, glutathione reductase, acylase I, cytochrome c oxidase, carboxylesterase, DNA polymerase, and catalase) that are involved in energy production or self-repair, ultimately leading to cell death. The interaction of ionic liquid with the bacterial cell wall has been presented in Fig. 15.4 (Fang et al. 2022).

15.4.3 Effects on Microbial Biofilms

ILs have been notorious to possess antibacterial capabilities, most notably broad spectrums of activity in opposition to both Gram-positive/negative bacteria, fungus, mycobacteria, etc. Moreover, the food industry relies heavily on food packaging to increase the safety and protection of the food. By incorporating substances like antimicrobial ionic liquids to lower the live count of bacteria, it is possible to create novel, inventive packaging materials with antimicrobial capabilities (Janani et al. 2020). As potential possibilities for fresh food packaging, 1-allyl-3-methylimidazolium chloride, biomolecules chiefly starch, lignin, and cellulose have been combined to create green antimicrobial composite films (Wu et al. 2009). The creation of gelatine-modified ionogel films with improved antibacterial and antioxidant properties involved the use of a bioactive ionic liquid called choline salicylate. The ionic liquid did not undergo any chemical modifications during the manufacturing process and instead served as a plasticizer in the film's structure. When compared to films without ionic liquid, the gelatin/ionic liquid-based films demonstrated good bacterial growth suppression. This might be as a result of cholinium salicylate's effective antibacterial action against *Bacillus subtilis*. In addition to demonstrating the potential to prolong the shelf life of red apples by avoiding air oxidation, the ionic liquid-loaded sheets also showed remarkable antioxidant and UV shielding capabilities as well as great mechanical strength. According to the results of a free radical scavenging assay, ionic liquid-loaded films with higher ionic liquid concentrations showed greater antioxidant capacity while films without ionic liquid did not produce any radical scavenging impact.

Antimicrobial poly(ionic liquids), created from polymerizable antimicrobial ionic liquids, combine the special qualities of antimicrobial ionic liquids with the inherent features of polymers. The antimicrobial poly(ionic liquids) appear to be the promising materials for designing antimicrobial membranes with improved processability, mechanical, and thermal stability, as well as higher permeation rates and separation capacity (Yan et al. 2019). In view of this context, antibacterial poly(ionic liquid)s membranes are manufactured via a 2-step synthetic technique: (1) Fabrication of IL monomers (1-vinyl-3- allylimidazolium bromide, 1-vinyl-3-butylimidazolium chloride, and 1-vinyl-3-hexylimidazolium bromide), and 2) the production of PILMs via fusion of IL monomers/co-monomers (styrene/acrylonitrile) with the implication of cross-linker (1,4-divinylbenzene). The PILMs demonstrated significant constancy in polar-protic/apolar-aprotic/polar-aprotic media, highlighting the extensive range of solvents in which they can be used. PILMs reduced the intensification of *Pseudomonas aeruginosa*/*Staphylococcus aureus* along with the antibacterial action reliant to the imidazolium ring moieties.

15.5 Applications of Ionic Liquids in Nanotechnology

15.5.1 Nanotechnology in Antimicrobial Research

The antimicrobial activity of nanoparticles has been given in Fig. 15.5. Numerous nanostructures have had their antimicrobial qualities studied. While some metals, like silver, zinc, and copper, have antibacterial mechanisms in their bulk form, other materials, like iron oxide, do not have antibacterial properties in their bulk form, but may have antibacterial properties in their nanoparticulate form. This distinction is crucial to understand the antibacterial properties of these nanostructures. From one nanoparticle to another, the antibacterial activity's mechanism varies.

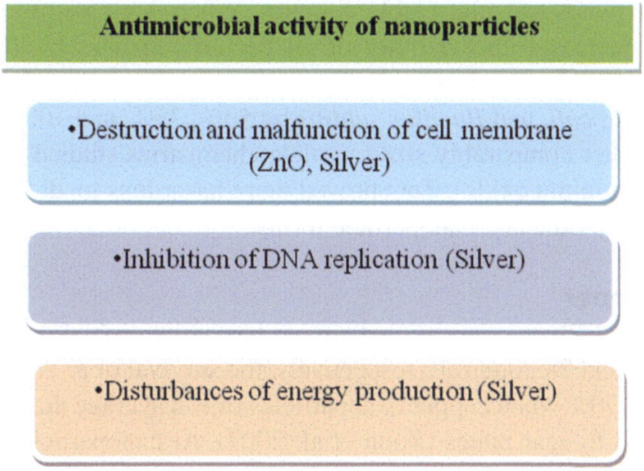

Fig. 15.5 The antimicrobial activity of nanoparticles (Karwowska 2017)

15.5.1.1 Silver

The development of silver nanoparticles was a natural and persuasive step given the history of using silver as an antibacterial agent. Since silver is a naturally antibacterial metal, it is believed that it has several antibacterial action mechanisms. In a study, on agar plates with nanoparticle concentrations of 75µg/mL or more, it was demonstrated that the growth of Gram-negative bacteria species (*Vibrio cholerae, E. coli, Salmonella typhi*, and *P. aeruginosa*) was inhibited by silver nanoparticles, which have an average diameter of 21 nm (Morones et al. 2005).

In an investigation contrasting the morphological characteristics of Gram-positive (*S. aureus*) and Gram-negative (*E. coli*) bacteria subjected to the same concentration of silver ions (10µg/mL silver nitrate) for 4–12 h, it was found that both types of bacteria displayed DNA condensation, cell membrane fragmentation, and damage to the cell wall (Feng et al. 2000). These traits are suggestive of injured and distressed microorganisms. Moreover, silver ions were found in the cytoplasm of every kind of bacteria. These findings provided more insight into how the metal ion component functions in silver's antibacterial action.

15.5.1.2 Zinc Oxide (ZnO)

ZnO has been shown to naturally reduce the activity of a wide range of (mostly Gram-positive) bacteria strains without the use of antibiotics (Söderberg et al. 1990). Implementation of nanotechnology has further enhanced the antibacterial properties of ZnO (Yamamoto 2001).

For instance, nanoparticles were found to have a stronger antibacterial effect than micronscale ZnO particles in a study that employed particle supplements to liquid cell suspensions to examine the antibacterial effect of both sizes (Jiang et al. 2009). ZnO micronscale particles and nanoparticles (average mean diameter = 60 nm, zeta potential under experimental circumstances = −5 mV) were put at a concentration of 20µg/mL to bacterial suspensions, cultured for 2 h, and then placed to agar plates to count viable colonies. When compared to controls, ZnO nanoparticles decreased the viability of all bacterial species by 100%, whereas micron-size ZnO particles decreased the colony counts of *Pseudomonas fluorescens*, *Escherichia coli*, and *Bacillus subtilis* by 50%, 75%, and 100%, respectively. Compared to other comparably sized particle chemistries studied (such as silicon dioxide and aluminium oxide), ZnO proved more hazardous to all kinds of bacteria in both the micronscale and nanoparticle forms.

15.5.1.3 Copper

Like conventional silver, conventional copper has antimicrobial qualities. At doses of 33.40μg/mL and 28.20μg/mL, respectively, the survival of *E. coli* and *B. subtilis* was reduced by 90% when copper nanoparticles (mean average diameter = 100 nm) were introduced to agar plates (Yoon et al. 2007). At nanoparticle concentrations more than 60μg/mL, the growth of both kinds of bacteria was totally stopped. In addition, somewhat smaller (mean average diameter = 40 nm) silver nanoparticles were evaluated in order to compare the antibacterial effects of the various metals. The amounts of silver nanoparticles required were greater than the required copper concentrations, at 58.4μg/mL and 32.12μg/mL, respectively, to inhibit *E. coli* and *B. subtilis* survival by 90%.

15.5.1.4 Iron Oxide

Based on a few researches on its effects on bacteria, ordinary iron oxide is not thought to be antibacterial; nonetheless, it has been shown to suppress bacterial activity. Iron oxide nanoparticles, for instance, with a mean average diameter of 9 nm and a zeta potential of −19.09 mV under experimental conditions, exhibited a reduction in *S. aureus* viability as demonstrated by a fall in the ratio of live to dead cells in a chain-like structure spanning 100–200 nm (Tran et al. 2010).

15.5.2 Synergy Between Ionic Liquids and Nanomaterials

Ionic liquids and nanoparticles can combine to generate a variety of hybrid forms, depending on how their intra- and intermolecular interactions are balanced. Ionic liquids, for instance, act as solvent media for colloidal dispersions (Ueno and Watanabe 2011), facilitating the dispersion of metal nanoparticles (Fonseca et al. 2006), nanostructured inorganic particles (Zhou and Antonietti 2003), graphene (Ravula et al. 2015), and carbon nanotubes (Shim and Kim 2009; Fukushima et al. 2003) as well as aiding in the dissolution of cellulose (Badgujar and Bhanage 2015) and dispersing the as-obtained nanocellulose (Li et al. 2012). Adding traditional stabilizers such as surfactants and/or polymers is not always necessary to maintain the stability of some colloidal particles in ionic liquids.

Presently, it is well recognized that the pharmacokinetics and biopharmaceutical properties of 95% of all novel prospective medicines are subpar. New potential for the diagnosis and treatment of many diseases has been made possible by the notion of NPs that were developed for the design of new drug delivery systems. In fact, it is conceivable to create novel approaches with a broad range of therapeutic applications while operating at submolecular scales. According to the current state of research, NPs can be created to enable medication delivery to specific sites and to enhance the pharmacokinetic profile of several substances with biological uses (Saji et al. 2010; Silva et al. 2016; Reis et al. 2008; Elkhodiry et al. 2016).

Numerous nano-based drug delivery technologies, including liposomes, fullerenes, carbon nanotubes, quantum dots, and other lipidic systems (such as solid lipid NPs, nanostructured lipid carriers, ethosomes, phytosomes, micelles, metallic NPs (e.g. gold and silver NPs), etc.), have been created. NPs can increase drug activity by shielding it from environmental deterioration, facilitating drug delivery and transportation and improving drug targeting by modifying surface charge with ligands (antibodies, nanobodies, aptamers, surfactants, polymers, and sugars, among others). A ligand may simply be a chain of inert molecules or it may contain a functional group like a thiol, amine, phosphine, or a biomolecule. NPs can also extend the time that the medicine is released over hours, days, or even weeks (Reis et al. 2014; Rosado et al. 2013).

It has become relevant to think about their combination as a way to achieve more effective hybrid systems that share the properties of both ILs and nanosystems (NS), taking into account the various advantageous properties that both materials present individually, particularly in their application in delivery systems. As a result, these novel materials exhibit extraordinary synergistic effects and are often more stable chemically and thermally (Ueno and Watanabe 2011). In quite a while, the equilibrium between intramolecular and intermolecular interactions has a significant impact on the structures that emerge from these diverse systems (Ghaedi and Vafaei 2017). For instance, by suspending NPs into ILs, which provide a protective barrier around the NPs and improve their chemical and thermal stability, stable colloidal dispersions or colloidal gels can be created.

Moreover, since IL characteristics can be adjusted to meet a particular demand, they might be created to enhance colloidal particle stabilization. The stability of metal NPs in terms of aggregation, as seen in iridium or gold NPs, is one benefit of using ILs as a reaction medium for NP production (Itoh et al. 2004). Moreover, under ambient, water-poor, or anhydrous circumstances, ILs make it easier to synthesize NP from highly polar starting materials, and their variable miscibility makes it simple to recycle the solvent and encapsulate the product (Zahmakıran and Özkar 2011). With remarkable experimental reproducibility, a simple one-pot technique for water-phase synthesis of gold nanoparticles was discovered in 2006. It makes use of the ecologically benign N-(2-hydroxyethyl)-N-methyl morpholinium tetrafluoroborate (Kim et al. 2006). Monodispersed and size-selective metal NPs were created as a result of the alcohol IL's reducing and stabilizing effects. ILs are salts and hence the solvation and stability of metal ions by ILs are higher than typical organic molecules. In fact, metallic NPs are frequently mixed with ILs for catalytic applications to increase their catalytic effectiveness and recycling capacity. Indeed, through intermolecular interactions, ILs can give NPs chemical protection and physicochemical property augmentation. They also increase NPs' durability and suitability for reuse in a variety of applications. Intermolecular interactions between

ILs and NPs act in unison to organize materials with various structures. These liquids may be covalently attached to the surface of the NPs or the NPs may be included in emulsions or prefabricated structures made of IL. While ILs are physically adsorbed on NPs, they can also be chemically bonded onto NP surfaces to create what are known as Supported Ionic Liquids nanoparticles. This process involves creating an immobilized layer on the NP surface, which improves the NPs' dispersibility and thermostability (Selvam et al. 2012).

It becomes obvious that the judicious pairing of ILs and NS may mark a significant advancement in pharmaceutics. By combining the desired properties of ILs, which can be correctly tailored, with the similarly distinctive characteristics of nanodevices, improved systems with greater and more tuned performance may be reached.

Further research must be done on the formulation of IL-NS combination materials, including the cautious preparation of various IL types, NS, and therefore new IL-NS delivery methods, without ignoring the interactions (both positive and negative) between ILs and the NS.

15.5.3 Innovative Nanotechnological Approaches for Antimicrobial Delivery

Ionic liquids have long been known to have antibacterial capabilities, most notably broad spectrums of activity against both Gram-positive and Gram-negative bacteria as well as fungus and mycobacteria. The combination of ionic liquids with other materials is known to enhance the intrinsic properties of pristine materials to destroy a wide range of microorganisms (Ye et al. 2012). The reviews that have already been written mostly concentrated on the structural characteristics of poly(ionic liquid)s (Muñoz-Bonilla and Fernández-García 2018) and ionic liquids, and they described alternate techniques for producing antibacterial materials based on ionic liquids. This chapter provides a thorough explanation of the antibacterial characteristics of materials based on ionic liquids.

A photocatalyst for the oxidation of 4-nitroaniline under ultraviolet light was created using benzimidazole-based ionic liquid coated on ZnO nanoparticles (ZnO@IL) (Rojas et al. 2020). According to the first-order kinetics of the 4-nitroaniline oxidation by ZnO@IL, a considerable deterioration took place. The ZnO@IL (150 microL, 5 mM) demonstrated antibacterial efficacy against bacteria (2.5105 CFU/mL) with microbiological inhibition of 43% and 40% against Gram-positive *S. aureus* and Gram-negative *E. coli*, respectively. The cell growth inhibition of ZnO@IL was linked to the imidazolium ring and the carboxylate group as integral reactive center that could generate the reactive oxygen species, interact with proteins (amide, carboxyl, phosphate, hydroxyl tails, and carbohydrate-related moieties), or intercalate between the purine and pyrimidine components in DNA of both types' bacteria. ZnO@IL might easily penetrate the cell, disrupt metabolic processes, impede respiration, tear down the cell wall, release components from the cell, and ultimately cause cell death (Fig. 15.6).

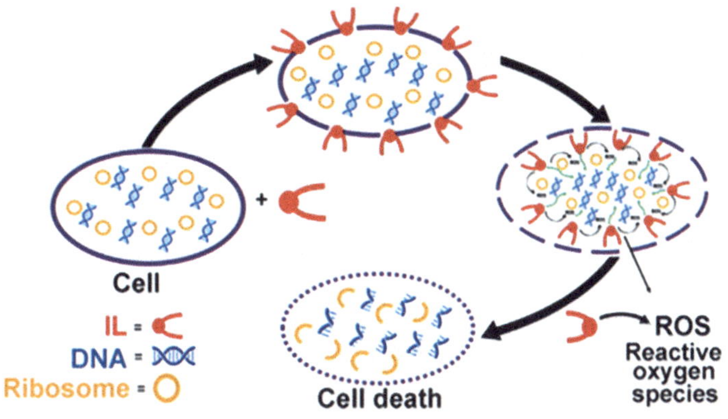

Fig. 15.6 The anticipated interactions amid ZnO@ILs and bacterial cells (Rojas et al. 2020)

15.6 Future Prospects and Challenges

Although novel prospects of the relevance of ILs have been stretched in various fields, ILs are used for skirmishing contagion. As ILs acquire anion exchangeability, the anions with diverse hydrophobicity can be customized to refrain the bacterial action. Antimicrobial ILs can also be employed as polymers/surface-active agents/membranes depending on how and where these are deployed. Besides, inclusive measures have been explored by the scientists toward conception and progression of manifold means instigating the production of compatible, non-toxic, and eco-friendly ILs. The ILs manufactured through green chemistry may be engaged in severance, refinement, and extraction of various bioactive components like amino acids, lipids, nucleic acids, fats, saponins, proteins, vitamins, oils, carotenoids, pharmaceuticals, etc. (Ventura et al. 2017). IL-based therapeutics have massive prospective to reinstate usual antibiotics in the combat of multidrug-resistant contagion. Well-deliberated and proscribed medical trials must be executed on stratagem of prospective therapeutic implication for the cure of pandemic diseases. Regardless of the accessible intricacy in receiving endorsement from the European Medicine Agency as well as US Food and Drug Administration, there have been commendable exertions to modify the biological action of IL-reliant antimicrobial compounds and to bestow these compounds as novel remedial agents in opposition to multidrug-resistant pathogens (Egorova et al. 2017; Moshikur et al. 2020; Shamshina et al. 2015).

Upon survey, it has been analysed that antimicrobial IL-reliant compounds have two foremost challengers: antimicrobial peptides as well as antimicrobial metal/metal oxide nanomaterials. Even though these challengers have stupendous characteristics, they have precincts that encumber their relevance. For illustration, antimicrobial peptides are soaring in their manufacturing outlay (Bommarius et al. 2010), probably lethal to mankind (Zhu et al. 2014), susceptible to ruthless ecological conditions like implicated medium, proteases as well as pH (Bahar and Ren 2013).

Further, these are mortified by various proteolytic enzymes and are prone to hemolysis owing to their amphipathicity as well as hydrophobicity (Mahlapuu et al. 2020). In addition, antimicrobial peptides lose their selectivity towards precise microbial stains when they are implicated in coating the materials (Ding et al. 2014). Moreover, the unfolding of a number of hefty antimicrobial peptides directs to a loss in their antimicrobial action (Bagheri et al. 2009). Metal/metal oxide nanomaterials are probably venomous to customary tissues. The form, composition, and size sway the cellular toxicity as well as antimicrobial characteristics of metal oxide and their allied nanomaterials (Makvandi et al. 2020), in turn confining their biomedical relevance. Opsonization, nonbiogenic creation, agglomeration, synergy with the monocytes and macrophages, in addition to DNA damage occurred by the production of imprudent oxygen moieties are supplementary disadvantages of antibacterial nanomaterials (Manimaran and Kannabiran 2017; Li et al. 2018). However, some of these disadvantages have been unravelled with latest emerald production techniques and surface amendments with aptamers, biomolecules, and peptides (Magala et al. 2017). Moreover, investigations are needed to resolve the cytotoxicity, biocompatibility as well as antimicrobial effectiveness of these green nanomaterials. Consequently, it emerges out that antimicrobial IL-derived compounds are appropriate antimicrobial agents of the future.

For the most part, antibacterial IL-reliant membranes are manufactured from vinyl-substituted ILs which are non-degradable. Thus, there is a rigorous inadequacy in terms of therapeutic prospects. To deal with this predicament, vinyl poly(ionic liquids) could be combined with eco-friendly man-made polymers such as polylactide, poly(butylene adipate), polyglycolide, polyorthoesters, polycaprolactone, poly(hydroxy butyrate), poly(lactide-coglycolide), or natural polymers such as carrageenan, gelatin, alginate, hyaluronic acid, chitosan, collagen, or gelatin to amplify their biodegradability (Zhang et al. 2017; Qian et al. 2019). Also, biomolecules may be constitutionally tailored with ILs (Elshaarawy et al. 2018; Bernard et al. 2018). Poly(ionic liquid)s comprising ecological moieties could further be primed, for instance, by means of polycondensation of 1,2-bis[N-(N′-hydroxylalkylimidazolium)]ethane salts with diverse diacid chlorides (terephthaloyl chloride, succinyl chloride, adipoyl chloride, and sebacoyl chloride) (Lee et al. 2011). Such exploitations can improve the biocompatibility of IL-reliant compounds. Additionally, investigations on the physicochemical as well as biological features of these biodegradable IL-composed compounds are obligatory to poise antimicrobial effectiveness with biocompatibility towards living cells.

15.7 Conclusion

The ILs have gained attention as potential designer and eco-friendly solvents that might be used in a variety of studies and industrial applications related to energy, health, and the environment. ILs have been viewed as an appealing alternative to traditional organic and inorganic solvents for applications based on biomolecules. They are important participants in projects involving drug delivery and drug discovery because of their wide chemical space and highly adjustable character. The implication of ILs as a novel approach to combat viruses could be made possible by the existing standstill in the medication discovery process. Moreover, the design of nanomaterials using ionic liquid is now one of the most developed and promising areas of nanotechnology because of the wide range of uses for these substances.

References

Abushaheen MA, Fatani AJ, Alosaimi M et al (2020) Antimicrobial resistance, mechanisms and its clinical significance. Dis Mon 66:100971

Ao M, Kim D (2013) Aggregation behavior of aqueous solutions of 1-dodecyl-3-methylimidazolium salts with different halide anions. J Chem Eng Data 58:1529–1534

Badgujar KC, Bhanage BM (2015) Factors governing dissolution process of lignocellulosic biomass in ionic liquid: current status, overview and challenges. Bioresour Technol 178:2–18

Bagheri M, Beyermann M, Dathe M (2009) Immobilization reduces the activity of surface-bound cationic antimicrobial peptides with no influence upon the activity spectrum. Antimicrob Agents Chemother 53:1132–1141

Bahar AA, Ren D (2013) Antimicrobial peptides. Pharmaceuticals 6:1543–1575

Bansal D, Malla N, Mahajan RC (2006) Drug resistance in amoebiasis. Indian J Med Res 123:115–118

Benedetto A, Ballone P (2018) Room-temperature ionic liquids and biomembranes: setting the stage for applications in pharmacology, biomedicine, and bionanotechnology. Langmuir 34:9579–9597

Bernard FL, Duczinski RB, Rojas MF et al (2018) Cellulose based poly (ionic liquids): tuning cation-anion interaction to improve carbon dioxide sorption. Fuel 211:76–86

Biswas K, Rao CEE (2007) Use of ionic liquids in the synthesis of nanocrystals and nanorods of semiconducting metal chalcogenides. Chem A Eur J 13:6123–6129

Bommarius B, Jenssen H, Elliott M et al (2010) Cost-effective expression and purification of antimicrobial and host defense peptides in *Escherichia coli*. Peptides 31:1957–1965

Busetti A, Crawford DE, Earle MJ et al (2010) Antimicrobial and antibiofilm activities of 1-alkylquinolinium bromide ionic liquids. Green Chem 12:420–425

Cao H, Alper H (2010) Palladium-catalyzed double carbonylation reactions of o-dihaloarenes with amines in phosphonium salt ionic liquids. Org Lett 12:4126–4129

Cieniecka-Rosłonkiewicz A, Pernak J, Kubis-Feder J et al (2005) Synthesis, anti-microbial activities and anti-electrostatic properties of phosphonium-based ionic liquids. Green Chem 7:855–862

Cole MR, Li M, El-Zahab B et al (2011) Design, synthesis, and biological evaluation of β-lactam antibiotic-based imidazolium-and pyridinium-type ionic liquids. Chem Biol Drug Des 78:33–41

Cuenca-Estrella M, Gomez-Lopez A, Mellado E et al (2003) Scopulariopsis brevicaulis, a fungal pathogen resistant to broad-spectrum antifungal agents. Antimicrob Agents Chemother 47:2339–2341

Cybulski J, Wisniewska A, Kulig-Adamiak A et al (2011) Mandelate and prolinate ionic liquids: synthesis, characterization, catalytic and biological activity. Tetrahedron Lett 52:1325–1328

Czekański Ł, Almeida TSD, Mota JP et al (2014) Synthesis of benzoazole ionic liquids and evaluation of their antimicrobial activity. Biomed Biopharm Res 11:227–235

Dake SA, Kulkarni RS, Kadam VN et al (2009) Phosphonium ionic liquid: a novel catalyst for benzyl halide oxidation. Synth Commun 39:3898–3904

Daschakraborty S, Biswas R (2014) Dielectric relaxation in ionic liquids: role of ion-ion and ion-dipole interactions, and effects of heterogeneity. J Chem Phys 140:014504

Davis JH Jr, Forrester KJ (1999) Thiazolium-ion based organic ionic liquids (OILs). 1, 2 novel OILs which promote the benzoin condensation. Tetrahedron Lett 40:1621–1622
de Almeida TS, Julio A, Caparica R et al (2015) Ionic liquids as solubility/permeation enhancers for topical formulations: skin permeation and cytotoxicity characterization. Toxicol Lett 2:S293
Ding Y, Wang W, Fan M et al (2014) Antimicrobial and anti-biofilm effect of Bac8c on major bacteria associated with dental caries and *Streptococcus mutans* biofilms. Peptides 52:61–67
Duman AN, Ozturk I, Tunçel A et al (2019) Synthesis of new water-soluble ionic liquids and their antibacterial profile against gram-positive and gram-negative bacteria. Heliyon 5:e02607
Egorova KS, Gordeev EG, Ananikov VP (2017) Biological activity of ionic liquids and their application in pharmaceutics and medicine. Chem Rev 117:7132–7189
Elkhodiry MA, Momah CC, Suwaidi SR et al (2016) Synergistic nanomedicine: passive, active, and ultrasound-triggered drug delivery in cancer treatment. J Nanosci Nanotechnol 16:1–18
Elshaarawy RF, Dechnik J, Hassan HM et al (2018) Novel high throughput mixed matrix membranes embracing poly ionic liquid-grafted biopolymer: fabrication, characterization, permeation and antifouling performance. J Mol Liq 266:484–494
Fahri F, Bacha K, Chiki FF et al (2020) Air pollution: new bio-based ionic liquids absorb both hydrophobic and hydrophilic volatile organic compounds with high efficiency. Environ Chem Lett 18:1403–1411
Fallon PG, Doenhoff MJ (1994) Drug-resistant schistosomiasis: resistance to praziquantel and oxamniquine induced in Schistosoma mansoni in mice is drug specific. Am J Trop Med Hyg 51:83–88
Fan A, Chuah GK, Jaenicke S (2012) Phosphonium ionic liquids as highly thermal stable and efficient phase transfer catalysts for solid–liquid Halex reactions. Catal Today 198:300–304
Fang Z, Zheng X, Li L et al (2022) Ionic liquids: emerging antimicrobial agents. Pharm Res 39:2391–2404
Feng QL, Wu J, Chen GQ et al (2000) A mechanistic study of the antibacterial effect of silver ions on *Escherichia coli* and *Staphylococcus aureus*. J Biomed Mater Res 52:662–668
Ferraz R, Silva D, Dias AR et al (2020) Synthesis and antibacterial activity of ionic liquids and organic salts based on penicillin G and amoxicillin hydrolysate derivatives against resistant bacteria. Pharmaceutics 12:221
Fishbain J, Peleg AY (2010) Treatment of Acinetobacter infections. Clin Infect Dis 51:79–84
Fonseca GS, Machado G, Teixeira SR et al (2006) Synthesis and characterization of catalytic iridium nanoparticles in imidazolium ionic liquids. J Colloid Interface Sci 301:193–204
Forero Doria O, Castro R, Gutierrez M et al (2018) Novel alkylimidazolium ionic liquids as an antibacterial alternative to pathogens of the skin and soft tissue infections. Molecules 23:2354
Frizzo CP, Bender CR, Salbego PR et al (2016) Thermodynamic properties of the aggregation behavior of a dicationic ionic liquid determined by different methods. Colloids Surf A Physicochem Eng 494:1–8
Fukushima T, Kosaka A, Ishimura Y et al (2003) Molecular ordering of organic molten salts triggered by single-walled carbon nanotubes. Science 300:2072–2074
Garcia MT, Ribosa I, Perez L et al (2014) Self-assembly and antimicrobial activity of long-chain amide-functionalized ionic liquids in aqueous solution. Colloids Surf B Biointerfaces 123:318–325
Gebbie MA, Smith AM, Dobbs HA et al (2017) Long range electrostatic forces in ionic liquids. Chem Commun 53:1214–1224
Ghaedi AM, Vafaei A (2017) Applications of artificial neural networks for adsorption removal of dyes from aqueous solution: a review. Adv Colloid Interface Sci 245:20–39
Ghandi K (2014) A review of ionic liquids, their limits and applications. Green Sustain Chem 4:44–53
Giernoth R (2010) Task-specific ionic liquids. Angew Chem Int Ed 49:2834–2839
Gilmore BF, Andrews GP, Borberly G et al (2013) Enhanced antimicrobial activities of 1-alkyl-3-methyl imidazolium ionic liquids based on silver or copper containing anions. N J Chem 37:873–876

Gomes JM, Silva SS, Reis RL (2019) Biocompatible ionic liquids: fundamental behaviours and applications. Chem Soc Rev 48:4317–4335

Hamilton KW, Fishman NO (2014) Antimicrobial stewardship interventions: thinking inside and outside the box. Infect Dis Clin North Am 28:301–313

Hayes R, Warr GG, Atkin R (2015) Structure and nanostructure in ionic liquids. Chem Rev 115:6357–6426

He B, Ou G, Zhou C et al (2013) Antimicrobial ionic liquids with fumarate anion. J Chem 2013:1–7

Hodyna D, Bardeau JF, Metelytsia L et al (2016) Efficient antimicrobial activity and reduced toxicity of 1-dodecyl-3-methylimidazolium tetrafluoroborate ionic liquid/β-cyclodextrin complex. Chem Eng J 284:1136–1145

Hoque J, Ghosh S, Paramanandham K, Haldar J (2019) Charge-switchable polymeric coating kills bacteria and prevents biofilm formation in vivo. ACS Appl Mater Interfaces 11:39150–39162

Hough WL, Smiglak M, Rodríguez H et al (2007) The third evolution of ionic liquids: active pharmaceutical ingredients. N J Chem 31:1429–1436

Howard SJ, Arendrup MC (2011) Acquired antifungal drug resistance in Aspergillus fumigatus: epidemiology and detection. Med Mycol J 49:S90–S95

Hutchings M, Truman A, Wilkinson B (2019) Antibiotics: past, present and future. Curr Opin Microbiol 51:72–80

Itoh H, Naka K, Chujo Y (2004) Synthesis of gold nanoparticles modified with ionic liquid based on the imidazolium cation. J Am Chem Soc 126:3026–3027

Iwai N, Nakayama K, Kitazume T (2011) Antibacterial activities of imidazolium, pyrrolidinium and piperidinium salts. Bioorg Med Chem Lett 21:1728–1730

Jamaledin R, Yiu CK, Zare EN et al (2020) Advances in antimicrobial microneedle patches for combating infections. Adv Mater 32:2002129

Janani N, Zare EN, Salimi F, Makvandi P (2020) Antibacterial tragacanth gum-based nanocomposite films carrying ascorbic acid antioxidant for bioactive food packaging. Carbohydr Polym 247:116678

Jiang W, Mashayekhi H, Xing B (2009) Bacterial toxicity comparison between nano-and micro-scaled oxide particles. Environ Pollut 157:1619–1625

Karwowska E (2017) Antibacterial potential of nanocomposite-based materials—a short review. Nanotechnol Rev 6:243–254

Kim KS, Choi S, Cha JH et al (2006) Facile one-pot synthesis of gold nanoparticles using alcohol ionic liquids. J Mater Chem 16:1315–1317

Kumar H, Katal A (2018) Spectroscopic and temperature dependent physicochemical studies on interactional behavior of phosphate salts in aqueous ionic liquid (1-butyl-3-methyl imidazolium tetrafluoroborate) solutions. J Mol Liq 251:273–285

Kumari P, Pillai VV, Benedetto A (2020) Mechanisms of action of ionic liquids on living cells: the state of the art. Biophys Rev 12:1187–1215

Lee M, Choi UH, Salas-de la Cruz D et al (2011) Imidazolium polyesters: structure–property relationships in thermal behavior, ionic conductivity, and morphology. Adv Funct Mater 21:708–717

Lei J, Sun L, Huang S et al (2019) The antimicrobial peptides and their potential clinical applications. Am J Transl Res 11:3919

Li J, Wei X, Wang Q et al (2012) Homogeneous isolation of nanocellulose from sugarcane bagasse by high pressure homogenization. Carbohydr Polym 90:1609–1613

Li S, Dong S, Xu W et al (2018) Antibacterial hydrogels. Adv Sci 5:1700527

Liu JF, Jiang GB, Jönsson JÅ (2005) Application of ionic liquids in analytical chemistry. TrAC Trends Anal Chem 24:20–27

Lu B, Zhou G, Xiao F et al (2020) Stimuli-responsive poly (ionic liquid) nanoparticles for controlled drug delivery. J Mater Chem B 8:7994–8001

Łuczak J, Jungnickel C, Łącka I et al (2010) Antimicrobial and surface activity of 1-alkyl-3-methylimidazolium derivatives. Green Chem 12:593–601

MacFarlane DR, Chong AL, Forsyth M et al (2018) New dimensions in salt–solvent mixtures: a 4th evolution of ionic liquids. Faraday Discuss 206:9–28

Magala P, Bocik WE, Majumdar A, Tolman JR (2017) Conformational dynamics modulate activation of the ubiquitin conjugating enzyme Ube2g2. ACS Omega 2:4581–4592
Magna L, Chauvin Y, Niccolai GP, Basset JM (2003) The importance of imidazolium substituents in the use of imidazolium-based room-temperature ionic liquids as solvents for palladium-catalyzed telomerization of butadiene with methanol. Organometallics 22:4418–4425
Mahlapuu M, Björn C, Ekblom J (2020) Antimicrobial peptides as therapeutic agents: opportunities and challenges. Crit Rev Biotechnol 40:978–992
Makvandi P, Wang CY, Zare EN et al (2020) Metal-based nanomaterials in biomedical applications: antimicrobial activity and cytotoxicity aspects. Adv Funct Mater 30:1910021
Manimaran M, Kannabiran K (2017) Actinomycetes-mediated biogenic synthesis of metal and metal oxide nanoparticles: progress and challenges. Lett Appl Microbiol 64:401–408
Melo CI, Bogel-Łukasik R, da Ponte MN, Bogel-Łukasik E (2013) Ammonium ionic liquids as green solvents for drugs. Fluid Ph Equilib 338:209–216
Miskiewicz A, Ceranowicz P, Szymczak M et al (2018) The use of liquids ionic fluids as pharmaceutically active substances helpful in combating nosocomial infections induced by Klebsiella Pneumoniae New Delhi strain, Acinetobacter Baumannii and Enterococcus species. Int J Mol Sci 19:2779
Moehring RW, Anderson DJ (2012) Antimicrobial stewardship as part of the infection prevention effort. Curr Infect Dis Rep 14:592–600
Morones JR, Elechiguerra JL, Camacho A et al (2005) The bactericidal effect of silver nanoparticles. Nanotechnology 16:2346
Moshikur RM, Chowdhury MR, Moniruzzaman M et al (2020) Biocompatible ionic liquids and their applications in pharmaceutics. Green Chem 22:8116–8139
Muñoz-Bonilla A, Fernández-García M (2018) Poly (ionic liquid) s as antimicrobial materials. Eur Polym J 105:135–149
Muzny CA, Schwebke JR (2013) The clinical spectrum of Trichomonas vaginalis infection and challenges to management. Sex Transm Infect 89:423–425
Obando D, Pantarat N, Handke R et al (2009) Synthesis, antifungal, haemolytic and cytotoxic activities of a series of bis (alkylpyridinium) alkanes. Bioorg Med Chem 17:6329–6339
Owens RC Jr (2008) Antimicrobial stewardship: concepts and strategies in the 21st century. Diagn Microbiol Infect Dis 61:110–128
Pendleton JN, Gilmore BF (2015) The antimicrobial potential of ionic liquids: a source of chemical diversity for infection and biofilm control. Int J Antimicrob Agents 46:131–139
Pernak J, Sobaszkiewicz K, Mirska I (2003) Anti-microbial activities of ionic liquids. Green Chem 5:52–56
Pernak J, Skrzypczak A, Lota G, Frackowiak E (2007) Synthesis and properties of trigeminal tricationic ionic liquids. Chem A Eur J 13:3106–3112
Pernak J, Borucka N, Walkiewicz F et al (2011) Synthesis, toxicity, biodegradability and physicochemical properties of 4-benzyl-4-methylmorpholinium-based ionic liquids. Green Chem 13:2901–2910
Popęda M, Płuciennik E, Bednarek AK (2014) Proteins in cancer resistance. Postępy Higieny i Medycyny Do'swiadczalnej 68:616–632
Prudêncio C, Vieira M, Van der Auweraer S, Ferraz R (2020) Recycling old antibiotics with ionic liquids. Antibiotics 9:578
Qi L, Cui J (2013) A schistosomiasis model with praziquantel resistance. Discret Dyn Nat Soc 2013:1–13
Qian L, Fan Y, Song H et al (2019) Poly (ionic liquid)/carboxymethyl chitosan complex-derived nitrogen and sulfur codoped porous carbon for high-performance supercapacitors. Ionics 25:4915–4924
Ravula S, Baker SN, Kamath G, Baker GA (2015) Ionic liquid-assisted exfoliation and dispersion: stripping graphene and its two-dimensional layered inorganic counterparts of their inhibitions. Nanoscale 7:4338–4353
Reis CP, Veiga FJ, Ribeiro AJ et al (2008) Nanoparticulate biopolymers deliver insulin orally eliciting pharmacological response. J Pharm Sci 97:5290–5305

Reis CP, Martinho N, Rosado C et al (2014) Design of polymeric nanoparticles and its applications as drug delivery systems for acne treatment. Drug Dev Ind Pharm 40:409–417

Rodero L, Mellado E, Rodriguez AC et al (2003) G484S amino acid substitution in lanosterol 14-α demethylase (ERG11) is related to fluconazole resistance in a recurrent Cryptococcus neoformans clinical isolate. Antimicrob Agents Chemother 47:3653–3656

Rojas LMG, Huerta-Aguilar CA, Tecuapa-Flores ED et al (2020) Why ionic liquids coated ZnO nanocomposites emerging as environmental remediates: enhanced photo-oxidation of 4-nitroaniline and encouraged antibacterial behavior. J Mol Liq 319:114107

Rosado C, Silva C, Reis CP (2013) Hydrocortisone-loaded poly (ε-caprolactone) nanoparticles for atopic dermatitis treatment. Pharm Dev Technol 18:710–718

Saji VS, Choe HC, Yeung KW (2010) Nanotechnology in biomedical applications: a review. Int J Nano Biomater 3:119–139

Samorì C (2011) Ionic liquids and their biological effects towards microorganisms. Curr Org Chem 15:1888–1904

Sekhon BS (2011) Ionic liquids: pharmaceutical and biotechnological applications. Asian J Pharm Biol Res 1:395–411

Selvam T, Machoke A, Schwieger W (2012) Supported ionic liquids on non-porous and porous inorganic materials—a topical review. Appl Catal A Gen 445:92–101

Shamshina JL, Kelley SP, Gurau G et al (2015) Chemistry: develop ionic liquid drugs. Nature 528:188–189

Shim Y, Kim HJ (2009) Solvation of carbon nanotubes in a room-temperature ionic liquid. ACS Nano 3:1693–1702

Silva CO, Molpeceres J, Batanero B et al (2016) Functionalized diterpene parvifloron D-loaded hybrid nanoparticles for targeted delivery in melanoma therapy. Ther Deliv 7:521–544

Singh V (2013) Antimicrobial resistance. In: Microbial pathogens and strategies for combating them: science, technology and education, vol 1. Formatex Research Center, Norristown, PA, pp 291–296

Söderberg TA, Sunzel B, Holm S et al (1990) Antibacterial effect of zinc oxide in vitro. Scand J Plast Reconstr Surg Hand Surg 24:193–197

Suppiah J, Zain RM, Nawi SH et al (2014) Drug-resistance associated mutations in polymerase (p) gene of hepatitis B virus isolated from malaysian HBV carriers. Hepat Mon 14:1–7

Sydow M, Owsianiak M, Framski G et al (2018) Biodiversity of soil bacteria exposed to sub-lethal concentrations of phosphonium-based ionic liquids: effects of toxicity and biodegradation. Ecotoxicol Environ Saf 147:157–164

Tran N, Mir A, Mallik D et al (2010) Bactericidal effect of iron oxide nanoparticles on *Staphylococcus aureus*. Int J Nanomedicine 15:277–283

Ueno K, Watanabe M (2011) From colloidal stability in ionic liquids to advanced soft materials using unique media. Langmuir 27:9105–9115

Ventura SPM, Silva FAE, Quental MV et al (2017) Ionic-liquid-mediated extraction and separation processes for bioactive compounds: past, present, and future trends. Chem Rev 117:6984–7052

Walkiewicz F, Materna K, Kropacz A et al (2010) Multifunctional long-alkyl-chain quaternary ammonium azolate based ionic liquids. New J Chem 34:2281–2289

Wang C, Luo H, Luo X et al (2010) Equimolar CO_2 capture by imidazolium-based ionic liquids and superbase systems. Green Chem 12:2019–2023

Wang H, Shi X, Yu D et al (2015) Antibacterial activity of geminized amphiphilic cationic homopolymers. Langmuir 31:13469–13477

Wang C, Qiao Y, Kang Y et al (2019) Bacteria-activated chlorin e6 ionic liquid based on cation and anion dual-mode antibacterial action for enhanced photodynamic efficacy. Biomater Sci 7:1399–1410

Wu RL, Wang XL, Li F et al (2009) Green composite films prepared from cellulose, starch and lignin in room-temperature ionic liquid. Bioresour Technol 100:2569–2574

Wutzler P (1997) Antiviral therapy of herpes simplex and varicellazoster virus infections. Intervirology 40:343–356

Yamamoto O (2001) Influence of particle size on the antibacterial activity of zinc oxide. Int J Inorg Mater 3:643–646

Yan X, Anguille S, Bendahan M, Moulin P (2019) Ionic liquids combined with membrane separation processes: a review. Sep Purif Technol 222:230–253

Ye Q, Gao T, Wan F et al (2012) Grafting poly (ionic liquid) brushes for anti-bacterial and anti-biofouling applications. J Mater Chem 22:13123–13131

Yoon KY, Byeon JH, Park JH, Hwang J (2007) Susceptibility constants of *Escherichia coli* and *Bacillus subtilis* to silver and copper nanoparticles. Sci Total Environ 373:572–575

Zafarani-Moattar MT, Asadzadeh B, Shahrisa A, Nazari MG (2014) Study of thermodynamic properties of l-serine in aqueous 1-carboxymethyl-3-methylimidazolium chloride solutions at 298.15 K. Fluid Ph Equilib 363:32–40

Zahmakıran M, Özkar S (2011) Metal nanoparticles in liquid phase catalysis; from recent advances to future goals. Nanoscale 3:3462–3481

Zakrewsky M, Lovejoy KS, Kern TL et al (2014) Ionic liquids as a class of materials for transdermal delivery and pathogen neutralization. Proc Natl Acad Sci 111:13313–13318

Zhang B, Sudre G, Quintard G et al (2017) Guar gum as biosourced building block to generate highly conductive and elastic ionogels with poly (ionic liquid) and ionic liquid. Carbohydr Polym 157:586–595

Zhao H, Xia S, Ma P (2005) Use of ionic liquids as 'green' solvents for extractions. J Chem Technol Biotechnol 80:1089–1096

Zhou Y, Antonietti M (2003) Synthesis of very small TiO_2 nanocrystals in a room-temperature ionic liquid and their self-assembly toward mesoporous spherical aggregates. J Am Chem Soc 125:14960–14961

Zhu X, Dong N, Wang Z et al (2014) Design of imperfectly amphipathic α-helical antimicrobial peptides with enhanced cell selectivity. Acta Biomater 10:244–257

Nanomaterials in the Development of Advanced Antimicrobial Agents

16

Neha Sharma, Mobina Kouser, and Monika Gupta

Abstract

Nanotechnology, the application of materials with atomic or molecular-scale dimensions, has grown in popularity for medical uses and is an extremely exclusive method for eradicating or suppressing the activity of several pathogens. Antibiotics and additives are being replaced with nanostructured materials (NSMs) in a variety of goods to impart a microbicidal effect. Due to the rising prevalence of antibiotic resistance, researchers are more focused on the therapeutic potential of nanoparticles (NPs) as antibacterial and antiviral agents. Different nanomaterials have evolved as innovative strategies to deal with microbial infections and the problems caused by conventional antimicrobials. Unique antimicrobial characteristics are exhibited by metal nanoparticles (MNPs), metal oxide nanoparticles (MeO-NPs), and other nanomaterials acting as a pivotal point for the growth of innovative medicinal devices. The small size of MNPs imparts the capacity to damage microbial cells through a variety of methods. Nanomaterials are established to be effective antimicrobial agents against a variety of microbial diseases. Nanomaterials offer an intriguing route to restrict microbial development before human infection, in contrast to antibiotics, which are used to treat illnesses and infections in patients. The focus of this chapter is to provide comprehensive overview of multiple nanomaterials-based antimicrobial agents such as MNPs, MeO-NPs, different nanocomposite, nanoemulsions, carbon-based nanomaterials, quantum dots, liposomes, micelles, etc. along with their synthesis strategies used, particle size & dimensions, particular targeted microbial strain, type of antimicrobial testing method used, reported MIC value, and antimicrobial action mechanisms.

N. Sharma · M. Kouser · M. Gupta (✉)
Department of Chemistry, University of Jammu, Jammu, India

M. Y. Wani et al. (eds.), *Nanotechnology Based Strategies for Combating Antimicrobial Resistance*, https://doi.org/10.1007/978-981-97-2023-1_16

Keywords

Nanomaterials · Antimicrobial agents · Liposomes · Quantum dots · Nanocomposite

16.1 Introduction

The importance of nanoparticles (NPs) or invisible particulate materials with a diameter of 1–100 nm (nm) in today's science cannot be underestimated (Song et al. 2021). Because of their enormous surface area to volume ratio, NPs are distinctive in many ways as delivery and antibacterial agents. NPs have been employed for human wellbeing in a variety of disciplines throughout the last few decades (Ingle et al. 2014). Due to their potential use in a variety of sectors, including medical science, drug delivery, electronics, waste water treatment etc., NPs have recently been the subject of significant scientific research (Sharmin et al. 2021). The distinctive qualities of the NP, such as its form, charge, size, high ratio of surface area to mass, and high reactivity, are the basis for its potential uses in a variety of sectors.

A group of illnesses or disorders that negatively impact human health and are brought on by pathogenic microbes (bacteria, viruses, fungi, protozoa, and parasites) are referred to as infectious diseases. Over the past few decades, the rise of infectious diseases and multidrug-resistant (MDR) strains has posed a threat to world health. On both living and nonliving surfaces, a vast variety of bacteria exist and thrive, which contributes to the emergence of infectious disease outbreaks, the prevalence of healthcare-associated infections, and the rise of MDR pathogens (Ribeiro et al. 2022). Antibiotics considerably reduce symptoms as well as signs linked with certain microorganisms and their existence. Because of widespread overconsumption and inappropriate dose, the therapeutic effect of conventional antimicrobial medicines is quickly degrading. The severe acute respiratory syndrome coronavirus 2 (SARS-CoV-2)-caused worldwide pandemic COVID-19 exacerbated the issue of MDR infections and increased the need for more potent antimicrobials. According to recent studies, bacteria have built resistance to β-lactam antibiotics by creating a new kind of β-lactamase (Fishovitz et al. 2014). A particular promising strategy to combat this kind of resistance built-up by microorganisms is NP-based therapy. As a result, conjugating antibiotic usage with NPs may increase inhibitory capability and reduce the chance that organisms will develop resistance (Sobhani et al. 2017). Nanotechnology is providing creative answers to the advancements in antimicrobial treatment, medication delivery, and the creation of improved materials (Arias et al. 2018). More precise drug targeting and administration, reduced toxicity while preserving therapeutic benefits, increased safety and biocompatibility, and quicker creation of novel safe medications are among the key objectives for research into nano-biotechnologies in drug delivery. In addition, the aims for NP entrapment of drugs are enhanced delivery to, or uptake by, target cells and/or a reduction in the toxicity of the free drug to non-target organs. Both circumstances will cause a rise in the therapeutic index, with the difference between dosages producing therapeutic efficacy (such as the killing of tumour cells) and toxicity to other organ systems. The physicochemical and biological behaviour of NPs may be changed dependent on their surroundings by appropriately functionalizing them with chemical/biological groups, resulting in more optimized applications (such as drug delivery). It is crucial to understand the antimicrobial mechanism of developed nanoparticles (NPs). Reactive oxygen species (ROS) generation has consistently been shown to be one of the cell cytotoxicity mechanisms for MNPs and MeO-NPs (An et al. 2009). The use of pharmacological combinations is a widespread clinical practice, effective in treating conditions like cancer, cardiovascular disease, microbial infections, and AIDS. Silver nanoparticles (AgNPs) are the main subject of most synergistic investigations. Gold (Au), copper (Cu), copper oxide (CuO), copper sulphide (CuS), iron (Fe), iron oxide (Fe_3O_4/Fe_2O_3), zinc (Zn), zinc oxide (ZnO), and platinum (Pt) are another MNPs which are explored in this approach for formulating antibiotic, antifungal, and antiviral substances (Faya et al. 2018).

In this chapter, we have provided an overview of the nanomaterials such as MNPs, metal-oxide, nanocomposite, nanoemulsions (NEs), nanocarbon, and others having antimicrobial activity.

16.2 Research Findings on Antimicrobial Nanomaterials

Over the past few decades, the emergence of MDR pathogens due to misuse and overuse of antimicrobials has left the pharmaceutical arsenal unarmed to fight against such severe microbial infections (Morsy et al. 2020). There are numerous antimicrobial medicines that can be used to combat infections, but they frequently have a number of drawbacks (Fadda et al. 2020; Grozav et al. 2021; Kumar and Kumar 2020). Infectious diseases constitute an increasing threat to public health and medical systems worldwide. A multisectoral effort is urgently needed to fulfil the Sustainable Development Goals. Methicillin-resistant Staphylococcus aureus (MRSA), which is a single bacterium, kills more Americans annually than emphysema, HIV/AIDS, Parkinson's disease, and homicide altogether (Spellberg et al. 2011). As a result, there is an acute requirement for the study and development of a new generation of advanced antimicrobial compounds. The scientific community has been concentrating on the investigation of nanomaterials, primarily metal-based (MNPs), in order to test their antibacterial capabilities and viability to remove pollution sources and diseases (Vimbela et al. 2017). They have allowed the development of novel drugs and materials based on the improvements made to the chemical, physical, and biological characteristics of NPs at the nanoscale in terms of their surface area, size, distribution, and shape (Singh et al. 2020; Buzea et al. 2007; Correa et al. 2020). Research demonstrates that the synthesis technique employed to produce the nanomaterials has a distinct impact on their antibacterial capabilities. Physical approaches often have the highest financial and energy expenditures (Dhand et al. 2015; Chandrasekaran et al. 2016). As a result, research has favoured chemical synthesis techniques that can produce a lot of nanomaterials in a short amount of time (Ingale and Chaudhari 2013).

16.2.1 Metal Nanoparticles as Alternative Antimicrobial Agents to Combat Multidrug Resistance

Among the various metals, silver, gold, copper, iron, zinc, cobalt, platinum, and palladium are among the most significant biomedical MNPs used in antimicrobial compositions. Saeede et al. synthesized AgNPs by using the A. atropatana extract as a reducing and capping agent. The prepared AgNPs have good physical and biological impacts. Additionally, AgNPs had antibacterial and anticancer properties and could more strongly induce the death of cells than necrosis in cancer cell lines (Dehghanizade et al. 2018). The effectiveness of AgPtNHs (silver-platinum nanohybrid) in combination with antibiotics against both Gram-positive and Gram-negative bacterial infections was tested. The three test pathogens such as E. coli, P. aeruginosa, and S. aureus were suppressed by AgPtNHs alone, but resilient antibacterial synergy was demonstrated when AgPtNHs were used in conjunction with antibiotics like rifampicin and novobiocin (Ranpariya et al. 2021). The enhanced antibacterial activities are created by the conjugation of MNPs and generic antimicrobial agents. Table 16.1 displays different MNPs as antimicrobial agents.

16.2.1.1 Mechanistic Mode of Action

It has been discovered that metal-based NPs have non-specific bacterial toxicity mechanisms because they do not attach to a specific receptor in the bacterial cell. This increases the spectrum of their antibacterial effect and makes it harder for germs to develop resistance to them. Most studies on the potency of metal-based NPs in bacteria have so far yielded favourable results for both Gram-positive and Gram-negative bacteria. The mechanism of action is poorly understood but the main reason is that the ion imbalance significantly disrupts the homeostasis of the microorganisms, impairing respiration, interrupting energy transmission, and ultimately causing cell death (Ribeiro et al. 2022).

16.2.2 Metal Oxide Nanoparticle-Based Antimicrobial Agents Against AMR

MeO-NPs are versatile because they have distinct physical and chemical properties related to their nanometer size. Recently, MeO-NPs have emerged as an intriguing alternative source of antibiotics for fighting microbes with high resistance. MeO-NPs not only have antibacterial qualities but also act as drug transporters, making it difficult for microbes to become resistant to them. As antimicrobial agents, some metal oxides are widely employed in healthcare, agriculture, and environmental protection. Potential antimicrobial agents are CoO, CaO, CuO, NiO, Cu_2O, MgO, TiO_2, Ag_2O, SiO_2, ZnO, and Fe_xO_y in their nanoforms. Their antimicrobial properties depend upon various factors. Similar to metal ions, MeO-NPs may use electrostatic interactions to damage the integrity of bacterial envelope as well as the potential of the cell membrane. CaONPs have a strong antibacterial potential and can neutralize microbial endotoxins. CaONPs are safe materials for both humans

Table 16.1 Different MNPs along with their synthesis strategies used, particle size and dimensions, microbial target, type of antimicrobial testing method used, and MIC are given

S. no.	Metal nanomaterials (MNPs)	Synthesis strategies used	Particle size and dimensions	Microbial target	Type of antimicrobial testing method used	Different strains tested	MIC (μg/mL)	Ref.
1	Ag NPs	Green synthesis using the extract of *Acacia rigidula* as an reducing and capping agent	22.46 nm	Antibactericidal activity in both gram-negative and gram-positive bacteria	Drop count method	1. Escherichia coli ATCC-11229 2. Pseudomonas aeruginosa ATCC 3. P. aeruginosa (clinical multidrug-resistant strain) 4. Bacillus subtilis ATCC	62.5, 15.6, 7.8, 0.5, respectively	Escárcega-González et al. (2018)
2	Fe – Ag1 bimetallic NPs with 19% Ag content	Reduction of silver ions by available zerovalent Fe NPs	Needle-shaped lepidocrocite nanostructures with size ranging between 10 and 20 nm	Antifungal	Microdilution method	*Candida albicans* (I and II), *Candida tropicalis, and Candida parapsilosis strains*	3.3 mg/L to 270 mg/L	Markova et al. (2013)

(continued)

Table 16.1 (continued)

S. no.	Metal nanomaterials (MNPs)	Synthesis strategies used	Particle size and dimensions	Microbial target	Type of antimicrobial testing method used	Different strains tested	MIC (μg/mL)	Ref.
3	Cu NPs	Modified polyol method by the reduction of copper acetate hydrate in the presence of Tween 80 by refluxing between 190° and 200 °C	Cu NPs crystallize in FCC structure with size <100 nm	Antibacterial and antifungal	Agar disk diffusion	*M. luteus, S. aureus, E. coli, K. pneumoniae, P. aeruginosa, A. falvus, A. niger, C. albicans*	(16 ± 0.781), (21 ± 1.612), (26 ± 0.985), (15 ± 0.689), (5 ± 1.136), (13 ± 0.577), (16 ± 0.354), (23 ± 1.258), respectively in mm	Ramyadevi et al. (2012)
4	Au NPs	Chemical reduction method	6–40 nm	Antibacterial	Diffusion method by Agar plates	*E. coli*	2.93 μg/mL	Shamaila et al. (2016)
5	Ni NPs	Green reduction method using extract of *O. sanctum*	12–36 nm	Antibacterial and antifungal		Bacteria: *S. typhi* *S. epidermidis* Fungi: *A. fumigatus* *A. niger*	25 ± 0.2 50 ± 0.5 50 ± 1.0 50 ± 0.8	Jeyaraj Pandian et al. (2016)
6	Co NPs	Green synthesis from the aqueous extract of *Ziziphora clinopodioides Lam*	29.08 nm	Antibacterial, antifungal, antioxidant, cytotoxicity	Agar diffusion tests and macrobroth dilution assay	Fungi: *C. albicans*	4 ± 0	Hou et al. (2020)

and animals, and because of their special structural and optical qualities, they may be employed as a possible drug delivery vehicle, such as for the administration of chemotherapy drugs. There are two types of copper oxides: cuprous oxide (Cu_2O) and cupric oxide (CuO). According to certain research, Cu_2O exhibits more antibacterial activity than CuO due to its increased capacity to produce poisonous Cu1. The ability of liberated Cu1 ions to interact with thiol groups from amino acid residues in proteins is what gives Cu_2O its bactericidal properties (Stanić and Tanasković 2020). Table 16.2 illustrates MeO-NPs as antimicrobial agents.

16.2.2.1 Mechanistic Mode of Action

MeO-NPs are widely sought-after as antibacterial agents due to their distinctive characteristics. But our knowledge of their molecular mechanisms remains in the beginning stages, and recent research has shown that the mechanism of the antibacterial activity depends on the type and chemistry of MeO-NP. The antimicrobial activity is primarily achieved through the generation of ROS. Other mechanisms of action include genotoxicity, photokilling, protein and enzyme malfunction, disruption of metal/metal ion equilibrium, and damage to cell membranes caused by electrostatic contact (Raghunath and Perumal 2017). MeO-NPs such as ZnO and TiO_2 cause the production of ROS such as superoxide anion (O_2^-), hydroxyl radicals ($OH^•$), hydrogen peroxide (H_2O_2), and organic hydroperoxides once they enter bacteria. These ROS are harmful to bacteria, damaging nearly all organic biomolecules like amino acids, carbohydrates, lipids, nucleic acids, and proteins ultimately leading to microbial death (Nel et al. 2006a). ROS primarily inhibit respiratory enzymes. In the electron transport chain, $O_2^{•-}$ causes broken iron-sulfur (Fe-S) clusters to release more ferrous ions, which reduces ATP synthesis. When treated with MeONPs, these electrostatic interactions cause morphological changes in bacterial cells that distort and destroy the bacterial cell membrane (Kohanski et al. 2007). When ZnO-NPs were present, ROS in fungus displayed an anticandidal activity via cytotoxicity and apoptotic cell death. ZnO-NPs are proven to prevent the development of fungi by obstructing cell function and hyphal production (Shoeb et al. 2013; He et al. 2011). Figure 16.1 shows an overview of antimicrobial mechanism by MeO-NPs.

16.2.3 Functionalized Quantum Dots as Antimicrobial Agents

QDs, a significant class of fluorescent labels for biological, biomedical, and biosensing applications, are semiconductor crystals with nanoscale dimensions (2–10 nm) and 200–10,000 atoms (Niemeyer 2001). QDs have demonstrated exceptional fluorescence features, including photostability, tunable emission spectra, and a high, specific bioactivity when coated and coupled to functional groups. This is due to their extremely tiny size. However, several studies also revealed that QDs had cellular toxicity (Neelgund et al. 2012). So many studies concentrated on figuring out how to make QDs less toxic while still improving their effectiveness as antibacterial agents. For instances, QDs when functionalized by polymers such as

Table 16.2 Different MeO-NPs along with their synthesis strategies used, particle size and dimensions, microbial target, type of antimicrobial testing method used, and MIC are given

S. no	Nanomaterial	Synthesis strategies used	Particle size and dimensions	Microbial target	Type of antimicrobial testing method used	MIC (minimum inhibitory concentration)	Reference
1	ZnO	Precipitation method using $Zn(NO_3)_2$ and NaOH	45 nm	*E. coli*, a Gram-negative bacterium	Paper disk diffusion assay	mm (*E. coli*) 27 (0.1)	Padmavathy and Vijayaraghavan (2008)
2	α-Fe_2O_3 NPs	Green synthesis using aqueous extract of *Laurus nobilis L.*	8.03 ± 8.99 nm, almost spherical in shape	Antifungal activity against *Aspergillus favus* and *Penicillium spinulosum*	Disk diffusion method	20 and 19 mm, respectively	Jamzad and Kamari Bidkorpeh (2020)
3	NiO NPs	Green synthesis using *Stevia* Leaf Broth	10–40 nm with average crystallite size 13.7 nm with almost spherical in shape	Gram-positive (*Bacillus subtilis* and *Streptococcus pneumonia*)	DPPH reduction assay	12 and 14 mm, respectively	Srihasam et al. (2020)
4	CuO NPs	Reflux condensation method	25 nm with flower-shaped nanostructure	Antifungal activity against *Rhizopus oryzae* and *Cladosporium carrionii*	Agar disk diffusion method	0.8 and 0.9, respectively (cm)	Mageshwari and Sathyamoorthy (2013)
5	CeO_2 NPs	Green synthesis using *O. Europaea* leaf extract	Each grain of size 24 nm was composed of a cluster of 6 nm crystallites	Antimicrobial activity against Gram-positive and Gram-negative bacteria	Disk diffusion method	10 and 19, respectively (mm)	Maqbool et al. (2016)

S. no	Nanomaterial	Synthesis strategies used	Particle size and dimensions	Microbial target	Type of antimicrobial testing method used	MIC (minimum inhibitory concentration)	Reference
6	Co_3O_4 NPs	Pyrolytic technique	Co_3O_4-I is square in shape with 10–25 nm Co_3O_4-II is hexagon in shape with 100–150 nm	*S. aureus and E. coli*	Agar diffusion method	128 μg/mL and 64 μg/mL, respectively for Co_3O_4-I 64 μg/mL and 32 μg/mL, respectively for Co_3O_4-II	Ghosh et al. (2014)
7	Chromium (III) oxide (Cr_2O_3) NPs	Solvothermal method	28.0481 nm	*Escherichia coli, Enterobactera erogenes, Klebsiella pneumonia, Serratiamarc escens*	Agar diffusion method	4, 12, 6, and 3, respectively (mm)	Muthu et al. (2022)

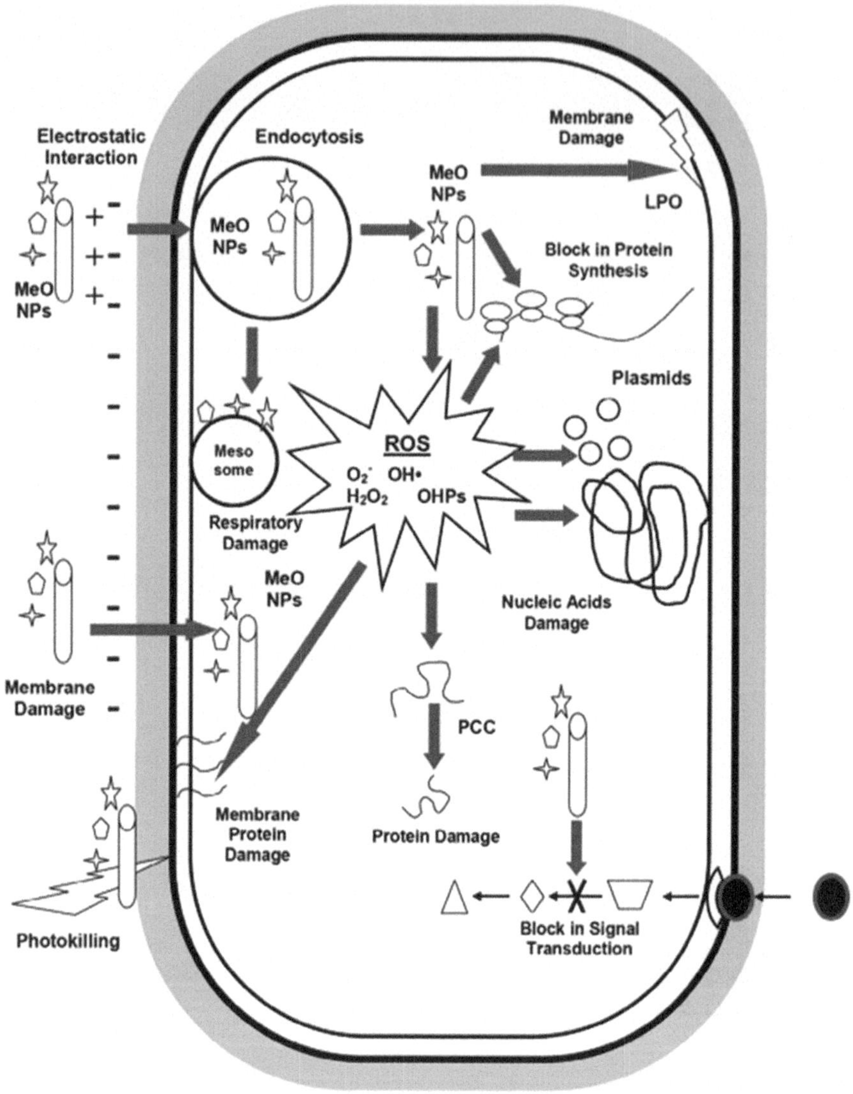

Fig. 16.1 Multiple mechanism of antibacterial action of MeO-NPs

poly(ethylene glycol), polyethyleneimine, and poly-L-lysine can also show enhanced antimicrobial activity. Graphene QDs GQDs having size in the range of 3–8 nm showed an antimicrobial activity against *Escherichia coli* and *Staphylococcus aureus*. In this case, GQDs were developed with a modest dose of H_2O_2. The H_2O_2 breakdown into the 3OH free radical could be catalyzed by the peroxidase-like activity generated from GQDs as shown in Fig. 16.2. The production of these free radicals also contributes to the inhibitory effects of GQDs on microorganisms,

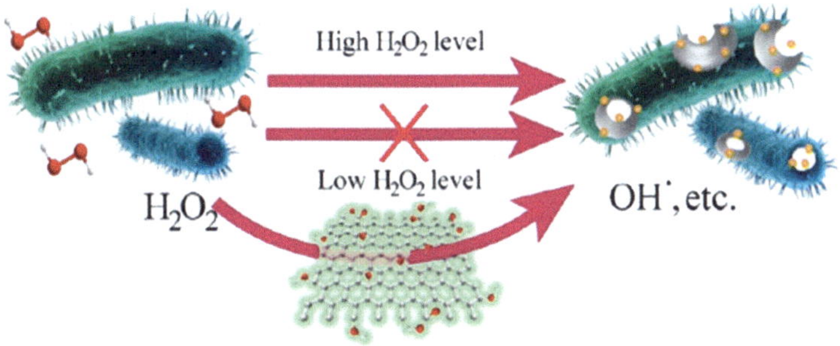

Fig. 16.2 How graphene quantum dots GQDs generate OH free radical to eradicate microorganism are shown

including the degradation of cell walls and membranes and the binding of nucleic materials (DNA and RNA) to prevent cell reproduction (Sun et al. 2014; Rajendiran et al. 2019). Table 16.3 represents different functionalized QDs as antimicrobial agents.

16.2.3.1 Mechanistic Mode of Action

The antimicrobial action by QDs generally proceed via three mechanistic pathways such as (1) cell membrane breakdown, (2) production of ROS, and (3) interaction with DNA/RNA to prevent cell growth. For instance, the surface binding of CdTe QDs and a ROS-dependent pathway are both involved in the antibacterial mode of action against *Escherichia coli*. As a result of QDs inserting into the cell membrane, heavy metal ions are released into the cells, decreasing superoxide dismutase (SOD) gene expression. This demonstrated how oxidative stress and damage to biomolecules in *E. coli* eventually result in cell death (Lu et al. 2008).

16.2.4 Nanoemulsions as Drug Carriers to Combat AMR

Now, one of the main causes of antimicrobial resistance (AMR) is the incorrect use of broad-spectrum antibiotics, particularly in hospitals. NEs, a lipid-based nanocarrier exhibit incredible potential towards improving the physiochemical and antibacterial properties of both hydrophilic and hydrophobic chemicals (Jafari et al. 2017; Pavoni et al. 2020). Antimicrobial NEs, composed of charged lipids and bioactive compounds, have high binding efficiency and electrostatic attraction. These nanodroplets, made from herbal plant oils, have broad biocidal activities against bacteria, providing a new generation of antimicrobials for Gram-positive and Gram-negative pathogens (Tayeb et al. 2022). Recently, NEs were created to improve the physicochemical and antibacterial properties of peppermint oil, and they demonstrated physiochemical stability and significant antimicrobial activity against *Escherichia coli*. In order to look into the antibacterial efficiency of these loaded NEs against carbapenem-resistant *Klebsiella pneumoniae*, researchers created clove essential

Table 16.3 Different functionalized quantum dots along with their synthesis strategies used, particle size and dimensions, microbial target, type of antimicrobial testing method used and MIC

S. no.	Functionalized quantum dots	Synthesis strategies used	Particle size and dimensions	Microbial target	Type of antimicrobial testing method used	MIC (minimum inhibitory concentration)	Reference
1	Polyamine-functionalized carbon quantum dots (CQD_{600})	Hydrothermal-assisted synthesis	Monodisperse nanocrystal with average diameter 2.02 nm	At pH 7.4 *S. aureus*, *E. coli*	Medium dilution method	346.8, 653.2, respectively, in μg/mL	Li et al. (2020)
2	Quantum dots functionalized with AP-CdSe; AP (antimicrobial peptides)	Solid-phase synthesis method	The average size is 14 nm	*E. coli, S. aureus, M. tetragenus, P. aeruginosa, B. subtilis*	Microtiter plate dilution method	6, 4, 24, 16, 24, respectively, in μg/mL	Li et al. (2021)
3	ZnO (QDs)	Wet chemical route	5–6 nm	*E. coli, C. albicans*	Broth microdilution assay	200 μg/mL	Preeti et al. (2020)
4	C (QDs) Carbon quantum dots	Refluxing carbon nanopowder in nitric acid	4–5 nm in average diameter	Both Gram-negative bacteria *E. coli* cells and on Gram-positive bacteria *B. subtilis* cells	Broth microdilution assay	64 μg/mL	Dong et al. (2017)
5	QCQD (Quaternary ammonium modified carbon quantum dots)	Green "one-pot" method	Particle size was about 5 nm	*S. epidermidis; S. aureus, L. monocytogenes; E. faecalis* respectively	Disk-diffusion test	5, 10, 20, respectively, in μg/mL	Zhao et al. (2020)

S. no.	Functionalized quantum dots	Synthesis strategies used	Particle size and dimensions	Microbial target	Type of antimicrobial testing method used	MIC (minimum inhibitory concentration)	Reference
6	Ag@S-GQDs (silver–sulfur doped graphene quantum dot)	One-step synthesis for the decoration of Ag NPs over sulfur-doped graphene quantum dots (S-GQDs)	Average diameter of S-GQDs and Ag NPs are about 4 nm and 22 ± 1.76 nm	*P. aeruginosa* and *S. aureus*	Agar well diffusion and macrodilution methods	70 and 35 μg/mL	Kadian et al. (2020)
7	ZnS quantum dots (QDs)	Green chemistry-mediated chemical precipitation reaction	Size range is 2–6 nm	*B. cereus, B. subtilis, K. pneumonia, E. coli, S. aureus, P. aeruginosa, S. epidermidis*	Killing kinetics assay	100, 75, 125, 100, 75, 125, 100, respectively, in μg/mL	Baruah et al. (2019)

oil-in-water (CEO-NEs) and water-in-oil-in-water NEs co-encapsulating CEO and meropenem (CEO-MEM-NEs). CEO-NEs (0.16% CEO MIC) and CEO–MEM-NEs (0.08% CEO and 1 μg/mL MEM MICs) have completely inactivated *K. pneumonia*. The synthesized NEs dramatically increased the antibacterial activity of CEO and MEM and have great potential as drug delivery systems of antimicrobial compounds (Cruz et al. 2005). Table 16.4 shows different reported NEs as antimicrobial agents.

16.2.4.1 Mechanistic Mode of Action

Bioactive phytochemicals in NEs inhibit broad-spectrum metabolic enzymes and drug efflux in drug-resistant bacteria (Krishnamoorthy et al. 2018). Antimicrobial

Table 16.4 Different nanoemulsion and their microbial target, type of antimicrobial testing method used, and MIC are given

S. no.	Nanoemulsions	Type of antimicrobial testing method used	Bacterial strains	MIC value	Reference
1	Nano-emulsion was prepared using essential oil from *C. viscosa* seeds	Agar well diffusion method	*E. coli, K. pneumonia, P. aeruginosa, S. aureus*	30, 30, 30, 25, respectively, in μg/mL	Krishnamoorthy et al. (2018)
2	PO nanoemulsion synthesized from mixture of peppermint oil (PO) with medium-chain triacylglycerol and stabilized with a food-grade biopolymer	Agar dilution method	*L. monocytogenes, S. aureus* ATCC 25923	0.5 (% v/v)	Liang et al. (2012)
3	Nanoemulsions containing Eucalyptus globulus oil having a size of approximately 76 nm	Macrodilution method	*P. aeruginosa PA01, C. albicans, C. tropicalis, C. glabrata*	114.125 mg/mL, 0.219 mg/mL, 0.885 mg/mL and 0.219 mg/mL	Quatrin et al. (2017)
4	Clove oil nanoemulsion	Broth microdilution method	*Escherichia coli* and *Bacillus cereus*	16 and 32 μg/mL, respectively	Shahavi et al. (2016)
5	Cinnamon oil nanoemulsion	Microbroth dilution assay	*Listeria monocytogenes* and *Salmonella* spp.	Average MIC values are 0.078% v/v and 0.039% v/v, respectively	Paudel et al. (2019)

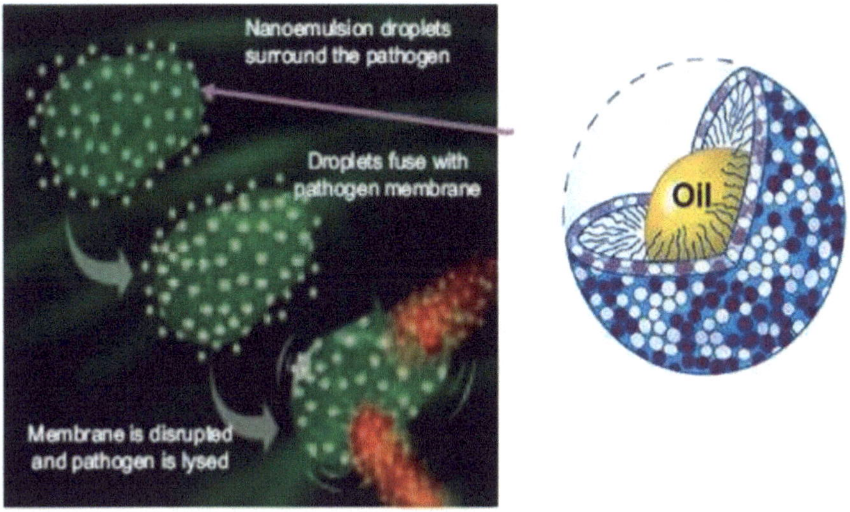

Fig. 16.3 Illustration of how nanoemulsions kill bacteria by physical mechanism

NEs are oil-in-water droplets with a size range of 20 to 200 nm. They consist of oil and water, stabilized by surfactants and alcohol. These droplets fuse with cell membrane lipid bilayers, increasing the fusion due to electrostatic attraction. The energy released from the oil and detergent emulsion destabilizes the pathogen's lipid membrane, leading to cell lysis and death as illustrated in Fig. 16.3. Nanoemulsions have nonspecific antimicrobial activity, allowing broad spectrum activity and limiting resistance generation. This makes them suitable for wound treatment and surface decontamination (Sinico et al. 2005; Zhao et al. 2023; Sonneville-Aubrun et al. 2004; Shams and Sahari 2016; Myc et al. 2003).

16.2.5 Liposomes as Efficient Drug Delivery Vehicles to Combat AMR

Liposomes are another promising nanocarrier that has been thoroughly investigated as a drug delivery system for AMR. Their capacity to encapsulate hydrophilic drugs inside the aqueous compartment and/or hydrophobic drugs inside the lipid bilayer sets them apart from other NPs, considerably expanding the variety of drugs that can be included (Cruz et al. 2005; Ruz et al. 2016). The schematic representation of the drug delivery ability of different class of drugs within various regions of liposomes is shown in Fig. 16.4.

Liposomal encapsulation has been shown in numerous studies to improve the stability and safety of antibiotics. It also produces more appropriate pharmacodynamic profiles and pharmacokinetic by extending the time that an antibiotic is in the bloodstream and allowing for targeted administration of the medication via a variety of routes to specific infection sites (Eleraky et al. 2020). The idea of mixing two

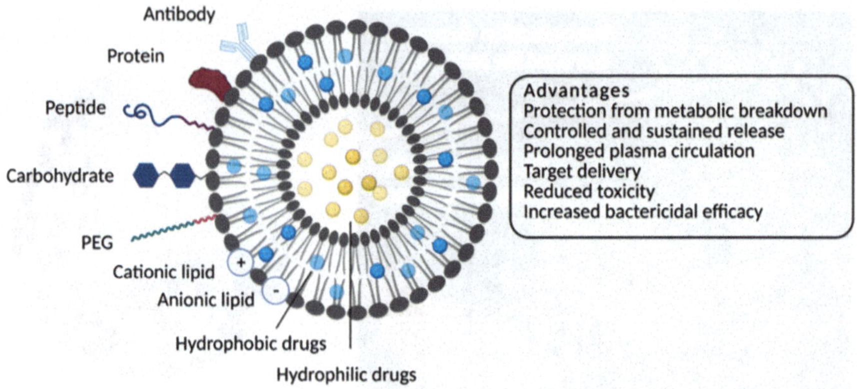

Fig. 16.4 Liposomes as efficient drug delivery vehicles to combat AMR

antibiotics inside the same liposomal formulation is being thought about in order to increase the effectiveness of therapies for lung infections caused by MDR bacteria. Wang et al. created and in vitro tested a dry powder inhaler liposomal formulation with colistin and ciprofloxacin against clinical isolates of *P. aeruginosa*. In this study, the investigators found that the two medicines worked well together, which makes them particularly appealing for MDR infections (Wang et al. 2018). For P. aeruginosa soft tissue infections, Price *et. al* created silver sulfadiazine encapsulated in liposomes. They found that one treatment greatly reduced bacteria CFUs compared to numerous applications of the free medication (Price et al. 1990). Bone infection therapy is additionally extremely difficult, demanding prolonged antibiotic use, and having a lower success rate. Keeping this mind, liposomal formulations offer an alternate technique for treating this kind of infection. A variety of antibiotics, including gentamicin, vancomycin, ceftazidime, and dicloxacillin, were integrated into liposomes using various lipid compositions, and the resulting formulations were tested against Methicillin-resistant *Staphylococcus aureus* (MRSA) bone infections. As a result, it is possible to produce liposomes that can specifically target the bone in this case (Ferreira et al. 2021). Thus, they can deliver the antibiotic directly to the infected site, tissue, or pathogen in a controlled and sustained manner, preventing premature enzymatic and immunological inactivation, limiting its distribution to healthy tissues, and reducing the possibility of toxic side effects. Table 16.5 explains different liposomes as antimicrobial agents.

16.2.5.1 Mechanistic Mode of Action

The mechanism of the improved antibacterial activity of liposomal formulations may be explained by a number of possibilities, including reduced electrostatic repulsion of liposomal antibiotics or protection of the drugs from bacterial enzymes (Mugabe et al. 2006). In our hypothesis, the fusing of these formulations with the bacterial outer membrane accounts for their increased antibacterial action.

Table 16.5 Different liposomes and their microbial target, type of antimicrobial testing method used, and MIC are given

S. no.	Liposomes	Type of antimicrobial testing method used	Microbial strains	MIC value	Reference
1	Liposome-encapsulated with gentamicin, an aminoglycoside antibiotic	Broth dilution technique	Two clinical strains (PA-1 and PA-136411) of *Pseudomonas aeruginosa*	512, 32 in mg/L, respectively	Rukholm et al. (2006)
2	Co-encapsulation of gallium with gentamicin in liposomes	Agar diffusion method	Highly antibiotic resistant strain of P. aeruginosa, i.e., PA-48912-1and PA-48913 growing in planktonic or biofilm community	4 and 2 in mg/L, respectively Lipo-Ga-GEN formulation is more effective than gentamicin	Halwani et al. (2008)
3	Liposome-enclosed ticarcillin	Iodometric assay	Strain A and strain B of *P. aeruginosa* Strain A was a B-lactamase-producing strain and strain B was non producing of B-lactamase	2% MIC 57.0 and 55.3 in μg/mL, respectively	Lagacé et al. (1991)
4	Liposome-enclosed tobramycin	Iodometric assay	Strain A and strain B of *P. aeruginosa*	2% MIC 48.5 and 53.6 in μg/mL, respectively	Lagacé et al. (1991)

16.2.6 Nanocomposites in Combating Antimicrobial Resistance

The nanocomposites are yet another promising method for eradicating MDR bacteria without inducing an outbreak of AMR. Smaller diameter NPs offer better antibacterial effects, according to some studies (Lu et al. 2013). Additionally, a smaller size stimulates the generation of ions that break down bacterial cell walls. H. Zheng and his coworkers created the La@GO nanocomposites of lanthanum hydroxide and graphene oxide to have a synergistic bactericidal impact on all tested resistant bacteria, i.e., *E. coli* (AMR), *L. crispatus* (AMR), *S. aureus* (AMR), and *P. aeruginosa* (AMR). More significantly, while conventional antibiotics and AgNPs cause a 16- to 64-fold rise in tolerance, a prolonged exposure of *E. coli* (AMR) to subminimum inhibitory concentrations of La@GO did not cause measurable secondary resistance. It is believed that a unique extracellular multitarget invasion killing mechanism involving lipid dephosphorylation, lipid peroxidation, and peptidoglycan breakdown prevents *E. coli* from evolving resistance to La@GO which has tolerant to 24 antibiotics (Zheng et al. 2019). Table 16.6 represents antimicrobial properties of different reported nanocomposites.

16.2.6.1 Mechanistic Mode of Action

Titania (TiO_2)-based antimicrobial nanocomposites have been the subject of current active research. Therefore, the mechanism behind the antimicrobial activity of TiO_2-based nanocomposite and most of the other nanocomposites involves rapid cell inactivation, decreased coenzyme-independent respiratory chains, reduced iron and phosphorous assimilation and transport, and reduced heme biosynthesis and degradation. These processes, as well as the significant changes in the cell wall, cause cell damage and death (Kubacka et al. 2014).

16.2.7 Carbon Nanomaterials as Antimicrobial Agents to Combat Multidrug Resistance

Employing carbon-based nanomaterials has the following characteristic features such as (1) they must be capable of eradicating or inhibiting the development of microorganisms at the intended place, (2) displays selective toxicity towards microorganisms while not damaging the nearby cells and tissue from harm, (3) after being diluted in bodily fluids, they need to be highly soluble and stable, (4) they should be very accessible and continue a particular activity for the desired length of time after arriving at the target spot, and (5) cost-effectiveness. In ongoing discussion on nanomaterials in the development of advanced antimicrobial agents, carbon nanotubes (CNTs) are especially crucial as they have the potential to be more cost-effective and effective than conventional antibiotic therapy in combating the rise of MDR microbes. CNTs additionally possess significant antibacterial activity. It has been demonstrated that when fixed to a surface in suspension or embedded in polymeric matrices to generate CNT-polymer composites, they can inhibit bacterial adherence and biofilm development (Maas 2016; Yick et al. 2015; Dong and Yang 2014; Vagos et al. 2020).

CNTs can be functionalized in order to enhance their association with microbial cells, thereby increase their antibacterial ability. Amine-functionalized multiwalled CNT (a-MWCNT) films significantly reduced *E. coli* optical density compared with poly(ε-caprolactone) (PCL) control ($p < 0.01$). The high bactericidal activity of a-MWCNT/PCL films seems to be associated with an extensive cell membrane lysis on the composite surface induced by quaternary ammonium groups (Teixeira-Santos et al. 2021). The use of CNTs in conjunction with antimicrobial medicines has drawn a lot of interest as a substitute for combating antibiotic resistance and developing fresh treatment possibilities. According to Qi et al., cefalexin was covalently immobilized in MWCNT/poly(ethylene glycol) (PEG) films, improving the antimicrobial and anti-adhesive properties of CNTs against Gram-positive and Gram-negative bacteria by preventing bacterial adhesion and lowering bacterial cell viability by 50% to 80% (Qi et al. 2012). Table 16.7 exhibits antimicrobial property of various carbon-based nanomaterials against particular microbial stains.

Table 16.6 Different nanocomposite along with their synthesis strategies used, particle size and dimensions, microbial target, type of antimicrobial testing method used, and MIC ae given

S. no.	Nanocomposite	Synthesis strategies used	Particle size and dimensions	Microbial target	Type of antimicrobial testing method used	MIC (minimum inhibitory concentration)	Reference
1	C-AuNp-Amp (chitosan-capped AuNps coupled with ampicillin)	Wet precipitation method	55–100 nm with spherical to ellipsoidal shape	*E. coli, S. aureus and K. mobilis*	The broth dilution assay	1.7, 1.8, and 1.3, respectively, in cm	Chamundeeswari et al. (2010)
2	ZnTe G4.NH_2 (amine terminated ZnTe/dendrimer nanocomposites)	Aqueous synthesis using 1:1 molar ratio of Zn^{2+}/Te^{2-} and dendrimer	<3.3 nm and 6.0 nm	Multidrug-resistant strains of *V. cholerae* and *E. coli*	Agar-diffusion assay	64 and 128 μg/mL, respectively	Ghosh et al. (2011)
3	ZnTe G3.5COOH (carboxyl terminated ZnTe/dendrimer nanocomposites)	Aqueous synthesis using 1:1 molar ratio of Zn^{2+}/Te^{2-} and dendrimer	>3.3 nm and 6.0 nm	Multidrug-resistant strains of *V. cholerae* and *E. coli*	Agar-diffusion assay	256 and 512 μg/mL, respectively	Ghosh et al. (2011)
4	CdO/ZnO-ESM	Bio-templated method from biowaste-eggshell membranes (ESM)	35.18 nm with both flower-like structure and spherical grains	*Klebsiella pneumonia, Pseudomonas aeruginosa, Salmonella enterica, Vibrio* sp	Agar well diffusion method	19, 18, 24, and 20, respectively in mm	Sundara Selvam et al. (2021)
5	GO@CS/ZnO	Graphene oxide were prepared by modified hammer's method followed by deposited by ZnO and chitosan	Deposition of ZnO NPs onto GO nanosheets with a range of 30 nm	Antimicrobial activity against *Staphylococcus aureus, Escherichia coli*	Broth and agar dilution methods	5 μg/mL and 2.5 μg/mL, respectively	Chowdhuri et al. (2015)

(continued)

Table 16.6 (continued)

S. no.	Nanocomposite	Synthesis strategies used	Particle size and dimensions	Microbial target	Type of antimicrobial testing method used	MIC (minimum inhibitory concentration)	Reference
6	Fe_3O_4 @ PEG–A Core-shell nanocomposite	Synthesized by relying on electrostatic forces and a reducing agent	20–25 nm	*E. coli, S. aureus, C. albicans*	Broth microdilution method	16, 16, and 8, respectively, in μg/mL	Zomorodian et al. (2018)
7	Ag-MgO nanocomposite	Seed growth method via the bioreduction route by the precipitation method	11.92 nm	*Escherichia coli*	Agar well diffusion	20 μg/mL	Ayinde et al. (2018)
8	15% $CdSAg_2S$	Chemical co-precipitation method	9.6 nm	*P. aeruinosa, E. coli* and *S. aureus*	Well diffusion method	80, 155, and 540, respectively, in μm	Iqbal et al. (2019)

16.2.7.1 Mechanistic Mode of Action

A number of mechanisms have been uncovered to explain the antibacterial action of carbon-based nanomaterials (CNMs) such as physical/mechanical destruction, oxidative stress (ROS-dependent and ROS-independent oxidative stress), photothermal effect, photocatalytic effect, lipid extraction effect, inhibition of bacterial metabolism, and isolation by wrapping (Xin et al. 2019). Oxidative stress is explained pictorially in Fig. 16.5. Physical/mechanical destruction can be explained as that the bacterial outer membrane or cell wall can be damaged physically or mechanically, resulting in dysfunction and cytoplasmic component leakage, which in turn results in bacteriostatic effects. The development of antibacterial drugs has been largely influenced by this pathway. For instance, when exposed to *Escherichia coli,* carboxylated nanodiamonds (NDs) exhibit an antibacterial effect through the destruction of the bacterial cell wall. This has been attributed to the CNMs attachment to the cell wall surface, which subsequently alters the protein structure of the bacterial cell wall (Chatterjee et al. 2015). Bacterial cells also have a lipid membrane which is essential for the uptake of nutrients and the elimination of toxic molecules. Bacterial cells treated with GO nanosheets have effectively been used to study the lipid extraction impact. The vigorous extraction of phospholipid molecules from the lipid membrane to the graphene nanosheets is demonstrated by molecular dynamics simulation. In order to bind to lipid molecules, GO nanosheets must have amphipathicity, which includes a hydrophobic plane, hydrophilic edges, and a large surface area. This property explains why GO nanosheets exhibit antibacterial activity through lipid extraction (Zhu et al. 2016).

16.2.8 Micelles as Drug Carriers to Combat Antimicrobial Resistance

Polymeric micelles (PMs) have also generated a lot of interest over the last 20 years, especially since they are powerful water-insoluble active principle solubilizers and efficient drug nanocarriers (Xu et al. 2013; Ghezzi et al. 2021). PMs represent a unique class of carriers formed from amphiphilic block or graft copolymers. These amphiphilic copolymers self-assemble at the nanometric level in aqueous conditions due to the thermodynamic incompatibility of the various hydrophilic and hydrophobic sequences. Different types of PMs were studied for the preparation of formulation with an antimicrobial activity to overcome the disadvantages coming from new drugs with the antimicrobial activity such as their low solubility and bioavailability, high toxicity, and inadequate release profile, limiting their clinical utilization (Tanase et al. 2021; Groo et al. 2018). Typically, the micelles should be small enough to avoid detection and destruction by the reticular endothelial system in order to have the best drug-loading efficiency throughout its circulation in blood. Free micelles have the lowest antimicrobial activity but when loaded with active pharmaceutical ingredient (API), 1-(5′-nitrobenzimidazole-2′-yl-sulphonyl-acetyl)-4-aryl-thiosemicarbazide, different trends were observed. It is studied that the free API exhibits enhanced antimicrobial activity against *S. aureus* (39 ± 0.6 mm)

Table 16.7 Different carbon-based nanomaterials along with their synthesis strategies used, particle size and dimensions, microbial target, type of antimicrobial testing method used, and MIC are given

S. no	Carbon nanomaterials	Synthesis strategies used	Particle size and dimensions	Microbial target	Type of antimicrobial testing method used	MIC (minimum inhibitory concentration)	Reference
1	Ag–RGO (Ag-doped reduced graphene oxide)	Facile green method using aqueous extract of the *C. esculenta* leaves as reducing agent	Average particle size of Ag is 17.2 nm over graphene sheets	*Staphylococcus aureus, Escherichia coli,* and *Candida albicans*	Agar well diffusion assay	12.5, 20 and 16, respectively, in μg/mL	Barua et al. (2014)
2	Mesoporous carbon nanospheres	Pyrolysis method employing agro-waste garlic peels	Average diameter of the nanospheres was 31 ± 6.3 nm	Gram-negative bacteria	Disc diffusion assay	480 ± 0.5 μg/mL	Roy et al. (2021)
3	f-CNT@AMB 2 Conjugates-Amphotericin B (AMB)	Chemical method synthesis	Outer average diameter was between 20 and 30 nm	Antifungal activity against *C. neoformans* ATCC 90112, *C. albicans* ATCC 90029, *C. parapsilosis* ATCC 90118	Broth microdilution susceptibility test	10 (1), 80 (8) and 10 (1), respectively, in (g/mL)	Benincasa et al. (2011)
4	Functionalized MWCNTs with lysine	Microwave-assisted synthesis	–	*A. niger, A. fumigatus, C. albicans, P. chrysogenum, S. cerevisiae*	Radial diffusion assay	12 ± 0.56, 9 ± 0.84, 11 ± 1.13, 10.8 ± 1.10, 9.86 ± 0.85	Zare-Zardini et al. (2013)
5	SWCNTs@ $mSiO_2$-TSD@Ag	TSD-mediated strategy	Ultrasmall particle diameter of Ag is 2.78 ± 0.70 nm which is smaller than the pore size of SWCNTs@$mSiO_2$ (~ 3.8 nm)	Multi-drug-resistant bacteria *E. coli, S. aureus*	Plate counting method	120 and 140, respectively, in μg/mL	Zhu et al. (2020)

S. no	Carbon nanomaterials	Synthesis strategies used	Particle size and dimensions	Microbial target	Type of antimicrobial testing method used	MIC (minimum inhibitory concentration)	Reference
6	THF/nC60	Synthesized using THF as a solvent	50 to 150 nm in diameter	Gram-positive bacterium *Bacillus subtilis*	Turbidity assay	0.08–0.10 in mg/L	Lyon et al. (2006)
7	son/nC60	Sonicating C60 dissolved in toluene with water	<200 nm in diameter but more dispersed	Gram-positive bacterium *Bacillus subtilis*	Turbidity assay	0.4–0.6 in mg/L	Lyon et al. (2006)
8	PVP/C60	In the presence of solubilizing agent such as PVP	<200 nm in diameter with more dispersity	Gram-positive bacterium *Bacillus subtilis*	Turbidity assay	0.6–1.0 in mg/L	Lyon et al. (2006)

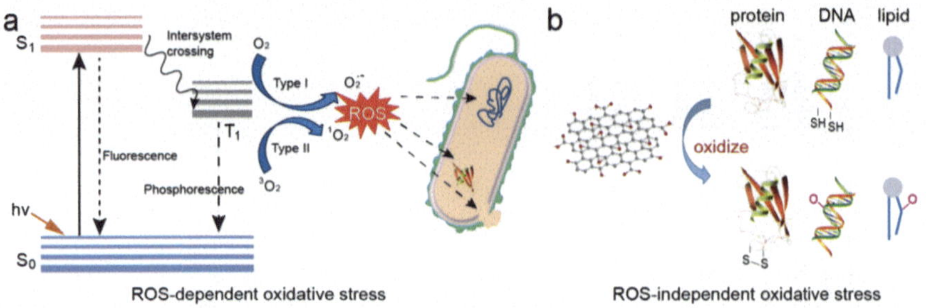

Fig. 16.5 Scheme for the ROS-dependent and ROS-independent antimicrobial action CNMs

and *E. coli* (25 ± 0.4 mm), both of which have inhibition diameters significantly larger than 15 mm. Additionally, *K. Pneumonias* showed a medium level of activity (14 0.5 mm), but *P. aeruginosa* and *P. gingivalis* showed essentially no activity. This trend is also evident in PMs that have APIs loaded. As a result, the antibacterial activity of Pluronic F127, a micelle, rises by raising the F127/API ratio (Popovici et al. 2022). Table 16.8 displays antimicrobial properties of different micelles.

16.2.8.1 Mechanistic Mode of Action

Apparently, micelles penetrated the microbe, damaged its membrane, and eventually burst it, exhibiting outstanding antimicrobial properties. As a matter of fact, microbes and micelles interact electrostatically, as the micelle corona's amino groups have positive charges, while the microbes have negative charges. Additionally, the phenyl groups in the antimicrobial polypeptide alter the permeability of microbe, thereby killing them. The adhesion and rupture mechanisms of antimicrobial micelles were supported by transmission electron microscopy (TEM) images (Jones et al. 2015). Figure 16.6 presents the schematic representation of mechanism of antimicrobial action of micelles.

16.2.9 Dendrimers-Based Nanoparticles as Drug Delivery Vehicles Against MDRB

The growth of nanomedicine to treat many ailments has made dendrimers an essential class of nanostructured carriers. Because of their structural variety and adaptability, dendrimers have been used to deliver genes and drugs in a number of ways. For instance, dendrimers with a hydrophilic periphery and a hydrophobic core can function as unimolecular micelles. This property has been utilized to ensnare and solubilize hydrophobic medications by trapping them inside the intramolecular cavity. Catalytic dendrimers are widely used as non-viral gene carriers. It has been claimed half-life of drugs can lengthen when dendrimers conjugated with drugs. For instance, when conjugated with polyamidoamine PAMAM dendrimer, the half-life of methotrexate is extended to 24 h from 24 min (Gupta and Nayak 2015). By reducing the frequency of drug administration and increasing pharmacological

Table 16.8 Various reported micelles with their respective microbial target, type of antimicrobial testing method used, and MIC are given

S. no	Micelles	Type of antimicrobial testing method used	Bacterial strains	MIC value	Reference
1	PLLA and peptide-based copolymer [PLLA31-b-poly(Phe24-stat-Lys36)] which can self-assemble into micelles	Broth microdilution method	Against both Gram-negative, *E. coli* and Gram-positive, bacteria *S. aureus*	40 μg/mL and 120 μg/mL	Xi et al. (2016)
2	Hydroxyl radical-generating ferrocene-loaded PCAE self-assemble to generate micelles under aqueous conditions	Agar well diffusion method	*E. coli* and *P. aeruginosa*	0.9 μg/mL	Park et al. (2016)
3	Triclosan loaded in PEG-g-PU-3 micelles with hydrophobic PCL core and hydrophilic PEG corona	Microplate broth dilution method and plate counting method	*S. aureus (ATCC29213) and E. coli (ATCC35218)*	0.03 in ppm	Su et al. (2018)
4	Hypocrellin A-loaded lipase sensitive polymer micelles (core-shell mPEG-PCL/HA) having average diameter as 87.6 ± 0.6 nm	Fluorescence microscopy	Methicillin-resistant *Staphylococcus aureus*	0.69 mg/L after light irradiation and in dark, it is 250 mg/L	Guo et al. (2020)
5	Piperacillin/tazobactam loaded micelles having semi-spherical morphology with a mean diameter of below 30 nm	Disc diffusion agar method	*P. aeruginosa* (0–29) strain	128 μg/mL	Morteza et al. (2019)

efficacy, the prolonged circulation half-life can increase patient compliance. Solubility of drugs also enhanced when conjugated with dendrimers. Drugs are more soluble when they are conjugated with dendrimers. For instance, when paclitaxel is conjugated with PAMAM dendrimer, its water solubility is increased by 9000-fold (Nel et al. 2006b).

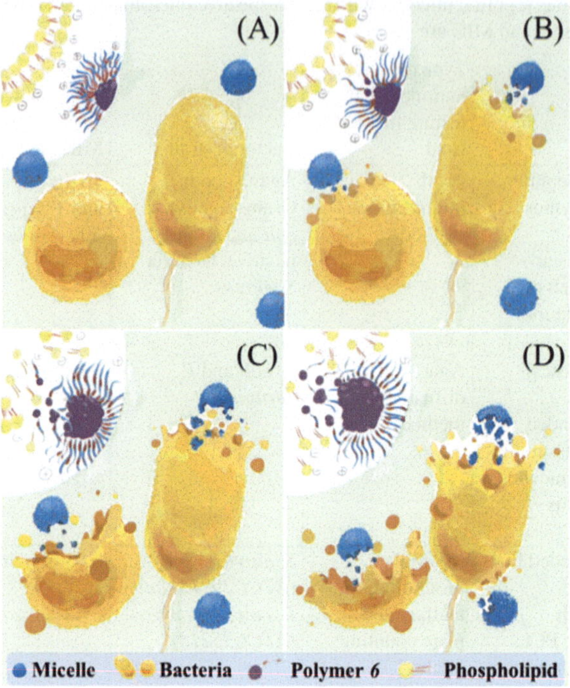

Fig. 16.6 Represent antimicrobial mechanism by micelles. (**a**) Electrostatic interaction between bacteria and micelles. (**b**) The antimicrobial micelles adhere to the bacteria. (**c**) The membrane of bacteria is pierced by the micelles. (**d**) Rupture of bacteria membrane

16.2.10 Mechanistic Mode of Action

Dendrimer biocides, with their high positive charge density, quickly interact with bacteria due to electrostatic interactions. They initially displace surface divalent ions and bind to negatively charged phospholipid membranes, causing a slight change in membrane permeability and neutralizing bacteria's surface charge. As more biocides are introduced, they can denature membrane proteins, penetrate phospholipid bilayers, and eventually lead to complete membrane disintegration, resulting in a bactericidal effect (Elder et al. 2006).

16.3 Limitation

The combination of antibiotic medication and nanotherapy is currently seen as a logical strategy to combat microbe resistance. However, other disadvantages, such as selectivity index, efficiency, and toxicity, require additional analysis. More research is required because the mechanism of action used by the NPs is not entirely known. The same effects as "traditional" particles (such as inflammation and lung cancer) may also be caused by NPs, although they may be more potent due to their larger surface area. However, NPs may potentially have new effects that were not previously recognized with larger particles, such as cardiovascular consequences, platelet aggregation, and mitochondrial damage. For example, MnO_2, Au, and carbon NPs can translocate to the brain from the nose and cause serious problems. ROS and oxidative stress have been found to be generated by MnO_2 NPs resulting from inhalation in the brain (Wang et al. 2022; Chen and Cooper 2002).

16.4 Conclusion and Future Outlook

MNPs, both in organic and inorganic form, due to their remarkable features give rise to a number of intriguing sectors in science and technology. In light of the current state of human health, comfort, and well-being, multifunctionality NP MNPs are warmly accepted by a number of industries, such as the health and personal care, textile and environmental sectors. MNPs implementation in the economic system on a wide scale is still a long way off, though. Therefore, in order to scale up and satisfy the expanding human demands, future studies should focus on developing unique, useful, and affordable approaches for manufacturing of these MNPs. MNPs with a number of other antibacterial and antifungal compounds have been successfully coupled. However, the synergistic use of MNPs in combination with other medicinal compounds is still restricted. There are only a few mechanistic investigations of MNP conjugates. Therefore, it seems that the combined action of MNPs and antimicrobial drugs is a key tactic for avoiding pathogenic resistance and impeding its development. It is possible to achieve effective antibacterial effects with lower medication concentrations, as in the majority of cases, the conjugation displayed lower cytotoxicity than each drug. Recent research investigations have examined cytotoxicity but additional study is required to establish the ideal balance between effect and toxicity. Last but not least, MNP provides an effective route for inhibiting microbial growth with reduced individual drug dosages, reduced cytotoxicity, and a wider range of antibacterial coverage. The loss of nanomaterials such as MNPs, MeO-NPs, etc. during the process of downstream processing can be hazardous to humans and have an impact on many ecosystems; thus, future research should focus on creating more effective methods for keeping nanomaterials. Also, in terms of exact mechanisms, antimicrobial properties of nanomaterials are still unknown. Microbicidal impact has been attributed to the formation of ROS or the onset of oxidative stress in some papers, while in others, metabolism cannot be linked to antimicrobial effect. Future research should thus take into account addressing the precise mechanism behind the antibacterial effect of NSMs.

References

An J, Zhang H, Zhang J, Zhao Y, Yuan X (2009) Preparation and antibacterial activity of electrospun chitosan/poly(ethylene oxide) membranes containing silver nanoparticles. Colloid Polym Sci 287:1425–1434

Arias L, Pessan J, Vieira A, Lima T, Delbem A, Monteiro D (2018) Iron oxide nanoparticles for biomedical applications: a perspective on synthesis, drugs, antimicrobial activity, and toxicity. Antibiotics 7(2):46

Ayinde WB, Gitari MW, Muchindu M, Samie A (2018) Biosynthesis of ultrasonically modified ag-MgO nanocomposite and its potential for antimicrobial activity. J Nanotechnol 2018:1

Barua S, Thakur S, Aidew L, Buragohain AK, Chattopadhyay P, Karak N (2014) One step preparation of a biocompatible, antimicrobial reduced graphene oxide–silver nanohybrid as a topical antimicrobial agent. RSC Adv 4(19):9777–9783

Baruah JM, Kalita S, Narayan J (2019) Green chemistry synthesis of biocompatible ZnS quantum dots (QDs): their application as potential thin films and antibacterial agent. Int Nano Lett 9:149–159

Benincasa M, Pacor S, Wu W, Prato M, Bianco A, Gennaro R (2011) Antifungal activity of amphotericin B conjugated to carbon nanotubes. ACS Nano 5(1):199–208

Buzea C, Pacheco II, Robbie K (2007) Nanomaterials and nanoparticles: sources and toxicity. Biointerphases 2:17–71

Chamundeeswari M, Sobhana SL, Jacob JP, Kumar MG, Devi MP, Sastry TP, Mandal AB (2010) Preparation, characterization and evaluation of a biopolymeric gold nanocomposite with antimicrobial activity. Biotechnol Appl Biochem 55(1):29–35

Chandrasekaran R, Gnanasekar S, Seetharaman P, Keppanan R, Arockiaswamy W, Sivaperumal S (2016) Formulation of Carica papaya latex-functionalized silver nanoparticles for its improved antibacterial and anticancer applications. J Mol Liq 219:232–238

Chatterjee A, Perevedentseva E, Jani M, Cheng CY, Ye YS, Chung PH, Cheng CL (2015) Antibacterial effect of ultrafine nanodiamond against gram-negative bacteria *Escherichia coli*. J Biomed Opt 20(5):051014–051014

Chen CZ, Cooper SL (2002) Interactions between dendrimer biocides and bacterial membranes. Biomaterials 23(16):3359–3368

Chowdhuri AR, Tripathy S, Chandra S, Roy S, Sahu SK (2015) A ZnO decorated chitosan–graphene oxide nanocomposite shows significantly enhanced antimicrobial activity with ROS generation. RSC Adv 5(61):49420–49428

Correa MG, Martínez FB, Vidal CP, Streitt C, Escrig J, de Dicastillo CL (2020) Antimicrobial metal-based nanoparticles: a review on their synthesis, types and antimicrobial action. Beilstein J Nanotechnol 11(1):1450–1469

Cruz MEM, Manuela GM, Bárbara M, Martins F, Luísa CM (2005) Liposomal superoxide Dismutases and their use in the treatment of experimental arthritis. Methods Enzymol 391:395–413

Dehghanizade S, Arasteh J, Mirzaie A (2018) Green synthesis of silver nanoparticles using Anthemis atropatana extract: characterization and in vitro biological activities. Artif Cells Nanomed Biotechnol 46(1):160–168

Dhand C, Dwivedi N, Loh XJ, Jie Ying AN, Verma NK, Beuerman RW, Lakshminarayanan R, Ramakrishna S (2015) Methods and strategies for the synthesis of diverse nanoparticles and their applications. RSC Adv 5:105003–105037

Dong X, Yang L (2014) Inhibitory effects of single-walled carbon nanotubes on biofilm formation from bacillus anthracis spores. Biofouling 30(10):1165–1174

Dong X, Awak MA, Tomlinson N, Tang Y, Sun YP, Yang L (2017) Antibacterial effects of carbon dots in combination with other antimicrobial reagents. PLoS One 12(9):0185324

Elder A, Gelein R, Silva V (2006) Translocation of inhaled ultrafine manganese oxide particles to the central nervous system. Environ Health Perspect 114:1172–1178

Eleraky NE, Allam A, Hassan SB, Omar MM (2020) Nanomedicine fight against antibacterial resistance: an overview of the recent pharmaceutical innovations. Pharmaceutics 12:142

Escárcega-González CE, Garza-Cervantes JA, Vázquez-Rodríguez A, Montelongo-Peralta LZ, Treviño-González MT, Díaz Barriga Castro E, Saucedo-Salazar EM, Chávez Morales RM, Regalado Soto DI, Treviño González FM, Carrazco Rosales JL, Cruz RV, Morones-Ramírez JR (2018) In vivo antimicrobial activity of silver nanoparticles produced via a green chemistry synthesis using *Acacia rigidula* as a reducing and capping agent. Int J Nanomedicine Volume 13:2349–2363

Fadda AA, Mohammed AR, Abdel-Galil E (2020) Synthesis and antimicrobial evaluation of some 4-quinolinylazo-N-pyrimidinyl benzenesulfonamide derivatives. Biointerface Res Appl Chem 10:4846–4852

Faya M, Kalhapure RS, Kumalo HM, Waddad AY, Omolo C, Govender T (2018) Conjugates and nano-delivery of antimicrobial peptides for enhancing therapeutic activity. J Drug Deliv Sci Technol 44:153–171

Ferreira M, Aguiar S, Bettencourt A, Gaspar MM (2021) Lipid-based nanosystems for targeting bone implant-associated infections: current approaches and future endeavors. Drug Deliv Transl Res 11:72–85

Fishovitz J, Hermoso JA, Chang M, Mobashery S (2014) Penicillin-binding protein 2a of methicillin-resistant Staphylococcus aureus. IUBMB Life 66(8):572–577

Ghezzi M, Pescina S, Padula C, Santi P, Del Favero E, Cantu L, Nicoli S (2021) Polymeric micelles in drug delivery: An insight of the techniques for their characterization and assessment in biorelevant conditions. J Control Release 332:312–336

Ghosh S, Ghosh D, Bag PK, Bhattacharya SC, Saha A (2011) Aqueous synthesis of ZnTe/dendrimer nanocomposites and their antimicrobial activity: implications in therapeutics. Nanoscale 3(3):1139–1148

Ghosh T, Dash SK, Chakraborty P, Guha A, Kawaguchi K, Roy S, Chattopadhyay T, Das D (2014) Preparation of antiferromagnetic Co_3O_4 nanoparticles from two different precursors by pyrolytic method: in vitro antimicrobial activity. RSC Adv 4(29):15022–15029

Groo AC, Matougui N, Umerska A, Saulnier P (2018) Reverse micelle-lipid nanocapsules: a novel strategy for drug delivery of the plectasin derivate AP138 antimicrobial peptide. Int J Nanomedicine 13:7565–7574

Grozav A, Fedoriv M, Chornous V, Yakovychuk N, Kemskyi S, Vovk M (2021) Synthesis and bioevaluation of 5-Chloro-4-(1,3-Oxazol-5-yl)-1H-Pyrrole-3-Carboxyamides as antimicrobial agents. Biointerface Res Appl Chem 11:10595–10606

Guo LY, Yan SZ, Tao X, Yang Q, Li Q, Wang TS, Yu SQ, Chen SL (2020) Evaluation of hypocrellin A-loaded lipase sensitive polymer micelles for intervening methicillin-resistant *Staphylococcus aureus* antibiotic-resistant bacterial infection. Mater Sci Eng C 106:110230

Gupta V, Nayak S (2015) Dendrimers: a review on synthetic approaches. J Appl Pharm Sci 5:117–122

Halwani M, Yebio B, Suntres ZE, Alipour M, Azghani AO, Omri A (2008) Co-encapsulation of gallium with gentamicin in liposomes enhances antimicrobial activity of gentamicin against *Pseudomonas aeruginosa*. J Antimicrob Chemother 62(6):1291–1297

He L, Liu Y, Mustapha A, Lin M (2011) Antifungal activity of zinc oxide nanoparticles against *Botrytis cinerea* and *Penicillium expansum*. Microbiol Res 166(3):207–215

Hou H, Mahdavi B, Paydarfard S, Zangeneh MM, Zangeneh A, Sadeghian N, Taslimi P, Erduran V, Sen F (2020) Retracted article: novel green synthesis and antioxidant, cytotoxicity, antimicrobial, antidiabetic, anticholinergics, and wound healing properties of cobalt nanoparticles containing *Ziziphora clinopodioides* Lam leaves extract. Sci Rep 10(1):12195

Ingale AG, Chaudhari AN (2013) Biogenic synthesis of nanoparticles and potential applications: an eco-friendly approach. J Nanomed Nanotechnol 4:1–7

Ingle AP, Duran N, Rai M (2014) Bioactivity, mechanism of action, and cytotoxicity of copper-based nanoparticles: a review. Appl Microbiol Biotechnol 98:1001–1009

Iqbal T, Ali F, Khalid NR, Tahir MB, Ijaz M (2019) Facile synthesis and antimicrobial activity of $CdS-Ag_2S$ nanocomposites. Bioorg Chem 90:103064

Jafari SM, Paximada P, Mandala I, Assadpour E, Mehrnia MA (2017) Nanoencapsulation technologies for the food and nutraceutical industries, pp 36–73

Jamzad M, Kamari Bidkorpeh M (2020) Green synthesis of iron oxide nanoparticles by the aqueous extract of *Laurus nobilis* L. leaves and evaluation of the antimicrobial activity. J Nanostruct Chem 10:193–201

Jeyaraj Pandian C, Palanivel R, Dhanasekaran S (2016) Screening antimicrobial activity of nickel nanoparticles synthesized using *Ocimum sanctum* leaf extract. J Nanoparticles 2016:1–13

Jones CH, Chen CK, Chen MF, Ravikrishnan A, Zhang HG, Gollakota A, Chung TC, Cheng C, Pfeifer BA (2015) PEGylated cationic polylactides for hybrid biosynthetic gene delivery. Mol Pharm 12:846–856

Kadian S, Manik G, Das N, Nehra P, Chauhan RP, Roy P (2020) Synthesis, characterization and investigation of synergistic antibacterial activity and cell viability of silver–sulfur doped graphene quantum dot (Ag@ S-GQDs) nanocomposites. J Mater Chem B 8(15):3028–3037

Kohanski MA, Dwyer DJ, Hayete B, Lawrence CA, Collins JJ (2007) A common mechanism of cellular death induced by bactericidal antibiotics. Cell 130(5):797–810

Krishnamoorthy R, Athinarayanan J, Periasamy VS, Adisa AR, Al-Shuniaber MA, Gassem MA, Alshatwi AA (2018) Antimicrobial activity of nanoemulsion on drug-resistant bacterial pathogens. Microb Pathog 120:85–96

Kubacka A, Diez MS, Rojo D, Bargiela R, Ciordia S, Zapico I, Albar JP, Barbas C, Martins dos Santos VA, Fernández-García M, Ferrer M (2014) Understanding the antimicrobial mechanism of TiO_2-based nanocomposite films in a pathogenic bacterium. Sci Rep 4(1):4134

Kumar A, Kumar A (2020) Design and synthesis of anti-convulsant and anti-bacterial activity of new hydrazone derivatives. Biointerface Res Appl Chem 10:5229–5236

Lagacé J, Dubreuil M, Montplaisir S (1991) Liposome-encapsulated antibiotics: preparation, drug release and antimicrobial activity against *Pseudomonas aeruginosa*. J Microencapsul 8(1):53–61

Li P, Yang X, Zhang X, Pan J, Tang W, Cao W, Zhou J, Gong X, Xing X (2020) Surface chemistry-dependent antibacterial and antibiofilm activities of polyamine-functionalized carbon quantum dots. J Mater Sci 55:6744–16757

Li W, Song P, Xin Y, Kuang Z, Liu Q, Ge F, Zhu L, Zhang X, Tao Y, Zhang W (2021) The effects of luminescent CdSe quantum dot-functionalized antimicrobial peptides nanoparticles on antibacterial activity and molecular mechanism. Int J Nanomedicine Volume 16:1849–1867

Liang R, Xu S, Shoemaker CF, Li Y, Zhong F, Huang Q (2012) Physical and antimicrobial properties of peppermint oil nanoemulsions. J Agric Food Chem 60(30):7548–7555

Lu Z, Li CM, Bao H, Qiao Y, Toh Y, Yang X (2008) Mechanism of antimicrobial activity of CdTe quantum dots. Langmuir 24(10):5445–5452

Lu Z, Rong K, Li J, Yang H, Chen R (2013) Size-dependent antibacterial activities of silver nanoparticles against oral anaerobic pathogenic bacteria. J Mater Sci Mater Med 24:1465–1471

Lyon DY, Adams LK, Falkner JC, Alvarez PJ (2006) Antibacterial activity of fullerene water suspensions: effects of preparation method and particle size. Environ Sci Technol 40(14):4360–4366

Maas M (2016) Carbon nanomaterials as antibacterial colloids. Materials 9(8):617

Mageshwari K, Sathyamoorthy R (2013) Flower-shaped CuO nanostructures: synthesis, characterization and antimicrobial activity. J Mater Sci Technol 29(10):909–914

Maqbool Q, Nazar M, Naz S, Hussain T, Jabeen N, Kausar R, Anwaar S, Abbas F, Jan T (2016) Antimicrobial potential of green synthesized CeO_2 nanoparticles from *Olea europaea* leaf extract. Int J Nanomedicine Volume 11:5015–5025

Markova Z, Siskova KM, Filip J, Cuda J, Kolar M, Safarova K, Medrik I, Zboril R (2013) Air stable magnetic bimetallic Fe–Ag nanoparticles for advanced antimicrobial treatment and phosphorus removal. Environ Sci Technol 47(10):5285–5293

Morsy MA, Ali EM, Kandeel M, Venugopala KN, Nair AB, Greish K, El-Daly M (2020) Screening and molecular docking of novel Benzothiazole derivatives as potential antimicrobial agents. Antibiotics 9(5):221

Morteza M, Roya S, Hamed H, Amir Z, Abolfazl A (2019) Synthesis and evaluation of polymeric micelle containing piperacillin/tazobactam for enhanced antibacterial activity. Drug Deliv 26(1):1292–1299

Mugabe C, Halwani M, Azghani AO, Lafrenie RM, Omri A (2006) Mechanism of enhanced activity of liposome-entrapped aminoglycosides against resistant strains of Pseudomonas aeruginosa. Antimicrob Agents Chemother 50(6):2016–2022

Muthu MS, Ajith P, Agnes J, Ramkumar R, Raja P, Anand DP (2022) Synthesis, characterizations and antibacterial studies of chromium trioxide nanoparticles. Int J Mod Trend Sci Technol 8:252–258

Myc A, Vanhecke T, Landers JJ, Hamouda T, Baker JR (2003) The fungicidal activity of novel nanoemulsion (X8W 60 PC) against clinically important yeast and filamentous fungi. Mycopathologia 155:195–201

Neelgund GM, Oki A, Luo Z (2012) Antimicrobial activity of CdS and Ag2S quantum dots immobilized on poly (amidoamine) grafted carbon nanotubes. Colloids Surf B Biointerfaces 100:215–221

Nel A, Xia T, Madler L, Li N (2006a) Toxic potential of materials at the nanolevel. Science 311(5761):622–627

Nel A, Xia T, Madler L (2006b) Toxic potential of materials at the nano level. Science 311:622–627

Niemeyer CM (2001) Nanoparticles, proteins, and nucleic acids: biotechnology meets materials science. Angew Chem Int Ed 40(22):4128–4158

Padmavathy N, Vijayaraghavan R (2008) Enhanced bioactivity of ZnO nanoparticles—an antimicrobial study. Sci Technol Adv Mater 9:035004

Park SC, Kim NH, Yang W, Nah JW, Jang MK, Lee D (2016) Polymeric micellar nanoplatforms for Fenton reaction as a new class of antibacterial agents. J Control Release 221:37–47

Paudel SK, Bhargava K, Kotturi H (2019) Antimicrobial activity of cinnamon oil nanoemulsion against *Listeria monocytogenes* and *Salmonella* spp. on melons. LWT 111:682–687

Pavoni L, Perinelli DR, Bonacucina G, Cespi M, Palmieri GF (2020) Nanomaterials 10:135–158

Popovici C, Popa M, Sunel V, Atanase LI, Ichim DL (2022) Drug delivery systems based on Pluronic micelles with antimicrobial activity. Polymers 14(15):3007

Preeti, Radhakrishnan VS, Mukherjee S, Mukherjee S, Singh SP, Prasad T (2020) ZnO quantum dots: broad spectrum microbicidal agent against multidrug resistant pathogens *E. coli* and *C albicans*. Front Nanotechnol 2:576342

Price CI, Horton JW, Baxter CR (1990) Topical liposomal delivery of antibiotics in soft tissue infection. J Surg Res 49:174–178

Qi X, Gunawan P, Xu R, Chang MW (2012) Cefalexin-immobilized multi-walled carbon nanotubes show strong antimicrobial and anti-adhesion properties. Chem Eng Sci 84:552–556

Quatrin PM, Verdi CM, de Souza ME, de Godoi SN, Klein B, Gundel A, Wagner R, de Almeida Vaucher R, Ourique AF, Santos RCV (2017) Antimicrobial and antibiofilm activities of nanoemulsions containing *Eucalyptus globulus* oil against *Pseudomonas aeruginosa* and *Candida* spp. Microb Pathog 112:230–242

Raghunath A, Perumal E (2017) Metal oxide nanoparticles as antimicrobial agents: a promise for the future. Int J Antimicrob Agents 49(2):137–152

Rajendiran K, Zhao Z, Pei DS, Fu A (2019) Antimicrobial activity and mechanism of functionalized quantum dots. Polymers 11(10):1670

Ramyadevi J, Jeyasubramanian K, Marikani A, Rajakumar G, Rahuman AA (2012) Synthesis and antimicrobial activity of copper nanoparticles. Mater Lett 71:114–116

Ranpariya B, Salunke G, Karmakar S, Babiya K, Sutar S, Kadoo N, Kumbhakar P, Ghosh S (2021) Antimicrobial synergy of silver-platinum nanohybrids with antibiotics. Front Microbiol 11:610968

Ribeiro AI, Dias AM, Zille A (2022) Synergistic effects between metal nanoparticles and commercial antimicrobial agents: a review. ACS Appl Nano Mater 5(3):3030–3064

Roy P, Bhat VS, Saha S, Sengupta D, Das S, Datta S, Hegde G (2021) Mesoporous carbon nanospheres derived from agro-waste as novel antimicrobial agents against gram-negative bacteria. Environ Sci Pollut Res 28:13552–13561

Rukholm G, Mugabe C, Azghani AO, Omri A (2006) Antibacterial activity of liposomal gentamicin against *Pseudomonas aeruginosa*: a time–kill study. Int J Antimicrob Agents 27(3):47–252

Ruz MEM, Simoes SI, Corvo ML, Martins MBF, Gaspar MM (2016) Drug delivery nanoparticles formulation and characterization. In: Formulation of NPDDS for macromolecules, vol 191, pp 35–49

Shahavi MH, Hosseini M, Jahanshahi M, Meyer RL, Darzi GN (2016) Clove oil nanoemulsion as an effective antibacterial agent: Taguchi optimization method. Desalin Water Treat 57(39):18379–18390

Shamaila S, Zafar N, Riaz S, Sharif R, Nazir J, Naseem S (2016) Gold nanoparticles: an efficient antimicrobial agent against enteric bacterial human pathogen. Nanmaterials 6(4):71

Shams N, Sahari MA (2016) Nanoemulsions: preparation, structure, functional properties and their antimicrobial effects. Appl Food Biotechnol 3:138–149

Sharmin S, Rahaman MM, Sarkar C, Atolani O, Islam MT, Adeyemi OS (2021) Nanoparticles as antimicrobial and antiviral agents: a literature-based perspective study. Heliyon 7(3):06456

Shoeb M, Singh BR, Khan JA, Khan W, Singh BN, Singh HB, Naqvi AH (2013) ROS-dependent anticandidal activity of zinc oxide nanoparticles synthesized by using egg albumen as a biotemplate. Adv Nat Sci Nanosci Nanotechnol 4(3):035015

Singh A, Gautam PK, Verma A, Singh V, Shivapriya PM, Shivalkar S, Sahoo AK, Samanta SK (2020) Green synthesis of metallic nanoparticles as effective alternatives to treat antibiotics resistant bacterial infections: a review. Biotechnol Rep 25:00427

Sinico C, De Logu A, Lai F, Valenti D, Manconi M, Loy G, Bonsignore L, Fadda AM (2005) Liposomal incorporation of *Artemisia arborescens* L. essential oil and in vitro antiviral activity. Eur J Pharm Biopharm 59(1):161–168

Sobhani Z, Samani SM, Montaseri H, Khezri E (2017) Nanoparticles of chitosan loaded ciprofloxacin: fabrication and antimicrobial activity. Adv Pharm Bull 7(3):427

Song X, Bayati P, Gupta M, Elahinia M, Haghshenas M (2021) Fracture of magnesium matrix nanocomposites—a review. Int J Lightweight Mater Manuf 4(1):67–98

Sonneville-Aubrun O, Simonnet JT, L'alloret F (2004) Nanoemulsions: a new vehicle for skincare products. Adv Colloid Interface Sci 108:145–149

Spellberg B, Blaser M, Guidos R, Boucher H, Bradley J (2011) Combating antimicrobial resistance: policy recommendations to save lives. Clin Infect Dis 52(Suppl. 5):397–S428

Srihasam S, Thyagarajan K, Korivi M, Lebaka VR, Mallem SPR (2020) Phytogenic generation of NiO nanoparticles using stevia leaf extract and evaluation of their in-vitro antioxidant and antimicrobial properties. Biomol Ther 10(1):89

Stanić V, Tanasković SB (2020) Antibacterial activity of metal oxide nanoparticles. In: Nanotoxicity, pp 241–274

Su Y, Zhao L, Meng F, Qiao Z, Yao Y, Luo J (2018) Triclosan loaded polyurethane micelles with pH and lipase sensitive properties for antibacterial applications and treatment of biofilms. Mater Sci Eng C 93:921–930

Sun H, Gao N, Dong K, Ren J, Qu X (2014) Graphene quantum dots-band-aids used for wound disinfection. ACS Nano 8(6):6202–6210

Sundara Selvam PS, Chinnadurai GS, Ganesan D, Perumal P, Kandan V (2021) Cadmium oxide-zinc oxide nanocomposites synthesized using waste eggshell membrane and its in-vitro assessments of the antimicrobial activities and minimum inhibitory concentration. J Inorg Organomet Polym Mater 31:816–835

Tanase MA, Raducan A, Oancea P, Ditu LM, Stan M, Petcu C, Scomoroscenco C, Ninciuleanu CM, Nistor CL, Cinteza LO (2021) Mixed Pluronic—Cremophor polymeric micelles as nanocarriers for poorly soluble antibiotics—the influence on the antibacterial activity. Pharmaceutics 13:435

Tayeb HH, Moqaddam SA, Hasaballah NH, Felimban RI (2022) Development of nanoemulsions for the delivery of hydrophobic and hydrophilic compounds against carbapenem-resistant Klebsiella pneumoniae. RSC Adv 12(40):26455–26462

Teixeira-Santos R, Gomes M, Gomes LC, Mergulhao FJ (2021) Antimicrobial and anti-adhesive properties of carbon nanotube-based surfaces for medical applications: a systematic review. iScience 24(1):102001

Vagos MR, Gomes M, Moreira JM, Soares OS, Pereira MF, Mergulhão FJ (2020) Carbon nanotube/poly (Dimethylsiloxane) composite materials to reduce bacterial adhesion. Antibiotics 9(8):434

Vimbela GV, Ngo SM, Fraze C, Yang L, Stout DA (2017) Antibacterial properties and toxicity from metallic nanomaterials. Int J Nanomedicine 12:3941–3965

Wang S, Yu S, Lin Y, Zou P, Chai G, Yu HH, Wickremasinghe H, Shetty N, Ling J, Li J (2018) Co-delivery of ciprofloxacin and Colistin in liposomal formulations with enhanced in vitro antimicrobial activities against multidrug resistant *Pseudomonas aeruginosa*. Pharm Res 35:187

Wang J, Li B, Qiu L, Qiao X, Yang H (2022) Dendrimer-based drug delivery systems: history, challenges, and latest developments. J Biol Eng 16(1):1–12

Xi Y, Song T, Tang S, Wang N, Du J (2016) Preparation and antibacterial mechanism insight of polypeptide-based micelles with excellent antibacterial activities. Biomacromolecules 17(12):3922–3930

Xin Q, Shah H, Nawaz A, Xie W, Akram MZ, Batool A, Tian L, Jan SU, Boddula R, Guo B, Liu Q (2019) Antibacterial carbon-based nanomaterials. Adv Mater 31(45):1804838

Xu W, Ling P, Zhang T (2013) Polymeric micelles, a promising drug delivery system to enhance bioavailability of poorly water-soluble drugs. J Drug Deliv 2013:340315

Yick S, Mai-Prochnow A, Levchenko I, Fang J, Bull MK, Bradbury M, Murphy AB, Ostrikov KK (2015) The effects of plasma treatment on bacterial biofilm formation on vertically-aligned carbon nanotube arrays. RSC Adv 5(7):5142–5148

Zare-Zardini H, Amiri A, Shanbedi M, Memarpoor-Yazdi M, Asoodeh A (2013) Studying of antifungal activity of functionalized multiwalled carbon nanotubes by microwave-assisted technique. Surf Interface Anal 45(3):51–755

Zhao C, Wu L, Wang X, Weng S, Ruan Z, Liu Q, Lin L, Lin X (2020) Quaternary ammonium carbon quantum dots as an antimicrobial agent against gram-positive bacteria for the treatment of MRSA-infected pneumonia in mice. Carbon 163:70–84

Zhao D, Ma Y, Wang W, Xiang Q (2023) Antibacterial activity and mechanism of cinnamon essential oil nanoemulsion against pseudomonas deceptionensis CM2. Heliyon 9(9):19582–19592

Zheng H, Ji Z, Roy KR, Gao M, Pan Y, Cai X, Wang L, Li W, Chang CH, Kaweeteerawat C, Chen C (2019) Engineered graphene oxide nanocomposite capable of preventing the evolution of antimicrobial resistance. ACS Nano 13(10):11488–11499

Zhu W, von dem Bussche A, Yi X, Qiu Y, Wang Z, Weston P, Hurt RH, Kane AB, Gao H (2016) Nanomechanical mechanism for lipid bilayer damage induced by carbon nanotubes confined in intracellular vesicles. Proc Natl Acad Sci 113(44):12374–12379

Zhu Y, Xu J, Wang Y, Chen C, Gu H, Chai Y, Wang Y (2020) Silver nanoparticles-decorated and mesoporous silica coated single-walled carbon nanotubes with an enhanced antibacterial activity for killing drug-resistant bacteria. Nano Res 13:389–400

Zomorodian K, Veisi H, Mousavi SM, Ataabadi MS, Yazdanpanah S, Bagheri J, Mehr AP, Hemmati S, Veisi H (2018) Modified magnetic nanoparticles by PEG-400-immobilized ag nanoparticles (Fe_3O_4@ PEG–ag) as a core/shell nanocomposite and evaluation of its antimicrobial activity. Int J Nanomedicine 13:3965–3973

Nanoparticles to Abate Antibiotic Resistance During the Management of Dental Diseases

17

C. Pushpalatha, Shwetha Venkataramana, Poornima Ramesh, B. M. Kavya, Shruthi Nagaraja, and K. Vineeth Kumar

Abstract

The intensive use or inappropriate selection of systemic antibiotic therapy is identified as a significant factor in the development of antibiotic-resistant bacterial strains. Antibiotic-resistant bacterial species are associated with a more aggressive disease process, poorer responses to therapy and demonstrate potential for increasing mortality rates. When bacteria become resistant to antibiotics, they oppose the activity of a particular antimicrobial agent to which it was previously susceptible. Increasing attention is being directed towards the management of antibiotic-resistant bacteria. In dentistry, nanoparticle (NP)-based treatments present great promise for treating antibiotic-resistant strains. NP-based antimicrobials have demonstrated their efficacy against both planktonic and biofilm infections. Nanostructured materials can be utilized for carrying antimicrobials, aid in the delivery of novel drugs, and may even have antimicrobial activity on their own. NPs such as silver (Ag) NPs, metal oxides (MeOs), bimetallic NPs,

C. Pushpalatha (✉)
Department of Pedodontics and Preventive Dentistry, Faculty of Dental Sciences, M.S. Ramaiah University of Applied Sciences, Bengaluru, Karnataka, India

S. Venkataramana
Department of Oral Medicine and Radiology, Faculty of Dental Sciences, M.S. Ramaiah University of Applied Sciences, Bengaluru, Karnataka, India

P. Ramesh · S. Nagaraja
Department of Conservative Dentistry and Endodontics, Faculty of Dental Sciences, M.S. Ramaiah University of Applied Sciences, Bengaluru, Karnataka, India

B. M. Kavya
Bowring and Lady Curzon Hospital, Bengaluru, Karnataka, India

K. V. Kumar
Department of Oral and Maxillo-Facial Surgery, Faculty of Dental Sciences, M.S. Ramaiah University of Applied Sciences, Bengaluru, Karnataka, India

M. Y. Wani et al. (eds.), *Nanotechnology Based Strategies for Combating Antimicrobial Resistance*, https://doi.org/10.1007/978-981-97-2023-1_17

zinc oxide, iron oxide, copper oxide, etc. have a potential toward drug resistance mechanisms in bacteria, inhibit biofilm formation or other important processes. Other approaches, such as combining plant-based antimicrobials and NPs to overcome toxicity issues, are being researched. This approach is also effective in inhibiting bacterial efflux pumps potential, formation of biofilms, interference of quorum sensing, and possibly plasmid curing. In this chapter, we will highlight present developments on NPs used to combat multidrug-resistant bacteria.

Keywords

Nanoparticles · Antimicrobial resistance · Metal nanoparticles · Oral biofilm

17.1 Introduction

The advent of antibiotics with the discovery of penicillin by Sir Alexander Fleming revolutionized modern medicine, saving countless lives (Fleming 2001). However, the prescription of antibiotics intended to treat only serious illnesses turned out to be a double-ended sword with the reporting of the first case of Methicillin-resistant *Staphylococcus aureus* in 1962 in the United Kingdom, followed by its prevalence steadily increasing worldwide (Sengupta et al. 2013). The subsequent updating and advancements in the development of novel antibiotics could barely keep pace with the emergence of bacterial resistance. One of the main contributing factors to the surge in antibiotic resistance was the misuse and over-prescription of antibiotics to treat illnesses not warranting its use. Apart from this obvious factor, genes conferring antibiotic-resistance to the bacteria suggest that the emergence of antibiotic resistance may even be a process resulting from an evolutionary consequence. This is confirmed by the presence of bacteria exhibiting resistance to antibiotics isolated from areas unimpeached by human intervention and having the least likelihood of prior exposure to antibiotics (Chattopadhyay and Grossart 2010). The emergence of antibiotic resistance possibly has a multi-factorial etiology, and the World Health Organization (WHO) recognized this as an imminent global health crisis, placing this among the top 10 threats to global health in 2019. Currently, antimicrobial resistance potentially causes the deaths of more than 700,000 people worldwide every year. The WHO has also predicted that this can possibly rise to ten million by 2050, meriting immediate remedial steps to curb this catastrophe (Aranda 2019). In February 2017, the WHO published a list of pathogens belonging to 12 bacterial families, which pose a great threat to mankind. These include the critical-priority bacteria, designated by the acronym ESKAPE, comprising of *Enterococcus faecium, Staphylococcus aureus, Klebsiella pneumoniae, Acinetobacter baumannii, Pseudomonas aeruginosa,* and *Enterobacter species,* that are considered highly resistant to antibiotics (Mulani et al. 2019). A policy package by the WHO in 2002 emphasized the need for ensuring rapid testing of antibiotic resistance, regulation on the sale of antibiotics, access to quality medical services, and the development of novel antimicrobials. In 2023, the European Centre for Disease Prevention and Control (ECDE) and the WHO/Europe jointly published a surveillance of antimicrobial resistance in Europe, that reported increases in resistance of last-line antibiotics including presence of carbapenem-resistant Acinetobacter species which if endemic will make eradication difficult. The reduction of antibiotic consumption and judicious prescription has been strongly advocated across global health agencies and the ECDE further emphasizes the need for the implementation of antimicrobial stewardship programs to curb antimicrobial resistance (World Health Organization 2023).

Prophylactic antibiotic prescription in dentistry should generally be limited to patients considered to be of high risk such as those suffering from congenital heart disease or endocarditis or therapeutically for the treatment of odontogenic and non-odontogenic infection with systemic involvement. Many epidemiological data suggest that more than 80% of antibiotic prescriptions in dentistry to prevent infection are unnecessary. Antibiotic prescription in dentistry accounts for about 10% of worldwide prescriptions (Oberoi et al. 2015). Moreover, strict compliance to the recommended course and duration of antibiotics, especially after initial alleviation of symptoms, is not certain, among most patients, further exacerbating the misuse of antibiotics in dentistry (Barone et al. 2020). A recent study also concluded that dentists might be influenced by patient's request for prescriptions, often leading to increased antibiotic prescribing habits (Al-Khatib and AlMohammad 2022). Most dental and periodontal diseases are best treated by local interventions. Contrary to the belief of many dentists, a systematic review in 2016 found no evidence to state that antibiotics help in pain relief in cases of irreversible pulpitis (Agnihotry et al. 2019). The potential benefits of antibiotics, side effects, and the risk of over-prescribing must be carefully considered in each clinical scenario to ensure judicious use of the same.

Apart from a preventive strategy to combat antimicrobial resistance, various therapeutic strategies have been proposed. The use of topical antibiotics has been proposed for several endodontic procedures and local drug delivery for periodontal treatment. Nanoparticle (NP)-based antimicrobials appear to offer a promising solution to managing antimicrobial resistance. Nanotechnology-based novel antimicrobial approach delivers antimicrobial agents that kills planktonic bacteria, biofilm structures as well as antibiotic-resistant species. NPs are known to easily penetrate biofilms and selectively target specific biofilm matrix constituents (Malaekeh-Nikouei et al. 2020). These NPs either have inherent therapeutic advantages that help in combating recalcitrant biofilms and evades antimicrobial resistance or maybe used as nanocarriers in precise delivery of antibiotics to their intended site of action or even triggered release of antibiotics. Depending on the type of NPs used, the mechanism of antimicrobial action may vary against various strains of bacteria. Novel nanoparticles combine various strategies such as photodynamic and photothermal therapy, antivirulence mechanisms, antibiotic release in response to stimuli, to provide a multimodal mechanism for bactericidal action, by causing physical damage to structural integrity of cell wall, enzymatic killing as well as generation of reactive oxygen species (ROS). Since NPs use multiple bactericidal strategies, the chances of developing drug resistance by bacteria are minimized as compared to conventional antibiotic therapy (Zhao et al. 2021). Novel nanoparticle-based antibacterial therapy has great potential in resisting oral bacterial biofilms and can act as an adjunct in inhibiting various oral infections such as dental caries, endodontic diseases, peri-implantitis, periodontitis, and maxillofacial infections (Qi et al. 2019). There is a need to comprehensively understand the structure, properties, and mechanisms of action of NPs and nanotechnology-based systems to effectively strategize its use in combating antimicrobial resistance. Contemporary research is directed towards the incorporation of NPs as a surface coating on implants, dental materials, and topical agents, for its antibacterial efficacy, potentially reducing the need for conventional antibiotic prescriptions. This chapter will offer insights into the advancements made with NP-based antibacterial applications towards controlling oral biofilm and oral diseases to tackle the issue of antimicrobial resistance.

17.2 Mechanism of Action of Nanoparticle-Based Antimicrobials Against Bacteria

The term "Nanoparticles" (NPs) refers to substances with a particle size range of 1–100 nm. They have a large surface area to volume ratio along with high reactivity offering unique and potent bactericidal characteristics and are thus being explored as an alternative to conventional antibiotics (Edmundson et al. 2013). It is considered that NPs can effectively prevent microbial drug resistance in certain cases. The exact mechanism of antimicrobial action of NPs is not clearly understood. It is proposed that NPs exhibit broad-spectrum antimicrobial properties against both Gram-positive and Gram-negative bacterial species including *Staph. aureus, E. coli, and P. aeruginosa* (Ramalingam et al. 2016). Various NPs exhibit different antimicrobial strategies such as disruption of bacterial membranes, biofilm formation hindrance, intracellular antibacterial effects by interaction with proteins and DNA, ROS generation, or alteration of bacterial cell membrane penetrability. These antibacterial mechanisms can occur simultaneously. This ensures that the likelihood of emergence of drug-resistant bacteria is way lesser since a microbe cannot possibly have multiple mutated genes (Wang et al. 2017). Furthermore, NPs can act as a medium or carrier of antibiotics for intracellular penetration and targeted action.

Contact with bacterial cells is imperative for the NPs to achieve their antibacterial action. The interactions can be in the form of electrostatic attraction, Van der Waal's forces, hydrophobic interactions, or receptor-ligand attachments, aiding the NPs to cross the bacterial cell membrane and interact with the cellular components, causing oxidative stress, damage enzymatic activity, and alter critical cellular metabolic processes (Li et al. 2012; Huh and Kwon 2011). Bacterial cells maintain the production and release of ROS in a balanced manner to maintain its structural integrity. However, in the presence of NPs, the production of ROS is exponentially increased, causing an imbalance and leading to oxidative stress. Different types of NPs produce different ROS by reducing oxygen molecules, which results in changes in cell membrane permeability and its damage (Cheloni et al. 2016). ROS alters the gene expression, facilitating bacterial cell apoptosis (Wu et al. 2011). The oxidative stress is a key factor in the alteration of components responsible for optimal

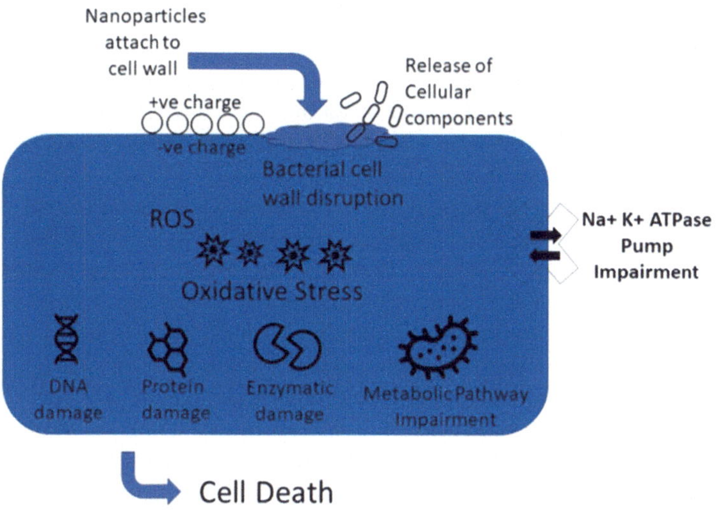

Fig. 17.1 Mechanism of action of nanoparticle-based antimicrobials against bacteria

morphology and physiological functioning of bacterial cells (Yu et al. 2014). It is proposed that metal ions released from NPs are responsible for some amount of weak antibacterial activity and act as carriers of antimicrobial substances (Hussein-Al-Ali et al. 2014). Non-oxidative mechanisms have also been hypothesized for the antibacterial activity of certain NPs although the exact processes are unclear (Leung et al. 2014). The diagrammatic representation of the antimicrobial action of nanoparticle-based antimicrobials is shown in Fig. 17.1.

17.2.1 Mechanisms Against Oral Biofilms and Planktonic Bacteria

Dental caries and periodontitis are diseases associated with polymicrobial biofilms, which are highly organized microbial communities enmeshed in a three-dimensional extracellular matrix (Karygianni et al. 2020). This structured assembly of the biofilms prevents and delays the penetration of conventional antibiotics. The biofilm contains highly varying species of microorganisms, which have different susceptibility influences on antibiotics, making them highly shielded against antibiotics (Stewart and Franklin 2008). To overcome this, increasing the dose of antibiotics leads to an increase in adverse effects and emergence of drug resistance. Achieving a minimal inhibitory concentration of antibiotics to tackle oral biofilms is highly challenging. In this regard, NPs have the unique ability to penetrate biofilms owing to their physical characteristics such as size, shape, surface properties, and interior structure. Contemporary research is directed towards exploring and modifying NPs for enhancing their antibiofilm activity. Various NP-based drug delivery systems have been developed such as lipid-based carriers, polymeric particles, carbon-based NPs, metal NPs, and combined antibacterial phototherapy, targeted at entry into biofilms (Malaekeh-Nikouei et al. 2020). There is ample literature evidence to demonstrate that the penetrative capability of NPs will also enable targeted delivery of conventional antibiotics inside the biofilm and shows promising biofilm eradication abilities. A study demonstrated enhanced antibiofilm activity against *Staph. aureus* biofilm by Clarithromycin-loaded solid lipid nanocarriers at lower drug concentrations (Sharma et al. 2016).

Antibacterial NPs are of particular interest for the eradication of oral biofilms since they can be considered for usage as topical agents to be left in situ on tooth surfaces or gingival sulcular areas. Certain metallic NPs such as Ag ions have been found to be effective against planktonic cells but are not penetrative into a biofilm (Chaw et al. 2005). Incorporating NPs into dental materials is being researched to combat biofilms. Zinc oxide NPs combined into composites have shown to combat *S. sobrinus* biofilm growth at low concentrations. Quaternary ammonium NPs demonstrated in-vitro growth inhibition of *S. mutans* by damaging bacterial cell wall (Yudovin-Farber et al. 2008). However, the challenge of not altering the physical and mechanical properties of composites by the incorporation of NPs remains a barrier and is a hot topic for current research.

The NP design may be modified to perform its triggered release in response to change in pH or hypoxia that are conducive for biofilm formation. The surface charge of the NPs also influences their antimicrobial and antibiofilm efficacies. A study by Li et al. in 2015 showed that positively charged NPs have greater biofilm penetration as compared to negatively charged or uncharged particles (Li et al. 2015). Similarly, hydrophobic cationic NPs exhibited a more robust mechanism of biofilm cytotoxicity (Duncan et al. 2015). However, pH-responsive NPs have been investigated against *S. mutans* biofilm and have shown positive results (Zhou et al. 2016). Metal and MeO-NPs disrupt the biofilm by inhibiting glucans production or by interfering with quorum sensing. The oxidative stress mediated by MeO-NPs also promotes biofilm exopolysaccharide matrix degradation of *S. mutans* (Dwivedi et al. 2014). Catalytic NPs comprising iron oxide NPs coated with dextran also result in biofilm disruption via pH-dependent free radical production (Cormode et al. 2018).

Numerous emerging technologies are being explored to suitably modify NPs for targeted drug delivery and oral biofilm disruption. However, off-target accumulation, systemic consequences by ingestion, tissue distribution, and clearance routes of these oral NPs need to be assessed to promote safety and efficacy of NP-based treatment modalities. Though it is unclear if the accumulation of NPs can cause systemic toxicity, augmenting the efficacy of NP delivery is of prime importance (Benoit et al. 2019).

17.3 Nanoparticle Properties Influencing Bacterial Action

NPs possess unique physical and chemical characteristics that profoundly influence their antimicrobial efficacy and potential. In order to exploit their usage to its optimal efficacy, it is imperative for us to understand the properties of NPs.

Smaller-sized NPs compared to large-sized NPs exhibit greater antibacterial efficacy and concurrently induce higher levels of toxicity due to an increased production of ROS that negatively affect the production of essential biomolecules (Karakoti et al. 2006). The shape of the NP also influences its action and efficacy. Although spherical NPs are the most common type, rods, cubes, sheets, plates, and tubes are also synthesized. The shape of NPs influences the stability of its internal structure that is in turn linked to the energy required to form oxygen vacancies, affecting bactericidal activity. In this context, nanocubes and rods are known to be more effective (Wang et al. 2014). A study done by Agarwal et al. in 2021 showed that the biofilm inhibition ability, especially for *E. coli* and *S. aureus* species, was better with nanocubes as compared to nanospheres. Although the exact mechanism of this improved antimicrobial activity by nanocubes is yet to be fully explored, it is said to be attributed to the distinct structural boundary of nanocubes that allows its edges to disrupt the biofilm (Agrawal et al. 2021). As discussed in the previous section, positively charged NPs exhibit greater antibacterial activity because they are able to alter the functioning of electron transport chain within the bacteria (Ivask et al. 2014). Moreover, a positively charged NP electrostatically attracts to the negatively charged bacterial cell wall, enhancing antimicrobial activity and damage to the bacterial membrane. Additionally, the interior architecture of the NPs may be modified such as in liposomes, micelles, and dendrimers to improve the antibiofilm properties of the nanosystems (Liu et al. 2019).

Drugs loaded in NP carriers are protected from detrimental chemical reactions, and the NPs help in targeting the drugs at a specific intended site of action. Nanocarriers minimize any systemic adverse reactions caused by the drugs as seen with conventional antibiotic therapy. The use of an appropriate NP carrier also enables sustained and controlled drug release and maintains the serum concentration of the drug at optimal levels, thus minimizing the risk of bacterial resistance (Wang et al. 2017). Combination therapy is also possible by packaging multiple drugs within the same NP. For example, studies have shown enhanced antibacterial efficacy by lipid-polymer hybrid NPs (Hadinoto et al. 2013).

17.4 Nanoparticle Classes for Antimicrobial Application

Based on their structural configurations and carrier system used for drug delivery, NPs are categorized as metal-based NPs such as Au, Ag, silica (SiO_2), zinc oxide (ZnO), nickel (Ni), magnesium oxide (MgO), copper oxide (CuO), titanium dioxide (TiO_2), iron oxide (Fe_2O_3), etc., lipid-based carriers such as liposomes and solid lipid NPs, polymeric particles, carbon-based NPs, and other modifications such as smart NPs. The following section will discuss various NP-based drug carrier systems and the role of each NP for antimicrobial applications.

17.4.1 Metal-Based Nanoparticles

Metals have been widely researched as antibacterial agents because of their historical usage. Metals such Ag, Au, Cu, and titanium are known to possess antimicrobial properties. These metals and MeO particles when used in conjunction with NPs exhibit a significant antimicrobial activity. Each metal-based NP has unique properties and has different mechanisms for antimicrobial effects. Each of them has to be further evaluated for its toxic effects and suitability for use in medicine and dentistry. The specific mechanism of different metal-based NPs is explained in detail in the sections below.

Metals with bactericidal activities selectively block certain metabolic pathways and eradicate multidrug-resistant microorganisms. Similar to antibiotics, metal compounds can distinguish between bacterial and mammalian targets, but this is only possible because cells have different metalloproteins and metal transport networks. Due to this distinction, metals can be used as effective, long-lasting antibacterial and biofilm-preventing materials (Gold et al. 2018). Metal may be transformed into NPs (size range: 1–100 nm) with strong, focused, and prolonged antibacterial action at low doses. Metallic NPs have significant antibacterial interactions with bacteria and biofilms due to their smaller size as compared to bacteria and high surface-to-volume ratio (Gold et al. 2018).

The metal NPs that have been subject to the most thorough investigations include AgNPs and AuNPs. CuONPs, ZnO NPs, TiO_2 NPs, and MgO NPs are examples of metallic oxide NPs that have shown antibacterial properties. Others include MnO_2 NPs, Fe_2O_3 NPs, and CaO NPs (Mba and Nweze 2021).

17.4.1.1 Pathways of Interaction of Metallic Nanoparticles with Bacterial Cells

17.4.1.1.1 Interactions at the Cell Membrane Level

Metal-based NPs can electrostatically interact with bacterial cell walls and interfere with the potential and integrity of the cell membrane by releasing divalent cations. The bacterial cell membrane is disrupted by the electrostatic interaction between the positive charge of the NPs and the negative charge of cellular components, which increases oxidative stress and damages bacterial proteins. A large volume of cytosolic water is released as a result of the damage in the cell membrane. As a result, ion imbalance and instability of the membrane respiration is hampered, energy transfer is disrupted, and finally, cell death occurs. The interaction of Ag, Au, ZnO, MgO, and TiO_2 NPs has proven this effect (Gold et al. 2018).

17.4.1.1.2 Attachment to Cytosolic Proteins

Metallic-based NPs interact with cytoplasmic DNA and enzymes to trigger an antibacterial response, resulting in reduced functions of the respiratory, metabolic, and ATP generation pathways. For instance, Ag binds to DNA and respiratory chain enzymes, preventing DNA replication and cell division. On the other hand, Au affects DNA via activating genes inside of the cell. As a result, the membrane's integrity is compromised, and the cell's cytoplasm accumulates ROS (Gold et al. 2018).

17.4.1.1.3 Release of Reactive Oxygen Species (ROS)

ROS induces severe oxidative stress and damage to the cell's macromolecules. These effects collectively result in lipid peroxidation, protein modification, enzyme inhibition, and RNA/DNA damage. By creating holes in the bacterial membrane under conditions of extreme oxidative stress, cells can be destroyed. Ag has been used to detect the production of the hydroxyl radical (OH). By producing H_2O_2 from glucose oxidase with an improved catalytic activity, Au, ZnO, and MgO show ROS production. When TiO_2 is exposed to light, both OH radicals and H_2O_2 are produced as ROS (Gold et al. 2018). Figure 17.2 gives the diagrammatic representation of the various pathways by which metal NPs interact with bacterial cells.

17.4.1.2 Production and Antimicrobial Actions of Metallic Nanoparticles

There are two ways to make metallic-based NPs such as top-down and bottom-up approaches. The top-down strategy is the method of reducing a bulk material to a size that can be manipulated at the nanoscale using ball milling or attrition. Contrarily, bottom-up methodologies employ a variety of strategies to construct a nanoparticle from unprocessed chemicals and physical surroundings to a final, fully functional product. While the bottom-up method takes a longer time, it allows for precise control over the chemical output and produces consistent particle shapes, sizes, and geometries with few defects. In contrast, the top-down method produces a wide size distribution, non-uniform particle geometries, and contains more impurities (Gold et al. 2018). Various approaches are used for the synthesis of Ag and Au nanohybrids such as nanoalloys, core shell NPs are shown in Fig. 17.3 (Loza et al. 2020).

17.4.1.2.1 Silver Nanoparticles

AgNPs are now regarded as the next generation of antibiotics due to their excellent efficiency in suppressing bacterial growth and reduced toxicity compared to other counterparts. Among all NPs, these are regarded as the most widely used NPs for commercial purposes (Mba and Nweze 2021).

Various steps of antimicrobial mechanism of Ag particles are as follows:

1. The first step in the cytotoxic mechanism of AgNPs often involves the adhesion and penetration of the microbial membrane surface.
2. Then, the release of Ag + ions results in detrimental effects on internal components via the generation of ROS and the activation of oxidative stress. These processes lead to the impairment of the Na+/K+ ATPase pump and signal transduction pathways.
3. At last, DNA (compounds containing phosphorus) interacts with Ag + ions and AgNPs, causing protein inactivation and consequent bacterial cell death.

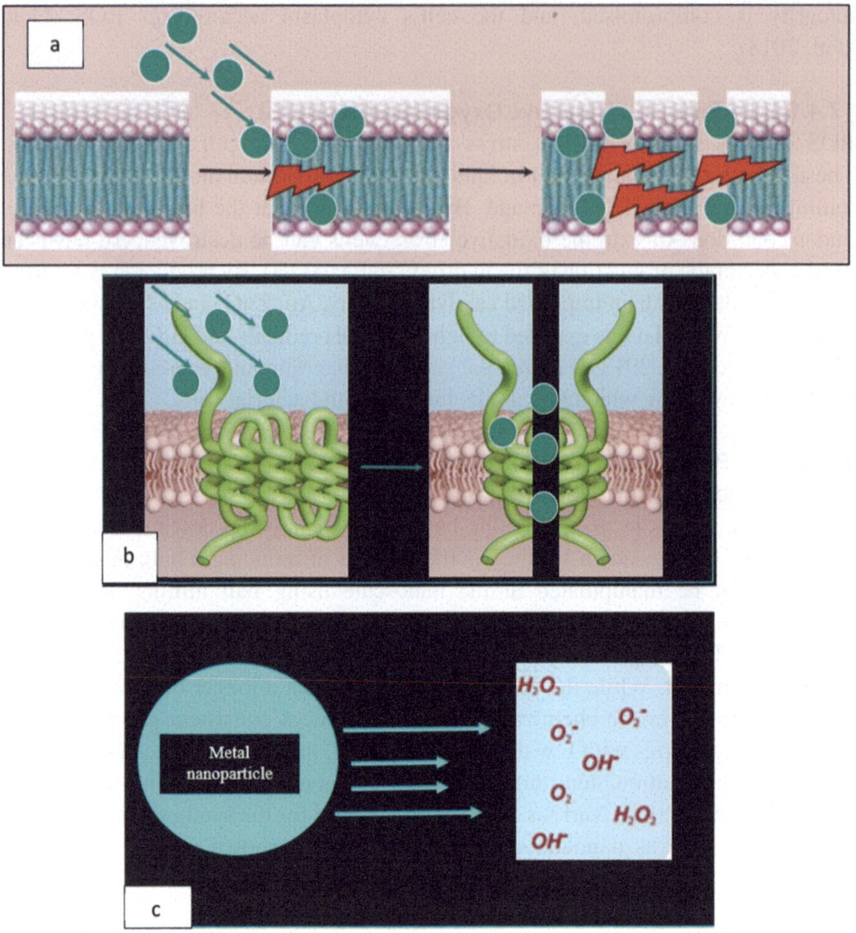

Fig. 17.2 Pathways of interaction of metallic nanoparticles with bacterial cells. (**a**) Attachment with cytosolic proteins. (**b**) Interaction with phospholipid layer. (**c**) Formation of ROS

The ions released from AgNPs interact with sulfur-containing components of the cell membrane, thus preventing the production of the cell wall proteins (Gatto and Bardi 2018).

Other elements influencing the release of Ag + ions include the following:

1. Size: Small AgNPs may easily enter cells, change their membrane architecture and structural integrity, increase permeability, and cause cell death as a result.
2. Variation in bacterial species: AgNP activity is influenced by the variations in cell wall composition, thickness, and organization among various bacterial species. Furthermore, the antibacterial efficacy of AgNPs is closely associated with the impairment of the cell wall and plasma membrane, as well as the inactivation

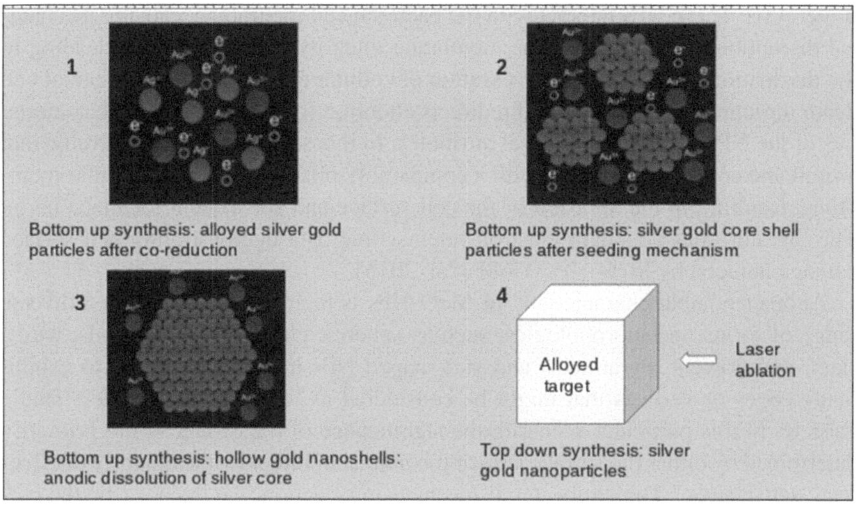

Fig. 17.3 Silver-gold system alloyed bimetallic NP synthesis (1). Bottom-up synthesis: results in co-reduced alloyed silver-gold NPs (2). Bottom-up synthesis: producing NPs with a silver-gold core and shell (3). Bottom-up synthesis: hollow gold nanoshells are produced due to anodic dissolution of the core silver (4). Top-down synthesis: using laser ablation, alloyed silver-gold NPs are created from an alloyed target. (Adapted under the Creative Commons CC BY license (Loza et al. 2020))

of proteins and peroxidation of membrane lipids. These processes result in the disruption of transport proteins and subsequent leaking of potassium ions.

17.4.1.2.2 Metal Oxides

Metallic elements have the capacity to react with oxygen, resulting in the formation of a MeO. MeOs are generated by the bonding of metal ions with oxides, leading to the production of a compact and tightly packed structure. Metal-oxide NPs have the potential to exhibit durable antibacterial properties and effectively inhibit biofilm formation by selectively targeting bacterial cells while sparing mammalian cells. AgO, CuO, Fe_2O_3, MgO, TiO_2, and ZnO are extensively investigated MeO-NPs in scientific research.

The antimicrobial properties of AgO and CuONPs have been shown via the release of Cu^{2+} and Ag^+ ions. Metal-organic frameworks (MOFs) display a wide range of physicochemical and functional properties, including magnetic, mechanical, electrical, and optical attributes. These features arise from the unique Lewis-dot topologies inherent to each MOF. MeO-NPs possess many advantageous characteristics that make them highly suitable for use as antimicrobial medicines. These include their exceptional durability, elevated stability, and less toxicity to mammalian cells, particularly when compared to organic NPs.

MeO-NPs possess nanoscale dimensions and exhibit variable surface chemistry, enabling them to elicit bacterial toxicity through multiple mechanisms, including lipid peroxidation, oxidative stress, cell membrane lysis, enzyme inhibition, and proteolysis. These NPs interact with the bacterial cell membrane, causing mechanical disruption. Consequently, the membrane integrity is compromised, leading to the dissipation of ion gradients, cessation of cellular metabolism, and eventual cell death through lysis. This method redirects attention from the chemical characteristics of the NPs to their mechanical attributes. In this scenario, it is conceivable that a rigid and coarse NP engages with a comparably pliable and fluidic cellular membrane, resulting in the abrasion of the cell surface and subsequent localized harm. The literature has abundant electron microscopic data demonstrating cell surface damage induced by MeO-NPs (Gold et al. 2018).

Another notable characteristic of MeO-NPs is their ability to exhibit a diverse range of forms and morphologies, such as spheres, plates, pyramids, rods, wires, stars, and flowers. Pyramidal- and star-shaped NPs have the potential to exhibit many edges or vertices that might be considered as "active sites" inside a single particle. In this particular scenario, the significance of the dosage is not primarily determined by either the surface area or the mass, but rather by the quantity of effective "active sites." The antibacterial mechanism of action is influenced by the creation of NPs and the alterations made to their surface. A range of methodologies have been used in the preparation of MeO-NPs, including sonochemical, reflux, biosynthesis, solvothermal methods, among others. The use of these synthesis techniques yields significant variations in the surface chemistry, size, and morphology of MeO-NPs. Additives serve to regulate the form, surface chemistry, and stability of a substance, while also enhancing its capacity to interact with certain cells. Positively charged NPs have the potential to engage with the surfaces of bacteria that carry a negative charge, such as *E. coli* and *S. aureus*. This interaction may lead to the breakdown of the cell wall or membrane (Gold et al. 2018).

17.4.1.2.3 Bimetallic Nanoparticles

Metallic NPs are often seen in either monometallic or bimetallic compositions. Monometallic NPs are comprised only of a single metal, exhibiting discernible physical and chemical characteristics. Bimetallic NPs, which are created by combining two distinct metals, have shown considerable potential in terms of their antibacterial effectiveness, exceeding that of NPs composed of a single metal. This enhanced efficacy may be attributed to the synergistic effects resulting from the integration of various metals, as well as the wide array of physiochemical characteristics exhibited by these NPs and their unique methods of action.

Bimetallic NPs have garnered significant interest over the last decade within the realms of science and technology due to their distinctive optical, electrical, magnetic, and catalytic characteristics. The achievement of desired qualities and performance can be attained by the careful selection of appropriate metal combinations and supports, together with the optimization of the composition of each metal type (Gatto and Bardi 2018; Arora et al. 2020).

Bimetallic nanostructures can be classified into two distinct categories such as mixed and segregated. These categories can be further subdivided based on the arrangement of atoms within the nanostructures. Specifically, the subcategories include alloy structures, intermetallic structures, subclusters (where two metals are

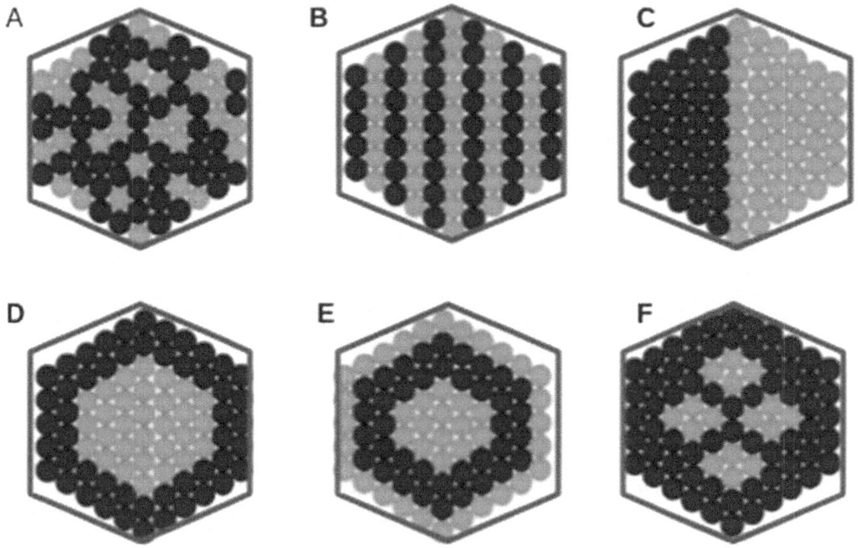

Fig. 17.4 Types of bimetallic nanoparticles: (**a**) alloyed, (**b**) intermetallic, (**c**) subclusters, (**d**) core–shell, (**e**) multishell core–shell, and (**f**) multiple core materials coated by a single shell. Yellow and purple spheres represent two different kinds of metal atoms. (Adapted from Srinoi P et al. (2018) under the Creative Commons CC BY license (Srinoi et al. 2018))

distributed separately with a shared interface), and core-shell structures (where a metal core is surrounded by a shell made of a different metal). A schematic representation of various types of bimetallic NPs is shown in Fig. 17.4. For instance, the combination of two distinct metals results in the formation of alloyed and intermetallic particles. These particles exhibit random and ordered arrangements of atoms, respectively. By carefully regulating the reduction rates of the two metal precursor components during synthesis, it is possible to achieve the appropriate size, shape, structure, morphology, and distribution of metals in the final bimetallic nanoparticles. Recent studies have investigated the antibacterial efficacy of several bimetallic NPs, such as Ag–Au, Ag–Cu, Fe–Ag, and Cu–Ni, which have a synergistic antibacterial impact (Gatto and Bardi 2018; Arora et al. 2020).

17.4.1.2.4 Zinc Oxide

Zn is a significant nutrient as it plays a critical role in the growth, development, and general health of animals. ZnONPs are often used in the inhibition of microbial proliferation, displaying efficacy against free-floating bacteria, and impeding the development of biofilms. These NPs can be produced by a range of techniques, spanning from environmentally friendly chemical practices to the use of sound waves (sonochemistry). The antibacterial action of the NPs may be partly attributable to two main factors such as the dissolution of metal ions and the formation of ROS. ZnO liberates Zn^{2+}in a liquid environment. These ions either adsorb onto the surface of bacteria or enter the bacterial cell. Once inside, they interact with functional groups present in proteins and nucleic acids, so impeding the enzyme activity and disrupting normal physiological processes (Dwivedi et al. 2014; Jiang et al. 2018).

In a study conducted by Sarwar et al. (2016), it was shown that nanosized ZnO induced notable oxidative stress in *Vibrio cholera*. The resultant effects included DNA destruction, protein leakage, membrane depolarization, and alterations in fluidity. In 2017, according to Ehsan and Sajjad, the antibacterial effectiveness of ZnONPs combined with antibiotics was observed against various bacterial strains, including *S. aureus, Proteus, Acinetobacter, P. aeruginosa, and E. coli*. These bacterial strains were initially resistant to antibiotics, but their sensitivity increased when exposed to the combination of NPs and antibiotics. Furthermore, it has been shown that these NPs have the ability to stimulate the generation of ROS even under conditions of low light intensity. This phenomenon might be attributed to the presence of surface flaws on the NPs (Sarwar et al. 2016).

A review published by Sirelkhatim et al. in 2015 indicated that ZnONPs exhibited a significant antimicrobial activity while maintaining lower toxicity levels towards mammalian cells, suggesting its potential use in combating microbial infections. ZnONPs have been shown to possess the ability to impede the development of biofilms and the generation of virulence factors that are reliant on quorum sensing in *P. aeruginosa* (Sirelkhatim et al. 2015).

17.4.1.2.5 Iron Oxide

Iron is one of the essential microelements in biological systems. It performs a number of significant tasks, including supporting the activity of various enzymes, including catalase, transport proteins, like haemoglobin, electron transport chain proteins, such as cytochromes, and DNA repair proteins. Regulatory proteins of *Salmonella enterica*, such as Fur, Fnr, NorR, SoxR, IscR, and NsrR, also contain iron (Gudkov et al. 2021). Iron is like a blade with two edges. Despite performing the aforementioned activities in living organisms, it is also capable of catalyzing the Fenton reaction, which damages proteins, lipids, and DNA. In this process, a hydroxyl radical and a Fe^{3+} ion is created, when the free Fe^{2+} ion interacts with hydrogen peroxide (H_2O_2). Fe^{2+} is regenerated as the first catalyst in the subsequent reaction of Fe^{3+} with the superoxide anion radical (O^{2-}), which results in the creation of molecular oxygen (O_2). Maintaining a very low amount of free iron ions inside cells is important to shield it from the harm brought on by the production of hydroxyl radicals. Both free iron ions and iron oxide nanoparticles (IONPs) have antibacterial capabilities; however, unlike free ions, IONPs have little to no harmful impact on human cells.

The production of IONPs may be accomplished by many methods. ROS production is one of the primary mechanisms of IONP toxicity. In turn, ROS have a genotoxic effect that damages DNA molecules. A reduction in the activity of the antioxidant system enzymes (SOD, catalase, and glutathione reductase) might result in an increase in ROS levels. Mercapto (-SH), amino (-NH), and carboxyl (-COOH) groups of proteins, particularly those in enzymes, can bind to metal ions, causing deactivation or partial inhibition. Additionally, as shown in SEM investigations on the cell wall of *S. aureus*, IONPs harm the integrity of bacterial cell walls. Antibiotic-resistant bacteria may express less antibiotic resistance genes (ARGs) when exposed to IONPs. The activity of F0/F1-ATPase, the rate of H^+ passage through the membrane, and the redox potential can all be affected by IONPs. For NPs with tiny diameters, a capability to prevent DNA replication by inactivating topoisomerase is disclosed. Electron microscopy was used to demonstrate that IONPs interacted directly with the *E. coli* cell wall. Additionally, IONPs can enter the cytoplasm, gather there, and damage cell walls and create vacuoles. In-vivo experiments with IONPs also demonstrated the highest level of antibacterial efficacy by inactivation of MRSA bacteria (Baptista et al. 2018).

A few of IONPs' drawbacks include their generally ineffective antibacterial activity against a variety of strains, as well as their inadequate biocompatibility and toxicity for several eukaryotic cell lines. By conjugating IONPs with other NPs like carbon nanotubes (CNTs), chitosan, polyethylene glycol, or antibiotics, altering their surface, using green production techniques, etc., it is possible to change their antibacterial activity (Baptista et al. 2018).

17.4.1.2.6 Copper Oxide

Due to their undeniable broad-spectrum bactericidal activity, low cost, physicochemical stability, biocompatibility properties, and compatibility with other materials to form polymers, copper NPs (CuNPs) or copper oxide NPs (CuONPs) have drawn significant attention as AgNPs substitutes in the medical field. Various types of NPs containing copper, such as CuNPs, CuONPs, and bimetallic NPs including copper, have been effectively synthesized. CuONPs have many advantages compared to organic antibacterial agents, including enhanced durability, consistent physical and chemical properties, and a relatively prolonged period of storage viability. CuONPs have physical properties that facilitate its integration with polymers, enabling the formation of composite materials using CuONPs (Wang et al. 2017; Cheloni et al. 2016).

CuNPs exhibit a higher rate of oxidation and worse stability compared to CuONPs. The production of CuNPs in an ambient environment is a significant challenge due to the inherent tendency of Cu to undergo oxidation upon exposure to atmospheric conditions, resulting in the formation of CuO species such as Cu_2O and CuO. The manufacture of CuNPs in the presence of polymers and surfactants is often necessary for the purpose of generating coatings on their surface. NPs that consist of copper in conjunction with another metal are often referred to as copper-containing bimetal NPs. The collaborative interaction between two metallic elements may have a synergistic effect, potentially enhancing their antimicrobial efficacy beyond that of a single metal (Wang et al. 2017; Cheloni et al. 2016).

A range of Cu-containing NPs may be integrated with biocompatible supporting materials that possess antibacterial properties and sustain their efficacy. Examples of such supporting materials include chitosan, cellulose polymers, hydrogels, and others. The antibacterial activity is positively associated with the concentration of CuONPs, and they also exhibit enhanced antibacterial efficacy against various bacterial and fungal strains. Chitosan has the potential to be integrated with

Cu-containing NPs for use in dental materials due to its biocompatible, biodegradable, and non-toxic nature. Chitosan has the potential to enhance the adhesion of copper to tooth surfaces and augment its antibiofilm properties via interactions with hydroxyapatite and the bacterial cell walls present on teeth. Chitosan has a significant efficacy in inhibiting the growth of Gram-negative bacteria, Gram-positive bacteria, and fungi. Additionally, it exhibits the ability to impede the adherence of oral microbes, hinder the formation of biofilms, and expedite the healing process of wounds and oral ulcers. In addition to their exceptional antibacterial properties, the CuNP-embedded hydrogels decreased inflammatory responses and accelerated angiogenesis in in-vivo models to hasten wound healing. Natural physical and chemical properties of Cu-containing NPs can be preserved while providing composites additional qualities that make them better suited for clinical applications. The achievement of this objective is realized by the amalgamation of Cu-containing NPs with other substances, resulting in the formation of composite materials (Wang et al. 2017; Cheloni et al. 2016).

In a broad sense, the deleterious effects of copper on microbial organisms manifest via the generation of ROS and the displacement or substitution of native cofactors within metalloproteins. Moreover, Cu plays a significant role in the innate immune response by promoting the generation of ROS during the phagocytic process inside immune cells, hence enhancing the bactericidal efficacy in bacterial engulfment (Wang et al. 2017; Cheloni et al. 2016).

Multiple studies have shown evidence that NPs possess enhanced antibacterial properties compared to conventional-sized materials, owing to their increased surface area and unique crystal structure. In addition, they have the ability to exert effect on many biological components inside microbial cells via unconventional mechanisms. The antibacterial activity of CuNPs is primarily influenced by physical characteristics such as size, shape, and environmental conditions, namely, temperature. Chemical factors to be considered include ambient conditions (such as pH level, moisture content, and composition of the medium), alterations resulting from the introduction of additional elements, and the extent of Cu oxidation (Wang et al. 2017; Cheloni et al. 2016).

The incorporation of NPs onto the surface or inside dental materials has the potential to enhance or maintain the mechanical properties of these materials. Additionally, this integration may provide diverse antibacterial capabilities to the material. Metalloproteinase inhibitors (MMPs) have the ability to impede the activity of metalloproteinases and exhibit several applications within the field of dentistry. The use of CuNPs has shown potential benefits in several dental applications such as dental adhesives, filling materials, implant coatings, and bracket coatings. CuNPs have the ability to effectively inhibit or eradicate a diverse array of oral pathogenic microorganisms, while mitigating the risk of microbial resistance. These NPs have promising antibacterial properties and furthermore demonstrate beneficial effects on oral tissues. The present focus of research is in exploring the potential of CuNPs in several domains, including dental filling materials, prosthetic devices and implant coatings, and oral antibacterial agents (Wang et al. 2017; Cheloni et al. 2016).

(NACP), calcium and phosphorus ions (Zhou et al. 2020). Poly(amidoamine) (PAMAM) are another set of polymeric particles with highly branched structures, which have tremendous potential to induce remineralization. A study published by Liang et al. in 2019 demonstrated that PAMAM with NACP has caries inhibiting properties, and these nanopolymers can be incorporated into composites and dentin adhesives to promote remineralization and prevent demineralization (Liang et al. 2019). NPs responsive to pH changes triggered by environmental changes are being explored for delivery of drugs, which disrupt the oral biofilm. Farnesol loaded NPs were found to be effective in reducing the biofilm virulence (Horev et al. 2015). NP systems based on dimethylaminoethyl methacrylate (DMAEMA) are also being studied, and it is found that composites with DMAEMA along with NPs of NCAP exhibited reduction in microorganisms within the oral biofilm (Zhu et al. 2022). Thus, NPs may be used in conjunction with dental materials for their antibacterial properties without significantly affecting their properties. Another example is the use of chitosan NPs with glass ionomer cements for anticaries property and prolonged fluoride release.

Fluoride-doped NCAP promotes dental remineralization and offers biomimetic strategies for the prevention of dental caries (Iafisco et al. 2018). Curcumin-doped poly(lactic-co-glycolic acid) NPs in dental adhesives demonstrated antibiofilm activity for *S mutans* and was recommended for bonding of orthodontic brackets to prevent white spot lesions (Ahmadi et al. 2020). Curcumin-loaded polymeric NPs have also been shown to reduce periodontal inflammation and tissue damage. Polymeric NPs composed of poly(ethylene glycol) (PEG) and polylactic-co-glycolic acid (PLGA) show antibiofilm activities against *S mutans* and are biodegradable and have potential for sustained release of drugs. Polymeric NPs loaded with chlorhexidine were effective in killing *S mutans* biofilms and were released in the specific acidic environments in response to pH changes (Zhou et al. 2016).

Zhang et al. in 2016 developed a hybrid drug carrier system by combining two distinct delivery platforms for enhanced antimicrobial functions. Antibiotics were loaded into polymeric NPs, which were embedded into a hydrogel network that can adhere to biological surfaces and was found to inhibit the formation of *E. coli*. The NPs allowed sustained release of antibiotics while the hydrogel network maintained tissue adhesion (Zhang et al. 2016).

Polymer-based drug carrier systems are being used as intrapocket drug delivery systems for the management of periodontal diseases due to their stability and ability to maintain optimal antibiotic concentration in the gingival crevicular fluid. Triclosan loaded into polymer-based NPs namely poly(D, L-lactide-co-glycolide), poly(D,L-lactide), and cellulose acetate phthalate has been investigated for the treatment of gingival inflammation and found a fast, sustained release of triclosan at targeted sites (Pinon-Segundo et al. 2005). Polymeric PolymP-n Active NPs combined with Ag and doxycycline showed substantial antimicrobial activity against numerous periodontal pathogens including *S. oralis, A. naeslundii, V. parvula, F. nucleatum, P. gingivalis, and Aa. Species* (Sánchez et al. 2019). Polymeric NPs are also of great use to deliver and increase local dose of agents that exhibit antibiofilm properties against *P. gingivalis* and *S. gordonii* as well.

Chitosan NPs have a positive charge, enabling attraction to the negatively charged cell wall of bacteria along with the mucoadhesive properties enabling a sustained release of drugs. Various in-vitro studies have proved the antibiofilm activities of chitosan including a broad spectrum of antibacterial activity against Gram-positive and Gram-negative bacteria and fungi. The use of chitosan NPs in the field of endodontics as an irrigant has been investigated, and a study showed that chitosan NPs removed smear layer as well as prevented recolonization of bacteria over radicular dentin (Jhaveri et al. 2021). Chitosan NPs along with fluoride and other polymeric NPs also demonstrated anticaries activity.

17.4.4 Lipid-Based Nanocarriers

Solid lipid nanoparticles (SLN) composed of triglycerides, glycerides, steroids, and waxes are advantageous as drug carrier systems due to their stability, targeted drug delivery, and controlled release. Structurally, they have a hydrophobic core of phospholipids with the drug dispersed in the core. In-vitro studies have shown that antibiotics and drugs loaded into SLN carriers exhibited antibiofilm activities against *S. epidermidis, S. aureus, and P. aeruginosa* (Bazzaz et al. 2016; Singh et al. 2014). A recent study done by Radaic et al. in 2022 investigated the potential of SLNs loaded with nisin and found that the system possesses significant antibacterial, antibiofilm, and anticancer properties. The study demonstrated the inhibition of *T. denticola*, a common oral pathogen, along with a disruption of oral biofilms and decreased cell viability of oral squamous cell carcinoma cells (Radaic et al. 2022). Although SLNs have many advantages, they have relatively limited space for efficient drug loading and can lead to premature drug expulsion. Nanostructured lipid carriers (NLC) and lipid drug conjugates were designed to overcome the challenges posed by SLNs. NLCs have higher drug loading capacity since they are composed of glycerides that have larger distance between fatty acid chains and the unstructured crystal, which enables higher amounts of drug accommodation (Mukherjee et al. 2009).

Liposomes are NPs consisting of an outer phospholipid bilayer with an inner hydrophilic core that is suitable for drug encapsulation. The phospholipid layer of liposomes attaches to the bacterial cell membrane that penetrates the peptidoglycans layer of bacteria, creating pores and transporting channels for sustained drug release and subsequent inhibition of bacterial growth. Yamakami and colleagues in 2016 studied the action liposome encapsulated nisin particles against *S. mutans* and found that they exhibited sustained inhibition of glucan synthesis by Streptococci and a potential to prevent dental caries (Yamakami et al. 2016). Liposomal carriers used in conjunction with amphotericin B elicited inhibition of Candida biofilms and substantial antifungal activities. Liposome carriers are effective for drugs used for pain management. Moraes et al. demonstrated the use of liposome loaded lignocaine/prilocaine for pain control during scaling and root planning procedures (Moraes et al. 2019).

17.4.5 Smart Nanoparticles

Smart material carrier systems have the unique ability to alter their properties to enable the release of antimicrobial drugs in response to environmental changes such as pH alterations, temperature, and light. Multiple drugs can be packaged in the same carrier using smart NPs. Apart from controlled release of therapeutic drugs, smart NPs have many advantages such as smaller particle size and larger surface area, triggered release of agents based on environmental factors, longer duration of action, improved pharmacokinetics, and decreased toxicity along with ability to combine different classes of drugs in the same carrier (Malaekeh-Nikouei et al. 2020).

However, pH-responsive drug delivery systems can be incorporated into dental materials or used directly for the prevention of dental caries. A study showed that Chlorhexidine-loaded NPs with a hydrogel-based carrier released chlorhexidine in acidic environments, exhibiting antibiofilm activity against *S mutans* (Peng et al. 2022). Chlorhexidine-loaded bioactive glass NPs are another example of pH-responsive smart NPs, showing anticaries activity. Chitosan NPs exhibit similar pH-sensitive antibacterial activity wherein drug degradation is prevented under physiological salivary pH conditions while acidic pH conditions trigger the release of drugs. Another novel smart NP system was developed for the treatment of periodontitis in diabetic patients, which uses responsiveness to multiple environmental changes including pH and enzyme activities as the trigger factors for drug release. The novel drug delivery system developed by Wang and colleagues initiated drug release in response to changes in pH and lipase activity within the oral microenvironment. The developed drug delivery systems exhibited significant antibacterial, anti-inflammatory, and antioxidant activities with osteogenic potential (Wang et al. 2022). Various smart resin composites, adhesives, resin infiltrates, and sealants are being developed for anticaries activities, which are pH-sensitive and promote remineralization. Efforts are also being made for the development of smart materials that can self-repair damaged tissues and potentially reverse dental caries and periodontitis (Yu et al. 2023). Another novel smart polymer strategy for antibiofilm activity against *S mutans* was developed that are pH-responsive and have charge switching potential in the acidic environment. The smart co-polymer becomes positively charged in acidic conditions that enables binding with negatively charged cell membrane, causing cell death of both planktonic and *S. mutans* biofilm (Bhat et al. 2022). Thus, smart NPs have potential for enabling targeted delivery of drugs to combat oral biofilms.

17.5 Nanoparticle-Based Therapeutics in Clinical Trials

Targeted and controlled drug delivery systems often use a diverse range of chemical, inorganic, polymeric, and metallic nanostructures. These nanostructures include dendrimers, micelles, SLNs, CNTs, and liposomes. For controlled and delayed drug release, low-soluble medicines in particular are enclosed in NPs. Several drugs, such as Caelyx®, Emend®, Mepact®, Abraxane®, Myocet®, and Rapamune®, have been commercially introduced for human use after the successful preclinical and clinical studies. The development of these innovative therapeutic agents presents an opportunity for the prospective utilization of peptides, nucleic acids (DNA and RNA), and genes as nanomedicines in the management of chronic ailments such as diabetes, cancer, atherosclerosis, myocardial ischemia, asthma, pulmonary tuberculosis, Parkinson's disease, Alzheimer's disease, and various other conditions. However, the size, shape, hydrophobicity, surface properties, as well as a number of other chemical and biophysical features affect the effectiveness of these pharmaceutical delivery techniques (Patra et al. 2018).

Efficient drug delivery strategies in biomedical applications are often attributed to nanoscale materials, which possess exceptional biocompatibility and biodegradability. When the size of particles is in the nanometer range, as opposed to the larger micron-sized particles often used in standard diagnostic procedures, the total surface area of the particles rises significantly, by many orders of magnitude. Nanoscale materials provide considerable potential for interacting with systems, mostly attributable to the heightened surface area that facilitates extensive contact with their environment. Nevertheless, the presence of toxicity concerns mandates further investigation to ensure optimal safety for the utilization of these pharmaceuticals within clinical environments (Patra et al. 2018).

One of the biggest threats to human health is the growth of pathogenic bacteria, which are drug resistant. Antibiotics attack bacteria's primary mechanisms of survival, preventing the formation of cell walls and the production of essential proteins, DNA, and RNA. Bacteria have the innate capacity to adapt quickly through mutation and DNA transfer in order to counter the threat posed by these antibiotics. A single microbe can pick up many drug-resistance genes from diverse organisms, creating a "superbug" that is resistant to a wide range of medications, and a bacterial "super-resistance" gene that renders them immune by causing enzymatic breakdown (Patra et al. 2018).

Table 17.1 summarizes the recent studies/clinical trials on NPs tested on organisms and their antimicrobial potential.

17.6 Nanotheranostic Potential

The potential to influence a disease's therapy through its diagnosis is known as theranosis (Jagtap et al. 2017). The goal of the nanotheranostic approach is to harness the biomedical applications of NPs to create novel diagnostic techniques that have the ability to alter the course of a disease's natural history by guiding treatment (Pedrosa et al. 2015).

With the development of antimicrobial resistance (AMR) in microbes that cause common infections like bacterial pneumonia, postoperative infections, and higher mortality infections like malaria, HIV, and tuberculosis that are getting harder to treat due to the rise in resistance, this novel approach can be used to address the major public health issues of our time (Inoue and Minghui 2017). Additionally,

Table 17.1 Summary of the recent studies/clinical trials on NPs tested on organisms and their antimicrobial potential

Author and year	Nanoparticle	Bacteria tested	Antimicrobial potential
Escárcega-González et al. (2018)	AgNPs	*E. coli, P. aeruginosa*, Clinically MDR strain of *P. aeruginosa and B. subtilis*	The antibacterial efficacy against the resistant microorganisms was evaluated in a mouse skin infection model. The optimal and secure use of novel psychoactive substances (NPS) as medicinal agents in animal models
Huq (2020)	AgNPs	Multidrug-resistant pathogenic microbes *S. aureus, P. aeruginosa,* and *E. coli*	The MIC values of *S. aureus* and *P. aeruginosa* were determined to be 100 μg/mL and 100 μg/mL, respectively. The minimum bactericidal concentrations (MBC) of *S. aureus* and *P. aeruginosa* were determined to be 200 μg/mL and 50 μg/mL, respectively
Akter and Huq (2020)	AgNPs	Drug-resistant microbes *P. aeruginosa, E. coli,* and *S. aureus*	The MIC values for *E. coli* and *S. aureus* were determined to be 6.25 and 50 μg/mL, respectively. Additionally, the minimum bactericidal concentration (MBC) values for *E. coli* and *S. aureus* were found to be 25 and 100 μg/mL, respectively
Dong et al. (2019)	AgNPs	Vibrio natriegens	The MIC and minimum bactericidal concentration (MBC) exhibited a dependence on the dosage administered. The minimum inhibitory concentration (MIC) spans a range of 1 to 11.5 μg/mL, whereas the minimum bactericidal concentration (MBC) falls within the range of 1.1 to 11.7 μg/mL
Hossain et al. (2019)	AgNPs	*Enteropathogenic E. coli, S. typhi, S. aureus, V. cholerae, E. faecalis, Hafnia alvei, Acninetobacter baumannii, E. coli* DH5α, *E. coli* K12 and *B. cereus*	The MIC for two different types of silver nanoparticles (bAgNPs) was determined to be 0.125 μg. The compound Et-bAgNPs exhibited the most potent antibacterial activity against *S. aureus* at a concentration of 60 μg during a 16-hour incubation period, resulting in an inhibition zone diameter (IZD) of 28 mm

(continued)

Table 17.1 (continued)

Author and year	Nanoparticle	Bacteria tested	Antimicrobial potential
Adebayo-Tayo et al. (2019)	AgNPs	*S. aureus, E. coli* 11,775, E. coli 35,218, *P. aeruginosa, Citrobacter* spp., *Salmonella typhi* 14,028, *B. cereus*	Effective antibacterial activity against all pathogens with IZD ranging from 1 to 21 mm exhibiting strong antibiofilm activity
Hu et al. (2019)	AgNPs	*S. aureus, B. cereus, S. enterica, P. aeruginosa, E. coli*	The MIC for Gram-positive bacteria is 16.12 μg/mL, whereas for Gram-negative bacteria it is 13.98 μg/mL. The cytotoxicity is seen to have a positive correlation with the concentration of NPs
Rodríguez-Serrano et al. (2020)	AgNPs	Uropathogenic *E. coli*	The MIC is 25 mg/mL. The concentration of 7.5 mg/L, which is below the minimum inhibitory concentration (MIC), was shown to effectively prevent around 97% of pathogenic biofilm development or cause disruption of 80% of mature biofilm
Hamouda et al. (2019)	AgNPs	*E. coli, B. cereus*	IZD of 22 mm and 20 mm. Mechanism: Damage to cell membrane, leakage of cytoplasm exterior to cell, internal diffusing of AgNPs to cell, cell disruption, disintegration, shrinking of protoplasm and, detachment of cellular membrane. Cytotoxicity: hemolytic activity showed that it was non-toxic to human RBC in low concentrations
Habibipour et al. (2019)	AgNPs	*P. aeruginosa*	Showed strong inhibition against biofilm formation at 0.1–0.5 mg/mL. Cytotoxicity: no significant toxicity against L929 cell line at 400 μg/mL
El-Sayed and El-Sayed (2020)	AgNPs	Multidrug-resistant *P. aeruginosa* and *S. aureus*	The results demonstrated a substantial inhibitory impact on *P. aeruginosa*, with a mean inhibition zone diameter of 22.4 mm. Additionally, MIC was determined to be 21.33 μg/mL

(continued)

Table 17.1 (continued)

Author and year	Nanoparticle	Bacteria tested	Antimicrobial potential
Wypij et al. (2018)	AgNPs	*P. aeruginosa, M. furfur, B. subtilis, E. coli, S. aureus,* and *K. pneumonia*	The MICs of the substances were found to be 16, 32, 64, 64, 256, and 26 μg/mL, respectively. Cytotoxicity assessment of in-vitro cytotoxic effects on mouse fibroblasts and cancer cells
Singh et al. (2018)	AuNPs	*P. aeruginosa, E. coli, S. aureus,* and *Vibrio parahaemolyticus*	The antibacterial activity was seen at concentrations of 15.3 ± 0.5 ppm, 11.7 ± 0.5 ppm, 14.7 ± 0.9 ppm, and 13.7 ± 1.1 ppm. The compound exhibited cytotoxic effects on human cervical cancer cells (HeLa) at a concentration of 10 μg/mL, but it did not have any inhibitory effects on human lung cancer cells (A549)
Folorunso et al. (2019)	AuNPs	*S. aureus, Clostridium, E. faecalis* and *K. pneumonia*	The observed phenomenon demonstrates a positive correlation between the concentration of the substance and its effectiveness in inhibiting the growth of microorganisms
Umar et al. (2019)	ZnONPs	*B. cereus, S. aureus, E. coli, K. pneumoniae,* and *S. typhi*	The observed antibacterial activity exhibited a significant dependence on the dosage administered. The investigation focused on the cytotoxicity of the MDA-MB231 and MCF-7 cell lines. The cytotoxic impact exhibited a dependence on the concentration
Rehman et al. (2019)	ZnONPs	*E. coli*	The MIC values for *E. coli* were found to be more than 8 mg/mL, while the minimum bactericidal concentration (MBC) values were greater than 16 mg/mL. In the case of *S. aureus*, the MIC values were 4 mg/mL, and the MBC values were 8 mg/mL
Ansari et al. (2020)	ZnONPs	*E. coli* and *S. aureus*	The MIC for *E. coli* was determined to be 125 μg/mL, whereas the MIC for *S. aureus* was found to be 62.5 μg/mL
Supraja et al. (2016)	ZnONPs	*Sphingobacterium* sp., *Acinetobacter* sp., *Ochrobactrum* sp.	The IZD are 3 mm, 1.7 mm, and 4 mm, respectively

(continued)

Table 17.1 (continued)

Author and year	Nanoparticle	Bacteria tested	Antimicrobial potential
Malaikozhundan et al. (2020)	ZnONPs	*E. faecalis, S. aureus, E. coli*, and *P. aeruginosa*	A more pronounced antibacterial efficacy of ws-ZnONPs was seen against *E. faecalis* and *S. aureus* at a concentration of 100 μg/mL. The growth of the biofilm formed by *E. faecalis* and *S. aureus* was significantly suppressed at a concentration of 100 μg/mL, in contrast to *E. coli* and *P. aeruginosa*. The results indicate that ws-ZnONPs exhibited the lowest LC50 value (9.65 μg/mL) in comparison to both uncoated ZnONPs (38.8 μg/mL) and leaf extract (13.06 μg/mL)
Ijaz et al. (2017)	CuONPs	*Klebsiella, E. coli, S. aureus*, and *B. subtilis*	Significant bactericidal activity of NPs against the diameters of inhibition zones for *Klebsiella* and *B. subtilis* were found to be 14 ± 0.05 mm and 15 ± 0.11 mm, respectively. At a concentration of 5 mg, the CuONPs exhibited significant antimicrobial activity against *S aureus, Klebsiella,* and *Bacillus subtilis*, resulting in inhibition zone diameters (IZD) of 10 ± 0.11 mm, 14 ± 0.05 mm, and 15 ± 0.11 mm, respectively
Arul Selvaraj et al. (2019)	CuONPs	*Proteus mirabilis, S. aureus*	There is a notable synergistic impact shown when β-lactam antibiotics are used in combination. The two species exhibited a significant decrease in biofilm development, with reductions of 85% and 93% observed, respectively
Thakur and Kumar (2019)	CuONPs	*Pseudomonas, Klebsiella, staphylococcus,* and *E. coli*	The IZD of the specimens are 11 mm, 12 mm, 8 mm, and 9 mm, respectively

according to Col and Brig (Das and Kataria 2010) and Tournier et al. (2019), AMR poses a risk for the potential spread of drug-resistant bacteria from hospitals and healthcare facilities. In order to identify AMR, ascertain antimicrobial susceptibility, find novel medications, and monitor treatment, it is required to deploy new diagnostic procedures (Burnham et al. 2017). In order to detect multiresistant microbes in tissues, monitor their treatment, and eradicate them using both pharmacological and physical methods, theranostic platforms have been explored recently (Xie et al. 2017; Vangara et al. 2018). Three main strategies have been employed in nanotheranostics: combining NPs with imaging agents, antimicrobial medications, and targeted ligands (Tonga et al. 2014).

17.6.1 Dental Diagnostics

The idea of nanobiosensing was created in an effort to enhance medical diagnostics. A biosensor may be described as an analytical device that integrates a biologically active element together with a suitable physical transducer, with the purpose of generating a detectable signal that is directly proportional to the quantity of chemical species present in a given sample. The enhancement of biorecognition and overall performance of bioreceptors was achieved by the integration of nanotubes, nanowires, and nanodots into the sensing assembly of nanobio receptors, resulting in the development of nanobio receptors. The use of NPs in oral hygiene is a growing area of research. Specifically, nanorobots are being included into mouthwash formulations to efficiently identify and eradicate pathogenic bacteria responsible for gingivitis and periodontitis (Rajendiran et al. 2021). By doing so, these nanorobots create an environment conducive to the growth of beneficial oral flora, promoting oral health. Moreover, the device would possess the ability to detect and identify food particles, tartar, and plaque, facilitating their elimination by rinsing (Shetty et al. 2013). This process would effectively contribute to the attainment of a hygienic and plaque-free oral cavity. The potential use of hollow spheres, core-shell structures, nanotubes, and nanocomposites as delivery systems for periodontal treatment can be seen in the foreseeable future. Nanorobots included in dentifrices are expected to continuously remove supra- and subgingival calculus. Additionally, they will act as a constant barrier against it by eliminating the germs that cause halitosis.

A promising and intriguing tool for the future is the use of platforms created by milli/microrobots to execute a specialized therapy in the event of the removal of biofilms in places where traditional medications cannot reach (Vikram Singh and Sitti 2016). The usage of biomimicry, in which nature is used as a model and mentor to build bioinspired goods so as to have more biocompatible platforms, is another fascinating strategy (Evangelopoulos et al. 2018). To accurately translate nanomedicine into clinical practice, this factor is crucial (Lagarce 2015). For the application of these instruments in clinical practice, toxicity and safety assessment platforms should also be considered. For the development of these methods for the detection and treatment of infectious disorders, it is necessary to evaluate the effects of theranostic applications on host autoimmunity/allergy as well as microbiota (Wypych and Marsland 2018).

17.7 Future Scope and Challenges of Current Research

Over the past 10 years, there has been an incredible rise in the use of nanomedicine. Nanotechnology-based pharmaceuticals are in the development to address many objectives, encompassing the enhancement of existing drugs, precise drug administration to particular anatomical sites, and eventually fostering greater patient adherence. Successful completion of both preclinical and clinical research using nanotechnology-based diagnostic and therapeutic approaches has been achieved. Nanomedicines have significant benefits over traditional therapies due to their ability to achieve controlled drug release, improved pharmacokinetics, and enhanced pharmacodynamics, leading to reduced toxicity and increased effectiveness. To achieve sustained drug delivery at a given place, an appropriate drug carrier is essential. Despite the apparent success of the majority of preclinical investigations, clinical translation remains a significant obstacle that necessitates numerous carefully planned clinical trials in numerous disease cohorts. Due to the inadequacy of existing regulations in guaranteeing the safety and quality of emerging nanomedicines, it is imperative to establish a distinct regulatory framework specifically tailored for nanosystems and nanomedicines in contrast to traditional pharmaceuticals. Furthermore, it is essential to develop suitable delivery methods quickly in order to fulfill present medical needs and advance earlier nanosystems. The current focus of nanosystems is shifting towards the delivery of peptides, nucleic acids, and genes, in addition to medication administration. However, it is crucial to conduct a comprehensive benefit-risk evaluation prior to granting authorization for human usage. The primary challenge associated with the translation of nanomedicines into clinical practice pertains to the adoption of a universal approach. While the preclinical stage involves conducting a comprehensive experimental approach, clinical trials need the establishment of dose standardization and a more in-depth examination of the toxicity profile of nanomedicines. Prior to being authorized for therapeutic applications in human illnesses, it is essential to assess the potential adverse consequences, both immediate and prolonged, of nanoencapsulation. This is due to its ability to extend the half-lives of drugs in circulation and enhance their retention inside the body (Halwani 2022).

In order for nanomedicines to be deemed suitable for clinical application, it is essential that they exhibit a significant advantage in terms of the risk-benefit ratio. The primary organs where nanomedicines accumulate after systemic administration are the liver, kidney, spleen, lungs, bone marrow, and lymph nodes. These organs need specific evaluation to assess any potential adverse effects. Based on the available research, it has been shown that nanomedicines possess cytotoxic qualities that have the potential to induce macrophage death and perhaps suppress the immune system. Hence, it is essential to tackle these concerns in order to safeguard the security and optimize the efficacy of nanomedicines within the realm of medical applications. Additional research is required to develop synthesis procedures. In the context of large-scale production, it is essential that the preparatory procedures used exhibit characteristics of simplicity, cost-effectiveness, and consistent reproducibility over several batches. Given the promise of nanotechnology as a tool for diagnosis and treatment, further clinical studies on an international scale are necessary (Halwani 2022).

The moral implications arise when considering the rigorous testing process that dental or medical nanoproducts undergo after extensive research and development. These tests encompass comprehensive preclinical in vitro evaluations, which examine the mechanical, toxicological, and immunological aspects associated with the particular product in question. Several reputable institutions, including the National Institute for Occupational Safety and Health and the U.S.A. Environmental Protection Agency, have issued recommendations pertaining to the assessment of hazards associated with NPs. The current decision-making procedure used by the ethical committee, which mostly relies on utilitarianism, is insufficient in addressing the fast-paced progress and uncertain future of nanotechnological advancements and products. This emphasizes the need for greater understanding of the science, along with a comparison of the risks, benefits, and ethical concerns at every stage of the development process (Halwani 2022).

The future holds the potential for the development of various materials that may be used in periodontal medication delivery systems. Pharmaceutical substances may be encapsulated inside biodegradable polymer nanospheres, therefore facilitating controlled drug release upon degradation of the nanospheres. A notable illustration of potential technological advancement may be seen in the recent development of Arestin, a product that incorporates tetracycline into microspheres to facilitate targeted medicine administration inside a periodontal pocket. Nanorobots are being included into mouthwash formulations to expedite the process of identifying and eliminating harmful bacteria associated with periodontitis and gingivitis, while preserving the beneficial oral microbiota (Halwani 2022).

The objective of managing infections in patients caused by antibiotic-resistant microorganisms may not be attainable within the foreseeable future, despite the promising potential of nanotechnology in the domains of diagnostics, prognostics, prediction, and treatment management. Recent discoveries and breakthroughs in NPs and nanotechnology have facilitated their use for commercial purposes in the detection and treatment of antibiotic-resistant infections. It is anticipated that when more and more nanotechnologies are commercially explored, this trend will continue to improve. Although we have numerous goals for nanodentistry, most of them are presently not feasible due to a multitude of challenges, including engineering, biological, and societal issues. The nanomolecular scale portion is extremely difficult to position and assemble accurately. In the field of nanodentistry, biologically compatible chemicals that are affordable, environmentally friendly, and morally acceptable are still far off (Halwani 2022).

References

Adebayo-Tayo B, Salaam A, Ajibade A (2019) Green synthesis of silver nanoparticle using *Oscillatoria* sp. extract, its antibacterial, antibiofilm potential and cytotoxicity activity. Heliyon 5(10):e02502

Agnihotry A, Thompson W, Fedorowicz Z, van Zuuren EJ, Sprakel J (2019) Antibiotic use for irreversible pulpitis. Cochrane Database Syst Rev 5:CD004969

Agrawal N, Mishra P, Ranjan R, Awasthi P, Srivastava A, Prasad D, Kohli E (2021) Nano-cubes over nano-spheres: shape dependent study of silver nanomaterial for biological applications. Bull Mater Sci 44(3):191

Ahmadi H, Haddadi-Asl V, Ghafari HA, Ghorbanzadeh R, Mazlum Y, Bahador A (2020) Shear bond strength, adhesive remnant index, and anti-biofilm effects of a photoexcited modified orthodontic adhesive containing curcumin doped poly lactic-co-glycolic acid nanoparticles: an ex-vivo biofilm model of *S. mutans* on the enamel slab bonded brackets. Photodiagnosis Photodyn Ther 30:101674

Akter S, Huq MA (2020) Biologically rapid synthesis of silver nanoparticles by *Sphingobium* sp. MAH-11T and their antibacterial activity and mechanisms investigation against drug-resistant pathogenic microbes. Artif Cells Nanomed Biotechnol 48(1):672–682

Al-Khatib A, AlMohammad RA (2022) Dentists' habits of antibiotic prescribing may be influenced by patient requests for prescriptions. Int J Dentistry 2022:1

Ansari MA, Murali M, Prasad D, Alzohairy MA, Almatroudi A, Alomary MN, Udayashankar AC, Singh SB, Asiri SM, Ashwini BS, Gowtham HG (2020) *Cinnamomum verum* bark extract mediated green synthesis of ZnO nanoparticles and their antibacterial potentiality. Biomol Ther 10(2):336

Aranda S (2019) Ten threats to global health in 2019. World Health Organization (WHO), Geneva, pp 1–8

Arora N, Thangavelu K, Karanikolos GN (2020) Bimetallic nanoparticles for antimicrobial applications. Front Chem 8:412

Arul Selvaraj RC, Rajendran M, Nagaiah HP (2019) Re-potentiation of β-lactam antibiotic by synergistic combination with biogenic copper oxide nanocubes against biofilm forming multidrug-resistant bacteria. Molecules 24(17):3055

Baptista PV, McCusker MP, Carvalho A, Ferreira DA, Mohan NM, Martins M, Fernandes AR (2018) Nano-strategies to fight multidrug resistant bacteria—"a Battle of the titans". Front Microbiol 9:1441

Barone A, Chatelain S, Derchi G, Di Spirito F, Martuscelli R, Porzio M, Sbordone L (2020) Antibiotic's effectiveness after erupted tooth extractions: a retrospective study. Oral Dis 26(5):967–973

Bazzaz BS, Khameneh B, Zarei H, Golmohammadzadeh S (2016) Antibacterial efficacy of rifampin loaded solid lipid nanoparticles against Staphylococcus epidermidis biofilm. Microb Pathog 93:137–144

Benoit DS, Sims KR Jr, Fraser D (2019) Nanoparticles for oral biofilm treatments. ACS Nano 13(5):4869–4875

Bhat R, Godovikova V, Flannagan SE, Li Y, Seseogullari-Dirihan R, González-Cabezas C, Kuroda K (2022) Targeting cariogenic *Streptococcus mutans* in oral biofilms with charge-switching smart antimicrobial polymers. ACS Biomater Sci Eng 9(1):318–328

Brown AN, Smith K, Samuels TA, Lu J, Obare SO, Scott ME (2012) Nanoparticles functionalized with ampicillin destroy multiple-antibiotic-resistant isolates of *Pseudomonas aeruginosa* and *Enterobacter aerogenes* and methicillin-resistant *Staphylococcus aureus*. Appl Environ Microbiol 78(8):2768–2774

Burnham CA, Leeds J, Nordmann P, O'Grady J, Patel J (2017) Diagnosing antimicrobial resistance. Nat Rev Microbiol 15(11):697–703

Chattopadhyay MK, Grossart HP (2010) Intractable, and here's why. BMJ 341:c6848

Chaw KC, Manimaran M, Tay FE (2005) Role of silver ions in destabilization of intermolecular adhesion forces measured by atomic force microscopy in Staphylococcus epidermidis biofilms. Antimicrob Agents Chemother 49(12):4853–4859

Cheloni G, Marti E, Slaveykova VI (2016) Interactive effects of copper oxide nanoparticles and light to green alga *Chlamydomonas reinhardtii*. Aquat Toxicol 170:120–128

Cormode DP, Gao L, Koo H (2018) Emerging biomedical applications of enzyme-like catalytic nanomaterials. Trends Biotechnol 36(1):15–29

Das S, Kataria VK (2010) Bioterrorism: a public health perspective. Med J Armed Forces India 66(3):255–260

Dong Y, Zhu H, Shen Y, Zhang W, Zhang L (2019) Antibacterial activity of silver nanoparticles of different particle size against *Vibrio natriegens*. PLoS One 14(9):e0222322

Duncan B, Li X, Landis RF, Kim ST, Gupta A, Wang LS, Ramanathan R, Tang R, Boerth JA, Rotello VM (2015) Nanoparticle-stabilized capsules for the treatment of bacterial biofilms. ACS Nano 9(8):7775–7782

Dwivedi S, Wahab R, Khan F, Mishra YK, Musarrat J, Al-Khedhairy AA (2014) Reactive oxygen species mediated bacterial biofilm inhibition via zinc oxide nanoparticles and their statistical determination. PLoS One 9(11):e111289

Edmundson M, Thanh NT, Song B (2013) Nanoparticles based stem cell tracking in regenerative medicine. Theranostics 3(8):573

El Sayed MT, El-Sayed AS (2020) Biocidal activity of metal nanoparticles synthesized by *Fusarium solani* against multidrug-resistant bacteria and mycotoxigenic fungi. J Microbiol Biotechnol 30(2):226

Escárcega-González CE, Garza-Cervantes JA, Vazquez-Rodríguez A, Montelongo-Peralta LZ, Treviño-Gonzalez MT, Díaz Barriga Castro E, Saucedo-Salazar EM, Chávez Morales RM, Regalado Soto DI, Treviño González FM, Carrazco Rosales JL (2018) In vivo antimicrobial activity of silver nanoparticles produced via a green chemistry synthesis using *Acacia rigidula* as a reducing and capping agent. Int J Nanomedicine 17:2349–2363

Evangelopoulos M, Parodi A, Martinez JO, Tasciotti E (2018) Trends towards biomimicry in theranostics. Nanomaterials 8(9):637

Fleming A (2001) On the antibacterial action of cultures of a penicillium, with special reference to their use in the isolation of B. influenzae. Bull World Health Organ 79:780–790

Folorunso A, Akintelu S, Oyebamiji AK, Ajayi S, Abiola B, Abdusalam I, Morakinyo A (2019) Biosynthesis, characterization and antimicrobial activity of gold nanoparticles from leaf extracts of *Annona muricata*. J Nanostruct Chem 9:111–117

Gatto F, Bardi G (2018) Metallic nanoparticles: general research approaches to immunological characterization. Nanomaterials 8(10):753

Gold K, Slay B, Knackstedt M, Gaharwar AK (2018) Antimicrobial activity of metal and metal-oxide based nanoparticles. Adv Therapeut 1(3):1700033

Gudkov SV, Burmistrov DE, Serov DA, Rebezov MB, Semenova AA, Lisitsyn AB (2021) Do iron oxide nanoparticles have significant antibacterial properties? Antibiotics 10(7):884

Habibipour R, Moradi-Haghgou L, Farmany A (2019) Green synthesis of AgNPs@ PPE and its *Pseudomonas aeruginosa* biofilm formation activity compared to pomegranate peel extract. Int J Nanomedicine 14:6891–6899

Hadinoto K, Sundaresan A, Cheow WS (2013) Lipid–polymer hybrid nanoparticles as a new generation therapeutic delivery platform: a review. Eur J Pharm Biopharm 85(3):427–443

Halwani AA (2022) Development of pharmaceutical nanomedicines: from the bench to the market. Pharmaceutics 14(1):106

Hamblin MR (2018) Fullerenes as photosensitizers in photodynamic therapy: pros and cons. Photochem Photobiol Sci 17(11):1515–1533

Hamouda RA, Hussein MH, Abo-Elmagd RA, Bawazir SS (2019) Synthesis and biological characterization of silver nanoparticles derived from the cyanobacterium *Oscillatoria limnetica*. Sci Rep 9(1):13071

Horev B, Klein MI, Hwang G, Li Y, Kim D, Koo H, Benoit DS (2015) pH-activated nanoparticles for controlled topical delivery of farnesol to disrupt oral biofilm virulence. ACS Nano 9(3):2390–2404

Hossain MM, Polash SA, Takikawa M, Shubhra RD, Saha T, Islam Z, Hossain S, Hasan MA, Takeoka S, Sarker SR (2019) Investigation of the antibacterial activity and in vivo cytotoxicity of biogenic silver nanoparticles as potent therapeutics. Front Bioeng Biotechnol 7:239

Hu X, Saravanakumar K, Jin T, Wang MH (2019) Mycosynthesis, characterization, anticancer and antibacterial activity of silver nanoparticles from endophytic fungus *Talaromyces purpureogenus*. Int J Nanomedicine 14:3427–3438

Huh AJ, Kwon YJ (2011) "Nanoantibiotics": a new paradigm for treating infectious diseases using nanomaterials in the antibiotics resistant era. J Control Release 156(2):128–145

Huq MA (2020) Green synthesis of silver nanoparticles using *Pseudoduganella eburnea* MAHUQ-39 and their antimicrobial mechanisms investigation against drug resistant human pathogens. Int J Mol Sci 21(4):1510

Hussein-Al-Ali SH, El Zowalaty ME, Hussein MZ, Geilich BM, Webster TJ (2014) Synthesis, characterization, and antimicrobial activity of an ampicillin-conjugated magnetic nanoantibiotic for medical applications. Int J Nanomedicine 9:3801–3814

Iafisco M, Degli Esposti L, Ramírez-Rodríguez GB, Carella F, Gómez-Morales J, Ionescu AC, Brambilla E, Tampieri A, Delgado-López JM (2018) Fluoride-doped amorphous calcium phosphate nanoparticles as a promising biomimetic material for dental remineralization. Sci Rep 8(1):17016

Ijaz F, Shahid S, Khan SA, Ahmad W, Zaman S (2017) Green synthesis of copper oxide nanoparticles using *Abutilon indicum* leaf extract: antimicrobial, antioxidant and photocatalytic dye degradation activity. Trop J Pharm Res 16(4):743–753

Inoue H, Minghui R (2017) Antimicrobial resistance: translating political commitment into national action. Bull World Health Organ 95(4):242

Ivask A, ElBadawy A, Kaweeteerawat C, Boren D, Fischer H, Ji Z, Chang CH, Liu R, Tolaymat T, Telesca D, Zink JI (2014) Toxicity mechanisms in Escherichia coli vary for silver nanoparticles and differ from ionic silver. ACS Nano 8(1):374–386

Jagtap P, Sritharan V, Gupta S (2017) Nanotheranostic approaches for management of bloodstream bacterial infections. Nanomedicine 13(1):329–341

Jhaveri J, Raichura Z, Khan T, Momin M, Omri A (2021) Chitosan nanoparticles-insight into properties, functionalization and applications in drug delivery and theranostics. Molecules 26(2):272

Jiang J, Pi J, Cai J (2018) The advancing of zinc oxide nanoparticles for biomedical applications. Bioinorg Chem Appl 2018:1

Karakoti AS, Hench LL, Seal S (2006) The potential toxicity of nanomaterials—the role of surfaces. JOM 58:77–82

Karygianni L, Ren Z, Koo H, Thurnheer T (2020) Biofilm matrixome: extracellular components in structured microbial communities. Trends Microbiol 28(8):668–681

Lagarce F (2015) Nanomedicines: are we lost in translation? Eur J Nanomed 7(2):77–78

Leung YH, Ng AM, Xu X, Shen Z, Gethings LA, Wong MT, Chan CM, Guo MY, Ng YH, Djurišić AB, Lee PK (2014) Mechanisms of antibacterial activity of MgO: non-ROS mediated toxicity of MgO nanoparticles towards *Escherichia coli*. Small 10(6):1171–1183

Li Y, Zhang W, Niu J, Chen Y (2012) Mechanism of photogenerated reactive oxygen species and correlation with the antibacterial properties of engineered metal-oxide nanoparticles. ACS Nano 6(6):5164–5173

Li X, Yeh YC, Giri K, Mout R, Landis RF, Prakash YS, Rotello VM (2015) Control of nanoparticle penetration into biofilms through surface design. Chem Commun 51(2):282–285

Liang K, Wang S, Tao S, Xiao S, Zhou H, Wang P, Cheng L, Zhou X, Weir MD, Oates TW, Li J (2019) Dental remineralization via poly (amido amine) and restorative materials containing calcium phosphate nanoparticles. Int J Oral Sci 11(2):15

Liu Y, Shi L, Su L, van der Mei HC, Jutte PC, Ren Y, Busscher HJ (2019) Nanotechnology-based antimicrobials and delivery systems for biofilm-infection control. Chem Soc Rev 48(2):428–446

Loza K, Heggen M, Epple M (2020) Synthesis, structure, properties, and applications of bimetallic nanoparticles of noble metals. Adv Funct Mater 30(21):1909260

Malaekeh-Nikouei B, Bazzaz BS, Mirhadi E, Tajani AS, Khameneh B (2020) The role of nanotechnology in combating biofilm-based antibiotic resistance. J Drug Deliv Sci Technol 60:101880

Malaikozhundan B, Vinodhini J, Kalanjiam MA, Vinotha V, Palanisamy S, Vijayakumar S, Vaseeharan B, Mariyappan A (2020) High synergistic antibacterial, antibiofilm, antidiabetic and antimetabolic activity of *Withania somnifera* leaf extract-assisted zinc oxide nanoparticle. Bioprocess Biosyst Eng 43:1533–1547

Mba IE, Nweze EI (2021) Nanoparticles as therapeutic options for treating multidrug-resistant bacteria: research progress, challenges, and prospects. World J Microbiol Biotechnol 37:1–30
Moraes GS, Santos IB, Pinto SC, Pochapski MT, Farago PV, Pilatti GL, Santos FA (2019) Liposomal anesthetic gel for pain control during periodontal therapy in adults: a placebo-controlled RCT. J Appl Oral Sci 28:e20190025
Mroz P, Tegos GP, Gali H, Wharton T, Sarna T, Hamblin MR (2007) Photodynamic therapy with fullerenes. Photochem Photobiol Sci 6(11):1139–1149
Mukherjee S, Ray S, Thakur RS (2009) Solid lipid nanoparticles: a modern formulation approach in drug delivery system. Indian J Pharm Sci 71(4):349
Mulani MS, Kamble EE, Kumkar SN, Tawre MS, Pardesi KR (2019) Emerging strategies to combat ESKAPE pathogens in the era of antimicrobial resistance: a review. Front Microbiol 10:539
Oberoi SS, Dhingra C, Sharma G, Sardana D (2015) Antibiotics in dental practice: how justified are we. Int Dent J 65(1):4–10
Patra JK, Das G, Fraceto LF, Campos EV, Rodriguez-Torres MD, Acosta-Torres LS, Diaz-Torres LA, Grillo R, Swamy MK, Sharma S, Habtemariam S (2018) Nano based drug delivery systems: recent developments and future prospects. J Nanobiotechnol 16(1):1–33
Pedrosa P, Vinhas R, Fernandes A, Baptista PV (2015) Gold nanotheranostics: proof-of-concept or clinical tool? Nanomaterials 5(4):1853–1879
Penders J, Stolzoff M, Hickey DJ, Andersson M, Webster TJ (2017) Shape-dependent antibacterial effects of non-cytotoxic gold nanoparticles. Int J Nanomedicine 29:2457–2468
Peng X, Han Q, Zhou X, Chen Y, Huang X, Guo X, Peng R, Wang H, Peng X, Cheng L (2022) Effect of pH-sensitive nanoparticles on inhibiting oral biofilms. Drug Deliv 29(1):561–573
Pinon-Segundo E, Ganem-Quintanar A, Alonso-Pérez V, Quintanar-Guerrero D (2005) Preparation and characterization of triclosan nanoparticles for periodontal treatment. Int J Pharm 294(1–2):217–232
Qi M, Chi M, Sun X, Xie X, Weir MD, Oates TW, Zhou Y, Wang L, Bai Y, Xu HH (2019) Novel nanomaterial-based antibacterial photodynamic therapies to combat oral bacterial biofilms and infectious diseases. Int J Nanomedicine 28:6937–6956
Radaic A, Malone E, Kamarajan P, Kapila YL (2022) Solid lipid nanoparticles loaded with Nisin (SLN-Nisin) are more effective than free Nisin as antimicrobial, Antibiofilm, and anticancer agents. J Biomed Nanotechnol 18(4):1227–1235
Rajendiran M, Trivedi HM, Chen D, Gajendrareddy P, Chen L (2021) Recent development of active ingredients in mouthwashes and toothpastes for periodontal diseases. Molecules 26(7):2001
Ramalingam B, Parandhaman T, Das SK (2016) Antibacterial effects of biosynthesized silver nanoparticles on surface ultrastructure and nanomechanical properties of gram-negative bacteria viz. *Escherichia coli* and *Pseudomonas aeruginosa*. ACS Appl Mater Interfaces 8(7):4963–4976
Rehman S, Jermy BR, Akhtar S, Borgio JF, Abdul Azeez S, Ravinayagam V, Al Jindan R, Alsalem ZH, Buhameid A, Gani A (2019) Isolation and characterization of a novel thermophile; *Bacillus haynesii*, applied for the green synthesis of ZnO nanoparticles. Artif Cells Nanomed Biotechnol 47(1):2072–2082
Rodríguez-Serrano C, Guzmán-Moreno J, Ángeles-Chávez C, Rodríguez-González V, Ortega-Sigala JJ, Ramírez-Santoyo RM, Vidales-Rodríguez LE (2020) Biosynthesis of silver nanoparticles by fusarium scirpi and its potential as antimicrobial agent against uropathogenic *Escherichia coli* biofilms. PLoS One 15(3):e0230275
Sánchez MC, Toledano-Osorio M, Bueno J, Figuero E, Toledano M, Medina-Castillo AL, Osorio R, Herrera D, Sanz M (2019) Antibacterial effects of polymeric PolymP-n active nanoparticles. An in vitro biofilm study. Dent Mater 35(1):156–168
Sarwar S, Chakraborti S, Bera S, Sheikh IA, Hoque KM, Chakrabarti P (2016) The antimicrobial activity of ZnO nanoparticles against vibrio cholerae: variation in response depends on biotype. Nanomedicine 12(6):1499–1509
Sengupta S, Chattopadhyay MK, Grossart HP (2013) The multifaceted roles of antibiotics and antibiotic resistance in nature. Front Microbiol 4:47

Sharma M, Gupta N, Gupta S (2016) Implications of designing clarithromycin loaded solid lipid nanoparticles on their pharmacokinetics, antibacterial activity and safety. RSC Adv 6(80):76621–76631

Shetty NJ, Swati P, David K (2013) Nanorobots: future in dentistry. Saudi Dent J 25(2):49–52

Singh B, Vuddanda PR, Vijayakumar MR, Kumar V, Saxena PS, Singh S (2014) Cefuroxime axetil loaded solid lipid nanoparticles for enhanced activity against *S. aureus* biofilm. Colloids Surf B Biointerfaces 121:92–98

Singh H, Du J, Singh P, Yi TH (2018) Ecofriendly synthesis of silver and gold nanoparticles by *Euphrasia officinalis* leaf extract and its biomedical applications. Artif Cells Nanomed Biotechnol 46(6):1163–1170

Sirelkhatim A, Mahmud S, Seeni A, Kaus NH, Ann LC, Bakhori SK, Hasan H, Mohamad D (2015) Review on zinc oxide nanoparticles: antibacterial activity and toxicity mechanism. Nano Lett 7:219–242

Srinoi P, Chen YT, Vittur V, Marquez MD, Lee TR (2018) Bimetallic nanoparticles: enhanced magnetic and optical properties for emerging biological applications. Appl Sci 8(7):1106

Stewart PS, Franklin MJ (2008) Physiological heterogeneity in biofilms. Nat Rev Microbiol 6(3):199–210

Supraja N, Prasad TN, Krishna TG, David E (2016) Synthesis, characterization, and evaluation of the antimicrobial efficacy of *Boswellia ovalifoliolata* stem bark-extract-mediated zinc oxide nanoparticles. Appl Nanosci 6:581–590

Thakur P, Kumar V (2019) Kinetics and thermodynamic studies for removal of methylene blue dye by biosynthesize copper oxide nanoparticles and its antibacterial activity. J Environ Health Sci Eng 17:367–376

Tonga GY, Moyano DF, Kim CS, Rotello VM (2014) Inorganic nanoparticles for therapeutic delivery: trials, tribulations and promise. Curr Opin Colloid Interface Sci 19(2):49–55

Tournier JN, Peyrefitte CN, Biot F, Merens A, Simon F (2019) The threat of bioterrorism. Lancet Infect Dis 19(1):18–19

Umar H, Kavaz D, Rizaner N (2019) Biosynthesis of zinc oxide nanoparticles using *Albizia lebbeck* stem bark, and evaluation of its antimicrobial, antioxidant, and cytotoxic activities on human breast cancer cell lines. Int J Nanomedicine 14:87–100

Vangara A, Pramanik A, Gao Y, Gates K, Begum S, Chandra RP (2018) Fluorescence resonance energy transfer based highly efficient theranostic nanoplatform for two-photon bioimaging and two-photon excited photodynamic therapy of multiple drug resistance bacteria. ACS Appl Bio Mater 1(2):298–309

Vikram Singh A, Sitti M (2016) Targeted drug delivery and imaging using mobile milli/microrobots: a promising future towards theranostic pharmaceutical design. Curr Pharm Des 22(11):1418–1428

Wang L, He H, Yu Y, Sun L, Liu S, Zhang C, He L (2014) Morphology-dependent bactericidal activities of ag/CeO2 catalysts against *Escherichia coli*. J Inorg Biochem 135:45–53

Wang L, Hu C, Shao L (2017) The antimicrobial activity of nanoparticles: present situation and prospects for the future. Int J Nanomedicine 14:1227–1249

Wang L, Li Y, Ren M, Wang X, Li L, Liu F, Lan Y, Yang S, Song J (2022) pH and lipase-responsive nanocarrier-mediated dual drug delivery system to treat periodontitis in diabetic rats. Bioactive Mater 18:254–266

World Health Organization (2023) Antimicrobial resistance surveillance in Europe 2023–2021 data. World Health Organization (WHO), Geneva

Wu B, Zhuang WQ, Sahu M, Biswas P, Tang YJ (2011) Cu-doped TiO2 nanoparticles enhance survival of Shewanella oneidensis MR-1 under ultraviolet light (UV) exposure. Sci Total Environ 409(21):4635–4639

Wypij M, Czarnecka J, Świecimska M, Dahm H, Rai M, Golinska P (2018) Synthesis, characterization and evaluation of antimicrobial and cytotoxic activities of biogenic silver nanoparticles synthesized from *Streptomyces xinghaiensis* OF1 strain. World J Microbiol Biotechnol 34:1–3

Wypych TP, Marsland BJ (2018) Antibiotics as instigators of microbial dysbiosis: implications for asthma and allergy. Trends Immunol 39(9):697–711

Xia MY, Xie Y, Yu CH, Chen GY, Li YH, Zhang T, Peng Q (2019) Graphene-based nanomaterials: the promising active agents for antibiotics-independent antibacterial applications. J Control Release 307:16–31

Xie X, Mao C, Liu X, Zhang Y, Cui Z, Yang X, Yeung KW, Pan H, Chu PK, Wu S (2017) Synergistic bacteria killing through photodynamic and physical actions of graphene oxide/ag/collagen coating. ACS Appl Mater Interfaces 9(31):26417–26428

Xiong MH, Bao Y, Yang XZ, Zhu YH, Wang J (2014) Delivery of antibiotics with polymeric particles. Adv Drug Deliv Rev 78:63–76

Yamakami K, Tsumori H, Shimizu Y, Sakurai Y, Nagatoshi K, Sonomoto K (2016) Cationic lipid content in liposome-encapsulated nisin improves sustainable bactericidal activity against *Streptococcus mutans*. Open Dent J 10:360

Yu J, Zhang W, Li Y, Wang G, Yang L, Jin J, Chen Q, Huang M (2014) Synthesis, characterization, antimicrobial activity and mechanism of a novel hydroxyapatite whisker/nano zinc oxide biomaterial. Biomed Mater 10(1):015001

Yu K, Zhang Q, Dai Z, Zhu M, Xiao L, Zhao Z, Bai Y, Zhang K (2023) Smart dental materials intelligently responding to oral pH to combat caries: a literature review. Polymers 15(12):2611

Yudovin-Farber I, Beyth N, Nyska A, Weiss EI, Golenser J, Domb AJ (2008) Surface characterization and biocompatibility of restorative resin containing nanoparticles. Biomacromolecules 9(11):3044–3050

Zhang Y, Zhang J, Chen M, Gong H, Thamphiwatana S, Eckmann L, Gao W, Zhang L (2016) A bioadhesive nanoparticle–hydrogel hybrid system for localized antimicrobial drug delivery. ACS Appl Mater Interfaces 8(28):18367–18374

Zhao Y, Chen L, Wang Y, Song X, Li K, Yan X, Yu L, He Z (2021) Nanomaterial-based strategies in antimicrobial applications: progress and perspectives. Nano Res 14:4417–4441

Zhou J, Horev B, Hwang G, Klein MI, Koo H, Benoit DS (2016) Characterization and optimization of pH-responsive polymer nanoparticles for drug delivery to oral biofilms. J Mater Chem B 4(18):3075–3085

Zhou W, Peng X, Zhou X, Weir MD, Melo MA, Tay FR, Imazato S, Oates TW, Cheng L, Xu HH (2020) In vitro evaluation of composite containing DMAHDM and calcium phosphate nanoparticles on recurrent caries inhibition at bovine enamel-restoration margins. Dent Mater 36(10):1343–1355

Zhu Z, Tang Z, Phillips JA, Yang R, Wang H, Tan W (2008) Regulation of singlet oxygen generation using single-walled carbon nanotubes. J Am Chem Soc 130(33):10856–10857

Zhu T, Huang Z, Shu X, Zhang C, Dong Z, Peng Q (2022) Functional nanomaterials and their potentials in antibacterial treatment of dental caries. Colloids Surf B Biointerfaces 6:112761

Nanocomposites of Metal Nanoparticles and Polymer as Platform of Alternative Approach in Combating Antimicrobial Resistance

18

Vesna Vodnik, Una Stamenović, and Ivana Vukoje

Abstract

As toxic microorganisms have developed resistance to each new antibiotic over the course of several decades, antimicrobial resistance (AMR) has evolved into a complicated public health problem of the present. It makes infections more difficult to cure and raises the risk of disease transmission. Faced with this reality and the necessity for action to find new sources of materials as an alternative to antibiotics and other antimicrobial agents, scientists have been directed toward the development of nanomaterials that exhibit these properties. Among them are the nanocomposites of metal nanoparticles (silver, copper, zinc, etc.) and polymers that provide high surface area, sorption capacity, and selectivity for target microbes. However, as significant as the development of new nanomaterials is, innovation without focusing on their potential side effects does not improve health outcomes. In this chapter, we will review our research and discuss new advancements for innovative nanocomposites in countering AMR and their cytotoxicity, covering the background of the recent literature.

Keywords

Antimicrobial resistance · Nanotechnology · Metal-polymer · nanocomposite · Cytotoxicity

V. Vodnik (✉) · U. Stamenović · I. Vukoje
Vinča Institute of Nuclear Sciences—National Institute of the Republic of Serbia, University of Belgrade, Belgrade, Serbia
e-mail: vodves@vin.bg.ac.rs; vodves@vinca.rs

M. Y. Wani et al. (eds.), *Nanotechnology Based Strategies for Combating Antimicrobial Resistance*, https://doi.org/10.1007/978-981-97-2023-1_18

18.1 Introduction

Antimicrobial resistance (AMR), as one of today's serious and complex public health problems, is the result of changes in the resistance of bacteria, viruses, fungi, and parasites over time to common drugs, which makes it difficult to treat the infections they cause and increases the risk of disease spreading (Jasovsky et al. 2016). The effectiveness of antimicrobials such as antibiotics, antivirals, antifungals, and anti-parasitic, widely used to prevent and treat infections, is now in jeopardy. A particularly urgent problem arises when bacteria develop resistance faster than new antibiotics can be developed. These microorganisms are referred to as superbugs. Of course, for many years, we have enjoyed the benefits of effective antibiotics to treat many common bacterial infections, but today's numbers show that more than one million people die each year from drug-resistant infections. It should also be noted that antibiotic resistance is not exclusively caused by antibiotics fighting infections. Substantial contributors include the release of non-metabolized antibiotics or their residues into the environment via manure/feces, inadequate sanitation/hygiene, and the use of antibiotics as growth promoters in poultry and livestock (Samreena et al. 2021). Regardless of numerous proposals and initiatives over the past few decades, the world has been unable to keep up, always moving in a vicious circle of misusage and facing the impossibility of finding the most appropriate approach to increase awareness when using antibiotics is in question. Infections caused by microbes that are prone to multiple drugs such as *Enterococcus faecium*, *Staphylococcus aureus*, *Klebsiella pneumoniae*, and *Pseudomonas aeruginosa* lead to approximately 700,000 annual deaths. By 2050, the yearly death rate is predicted to surpass cancer-related deaths and reach ten million if an adequate action plan is not put into place (Kraker et al. 2016). Regarding this, methicillin-resistant *Staphylococcus aureus* is one of the most dangerous resistant bacteria, which can easily be infected during prolonged hospital stays and unhygienic medical conditions, causing local skin infections, pneumonia, pericarditis, sepsis, and death (Turner et al. 2019). Therefore, monitoring systems should adopt a worldwide strategy that incorporates interdisciplinary and collaborative efforts to provide the best possible health for people, animals, and the environment. Faced with this reality, when the discovery of novel antibiotics is constrained by the time and complexities, the need for action is imperative. It is becoming clear that there is an urge to invest time and effort into the development and implementation of new types of materials that could successfully help antibiotics in their inevitable combat against microbes. Fortunately, advances in nanomaterial applications have given new hope and opened new doors for combating this problem. Among them, metal nanoparticles (MNPs) of silver (Ag), copper (Cu), zinc oxide (ZnO), iron oxide (Fe_3O_4), magnesium oxide (MgO), etc. have been found to be effective against multidrug-resistant bacteria (Mishraa et al. 2022; Vodnik and Bogdanović 2019; Bogdanović et al. 2014; Vukoje et al. 2017, 2019; Nejabatdoust et al. 2019; Gabrielyan et al. 2019; Nguyen et al. 2018). The physicochemical properties, including size, shape, surface chemistry, reactivity, stability, and agglomeration/aggregation, are essential factors for their antimicrobial use. Moreover, the advantage related to use of MNPs with antibacterial activity regards their intrinsic properties, such as the magnetic separation of Fe_3O_4, the photocatalytic activity of ZnO, and the plasmonic properties of Ag. Their better antimicrobial efficacies could be reached by surface functionalization or incorporation in polymer matrices, enabling them to execute multiple novel bactericidal pathways to achieve the antimicrobial activity. Their combination with polymers in the form of metal-polymer nanocomposites (MPNCs) can overcome the antibiotic-resistance mechanisms owing to the unique properties of both counterparts, which have an impact on the resulting synergistic effects (Vodnik and Bogdanović 2019; Bogdanović et al. 2015, 2018; Stamenović et al. 2021). Thus, these MPNCs are able to bind and disrupt the bacterial membrane, causing leakage of cytoplasmic components (Bogdanović et al. 2015). Also, many researchers have shown that if they combine with antibiotics, they can disrupt microbial strains at much lower concentrations (Herrera et al. 2021). Their application provides a potential substitute for existing antimicrobials and can be used as a starting point for investigating innovative materials in this area. Unfortunately, there is another side to using these nanocomposites—potentially negative impact on the environment and toxicity to humans. We are aware of the fact that their cytotoxicity has not been fully tested. Therefore, as vital as the development of new antimicrobial materials is, innovation without focusing on access to toxicity does not improve health outcomes. As a result, progress in this field must be accelerated, and more research must be conducted to create safe MPNCs for commercial applications. In this chapter, we performed a thorough analysis and critical assessment of previously published results in the growing field of exploring MPNCs in combating AMR. Moreover, MPNCs mode of action and microbe-resistant mechanisms which determine the final effectiveness of combined MNPs antimicrobial treatment with polymers and their impact on the resulting synergistic effects are also discussed. All of this research provides a new look and perspective on this issue, offers a useful resource for further complementary investigations, and helps to navigate further approaches in combating AMR.

18.2 The Problem of Antibiotic Resistance and Demand for Antimicrobial Nanomaterials

As already mentioned, even with the widespread, continuous development and commercialization of antibiotics, microorganisms still find a way to cause millions of deaths each year (Samreena et al. 2021). However, we must keep in mind that the evolution of resistance in bacteria is a natural process that cannot be avoided. Resistance develops genetically, either through mutation or by the acquisition of new resistance genes via genetic exchange processes (Davis et al. 2018). No matter how much effort is put into controlling these processes, the evolution of antibiotic resistance is unstoppable. Excessive and inconsiderate use of antibiotics speeds up nature's course. AMR is therefore an inevitable phenomenon that decreases the efficacy of modern medicine and impacts individuals all over the world. So far, three broad categories of prevention, diagnosis, and therapy of infectious diseases have been defined. The first step is to prevent antibiotics from being easily accessible both locally and globally. Second, conservation efforts such as appropriate prescribing and surveillance should be implemented to reduce the use of antimicrobials. The third phase is to encourage new biotechnological advances and development in infection control technologies (Mendelson and Matsoso 2015). With the development of nanotechnology in recent decades, we hope to combat this problem by developing and implementing new types of materials that could successfully assist antibiotics in their inevitable fight against microbes.

18.3 Why Nanocomposites of Metal Nanoparticles and Polymers?

One of the important impacts of nanotechnology comes from the development and preparation of nanocomposite-based antimicrobial materials that rely on the interactions of MNPs (metals or metal oxides) and polymers (synthetic or biopolymers), and their proper combination for the goal-orienting performance of the resulting MPNCs. In this regard, different types of nanocomposites are progressively being investigated in an attempt to develop an efficient method for regulating the high mortality rate caused by the evolution and emergence of multidrug-resistant pathogens (Bogdanović et al. 2015; Colino et al. 2021; Zare and Shabani 2016). Because microbes are less likely to develop resistance to nanomaterials, certain metals, metal oxides, metal halides, and bimetals have been found to exhibit various actions against them in a nanoparticulate form (Dizaj et al. 2014; Hemeg 2017). Their potential to interact with multidrug-resistant microorganism and correspondingly prevent biofilm formation has been projected as a new line of defense against them. The unique size-dependent properties of this NPs and its large effective surface, with a high percentage of unsaturated surface atoms that tend to interact with the surrounding media, are exploited widely in various areas and have occupied prime importance in the medical field—they are being utilized in diagnostic and therapeutic tools to better understand, detect, and treat human diseases (Dizaj et al. 2014; Hemeg 2017). Their high surface reactivity is attributed to their small particle size and consequently better uptake by cells and tissues via endocytosis and other pathways with respect to the classical treatments. The cellular uptake is also influenced by the shape of the NPs in the following order: triangular > rods > spheres > cylinders > cubes, which is attributed to the number of NPs facets interacting with the microbe components (Albanese et al. 2012). Thus, triangular NPs have more facets than spherical- or rod-shaped, and thus, it cause more cellular damage (Makabenta et al. 2021). In addition to size and shape, the surface charge of these NPs (positively charged NPs surface can enhance antibacterial activity) plays a pivotal role in reducing microbial resistance by disrupting the potential and integrity of their membranes, adhesion to the cell surface, causing lipid and protein damage, preventing the growth of biofilms, generating reactive oxygen species (ROS), improving host immune responses, ribosome destabilization, and intercalation between DNA bases (Mishraa et al. 2022; Guerrero Correa et al. 2020). The antimicrobial mechanism of different MNPs varies from particle to particle, and it is not fully understood, but it is clear that metal oxide NPs generate ROS, while other MNPs exhibit the damaging properties mentioned above, related to their physical structure and metal ion release. Among the MNPs, silver is the most widely studied NPs and probably the most used nanofiller in MPNCs due to its distinctive physical, chemical, and biological properties and ability to anchor easily to polymer matrices (Bruna et al. 2021; Akter et al. 2018). They have established themselves as a promising alternative to combat different microorganisms and have good contact with their membrane, leading to a dissipation of the proton motive force and blocking oxidative phosphorylation (Lok et al. 2006). Like other MNPs (Cu, CuO, ZnO, Au, etc.), they have an affinity for sulfur and phosphorus from the cell membrane, interact with many vital enzymes, and modulate microbial signal transduction (Shrivastava et al. 2007). Beside Ag NPs, Cu NPs exhibit unique characteristics that make them equally suitable for antimicrobial activities (Bogdanović et al. 2018; Ermini and Voliani 2021). They are cheap and easily accessible, generating oxidative stress and causing the disassembly of viruses or bacterial membranes. On the other hand, the use of MNPs also has some limitations, such as the development of microbial resistance towards MNPs owing to their repeated exposure, which could be solved by their surface functionalization. Second, the instability of the MNPs, which is reflected in their oxidation and agglomeration, as well as the uncontrolled release of metal ions in the surroundings, occurs as a consequence of their high reactivity. This problem can be overcome by inserting them into polymer matrices, forming MPNCs, and also by reducing their toxicity for safer cellular uptake and environmental release. In such structures, polymer chains act as "bridges" between MNPs, providing support by limiting the possibility of oxidation, aggregation/agglomeration, and the production of thermodynamically favorable materials (Bogdanović et al. 2015). When incorporated into a polymer matrix, MNPs remain as individual nanoparticles or smaller clusters, ensuring better contact with microorganisms and consistent antimicrobial performance. In this case, MPNCs can provide a controlled and sustained release of metal ions over time. This slow release ensures a prolonged and effective antimicrobial action, which reduces the risk associated with sudden bursts of high ion concentrations, making them suitable for applications requiring long-term antimicrobial protection. To reach the appropriate level of antibacterial activity, the concentration of MNPs in the polymer matrix can be changed. This adaptability allows MPNCs to be tailored for specific purposes, ranging from weak antibacterial effects to strong bactericidal properties. But it should be undertaken with caution, since MNPs have the potential to be cytotoxic to human cells and cause environmental hazards. Their selection and concentration in a given application should be carefully tuned to balance antimicrobial efficacy and potential safety issues. On the other side, by using MNPs in a nanocomposite form, the overall amount of metal required can be reduced while maintaining the same level of the antimicrobial activity. This makes the nanocomposite more cost-effective compared to using pure MNPs. If they are incorporated into various types of polymers, this versatility widens the range of potential applications for MPNCs. Also, they typically have better stability during storage and handling, preventing oxidation or degradation of the MNPs and extending their shelf-life. Thus, developing MPNCs materials that take advantage of the hosts—polymers and dispersed MNPs—is one viable way to improve their capacity, selectivity, stability, and application, as well as exploit the synergistic effect of the components (Vodnik and Bogdanović 2019; Bogdanović et al. 2015; Naseem and Waseem 2022). These materials also have the potential to reduce MNPs discharge into the environment and their toxicity, have a balance of hydrophobic and hydrophilic groups, and finally, have biocompatibility. Therefore, the choice of non-toxic polymer and type, as well as the size and shape of NPs, plays important roles in determining cellular uptake, mode of action against selected microbes, and potential toxicity.

The most extensively used polymeric hosts for MNPs are conducting polymers (polyaniline (PANI), polypyrrole (PPy) (Bogdanović et al. 2015; Stamenović et al. 2021; Mašojević et al. 2023), or other polymers—polythiophene (PTh) (Karimi et al. 2021), polymethylmethacrylate (PMMA) (Di Mauro et al. 2017), macroporous copolymers, (Vukoje et al. 2014a, 2017, 2019) and new types of polymeric hosts—biopolymers such as polysaccharides, chitosan, alginate, and cellulose (Vukoje et al. 2022; Bagavathy et al. 2019; Booshehri et al. 2015; Devi et al. 2016). Conducting polymers are suitable hosts for MNPs due to their high electrical conductivity, good environmental stability, non-toxicity, porous structure, water insolubility, ease of preparation, and relatively low cost. Their large surface area is rich in heteroatoms from their functional groups, such as nitrogen (N), carbon (C), and sulfur (S), which can actively take part as antimicrobial hotspots. The presence of a significant number of amine and imine functional groups in PANI and PPy with –N+ opens up the possibility of their use in the adsorption processes (Bogdanović et al. 2015; Mašojević et al. 2023; Della Pina et al. 2020).Macroporous copolymers, due to their morphological properties and possibility to react with properly selected modifiers with the ability to reduce Ag^+ ions to the metal state and immobilize formed AgNPs, are also good candidates for antimicrobial-oriented materials (Vukoje et al. 2014a, 2017, 2019). Biopolymers have risen to the top of research surveys because of their non-toxicity, low cost, high efficiency, ease of degradation, eco-friendly nature, availability, and biodegradability. Chitosan, for example, is a biodegradable and antibacterial biopolymer, and its amino and hydroxyl groups are active sites for antimicrobial action, whereas densely arranged hydrogen-bonded molecules contribute to cellulose's chemical stability, mechanical strength, sound swelling resistance, and hydrophilicity (Nasrollahzadeh et al. 2021). Antimicrobial textiles are another attractive research area where challenges regarding AMR should be overcome. Cotton, polyester, and polyamide fabrics are used as polymer carriers for AgNPs in order to obtain nanocomposite textile materials with stable and persistent antimicrobial properties (Radetić et al. 2008; Ilić et al. 2009, 2010; Vukoje et al. 2014b). A different combination of MNPs and polymers, together with their characteristics, and antimicrobial activities are briefly summarized in Table 18.1.

Table 18.1 Different MPNCs, their characteristics, antimicrobial activities, and toxicity

Nanocomposite.	MNPs size and shape, polymer size/morphology, MNPs content	Tested microbes	Microbiological analysis: method (Colony forming units—CFU, disk diffusion assay, etc.) incubation time, *t*, pH	*R* (Reduction, %), MIC (μg/mL), MBC (μg/mL), MPNCs concentration	Toxicity assay (threshold) μg/mL
Ag-poly(GMA-co-EGDMA) (Vukoje et al. 2014a, 2017, 2019)	Ag spheres, 6.6 nm, 10 wt% Ag; Ag spheres, 12.1 nm, 5 wt% Ag	*E. coli*	CFU, % reduction, 2 h contact, 37 °C	12.5 ppm: 96.4 96.3	–
Ag-PANI (Shaban et al. 2018)	Ag spheres, 20 nm, PANI microgranules	*E. coli, S. aureus, B. subtilis, Salmonella*	Disk diffusion assay, pH 7, 37 °C	400 ppm: 18.0 mm 20.0 mm 19.0 mm 22.5 nm	–
Ag-PANI/PVP (Stamenović et al. 2021)	Ag spheres, 7 ± 2.3 nm, granular PANI, 43.8 wt% Ag; Ag triangles, 25–140 nm, wrinkled PANI 44.4 wt% Ag	*E. coli* *S. aureus* *C. albicans*	CFU, % reduction, 1–4 h contact, 37 °C	20 ppm, 1 h, 98.2–99.0% *E. coli;* 20 ppm, 4 h, 99.2% *S. aureus;* 20 ppm, 4 h, 91.1% *C. albicans*	>5
Ag-PPy (Liu et al. 2015)	Ag spheres, 15–40 nm, PPy nanowires	*E. coli*	Disk diffusion assay, 37 °C	15.4 mm	–
Ag-polyester Ag-polyamide (Radetić et al. 2008)	Ag spheres, 10 nm, ripple-like structure of fibers	*E. coli* *S. aureus* *E. coli* *S. aureus*	CFU, % reduction, 24 h contact pH 7.2 37 °C	10 ppm: 99.2 85.7 99.8 99.9	–

(continued)

Table 18.1 (continued)

Nanocomposite.	MNPs size and shape, polymer size/morphology, MNPs content	Tested microbes	Microbiological analysis: method (Colony forming units—CFU, disk diffusion assay, etc.) incubation time, *t*, pH	*R* (Reduction, %), MIC (μg/mL), MBC (μg/mL), MPNCs concentration	Toxicity assay (threshold) μg/mL
Ag-cotton (Ilić et al. 2009)	Ag spheres, 10 nm	*E. coli* *S. aureus* *C. albicans*	% reduction, 24 h contact pH 7.2, 37 °C	10 ppm: 99.7 99.9 99.9	–
Ag-chitosan-cotton (Vukoje et al. 2014b)	Ag nanoplates, 40–60 nm, thickness 5 nm 45.7 μg of Ag per 1 g of fabric	*E. coli* *S. aureus* *C. albicans*	% reduction, 24 h contact pH 7.2, 37 °C	99.9 99.6 99.9	–
Ag-PPy-cotton, Ag-PANI-cotton (Marakova et al. 2017)	11 wt% Ag clusters (50–500 nm) 4 wt% Ag clusters	*E. coli* *S. aureus*	Disk diffusion assay, 37 °C, 24 h	MIC 10 μg/mL, 85% (*E. coli*), MIC 200 μg/mL, 95% (*S. aureus*) No growth under sample with 0.05 M $AgNO_3$	Absence of cytotoxicity at extract concentrations up to 25% regardless Ag content
Ag-glucose, sucrose, dextran (Vukoje et al. 2022)	Ag spheres, 15–30 nm 10 ppm Ag	*E. coli* *S. aureus*	% reduction, 37 °C, 24 h	99.9 99.9	>2.5 ppm MPNCs
Ag-chitosan (Sharma 2017)	162.3 ± 35 nm spherical MPNCs particles, 1.5% (w/w) Ag	*E. coli* *E. aerogenes* *P. aeruginosa* *B. cereus* *E. faecalis* *E. coli* *E. aerogenes* *P. aeruginosa* *B. cereus* *E. faecalis*	MIC, 37 °C, 12 h MBC, 37 °C, 12 h	192.07 172.10 193.32 274.56 232.66 204.00 308.05 258.80 367.19 378.00	–

Cu-PANI (Bogdanović et al. 2015, 2018)	Cu spheres, 6 nm, dendritic PANI structure, 14.3 wt% Cu	*E. coli* *S. aureus* C. *albicans*	CFU, % reduction, 2 h contact	1 ppm, <10 CFU, 99.9% for all tested microbes	>0.7 ppm
Cu-PPy (Mašojević et al. 2023)	Cu spheres 25 nm, granular PPy, 9.45 wt% Cu	*E. coli* *S. aureus*	CFU, % reduction, 24 h contact	10–20 ppm: 99.9% *E. coli* and *S. aureus;* 2 ppm: 99.17% *S. aureus*	>100 ppm PPy
CuO-cellulose (Booshehri et al. 2015)	MPNCs 700 nm/300 nm/30 nm lenght/width/thickness manoleave shape, CuO nanorods, diameter 5 nm, 25 wt% CuO	*E. coli* *S. aureus*	Plate counting method	180 min contact Total disinfection 150 min contact Total disinfection	–
ZnO-PANI (Mohsen et al. 2018)	61.6 nm MPNCs size, 8 wt% ZnO	*E. coli* *S. aureus*	Disk diffusion assay	13.0 nm 16.0 nm	–
ZnO-Ag-PANI (Hou et al. 2016)	ZnO spheres, 25–30 nm, 2 wt% Ag	*E. coli* *S. aureus* C. *albicans* *E. coli* *S. aureus* C. *albicans* *E. coli* *S. aureus* C. *albicans*	MIC 37 °C, 24 h MBC 37 °C, 24 h Disk diffusion assay, 37 °C, 24 h	10 10 5 25 25 10 23.8 mm 22.4 mm 25.4 mm	According to HG/T3794–2005 standard, if the MICs of inorganic antimicrobial <800 ppm, it is identified as qualified

(continued)

Table 18.1 (continued)

Nanocomposite.	MNPs size and shape, polymer size/morphology, MNPs content	Tested microbes	Microbiological analysis: method (Colony forming units—CFU, disk diffusion assay, etc.) incubation time, *t*, pH	*R* (Reduction, %), MIC (μg/mL), MBC (μg/mL), MPNCs concentration	Toxicity assay (threshold) μg/mL
ZnO-chitosan (Bagavathy et al. 2019)	10 μm diameter MPNCs polymorphological shape, average particle size 259.6 nm	*E. coli* *Bacilus* *E. coli* *Bacilus*	Disk diffusion assay	100 μL: 3 mm 4 mm 120 μL: 4 mm 2 mm	–
Pd-PPy (Salabat et al. 2015)	Pd spheres, 21.5 nm	*S. aureus* *Bacillus* *E. coli* *Klebsiella* *S. aureus*	Disk diffusion assay, 37 °C, 24 h MIC, 37 °C, 90 min MBC 37 °C, 90 min	100 mg/mL: 23 mm 10 mm 16 mm 14 mm 5.78 mg/mL 23.12 mg/mL	–
Fe_3O_4-PPy (Guo et al. 2021)	Fe_3O_4 spheres, 185 ± 15 nm; MPNCs size 293 ± 27 nm	*E. coli* *S. aureus*	Magnetic adsorption 10 min; interaction10 min	100 ppm, MPNCs >50%	>100 ppm MPNCs

18.4 Design of MPNCs, Their Antimicrobial Mechanism, and Efficacy

The objective of antimicrobial agents is to prevent the growth and proliferation of harmful microorganisms, ultimately aiding in the control and treatment of infections and diseases. The antimicrobial activity can be found in a wide range of substances, including natural compounds, synthetic chemicals, and engineered materials like nanocomposites, as discussed earlier. Research and development of new antimicrobial agents, including those found in nanocomposites, continue to be a priority to address emerging microbial threats and combat infectious diseases effectively. When designed with specific NPs, nanocomposites can exhibit strong antimicrobial activity against a wide range of microorganisms. It is important to note that this activity can vary depending on the type of NPs used, their size, shapes, and concentration, the nature of the matrix material, and the targeted microorganisms. In the case of MNPs such as Ag, Cu, Pt, Pd, Mn, ZnO, Fe_3O_4, and MgO, it involves several key processes that contribute to their effectiveness against a wide range of microorganisms, including bacteria, viruses, and fungi. It can be stated that the primary mode of action of each particular NPs is based on the release of ions from the nanoparticle's surface (Bogdanović et al. 2014, 2015, 2018; Guerrero Correa et al. 2020). The released ions interact with the cell membrane of microorganisms, and penetrate the cell membrane, disrupting its integrity and causing damage. Inside the microbial cell, ions can also trigger the production of ROS through various mechanisms as highly reactive molecules that can cause oxidative stress, leading to damage to cellular components such as proteins, lipids, and DNA (Guerrero Correa et al. 2020). The ions can also directly attach to microbial DNA, leading to DNA damage and interfering with essential cellular processes, including replication and transcription (Bogdanović et al. 2014). Furthermore, interaction with microbial proteins might result in the inactivation or denaturation of vital enzymes and proteins necessary for the microorganism's survival. MNPs have the ability to inhibit or prevent the production of biofilms, which are protective coatings generated by microbes to withstand antimicrobial agents and the host's immunological response. A schematic view of the interaction of MNPs and polymers with the microbe and the different types of damage that occur as a result of that interaction is given in Fig. 18.1.

Considering CuNPs, which have several advantages compared to other metallic NPs (cheaper and easier synthesis), they exert a very fast toxic action on microbial cells, since they dissolve faster than other MNPs by releasing ions in the surroundings (Bogdanović et al. 2014; Ermini and Voliani 2021). Bogdanović et al., showed that spherical and bare CuNPs (diameter of 5 nm, concentration of 32 ppm) effectively reduce more than 98% of *E. coli, S. Aureus,* and *C. albicans*, limit the formation of new ones, and inflict severe membrane damage only within 2 h of contact (Bogdanović et al. 2014). Among the tested strains, *E. coli* cells have the least resistance to them and suffer the most cell wall damage, ranging from grooves to pore-like lesions in the outer membrane as well as the collapse of the cell structure at the apical ends. It was discovered that the large active surface of CuNPs, as well as the

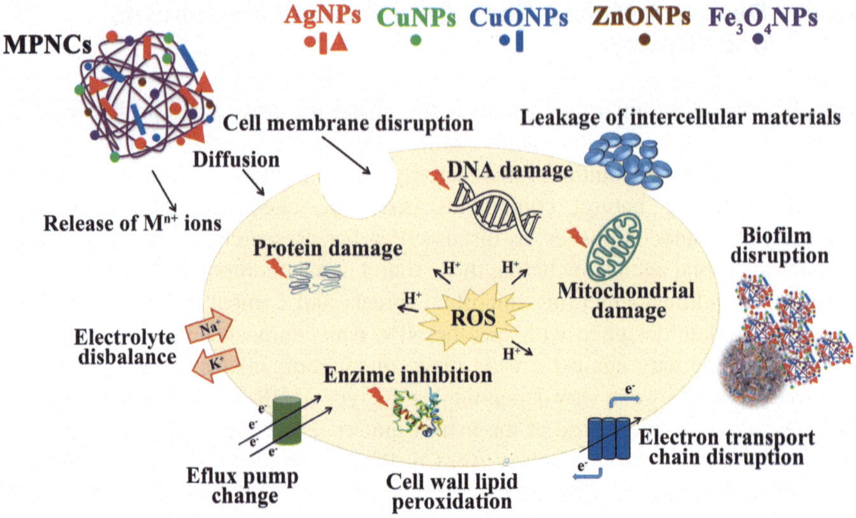

Fig. 18.1 Schematic representation of the interaction of MNPs and polymers with the microbe and its different types of damages

presence of highly reactive crystallographic planes (111) on them, allows for direct interaction with the cell membrane, oxidation of Cu on its surface with the release of ions, and reaction with certain functional groups inside the cell, resulting in the disruption of its biochemical functions.

The spherical and bare AgNPs (6 nm) also showed maximal bacterial reduction (99.9%) for both bacteria, *E. coli* and *S. aureus*, practically for all samples in the examined concentration range (0.5–10.0 ppm), if the exposure duration is sufficient (24 h) (Vukoje et al. 2022). However, even after 24 h, 10^2–10^3 CFU/mL of both bacteria remain viable at the lowest tested concentration of AgNPs (0.5 ppm), indicating that this concentration of silver is insufficient to inactivate the entire initial concentration of bacteria inoculum (10^6–10^7 CFU/mL). These NPs display non-toxic behavior (cell viability of pre-osteoblast cells >70%) for concentrations smaller than 0.6 ppm. Other examples of MNPs also show excellent properties towards selected microbes, depending on NPs size, concentration, and time of exposure to them (Mba and Nweze 2021). Moreover, it is important to note that MNPs, unlike conventional antibiotics, are less likely to induce microbial resistance due to their diverse mechanisms of action. This makes them useful alternatives in an era of rising antibiotic resistance. Although MNPs are highly effective antimicrobial agents, their incorporation into polymers to form nanocomposites offers several advantages over using pure MNPs, as already mentioned, including a stabilizing environment that hinders the agglomeration and oxidation of MNPs.

Conducting polymers PANI and PPy as multilateral materials have the ability to tune properties for specific applications in electronics, sensors, batteries, and capacity to act as a antimicrobial agents, which makes them fascinating and versatile materials in the realm of advanced materials (Maruthapandi et al. 2022; Zare et al.

2019). They are suitable hosts for various MNPs and their antimicrobial activity. When CuNPs are combined with PANI to form a Cu-PANI MPNCs, it makes them more stable and effective against microbes (Bogdanović et al. 2015, 2018). The authors showed that this composite has a symmetrical dendritic structure with mutually parallel branches consisting of PANI nanofibers and equally distributed spherical CuNPs with a diameter of 6 nm (Bogdanović et al. 2015, 2018). The Cu-PANI MPNCs revealed exceptional antimicrobial properties against *E. coli*, *S. aureus*, and *C. albicans,* both in reduction and prevention of their growth, which are significantly better than the activity of bare CuNPs and pure PANI for a much lower tested composite concentration and incubation time with microbes. In particular, unlike CuNPs and pure PANI, the Cu-PANI entirely prevented the growth of all three microorganisms after 1 and 2 h of contact at concentrations of 1 ppm (Table 18.1). The results demonstrate that Cu-PANI can be a swift and efficient antibacterial agent. Moreover, the atomic force microscopy (AFM) analysis revealed that this nanocomposite caused significant damage to the cell walls of all tested microorganisms, considerably more than the CuNPs alone, with *E. coli* cells suffering the most damage, with no surviving colonies observed. An example of the different stages of *E. coli* cell damage, due to interaction with Cu-PANI nanocomposite (1, 2, and 3 h of contact) in comparison with an untreated, smooth *E. coli* strain (a) can be seen in Fig. 18.2. Although the particular mechanism of their antibacterial effect is

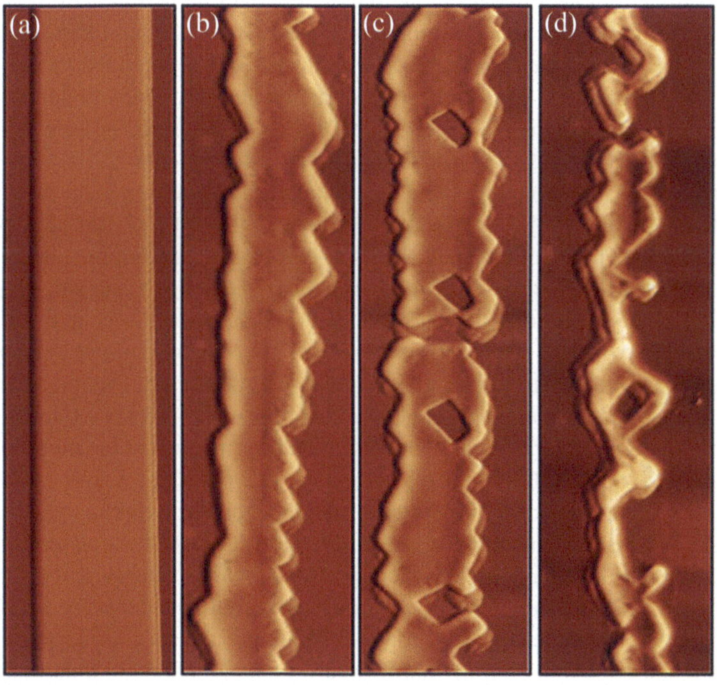

Fig. 18.2 AFM images of *E. coli* before (**a**) and after incubation with Cu-PANI nanocomposite (20 ppm) for (**b**) 1 h, (**c**) 2 h, and (**d**) 3 h

undetermined, many conclusions have been drawn due to the synergistic effect of CuNPs and PANI chains, in the nanocomposite, as well as its morphology—large areas of direct physical contact with microbes, which prevent their nutrition, play a key role in obtaining increased antibacterial efficacy. In addition, PANI possess a positively charged backbone, that can interact with the negatively charged cell surfaces of microorganisms through electrostatic forces together with positive charged ions released from CuNPs. Thus, the coalescence of metal cations and PANI polarons has the potential to increase the attraction and adherence to microbial cell surfaces, enhancing the disruption of cell membranes. Based on the findings of the Cu-PANI cytotoxic properties as well as an evaluation of the cellular oxidative state, the researchers emphasized the significance of exercising caution while using doses of Cu-PANI nanocomposite exceeding 0.7 ppm. Regardless of such small concentrations at the biosafety level, nanocomposite showed efficient bactericidal and fungicidal activities even at its low concentrations ($\leq$1 ppm) within the incubation time of 2 h (taking into account that the amount of Cu in the nanocomposite was 14.3 wt%).

When it comes to combinations of Ag and PANI in the work of Shaban et al., where the average size of Ag crystallites was 20 nm, the Ag/PANI MPNCs composite concentration of 400 ppm resulted in the following inhibitory effects on four bacterial growths: *Salmonella* exhibited the largest inhibition zone with a diameter of 22.5 mm, followed by *S. aureus* with 20.0 mm, *B. subtilis* with 19.0 mm, and *E. coli* with 18.0 mm (Shaban et al. 2018). The inhibition zones of all bacteria increase with increasing the active material from 200 to 400 ppm that inhibits the growth of bacteria, while they decrease with further increasing the dose concentrations to 600 ppm due to the aggregation of nanomaterials, which greatly decrease their antibacterial activities. As well as the Cu-PANI composite, the combination of AgNPs with PANI in this work, led to their synergistic effect on tested bacteria and the electrostatic attraction with their functional groups. The spherical morphology of Ag/PANI composites also played a vital role in the antibacterial activity since their nanoporous particles can gain access to cells easily through membrane pores. Another tested MPNC is Ag–PANI/PVP, synthesized through the in situ integration of AgNPs in polymer during the oxidative polymerization of aniline in the presence of polyvinylpyrrolidone (PVP), whose role was not only to facilitate the polymerization process but also to reduce the possibility of AgNPs aggregation and macroscopic precipitation (Stamenović et al. 2021). In this work, two MPNCs were tested with a comparable silver content of 40 wt% but differ in terms of the morphological characteristics of the AgNPs (spheres with a diameter of 7.0 ± 2.3 nm/triangles with an edge length ranging from 25 to 140 nm) and the granular/wrinkling pattern of the polymer. The assessment of the antimicrobial capabilities of the composites and their potential utilization as a potent tool against *E. coli*, *S. aureus*, and *C. albicans* has revealed a significant suppression of microbial proliferation exceeding 90% at relatively low composite concentrations of 5 ppm and an exposure duration of 4 h. A comprehensive evaluation of the biosafety of these MPNCs, focusing on the assessment of genotoxicity and redox modifying activity using human peripheral blood cells as a representative model system, indicated that MPNCs do not exhibit any observable ability to cause DNA damage, but they have prooxidant and cytostatic effects on human peripheral blood cells. In the study by Mohsen et al., the antibacterial properties of ZnONPs (crystallite size of 46.7 nm), PANI, and PANI/ZnO MPNCs (spherical crystallite structure of 61.6 nm) were tested (Mohsen et al. 2018). The authors showed that the antibacterial efficacy of ZnONPs and PANI/ZnO composites against *S. aureus* is higher compared to *E. coli*, while PANI exhibited an inhibition zone diameter of only 9 mm/mg against *E. coli* and indicated inadequate action against *S. aureus* (Table 18.1). Furthermore, the level of activity exhibits an upward trend as the ZnO content in MPNCs is raised, ultimately reaching its peak efficiency at a ZnO weight percentage of 8%. The most favorable outcomes are achieved for MPNCs in which the synergistic effect of PANI and MNPs becomes prominent, but the primary factor contributing to the antibacterial properties of these MPNCs is the presence of ZnONPs, which effectively hinder bacterial development through their photocatalytic capabilities. In another study, Hou et al. focused on the investigation of the antibacterial activity of Ag-doped ZnO/PANI composite (Hou et al. 2016). The authors introduced Ag as a dopant in ZnO to enhance its photocatalytic performance, resulting in the formation of AZO ($Ag_{0.02}Zn_{0.98}O_{0.99}$). Based on the findings of the morphological investigation, they showed that ZnO particles, with an average size ranging from 25 to 30 nm, have a uniform distribution, conversely, AZO particles display a tendency to agglomerate. The antimicrobial activity of the AZO/PANI MPNCs against *E. coli*, *S. aureus*, and *C. albicans* is more significant compared to pure PANI, and the minimum inhibitory concentrations (MICs)/minimum bactericidal concentrations (MBCs) are 10/25 mg/L, 10/25 mg/L, and 5/10 mg/L, respectively. These values are below the standard threshold of 800 mg/L, indicating that the composite can be considered effective according to the established criteria (Table 18.1). The process clarifies that the application of PANI coating in conjunction with AZO has the potential to stabilize the produced e^- and h^+ species originating from AZO, which subsequently undergo a reaction with adsorbed oxygen (O_2) and water (H_2O), resulting in the formation of ROS. These ROS have the ability to cause damage to the microbial cell membrane and other intramolecular components.

The synergistic interaction of electrostatic forces by charged polypyrrole, another host for MNPs, together with their permeation through the cell wall, also leads to the leakage of cytoplasmic components and increases the antimicrobial abilities or favors the antibiofilm activity of the resulting material, such as Cu/PPy, Ag/PPy, Fe_3O_4/PPy, and Pd/PPy nanocomposites (Mašojević et al. 2023; Liu et al. 2015; Guo et al. 2021; Salabat et al. 2015). The antibacterial abilities of Ag/PPy nanowires with spherical AgNPs (15–40 nm) against *E. coli* were achieved by assessing the resulting diameter of the inhibitory zone around 15.4 mm (Table 18.1) (Liu et al. 2015). Testing the efficacy of Cu/PPy composite (spherical CuNPs with an average diameter of ~25 nm and granular PPy), against *E. coli* and *S. aureus* showed that 24-h exposure of 2 ppm of Cu/PPy to *S. aureus*, has a near-complete bactericidal effect, resulting in a 99.17% inhibition of bacterial growth, while concentrations of 10 and 20 ppm, reached a cidal value of 99.99% for *S. aureus* and *E. coli*, respectively (Mašojević et al. 2023). In this NCs, the presence of the polymer matrix facilitates a significant contact area between the MPNCs and pathogens, enabling more efficient interaction between CuNPs and bacteria, particularly along their active (111) planes. The communication between bacteria and their surrounding medium is impeded, thereby inhibiting the nutritional process of the bacteria. The presence of safe and environmentally acceptable quantities of Cu/PPy suggests that it is a viable option as an antibacterial agent with potential for further application. In another work, authors demonstrated the sterilizing capabilities of Fe_3O_4/PPy composite and removal of over 50% of *E. coli* and *S. aureus* with a composite concentration of 100 μg/mL during a period of 10 min of interaction and subsequent magnetic adsorption (Guo et al. 2021). The present study offers compelling experimental evidence that Fe_3O_4/PPy MPNCs possess significant promise as durable and recyclable nanocomposite materials for the purposes of bacteria adsorption and photothermal sterilization. A wide range of antibacterial activity with highly favorable outcomes was also observed for Pd-PPy composites (Salabat et al. 2015). Its zones of inhibition were found to be 23 ± 0.5 mm, 10 ± 0.25 mm, 16 ± 0.3 mm, and 14 ± 0.6 mm for *S. aureus*, *Bacillus* sp., *E. coli*, and *Klebsiella* sp., respectively. Also, its MIC and MBC findings, conducted on *S. aureus,* indicate that the MIC value was 5.78 mg/mL, while the MBC result indicated cell destruction at a concentration of 23.12 mg/mL. The modification of macroporous copolymers like poly(GMA-*co*-EGDMA) with molecules containing free $–NH_2$ groups allows the production of carriers that can mediate the in situ synthesis of AgNPs as another kind of antimicrobial agents (Vukoje et al. 2014a, 2017, 2019). The benefits of in situ synthesis of AgNPs are evident in the considerable uniformity of the resulting particles, as well as the ability to regulate particle size through the length of the diamine hydrocarbon chain. Significantly, high percentages of microbial reduction of *E. coli* were achieved using Ag/GMA-*co*-EGDMA MPNCS even when the concentration of silver was as low as 12.5 ppm (Table 18.1).

The approach used for implementing biomolecules as hosts for MNPs involves the employing of various carbohydrate molecules, proteins, peptides, or DNA to stabilize and regulate the formation of MNPs with different sizes, shapes, and features with a tendency to be used as antimicrobial agents (Kukushina et al. 2021). These materials provide ecologically sustainable alternatives for the synthesis of MNPs and their applications. For instance, the enhanced long-term stability of AgNPs (15–30 nm in size) was achieved when they were prepared using the reducing, stabilizing, and coating abilities of glucose, sucrose, and dextran due to the positive correlation with the molecular weight of these molecules (Vukoje et al. 2022). Based on the obtained results, maximum bacterial reduction (99.9%) for *E. coli* and *S. aureus* is achieved for all samples when AgNPs concentration is 10.0 ppm if the exposure time is sufficiently long (24 h). The Ag/dextran MPNCs exhibited the broadest spectrum of non-toxic behavior, with a concentration range of up to 2.5 ppm. This concentration range is almost four times greater than that reported for bare AgNPs (0.6 ppm). Chitosan (CS) is a derivative of the natural polymer chitin, which is the second most common polysaccharide after cellulose, and therefore, it is widely used as a host for MPNs (Yadav et al. 2019). According to Sharma S., MIC and MBC values for Ag/CS MPNCs towards *E. coli*, *E. aerogenes*, *P. aeruginosa*, *B. cereus*, and *E. faecalis* are presented in Table 18.1, demonstrating the synergistic action of CS nanocarriers and impregnated AgNPs (Sharma 2017). Potara et al. proposed two potential mechanisms for the synergistic effect observed in the Ag/CS MPNCs (Potara et al. 2011). First, the authors suggest that the application of CS on the surface of the AgNPs results in steric stabilization, preventing or reducing aggregation. This steric stabilization increases the effective concentration of particles available for interaction with the cellular surface. Second, the positive charge introduced by the CS may enhance the binding of the particles to negatively charged sites on the cell surface.

When discussing the use of biomaterials as viable substrates for MNPs, it is imperative to acknowledge the significance of cotton fabric. Cotton is one of the most widely used natural fabrics in the world due to its numerous favorable qualities and versatility. But, under certain temperature and humidity conditions, it can easily support the growth of different microbes. The reason for this phenomenon is attributed to the capacity of cotton fibers to absorb and hold moisture, thereby establishing a favorable environment that promotes the proliferation of microorganisms. Consequently, some textile products are treated with antimicrobial agents during manufacturing to inhibit the growth of microbes. These treatments can be effective in reducing microbial colonization, but their effectiveness might vary over time and with use. For instance, Vukoje et al. investigated the influence of Ag nanoplates with triangular and truncated-triangular shapes on the antimicrobial activity and color of cotton fabrics pretreated with CS (Vukoje et al. 2014b). These nanoplates have an edge length ranging from 40 to 60 nm and exhibit a consistent thickness of approximately 5 nm. Based on the results, CS-treated cotton fabric impregnated with these MNPs provides maximum microbial reduction in the cases of *E. coli* and *C. albicans* ($R = 99.9\%$) and a slightly smaller microbial reduction in the case of *S. aureus* ($R = 99.6\%$). The authors suggested that the improved antimicrobial activity of CS-treated cotton fabric loaded with AgNPs compared to untreated fabric was attributed to the enhanced binding of AgNPs to free amino and hydroxy groups of CS and, therefore, to a higher content of deposited AgNPs available to interact with microbes. Interesting research was published by Marakova et al., using cotton fabric to deposit conducting polymers (PANI or PPy) and their reduction abilities to form AgNPs on them (Marakova et al. 2017). The PPy-coated cotton, with and without deposited AgNPs, exhibited notable antibacterial activity against *S. aureus* and *E. coli,* together with minimal cytotoxicity, suggesting that this material holds significant potential for diverse applications. Also, the samples coated with PANI exhibited reduced antibacterial efficacy (Table 18.1) and increased cytotoxicity in comparison to materials based on PPy.

18.5 Challenges in the Application of Metal-Polymer Nanocomposite as the Next Generation of Antimicrobials and Their Future Perspective

As we have seen before, society faces serious challenges, and the development of new nanomaterials can offer solutions directly or indirectly in various aspects of health. They are also expected to support our health and well-being through the development of antimicrobial materials. Overall, MPNCs add a new dimension to the design of these materials, potentially leading to the next generation of them. However, along with the benefits of MPNCs research and development come potential risks. Despite their rapid development, their ease of use, sensitivity to various microbes, toxicity, and side effects in biological systems and the environment are less known. The evaluation of their toxicity can be done by considering their interaction with living cells and additional analysis of their interactions with tissues and organs, and/or accumulation in them. Therefore, the ability to assess and mitigate these risks is essential to ensuring their safety. Their use in this area is determined by the chemical nature of the constituents, the optimal dose, short-term and long-term exposure to targeted microbes, efficiency, high economic value, and the possibility of safely removing them from the environment. First of all, it is important that novel MPNCs be biocompatible to reduce health hazards. Other important steps are found in optimizing some parameters during their design, such as (a) stability, (b) functionality, (c) reactivity, (d) retention time, (e) toxicity/distribution in the living body, (f) environmental release, (g) easy separation after use, and (h) reuse. The next step is standardization of the experimental conditions, i.e., development of guidelines for the antimicrobial and antibiofilm evaluation of MPNCs by international regulatory bodies. The current situation indicates a lack of reproducibility and comparability of data from different research studies, as well as detailed insights into methods used for assessing antimicrobial efficiency, which are only partially elaborated in terms of standardization processes. Therefore, it is necessary to define protocols with minimum reporting standards related to these activities, so that the results of different studies can be compared. Other important steps are found in the safety and toxicity profiles of MPNCs systems. Importantly, before their efficient use as antimicrobial agents occurs, clear insights into regarding their activity in complex biological media, i.e., interaction with biological fluids (proteins, lipids, plasma, serum, electrolytes, and metabolites), as well as their non-toxic doses, are crucial. Next, their financial implications and large-scale application, from laboratory to industrial, should be addressed before their implementation. Finally, their financial implications and broad application, from laboratory to industrial, should be addressed, since their benefits do not exist until they are commercialized and implemented at scale.

18.6 Conclusion

With the current state of the art, nanotechnology exhibits a broad platform that enables many novel approaches to create nanocomposite-based antimicrobial materials that rely on interactions between MNPs and polymers. These nanocomposites have already brought many benefits to society, and more are yet to be discovered, but they also come with potential risks. Their interdisciplinary research evaluating distinct aspects and the ability to mitigate these risks is necessary to ensure that they are safe and can be used as an alternative in the fight against AMR. Gaps remain in our understanding of their interaction with humans and the environment, so studies on the toxicity and biocompatibility of various combinations of metal NPs and polymers are needed for translation into practical applications. Moreover, for their realization as effective next-generation antimicrobials, precisely defined protocols must be addressed. In the years to come, there remains hope that these nanomaterials will overcome the antibiotic resistance crisis.

Acknowledgement The research was funded by the Ministry of Education, Science and Technological Development of the Republic of Serbia (contract no. 451-03-47/2023-01/200017).

References

Akter M, Sikder MT, Rahman MM et al (2018) A systematic review on silver nanoparticles-induced cytotoxicity: physicochemical properties and perspectives. J Adv Res 9:1–16

Albanese A, Tang PS, Chan WCW (2012) The effect of nanoparticle size, shape, and surface chemistry on biological systems. Annu Rev Biomed Eng 14:1–16

Bagavathy MS, Perachiselvi M, Feiona TA et al (2019) Adsorption and antibacterial studies using encapsulated chitosan/ZnO nanocomposite. Res J Life Sci Bioinform Pharm Chem Sci 5:735–755

Bogdanović U, Lazić V, Vodnik V et al (2014) Copper nanoparticles with high antimicrobial activity. Mater Lett 128:75–78

Bogdanović U, Vodnik VV, Mitrić M et al (2015) Nanomaterial with high antimicrobial efficacy—copper/polyaniline nanocomposite. ACS Appl Mater Interfaces 7:1955–1966

Bogdanović U, Dimitrijević S, Škapin SD et al (2018) Copper-polyaniline nanocomposite: role of physicochemical properties on the antimicrobial activity and genotoxicity evaluation. Mater Sci Eng C 93:49–60

Booshehri AY, Wang R, Xu R (2015) Simple method of deposition of CuO nanoparticles on a cellulose paper and its antibacterial activity. Chem Eng J 262:999–1008

Bruna T, Maldonado-Bravo F, Jara P et al (2021) Silver nanoparticles and their antibacterial applications. Int J Mol Sci 22:7202

Colino CI, Lanao JM, Gutierrez-Millan C (2021) Recent advances in functionalized nanomaterials for the diagnosis and treatment of bacterial infections. Mater Sci Eng C 121:111843

Davis M, Whittaker A, Lindgren M et al (2018) Understanding media publics and the antimicrobial resistance crisis. Glob Public Health 13:1158–1168

Della Pina C, De Gregorio MA, Dellavedova P et al (2020) Polyanilines as new sorbents for hydrocarbons removal from aqueous solutions. Materials 13:2161

Devi GK, Kumar PS, Kumar KS (2016) Green synthesis of novel silver nanocomposite hydrogel based on sodium alginate as an efficient biosorbent for the dye wastewater treatment: prediction of isotherm and kinetic parameters. Desalin Water Treat 57:27686–27699

Di Mauro A, Cantarella M, Nicotra G et al (2017) Novel synthesis of ZnO/PMMA nanocomposites for photocatalytic applications. Sci Rep 7:40895

Dizaj SM, Lotfipour F, Barzegar-Jalali M et al (2014) Antimicrobial activity of the metals and metal oxide nanoparticles. Mater Sci Eng C 44:278–284

Ermini ML, Voliani V (2021) Antimicrobial nano-agents: the copper age. ACS Nano 15:6008–6029

Gabrielyan L, Hakobyan L, Horhannisyan A et al (2019) Effects of iron oxide (Fe_3O_4) nanoparticles on *E. coli* antibiotic resistant strains. J Appl Microbiol 126:1108–1116

Guerrero Correa M, Martínez FB, Vidal CP et al (2020) Antimicrobial metal-based nanoparticles: a review on their synthesis, types and antimicrobial action. Beilstein J Nanotechnol 11:1450–1469

Guo N, Cang F, Wang Z et al (2021) Magnetism and NIR dual response polypirrole-coated Fe_3O_4 nanoparticles for bacteria removal and inactivation. Mater Sci Eng C Mater Biol Appl 126:112143

Hemeg HA (2017) Nanomaterials for alternative antibacterial therapy. Int J Nanomedicine 12:8211–8225

Herrera G, Pena-Bahamonde J, Paudel S et al (2021) The role of nanomaterials and antibiotics in microbial resistance and environmental impact: an overview. Curr Opin Chem Eng 33:100707

Hou Y, Feng J, Wang Y et al (2016) Enhanced antibacterial activity of Ag-doped ZnO/polyaniline nanocomposites. J Mater Sci Mater Electron 27:6615–6622

Ilić V, Šaponjić Z, Vodnik V et al (2009) The influence of silver content on antimicrobial activity and color of cotton fabrics functionalized with ag nanoparticles. Carbohydr Polym 78:564–569

Ilić V, Šaponjić Z, Vodnik V et al (2010) Bactericidal efficiency of silver nanoparticles deposited onto radio frequency plasma pretreated polyester fabrics. Ind Eng Chem Res 49:7287–7293

Jasovsky D, Littmann J, Zorzetand A et al (2016) Antimicrobial resistance-a threat to the world's sustainable development. Ups J Med Sci 121:159–164

Karimi Z, Khalili R, Zazouli MA (2021) Surface modified polythiophene/Al_2O_3 and polyaniline/Al_2O_3 nanocomposites using poly(vinylalcohol) for the removal of heavy metal ions from water: kinetics, thermodynamic and isotherm studies. Water Sci Technol 84:182–199

Kraker MEA, Stewardson AJ, Harbarth S (2016) Will 10 million people die a year due to antimicrobial resistance by 2050? PLoS Med 13:1002184

Kukushina E, Hossain SI, Sporttelli MC et al (2021) Ag-based synergistic antimicrobial composites. A critical review. Nanomaterials 11:1687

Liu F, Yuan Y, Li L et al (2015) Synthesis of polypirrole nanocomposite decorated with silver nanoparticles with electrocatalysis and antibacterial property. Compos Part B Eng 69:232–236

Lok CN, Ho CM, Chen R et al (2006) Proteomic analysis of the mode of antibacterial action of silver nanoparticles. J Proteome Res 5:916–924

Makabenta JMV, Nabawy A, Li CH et al (2021) Nanomaterial-based therapeutics for antibiotic-resistant bacterial infections. Nat Rev Microbiol 19:23–36

Marakova N, Humpolicek P, Kasparkova V et al (2017) Antimicrobial activity and cytotoxicity of cotton fabric coated with conducting polymers, polyaniline or polypyrrole, and with deposited silver nanoparticles. Appl Surf Sci 396:169–176

Maruthapandi M, Saravanan A, Gupta A et al (2022) Antimicrobial activities of conducting polymers and their composites. Macromolecules 2:78–99

Mašojević D, Stamenović U, Otoničar M et al (2023) Evaluation of the antibacterial effectiveness of novel copper/polypirrole nanocomposite. Mater Lett 338:134051

Mba IE, Nweze EI (2021) Nanoparticles as therapeutic options for treating multidrug-resistant bacteria: research progress, challenges, and prospects. World J Microbiol Biotechnol 37:108

Mendelson M, Matsoso MP (2015) The World Health Organization global action plan for antimicrobial resistance. S Afr Med J 105:325

Mishraa A, Pradhana D, Haldera J et al (2022) Metal nanoparticles against multi-drug-resistance bacteria. J Inorg Biochem 237:111938

Mohsen R, Morsi SMM, Selim MM et al (2018) Electrical, thermal, morphological, and antibacterial studies of synthesized polyaniline/zinc oxide nanocomposites. Polym Bull 76:1–21

Naseem T, Waseem M (2022) A comprehensive review on the role of some important nanocomposites for antimicrobial and wastewater applications. Int J Environ Sci Technol 19:2221–2246

Nasrollahzadeh M, Sajjadi M, Iravani S et al (2021) Starch, cellulose, pectin, gum, alginate, chitin and chitosan derived (nano) materials for sustainable water treatment: a review. Carbohydr Polym 251:116986

Nejabatdoust A, Zamani H, Salehzadeh A (2019) Functionalization of ZnO nanoparticles by glutamic acid and conjugation with thiosemicarbazide alters expression of efflux pump genes in multiple drug-resistant *Staphylococcus aureus* strains. Microb Drug Resist 25:966–974

Nguyen N-YT, Gruelling N, Wetteland CL et al (2018) Antimicrobial activities and mechanisms of magnesium oxide nanoparticles (nMgO) against pathogenic bacteria, yeasts, and biofilms. Sci Rep 8:16260

Potara M, Jakab E, Damert A et al (2011) Synergistic antibacterial activity of chitosan-silver nanocomposites on *Staphylococcus aureus*. Nanotechnology 22:135101

Radetić M, Ilić V, Vodnik V et al (2008) Antibacterial effect of silver nanoparticles deposited on corona-treated polyester and polyamide fabrics. Polym Adv Technol 19:1816–1821

Salabat A, Mirhoseini F, Mahdieh M et al (2015) A novel nanotube-shaped polypirrole-Pd composite prepared using reverse microemulsion polymerization and its evaluation as an antibacterial agent. N J Chem 39:4109–4114

Samreena, Ahmad I, Malak HA, Abulreesh HH (2021) Environmental antimicrobial resistance and its drivers: a potential threat to public health. J Glob Antimicrob 27:101–111

Shaban M, Rabia M, Fathallah W et al (2018) Preparation and characterization of polyaniline and ag/polyaniline composite nanoporous particles and their antimicrobial activities. J Polym Environ 26:434–442

Sharma S (2017) Enhanced antibacterial efficacy of silver nanoparticles immobilized in a chitosan nanocarrier. Int J Biol Macromol 104:1740–1745

Shrivastava S, Bera T, Roy A et al (2007) Retracted: characterization of enhanced antibacterial effects of novel silver nanoparticles. Nanotechnology 18:225103

Stamenović U, Davidović S, Petrović S et al (2021) Antimicrobial and biological effects of polyaniline/polyvinylpyrrolidone nanocomposites loaded with silver nanospheres/triangles. New J Chem 45:12711–12720

Turner NA, Sharma-Kuinkel BK, Maskarinec SA et al (2019) Methicillin-resistant *Staphylococcus aureus*: an overview of basic and clinical research. Nat Rev Microbiol 17:203–218

Vodnik VV, Bogdanović U (2019) Metal nanoparticles and their composites: a promising multifunctional nanomaterial for biomedical and related applications. In: Grumezescu V, Grumezescu AM (eds) Materials for biomedical engineering: inorganic micro and nanostructures. Elsevier, Amsterdam, pp 397–427

Vukoje I, Džunuzović E, Vodnik V et al (2014a) Synthesis, characterization, and antimicrobial activity of poly(GMA-co-EGDMA) polymer decorated with silver nanoparticles. J Mater Sci 49:6838–6844

Vukoje I, Lazić V, Vodnik V et al (2014b) The influence of triangular silver nanoplates on antimicrobial activity and color of cotton fabrics pretreated with chitosan. J Mater Sci 49:4453–4460

Vukoje I, Džunuzović E, Lončarević D et al (2017) Synthesis, characterization, and antimicrobial activity of silver nanoparticles on poly(GMA-co-EGDMA) polymer support. Polym Compos 38:1206–1214

Vukoje I, Džunuzović E, Dimitrijević S et al (2019) Size-dependant antibacterial properties of ag nanoparticles supported by amino-functionalized poly(GMA-co-EGDMA) polymer. Polym Compos 40:2901–2907

Vukoje I, Lazić V, Sredojević D et al (2022) Influence of glucose, sucrose, and dextran coatings in the stability and toxicity of silver nanoparticles. Int J Biol Macromol 194:461–469

Yadav M, Goswami P, Paritosh K et al (2019) Seafood waste: a source for preparation of commercially employable chitin/chitosan materials. Bioresour Bioprocess 6:8

Zare Y, Shabani I (2016) Polymer/metal nanocomposites for biomedical applications. Mater Sci Eng C 60:195–203

Zare EN, Makvandi P, Ashtari B et al (2019) Progress in conductive polyaniline-based nanocomposites for biomedical applications: a review. J Med Chem 63:1–22

Green-Synthesized Nanoparticles to Combat Multidrug-Resistant Bacteria

19

Rutuja S. Palaskar, Darshana S. Dhokane, and Balaprasad G. Ankamwar

Abstract

Antibiotic resistance has rapidly climbed in recent years, threatening millions of lives worldwide and presenting a serious threat to community health. Since resistance has spread widely and many different species of bacteria share resistance genes, multidrug-resistant (MDR) microorganisms have come to light. This led to international efforts to develop new and improved antibacterial materials to treat bacteria with multidrug resistance. Global interest is currently growing in the utilization of innovative biomaterials made of nanoparticles (NPs) to completely achieve this accomplishment. For the treatment of drug-resistant infections, NPs may prove to be a vitally important therapeutic option. The metals and metal oxide NPs seem to have the most potential of all classes of NPs and have grabbed the curiosity of numerous researchers. Due to their special physicochemical characteristics, metal and metal oxide NPs such as gold, silver, iron oxide, copper, zinc oxide, and titanium dioxide. Have demonstrated antibacterial effects at very low concentrations. Synthesis of NPs by physical and chemical procedures can result in environmental contamination, health risks to humans, technological issues, high costs, and ineffectiveness. To address these issues, emerging in popularity are biological methods that use safe, non-toxic substances during the synthesis process. The process known as "green synthesis" produces NPs by using biological sources, which act as reducing and capping agents. There are numerous methods by which these NPs fight bacteria, including interference with bacterial biofilms, disruption of cell membranes, and generation of reactive oxygen species (ROS). These mechanisms lessen the possibility of resis-

Rutuja S. Palaskar and Darshana S. Dhokane have contributed equally to this work.

R. S. Palaskar · D. S. Dhokane · B. G. Ankamwar (✉)
Bio-Inspired Materials Research Laboratory, Department of Chemistry, Savitribai Phule Pune University (Formerly University of Pune), Ganeshkhind, Pune, India

M. Y. Wani et al. (eds.), *Nanotechnology Based Strategies for Combating Antimicrobial Resistance*, https://doi.org/10.1007/978-981-97-2023-1_19

tance developing while preventing bacterial growth. NPs produced using green synthesis offer a potentially effective method for battling bacteria with multiple MDRs in addition to the weapons of antimicrobial agents due to their environmentally friendly production practices, strong antibacterial characteristics, and low toxicity.

Keywords

Green synthesis · Nanoparticles · Antibiotic · Antibacterial · Reactive oxygen species

19.1 Introduction

One of the most significant and outstanding breakthroughs in the medical sector is the discovery of antibiotics. They are now required in all critical care procedures, including surgery, organ transplantation, and the management of several bacterial illnesses. Antibiotic resistance in bacterial species unfortunately results from the widespread use of antibiotics, especially when they are misused. The therapeutic success is currently under threat due to the enormous rise in bacterial resistance to antibiotics. According to a World Health Organization (WHO) assessment, resistance to antibiotics has come to light as one of the biggest worries to public health in the twenty-first century (Singh et al. 2020). Ineffective antimicrobials against rapidly evolving resistant bacterial strains are responsible for over 700,000 deaths annually, according to a WHO report from 2019, putting immune-compromised individuals (such as cancer patients and others) at an even higher risk of unfavorable outcomes and treatment complications. This supports the need for ongoing surveillance in medical settings and the development of creative strategies for better management of this intricate, multifaceted challenge, namely combating/eradicating the MDR microorganisms without unfavorable, long-term treatment effects (Arshad et al. 2022). Reports on the resistance to the majority of commercially available antibiotics are published, ranging from penicillin (1928) to ceftaroline (2010). Even more unexpectedly, reports of antibiotic resistance began to surface even before several of these drugs were officially released on the market. Ceftaroline, levofloxacin, and penicillin serve as examples. Some of these infections develop antibiotic resistance, and they are sometimes referred to as "superbugs" because of this. In the developed world, dangerous risks from drug-resistant *Enterobacteriaceae, Mycobacterium TB, Klebsiella pneumoniae, and Neisseria gonorrhoeae* are present (Khan et al. 2016).

In general, NPs are unusual because they differ from their bulk counterparts in terms of their optical, mechanical, catalytic, conductivity, and electrical characteristics. Due to the wide range of uses they exhibit, including use in medicine, the synthesis of customized NPs and nanomaterials has therefore been carried out on a massive scale (Gharpure et al. 2019). Various pathogenic microorganisms and bacterial drug resistance have grown to be major challenges in recent years. The search for novel antibacterial drugs has so far entailed several research projects. Due to their high surface area-to-volume ratio and distinctive chemical complexity, nanoscale materials have recently been recognized as new antibacterial agents. For the goal of creating novel materials at the nanoscale level, nanotechnology is a discipline that is expanding quickly. This field is anticipated to create a new platform for employing nanomaterials to treat and prevent illness. Metallic NPs emerged as the most intriguing technique among the most promising nanomaterials with antibacterial capabilities (Das et al. 2017). Since they are less harmful to the environment and require a straightforward approach, green solutions utilizing biological entities have drawn a big of interest throughout the past few years in order to reduce the usage of hazardous chemicals during the manufacturing of metal NPs. In terms of environmentally friendly approaches, plant-mediated NP synthesis is superior to microbial ones due to its low cost, quick synthesis, increased stability, and wide range of biological activities caused by the presence of diverse phytocompounds. Pt, Au, Ag, Cu, and Zn are used in the environmentally friendly way of making metal NPs, and Au stands out for having a special, controllable surface plasmon resonance (SPR) (Vijayakumar et al. 2017). Metal or nonmetal NPs such as silver (Ag), gold (Au), copper (Cu), bismuth (Bi), and selenium (Se), as well as metal oxide NPs such as ZnO, TiO_2, CaO, MgO, Fe_2O_3, or Al_2O_3, are among the nanostructured antibacterial materials. The majorly these nanostructures exhibit inherent antimicrobial properties through no specific activity, which may prevent the emergence of bacterial resistance (Hochvaldová et al. 2022). Infections caused by biofilms that are resistant to conventional antibiotics and persistent MDR bacteria can be treated using NPs. To maximize the therapeutic effect and reduce host toxicity, NPs can be engineered with specialized surface functionalities. NPs' multimodal antibacterial actions have the potential to greatly slow down or stop the emergence of drug resistance, offering an effective solution to the problems associated with the post-antibiotic period (Karnwal et al. 2023).

19.2 The Burden of Antibiotic Resistance and the Need for Antibacterial Nanomaterials

The WHO has identified three major worldwide public health challenges for the twenty-first century, with antimicrobial drug resistance being one of them. *Enterobacter, Staphylococcus, Klebsiella, Actinobacter, Pseudomonas,* and *Enterobacter,* collectively known as the ESKAPE group, are the most dangerous and have caused the most concern. They are all associated with a high fatality rate. The WHO's figures show that overuse of antibiotics is to blame for about 80% of MDR microbial organisms and these infections have significant harmful consequences. Due to the growth of MDR bacteria and other resistant infections, there are currently few preventive and therapeutic options. Alternative therapy approaches are required for microbial infections. The need for improved treatment alternatives stems from the failure of the majority of antibiotics (Morgan et al. 2011). Consequently, the increase in MDR and heavily drug-resistant (DR) bacteria continues to be a major public health challenge since it is connected to high rates of mortality, morbidity, and treatment costs (Yang et al. 2021). This is exacerbated by the ability of a variety of bacterial species to produce biofilms, which have been associated with 65–80% of human infections (Mba and Nweze 2021). When it comes to managing infected patients, the problem of resistance frequently leaves practitioners without a solid substitute. The situation has also been made worse by the appearance of super bacteria (bacteria resistant to practically all antibiotics) and MDR. This is generally attributed to the overuse of antibiotics, which favors the development and evolution of strains with diverse genotypic and phenotypic characteristics (Wang et al. 2017). Microbes' reaction to antimicrobial attack serves as a key example of adaptation and the apex of evolution. According to studies, microbial infections cause roughly three million fatalities each year in developing nations. Despite the enormous advancements in diagnosis and treatments that have been made over the years, it nevertheless results in over ten million fatalities annually, especially in tropical nations. In medicine, antimicrobial drug resistance has grown to be a significant global public health issue. The annual economic cost of MDR microorganisms is estimated to be $20 billion in the USA alone (Mba and Nweze 2021).

Numerous mechanisms have been developed by microbes to combat antibiotics (Fig. 19.1a, b). About 60–70% of today's antibiotics were ineffective against intracellular infections because of their poor permeability and low intracellular retention. Cell penetration is challenging with popular antibiotics such as beta-lactams and aminoglycosides due to their hydrophilic nature (Slavin et al. 2017). In Gram-negative bacteria, aminoglycoside antimicrobial agents pass through porn channels before being actively carried inside the cell, after which they bind to the 30S ribosomal subunit irrevocably and inhibit the synthesis of protein. Beta-lactam antibacterial agents, however, bind to penicillin-binding proteins, preventing the synthesis of cell wall components such as peptidoglycan and rendering autolytic inhibitors of enzymes.

It is likely that NPs are more efficient when paired with antibacterial agents simply due to facile for the NPs to penetrate the cell and easily break down the cell wall. The opposite is also true; when NPs tear down the cell wall, antibiotics specially aminoglycosides, whose mechanistic action does not include breakdown of cell wall, find it simpler to enter the cell. Beta-lactam antibiotics and aminoglycosides both include amino and hydroxyl groups, which potentially engage as targets of NPs (Railean-Plugaru et al. 2016). The antibacterial effect of antibiotic-conjugated NPs is greater than that of the NP alone and the antibiotic or demonstrating a synergistic effect and suggesting that NPs and antibiotics employ various antibiotic pathways (Slavin et al. 2017). Because NPs not only operate as transporters for natural antibiotics and antimicrobials, but they also actively combat bacteria, and they offer the most efficient way to treat MDR bacteria. Additionally, when they are attached to their surface or encased within a structure, loaded NPs have the ability to convey a range of treatments to the site of infection in a way that is both secure and efficient (Railean-Plugaru et al. 2016).

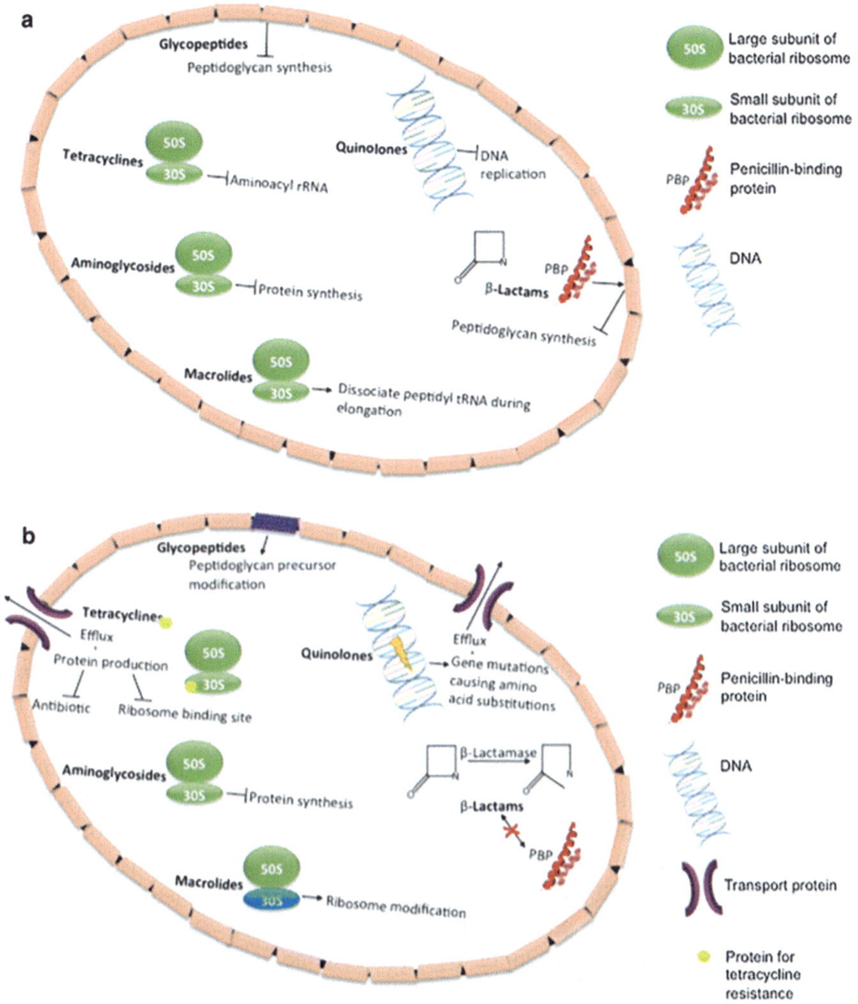

Fig. 19.1 Resistance against some antibiotic classes and their mechanisms. (**a**) A representation explaining the antibiotic mechanisms underlying β-lactams (e.g., penicillin), macrolides (e.g., azithromycin), tetracyclines (e.g., tetracycline), quinolones (e.g., ciprofloxacin), aminoglycosides (e.g., amikacin), and glycopeptides (e.g., vancomycin, and teicoplanin), (**b**) Bacteria produce antibiotic resistance mechanisms (Slavin et al. 2017)

19.3 Strategies to Combat Bacteria

ROS is a metabolic byproduct of oxidation that is produced inside the cells. These byproducts have a role in the differentiation, signaling, viability, and death of cells. Oxidative stress is induced by the accumulation of ROS at hazardous levels (Fig. 19.2). The bonding of protein thiols with hydroxyl radicals and superoxide, which inactivates membrane receptors, is arguably the most recognized way that

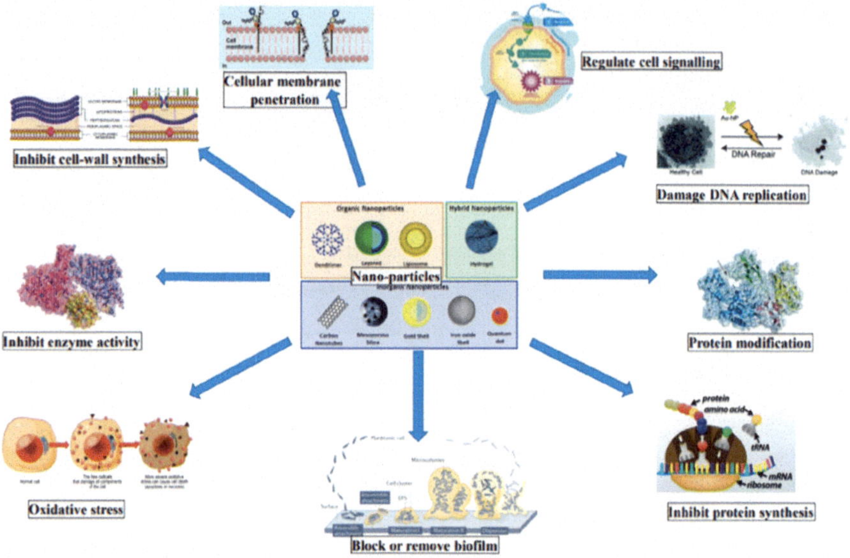

Fig. 19.2 NPs employ a variety of strategies to combat bacteria (Karnwal et al. 2023)

ROS can induce toxicity to cells. NPs can produce ROS through a variety of methods, including attachment to intracellular organelles, biomolecule oxidations using NADPH oxidase, and direct ROS generation (Karnwal et al. 2023). The main antibacterial method exploited by these NPs (photodynamic treatment) is formation of ROS, which is due to the inherent photocatalytic activity of several metal-based NPs (Gehring et al. 2016). The DNA and internal proteins of *B. subtilis* and *E. coli* are harmed as a result of this process, which also produces ROS. Resistance to antibiotic *E. coli* and *S. aureus* was subjected to tests using Ag-ZnO nanocomposites, which also showed antibacterial activity (Oves et al. 2019; Ananda et al. 2019). When these processes came together, a series of events occurred that had bactericidal effects, including the suppression of DNA replication, the dislocation of protein function, the dripping of cellular macromolecules, and the breakdown of cell membranes (Lemire et al. 2013). Negatively charged cell walls of Gram-positive and Gram-negative bacteria and positively charged NPs interact electrostatically. NPs interact strongly with the bacterial membrane, leading to penetration into the cell membrane, disrupting the integrity of the bacterial membrane (Punniyakotti et al. 2020). Finding a route through the physical barrier that biofilms provide is essential for their removal. In general, cationic NPs and uncharged NPs with diameters under 350 nm diffuse easily through the pores of the biofilm matrix (Qi et al. 2019). Bacterial survival and function are dependent on intracellular signaling and cell homeostasis. These issues include DNA damage, modifications to protein synthesis, and changes in gene expression (Karnwal et al. 2023).

19.4 Nanoparticles: Antibacterial Approach

19.4.1 Silver Nanoparticles (Ag NPs)

Ag NPs are on the route to become an amazing medication in the age of nanotechnology, tackling the issue of MDR acquired by many pathogenic bacteria (Kumari et al. 2017). Antibacterial and antibiofilm actions of metal NPs, particularly Ag NPs, against bacterial and fungi pathogens have been studied. There are several findings on Ag NPs' environmentally friendly manufacture and antibacterial properties (Qais et al. 2020). The suggested plant-mediated synthesis technique is a low-cost method capable of synthesizing Ag NPs with noticeably better killing of MDR bacteria (Yousef et al. 2022). By characterizing NPs, the current study concentrated on environmentally friendly Ag NPs, which have a greater ability to produce effective antibacterial outcomes than chemically synthesized Ag NPs and hence have a large promise for the development of innovative antibacterial medications (Yousef et al. 2022). Gram-positive and Gram-negative organisms were both resistant to the antibacterial effects of the Ag NPs synthesized from citrus and orange peels. Ag NP toxicity is influenced by their size, surface charge, solution chemistry, shape, and solubility (Kirtiwar et al. 2019). It is believed that larger surface areas result from smaller particle sizes, which in turn increases antibacterial activity. According to this study, bigger particles have greater antibacterial activity than smaller particles, which is counter to previous studies (Kirtiwar et al. 2019).

Mimusops elengi L. leaf extracts were utilized for phyto-synthesize of Ag NPs. P. Prakash et al. evaluated a distinct zone of inhibition toward *Klebsiella pneumonia, Micrococcus luteus*, and *Staphylococcus aureus* and demonstrated the exceptional antibacterial effectiveness of the produced Ag NPs against MDR clinical isolates (Prakash et al. 2013). Gopinath et al. fabricated spherical-shaped Ag NPs with a size variance of 16–28 nm from the fruit bodies of the medicinal plant *Tribulus terrestris L.* These green-produced Ag NPs were tested against MDR bacteria, including *Staphylococcus aureus, Pseudomonas aeruginosa, Escherichia coli, Bacillus subtilis, and Streptococcus pyogens* and were found to have bactericidal action (Gopinath et al. 2015). Mysoon M. Al-Ansari shows that combining Ag NPs with layers of *T. procumbens* bioactive chemicals targets the bacteria's cell membrane and causes both internal and exterior cellular harms. Compared to bigger Ag NPs, Ag NPs of modest size (20–40 nm) readily bind to bacterial surfaces. Smaller particles can easily interact with cellular organelles. MDR bacterial isolates are susceptible to the antimicrobial effects of Ag NPs made from *T. procumbens*' inflorescence extracts. Therefore, more research may be done to create defenses against bacterial infections of the urinary system that are resistant to numerous medications (Al-Ansari et al. 2020). Das et al. illustrate an effective green approach for synthesizing Ag NPs utilizing *O. gratissimum* leaf extract as the bio-reductant. MDR *E. coli* and *S. aureus* strains were used as test subjects to evaluate their antibacterial efficacy. In both kinds of strains, Ag NPs were also discovered to be a potent inhibitor of the development of biofilms. The cell membrane is harmed by the bacterial

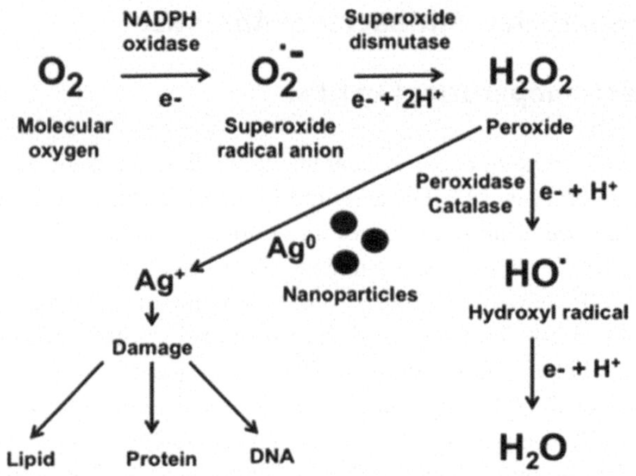

Fig. 19.3 Diagram illustrating the function of NPs in the production of ROS (Slavin et al. 2017)

death process, which is followed by ROS production. This nanomaterial provides an additional alternative for treating bacterial infections that are resistant to many drugs. Consequently, these biogenic Ag NPs can be used to treat superbugs in a variety of medical settings (Das et al. 2017). Ag NPs were produced using *Cinnamomum tamala* (Tejpata) leaf extracts. This finding might help scientists create new therapeutic drugs to fight off germs that are resistant to antibiotics. The creation of green NPs might be a crucial research tool for discovering therapeutic ingredients (Al Mashud et al. 2022). Ag NPs have been used to achieve this goal and have proven to be highly efficient against *A. baumannii, P. aeruginosa, Enterobacteriaceae*, and MDR *S. aureus* (Mateo and Jiménez 2022). Interactions at many levels are part of the Ag NPs' mode of action. They alter the permeability of the plasma membrane, which releases the intracellular contents and results in cell death. They decrease ribosomal subunit expression, inhibit DNA replication, and deactivate proteins and enzymes. They cause membrane-bound respiratory enzymes to act differently, which results in the production of ROS and oxidative damage to cell components (Fig. 19.3) (Mateo and Jiménez 2022).

Ag NPs are internalized by cells, and then, Ag NPs interact with proteins and DNA, influencing the essential and critical activities of the cells. Ag NPs have been shown to have antibacterial action by a process that involves the release of silver ions from the NPs, which negatively impacts the functions of proteins and DNA (Anupam et al. 2019).

19.4.2 Zinc Oxide Nanoparticles (ZnO NPs)

Plant extracts are being utilized more frequently to create ZnO NPs because they are simple, inexpensive, eco-friendly, and non-toxic. Using a plant extract as a capping/reducing agent and then characterizing the results, an affordable biosynthetic method of ZnO synthesis was defined (Saee et al. 2023). ZnO NP synthesis from different plant species such as *Aristolochia indica* (Steffy et al. 2018), *Cassia siamea* (Ahamad et al. 2023), *Bixa orellana (*Gharpure et al. 2022a*)*, *Prunus dulcis* (Saee et al. 2023), *Albizia lebbeck* (Gharpure and Ankamwar 2023), and *Lactuca virosa (*Gharpure et al. 2022b*)* is used to combat MDR bacteria. The clinically isolated MDR strains from diabetic foot ulcer of bacterial strains *S. aureus ATCC 29213, E. faecalis ATCC 29212, P. aeruginosa ATCC 27853, and E. coli ATCC 25922* were all effectively killed by green-synthesized ZnO NPs (Steffy et al. 2018). The bio-formulated ZnO NPs demonstrated an effective synergistic effect with colistin against the relevant species, demonstrating the possible utilization of ZnO NPs in conjunction with colistin as an efficient antibacterial agent toward the MDR pathogens, i.e., *P. aeruginosa* and *A. baumannii*. The mixture of colistin and ZnO NPs may also be employed as a source for developing efficient coatings for utilization in intensive care units, allowing for the efficient management of hospital gain illnesses (Yassin et al. 2022). MRSA and VRE strains were utilized to evaluate the antibacterial activity of ZnO NPs, and the outcome revealed anti-MRSA activity at 400 and 625 μg/mL, respectively. Therefore, these biogenic ZnO NPs show a significant antibacterial activity against microorganisms linked to various major systemic infections (Morowvat et al. 2023). According to Lakshmi et al., the synthesized NPs performed effectively against Gram-negative and Gram-positive bacteria. Thus, they demonstrated that biosynthesized NPs made from agricultural waste materials are the best for their eco-friendly products when compared to NPs produced using chemicals. Additionally, following extensive clinical research and biocompatibility evaluation, ZnO NPs were tested as a potential replacement for commercial antibiotics (Jothi Lakshmi et al. 2022). The respiratory chain and cell division are the targets of the antibacterial action of green-produced *Cassia siamea* (L.)-produced ZnO NPs (ZnO@Cs-NPs), further enhancing their bactericidal efficacy. ZnO NP suspensions result in higher levels of ROS production. These reactive species involve hydroxide (OH), hydrogen peroxide (H_2O_2), and superoxide anion (O_2). The primary mechanism underlying ZnO NPs' antibacterial action is ROS production (Fig. 19.4) (Ahamad et al. 2023). Numerous studies show that the interactions of ZnO NPs with different bacteria affect the structure of the cellular membrane, which ultimately results in cytoplasmic spillage and cell death (Gharpure et al. 2022c).

ZnO NPs fabricated using various plant parts of *Bixa orellana* were explored to study the antibacterial activity. FESEM and TEM images show the size of L-ZnO, S-ZnO, and Sc-ZnO NPs (L-ZnO is ZnO NP synthesis from leaf extract, S-ZnO is ZnO NP synthesis from seed extract, and Sc-ZnO is ZnO NP synthesis from seed coat extract) (Figs. 19.5 and 19.6). Figure 19.6a depicts the presence of spherical particles in L-ZnO, which has a size range of 169–259 nm; Fig. 19.6b depicts almond-like morphologies in S-ZnO, which has a size range of 304–465 nm, and Fig. 19.6c depicts spherical particles in Sc-ZnO, which has a size range of 278–654 nm. *E. coli, P. aeruginosa,* and *S. aureus, B. subtilis* microorganisms were utilized to analyze the antibacterial activities of ZnO NPs produced before and after calcination of *B. orellana* leaf, seed, and seed coat extracts. Only *S. aureus*

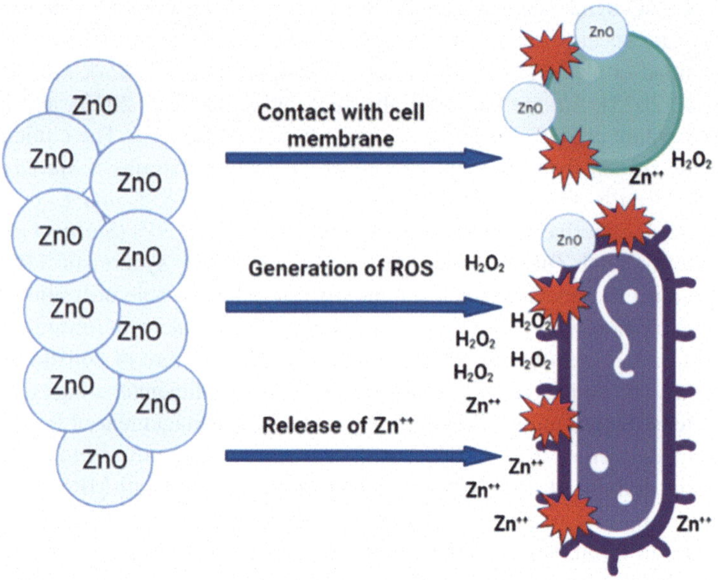

Fig. 19.4 Mechanisms of ZnO NPs' antibacterial efficiency against bacterial pathogens are shown schematically in many ways, involve ROS generation, Zn^{2+} release, incorporation of ZnO NPs into bacteria, and electrostatic interactions (Ahamad et al. 2023)

and *B. subtilis* were shown to be resistant to the antibacterial effects of L-ZnO, S-ZnO was only effective and adverse to *E. coli*, and Sc-ZnO was at best effective against *S. aureus* prior to calcination. Table 19.1 shows the zones of inhibition of L-ZnO, S-ZnO, and Sc-ZnO uncalcined at various concentrations (Gharpure et al. 2022a).

ZnO NPs produced using biological synthesis from plant extracts of ginger (*Zingiber officinale*), garlic (*Allium sativum*), and lemon (*Citrus lemon*) have antimicrobial activity against Gram-positive and Gram-negative bacteria, including MRSA, *Enterococcus* sp., *Staphylococcus* sp., and *E. coli, Acinetobacter* sp., *Klebsiella* sp., and *Pseudomonas* sp. Results can be used as an alternate treatment or remedy for bacteria that are resistant to many drugs in a variety of medical and biological applications (AbdElrahman et al. 2023).

19.4.3 Titanium Dioxide Nanoparticles (TiO_2 NPs)

TiO_2 is currently one of the metal oxide NPs that are most frequently employed in the fields of medicine and others. When exposed to UV radiation, TiO_2, which is known to be biologically inactive and harmless at the microscale, develops photocatalytic characteristics. The semiconductor TiO_2 has a conduction band with devoid energy levels and an electron-filled valence band. An energy hole separates

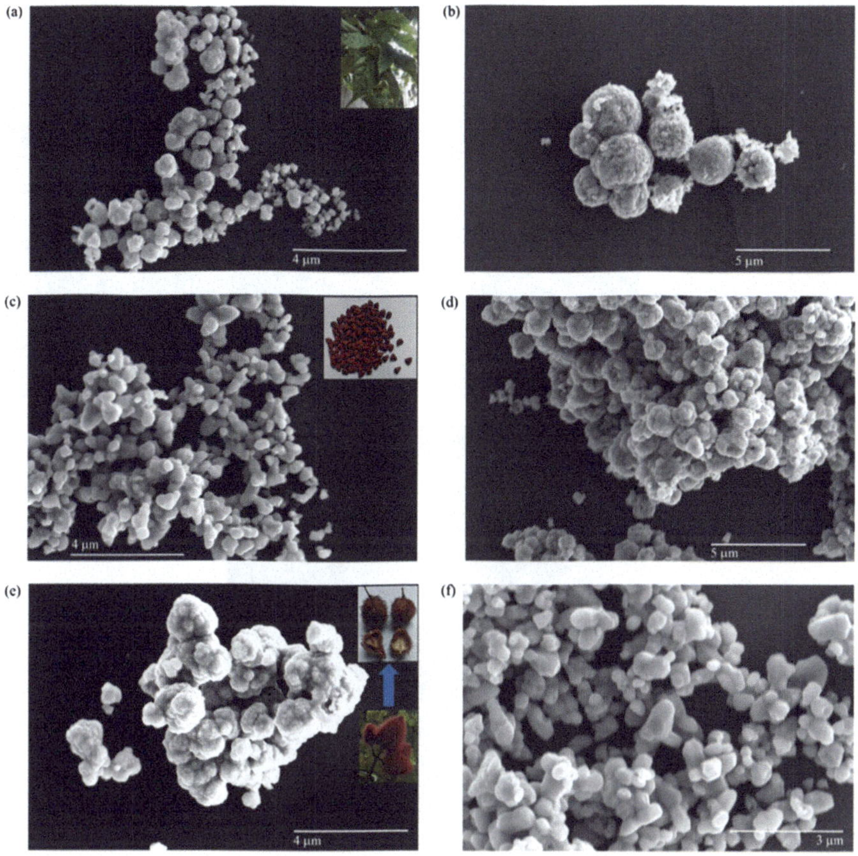

Fig. 19.5 Different FESEM magnifications were used to analyze the morphology of L-ZnO (**a**, **b**) with a visual showing a *Bixa orellana* leaf, S-ZnO (**c**, **d**), with a visual showing a *Bixa orellana* seed, and Sc-ZnO (**e**, **f**), with a visual showing a *Bixa orellana* seed coat (Gharpure et al. 2022a)

these bands from one another. Electrons in the valence band must travel to the conduction band for TiO_2 to have photocatalytic properties. For the electron transfer to occur, the photon energy that the semiconductor absorbs should be higher than the energy level of the hole that exists in between the two bands. UV light leads to valence band electrons in TiO_2 to migrate into the conduction band, where they produce conduction band electrons and valence band holes. When atmospheric H_2O and O_2 are combined with these active electrons and holes, ROS such as superoxide ions and hydroxyl radicals are produced. Photocatalytic oxidation of NPs generates H_2O_2 and singlet oxygen in addition to hydroxyl and superoxide radicals. These radicals have been proven to be highly reactive and destroy organic substances. These characteristics led to the usage of TiO_2 NPs in antibacterial applications (Blake et al. 1999; Hoffmann et al. 1995; Jacoby et al. 1998; Mills and Le Hunte 1997).

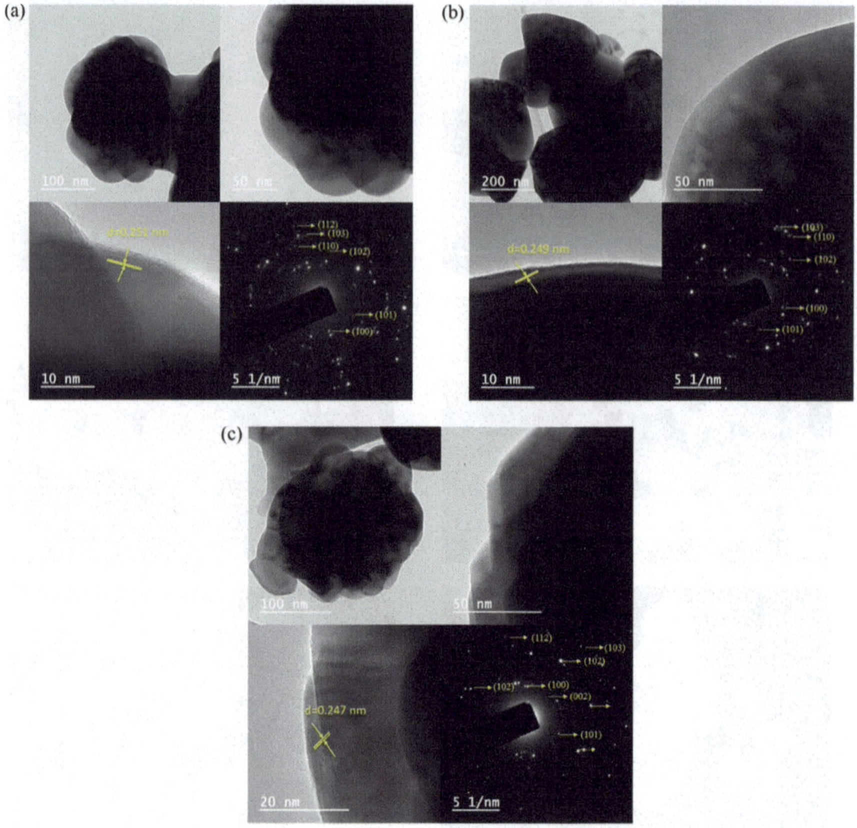

Fig. 19.6 TEM pictures include L-ZnO, S-ZnO, and Sc-ZnO NPs, as well as lattice fringes and SAED patterns (Gharpure et al. 2022a)

Table 19.1 L-ZnO, S-ZnO, and Sc-ZnO uncalcined have antibacterial activity against a number of bacteria (Gharpure et al. 2022a)

		Concentration of ZnO NPs (mg/mL)				
NPs	Micro-organism	10	5	2.5	1.25	0.625
L-ZnO	*E. coli*					
	S. aureus	19 mm	11 mm	9 mm		
	P. aeruginosa					
	B. subtilis	14 mm	10 mm			
S-ZnO	*E. coli*	8 mm				
	S. aureus					
	P. aeruginosa					
	B. subtilis					
Sc-ZnO	*E. coli*					
	S. aureus	10 mm				
	P. aeruginosa					
	B. subtilis					

TiO_2 NPs were biosynthesized using a leaf extract from *Trigonella foenum-graecum*, which generated spherical NPs with sizes ranging from 20 to 90 nm. Their antibacterial activity has been assessed using the conventional disk diffusion method (Subhapriya and Gomathipriya 2018). TiO_2 NP green production with *Glycyrrhiza glabra* root extract has also been described in detail (Mills and Le Hunte 1997). In vivo toxicity experiments utilizing zebrafish embryonic model (*Danio rerio*) are also performed. Antibacterial activities against *S. aureus* and *K. pneumonia* were studied. The healthy adult fish embryos were exposed to several NP variants without exhibiting any recognizable abnormalities at any stage of development (Mills and Le Hunte 1997). By eliminating the use of harmful chemicals during procedures, TiO_2 NP fabrication using microorganisms, such as fungi, yeasts, and bacteria, also satisfies the demands and the constantly expanding technical demand toward eco-friendly techniques. Bio-reducing, capping, and stabilizing features of the metabolites produced by microorganisms enhance production efficiency. The primary substance that served as a stabilizing and capping agent during the production of TiO_2 NPs from green materials was glycyl L-proline, one of *Aeromonas hydrophila* bacteria's most prevalent metabolites. In addition, interest in using fungus to create metal oxide NPs has grown during the past few years. Additionally, fungi enzymes and other metabolites have the inherent ability to extract ionic or elemental state metals from their related salts (Rajakumar et al. 2012; Jayaseelan et al. 2013; Bansal et al. 2005). Every year, bacterial infections, which can result in serious illnesses in humans, are caused by dangerous bacteria such as *S. aureus, A. baumannii, B. cepacia, K. pneumoniae, C. difficile, E. coli, P. aeruginosa, M. tuberculosis, and N. gonorrhea* (López de Dicastillo et al. 2021).

19.4.3.1 Antimicrobial Mechanism of TiO_2 NPs

TiO_2 NPs have an antibacterial action that is photo-dependent and linked to the generation of free radicals. These free radicals induce peroxidation, which has an impact on bacterial LPS, peptidoglycans, and the membrane bilayer. The exterior membrane of *E. coli* was severely damaged by ultraviolet (UV) rays and/or TiO_2 NPs when their combined antibacterial effects were studied (Jiang et al. 2010; Wu et al. 2010; Liu et al. 2010). TiO_2 has a large band gap of 3.2 eV, which under UV illumination may lead to the creation of electron-hole pairs having high energy. Its main photocatalytic feature is light with a wavelength of 385 nm (Aileni et al. 2018). Catalases and superoxide dismutase are two examples of the enzymatic antioxidant defense mechanisms found in bacteria, which counteract the effects of ROS radicals and lipid peroxidation in addition to antioxidants such as ascorbic acid, carotene, and tocopherol. When these systems are breached, series of redox reactions may lead to the modification of several crucial components (cell wall, DNA, etc.) and metabolic pathways, ultimately resulting in cell death (Kiwi and Sami 2018). The chemical composition of the cell wall depends on the type of microbe; for example, in fungus and yeast, chitin and polysaccharides contribute in the majority of the cell wall compositions (Gow et al. 2017). The majority of TiO_2 NP research concentrated on losing membrane integrity caused by the oxidation of phospholipids, which is induced by ROS such as H_2O_2 and hydroxyl radicals. The leaking of cellular content was due to an increase in membrane fluidity. Gram-positive bacteria have one layer of the membrane that is shielded by numerous layers of peptidoglycan, as opposed to Gram-negative bacteria, which have two layers of membranes, an inner and an outer membrane with a thin layer of peptidoglycan in between. Because there is no peptidoglycan protective coating, as there is in Gram-positive bacteria, the outer membrane is highly susceptible to mechanical rupture. We may infer that the efficiency of bacterial inactivation is mostly influenced by the cell walls' ability to withstand damage and the level of ROS formation (Pavlović et al. 2019; Khezerlou et al. 2018; Zainab et al. 2019; Zhu et al. 2018). Microbes exposed to the photocatalytic activity of TiO_2 NP demonstrated cell inactivation at the regulatory network and signaling levels, a significant reduction in respiratory chain activity, and resistance in the capacity to transfer and absorb iron and phosphorus. The significant cell wall and membrane changes caused by these mechanisms were the primary explanations for the biolethal activity of TiO_2 NPs (López de Dicastillo et al. 2021).

Metal oxide NPs produced using green synthesis exhibited comparable antibacterial efficacy at very low concentrations against both Gram-positive and Gram-negative bacteria. The ability of metal oxide NPs to exhibit antibacterial activity against a number of common bacterial agents enhanced our optimism that we would be able to employ them as a substitute antibacterial agent in the therapy of infectious disorders (Akhter et al. 2019). The strongest antibacterial activity against the five chosen MDR isolates was demonstrated by Ag NPs, followed by TiO_2 NPs, out of the four distinct biosynthesized metal NPs that were investigated (Alsamman et al. 2023).

The antimicrobial activity of various compositions of NPs against different microbial species is summarized in Table 19.2.

Table 19.2 Different compositions of NPs with their antibacterial activity

Sr. no	NPs	Source	Size of NPs	Antibacterial resistance against	Zone of inhibition (mm)	MIC (μg/mL)	Reference
1	Silver NPs	*Mimusops elengi, L., leaf extracts*	55–83 nm	*Klebsiella pneumonia*	*Control = 00* *5 μg = 8* 10 μg = 18	–	Prakash et al. (2013)
				Staphylococcus aureus	Control = 00 5 μg = 6.75 10 μg = 8 15 μg = 10		
				Micrococcus luteus	Control = 00 5 μg = 7 10 μg = 8.25 15 μg = 11		
2	Silver NPs	*Tribulus terrestris L.*	16–28 nm	*Staphylococcus aureus*	–	11	Gopinath et al. (2015)
				Escherichia coli		8.25	
				Streptococcus pyogens		10.75	
				Pseudomonas aeruginosa		7.5	
				Bacillus subtilis		11.5	
				Proteus vulgaris		8.75	
3	Silver NPs	*Tridax procumbens*	20–40 nm	*S. saprophyticus*	21.0 ± 1.7	–	Al-Ansari et al. (2020)
				P. aeruginosa	18.0 ± 1.3		
				K. pneumoniae	18.0 ± 0. 09		
				E. coli	17.0 ± 1.70		
4	Silver NPs	*Ocimum gratissimum leaf extract*	31 nm	*E. coli*	–	4	Das et al. (2017)
5	Silver NPs	*Cinnamomum tamala (Tejpata)*	25–30 nm	*P. aeruginosa*	7.67 ± 0.577	–	Al Mashud et al. (2022)

(continued)

Table 19.2 (continued)

Sr. no	NPs	Source	Size of NPs	Antibacterial resistance against	Zone of inhibition (mm)	MIC (μg/mL)	Reference
6	Zinc oxide NPs	*Aristolochia indica*	22.5 nm	*S. aureus*	–	200	Steffy et al. (2018)
				P. aeruginosa		100	
				E. coli		100	
7	Titanium dioxide NPs	*Trigonella foenum-graecum*	20 nm	*Enterococcus faecalis*	11.4 ± 0.5	–	Mills and Le Hunte (1997)
				Staphylococcus aureus	11.2 ± 0.4		
				Streptococcus faecalis	11.6 ± 0.3		
				Bacillus subtilis	10.5 ± 0.4		
				Yersinia enterocolitica	10.6 ± 0.5		
				Proteus vulgaris	8.5 ± 0.4		
				Escherichia coli	8.5 ± 0.4		
				Pseudomonas aeruginosa	10.3 ± 0.4		
				Klebsiella pneumoniae	10.2 ± 0.8		
				Candida albicans	10.2 ± 0.5		
8	Titanium dioxide NPs	*Aloe barbadensis mill*	20 nm	*Pseudomonas aeruginosa PAO1*	11 and 14	31.25	Rajkumari et al. (2019)
9	Titanium dioxide NPs	*Psidium guajava*	32.58 nm	*A. hydrophila*	17	–	Santhoshkumar et al. (2014)
				P. mirabilis	20		
				E. coli	23		
				S. aureus	25		
				P. aeruginosa	19		
10	Titanium dioxide NPs	*Azadirachta indica* leaf extract	25–87 nm	*S. aureus*	–	20.83	Thakur et al. (2019)
				K. pneumoniae		16.66	
				B. subtilis		25	
				S. typhi		10.42	
				E. coli		10.42	

19.5 Conclusion

In the field of antimicrobial research, using green-synthesized NPs to fight MDR bacteria is a viable and sustainable strategy. These environmentally friendly processes and biocompatible materials were used to create the NPs, which have a number of benefits in the battle against diseases with medication resistance. Importantly, the green synthesis techniques are economical and environmentally friendly since they frequently use plant extracts, microbes, or other natural sources. They adhere to the principles of green chemistry by reducing the use of hazardous chemicals and energy-intensive procedures. Moreover, the green NPs have outstanding antibacterial qualities. Due to their large surface area and small size, they engage with bacterial cells more effectively and therefore exert potent antimicrobial activity against a variety of strains. Additionally, they frequently use numerous mechanisms in their mode of action, which minimizes the chance that resistance will arise. In addition, compared to conventional antibiotics, these NPs are often biocompatible and provide less health concerns to people. This reduces their potential for adverse effects and decreases toxicity, making them a safer alternative for therapeutic applications. A broad range of NPs with various characteristics is also produced by the variety of natural sources accessible for green synthesis, allowing modification for particular applications and bacterial strains. This adaptability is a crucial tool in the continuing fight with microorganisms that resist many drugs. Transforming this technology from the laboratory to clinical use requires extensive safety and effectiveness testing, regulatory clearances, and standardization of synthesis procedures. Furthermore, it is essential to understand the abiding effects of NP release in the ecosystem. In conclusion, green-synthesized NPs provide a viable and effective option in the struggle against bacteria that are resistant to a variety of drugs. These NPs have the potential to change the area of antimicrobial medicine, addressing one of the most significant global health issues.

19.6 Future Aspects

For the future of biotechnology and health care, the use of green-synthesized NPs to battle MDR bacteria holds significant potential. Green synthesis, which includes creating NPs from natural materials such as plant extracts and microbes, has a number of benefits, including the possibility of scalable manufacturing. NPs created through green synthesis are constantly being improved in terms of their antibacterial capabilities. This involves simplifying the manufacturing procedure, modifying surface changes, and altering NP size and shape to improve their efficacy against MDR bacteria. An important area of focus should be designing NPs with precise targeting abilities. NPs functionalized with ligands or antibodies can help to ensure that they attach to bacteria only, causing less harm to healthy cells and increasing therapy effectiveness. It may be possible to circumvent bacterial resistance by combining conventional antibiotics with green-synthesized NPs, which may have a synergistic impact. This strategy can improve the efficiency of already available antibiotics and lessen the possibility of the emergence of new resistance. In order to defend themselves from antibiotics, MDR bacteria frequently create protective biofilms. In order to break up these biofilms, green-synthesized NPs may be used, making bacteria easier to therapy. To tackle MDR bacteria from many sides, researchers are looking into the possibility of integrating green-synthesized NPs with other treatment modalities, such as phage therapy or immunotherapy. In conclusion, green-synthesized NPs present a viable way to tackle the worldwide problem of MDR microorganisms. To make these NPs a useful weapon in the struggle against drug-resistant diseases, further research will be carried on to enhance them, investigate new applications, and overcome regulatory obstacles.

Acknowledgements Balaprasad Ankamwar thanks Inter-University Accelerator Centre, Aruna Asaf Ali Marg, New Delhi-110067, India, for the Research Project with financial support (Ref. IUAC/XIII.7/UFR-70316, Dated July 12, 2023).

References

AbdElrahman TK, Gebreel HMA, Youssef HA (2023) Assessing the effectiveness of green synthesized zinc oxide nanoparticles in controlling multidrug-resistant clinical bacteria. Indian J Microbiol 63(1):65–72. https://doi.org/10.1007/s12088-022-01048-3

Ahamad KM, Lone SA, Shahid M, Zeyad MT, Syed A, Ehtram A, Elgorban AM, Verma M, Danish M (2023) Phytogenically synthesized zinc oxide nanoparticles (ZnO-NPs) potentially inhibit the bacterial pathogens: in vitro studies. Toxics 11(5):452. https://doi.org/10.3390/toxics11050452

Aileni K, Kambalyal P, Shanmugasundaram K, Rajesh V, Donthula S, Patil S (2018) Comparative evaluation of antimicrobial efficacy of silver, titanium dioxide and zinc oxide nanoparticles against *Streptococcus mutans*. Pesquisa Brasileira em Odontopediatria e Clínica Integrada 18(1):4150. https://doi.org/10.4034/PBOCI.2018.181.88

Akhter SMH, Mohammad F, Ahmad S (2019) *Terminalia belerica* mediated green synthesis of nanoparticles of copper, iron and zinc metal oxides as the alternate antibacterial agents against some common pathogens. BioNanoScience 9:365–372. https://doi.org/10.1007/s12668-019-0601-4

Al Mashud MA, Moinuzzaman M, Hossain MS, Ahmed S, Ahsan G, Reza A, Anwar RR, Uddin MH, Momin MA, Hena Mostofa Jamal MA (2022) Green synthesis of silver nanoparticles using *Cinnamomum tamala* (Tejpata) leaf and their potential application to control multidrug resistant *Pseudomonas aeruginosa* isolated from hospital drainage water. Heliyon 8(7):e09920. https://doi.org/10.1016/j.heliyon.2022.e09920

Al-Ansari MM, Dhasarathan P, Ranjitsingh AJ, Al Humaid LA (2020) Challenging multidrug-resistant urinary tract bacterial isolates via bio-inspired synthesis of silver nanoparticles using the inflorescence extracts of *Tridax procumbens*. J King Saud Univ Sci 32:3145–3152. https://doi.org/10.1016/j.jksus.2020.08.028

Alsamman AM, Khedr M, Kabary HA, El-Sehrawy MH (2023) Elimination of pathogenic multidrug resistant isolates through different metal oxide nanoparticles synthesized from organic plant and microbial sources. Microb Pathog 178:106055. https://doi.org/10.1016/j.micpath.2023.106055

Ananda AP, Manukumar HM, Krishnamurthy NB, Nagendra BS, Savitha KR (2019) Assessment of antibacterial efficacy of a biocompatible nanoparticle PC@AgNPs against *Staphylococcus aureus*. Microb Pathog 126:27–39. https://doi.org/10.1016/j.micpath.2018.10.029

Anupam R, Onur B, Sudip S, Amit KM, Deniz Y (2019) Green synthesis of silver nanoparticles: biomolecule-nanoparticle organizations targeting antimicrobial activity. RSC Adv 9:2673–2702. https://doi.org/10.1039/C8RA08982E

Arshad H, Misbah S, Pasha U, Sadaf S (2022) Synthesis of Aloe vera-conjugated silver nanoparticles for use against multidrug-resistant microorganisms. Electron J Biotechnol 55:55–64. https://doi.org/10.1016/j.ejbt.2021.11.003

Bansal V, Rautaray D, Bharde A et al (2005) Fungus-mediated biosynthesis of silica and titania particles. J Mater Chem 15(26):2583. https://doi.org/10.1039/b503008k

Blake DM, Maness PC, Huang Z, Wolfrum EJ, Huang J, Jacoby WA (1999) Application of the photocatalytic chemistry of titanium dioxide to disinfection and the killing of cancer cells. Sep Purif Rev 28(1):1–50. https://doi.org/10.1080/03602549909351643

Das B, Dash SK, Mandal D, Ghosh TK, Chattopadhyay S, Tripathy S, Das S, Dey SK, Das D, Roy S (2017) Green synthesized silver nanoparticles destroy multidrug resistant bacteria via reactive oxygen species mediated membrane damage. Arab J Chem 10:862–876. https://doi.org/10.1016/j.arabjc.2015.08.008

Gehring J, Trepka B, Klinkenberg N, Bronner H, Schleheck D, Polarz S (2016) Sunlight-triggered nanoparticle synergy: teamwork of reactive oxygen species and nitric oxide released from mesoporous *Organo silica* with advanced antibacterial activity. J Am Chem Soc 138(9):3076–3084. https://doi.org/10.1021/jacs.5b12073

Gharpure S, Ankamwar B (2023) *Albiziz lebbeck* mediated ZnO photosynthesis and their nanoantibacterial and biocompatibility studies. Appl Nanosci 13(6):4497–4513. https://doi.org/10.1007/s13204-022-02725-x

Gharpure S, Kirtiwar S, Palwe S, Akash A, Ankamwar B (2019) Non-antibacterial as well as non-anticancer activity of flower extract and its biogenous silver nanoparticles. Nanotechnology 30(19):195701. https://doi.org/10.1088/1361-6528/ab011a

Gharpure S, Yadwade R, Ankamwar B (2022a) Non-antimicrobial and non-anticancer properties of ZnO nanoparticles biosynthesized using different plant parts of *Bixa orellana*. ACS Omega 7(2):1914–1933. https://doi.org/10.1021/acsomega.1c05324

Gharpure S, Yadwade R, Ankamwar B (2022b) *Lactuca virosa* leaf-mediated biosynthesis of zinc oxide nanoparticles and estimation of antimicrobial and anticancer activities. Chem Lett 51(7):739–743. https://doi.org/10.1246/cl.220071

Gharpure S, Yadwade R, Chakraborty B, Makar R, Chavhan P, Kamble S, Pawar P, Ankamwar B (2022c) Bioactive properties of ZnO nanoparticles synthesized using *Cocos nucifera* leaves. 3 Biotech 12(2):45. https://doi.org/10.1007/s13205-022-03110-9

Gopinath V, Priyadarshini S, Venkatkumar G, Muthupandian S, Mubarak D (2015) *Tribulus terrestris* leaf mediated biosynthesis of stable antibacterial silver nanoparticles. Pharm Nanotechnol 3:26–34. https://doi.org/10.2174/2211738503666150626160843

Gow NAR, Latge JP, Munro CA (2017) The fungal cell wall: structure, biosynthesis, and function. Microbiol Spectr 5(3). https://doi.org/10.1128/microbiolspec.FUNK-0035-2016

Hochvaldová L, Večeřová R, Kolář M, Prucek R, Kvítek L, Lapčík L, Panáček A (2022) Antibacterial nanomaterials: upcoming hope to overcome antibiotic resistance crisis. Nanotechnol Rev 11(1):1115–1142. https://doi.org/10.1515/ntrev-2022-0059

Hoffmann MR, Martin ST, Choi W, Bahnemann DW (1995) Environmental applications of semiconductor photocatalysis. Chem Rev 95:69–96. https://doi.org/10.1021/cr00033a004

Jacoby WA, Maness PC, Wolfrum EJ, Blake DM, Fennel JA (1998) Mineralization of bacterial cell mass on a photocatalytic surface in air. Environ Sci Technol 32(17):2650–2653. https://doi.org/10.1021/es980036f

Jayaseelan C, Rahuman AA, Roopan SM, Kirthi AV, Venkatesan J, Kim SK, Iyappan M, Siva C (2013) Biological approach to synthesize TiO2 nanoparticles using *Aeromonas hydrophila* and its antibacterial activity. Spectrochim Acta A Mol Biomol Spectrosc 107:82–89. https://doi.org/10.1016/j.saa.2012.12.083

Jiang X, Yang L, Liu P, Li X, Shen J (2010) The photocatalytic and antibacterial activities of neodymium and iodine doped TiO(2) nanoparticles. Colloids Surf B Biointerfaces 79(1):69–74. https://doi.org/10.1016/j.colsurfb.2010.03.031

Jothi Lakshmi N, Anandakumar S, Sampathkumar V (2022) Green synthesis and characterization of ZnO nanoparticles from Manihot (cassava) peel and their antibacterial study. Global NEST J 25(3):56–62. https://doi.org/10.30955/gnj.004541

Karnwal A, Kumar G, Pant G, Hossain K, Ahmad A, Alshammari MB (2023) Perspectives on usage of functional nanomaterials in antimicrobial therapy for antibiotic-resistant bacterial infections. ACS Omega 8(15):13492–13508. https://doi.org/10.1021/acsomega.3c00110

Khan ST, Musarrat J, Al-Khedhairy AA (2016) Countering drug resistance, infectious diseases, and sepsis using metal and metal oxides nanoparticles: current status. Colloids Surf B Biointerfaces 146:70–83. https://doi.org/10.1016/j.colsurfb.2016.05.046

Khezerlou A, Alizadeh-Sani M, Azizi-Lalabadi M, Ehsani A (2018) Nanoparticles and their antimicrobial properties against pathogens including bacteria, fungi, parasites and viruses. Microb Pathog 123:505–526. https://doi.org/10.1016/j.micpath.2018.08.008

Kirtiwar S, Gharpure S, Balaprasad A (2019) Effect of nutrient media on antibacterial activity of silver nanoparticles synthesized using *Neolamarckia cadamba*. J Nanosci Nanotechnol 19(4):1923–1933. https://doi.org/10.1166/jnn.2019.16117

Kiwi J, Sami R (2018) Mechanisms of the antibacterial effects of TiO2–FeOx under solar or visible light: Schottky barriers versus surface Plasmon resonance. Coatings 8(11):391. https://doi.org/10.3390/coatings8110391

Kumari M, Pandey S, Giri VP, Bhattacharya A, Shukla R, Mishra A, Nautiyal CS (2017) Tailoring shape and size of biogenic silver nanoparticles to enhance antimicrobial efficacy against MDR bacteria. Microb Pathog 105:346–355. https://doi.org/10.1016/j.micpath.2016.11.012

Lemire JA, Harrison JJ, Turner RJ (2013) Antimicrobial activity of metals: mechanisms, molecular targets and applications. Nat Rev Microbiol 11(6):371–384. https://doi.org/10.1038/nrmicro3028

Liu P, Duan W, Wang Q, Li X (2010) The damage of outer membrane of *Escherichia coli* in the presence of TiO_2 combined with UV light. Colloids Surf B Biointerfaces 78(2):171–176. https://doi.org/10.1016/j.colsurfb.2010.02.024

López de Dicastillo C, Guerrero Correa M, Martínez B, Streitt C, José Galotto M (2021) Antimicrobial effect of titanium dioxide nanoparticles. Intech Open, London. https://doi.org/10.5772/intechopen.90891

Mateo EM, Jiménez M (2022) Silver nanoparticle-based therapy: can it be useful to combat multidrug resistant bacteria. Antibiotics 11(9):1205. https://doi.org/10.3390/antibiotics11091205

Mba IE, Nweze EI (2021) Nanoparticles as therapeutic options for treating multidrug-resistant bacteria: research progress, challenges, and prospects. World J Microbiol Biotechnol 37(6):108. https://doi.org/10.1007/s11274-021-03070-x

Mills A, Le Hunte S (1997) An overview of semiconductor photocatalysis. J Photochem Photobiol A 108:1–35. https://doi.org/10.1016/S1010-6030(97)00118-4

Morgan DJ, Okeke IN, Laxminarayan R, Perencevich EN, Weisenberg S (2011) Non-prescription antimicrobial use worldwide: a systematic review. Lancet Infect Dis 11(9):692–701. https://doi.org/10.1016/S1473-3099(11)70054-8

Morowvat MH, Kazemi K, Jaberi MA, Amini A, Gholami A (2023) Biosynthesis and antimicrobial evaluation of zinc oxide nanoparticles using *Chlorella vulgaris* biomass against multidrug-resistant pathogens. Materials 16(2):842. https://doi.org/10.3390/ma16020842

Oves M, Rauf MA, Hussain A, Qari HA, Khan AAP, Muhammad P, Rehman MT, Alajmi MF, Ismail IIM (2019) Antibacterial silver nanomaterial synthesis from *Mesoflavibacter zeaxanthinifaciens* and targeting biofilm formation. Front Pharmacol 10:801. https://doi.org/10.3389/fphar.2019.00801

Pavlović VP, Vujančević JD, Mašković P, Ćirković J, Papan JM, Kosanović D, Dramićanin MD, Petrović PB, Vlahović B, Pavlović VB (2019) Structure and enhanced antimicrobial activity of mechanically activated nano TiO_2. J Am Ceram 102:7735–7745. https://doi.org/10.1111/jace.16668

Prakash P, Gnanaprakasam P, Emmanuel R, Arokiyaraj S, Saravanan M (2013) Green synthesis of silver nanoparticles from leaf extract of *Mimusops elengi*, Linn. for enhanced antibacterial activity against multi drug resistant clinical isolates. Colloids Surf B Biointerfaces 108:255–259. https://doi.org/10.1016/j.colsurfb.2013.03.017

Punniyakotti P, Panneerselvam P, Perumal D, Aruliah R, Angaiah S (2020) Anti-bacterial and anti-biofilm properties of green synthesized copper nanoparticles from *Cardiospermum hali-*

cacabum leaf extract. Bioprocess Biosyst Eng 43(9):1649–1657. https://doi.org/10.1007/s00449-020-02357-x

Qais FA, Shafiq A, Ahmad I, Husain FM, Khan RA, Hassan I (2020) Green synthesis of silver nanoparticles using *Carum copticum*: assessment of its quorum sensing and biofilm inhibitory potential against gram negative bacterial pathogens. Microb Pathog 144:104172. https://doi.org/10.1016/j.micpath.2020.104172

Qi M, Chi M, Sun X, Xie X, Weir MD, Oates TW, Zhou Y, Wang L, Bai Y, Xu HH (2019) Novel nanomaterial-based antibacterial photodynamic therapies to combat oral bacterial biofilms and infectious diseases. Int J Nanomedicine 14:6937–6956. https://doi.org/10.2147/IJN.S212807

Railean-Plugaru V, Pomastowski P, Rafinska K, Wypij M, Kupczyk W, Dahm H, Jackowski M, Buszewski B (2016) Antimicrobial properties of biosynthesized silver nanoparticles studied by flow cytometry and related techniques. Electrophoresis 37(56):752761. https://doi.org/10.1002/elps.201500507

Rajakumar G, Rahuman AA, Roopan SM, Khanna VG, Elango G, Kamaraj C, Zahir AA, Velayutham K (2012) Fungus-mediated biosynthesis and characterization of TiO_2 nanoparticles and their activity against pathogenic bacteria. Spectrochim Acta A Mol Biomol Spectrosc 91:23–29. https://doi.org/10.1016/j.saa.2012.01.011

Rajkumari J, Magdalane CM, Siddhardha B, Madhava J, Ramalingam G, Al-Dhabi NA, Arasu MV, Ghilan AKM, Duraipandiayan V, Kaviyarasu K (2019) Synthesis of titanium oxide nanoparticles using Aloe barbadensis mill and evaluation of its antibiofilm potential against *Pseudomonas aeruginosa* PAO1. Photochem Photobiol 201:111667. https://doi.org/10.1016/j.jphotobiol.2019.111667

Saee G, Rachana Y, Shuana M, Balaprasad A (2023) *Prunus dulcis*-mediated zinc oxide nanoparticles: synthesis, characterization, and an evaluation of their biocidal properties. Phys Scr 98:105004. https://doi.org/10.1088/1402-4896/acf3a7

Santhoshkumar T, Rahuman AA, Jayaseelan C, Rajakumar G, Marimuthu S, Kirthi AV, Velayutham K, Thomas J, Venkatesan J, Kim SK (2014) Green synthesis of titanium dioxide nanoparticles using *Psidium guajava* extract and its antibacterial and antioxidant properties. Asian Pac J Trop Med 7(12):968–976. https://doi.org/10.1016/S1995-7645(14)60171-1

Singh A, Gautam PK, Verma A, Singh V, Shivapriya PM, Shivalkar S, Sahoo AK, Samanta SK (2020) Green synthesis of metallic nanoparticles as effective alternatives to treat antibiotics resistant bacterial infections: a review. Biotechnol Rep 31(25):e00427. https://doi.org/10.1016/j.btre.2020.e00427

Slavin YN, Asnis J, Häfeli UO, Bach H (2017) Metal nanoparticles: understanding the mechanisms behind antibacterial activity. J Nanobiotechnol 15(1):65. https://doi.org/10.1186/s12951-017-0308-z

Steffy K, Shanthi G, Maroky AS, Selvakumar S (2018) Enhanced antibacterial effects of green synthesized ZnO NPs using *Aristolochia indica* against multi-drug resistant bacterial pathogens from diabetic foot ulcer. J Infect Public Health 11(4):463471. https://doi.org/10.1016/j.jiph.2017.10.006

Subhapriya S, Gomathipriya P (2018) Green synthesis of titanium dioxide (TiO_2) nanoparticles by *Trigonella foenum-graecum* extract and its antimicrobial properties. Microb Pathog 116:215–220. https://doi.org/10.1016/j.micpath.2018.01.027

Thakur B, Kumar A, Kumar D (2019) Green synthesis of titanium dioxide nanoparticles using *Azadirachta indica* leaf extract and evaluation of their antibacterial activity. S Afr J Bot 124:223–227. https://doi.org/10.1016/j.sajb.2019.05.024

Vijayakumar S, Vaseeharan B, Malaikozhundan B, Gopi N, Ekambaram P, Pachaiappan R, Velusamy P, Murugan K, Benelli G, Suresh K, Uriyanarayana Moorthy M (2017) Therapeutic effects of gold nanoparticles synthesized using Musa paradisiaca peel extract against multiple antibiotic resistant enterococcus faecalis biofilms and human lung cancer cells (A549). Microb Pathog 102:173–183. https://doi.org/10.1016/j.micpath.2016.11.029

Wang L, Hu C, Shao L (2017) The antimicrobial activity of nanoparticles: present situation and prospects for the future. Int J Nanomedicine 12:1227–1249. https://doi.org/10.2147/IJN.S121956

Wu P, Xie R, Imlay K, Shang JK (2010) Visible-light-induced bactericidal activity of titanium dioxide codoped with nitrogen and silver. Environ Sci Technol 44(18):6992–6997. https://doi.org/10.1021/es101343c

Yang X, Ye W, Qi Y, Ying Y, Xia Z (2021) Overcoming multidrug resistance in bacteria through antibiotics delivery in surface-engineered nano-cargos: recent developments for future nano-antibiotics. Front Bioeng Biotechnol 9:696514. https://doi.org/10.3389/fbioe.2021.696514

Yassin MT, Al-Askar AA, Maniah K, Al-Otibi FO (2022) Green synthesis of zinc oxide nano-crystals utilizing *Origanum majorana* leaf extract and their synergistic patterns with Colistin against multidrug-resistant bacterial strains. Crystals 12(11):1513. https://doi.org/10.3390/cryst12111513

Yousef A, Abu Elghait M, Barghoth MG, Elazzazy AM, Desouky SE (2022) Fighting multidrug-resistant *Enterococcus faecalis* via interfering with virulence factors using green synthesized nanoparticles. Microb Pathog 173(Pt A):105842. https://doi.org/10.1016/j.micpath.2022.105842

Zainab NJ, Olfat AM, Faisa LA (2019) Studying the effect of synthesized nano-titanium dioxide via two phases on the *Pseudomonas aeruginosa* and *Portus* bacteria as antimicrobial agents. Int J Nanoelectron Mater 12(3):329338. http://dspace.unimap.edu.my/xmlui/handle/123456789/61117i

Zhu Z, Huanhuan C, Sun DW (2018) Titanium dioxide (TiO_2) photocatalysis technology for non-thermal inactivation of microorganisms in foods. Trends Food Sci Technol 75:23. https://doi.org/10.1016/j.tifs.2018.02.018

Antimicrobial Nanofibers to Fight Multidrug-Resistant Bacteria

20

Carolina A. M. Ferreira, Sara F. C. Guerreiro, Tatiana Padrão, Nuno M. F. Alves, and Juliana R. Dias

C. A. M. Ferreira
CDRSP—Centro para o Desenvolvimento Rápido e Sustentado do Produto, Instituto Politécnico de Leiria, Marinha Grande, Leiria, Portugal

ICBAS—Instituto de Ciências Biomédicas Abel Salazar, Universidade do Porto, Porto, Portugal

CECA—Centro de Estudos de Ciência Animal, Instituto de Ciências, Tecnologias e Agroambiente (ICETA) da Universidade do Porto, Praça Gomes Teixeira, Apartado, Porto, Portugal

MARE—Centro de Ciências do Mar e do Ambiente, Instituto Politécnico de Leiria, Peniche, Portugal

S. F. C. Guerreiro
CDRSP—Centro para o Desenvolvimento Rápido e Sustentado do Produto, Instituto Politécnico de Leiria, Marinha Grande, Leiria, Portugal

i3S—Instituto de Investigação e Inovação em Saúde, Universidade do Porto, Porto, Portugal

Departamento de Física Médica, Instituto Português de Oncologia (IPO-Porto), Porto, Portugal

FEUP—Faculdade de Engenharia, Universidade do Porto, Porto, Portugal

T. Padrão
CDRSP—Centro para o Desenvolvimento Rápido e Sustentado do Produto, Instituto Politécnico de Leiria, Marinha Grande, Leiria, Portugal

i3S—Instituto de Investigação e Inovação em Saúde, Universidade do Porto, Porto, Portugal

FEUP—Faculdade de Engenharia, Universidade do Porto, Porto, Portugal

INEB—Instituto de Engenharia Biomédica, Universidade do Porto, Porto, Portugal

N. M. F. Alves · J. R. Dias (✉)
CDRSP—Centro para o Desenvolvimento Rápido e Sustentado do Produto, Instituto Politécnico de Leiria, Marinha Grande, Leiria, Portugal
e-mail: juliana.dias@ipleiria.pt

M. Y. Wani et al. (eds.), *Nanotechnology Based Strategies for Combating Antimicrobial Resistance*, https://doi.org/10.1007/978-981-97-2023-1_20

Abstract

Multidrug-resistant bacteria are among the most challenges in healthcare systems. According to the World Health Organization, these bacteria that belong to the critical and high-priority bacteria groups are resistant to carbapenem or multiple antibiotics, respectively. They are commonly associated with bone, skin, and soft tissue infections, and current treatment strategies involve the use of oral and intravenous multiple antibiotics. Consequently, infected injured sites benefit from combined antimicrobial and regenerative action during the tissues' repair. The high demand for effective solutions led to the development of biocompatible and antibacterial polymeric membranes for local therapy. These highly porous membranes are composed by nanofibers' network providing a matrix for cellular proliferation and local release of antimicrobial agents. Nowadays, there are several techniques to produce nanofibers like template synthesis, self-assembly, freeze-drying, melt-blowing, electrospinning, among others. Electrospinning emerges in the production of nanofibers with intrinsic characteristics. The high surface area-to-volume ratio and the nanometric diameters allow mimicking innate tissues increasing biomedical fields' attention. Therefore, electrospun nanofibers become favoured as drug delivery systems, improving drugs' dissolution rates while maximizing their action. Additionally, they permits the loading several drugs in higher concentrations and control their release rate without associated toxicity. Hence, antibiotics drugs have been the most common encapsulated medicinal compounds using numerous polymers combinations as carriers. Therefore, the versatility of electrospinning led to advanced approaches which combine complex strategies to locally provide multiple antimicrobial agents through the same platform and consider different release profiles. Overall, further developments should continue to complex delivery systems to meet clinical requirements.

Keywords

Multidrug-resistant bacteria · Antimicrobial nanofibers · Antibiotics · Polymers · Antimicrobial agents

20.1 Carolina A. M. Ferreira, Sara F. C. Guerreiro, and Tatiana Padrão have contributed equally to this work.Introduction

Multidrug-resistant bacteria have been identified as a leading cause of death worldwide, and drug resistance was estimated to be responsible for 1.3 deaths in 2019 (Murray et al. 2023). The World Health Organization's (WHO) list of high-priority pathogens includes *Acinetobacter baumannii* (*A. baumannii*), *Enterococcus faecium* (*E. faecium*), *Enterobacter* spp., *Klebsiella pneumoniae* (*K. pneumoniae*), *Pseudomonas aeruginosa* (*P. aeruginosa*), and *Staphylococcus aureus* (*S. aureus*) (World Health Organization 2017). However, due to its high mortality and prevalence, the concerns are extended to other strains such as *Escherichia coli* (*E. coli*), which was included in the list of the six leading pathogens (Murray et al. 2023). These bacteria can cause bloodstream infections, lower respiratory infections, intra-abdominal infections, urinary infections, tuberculosis, and skin infections (Murray et al. 2023; Ding et al. 2022). From local to generalized infections, the six pathogens named ESKAPEE were responsible for 9.29 k deaths in 2019 (Murray et al. 2023). The high mortality is associated with the lack of efficient treatments due to the drug resistance acquired as the result of inadequate and long-term use of antibiotics (Parmanik et al. 2023).

In fact, the treatment of infections caused by ESKAPEE bacteria focuses on the administration of antibiotics, which most of the time prove to be ineffective due to the resistance acquired by the bacteria over time (Parmanik et al. 2023; Pulingam et al. 2022). Therefore, the search for solutions in the health field has led researchers to explore new organic and inorganic materials with antibacterial activity (Saidin et al. 2021). Most of these materials are effective mainly against ESKAPEE bacteria such as *S. aureus* and *E. coli* (Chen et al. 2015; Fumakia and Ho 2016). Considering these findings, the alternatives for the treatment of infections caused by multi-resistant bacteria are increasing and novel antimicrobial materials are being explored (Ding et al. 2019). At the same time, the schedule and routes of administration of antibacterial agents also play a key role in the effectiveness of treatments and novel solutions have been explored (Landersdorfer et al. 2023). Intravenous administration or oral administration has been the favored routes of administration; however, the recent advances in science have opened the door to nanotechnology and to new related approaches through delivery systems for drug-controlled release (Ding et al. 2022; Landersdorfer et al. 2023). Among the various drug-delivery systems, this chapter thoroughly explores the potential of nanofibers as a promising solution in the fight against multidrug-resistant bacteria (Topcu et al. 2021).

In this context, nanofibers produced by electrospinning stand out due to their remarkable physical and chemical properties, which allow them to compete with other nanofiber production methods (Gao et al. 2019; Merzougui et al. 2022). These structures provide a surface for interaction with antimicrobial agents, enhancing their efficacy with diameters that range from a few nanometers to a few micrometers and a high surface area-to-volume ratio (Song et al. 2017). In addition, the versatility in the selection of materials allows electrospun fibers to be customized to meet specific requirements, such as the controlled release of drugs (Gaydhane et al. 2023). These characteristics provide nanofibers a significant advantage over other conventional structures, particularly for the treatment of local infections. In this chapter, the principles behind the electrospinning process will be covered in detail, as well as the materials used for the production of nanofibers, the antibacterial agents incorporated, and the functionalization strategies adopted to give or enhance antimicrobial properties of polymeric nanofibers. The recent advances that demonstrate the potential of nanofibers in combating challenging infections will also be discussed in this chapter.

20.2 Multidrug-Resistant Bacteria: General Overview

Multidrug-resistant bacteria have emerged as a significant global concern in healthcare systems, posing a serious threat to public health nowadays. These resilient pathogens, which include *A. baumannii*, *P. aeruginosa*, *K. pneumoniae*, *E. coli,* and *S. aureus*, are resistant to multiple antibiotics comprising carbapenems and third-generation cephalosporins and can cause severe and often deadly infections (World Health Organization 2017). Traditional antibiotics are losing their efficacy as bacteria develop resistance mechanisms, resulting from genetic and physiological factors transmitted both vertically within species and horizontally across different species (Murray et al. 2023). Each of these bacterial species presents unique challenges in the healthcare setting, and traditional treatment options often fall short in the face of their advancing resistance. A promising strategy in the fight against these bacteria has been to explore the field of nanofiber technology, since these nanoscale materials possess unique properties that make them excellent candidates for fighting multidrug-resistant bacterial infections. This section will provide an overview of multidrug-resistant bacteria including their main characteristics, prevalence, related infections, and how nanofibers can play a crucial role in the fight against multidrug-resistant bacteria.

20.2.1 *Acinetobacter baumannii*

Acinetobacter species are commonly distributed in the environment and readily contaminate the hospital environment, particularly *A. baumannii*, as it has a prevalence of cross-contamination in nosocomial infections and a relatively extended lifetime on human hands (Santajit and Indrawattana 2016). It is a gram-negative coccobacillus that can infect the urinary and respiratory systems among other locations (Santajit and Indrawattana 2016; Peleg et al. 2008). The WHO confirmed that *A. baumannii* is one of the most critical of the ESKAPE microorganisms that effectively escape the effects of antibacterial drugs (Boucher et al. 2009). Nowadays, it shows resistance to most first-line antibiotics, which is a serious problem, especially in intensive care units and surgical rooms (Vázquez-López et al. 2020). In recent years, the development of engineered nanofiber carriers to combat different types of ESKAPEE bacteria has been reported (Elsherbiny et al. 2022; Snetkov et al. 2022). Despite being a less frequently studied bacterium, some works have shown the potential of the nanofibers developed to fight this type of bacteria. *A. baumannii* bacterial growth was inhibited with the use of dual antibiotics (colistin/meropenem)—loaded chitosan (CS)/poly (vinyl alcohol) (PVA) nanofibers (Alizadeh et al. 2023). Moreover, electrospun nanofibers from cellulose acetate and lignin loaded with copper (Cu) complexes were active against *A. baumannii* to control diaper dermatitis (Elsherbiny et al. 2022).

20.2.2 *Pseudomonas aeruginosa*

P. aeruginosa is a gram-negative opportunistic bacterium responsible for severe infections in burns, ulcers, surgery wounds, and traumatic injuries (Ashfaq et al. 2017). Moreover, this bacterium can acquire several antibiotic resistance strategies, which are extended to different antibiotic classes such as piperacillin-tazobactam, ceftazidime, cefepime, aztreonam, meropenem, imipenem-cilastatin, ciprofloxacin (CIP), and levofloxacin (Kunz Coyne et al. 2022). In 2019, the global mortality rate per 100,000 population of multidrug-resistant *P. aeruginosa* was 22. 400 related to skin and subcutaneous bacterial infections, being one of the five leading deadly pathogens (Ikuta et al. 2022). Consequently, *P. aeruginosa* has been placed among the priority pathogens according to the WHO and the Center for Disease Control (CDC) since its treatment is challenging and it is responsible for life-threatening infections (Wood et al. 2023). For this reason, developing innovative antimicrobial wound dressings is urgently needed. Recent advances in nanotechnology and materials engineering have led to the development of advanced antimicrobial electrospun wound dressings and drug delivery systems (DDSs). Considering the context, several authors have developed antimicrobial nanofibers capable to fight *P. aeruginosa* (Kunz Coyne et al. 2022; Khosravimelal et al. 2021; Varshosaz et al. 2020; Sarhan et al. 2021).

20.2.3 *Klebsiella pneumoniae*

K. pneumoniae is a gram-negative bacteria that can cause different infections, such as pneumonia, gastrointestinal infections, urinary tract, bloodstream infections, and wound or surgical site infections (Assoni et al. 2021). Besides, *K. pneumoniae* is linked to community-associated infections, including liver abscesses, endophthalmitis, and meningitis (Assoni et al. 2021). Additionally, infections by *K. pneumoniae* are common in long-term hospital stays, affecting infants, older people, and immunocompromised patients who require the use of ventilators and catheters (Fayez et al. 2021). One of the reasons for *K. pneumoniae* to be considered multidrug-resistant bacteria is because of its ability to form a biofilm, which allows it to attach to tissues and abiotic surfaces. Infections originating from biofilm-forming *K. pneumoniae* strains are more difficult to treat than other strains, since biofilms delay the bioavailability of antibiotics and restrict host immune defense (Assoni et al. 2021). However, the literature refers to several studies of nanofibers with antimicrobial activity to fight *K. pneumoniae* (Snetkov et al. 2022; Ikuta et al. 2022; Saikaew and Intasanta 2021; Saleemi et al. 2020).

20.2.4 *Escherichia coli*

A member of the *Enterobacteriaceae* bacterial family, *E. coli,* is one of the most frequent causes of several common bacterial infections in humans and animals. This gram-negative bacterium is the leading cause of enteritis, urinary tract infection, septicemia, neonatal meningitis, among other clinical infections (Allocati et al. 2013). Besides being an important member of the gastrointestinal tract of humans from birth, *E. coli* species contains various pathotypes that cause the aforementioned infections (Kaper et al. 2004), being responsible for more than 500,000 deaths in 2019, and is the leading cause of death in 37 countries (Ikuta et al. 2022). Several multidrug-resistant high-risk bacteria, such as some *E. coli* species, have developed in the last years, mostly as a result of the increasing selection associated with antibiotic usage (Paitan 2018). The development of nanofibers based on different materials and incorporating various drugs has been explored and tested against different types of *E. coli* species (Aytac et al. 2014; Kim and Lee 2014). Core-shell nanofibers containing vancomycin and imipenem/cilastatin were developed and tested against *E. coli* inducing its inhibition (Davani et al. 2021).

20.2.5 *Staphylococcus aureus*

S. aureus is a gram-positive bacterium with cells organized in characteristic clusters. It is a commensal organism that often resides asymptomatically in the skin and mucosa of mammals. Discovered in the 1880s, it is known as an opportunistic human pathogen, causing numerous diseases, from minor skin infections to severe bacteremia and necrotizing pneumonia, being the major cause of nosocomial and community infections (Abdelbary et al. 2017). A systematic analysis for the Global Burden of Disease Study in 2019, published in the Lancet, found that *S. aureus* was the common bacterial cause of death in 135 countries and was also responsible for most of the deaths among people older than 15 years worldwide (Ikuta et al. 2022). In the pre-antibiotic era, *S. aureus* infections were highly deadly. The introduction of penicillin in the 1940s improved outcomes, but penicillin-resistant strains emerged as early as 1942. By the late 1960s, over 80% of *S. aureus* in hospitals and communities were penicillin-resistant. Methicillin, designed to overcome the resistance, faced a similar fate since the 1960s, with *methicillin-resistant Staphylococcus aureus* (MRSA) strains becoming a global concern since the mid-1990s (Abdelbary et al. 2017). Nanofibers incorporating different antimicrobial agents have been explored against *S. aureus* in different studies (Fathi et al. 2020). Nano-CS/*Eucalyptus* oil/cellulose acetate nanofibers reduced 87% of *S. aureus* colony-forming units (Elbhnsawi et al. 2023).

20.3 Antibacterial Nanofiber Processing

Antibacterial nanofibers have been investigated as a promising alternative to conventional approaches for local control of bacterial infections (Albright et al. 2018). These fibrillar structures are typically composed of polymeric materials and can incorporate antibacterial agents to create composites with strong antimicrobial performance (Parham et al. 2022). Several techniques have been used to produce nanofibers including drawing, template synthesis, self-assembly, conventional electrospinning, and melt electrospinning (Alghoraibi and Alomari 2018). Conventional electrospinning has proven to be the most versatile, easy to scale up, and offers the highest surface area-to-volume ratio (Shetty et al. 2022). Other methods using polymer solutions present disadvantages compared to electrospinning: Template synthesis requires a mold to produce nanofibers of specific diameters, and self-assembly produces nanofibers from smaller molecules whose final shape is difficult to predict (Alghoraibi and Alomari 2018). In both cases, electrospinning proved to be a faster and more customized technique for obtaining nanofibers with reduced diameters (Alghoraibi and Alomari 2018). Moreover, solution electrospinning presents the same advantages compared to other variants of electrospinning such as melt electrospinning or nanofiber-producing techniques such as drawing, both involving polymer melt. The use of melt polymer production techniques avoids hazardous solvents; however, the nanofibers are discontinuous, and their diameter is typically larger than the nanofibers produced by solution-based methods (Bubakir et al. 2019). Therefore, the processing technique defines the nanofiber morphology namely in terms of diameter and spatial orientation, which influence the mechanical properties and the efficacy of the antibacterial behavior of the structure (De Cesare et al. 2019; Ji et al. 2021) (Fig. 20.1). Electrospinning offers continuous nanofibers with diameters ranging from 100 nm to 1 μm, and it is an easily scalable process with the capability to include antibacterial agents (Parham et al. 2022; Alghoraibi and

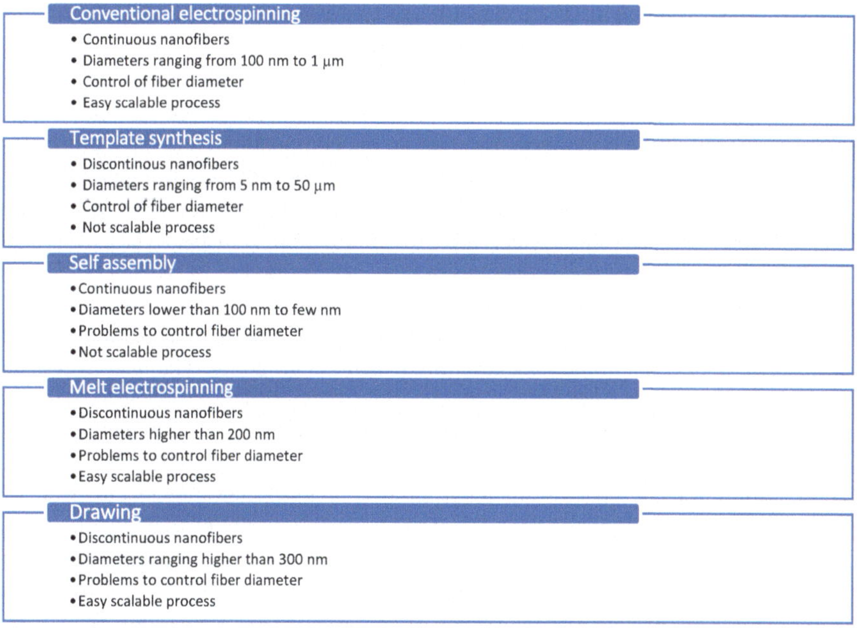

Fig. 20.1 Comparison of nanofiber characteristics produced by different techniques (Alghoraibi and Alomari 2018; Rasouli and Barhoum 2019; Dias et al. 2020)

Alomari 2018; Yang et al. 2018). Considering the antibacterial context, the potential applications of electrospun nanofibers are extended from wound healing to air filtration since they work as a barrier for bacteria intrusion while allowing gas exchanges, which are essential to promote cellular proliferation.

Electrospinning uses a high-voltage source to create an electric field that attracts a polymer solution in which the polymer is dissolved in a suitable solvent. A conventional electrospinning setup is composed of several components: a high-voltage source that creates the electrical field; a syringe pump that propels the solution to the tip of a needle; and a metal collector that collects the nanofibers (Gao et al. 2019). The polymer solution is continuously injected from a syringe pump, and drops emerge from the tip of the needle. Due to the attractive forces created, the droplets are stretched to form a jet that is deposited in the collector, which is located on the opposite side of the needle. The electrospun nanofibers deposited in the collector create a porous network that forms the final membrane (Gao et al. 2019). According to the type of collector used, the spatial organization of nanofibers is changed from random to oriented nanofibers; consequently, it is possible to mimic biological structures such as extracellular matrix (Parham et al. 2022; Yang et al. 2018). Besides, the morphology of electrospun nanofibers is influenced by the production parameters considered during the process, and these include applied voltage, flow rate, and distance between the needle and the collector. Additionally, properties of the polymer solution such as polymer concentration, molecular weight, viscosity, and conductivity and environmental conditions such as temperature and humidity can also impact the final morphology (Topcu et al. 2021; Alghoraibi and Alomari 2018; Abadi et al. 2022; Rasouli and Barhoum 2019; Dias et al. 2020).

The optimization of processing, production, and environmental parameters is crucial to achieve uniform fiber diameters, bead-free, and with specific biological and mechanical properties (Yang et al. 2018). For this reason, the nanofiber morphology, as well as the polymeric materials in the composition, and the methodologies used to incorporate the nanoparticles affect the antibacterial performance (Topcu et al. 2021). The porous structure of electrospun meshes and the reduced fiber diameter increase the surface area for interaction with bacteria and consequently the exposure to the antibacterial agents incorporated in the fibers (Shetty et al. 2022). Therefore, antibacterial agents delivered through electrospun nanofibers are typically more available compared to films or other similar structures that encapsulate the agents in the bulk structure (Shetty et al. 2022). The nanofibers can be composed of one or more polymeric materials with intrinsic antibacterial activity or, in the alternative, use polymers in the structure that contain antibacterial agents incorporated either in situ or after processing (Parham et al. 2022; Shetty et al. 2022). Different techniques for the incorporation of antibacterial agents in electrospun nanofibers have been explored including in situ methods such as blending, coaxial, and emulsion or after-processing methods such as surface adsorption (Buzgo et al. 2018). Considering in situ techniques, blending has been tested for a wide range of antibacterial agents (Zhou et al. 2022). During blending, these antibacterial agents are mixed with the polymer solution before the electrospinning process. As a result, the active ingredients are encapsulated and have a prolonged release profile (Buzgo et al. 2018). Also, coaxial electrospinning has been explored to obtain composite nanofibers with a core-shell structure (Lan et al. 2021; He et al. 2017). The two polymers are incorporated by using coaxially placed needles (Buzgo et al. 2018). With this strategy, different properties can be achieved between the core and the shell as an increase in mechanical performance in the core and antimicrobial activity in the shell (Huang et al. 2017). Core-shell nanofibers can also be produced using a single needle system by emulsion when two polymeric solutions in different liquid phases are combined (Buzgo et al. 2018). Besides, after-processing methods have been explored to incorporate antibacterial agents on nanofibers' surface (Shin et al. 2018; Pereira et al. 2020). Surface adsorption requires the electrospun nanofibers to be immersed in a solution of a dispersed antibacterial agent to create the composite material (Shetty et al. 2022). In this case, the release of the antibacterial agents is fast, which is suitable for short-term drug delivery (Buzgo et al. 2018).

The highly customized capacity of electrospinning allows several design modifications and set-up configurations including different collector types, needle types, multiple needles and needless systems, air-jet assisted, and centrifugal forces (Gao et al. 2019; Alghoraibi and Alomari 2018; Abadi et al. 2022; Buzgo et al. 2018; Huang et al. 2017). Therefore, the versatility of the electrospinning technique is the key to the complexity of structures possible to develop (Abadi et al. 2022). Electrospun nanofibers have advantages that proved to be more effective for antibacterial solutions than micro- or macrostructures (Yang et al. 2018). Apart from functionalization and their potential surface properties, the advantages of electrospun nanofibers include low weight, high flexibility and malleability, high surface area-to-volume ratio, and porous structure with space for cellular proliferation (Yang et al. 2018). These characteristics are valuable for electrospun nanofibers in tissue engineering applications that combine the need for tissue regeneration in injured areas with the control of bacterial infections.

20.4 Current Approaches of Nanofibers as Drug Delivery Systems

Currently, there are two approaches using electrospun nanofibers to combat multidrug-resistant bacteria, namely (1) conventional and (2) advanced approaches. Conventional approaches explore the incorporation of commercial antibiotics and organic and inorganic agents into the nanofibers. However, advanced approaches explore the use of multi-agent to combat co-infections.

20.4.1 Conventional Approaches

20.4.1.1 Commercial Antibiotics

The current approaches focus on nanofibers as DDSs due to their characteristics such as high surface-to-volume ratio and the possibility of drugs' entrapment in higher concentrations without associated toxicity (Tabakoglu et al. 2022). This approach allows to study and control drugs' release kinetics in the infected sites (Kajdič et al. 2019). The selection of appropriate materials for biomedical applications is crucial. For that reason, the most used drugs entrapped in nanofibers are antibiotics authorized by the United States Food and Drug Administration (FDA). Many antibiotics have been used in nanofibers, including rifampicin, vancomycin, gentamicin, metronidazole, amoxicillin, CIP, minocycline and doxycycline, and linezolid (Lopes Gama et al. 2023).

Rifampicin (RIF) is a common antibiotic that inhibits bacterial RNA polymerase, providing a bactericidal action against a broad spectrum of bacteria (Lopes Gama et al. 2023; Gruppuso et al. 2022). In this context, RIF was loaded at 0.1% w/v in polycaprolactone (PCL) electrospun meshes for ureteral regeneration. Additionally, the authors treated the PCL/RIF electrospun meshes with air plasma to support the adhesion and proliferation of urotheliocytes. Furthermore, this treatment reduced statistically the antimicrobial activity of the meshes against the bacteria of the ESKAPEE group compared with meshes without treatment. Moreover, the amount of RIF released by PCL/RIF and PCL/RIF/plasma was above the minimum inhibitory concentration (MIC) and minimum bactericidal concentration (MBC) of the antibiotic rifampicin alone. This reveals that the antimicrobial efficacy of electrospun meshes is higher than the antibiotic alone, suggesting that bacteria probably tend to be in contact with higher concentrations of antibiotic (Gruppuso et al. 2022). In fact, these similar findings were observed in another study. In the work of Fathi et al. (Fathi et al. 2020) alginate (ALG) 4 wt%, poly (ethylene oxide) (PEO) 4 wt%, and a copolymer surfactant, pluronic F-127 wt%, were used to produce nanofibers loaded with the antibiotic vancomycin (VAN) to control *Staphylococcus aureus*-induced skin infections. The in vivo antibacterial activity of VAN-loaded nanofibers was evaluated using an MRSA-infected skin abrasion model to represent chronic wound infection. After 24 h of treatment, the percentages of bacterial count in the infected abrasions were dramatically reduced in the treatment using vancomycin-loaded nanofibers or free drug solution, as compared to the control (untreated) group ($p < 0.01$). Furthermore, after 48 h and 72 h, both treatments continued to show a highly significant reduction of bacterial counts; additionally, bacterial counts were significantly decreased by the vancomycin-loaded nanofibers as compared to the free vancomycin solution ($0.05 > p \geq 0.01$) (Fathi et al. 2020).

The use of broad-spectrum aminoglycoside antibiotics, such as gentamycin (GEN), is already documented in the literature using nanofibers (Chen et al. 2017; Seo et al. 2022; Hadisi et al. 2020). Hadisi et al. (2020) produced an electrospun nano-biocomposite scaffold composed of silk fibroin (SF), hardystonite (HT—a calcium zinc silica), and GEN in different concentrations (1, 3, and 6 wt%) to combat implant-related bacterial infections and postsurgical complications. The results reveal that the addition of GEN in the SF-HT scaffolds significantly increased the zone of inhibition (ZOI) for *E. coli*, in the order of SF-HT-6GEN (2.35 ± 0.1 mm) > SF-HT-3GEN (1.22 ± 0.09 mm) > SFHT-1GEN (0.98 ± 0.1 mm). Similar results were found for *S. aureus* with the addition of GEN; the ZOI increased in the order of SF-HT-6GEN (2.87 ± 0.17 mm) > SF-HT-3GEN

(1.99 ± 0.09 mm) > SF-HT-1GEN (1.33 ± 0.11 mm). The antimicrobial activity of GEN-loaded nanofibers is dependent on antibiotic concentration and higher potency for gram-positive bacteria (Hadisi et al. 2020).

The use of nanofibers loaded with antibiotics can be applied to treat several types of infections. Bacterial vaginosis (BV), which is triggered by anaerobes, such as *Gardnerella vaginitis* and other associated pathogens forms biofilms. In this context, PCL/polyvinyl alcohol (PVA) + 0.5 w/v% of metronidazole (MTZ) + 0.1% to 0.5 v/v of sodium cocoamphoacetate (surfactant used in personal care products) electrospun nanofibers were produced as a novel mucoadhesive patch to manage BV (Ilomuanya et al. 2023). This antibiotic is a nitroimidazole drug responsible for inhibiting the DNA repair enzymes, especially in anaerobes and protozoa (El-Newehy et al. 2012). In this study, the authors also incorporated *Lactobacillus* spp. in PCL/PVA solution to increase the culture in situ for bacteria clearance (Ilomuanya et al. 2023). The results show that increasing the sodium cocoamphoacetate (CAP) concentration had led to increased mucoadhesion, in which F6 (PCL/PVA + 0.5 wt% MTZ + 0.5 v/v CAP) showed the highest mucoadhesion, as compared to the F7 (PCL/PVA + 0.5 wt% MTZ + 0.25 v/v CAP +150 × 10^6 *Lactobacilli* spp./mL) and F8 (PCL/PVA + 0.5 wt% MTZ + 0.1 v/v CAP +150 × 10^6 *Lactobacilli* spp./mL) meshes with 57.86 ± 0.95%, as compared to 42.67 ± 1.22% and 50.89 ± 1.01%, respectively. Moreover, F6 mesh had a stabilized structure by CAP allowing a sustained release of MTZ, ranging from 14.8502 μg/cm^2 at 5 min to 514.937 μg/cm^2 at 6 h. The viability of the *Lactobacilli* culture in electrospun meshes was assessed 30 days post-electrospinning, revealing a 150 and 143 CFU/mL *Lactobacillus* spp. in F7 and F8, respectively, compared with 198, and 214 CFU/mL in the emulsion prior to electrospinning. The safety assessments performed in female rats indicated that only F8 exhibited an absence of atrophy and inflammation in the vaginal epithelium of the rats, as compared to the control group with no treatment. However, F6 had a protection layer against invading pathogens and against water loss in the stratified squamous epithelium. The developed electrospun meshes demonstrate adequate tensile strength and mucoadhesive properties. The structure presents potential for intravaginal delivery of Lactobacilli spp. and sustainable release of MTZ, which are facilitated by the addition of CAP. Nevertheless, in order to be safe on vaginal tissues, CAP concentration should be reduced.

Amoxicillin (AMX) belongs to the penicillin class and is used for the treatment of infections caused by gram-positive bacteria. Nonetheless, AMX has a low metabolic rate in the human body, and around 80–90% is unmetabolized (Qutob et al. 2022). Consequently, it is excreted by the body, ending up in aquatic ecosystems and at risk of accumulating in the environment if not properly treated. One strategy to reduce these drawbacks, while maintaining AMX's activity, its loading AMX's into electrospun nanofibers. Ajalloueia et al. (2022) compared the pullulan (Pull—a naturally derived hydrophilic biopolymer) and poly(lactic-co-glycolic) acid (PLGA—a synthetic hydrophobic biopolymer) nanofibers as two potential carriers of an AMX delivery, to topical skin delivery applications. Additionally, the authors tested a multilayer structure to reduce antibiotic consumption and increase storage time. AMX was loaded into both PLGA and Pull electrospun nanofibers (mono-layers) with final concentrations of 1, 2, or 5 w/w%, while for the fabrication of the multi-layered structure (PLGA-Pull-PLGA), AMX was added only to the Pull layer or to all three layers. In antibacterial activity assays, loading AMX into Pull (AMX/Pull) nanofibers not only preserved the antibiotic's efficacy but also resulted in comparable performance for both *E. coli* and *S. aureus* compared to the normal AMX disk, whereas in PLGA nanofibers to show a similar AMX disk (20 μg) efficacy it requires 50 μg of AMX encapsulated in nanofibers to inhibit both bacterial growths. The efficacy of AMX in Pull and PLGA nanofibers was evaluated over time, namely 6 weeks (15, 20, 25, and 40 μg), 4 months (20 μg), and 8 months (20 μg) of storage. Interestingly, for AMX/Pull, the inhibition zone for samples at 4-month and 8-month storage times was similar to the inhibition zone of 6-week storage against *E. coli* and *S. aureus* (Fig. 20.2). These findings were comparable to those obtained with regular paper disks containing AMX. The results of in vitro release indicated that in all conditions most of the AMX (>90%) is released within the first 10 min from mono-layer AMX/Pull electrospun meshes. In order to extend the release, a three-layered sandwich structure was produced, in which the AMX/Pull electrospun mesh was placed between two PLGA electrospun layers. This way, a sustained release was observed, where only 40% of the AMX was released after 4 days. This property is advantageous in applications such as wound-healing dressings, where a mix of burst and sustained release is desired to promptly protect the tissue after injury and prophylactic effect during regeneration is needed (Ferreira et al. 2021a). According to the in vivo assays, the re-epithelization was achieved, but without statistical differences with the control, which uses an antibiotic in the open wound with >2000 times higher concentration than the AMX in nanofibers. Overall, findings show that Pull nanofibers may be used as an efficient carrier for long-term antibiotic delivery and this product efficacy is guaranteed to be used commercially as a shelf product for different local antibacterial applications (Ajalloueian et al. 2022).

Ciprofloxacin (CIP) is a second-generation fluoroquinolone antibiotic that restrains cell division by inhibiting DNA gyrase and topoisomerase IV (type II topoisomerases), which are necessary for bacterial DNA separation (Kanth et al. 2023). Furthermore, CIP is the world's fifth-most-produced generic antibiotic, and for that reason, it is commonly used in electrospun nanofibers to treat diverse infections (Teno et al. 2022). Polyvinylpyrrolidone (PVP) was used to load CIP in films and nanofiber meshes to evaluate their performance as DDSs in a wound infection model based on ex vivo human skin (Rancan et al. 2019). The PVP films were produced through solvent casting method starting from aqueous solutions of PVP (3% w/v), while PVP electrospun meshes were obtained from PVP solution (25% w/v) in acetic acid (30% v/v) and different concentrations of CIP (2.2, 44, and 88 mM). The CIP release kinetics is different in both systems, in which PVP films release faster high amounts of CIP (approximately 12 μg after only 2 h), than nanofiber, in which the concentration release is kept constant after 6 h. Moreover, films maintained the reached drug concentration in the wound over 24 h. On the contrary, in

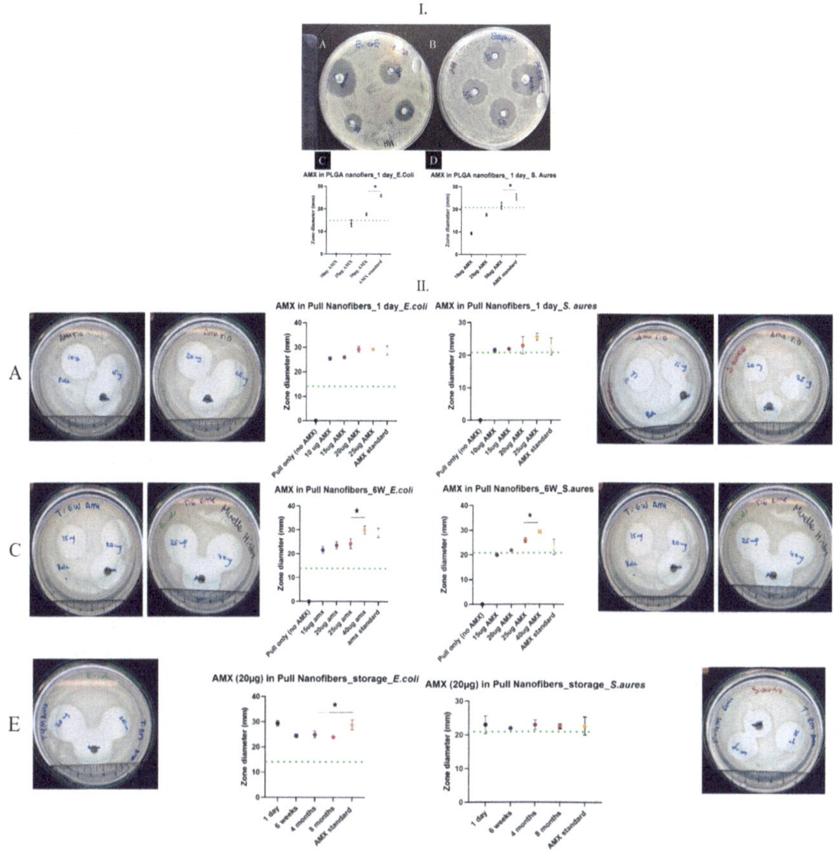

Fig. 20.2 Disk diffusion study of AMX/PLGA samples with different AMX concentrations against *S. aureus* and *E. coli*, 1 day after fabrication (**I—a**, **b**) compared with the standard AMX disk (**I—c**, **d** -*$p < 0.05$). The green broken line represents the breaking point. Disk diffusion assay comparing (*$p < 0.05$) the inhibition zone of mono-layer samples of AMX/Pull with different AMX concentrations against *S. aureus* and *E. coli* at time points of 1 day, 6 weeks, 4 months, and 8 months post-fabrication. (**II**). (Images reproduced with permission from (Ajalloueian et al. 2022))

samples treated with nanofiber mesh, the maximal concentration was maintained for a shorter time range (6–16 h), suggesting that a reduction of local drug concentration in the wound is due to diffusion to the adjacent tissue. The authors admit that for the treatment of biofilm infections, high antibiotic concentrations are required in PVP films. On the other side, the more controlled delivery profile of the nanofiber meshes might be more beneficial for the prevention of wound infections. The release profile is correlated with the antibacterial activity. Thus, PVP films were capable of inhibiting *P. aeruginosa in* both planktonic and biofilm when loaded with 500 µg of CIP, and the two lower concentrations resulted in a 3-log reduction of bacteria concentration with respect to the untreated controls. However, for the same CIP dose PVP nanofibers have registered counts, corresponding to a maximal 3-log reduction achieved by the highest concentration. The antibacterial activity of PVP foils was explained by the superior percentage of residual acetic acid (1.4%) than nanofiber meshes (0.1%). The findings highlight the importance of drug delivery systems with a prolonged release profile, such as nanofibers, in the treatment of *P. aeruginosa* wound infections (Rancan et al. 2019).

Minocycline (MIN) and doxycycline (DOX) are antibiotics belonging to tetracyclines' class. These drugs prevent aminoacyl-tRNA molecules from attaching to the ribosomal acceptor sites, hence preventing protein production in bacterial ribosomes (S30) (Angelette et al. 2023). Minocycline has a variety of activities that may be useful for non-infectious diseases in addition to its antibacterial activity. Allied with antibacterial activity, MIN has been used also to promote bone formation (Ma et al. 2020). Hence, MIN-PLGA electrospun membranes were produced using several MIN concentrations (1, 2, and 3 wt%, relative to the polymer weight) to evaluate their effect on osteogenesis in vitro and in vivo using a rat model for periodontitis. Periodontitis is a serious gum infection by oral pathogens that damages and destroys the bone that supports the teeth (Verma et al. 2012; Mirzaeei et al. 2021). This work demonstrated that MIN-PLGA, especially the mesh containing 2% MIN, promotes osteoblast proliferation. Despite an initial burst release of MIN, the MIN-PLGA electrospun meshes showed an adequate rate and a sustained release in vitro, which improved osteogenesis after periodontitis in vivo (Ma et al. 2020). In another study, DOX was used in polybutylene adipate-co-terephthalate (PBAT)/gelatin nanofiber that was posterior grafted with arginylglycylaspartic acid (RGD) peptide to act as wound dressing (Varshosaz et al. 2020). In the mentioned study, the DOX concentration was changed between 0.4, 1, and 1.6 w/w% of the polymer weight, PBAT concentration was studied between 17 and 23 w/v%, and gelatin concentration was fixed at 50 w/v%. From morphological analysis of DOX-PBAT/GEL meshes, the authors have shown that increasing DOX concentration will increase the mesh porosity and as a consequence was observed that DOX release was also increased, in two distinct kinetics profiles: burst release and sustainable release. The amount of matrix metalloproteinase (MMP) obtained from the culture supernatant of L929 cells confirms that DOX-PBAT/GEL inhibits MMPs more significantly than PBAT/gelatin nanofibers. DOX-loaded nanofibers' antibacterial activity was also investigated against *S. aureus* and *P. aeruginosa.* While the PBAT/GEL nanofibers had no antibacterial impact, the DOX-PBAT/GEL nanofibers had similar antibiotic effects to the free DOX. However, RGD-DOX-PBAT/GEL antibacterial activity was significantly less than the DOX-PBAT/GEL nanofibers ($p < 0.05$). In in vivo assays, when compared to the control group, the DOX-loaded nanofibers coupled with RGD (40 μg/100 mg of nanofibers) had a substantial impact on wound healing. The formation of collagen was verified and the re-epithelialization without inflammation, while granulation tissue was less than other conditions after 3 days of treatment. PBAT/GEL nanofibers grafted with RGD peptides and loaded with DOX are promising materials for wound healing (Varshosaz et al. 2020).

Other antibiotics than tetracyclines' class capable of inhibiting protein synthesis are the oxazolidinones, such as linezolid (LZD), which affects both the 30S and 50S ribosomal subunits. LZD is used to treat infections triggered by gram-positives, such as *S. aureus* hospital-acquired pneumonia, MRSA, vancomycin-resistant *E.s faecium* (VREF), *Streptococcus pyogenes* complicated skin infections, and community-acquired pneumonia caused by *Streptococcus pneumoniae* (Hashemian et al. 2018). In Boncu et al.'s (Eren Boncu et al. 2020) work, LZD-loaded electrospun PLGA and PCL fiber meshes with controlled drug release were developed for the treatment and prevention of prosthesis-related infections. The main advantage of using PLGA and PCL electrospun meshes instead of poly(methyl methacrylate) (PMMA) bone cement containing antibiotics is that the biodegradability of polymers does not require additional surgeries to remove the implant that is no longer functional. To produce electrospun meshes, different formulations were prepared with different PLGA:PCL ratios (ranging from 100:0 to 50:50) and different LZD concentrations (5, 10, or 20 w/w%). Results demonstrated that increasing the LZD concentration from 5 to 10% resulted in a higher burst effect and accelerated in vitro drug release. However, increasing the amount of LZD from 10 to 20% did not show significant differences in release. The inhibition zone was measured on the first and last days of the activity period and the total activity period of the formulations registered against MRSA. The results show that the PLGA fibers with 5% of LZD present an efficacy that lasted for 11 days, but increasing the amount of LZD to 10% the efficacy was extended to 19 days. The higher antibacterial efficiency was observed for the PLGA fibers with 20% of LZD formulation on the 28th day. For this reason, this formulation was selected for the in vivo tests in MRSA-infected rats with tibial fractures. The findings demonstrated that the treatment and prophylaxis groups treated with electrospun meshes were more effective than the untreated control group and intraperitoneal LZD administration two times a day for 28 days. In the histopathological assay for prophylactic and therapeutic use of electrospun meshes with 20% of LZD was observed less severe *in vivo* infection and faster healing in infected rats with prosthesis-stabilized fractures compared to both the untreated control group and intraperitoneal group where the commercial preparation was intraperitoneally applied twice a day. These were confirmed by X-ray images of bone fractures. This work reveals that the treatment efficiency will increase since the one-time administration of fiber eliminates the need for two injections per day. Moreover, reducing the antibiotic dose by 37-fold in fiber mat treatment compared to conventional treatment prevents antibiotic resistance (Eren Boncu et al. 2020).

In relationship to other antibiotic carries, the abovementioned studies using nanofibers have demonstrated a high loading capacity and high antibacterial encapsulation efficiency, with a biphasic release kinetics (burst and stationary), which is a major asset to combat most infections and low systemic toxicity. Regarding the promising results in *in vivo* models described in this section, few studies reach the preclinical and clinical trials, due to regulatory hurdles. Nevertheless, a randomized clinical trial (NCT03690960) has produced a triple antibiotic mixture (TAP) with CIP, MET, and MIN loaded into electrospun meshes to combat infected root dentine and promote endodontic regeneration. Despite the clinical trial are already finished, the outputs are not available yet (Cochrane Library 2018). Similarly, a clinical study using PCL nanofibers containing MET to treat periodontal diseases was performed (Verma et al. 2012). In contrast to the control group, which had values of 50%, 71%, and 71% for probing depth (PD), plaque index (PI), and gingival index (GI), respectively, the group treated with nanofibers containing MET shows higher reductions in PD, PI, and GI with 68%, 85%, and 90%, respectively ($p < 0.01$). The incorporation of MET into the electrospun fibers demonstrated higher efficiency to treat periodontal disease (Rancan et al. 2019).

The number of antibiotics that are effective against harmful bacteria has decreased during the past 10 years (He et al. 2022). This scenario highlights the need to explore new approaches and antimicrobial agents, where electrospun nanofibers have demonstrated that they can play a prominent role.

20.4.1.2 Organic Agents

The high rate of deaths associated with infections triggered by multidrug-resistant bacteria has led scientists to discover innovative compounds with novel mechanisms of action or to find new approaches to combat diseases caused by drug-resistant bacteria (Álvarez-Martínez et al. 2020). Organic compounds represent a potential source of bioactive molecules with remarkable antibacterial properties that might be useful in combat against bacterial infections (Pilehvar-Soltanahmadi et al. 2017). Plant and seaweed-based extracts and their oils, as well as other molecules from animals, bacteria, and fungi, are (natural) organic compounds that are being exploited to be used in nanotechnologies, such as electrospun nanofibers (Álvarez-Martínez et al. 2020; Guerreiro et al. 2021; Ferreira et al. 2021b). Plant-based compounds are the most used in the development of new approaches, as a result of their endurance to extreme environmental conditions, and years of adaptation to defend against predators, abiotic stress, and infections (Álvarez-Martínez et al. 2020). Recently, Snetkov and collaborators (Snetkov et al. 2022) developed curcumin-loaded nanofibers against multi-resistant bacteria of the ESKAPE group. Curcumin is a bioactive curcuminoid polyphenol that is isolated from the rhizomes of Curcuma longa (Merrell et al. 2009). Curcumin was added directly to the spinning polymer solution (hyaluronic acid sodium salt—HA-T) with the weight ratios of biopolymer to curcumin as follows: 1:0 (HA_0), 25:1 (HA_25), 15:1 (HA_15), and 5:1(HA_5). Based on the antimicrobial activity using the agar disk diffusion assay, increasing the amount of curcumin in nanofibers resulted in bigger inhibition halos for all control conditions and bacteria, with the exception of *K. pneumoniae* and *A. baumannii*. Moreover, gram-positives were more susceptible to curcumin-loaded nanofibers than gram-negatives. The median ZOI diameters for gram-positive bacteria were between 9.0 and 19.0 mm, while those for gram-negative bacteria ranged from 6.0 to 14.0 mm. However, MIC values revealed that curcumin-loaded nanofibers need lower concentrations to kill most of the bacteria in the study in relation to other literature studies using the same bioactive compound. Indeed, to fight *K. pneumoniae* and *A. baumannii* strains were required more than 2000 μg/mL and 5000 μg/mL, respectively. While in the study of Snetkov and collaborators (Assoni et al. 2021), the MIC values were significantly lower (360 μg/mL). Another plant-derived component often employed in nanofiber composition is *Calendula officinalis* (CO) extract. Chitosan/polyethylene oxide (CS/PEO) nanofibers containing CO extract were studied to prevent wound infection (Kharat et al. 2021) (Fig. 20.3). Quantitative antibacterial activity was evaluated in two strains of bacteria, *S. aureus* and *E. coli,* by disk diffusion and colony counting assay. The disk diffusion data indicated that, as predicted, the ZOI diameters were higher than on control condition CS/PEO,

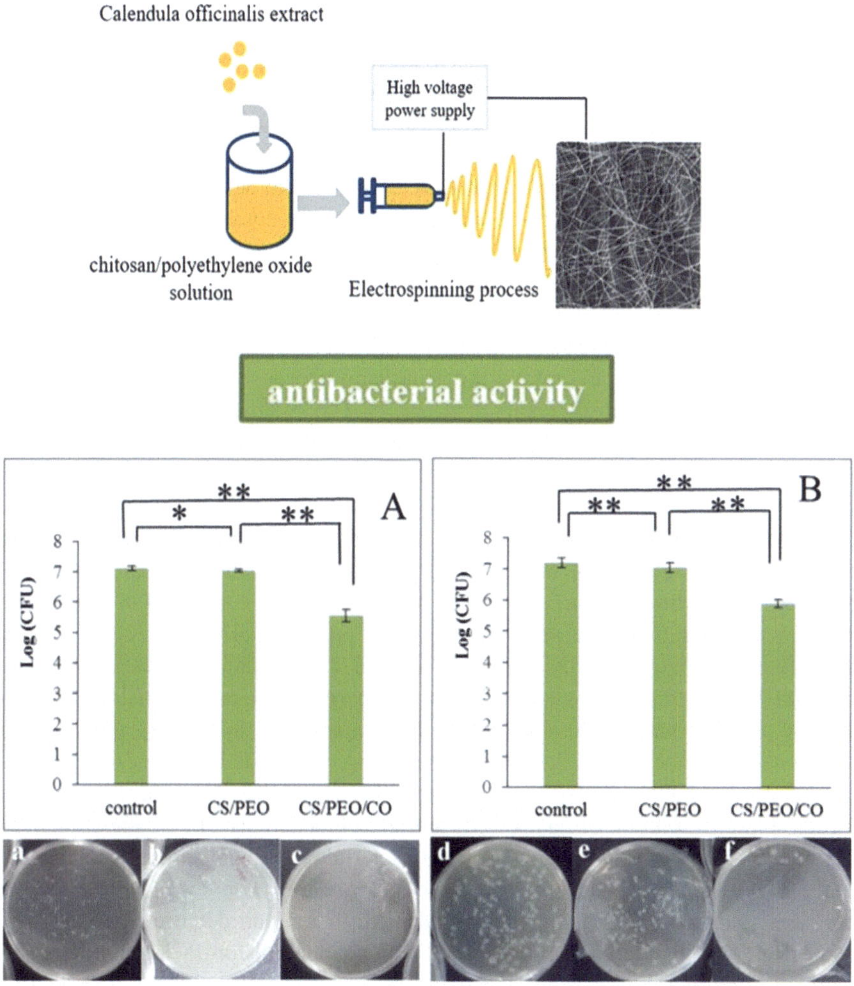

Fig. 20.3 Antibacterial activity of the CS/PEO and CS/PEO/CO dressings against (**a**) *S. aureus* [(a) untreated control, (b) CS/PEO nanofibers, and (c) CS/PEO/CO nanofibers] and (**b**) *E. coli* [(d) untreated control, (e) CS/PEO, and (f) CS/PEO/CO nanofibers] on 7×10^8 CFU/mL bacteria. Control is untreated samples. All experiments were done in triplicate, and data are analyzed by Student's *t*-test; $*p < 0.05$ and $**p < 0.01$. As shown, nanofibers containing CO extract exhibited a significant reduction in the number of bacterial colonies compared with CS/PEO control. (Figures reproduced with permission from (Kharat et al. 2021))

where *E. coli* and *S. aureus* ZOI registered 20.9 ± 1.1 (mm) and 22.0 ± 0.4 (mm), respectively, with no statistically significant differences. In terms of colony counts, CO extract-loaded nanofibers exhibited a 96.9% reduction in *S. aureus* and a 94.8% reduction in *E. coli* colony counts. However, control CS/PEO nanofibers showed a 14.9% reduction in *S. aureus* and a 30.0% reduction in *E. coli* colony count. The authors attributed the antimicrobial effect of CO extract to the presence of phenolic compounds and flavonoids (Kharat et al. 2021). Correspondingly, in the work of Hadipour-Goudarz et al. (2023) the superior antibacterial activity of acorn extract loaded in chitosan/gelatin/polyvinyl alcohol (CS/GEL/PVA) nanofibers was related to the presence of bioactive compounds in acorn extract such as flavonoids, tannins, polyphenols, and alkaloids. This allowed acorn-loaded nanofibers to reach an inhibition of 90% against *S. aureus* (1×10^4 CFU/mL) and 80% against *E. coli* (9×10^4 CFU/mL). Seaweed-derived compounds due to the plentiful structures that possess also have shown similar promising antimicrobial activity. In a study, electrospun meshes of GEL/CS doped with a phlorotannin-enriched (marine polyphenol) extract from seaweed *Undaria pinnatifida* demonstrated higher potency against *P. aeruginosa* over *S. aureus* through disk-diffusion assays. Furthermore, the meshes with higher amounts of phlorotannin-enrich extract showed significantly greater inhibition compared to other conditions (Ferreira et al. 2021a).

Essential oils (EOs) are among the promising approaches for the treatment of ESKAPEE-related biofilms (Panda et al. 2022). The volatile compounds, polyphenols, and flavonoids among other secondary metabolites found in plant-derived EOs have the potential to exhibit wide antibacterial and antibiofilm action against pathogenic organisms, such as the ESKAPE bacteria (Panda et al. 2022). Nevertheless, plant EOs present some drawbacks such as solubility. A strategy to overcome this issue is the formulation of nanoemulsions, in which small oil droplets are surrounded by a stabilizing surfactant. The nanoemulsion is composed of one single phase with no remarkable separation or precipitation (El-Naggar et al. 2023). Therefore, El-Naggar et al. (2023) developed electrospun meshes using PCL blended with different concentrations of peppermint oil nanoemulsion (PO) to explore their antimicrobial and antibiofilm capabilities against various human pathogenic microbes (Fig. 20.4). PO nanoemulsion was added to 10 wt% PCL solution, in different amounts (1, 2, 3, and 4 mL (PCL-1 to PCL-4)) until it reached an electrospinning solution of 20 mL. The antimicrobial activity of PO nanoemulsion-PCL nanofibers was tested against several bacterial and fungal species encompassing *A. baumannii*, *P. aeruginosa*, *S. aureus*, *Listeria monocytogenes*, *Candida albicans*, and *Aspergillus niger* through disk diffusion and determination of MIC and MBC. The obtained data proved that the PO nanoemulsion-PCL nanofibers had increased antimicrobial activity and registered greater ZOI diameters with PCL-4 against *A. baumannii*, *P. aeruginosa*, *S. aureus*, *L. monocytogenes*, *C. albicans*, and *A. niger* that measured 23.2, 25.4, 22.7, 21.5, 19.2, and 18 mm, respectively. These results were corroborated by the growth curve of pathogens before and after subjecting to the effective dose (200 μg/mL) of explored nanofibers. Remarkably, both PO oil nanoemulsion and PCL nanofiber exhibited the lowest levels of antimicrobial activity (El-Naggar et al. 2023). Similarly, the values of MIC and MBC demonstrated that among the PO nanoemulsion-PCL nanofibers, the PCL-4 presented the

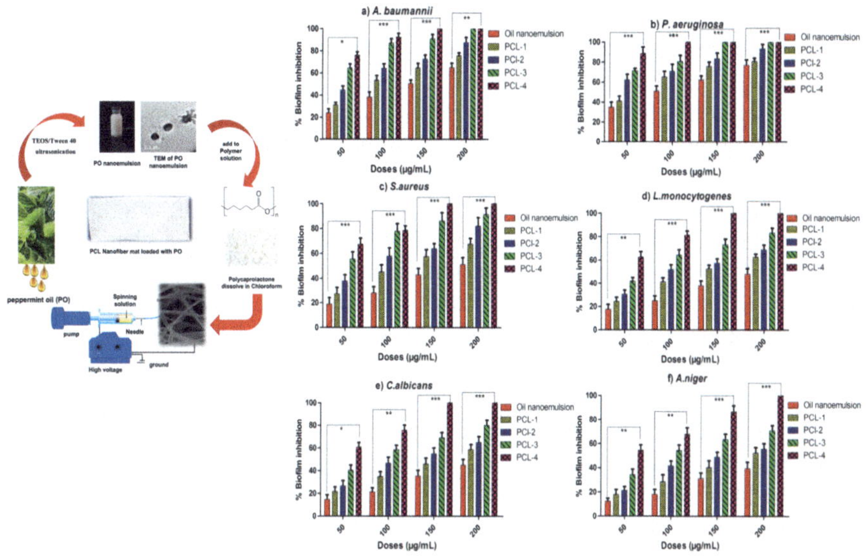

Fig. 20.4 Reduction in the biofilm formation of (**a**) *A. baumannii*, (**b**) *P. aeruginosa*, (**c**) *S. aureus*, (**d**) *L. monocytogenes*, (**e**) *C. albicans*, and (**f**) *A. niger* in the presence of 50, 100, 150, and 200 µg/mL concentrations of nanofibers. The error bars epitomize SD of three independent trials. Two-way analysis of variance (ANOVA) states; * indicates weak correlation ($p \leq 0.05$); ** indicates moderate correlation ($p \leq 0.01$); *** indicates strong correlation ($p \leq 0.001$). (Reproduced with permission from (El-Naggar et al. 2023))

highest potency against all the pathogens. The estimated inhibitory concentrations (MIC and MBC) of PCL-4 were 50 and 75 µg/mL, 75 and 100 µg/mL, 75 and 100 µg/mL, and 100 and 125 µg/mL against *A. baumannii*, *P. aeruginosa*, *L. monocytogenes*, *C. albicans*, and *A. niger*, respectively. Results indicated that 125 µg/mL of PCL-4 was the MIC and MBC against all tested microbial pathogens, whereas PCL nanofiber alone required >200 µg/mL. The authors affirm that PO nanoemulsion-PCL nanofiber antimicrobial activity is related to the capacity to interact with microbial cell membranes by the generation of reactive oxygen species (ROS), which can destabilize microorganism cell membranes by interfering with critical processes, such as reproduction (El-Naggar et al. 2023). Hence, protein leakage from microbial cell membranes was evaluated with several exposure times (3–24 h) of 200 µg/mL of PCL-2, PCL-3, and PCL-4 nanofibers. Protein release from inside bacterial and fungal cells increased with increasing PO nanoemulsion concentrations across 3-h and 24-h exposure intervals. Thus, protein spillage of PCL-4. was 348, 392, 336, 284, 257, and 143 µg/mL for *A. baumannii*, *P. aeruginosa*, *S. aureus*, *L. monocytogenes*, *C. albicans*, and *A. niger*, correspondingly. The antibiofilm activities of nanofibers were explored, which using 200 µg/mL of PCL-4 reached complete biofilm inhibition of all tested microbes. To conclude, the nanofibers of PCL loaded with PO nanoemulsion reveal a potent antimicrobial activity, better than PO nanoemulsion alone, revealing an advantageous approach to fight multidrug-resistant bacteria. Despite highlighting the PO nanoemulsion in this section, other EOs such as thyme oil, lavender oil, eugenol oil, and tea tree oil are being loaded in nanofibers with antibacterial efficacy against *E. coli*, *K. pneumoniae*, *S. aureus* strains, and other pathogenic organisms was also demonstrated outstanding results (Yue et al. 2021; Shi et al. 2022; Kwon et al. 2023). Nevertheless, a broad spectrum of other different EOs continues under investigation, and others used before in traditional medicine are demonstrating effective antimicrobial activity when applied in nanofibers (Elbhnsawi et al. 2023; Ustundag and Piskin 2023; Akturk et al. 2023).

From animal-derived compounds, CS is the most used polysaccharide in the production of electrospun meshes to enhance their antimicrobial activities, due to its polycationic nature (Ibrahim et al. 2023). This polycationic nature allows CS to permeabilize prokaryotic cell walls by creating ionic complexes with the negative charges present on the phospholipids and lipopolysaccharides found in gram-negative bacteria's outer and inner membranes, as well as on the teichoic acids connected to the peptidoglycan found in gram-positive bacteria's cell walls (Keirouz et al. 2020). However, to overcome the reduced spinnability of CS, it should be blended with other polymers, either natural or synthetic such as gelatin, alginate, collagen, zein, silk fibroin, PEO, PVA, poly(lactic acid) (PLA), PCL, and others to resolve the lower spinnability (Kharat et al. 2021; Hadipour-Goudarzi et al. 2023; Ibrahim et al. 2023). Based on this, Wu et al. (Wu et al. 2022) evaluated the antimicrobial activity of quaternized CS-PVA electrospun meshes. For that, 10 w/w% PVA solution, 3 w/w% CS solution, and 3 w/w% quaternized CS (HTCC) were used, which were mixed at different ratios PVA: CS (90/10–50/50) and PVA:HTCC ratio (90/10–50/50). In terms of antibacterial performance, the PVA-CS nanofiber mesh synthesized at a mass ratio of 7:3 recorded the highest antimicrobial activity, with an antibacterial efficacy (AE) of 73.82% against *E. coli*. However, in this study, AE was not directly proportional to the amount of CS. It was observed that the AE of the PVA-HTCC nanofiber meshes was significantly higher than that of the PVA-CS nanofiber mesh independent of the ratio of PVA to CS. Moreover, as the mass ratio of PVA to HTCC changed from 9:1 to 7:3, the AE of the PVA-HTCC nanofiber mesh increased. The maximal AE (~90%) was reached using the PVA-HTCC nanofiber mesh prepared at a 7:3 mass ratio (Wu et al. 2022). The results reported in this study suggested that the PVA-HTCC nanofiber meshes can be a promising material for biomedical applications such as wound healing and drug delivery. Regarding, the antimicrobial activities of CS, other natural polysaccharides are often used in electrospun mesh formulations such as soy, silk fibroin (SF), gelatin, cellulose, keratin, starch, and alginate (Chee and Nugent 2019). Notwithstanding some of the abovementioned polysaccharides do not present intrinsic antimicrobial activity; however, due to their functional groups they can act synergistically with other nanofiber polymers and show different methods of interactions with pathogenic organisms or particles/contaminants (interception, diffusion, impaction, sieving), which aid to inhibit or kill bacteria (Guerreiro et al. 2022). Other current approaches to combat multidrug-resistant bacteria, which have gained attention to use in nanofiber combined with other organic compounds, are the use of antimicrobial peptides (AMPs).

Antimicrobial peptides are short-cationic peptides with an amphipathic structure that inhibits the growth of bacteria, fungi, and viruses by changing membrane permeability and destroying it (Afshar et al. 2021). In this context, electrospun meshes were prepared with CS and PEO (30 mg/mL solution) containing different concentrations (1, 3, and 5 mg/mL) of an AMP (NP10). The antimicrobial activity was assessed by disk diffusion after 1 and 7 days against *S. aureus* and *E. coli*. The overall results demonstrated that CS-PEO-0.5%NP10 nanofiber mesh had the best morphology and antibacterial activity against both strains. The findings demonstrate that the antibacterial activity increases with time. Meanwhile, the results showed that the area of inhibition of *E. coli* was usually more extensive than that of *S. aureus,* which is attributed to the effect of the CS-PEO-NP10 nanofiber on E. coli *membrane*. In the in vitro release experiment, NP10 was released from the CS-PEO-0.5%NP10 nanofiber mesh in a burst first and then slowly and continuously, being an asset for the clinical application of AMPs and treatment of wound infections (Yu et al. 2021). Similarly, another AMP was conjugated with the natural polymers CS and SF to produce antibacterial wound dressing. In this study, was produced an antibacterial CS/SF electrospun nanofiber containing different concentrations of CM11 peptide. CM11 is a short cecropin-melittin-derived cationic AMP that is comprised of 11 amino acid residues (WKLFKKILKVL). Several investigations have demonstrated the antibacterial activity of this peptide against a wide range of gram-positive and gram-negative organisms that are resistant to antibiotics (Moghaddam et al. 2012, 2014; Moravej et al. 2019). Considering this, the study's findings demonstrated the dose-dependent antibacterial efficacy of 32 μg/mL of CM11 peptide-loaded CS/SF bilayer against conventional strains of *S. aureus*, *E. coli*, and *P. aeruginosa* as well as their multidrug-resistant strains isolated from burn patients (Fig. 20.5). Representing a promising candidate for further clinical investigations (Khosravimelal et al. 2021), these results are corroborated by the work of Moghaddam et al., (Moosazadeh Moghaddam et al. 2023) which used the same AMP. CM11, using an antibacterial human amniotic membrane/SF(AM/SF) electrospun nanofiber. The electrospun mesh containing CM11 (16 or 32 μg/mL (AM/SF/P16 and AM/SF/P32)) exhibits antimicrobial activity, although it shows a greater effect of the peptide against *S. aureus* compared to *P. aeruginosa*. Additionally, a comparison between the diameters of the halos around the disks revealed that the electrospun mesh with a higher concentration of CM11 exerts a greater growth inhibitory effect (Moosazadeh Moghaddam et al. 2023). AM/SF/CM11 electrospun meshes were regarded as an adequate option for antibacterial approaches that are commonly used to treat infection-prone lesions such as burns. Nevertheless, other AMPs already were used in nanofibers with similar results against *P. aeruginosa*, MRSA *S. aureus, A. baumannii* (Weishaupt et al. 2020; Chaiarwut et al. 2021).

Clinical practice is facing a significant problem due to the rising number of skin wound infections, particularly in cases when traditional antibiotic treatment is ineffective (Fayez et al. 2021). Within this framework, bacteriophages have surfaced as potentially effective substitutes for treating bacteria that are resistant to antibiotics. Bacteriophages are bacteria-infecting viruses that can target specific bacterial species or strains (Zargari et al. 2022; Kielholz et al. 2023). In the study of Kielholz et al. (2023), bacteriophage-loaded electrospun fiber meshes were developed as

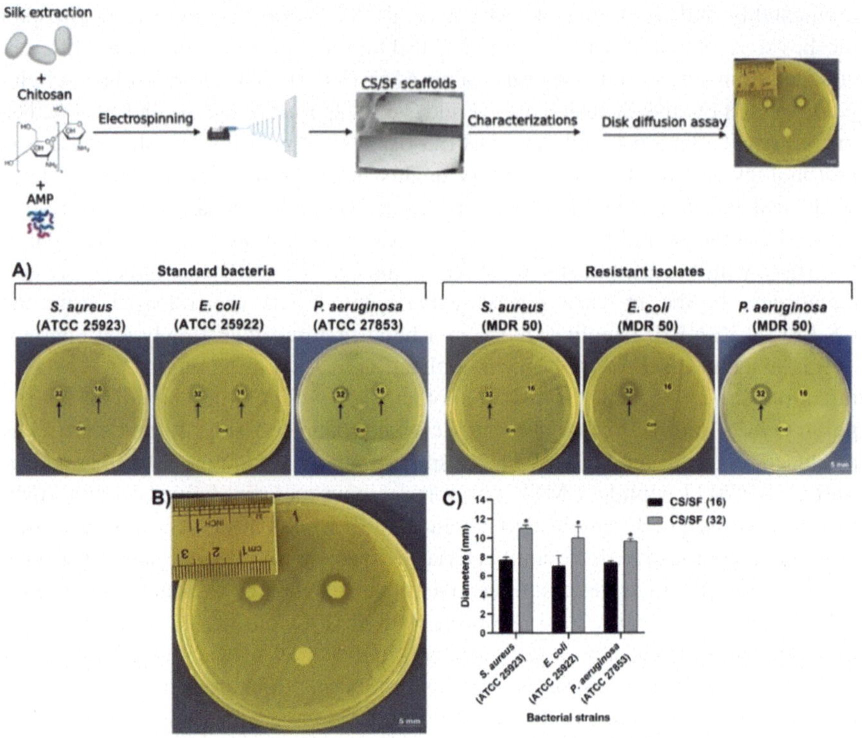

Fig. 20.5 Assessment of antibacterial activity peptide-loaded CS/FS by disk diffusion assay. (**a**) The antibacterial activity peptide-loaded CS/SF (16 and 32 μg/mL peptide, respectively) bilayers compared with control for standard strains of *S. aureus*, *E. coli*, and *P. aeruginosa* in MHA plates and resistant strains. (**b**) The diameter of the growth inhibition zone recorded by a ruler. (**c**) Diameters around peptide-loaded CS/SF (16 μg/mL) disks are compared to peptide-loaded CS/SF (32 μg/mL). There are significant differences in the electrospun meshes containing higher concentrations of CM11 peptide (32 μg/mL) with peptide-loaded CS/SF containing 16 μg/mL peptide in both standard strains ($p < 0.05$). Results from resistant strains and all controls are excluded as no diameters have formed and there is nothing to compare. (Reproduced with permission from (Khosravimelal et al. 2021))

next-generation wound dressings for the treatment of infected wounds. The authors used a coaxial electrospinning approach, creating fibers with a protective poly(vinylpolypyrrolidone) (PVP) polymer shell, enveloping bacteriophages in the core (EBHT bacteriophages targeting *S. aureus*, or JG004 phages inhibiting *P. aeruginosa*) (Fig. 20.6). The antimicrobial activity of the bacteriophage-loaded electrospun meshes was tested directly after electrospinning (fresh) and after 1, 2, and 4 weeks of storage at either room temperature (RT), 5 °C, or −20 °C. Bacteriophage fibers showed zones of inhibition after 1 week of storage in each condition tested. However, EBHT electrospun meshes displayed a higher ZOI diameter at RT (8.3 ± 0.3 mm) at both cooler conditions (8.9 ± 0.3 mm), than JG004-loaded fibers with significantly smaller diameters especially in cooler conditions (7.5 ± 0.3 mm

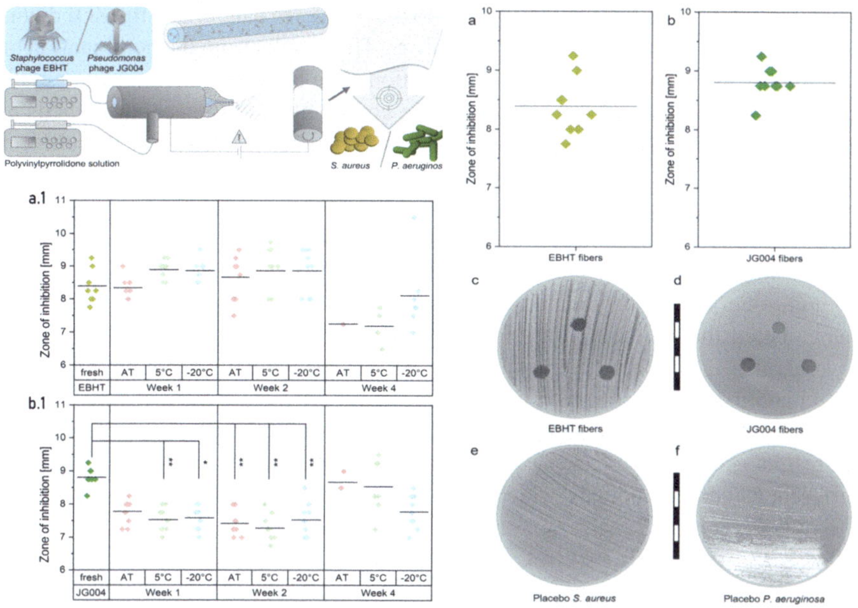

Fig. 20.6 Inhibition zone diameters of EBHT (**a**) and JG004 fiber samples (**b**) against *S. aureus* and *P. aeruginosa*, respectively. Representative photographs of agar plates incubated with either bacteriophage fibers (**c**, **d**) or placebo fibers (**e**, **f**). Scale bars represent 5 cm with 1-cm intervals. Zone of inhibition diameters of EBHT fibers (**a.1**) and JG004 fibers (**b.1**) directly after electrospinning (fresh) and after 1, 2, and 4 weeks of storage at either ambient temperature (AT, red), 5 °C (green), or −20 °C (blue). (Reproduced with permission from (Kielholz et al. 2023))

at 5 °C and 7.6 ± 0.3 mm at − 20 °C). Also, the same condition had significantly reduced activity compared to fresh samples after 2 weeks. Interestingly, after 4 weeks of storage, most bacteriophage-loaded fiber samples stored at RT were incapable of inhibiting their target bacteria. The results revealed that some remaining bacteriophage-loaded fibers continue to be active against pathogens when stored at cold conditions, with superior performance for fibers stored at −20 °C, as verified in literature (Koo et al. 2016). The authors justified as interindividual variability of EBHT and JG004 bacteriophages and also related pH and osmolality (Kielholz et al. 2023). Similar results were obtained in another work, where PCL nanofiber loaded with Mycobacteriophage Fulbright when stored at −20 °C, the bacteriophage remains viable for up to 11 months in the fiber (Kotturi et al. 2022). In terms of antimicrobial activity, the results indicated that the PCL_Fulbright electrospun meshes only reached 62.93% inhibition, after 24 h, when bacteriophages are released into the media and lysed the host cells successfully. The authors also indicated that bacteriophage activity decreased with higher temperatures, and highly acidic and highly basic environments are due to the denaturation of phage and host proteins (Kotturi et al. 2022). Similarly, when comparing *Aloe vera*/PVA nanofibers carrying *Pseudomonas* sp. bacteriophage to the *Aloe vera* nanofibers, the ZOI diameter was higher in phage-*Aloe vera* nanofibers. This results from the combination of ingredients, which allows the growth inhibition zone to be increased (Zargari et al. 2022). Regarding antimicrobial activity, the employment of phages in biomedical applications still requires some additional research before reaching the clinical stage, mainly related to the stability of phages over time.

In summary, great progress has been achieved in the development of organic-based antimicrobial nanofibers. The combination of innovative antimicrobial compounds and biopolymers could offer a diverse range of options for the next generation of non-toxic and biodegradable antimicrobial nanofibers. However, it should be remembered that to ensure antibacterial activity when using organic compounds, several other factors must be methodically assessed, essentially, the immobilization onto nanofibers and estimation of the longer-term efficacy of these nature-derived materials. Additionally, it is important to understand that higher antimicrobial activity results using extracts and oils are generally derived from a synergetic effect of different natural compounds, which can be reduced through the purification process.

20.4.1.3 Inorganic Agents

The use of inorganic nanoparticles (NPs) is another strategy for fighting the ESKAPEE bacteria. Inorganic NPs composed of transition metals such as silver (Ag), copper (Cu), zinc (Zn), and their oxides are some of the most significant inorganic NPs since they can provide a broad-spectrum bactericidal activity and act as active reservoirs, in contrast to the burst release often observed with the use of antibiotics (Ditaranto et al. 2018; Lemire et al. 2013). Moreover, they present long shelf life, drug stability, and safety to human health even in concentrations that are already efficient to inhibit bacterial growth of various microorganisms. Inorganic NPs present antibacterial activity through multiple potential mechanisms of action, including bacteria cell wall and membrane alterations, oxidative stress induction caused by an intracellular generation of reactive oxygen species (ROS), essential proteins/enzymes and DNA inhibition, metabolic pathway disruption, and metal ion intracellular accumulation (Slavin et al. 2017). These alterations lead to bacterial damage resulting in bacterial death. Although these are the most reported mechanisms of action for antibacterial activity, there are other very important factors in the interaction of NPs with bacteria (Gold et al. 2018). Physicochemical properties such as the zeta potential, size, and shape of inorganic NPs can also affect their capacity to damage bacteria. Smaller NPs generally exhibit greater antibacterial activity due to a higher surface area-to-volume ratio compared to larger NPs (Dizaj et al. 2014; Linklater et al. 2020). This leads to enhanced ROS generation, damaging essential biomolecules such as DNA, proteins, and lipids. However, some studies indicate that larger NPs can be more effective, suggesting that size is not the sole determinant of toxicity. The shape of NPs also influences the antibacterial activity, although the spherical shape is most reported, other shapes such as sheets, plates, tube cubes, and rods have been studied. Nanocubes and rods seem to be more effective than other shapes, possibly due to the exposed planes and the oxidation levels of metals. Cerium dioxide (CeO_2) nanosheets exhibited stronger antibacterial activity than

CeO_2 NPs. Both were synthesized similarly, but nanosheets had a larger surface area and higher oxygen vacancy concentration, resulting in enhanced ROS generation and greater antibacterial efficacy (Abbas et al. 2016). The presence of defects, corners, or edges in the nanoparticle structure acts as abrasive elements that increase the antibacterial activity (Pelletier et al. 2010). The charge of NPs displayed an important factor in the antibacterial activity of materials. Positively charged NPs have been shown to increase antibacterial activity because they are electrostatically attracted to the negatively charged bacterial cell wall, increasing their efficacy.

Also, the antibacterial activity of the NPs is influenced by the structure of bacteria. Based on their cell wall structure, bacteria are categorized as gram-positive and gram-negative bacteria. Gram-positive have a thick peptidoglycan layer, while gram-negative bacteria possess a thin peptidoglycan layer along with an outer lipopolysaccharide membrane (Dörr et al. 2019). Several studies suggest that gram-negative bacteria may be more susceptible to nanoparticle effects due to these structural differences. Gram-positive bacteria have thicker peptidoglycan layers that may act as a protective barrier against nanoparticle entrance and ion release into the cell.

For this purpose, several metal and/or metal oxide NPs have been combined with encapsulation/stabilization into electrospun fibers. The strategy of using reduced amounts of metal combined with encapsulation is expected to enable the production of environmentally sustainable materials, non-toxic to human health, and have reduced nanotoxicology problems associated. In this section, different studies incorporating different types of metals or metal oxides in electrospun fibers and their ability to inhibit the growth of several ESKAPEE bacteria will be addressed.

There are two general approaches to incorporate inorganic NPs into polymers, in situ and ex situ. In situ refers to the formation of NPs inside a polymer matrix by decomposition processes to reduce the size of NPs to less than 100 nanometers. Ex situ refers to the production of NPs before incorporating the polymer matrices (León-Buitimea et al. 2022).

A study performed by Rouhi and Ebrahimi (2023) used the in situ and ex situ methods to investigate the antibacterial and wound healing properties of PCL electrospun nanofibers loaded with Ag. In the in situ method, PCL was mixed with a solvent blend with reducing agents and silver nitrate ($AgNO_3$) was added in different concentrations of 0.5, 1, and 1.5 wt%. Electrospinning was performed after 7 days, once Ag NPs had formed. In the ex situ method, PCL and Ag NPs (<100 nm) were mixed in the solvent blend before electrospinning. The antibacterial activity of PCL-Ag nanofibers was assessed by AATCC100 standard test *"Test method for Antibacterial Finishes on Textile Materials."* The results showed a reduction in the number of bacterial colonies for both *S. aureus* and *E. coli,* with more effectiveness on *S. aureus*. Ag NPs showed higher destructive ability against *S. aureus* due to their sensitive structure characteristic of gram-positive bacteria. The electrospun nanofibers prepared using the in situ method exhibited higher antibacterial activity (in situ 0.5 wt%—50%; 1.5 wt%—64% of *E. coli* reduction; 0.5 wt%—54%; 1.5 wt%—70% of *S. aureus* reduction); compared to those prepared with ex situ method (ex situ 1.5%—22 wt%; of *E. coli* reduction; 1%—27 wt%; of *S. aureus* reduction). These differences are attributed to the smaller size of Ag NPs formed in the in situ method, leading to more Ag NPs on the surface of the fibers, which are known to be effective against bacteria (Rouhi and Ebrahimi 2023). Although the antibacterial results did not show a high antibacterial efficacy, since there was no reduction of more than 1 log for any of the tested conditions, the authors consider that PCL-Ag electrospun nanofiber mats produced by in situ method are a good alternative for antibacterial and medical applications.

The antibacterial activity of Ag was shown in the nineteenth century, and it is commonly used for skin treatment, food applications, and coatings for orthopedic and vascular devices (Prabhu and Poulose 2012). However, the use of Ag presents a considerable problem due to its nonessential nature to the host and potential to cause toxic effects. A strategy employed by many researchers to mitigate Ag NP cytotoxicity is using polymers as capping agents or for controlled release (Abdallah et al. 2020; Panthi et al. 2020). Spagnol et al. (2018) used PVA as capping agents in Ag NPs, which were tested against *S. aureus, E. coli, and P. aeruginosa* and showed the ability to inhibit bacterial growth (Spagnol et al. 2018).

Karuppannan et al. (2021) synthesized copper oxide (CuO) NPs and infused them into PCL/gelatin nanofibers in the ratio of 1, 8, and 4 wt%, respectively, as an antibacterial wound dressing. The produced nanofibers with a mean fiber diameter of 135.3 ± 11.8 nm showed an in vitro initial burst release of CuO NPs followed by a sustained release up to 48 h. The antibacterial potential of these nanofibers was observed against wound pathogens such as S. *aureus*, MRSA, *Pseudomonas aeruginosa*, and *E. coli*. The antibacterial activity was assessed by Kirby-Bauer disk diffusion assay showing inhibition zones between 30.5 and 50 mm for the tested bacteria and 10 × 10 mm nanofiber membrane (Karuppannan et al. 2021; Ramalingam et al. 2019). Also, the antibacterial activity was evaluated using bacterial cell viability assay under proliferating conditions, which showed a reduction of more than 4-log in the number of colony-forming units of *S. aureus* and MRSA and more than 5-log for gram-negative bacteria (Karuppannan et al. 2021).

Zinc oxide (ZnO) NPs have been extensively used due to their well-known thermal stability and antimicrobial agent (Fan and Lu 2005). According to the Artifon et al. (2019) study, different ZnO NP contents (0, 10, 30, and 50 w/w%) were used in polyetherimide (PEI) suspensions (15 w/v%) through electrospinning process to create electrospun composite fibers. This study was carried out under harsh environmental conditions including laser ablation and solvent corrosion to assess antibacterial effectiveness, morphology changes, and contact angle, among other properties. Antibacterial results evaluated using an adapted version of the Japanese Industrial Standard (JIS) Z 2801 demonstrated an inhibition rate of up to 100% for *S. aureus* and 99% for *E. coli* after 24 h. For this purpose, 60 μL of inoculum was applied to the fibers' surface of 40 × 40 mm^2 and a 24 × 24 mm^2 coverslip was placed over the microbial suspension. After the recovery of the bacteria and the subsequent process of serial dilutions, the solution followed the conventional plate counting (Artifon et al. 2019). Composite fibers with a ZnO/PEI ratio of 50% demonstrated notable antibacterial efficacy against both bacteria. The simulated harsh conditions used in this study had no significant impact on bacterial inhibition and the morphological properties of fibers (Fig. 20.7a). Additionally, electrospun fibers with different loadings of ZnO NPs have also been shown to be effective against other types of ESKAPE bacteria including the gram-negative *K. pneumoniae* (Lee 2009; Abdelhafiz et al. 2022; Geetha et al. 2022; Dural Erem et al. 2011), *A. baumannii,* and *P. aeruginosa* (Ahmed et al. 2018; Salam et al. 2021; Mahamuni-Badiger et al. 2020). Pranjali et al. (Mahamuni-Badiger et al. 2020) reported the antibacterial and antibiofilm activity of microfibers composed of ZnO NPs (1, 3, and 5 wt%) with poly(3-hydroxybutyrate-co-3-hydroxyvalerate) (PHBV)/PEO in a 4:1 ratio, for pathogenic bacteria gram-positive *S. aureus* and gram-negative *P. aeruginosa.* The antibacterial activity assessed by disk diffusion assay showed that

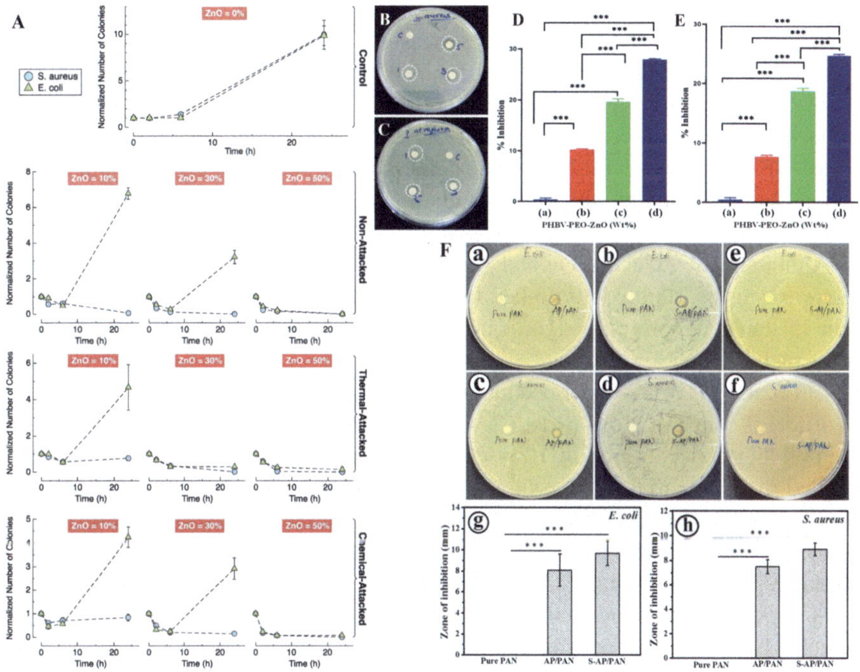

Fig. 20.7 (**a**) Microbial behavior of bacteria *S. aureus* and *E. coli* in contact with composite fibers in the absence and in the presence of ZnO (non-attacked, thermal attacked, and chemical attacked samples). Adapted from (Artifon et al. 2019). Antibacterial activity of neat PHBV/PEO and PHBV/PEO/ZnO (1, 3, and 5%) microfibers against pathogenic bacteria (**b**) *S. aureus* and (**c**) *P. aeruginosa*. Antibiofilm activity of (a) neat PHBV/PEO, (b) PHBV/PEO/ZnO (1 wt%), (c) PHBV/PEO/ZnO (3 wt%), and (d) PHBV/PEO/ZnO microfibers (5 wt%) against (**d**) *S. aureus* and (**e**) *P. aeruginosa*. Adapted from (Mahamuni-Badiger et al. 2020). (**f**) Disk diffusion test on *E. coli* (a, b) and *S. aureus* (c, d) by pure PAN, AP/PAN, and S-AP/PAN. Panels (e) and (f) represent disk diffusion test on *E. coli* and *S. aureus*, respectively, by pure PAN and S-AP/PAN in dark condition. Zone of inhibition of AP/PAN and S-AP/PAN on *E. coli* (g) and S. aureus (h) compared to Pure PAN statistically assessed by one-way ANOVA using SPSS software. $p < 0.05$ considered significant and represented as $^{*}p < 0.05$, $^{**}p < 0.01$ and $^{***}p < 0.001$. (Adapted from (Panthi et al. 2020))

PHBV/PEO/ZnO (5%) microfibers have a greater antibacterial activity than the other microfibers against *S. aureus* (inhibition zone—4 mm) as compared with *P. aeruginosa* (inhibition zone—3 mm) (Fig. 20.7b, c)). Similar results were also shown for antibiofilm activity of neat PHBV/PEO, PHBV/PEO/ZnO (1, 3, and 5 wt%) microfibers against the same bacteria, indicating that PHBV/PEO/ZnO (5 wt%) showed the maximum percentage of biofilm inhibition (28.17%) in *S. aureus* as compared with *P. aeruginosa* (24.51%) (Fig. 20.7d, e). The percentage of biofilm inhibition was increased with the increase in ZnO content in the PHBV/PEO microfibers (Mahamuni-Badiger et al. 2020).

The PLA electrospun fibers with different concentrations of titanium dioxide (TiO_2) were reported, and their bactericidal properties were determined. Toniatto et al. (2017) incorporated TiO_2 NPs (1, 3, and 5% wt) into PLA electrospun fibers (9 wt%) and *S. aureus* bacterial strain was used to study the antibacterial performance of fabricated fibers. The authors measured the absorbance (490 nm) after 6 h and 24 h of PLA/TiO_2 nanofiber (5 x 5 mm^2) incubation. At 6 h, higher TiO_2 concentrations (3 and 5 wt%) exhibited significant bacteria reduction. However, the bactericidal effect increased from 30% to 60% after 24 h, particularly in PLA/5 wt% TiO_2 (Toniatto et al. 2017). The authors suggest that this effect on nanofibers is notably influenced by ultraviolet irradiation activation of TiO_2 NPs, which show potential as contact-killing antimicrobial materials.

Although ZnO, CuO, TiO_2, and Ag show effectiveness against a wide range of microorganisms, the toxicity linked to heavy metals and their potential accumulation in the body presents significant risks, which leads to the need to explore other metals.

Liu et al. (2021a) incorporated magnesium oxide (MgO) NPs (~50 nm diameter), in different concentrations (0.1 to 2 w/v%), on PCL/GEL nanofiber membranes at the polymer concentration of 10 w/v% with a ratio of 8:2, by electrospinning and studied their potential for dressing infected wounds. The antibacterial activity of MgO-incorporated electrospun membranes was effective against gram-positive *S. aureus* and *S. epidermidis* and gram-negative *E. coli* with the ability to inhibit bacterial growth is proportional to the MgO NP concentration. The membrane with 0.5% MgO inhibited more than 98% *E. coli*, 90% *S. aureus,* and 94% *S. epidermidis* in vitro, promoted cell growth, and accelerated wound healing in a rat model (Liu et al. 2021a). In another study, the same authors investigated the potential of membranes with MgO NPs, changing the polymeric matrix to PLA/GEL. This study also demonstrated the membrane's ability to inhibit the growth of *S. aureus* and *E. coli,* in a dose-dependent manner, besides its osteogenic potential and periodontal regeneration through an in vivo rat model (Fig. 20.8a, b) (Liu et al. 2020). MgO NPs were also used in the development of a coaxial electrospun nanofiber to guide bone regeneration by combining PCL with a parathyroid hormone (PTH) in the core layer and MgO NPs (diameter <50 nm) with PCL in the shell layer (PCL: MgO NPs (1:0.3)) (Dong et al. 2023). The antibacterial activity was evaluated after 24 h using MTT assay. MgO-PCL/PTH-PCL membranes exhibited bacterial death of 95% for *S. aureus* and 89% *E. coli* due to the MgO NP release.

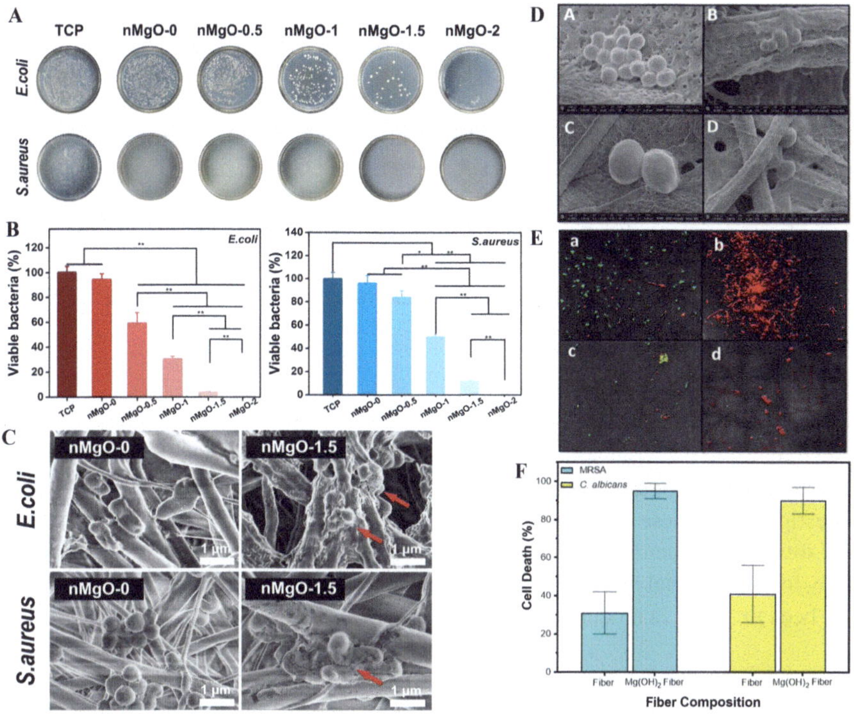

Fig. 20.8 Antibacterial activity of tissue culture plate (TCP) and MgO NP-incorporated membranes against *E. coli* and *S. aureus*. (**a**) Digital picture of agar plates with CFUs (Liu et al. 2020); (**b**) quantitative viable bacterial determined from corresponding CFUs (Liu et al. 2020). (**c**) SEM images of bacteria after 24 h of incubation with MgO-0 and MgO-1.5 NPs. Statistical analysis was performed by the one-way ANOVA followed by Tukey's multiple comparison test, $*p < 0.05$, $**p < 0.01$. Adapted from (Liu et al. 2020). (**d**) High-resolution scanning electron microscopy images of PEO/PCL bandage fibers and (**e**) confocal laser scanning microscopy, representing the viability of MRSA and *C. albicans* on control and $Mg(OH)_2$ nanosheet fibers, respectively (Truskewycz et al. 2021). Control fibers with (A,a) MRSA and (C,c) *C. albicans*; $Mg(OH)_2$ nanosheets with (B,b) MRSA and (D, d) *C. albicans* (Truskewycz et al. 2021). (**f**) Percentage of MRSA and C. albicans cell death following exposure to the electrospun bandages with and without $Mg(OH)_2$ incorporation. (Adapted from (Truskewycz et al. 2021))

Fluorescent magnesium hydroxide ($Mg(OH)_2$) nanosheets were integrated into biocompatible electrospun fibers and revealed the bactericidal and fungicidal capacity against relevant bacteria and fungi. $Mg(OH)_2$ nanosheets were added (1% in methanol) to the PEO/PCL solution at 20 and 5 w/v%, respectively, in a ratio of 2:1. The antimicrobial activity was assessed through an 18 h incubation of 5 × 5 mm electrospun membranes with MRSA and *C. albicans* (Truskewycz et al. 2021). The cell viability of the adhered bacteria to the membranes was evaluated using a live/dead Baclight bacterial viability kit and imaged with confocal microscopy (Fig. 20.8d–f). The integration of $Mg(OH)_2$ nanosheets in the electrospun fibers caused irreversible damage to microbial cells with 95% of cell death for MRSA and 90% for fungal strain (Truskewycz et al. 2021).

Recently, cerium oxide (CeO) NPs have been used for applications in wound healing due to the antioxidant properties and ROS scavenging activity conferred by crystal structure alternations between Ce III and Ce IV oxidation states (Rather et al. 2018). CeO NPs were synthesized and incorporated (25 v/v% of the 20 mM CeO NP solution) into the PCL/gelatin nanofibers at 10 and 20 w/v%, respectively, and showed antibacterial activity against *P. aeruginosa* and *S. aureus* (Fei et al. 2021). The antibacterial properties of Ce NPs were also verified in different electrospun nanofibers showing effective toxicity against both gram-positive (*S. aureus*) and gram-negative (*E. coli, K. pneumoniae, and P. aeruginosa*) bacterial strains showing a concentration-dependent effect of NPs (Unnithan et al. 2014; Zamani et al. 2021).

20.4.2 Advanced Approaches

When multiple bacterial strains or species coexist within a host, we are facing a co-infection that introduces distinctive complexities in clinical practice and scientific investigations. These instances can intricately entangle the processes of diagnosis and treatment since diverse bacteria may interact synergistically or respond differently to antibacterial therapies. Based on this, the search for multi-agent approaches has been growing in recent years.

20.4.2.1 Multi-Agent

Chronic wounds comprise traumatic, pressure, and ischemic ulcers that represent a serious public health problem without an effective solution and for which therapies based on a single antibacterial agent may be insufficient (Maslova et al. 2021; Kvich et al. 2020; Johani et al. 2017). The characteristics of the wound site, including the anatomical location, microbiome composition, and moisture content, play a key role in determining the range of bacteria observed in the wound bed (Sachdeva et al. 2022). In this context, multiple bacterial species can simultaneously affect the same injured site; therefore, antibacterial formulations combining different agents can be advantageous to provide a synergetic therapeutic approach (Jolivet-Gougeon and Bonnaure-Mallet 2019). For advanced chronic wounds, co-infections caused by multiple bacterial species can be found under a suitable environment for the proliferation of such microorganisms (Johani et al. 2017; Fischer et al. 2021). An example of competitive bacterial coexistence that evolves to cooperative action of two bacterial species is the adaptation and enhancement of antibiotic resistance by *S. aureus* during coexistence with *P. aeruginosa* as observed in patients suffering from cystic fibrosis (Briaud et al. 2019, 2020). Several research studies evaluate the efficiency of multiple strategies to fight co-infections, and promising approaches include the use of electrospun nanofibers as carriers as explored by Kielholz et al. (2023). In this study, specific bacteriophages selectively target bacteria and destroy their immune system or unbalance their activity. Electrospun nanofibers of PVP (14 w/v%) incorporating *Staphylococcus* bacteriophage EBHT and *Pseudomonas* phage JG004 were developed for antibacterial effect against *S. aureus* and *P. aeruginosa*, respectively (Kielholz et al. 2023). In this situation, co-infection benefits from the synergetic effect of nanofibers based on multi-antibacterial agents, which can promote collaborative action. Based on this, Federico et al. (2022) also explored a dual approach and evaluated the performance of graphene oxide (GO, 0.02 and 0.03 w/v%) incorporated in electrospinning solution of PVA (5 w/v%) with CIP (loading of 1.8 w/w%) (Federico et al. 2022). After three cycles of UV irradiation, antibacterial activity against *P. aeruginosa* and *S. aureus* electrospun nanofibers including both CIP and GO shown to have more than 1.5 and 2 log reduction compared to single-agent versions without CIP with 1.5 and 0.5 log reduction, respectively (Fig. 20.9). Therefore, the high prevalence of multi-resistant bacteria combined with scenarios for co-existence highlights the importance of developing complex nanofibers based on multiple agents able to act under different antibacterial mechanisms and reach a broad spectrum of bacteria (Puca et al. 2021).

In fact, different strategies can be adopted to develop more effective antibacterial nanofibers based on a multi-agent approach. However, the multiple active agents can be incorporated within the nanofibers and the polymeric material itself does not contribute to the antibacterial activity, apart from the physical barrier offered (Merzougui et al. 2022). In this situation, the antibacterial effect is primarily derived from the active agents present in the nanofibers (Sethuram and Thomas 2023). However, the polymeric materials of electrospun nanofiber may have intrinsic bioactivity enhanced by antibacterial polymers such as CS (Ardila et al. 2016; Kalantari et al. 2019). In both situations, polymeric nanofibers serve as a delivery system of antibacterial agents providing a platform for their controlled release and contact with bacteria (Homaeigohar and Boccaccini 2020). The incorporation of multiple antibacterial agents in the nanofiber is the first step to obtaining a dual-functional structure with favorable wound healing and antimicrobial-promoting characteristics, which is a compelling solution for tissue engineering applications (Homaeigohar and Boccaccini 2020). Multi-agent approaches include the simultaneous usage of

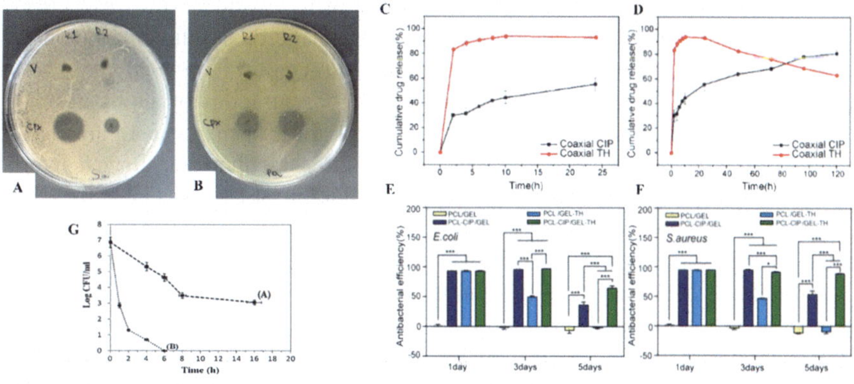

Fig. 20.9 (**a**, **b**) Antimicrobial activity was determined by agar diffusion for *S. aureus* (a) and *P. aeruginosa* (b) without CPX (V) or with CPX (1.8% w/w, CPX load of approximately 15 μg) (CPX). R1 and R2 are duplicates. Adapted from (Federico et al. 2022); (**c**–**f**) adapted from in vitro drug release of CIP and TH after 24 h and 120 h. Antibacterial effects of different nanofibers by CCK-8 assay ($n = 3$). * $-p < 0.05$; ** $p < 0.01$; *** $p < 0.001$ (Lin et al. 2022); (**g**) inhibitory zone against *Staphylococcus aureus* and *Escherichia coli* for different contents of drug-loaded nanofibers. (Adapted from (Hu et al. 2018))

different metallic and metallic oxide NPs, natural polymers, and antibiotics, all with different targets and distinct active mechanisms to achieve a broad spectrum of bacteria and maximize the antibacterial effect (Merzougui et al. 2022). The recent studies conducted in this domain concerning multi-resistant bacteria highlight the inhibition rates achieved through the incorporation of antibacterial agents within electrospun nanofibers (Topcu et al. 2021).

Currently, the treatment of infections caused by multi-resistant bacteria involves as a clinical gold standard the intravenous administration of multiple antibiotics to increase the spectrum of bacterial species covered (Dodds 2017). In this way, nanofibers appear as an alternative to the administration of conventional therapy by offering a vehicle for local drug delivery while secondary effects are minimized (Topcu et al. 2021). For this reason, nanofibers can incorporate different antibiotics simultaneously and be released under controlled conditions, which are achieved based on the development of advanced electrospun structures (Topcu et al. 2021; Homaeigohar and Boccaccini 2020). To this end, Lin et al. (Lin et al. 2022) developed an electrospun wound dressing using coaxial production for simultaneous incorporation of CIP and tetracycline hydrochloride (TH) on PCL (9 w/v%) and gelatin (10 w/v%), respectively, with a total of 20 mg of antibiotics loaded per mesh (Lin et al. 2022). The quick dissolution of gelatin allows a faster release of TH (1 w/w%) with 83.51% released in 2 h, compared to CIP (1 w/w%) incorporated on PCL which required 5 days to approximately achieve the same released amount (Fig. 20.9). Therefore, the initial antibacterial efficacy of electrospun nanofiber against *E. coli* and *S. aureus* is due to the release of large concentrations of TH and complemented and maintained over time by the prolonged action of CIP (Lin et al. 2022). After 5 days, the antibacterial effect of electrospun nanofibers was still over 50% and higher against *S. aureus* than *E. coli* (Lin et al. 2022). Also, Alizadeh et al. (Alizadeh et al. 2023) showed the synergetic effect of colistin and meropenem incorporated on an electrospinning solution of PVA (12 wt%) and CS (3 wt%) prepared in a ratio of 1:3, respectively (Alizadeh et al. 2023). The antibacterial activity of the dual antibiotic nanofibers tested using colony count and disk diffusion methods against *A. baumannii*, *P. aeruginosa,* and *E. coli* was compared for different concentrations of antibiotics. The results show that 128 μg/mL of colistin and 256 μg/mL of meropenem are slowly released from the nanofibers over 7 days. This multi-agent approach showed significant efficacy against the extensively drug-resistant clinical isolates (17×10^6 CFU/mL) and ATCC 19606 (1.7×10^6 CFU/mL) of *A. baumannii* compared to the control situation ($>30 \times 10^6$ CFU/mL) (Alizadeh et al. 2023).

Jatoi et al. (2019) explored the antibacterial activity of composite NPs of TiO_2 and Ag incorporated in electrospun nanofibers of cellulose acetate (17 wt%). The antibacterial activity was tested against *E. coli* and *S. aureus* (Jatoi et al. 2019). The capacity of TiO_2 to react with water and oxygen molecules and generate oxidative species that damage bacteria and DNA strains has been reported in the literature (Chen et al. 2019; Wang et al. 2020). However, for single TiO_2 NPs incorporated in cellulose acetate nanofibers, a hollow width was not observed (Jatoi et al. 2019). Therefore, Ag NPs were simultaneously incorporated to enhance the antibacterial effect of TiO_2 and to provide an effective solution based on synergetic the effect of both TiO_2 and Ag NPs (Jatoi et al. 2019; Soo et al. 2020). The affinity of Ag^+ with bacteria cell wall, which leads to their accumulation and consequent structural damage, is a complementary action mechanism that contributes to enhancing the antibacterial activity of the nanofibers as proved by the diffusion method. The results showed a halo width between 1.73 and 1.80 mm, for *E. coli* and *S. aureus*, respectively (Jatoi et al. 2019). The simultaneous incorporation of metallic NPs proved to be effective on *E. coli* and *S. aureus* inactivation or killing including electrospun nanofibers incorporating iron (Fe), Zn/ZnO, Ag, and Cu/CuO (Jatoi et al. 2019; Soo et al. 2020; Yasin et al. 2018; Ashfaq et al. 2016; Kharaghani et al. 2018; Sekar et al. 2019; Jia et al. 2019). Sekar et al. (2019) developed electrospun nanofibers of PVA (8.5 w/v%) incorporating Fe-doped Zn NPs (in different weights of 0, 4, 8, and 12 wt%) to create an antibacterial and biocompatible material for biomedical applications. For the highest concentration of Fe-doped Zn NPs, a halo of inhibition of 14 ± 0.4 mm and 19 ± 0.3 mm was measured for *E. coli* and *S. aureus*, respectively (Sekar et al. 2019). Therefore, the solution developed was more effective on gram-positive than gram-negative bacteria. Besides, Kharaghani et al. (2018) evaluated the synergetic effect of Ag and Cu nanoparticles (0.5 g/L) incorporated by reduction on PVA/CS nanofibers with a polymer weight ratio of 8:2 (Kharaghani et al. 2018). After evaluation through disk diffusion, it was possible to observe an enhancement of the antibacterial activity of nanofibers incorporating both Ag and Cu NPs compared to these NPs alone against *E. coli* (Kharaghani et al. 2018). In this situation, the PVA/CS nanofibers incorporating Ag NP alone are shown to have a higher antibacterial effect than PVA/CS nanofibers with CuO NPs for the same concentrations and size of particles. This conclusion agrees with the previous results reported in the literature as published by Singh et al. (2013), where the antibacterial activity of Cu and Ag incorporated in non-electrospun nanofibers of activated carbon was evaluated (Singh et al. 2013). The study was conducted under 72 h, and it was shown that for long-term application Ag is more effective than Cu against *E. coli* and as effective as Cu against *S. aureus* (Singh et al. 2013). This study shows the favorable antibacterial effect of both metal-based NPs for short-term applications while highlighting the different performance of both Ag and Cu as antibacterial agents for long-term usage. More recently, it was proven that combining Ag and Cu can have benefits due to the synergetic effects (Kharaghani et al. 2018). The synergetic effect of Ag combined with ZnO was also evaluated after incorporation in PVP/PCL nanofibers with a total polymer concentration of 12 wt% in a ratio of 1:3, respectively (Hu et al. 2018). The nanofibers were characterized after 20 h in contact with *S. aureus* and *E. coli*. In this situation, the nanofibers containing Ag proved to be more effective than nanofibers containing only ZnO (Fig. 20.9). However, the maximum antibacterial activity was observed for PVP/PCL nanofibers with 0.25% of ZnO and 0.75% of Ag (Hu et al. 2018). Besides, Jiang et al. (2022) observed the synergetic effects of Ag and silicon dioxide (SiO_2) encapsulated in polyacrylonitrile (PAN) nanofibers and polyquaternary ammonium salt (PMT) (Jiang et al. 2022). To prepare the electrospinning solution, 1.2 g PAN and 8.5 mg $AgNO_3$ were added to 10 mL DMF. Then, the compounds to perform PMT reactions were also added to the electrospinning solution as described before (Jiang et al. 2022). For the nanofibers incorporating PMT and Ag/SiO_2 separately, the nanofibers including PMT proved to be more effective against *E. coli* and as effective as Ag and SiO_2 against

S. aureus. After combining both PMT and Ag/SiO_2, there was a significant reduction of 448 and 192 μg/mL in the MIC for *E. coli* and *S. aureus*, respectively, compared to PMT alone (Jiang et al. 2022).

However, the intrinsic antibacterial properties of the polymers have been explored since the polymeric material used to create the nanofibers may possess inherent antibacterial activity. In this case, the polymeric nanofibers function as both a structural and bioactive role in antibacterial activity and can be combined with other antibacterial agents. Polymeric materials with intrinsic antibacterial activity such as CS have been widely explored to produce electrospun nanofibers with antibacterial activity and enhance their action by incorporating additional antibacterial agents (Antunes et al. 2015). Hang et al. (2010) showed that electrospun nanofibers of PVA (12 wt%) and CS (4 wt%) with proven antibacterial behavior validated through the CFU method can enhance their activity against *E. coli* after incorporation of silver nitrate (1 wt%) (Hang et al. 2010). This approach has been reported by others and focuses on the incorporation of positively charged molecules to increase the positive charge of the structure and its electric potential (Slavin et al. 2017). Consequently, the positively charged nanofibers allow the binding with the bacteria wall, which leads to changes in bacteria wall potential with severe impact on their normal activity (Slavin et al. 2017). Also, Liu et al. (2021b) developed blended PVA (10 w/w%) and CS (2 w/w%) nanofibers to provide target delivery of moxifloxacin (3.6 mg/mL) (Liu et al. 2021b). In this case, the diffusion method used to evaluate the performance of nanofibers did not show the antibacterial effect of CS in this work (Liu et al. 2021b). Other studies using inhibition zone methodology report the same inactivity of CS (Bakhsheshi-Rad et al. 2020). Consequently, depending on the validation test chosen to assess antibacterial activity, the results may differ; therefore, it is crucial to follow the standard guidelines defined for each situation (ASTM E2922 2023; ASTM E2149 2023; ISO 20776-2:2021(en) 2023). Since the disk diffusion method is more suitable for the validation of antibacterial agents that are released than when they act by direct contact, the antibacterial activity of electrospun nanofibers would benefit from cross-validation with other methods such as broth and agar dilution (Perez-Gavilan et al. 2021; Balouiri et al. 2016). The most frequent practice for the development of multi-agent nanofibers includes the simultaneous use of antibacterial agents from different families such as drugs, metal-based nanoparticles, and polymers with distinct action mechanisms that require diversity in the validation methods adopted (Balouiri et al. 2016).

The wide range of possibilities for the functionalization of electrospun nanofibers opens doors for the development of more complex multi-agent systems composed of multi-materials and several layers with different release rates. The versatility of the technique makes it possible to combine more than one active agent in electrospun fibers. This wide range of options makes it possible to explore new synergies between antibacterial compounds and therefore cover a greater number of possible interactions with pathogens (Abadi et al. 2022).

Table 20.1 provides a sum up of several studies developed including antibacterial agents into electrospun nanofibers to provide local and effective solution to mono- or co-infections.

Table 20.1 List of potential electrospun nanofibers to inhibit multidrug-resistant bacteria

Multidrug-resistant bacteria	Nanofibers materials	Type of antimicrobial agent	Antimicrobial agent	Antibiotic class and inhibition mechanism	Application	Ref.
Acinetobacter baumannii	Chitosan and PVA	Antibiotic	Colistin/ meropenem	Carbapenems: Disruption of bacterial cell membrane/inhibition of cell wall synthesis	Infected wounds	Alizadeh et al. (2023)
Pseudomonas aeruginosa	Cellulose acetate and lignin polymers	Inorganic agent	Copper complexes	n.a	Diaper dermatitis	Elsherbiny et al. (2022)
	Chitin nanofibers sheets	Inorganic agent	Silver nanoparticles	n.a	Antimicrobial biomaterials	Nguyen et al. (2014), Huang et al. (2022), Alven et al. (2021) and Liu et al. (2021a)
	Polybutylene adipate-co-terephthalate (PBAT) and gelatin	Antibiotic	Doxycycline	Tetracycline: Inhibition of protein synthesis (S30)	Antimicrobial wound dressing	Varshosaz et al. (2020)
	Gelatin and chitosan	Organic agent	Phlorotannin extract	n.a	Antimicrobial wound dressing	Ferreira et al. (2021a)
	PVP	Antibiotic	Ciprofloxacin	Fluoroquinolones: inhibition of DNA topoisomerases	DDSs in a wound infection model	Rancan et al. (2019)
Klebsiella pneumoniae	PCL	Antibiotic	Rifampicin	mRNA synthesis	Ureteral regeneration	Gruppuso et al. (2022)
	Polyacrylonitrile blended with polyvinylidene fluoride	Multi-agent	*Garcinia mangostana L.* extract and silver nanoparticles	n.a	Antimicrobial air filters	Snetkov et al. (2022), Ikuta et al. (2022), Saikaew and Intasanta (2021) and Saleemi et al. (2020)

(continued)

Table 1 (continued)

Multidrug-resistant bacteria	Nanofibers materials	Type of antimicrobial agent	Antimicrobial agent	Antibiotic class and inhibition mechanism	Application	Ref.
Escherichia coli	PVA	Organic agent	Allyl isothiocyanate (AITC) and β-cyclodextrin complex	n.a	Biomedical applications	Aytac et al. (2014) and Kim and Lee (2014)
	PVP and gelatin PEO and chitosan (core-shell)	Antibiotic	Vancomycin and imipenem/cilastatin	No lactam glycopeptide/ carbapenems: Inhibition of cell wall synthesis	Diabetic foot ulcer	Davani et al. (2021)
	Silk fibroin and hardystonite	Antibiotic	Gentamicin	Amino-glycoside: inhibition of protein synthesis (S30)	Combat implant-related infections	Hadisi et al. (2020)
	PCL and gelatin	Inorganic agent	Magnesium oxide nanoparticles	n.a	Infected wounds	Nguyen et al. (2014), Huang et al. (2022), Alven et al. (2021) and Liu et al. (2021a)
Staphylococcus aureus	Cellulose acetate	Organic agent	Eucalyptus oil and chitosan nanoparticles	n.a	Wound healing activities	Elbhnsawi et al. (2023)
	PEO and alginate	Antibiotic	Vancomycin	No lactam glycopeptide: Inhibition of cell wall synthesis	Skin infections	Fathi et al. (2020)
	PCL and gelatin	Inorganic agent	Copper oxide nanoparticles	n.a	Wound dressing	Karuppannan et al. (2021) and Ramalingam et al. (2019)
	PLGA and PULL	Antibiotic	Amoxicillin	Aminopenicillins: Inhibition of cell wall synthesis	Long-term antibiotic delivery	Ajalloueian et al. (2022)

20.5 Conclusion

The multidrug-resistant bacteria represent a global health challenge. These types of bacteria have developed resistance to multiple antibiotics, increasing the difficulty to treat infections. In fact, this resistance is a consequence of overuse and misuse of antibiotics, as well as inadequate infection treatments. Multidrug-resistant bacteria can lead to prolonged illnesses, increased healthcare costs, and even life-threatening conditions when effective treatments are limited. Based on this, the demand for new therapeutic agents and new therapeutic approaches is needed. The use of electrospun nanofibers for local treatment has been growing due to the ability to incorporate different antibacterial agents, single or multi-agents, and due to the ability to provide a sustainable and controlled antibacterial release. Indeed, electrospun nanofibers have emerged as a promising alternative. They can offer not only innovative treatment methods but also include new antibacterial agents that bacteria have not developed resistance to yet.

Acknowledgements This work was financially supported by the Fundação para a Ciência e a Tecnologia FCT/MCTES (PIDDAC) through the following projects: UIDB/04044/2020; UIDP/04044/2020; Associate Laboratory ARISE LA/P/0112/2020; PTCentroDiH (03/C16-i03/2022-768), "OP-SMARTherapy, 2022.04238.PTDC," and "InnovaBIOMAS, 2022.10564. PTDC, INOV.AM—Inovação em Fabricação Aditiva, 02-C05-i01.01-2022. PC644865234-00000004 (WP14)." This study was also supported by the European Union through PREDICTOS "PREDICTOS, 101079372." This work is, also, supported by the Fundação para a Ciência e Tecnologia (FCT) through the grants awarded to Carolina Ferreira (2021.04541.BD), to Sara Guerreiro (2021.05893.BD), and to Tatiana Padrão (2020.09198.BD).

References

Abadi B, Goshtasbi N, Bolourian S, Tahsili J, Adeli-Sardou M, Forootanfar H (2022) Electrospun hybrid nanofibers: fabrication, characterization, and biomedical applications. Front Bioeng Biotechnol 10:986975

Abbas F, Iqbal J, Jan T, Badshah N, Mansoor Q, Ismail M (2016) Structural, morphological, Raman, optical, magnetic, and antibacterial characteristics of CeO2 nanostructures. Int J Miner Metall Mater 23:102–108

Abdallah OM, El-Baghdady KZ, Khalil MMH, El Borhamy MI, Meligi GA (2020) Antibacterial, antibiofilm and cytotoxic activities of biogenic polyvinyl alcohol-silver and chitosan-silver nanocomposites. J Polym Res 27:1–9

Abdelbary MMH, Basset P, Blanc DS, Feil EJ (2017) The evolution and dynamics of methicillin-resistant *Staphylococcus aureus*. In: Genetics and evolution of infectious diseases, 2nd edn. Elsevier, Amsterdam, pp 553–572

Abdelhafiz M, Shalaby ASA, Hussein AK (2022) Preparation and characterization of bioactive polyvinylpyrrolidone film via electrospinning technique. Microsc Res Tech 85:3347–3355

Afshar A, Yuca E, Wisdom C, Alenezi H, Ahmed J, Tamerler C, Edirisinghe M (2021) Next-generation antimicrobial peptides (AMPs) incorporated nanofibre wound dressings. Med Devices Sens 4:1–11

Ahmed R, Tariq M, Ali I, Asghar R, Khanam PN, Augustine R, Hasan A (2018) Novel electrospun chitosan/polyvinyl alcohol/zinc oxide nanofibrous mats with antibacterial and antioxidant properties for diabetic wound healing. Int J Biol Macromol 120:385–393

Ajalloueian F, Asgari S, Guerra PR, Chamorro CI, Ilchenco O, Piqueras S, Fossum M, Boisen A (2022) Amoxicillin-loaded multilayer pullulan-based nanofibers maintain long-term antibacterial properties with tunable release profile for topical skin delivery applications. Int J Biol Macromol 215:413–423

Akturk A, Kasikci FN, Dikmetas DN, Karbancioglu-Guler F, Erol-Taygun M (2023) Hypericum perforatum oil and vitamin a palmitate-loaded gelatin nanofibers cross-linked by tannic acid as wound dressings. ACS Omega 8:24023–24031

Albright V, Xu M, Palanisamy A, Cheng J, Stack M, Zhang B, Jayaraman A, Sukhishvili SA, Wang H, Albright V, Palanisamy A, Sukhishvili SA, Xu M, Cheng J, Zhang B, Wang H, Stack M, Jayaraman A (2018) Micelle-coated, hierarchically structured nanofibers with dual-release capability for accelerated wound healing and infection control. Adv Healthc Mater 7:1800132. https://doi.org/10.1002/adhm.201800132

Alghoraibi I, Alomari S (2018) Different methods for nanofiber design and fabrication. In: Handbook of nanofibers, pp 1–46. https://doi.org/10.1007/978-3-319-42789-8_11-2

Alizadeh S, Farshi P, Farahmandian N, Ahovan ZA, Hashemi A, Majidi M, Azadbakht A, Darestanifarahani M, Sepehr KS, Kundu SC, Gholipourmalekabadi M (2023) Synergetic dual antibiotics-loaded chitosan/poly (vinyl alcohol) nanofibers with sustained antibacterial delivery for treatment of XDR bacteria-infected wounds. Int J Biol Macromol 229:22–34

Allocati N, Masulli M, Alexeyev MF, Di Ilio C (2013) Escherichia coli in Europe: an overview. Int J Environ Res Public Health 10:6235–6254

Álvarez-Martínez FJ, Barrajón-Catalán E, Micol V (2020) Tackling antibiotic resistance with compounds of natural origin: a comprehensive review. Biomedicine 8:405

Alven S, Buyana B, Feketshane Z, Aderibigbe BA (2021) Electrospun nanofibers/nanofibrous scaffolds loaded with silver nanoparticles as effective antibacterial wound dressing materials. Pharmaceutics 13:964

Angelette AL, Rando LL, Wadhwa RD, Barras AA, Delacroix BM, Talbot NC, Ahmadzadeh S, Shekoohi S, Cornett EM, Kaye AM, Kaye AD (2023) Tetracycline-, doxycycline-, minocycline-induced Pseudotumor cerebri and esophageal perforation. Adv Ther 40:1366–1378

Antunes BP, Moreira AF, Gaspar VM, Correia IJ (2015) Chitosan/arginine–chitosan polymer blends for assembly of nanofibrous membranes for wound regeneration. Carbohydr Polym 130:104–112

Ardila N, Medina N, Arkoun M, Heuzey MC, Ajji A, Panchal CJ (2016) Chitosan–bacterial nanocellulose nanofibrous structures for potential wound dressing applications. Cellul 23:3089–3104. https://doi.org/10.1007/s10570-016-1022-y

Artifon W, Pasini SM, Valério A, González SYG, de Ulson SM, de Souza AAU (2019) Harsh environment resistant-antibacterial zinc oxide/Polyetherimide electrospun composite scaffolds. Mater Sci Eng C 103:109859

Ashfaq M, Verma N, Khan S (2016) Copper/zinc bimetal nanoparticles-dispersed carbon nanofibers: a novel potential antibiotic material. Mater Sci Eng C 59:938–947

Ashfaq M, Verma N, Khan S (2017) Highly effective cu/Zn-carbon micro/nanofiber-polymer nanocomposite-based wound dressing biomaterial against the *P. aeruginosa* multi- and extensively drug-resistant strains. Mater Sci Eng C 77:630–641

Assoni L, Girardello R, Converso TR, Darrieux M (2021) Current stage in the development of *Klebsiella pneumoniae* vaccines. Infect Dis Ther 10:2157–2175. https://doi.org/10.1007/s40121-021-00533-4

ASTM E2149 (2023) Standard test method for determining the antimicrobial activity of antimicrobial agents under dynamic contact conditions [Internet]. https://www.astm.org/e2149-20.html

ASTM E2922 (2023) Standard Guide for the use of standard test methods and practices for evaluating antibacterial activity on textiles. https://www.astm.org/e2922-15.html

Aytac Z, Dogan SY, Tekinay T, Uyar T (2014) Release and antibacterial activity of allyl isothiocyanate/β-cyclodextrin complex encapsulated in electrospun nanofibers. Colloids Surf B Biointerfaces 120:125–131

Bakhsheshi-Rad HR, Hadisi Z, Ismail AF, Aziz M, Akbari M, Berto F, Chen XB (2020) In vitro and in vivo evaluation of chitosan-alginate/gentamicin wound dressing nanofibrous with high antibacterial performance. Polym Test 82:106298

Balouiri M, Sadiki M, Ibnsouda SK (2016) Methods for in vitro evaluating antimicrobial activity: a review. J Pharm Anal 6:71

Boucher HW, Talbot GH, Bradley JS, Edwards JE, Gilbert D, Rice LB, Scheld M, Spellberg B, Bartlett J (2009) Bad bugs, no drugs: no ESKAPE! An update from the Infectious Diseases Society of America. Clin Infect Dis 48:1–12

Briaud P, Camus L, Bastien S, Doléans-Jordheim A, Vandenesch F, Moreau K (2019) Coexistence with *Pseudomonas aeruginosa* alters *Staphylococcus aureus* transcriptome, antibiotic resistance and internalization into epithelial cells. Sci Rep 9:1–14. https://www.nature.com/articles/s41598-019-52975-z

Briaud P, Bastien S, Camus L, Boyadjian M, Reix P, Mainguy C, Vandenesch F, Doléans-Jordheim A, Moreau K (2020) Impact of coexistence phenotype between *Staphylococcus aureus* and *Pseudomonas aeruginosa* isolates on clinical outcomes among cystic fibrosis patients. Front Cell Infect Microbiol 10:527225

Bubakir MM, Li H, Barhoum A, Yang W (2019) Advances in melt electrospinning technique. In: Handbook of nanofibers, pp 125–156. https://doi.org/10.1007/978-3-319-53655-2_8

Buzgo M, Mickova A, Doupnik M, Rampichova M (2018) Blend electrospinning, coaxial electrospinning, and emulsion electrospinning techniques. In: Core-shell nanostructures for drug delivery and theranostics: challenges, strategies and prospects for novel carrier systems, pp 325–347

Chaiarwut S, Ekabutr P, Chuysinuan P, Chanamuangkon T, Supaphol P (2021) Surface immobilization of PCL electrospun nanofibers with pexiganan for wound dressing. J Polym Res 28:344

Chee BS, Nugent M (2019) Electrospun natural polysaccharide for biomedical application. In: Natural polysaccharides in drug delivery and biomedical applications. Elsevier, Amsterdam, pp 589–615

Chen WY, Chang HY, Lu JK, Huang YC, Harroun SG, Tseng YT, Li YJ, Huang CC, Chang HT (2015) Self-assembly of antimicrobial peptides on gold nanodots: against multidrug-resistant bacteria and wound-healing application. Adv Funct Mater 25:7189–7199. https://doi.org/10.1002/adfm.201503248

Chen S, Ge L, Mueller A, Carlson MA, Teusink MJ, Shuler FD, Xie J (2017) Twisting electrospun nanofiber fine strips into functional sutures for sustained co-delivery of gentamicin and silver. Nanomedicine 13:1435–1445

Chen KN, Sari FNI, Ting JM (2019) Multifunctional TiO2/polyacrylonitrile nanofibers for high efficiency PM2.5 capture, UV filter, and anti-bacteria activity. Appl Surf Sci 493:157–164

Cochrane Library (2018) Antimicrobial effect of modified antibiotic nanofibers for regenerative endodontics procedures [Internet]. https://classic.clinicaltrials.gov/ct2/show/NCT03690960?term=nanofiber&draw=2&rank=2

Davani F, Alishahi M, Sabzi M, Khorram M, Arastehfar A, Zomorodian K (2021) Dual drug delivery of vancomycin and imipenem/cilastatin by coaxial nanofibers for treatment of diabetic foot ulcer infections. Mater Sci Eng C 123:111975

De Cesare F, Di Mattia E, Zussman E, Macagnano A (2019) A study on the dependence of bacteria adhesion on the polymer nanofibre diameter. Environ Sci Nano 6:778–797. https://pubs.rsc.org/en/content/articlehtml/2019/en/c8en01237g

Dias JR, Alves AIF, Marzia-Ferreira CA, Alves NM (2020) Electrospinning fabrication strategies. In: Electrospun materials and their allied applications, pp 1–52. https://doi.org/10.1002/9781119655039.ch1

Ding X, Wang A, Tong W, Xu F-J, Ding X, Wang A, Tong W, Xu F (2019) Biodegradable antibacterial polymeric nanosystems: a new hope to cope with multidrug-resistant bacteria. Small 15:1900999. https://doi.org/10.1002/smll.201900999

Ding X, Tang Q, Xu Z, Xu Y, Zhang H, Zheng D, Wang S, Tan Q, Maitz J, Maitz PK, Yin S, Wang Y, Chen J (2022) Challenges and innovations in treating chronic and acute wound infections: from basic science to clinical practice. Burns Trauma 10:14

Ditaranto N, Basoli F, Trombetta M, Cioffi N, Rainer A (2018) Electrospun nanomaterials implementing antibacterial inorganic nanophases. Appl Sci 8:1643

Dizaj SM, Lotfipour F, Barzegar-Jalali M, Zarrintan MH, Adibkia K (2014) Antimicrobial activity of the metals and metal oxide nanoparticles. Mater Sci Eng C 44:278–284

Dodds DR (2017) Antibiotic resistance: a current epilogue. Biochem Pharmacol 134:139–146

Dong Y, Yao L, Cai L, Jin M, Forouzanfar T, Wu L, Liu J, Wu G (2023) Antimicrobial and pro-osteogenic coaxially electrospun magnesium oxide nanoparticles-Polycaprolactone/parathyroid hormone-Polycaprolactone composite barrier membrane for guided bone regeneration. Int J Nanomedicine 18:369–383

Dörr T, Moynihan PJ, Mayer C (2019) Editorial: bacterial cell wall structure and dynamics. Front Microbiol 10:2051. https://doi.org/10.3389/fmicb.2019.02051/full

Dural Erem A, Ozcan G, Skrifvars M (2011) Antibacterial activity of PA6/ZnO nanocomposite fibers. Text Res J 81:1638–1646

Elbhnsawi NA, Elwakil BH, Hassanin AH, Shehata N, Elshewemi SS, Hagar M, Olama ZA (2023) Nano-chitosan/eucalyptus oil/cellulose acetate nanofibers: manufacturing, antibacterial and wound healing activities. Membranes (Basel) 13:604. https://www.mdpi.com/2077-0375/13/6/604

El-Naggar ME, Abdelgawad AM, Abdel-Sattar R, Gibriel AA, Hemdan BA (2023) Potential antimicrobial and antibiofilm efficacy of essential oil nanoemulsion loaded polycaprolactone nanofibrous dermal patches. Eur Polym J 184:111782

El-Newehy MH, Al-Deyab SS, Kenawy ER, Abdel-Megeed A (2012) Fabrication of electrospun antimicrobial nanofibers containing metronidazole using nanospider technology. Fibers Polym 13:709–717

Elsherbiny DA, Abdelgawad AM, El-Naggar ME, Hemdan BA, Ghazanfari S, Jockenhövel S, Rojas OJ (2022) Bioactive tri-component nanofibers from cellulose acetate/lignin//N--vanillidene-phenylthiazole copper-(II) complex for potential diaper dermatitis control. Int J Biol Macromol 205:703–718

Eren Boncu T, Uskudar Guclu A, Catma MF, Savaser A, Gokce A, Ozdemir N (2020) In vitro and in vivo evaluation of linezolid loaded electrospun PLGA and PLGA/PCL fiber mats for prophylaxis and treatment of MRSA induced prosthetic infections. Int J Pharm 573:118758

Fan Z, Lu JG (2005) Zinc oxide nanostructures: synthesis and properties. J Nanosci Nanotechnol 5:1561–1573

Fathi HA, Abdelkader A, AbdelKarim MS, Abdelaziz AA, El Mokhtar MA, Allam A, Fetih G, El Badry M, Elsabahy M (2020) Electrospun vancomycin-loaded nanofibers for management of methicillin-resistant *Staphylococcus aureus*-induced skin infections. Int J Pharm 586:119620. https://doi.org/10.1016/j.ijpharm.2020.119620

Fayez MS, Hakim TA, Agwa MM, Abdelmoteleb M, Aly RG, Montaser NN, Abdelsattar AS, Rezk N, El-Shibiny A (2021) Topically applied bacteriophage to control multi-drug resistant *Klebsiella pneumoniae* infected wound in a rat model. Antibiotics 10:1048

Federico S, Catania V, Palumbo FS, Fiorica C, Schillaci D, Pitarresi G, Giammona G (2022) Photothermal nanofibrillar membrane based on hyaluronic acid and graphene oxide to treat *Staphylococcus aureus* and *Pseudomonas aeruginosa* infected wounds. Int J Biol Macromol 214:470–479

Fei Y, Huang Q, Hu Z, Yang X, Yang B, Liu S (2021) Biomimetic cerium oxide loaded gelatin PCL nanosystems for wound dressing on cutaneous care management of multidrug-resistant bacterial wound healing. J Clust Sci 32:1289–1298

Ferreira CAM, Januário AP, Félix R, Alves N, Lemos MFL, Dias JR (2021a) Multifunctional gelatin/chitosan electrospun wound dressing dopped with *Undaria pinnatifida* phlorotannin-enriched extract for skin regeneration. Pharmaceutics 13:2152

Ferreira CAM, Félix R, Félix C, Januário AP, Alves N, Novais SC, Dias JR, Lemos MFL (2021b) A biorefinery approach to the biomass of the seaweed *Undaria pinnatifida* (Harvey Suringar, 1873): obtaining phlorotannins-enriched extracts for wound healing. Biomolecules 11:461. https://www.mdpi.com/2218-273X/11/3/461/htm

Fischer AJ, Singh SB, LaMarche MM, Maakestad LJ, Kienenberger ZE, Peña TA, Stoltz DA, Limoli DH (2021) Sustained coinfections with *Staphylococcus aureus* and *Pseudomonas aeruginosa* in cystic fibrosis. Am J Respir Crit Care Med 203:328–338. https://pubmed.ncbi.nlm.nih.gov/32750253/

Fumakia M, Ho EA (2016) Nanoparticles encapsulated with LL37 and Serpin A1 promotes wound healing and synergistically enhances antibacterial activity. Mol Pharm 13:2318–2331. https://pubmed.ncbi.nlm.nih.gov/27182713/

Gao X, Han S, Zhang R, Liu G, Wu J (2019) Progress in electrospun composite nanofibers: composition, performance and applications for tissue engineering. J Mater Chem B 7:7075–7089

Gaydhane MK, Sharma CS, Majumdar S (2023) Electrospun nanofibres in drug delivery: advances in controlled release strategies. RSC Adv 13:7312–7328. https://pubs.rsc.org/en/content/articlehtml/2023/ra/d2ra06023j

Geetha K, Sivasangari D, Kim H-S, Murugadoss G, Kathalingam A (2022) Electrospun nanofibrous ZnO/PVA/PVP composite films for efficient antimicrobial face masks. Ceram Int 48:29197–29204

Gold K, Slay B, Knackstedt M, Gaharwar AK (2018) Antimicrobial activity of metal and metal-oxide based nanoparticles. Adv Ther (Weinh) 1:1–15

Gruppuso M, Guagnini B, Musciacchio L, Bellemo F, Turco G, Porrelli D (2022) Tuning the drug release from antibacterial Polycaprolactone/rifampicin-based core-shell electrospun membranes: a proof of concept. ACS Appl Mater Interfaces 14:27599–27612

Guerreiro SF, JF AV, Dias JR, Guerreiro NS AFC, JF AV, Dias JR, Alves N (2021) Box-Behnken design a key tool to achieve optimized PCL/gelatin electrospun mesh. Macromol Mater Eng 306:2000678. https://doi.org/10.1002/mame.202000678

Guerreiro SFC, Ferreira CAM, Valente JFA, Patrício TMF, Alves NMF, Dias JR (2022) Electrospun-based membranes as a key tool to prevent respiratory infections. Polymers (Basel) 14:3787

Hadipour-Goudarzi E, Hemmatinejad N, Shokrgozar MA (2023) Fabrication of acorn-loaded chitosan/gelatin nanofibrous web to increase antibacterial activity for wound-healing applications. Fibers Polym 24:893–905

Hadisi Z, Bakhsheshi-Rad HR, Walsh T, Dehghan MM, Farzad-Mohajeri S, Gholami H, Diyanoush A, Pagan E, Akbari M (2020) In vitro and in vivo evaluation of silk fibroin-hardystonite-gentamicin nanofibrous scaffold for tissue engineering applications. Polym Test 91:106698

Hang AT, Tae B, Park JS (2010) Non-woven mats of poly(vinyl alcohol)/chitosan blends containing silver nanoparticles: fabrication and characterization. Carbohydr Polym 82:472–479

Hashemian SMR, Farhadi T, Ganjparvar M (2018) Linezolid: a review of its properties, function, and use in critical care. Drug Des Devel Ther 12:1759–1767

He M, Jiang H, Wang R, Xie Y, Zhao C (2017) Fabrication of metronidazole loaded poly (ε-caprolactone)/zein core/shell nanofiber membranes via coaxial electrospinning for guided tissue regeneration. J Colloid Interface Sci 490:270–278

He J, Hong M, Xie W, Chen Z, Chen D, Xie S (2022) Progress and prospects of nanomaterials against resistant bacteria. J Control Release 351:301–323

Homaeigohar S, Boccaccini AR (2020) Antibacterial biohybrid nanofibers for wound dressings. Acta Biomater 107:25–49

Hu M, Li C, Li X, Zhou M, Sun J, Sheng F, Shi S, Lu L (2018) Zinc oxide/silver bimetallic nanoencapsulated in PVP/PCL nanofibres for improved antibacterial activity. Artif Cells Nanomed Biotechnol 46:1248–1257. https://pubmed.ncbi.nlm.nih.gov/28826242/

Huang ZX, Wu JW, Wong SC, Qu JP, Srivatsan TS (2017) The technique of electrospinning for manufacturing core-shell nanofibers. Mater Manuf Process 33:202–219. https://doi.org/10.1080/10426914.2017.1303144

Huang L, Yu L, Yin X, Lin Y, Xu Y, Niu Y (2022) Silver nanoparticles with vanadium oxide nanowires loaded into electrospun dressings for efficient healing of bacterium-infected wounds. J Colloid Interface Sci 622:117–125

Ibrahim MA, Alhalafi MH, Emam E-AM, Ibrahim H, Mosaad RM (2023) A review of chitosan and chitosan nanofiber: preparation, characterization, and its potential applications. Polymers (Basel) 15:2820

Ikuta KS, Swetschinski LR, Robles Aguilar G, Sharara F, Mestrovic T, Gray AP, Davis Weaver N, Wool EE, Han C, Gershberg Hayoon A, Aali A, Abate SM, Abbasi-Kangevari M, Abbasi-Kangevari Z, Abd-Elsalam S, Abebe G, Abedi A, Abhari AP et al (2022) Global mortality associated with 33 bacterial pathogens in 2019: a systematic analysis for the global burden of disease study 2019. Lancet 400:2221–2248

Ilomuanya MO, Bassey PO, Ogundemuren DA, Ubani-Ukoma UN, Tsamis A, Fan Y, Michalakis K, Angsantikul P, Usman A, Amenaghawon AN (2023) Development of mucoadhesive electrospun scaffolds for intravaginal delivery of *Lactobacilli* spp., a Tenside, and metronidazole for the management of bacterial vaginosis. Pharmaceutics 15:1263

ISO 20776-2:2021(en) (2023) Clinical laboratory testing and in vitro diagnostic test systems—Susceptibility testing of infectious agents and evaluation of performance of antimicrobial susceptibility test devices—Part 2: evaluation of performance of antimicrobes [Internet]. https://www.iso.org/obp/ui/#iso:std:iso:20776:-2:ed-2:v1:en

Jatoi AW, Kim IS, Ni QQ (2019) Cellulose acetate nanofibers embedded with AgNPs anchored TiO2 nanoparticles for long term excellent antibacterial applications. Carbohydr Polym 207:640–649

Ji SM, Tiwari AP, Oh HJ, Kim HY (2021) ZnO/ag nanoparticles incorporated multifunctional parallel side by side nanofibers for air filtration with enhanced removing organic contaminants and antibacterial properties. Colloids Surf A Physicochem Eng Asp 621:126564

Jia L, Huang X, Liang H, Tao Q (2019) Enhanced hydrophilic and antibacterial efficiencies by the synergetic effect TiO2 nanofiber and graphene oxide in cellulose acetate nanofibers. Int J Biol Macromol 132:1039–1043

Jiang C, Ling Z, Xu Y, Bao J, Feng L, Cheng H, Zhao W, Zhao C (2022) Long-term, synergistic and high-efficient antibacterial polyacrylonitrile nanofibrous membrane prepared by "one-pot" electrospinning process. J Colloid Interface Sci 609:718–733

Johani K, Malone M, Jensen S, Gosbell I, Dickson H, Hu H, Vickery K (2017) Microscopy visualisation confirms multi-species biofilms are ubiquitous in diabetic foot ulcers. Int Wound J 14:1160–1169. https://doi.org/10.1111/iwj.12777

Jolivet-Gougeon A, Bonnaure-Mallet M (2019) Bacterial persistence in biofilms and antibiotics. In: Antibiotic drug resistance, pp 181–210. https://doi.org/10.1002/9781119282549.ch9

Kajdič S, Planinšek O, Gašperlin M, Kocbek P (2019) Electrospun nanofibers for customized drug-delivery systems. J Drug Deliv Sci Technol 51:672–681

Kalantari K, Afifi AM, Jahangirian H, Webster TJ (2019) Biomedical applications of chitosan electrospun nanofibers as a green polymer—review. Carbohydr Polym 207:588–600

Kanth S, Malgar Puttaiahgowda Y, Gupta S, Swathi T (2023) Recent advancements and perspective of ciprofloxacin-based antimicrobial polymers. J Biomater Sci Polym Ed 34:918–949

Kaper JB, Nataro JP, Mobley HLT (2004) Pathogenic *Escherichia coli*. Nat Rev Microbiol 2:123–140

Karuppannan SK, Ramalingam R, Khalith SBM, Musthafa SA, Dowlath MJH, Munuswamy-Ramanujam G, Arunachalam KD (2021) Copper oxide nanoparticles infused electrospun polycaprolactone/gelatin scaffold as an antibacterial wound dressing. Mater Lett 294:129787

Keirouz A, Radacsi N, Ren Q, Dommann A, Beldi G, Maniura-Weber K, Rossi RM, Fortunato G (2020) Nylon-6/chitosan core/shell antimicrobial nanofibers for the prevention of mesh-associated surgical site infection. J Nanobiotechnol 18:1–17

Kharaghani D, Khan MQ, Tamada Y, Ogasawara H, Inoue Y, Saito Y, Hashmi M, Kim IS (2018) Fabrication of electrospun antibacterial PVA/Cs nanofibers loaded with CuNPs and AgNPs by an in-situ method. Polym Test 72:315–321

Kharat Z, Amiri Goushki M, Sarvian N, Asad S, Dehghan MM, Kabiri M (2021) Chitosan/PEO nanofibers containing *Calendula officinalis* extract: preparation, characterization, in vitro and in vivo evaluation for wound healing applications. Int J Pharm 609:121132

Khosravimelal S, Chizari M, Farhadihosseinabadi B, Moosazadeh Moghaddam M, Gholipourmalekabadi M (2021) Fabrication and characterization of an antibacterial chitosan/silk fibroin electrospun nanofiber loaded with a cationic peptide for wound-dressing application. J Mater Sci Mater Med 32:114

Kielholz T, Rohde F, Jung N, Windbergs M (2023) Bacteriophage-loaded functional nanofibers for treatment of *P. aeruginosa* and *S. aureus* wound infections. Sci Rep 13:1–13. https://doi.org/10.1038/s41598-023-35364-5

Kim SS, Lee J (2014) Antibacterial activity of polyacrylonitrile–chitosan electrospun nanofibers. Carbohydr Polym 102:231–237

Koo CKW, Senecal K, Senecal A, Nugen SR (2016) Dehydration of bacteriophages in electrospun nanofibers: effect of excipients in polymeric solutions. Nanotechnology 27:485102

Kotturi H, Lopez-Davis C, Nikfarjam S, Kedy C, Byrne M, Barot V, Khandaker M (2022) Incorporation of Mycobacteriophage Fulbright into Polycaprolactone electrospun nanofiber wound dressing. Polymers (Basel) 14:1948

Kunz Coyne AJ, El Ghali A, Holger D, Rebold N, Rybak MJ (2022) Therapeutic strategies for emerging multidrug-resistant *Pseudomonas aeruginosa*. Infect Dis Ther 11:661

Kvich L, Burmølle M, Bjarnsholt T, Lichtenberg M (2020) Do mixed-species biofilms dominate in chronic infections?–need for in situ visualization of bacterial organization. Front Cell Infect Microbiol 10:560993

Kwon S, Yang H, Lee S (2023) Core–shell nanofibers as carrier systems for sustained delivery of tea tree oil. Text Res J 93

Lan X, Liu Y, Wang Y, Tian F, Miao X, Wang H, Tang Y (2021) Coaxial electrospun PVA/PCL nanofibers with dual release of tea polyphenols and ε-poly (L-lysine) as antioxidant and antibacterial wound dressing materials. Int J Pharm 601:120525

Landersdorfer CB, Gwee A, Nation RL (2023) Clinical pharmacological considerations in an early intravenous to oral antibiotic switch: are barriers real or simply perceived? Clin Microbiol Infect 29:1120–1125

Lee S (2009) Multifunctionality of layered fabric systems based on electrospun polyurethane/zinc oxide nanocomposite fibers. J Appl Polym Sci 114:3652–3658

Lemire JA, Harrison JJ, Turner RJ (2013) Antimicrobial activity of metals: mechanisms, molecular targets and applications. Nat Rev Microbiol 11:371–384. https://www.nature.com/articles/nrmicro3028

León-Buitimea A, Garza-Cárdenas CR, Román-García MF, Ramírez-Díaz CA, Ulloa-Ramírez M, Morones-Ramírez JR (2022) Nanomaterials-based combinatorial therapy as a strategy to combat antibiotic resistance. Antibiotics 11:794

Lin M, Liu Y, Gao J, Wang D, Xia D, Liang C, Li N, Xu R (2022) Synergistic effect of co-delivering ciprofloxacin and tetracycline hydrochloride for promoted wound healing by utilizing coaxial PCL/gelatin nanofiber membrane. Int J Mol Sci 23:1895. https://www.mdpi.com/1422-0067/23/3/1895/htm

Linklater DP, Baulin VA, Le Guével X, Fleury J, Hanssen E, Nguyen THP, Juodkazis S, Bryant G, Crawford RJ, Stoodley P (2020) Antibacterial action of nanoparticles by lethal stretching of bacterial cell membranes. Adv Mater 32:2005679

Liu X, He X, Jin D, Wu S, Wang H, Yin M, Aldalbahi A, El-Newehy M, Mo X, Wu J (2020) A biodegradable multifunctional nanofibrous membrane for periodontal tissue regeneration. Acta Biomater 108:207–222

Liu M, Wang X, Li H, Xia C, Liu Z, Liu J, Yin A, Lou X, Wang H, Mo X, Wu J (2021a) Magnesium oxide-incorporated electrospun membranes inhibit bacterial infections and promote the healing process of infected wounds. J Mater Chem B 9:3727–3744. https://pubs.rsc.org/en/content/articlehtml/2021/tb/d1tb00217a

Liu Q, Ouyang WC, Zhou XH, Jin T, Wu ZW (2021b) Antibacterial activity and drug loading of moxifloxacin-loaded poly(vinyl alcohol)/chitosan electrospun nanofibers. Front Mater 8:643428

Lopes Gama E, Silva G, De Souza S, De Bustamante MM, Lopes Dias M, Machado Costa A, Rossi AM, Paula A, Matos S, Santos-Oliveira R, Ricci-Junior E, Journal A (2023) Antibiotics-loaded nanofibers fabricated by electrospinning for the treatment of bone infections. Arab J Chem 16:104392

Ma Y, Song J, Almassri HNS, Zhang D, Zhang T, Cheng Y, Wu X (2020) Minocycline-loaded PLGA electrospun membrane prevents alveolar bone loss in experimental periodontitis. Drug Deliv 27:151–160

Mahamuni-Badiger PP, Patil PM, Patel PR, Dhanavade MJ, Badiger MV, Marathe YN, Bohara RA (2020) Electrospun poly (3-hydroxybutyrate-co-3-hydroxyvalerate)/polyethylene oxide (PEO) microfibers reinforced with ZnO nanocrystals for antibacterial and antibiofilm wound dressing applications. New J Chem 44:9754–9766

Maslova E, Eisaiankhongi L, Sjöberg F, McCarthy RR (2021) Burns and biofilms: priority pathogens and in vivo models. NPJ Biofilms Microbiomes 7:73

Merrell JG, McLaughlin SW, Tie L, Laurencin CT, Chen AF, Nair LS (2009) Curcumin-loaded poly(ε-caprolactone) nanofibres: diabetic wound dressing with anti-oxidant and anti-inflammatory properties. Clin Exp Pharmacol Physiol 36:1149–1156

Merzougui C, Miao F, Liao Z, Wang L, Wei Y, Huang D (2022) Electrospun nanofibers with antibacterial properties for wound dressings. Biomater Sci Polym Ed 33:2165–2183. https://doi.org/10.1080/09205063.2022.2099662

Mirzaeei S, Mansurian M, Asare-Addo K, Nokhodchi A (2021) Metronidazole- and amoxicillin-loaded PLGA and PCL nanofibers as potential drug delivery systems for the treatment of periodontitis: in vitro and in vivo evaluations. Biomedicine 9:975

Moghaddam MM, Abolhassani F, Babavalian H, Mirnejad R, Azizi Barjini K, Amani J (2012) Comparison of in vitro antibacterial activities of two cationic peptides CM15 and CM11 against five pathogenic bacteria: *Pseudomonas aeruginosa, Staphylococcus aureus, Vibrio cholerae, Acinetobacter baumannii, and Escherichia coli*. Probiotics Antimicrob Proteins 4:133–139

Moghaddam MM, Barjini KA, Ramandi MF, Amani J (2014) Investigation of the antibacterial activity of a short cationic peptide against multidrug-resistant *Klebsiella pneumoniae* and *Salmonella typhimurium* strains and its cytotoxicity on eukaryotic cells. World J Microbiol Biotechnol 30:1533–1540

Moosazadeh Moghaddam M, Farhadie B, Mirnejad R, Kooshki H (2023) Evaluation of an antibacterial peptide-loaded amniotic membrane/silk fibroin electrospun nanofiber in wound healing. Int Wound J 20:3443

Moravej H, Fasihi-Ramandi M, Moghaddam MM, Mirnejad R (2019) Cytotoxicity and antibacterial effect of Trp-substituted CM11 cationic peptide against drug-resistant isolates of *Brucella melitensis* alone and in combination with recommended antibiotics. Int J Pept Res Ther 25:235–245

Murray CJ, Ikuta KS, Sharara F, Swetschinski L, Robles Aguilar G, Gray A, Han C, Bisignano C, Rao P, Wool E, Johnson SC, Browne AJ, Chipeta MG, Fell F, Hackett S, Haines-Woodhouse G, Kashef Hamadani BH, Kumaran EAP et al (2023) Global burden of bacterial antimicrobial resistance in 2019: a systematic analysis. Lancet 399:629–655. http://www.thelancet.com/article/S0140673621027240/fulltext

Nguyen VQ, Ishihara M, Kinoda J, Hattori H, Nakamura S, Ono T, Miyahira Y, Matsui T (2014) Development of antimicrobial biomaterials produced from chitin-nanofiber sheet/silver nanoparticle composites. J Nanobiotechnol 12:1–9

Paitan Y (2018) Current trends in antimicrobial resistance of *Escherichia coli*. In: *Escherichia coli*, a versatile pathogen, pp 181–211

Panda SK, Buroni S, Swain SS, Bonacorsi A, da Fonseca Amorim EA, Kulshrestha M, da Silva LCN, Tiwari V (2022) Recent advances to combat ESKAPE pathogens with special reference to essential oils. Front Microbiol 13:1029098

Panthi G, Hassan MM, Kuk Y-S, Kim JY, Chung H-J, Hong S-T, Park M (2020) Enhanced antibacterial property of sulfate-doped Ag3PO4 nanoparticles supported on PAN electrospun nanofibers. Molecules 25:1411

Parham S, Kharazi AZ, Bakhsheshi-Rad HR, Kharaziha M, Ismail AF, Sharif S, Razzaghi M, RamaKrishna S, Berto F (2022) Antimicrobial synthetic and natural polymeric nanofibers as wound dressing: a review. Adv Eng Mater 24:2101460. https://doi.org/10.1002/adem.202101460

Parmanik A, Das S, Kar B, Bose A, Dwivedi GR, Pandey MM (2023) Current treatment strategies against multidrug-resistant bacteria: a review. Curr Microbiol 79:388. https://doi.org/10.1007/s00284-022-03061-7

Peleg AY, Seifert H, Paterson DL (2008) *Acinetobacter baumannii*: emergence of a successful pathogen. Clin Microbiol Rev 21:538–582

Pelletier DA, Suresh AK, Holton GA, McKeown CK, Wang W, Gu B, Mortensen NP, Allison DP, Joy DC, Allison MR (2010) Effects of engineered cerium oxide nanoparticles on bacterial growth and viability. Appl Environ Microbiol 76:7981–7989

Pereira AGB, Fajardo AR, Gerola AP, Rodrigues JHS, Nakamura CV, Muniz EC, Lo HY (2020) First report of electrospun cellulose acetate nanofibers mats with chitin and chitosan nanowhiskers: fabrication, characterization, and antibacterial activity. Carbohydr Polym 250:116954

Perez-Gavilan A, de Castro JV, Arana A, Merino S, Retolaza A, Alves SA, Francone A, Kehagias N, Sotomayor-Torres CM, Cocina D, Mortera R, Crapanzano S, Pelegrín CJ, Garrigos MC, Jiménez A, Galindo B, Araque MC, Dykeman D et al (2021) Antibacterial activity testing methods for hydrophobic patterned surfaces. Sci Rep 11:1–10. https://www.nature.com/articles/s41598-021-85995-9

Pilehvar-Soltanahmadi Y, Dadashpour M, Mohajeri A, Fattahi A, Sheervalilou R, Zarghami N (2017) An overview on application of natural substances incorporated with electrospun nanofibrous scaffolds to development of innovative wound dressings. Mini Rev Med Chem 18:414–427

Prabhu S, Poulose EK (2012) Silver nanoparticles: mechanism of antimicrobial action, synthesis, medical applications, and toxicity effects. Int Nano Lett 2:1–10. https://doi.org/10.1186/2228-5326-2-32

Puca V, Marulli RZ, Grande R, Vitale I, Niro A, Molinaro G, Prezioso S, Muraro R, Di Giovanni P (2021) Microbial species isolated from infected wounds and antimicrobial resistance analysis: data emerging from a three-years retrospective study. Antibiotics 10:1162

Pulingam T, Parumasivam T, Gazzali AM, Sulaiman AM, Chee JY, Lakshmanan M, Chin CF, Sudesh K (2022) Antimicrobial resistance: prevalence, economic burden, mechanisms of resistance and strategies to overcome. Eur J Pharm Sci 170:106103

Qutob M, Shakeel F, Alam P, Alshehri S, Ghoneim MM, Rafatullah M (2022) A review of radical and non-radical degradation of amoxicillin by using different oxidation process systems. Environ Res 214:113833

Ramalingam R, Dhand C, Leung CM, Ong ST, Annamalai SK, Kamruddin M, Verma NK, Ramakrishna S, Lakshminarayanan R, Arunachalam KD (2019) Antimicrobial properties and biocompatibility of electrospun poly-ε-caprolactone fibrous mats containing *Gymnema sylvestre* leaf extract. Mater Sci Eng C 98:503–514

Rancan C, Jurisch B-P, Vogt B, Schaudinn. (2019) Evaluation of drug delivery and efficacy of ciprofloxacin-loaded povidone foils and nanofiber mats in a wound-infection model based on ex vivo human skin. Pharmaceutics 11:527

Rasouli R, Barhoum A (2019) Handbook of nanofibers, pp 733–774

Rather HA, Thakore R, Singh R, Jhala D, Singh S, Vasita R (2018) Antioxidative study of cerium oxide nanoparticle functionalised PCL-gelatin electrospun fibers for wound healing application. Bioact Mater 3:201–211

Rouhi H, Ebrahimi NG (2023) Antibacterial and in vivo studies of poly (ε-caprolactone)-silver electrospun nanofibers: effect of preparation methods on the properties. Int J Polym Mater Polym Biomater 72:460–471

Sachdeva C, Satyamoorthy K, Murali TS (2022) Microbial interplay in skin and chronic wounds. Curr Clin Microbiol Rep 9:21–31. https://doi.org/10.1007/s40588-022-00180-4

Saidin S, Jumat MA, Mohd Amin NAA, Saleh Al-Hammadi AS (2021) Organic and inorganic antibacterial approaches in combating bacterial infection for biomedical application. Mater Sci Eng C 118:111382

Saikaew R, Intasanta V (2021) Versatile nanofibrous filters against fine particulates and bioaerosols containing tuberculosis and virus: multifunctions and scalable processing. Sep Purif Technol 275:119171

Salam A, Hassan T, Jabri T, Riaz S, Khan A, Iqbal KM, Khan SU, Wasim M, Shah MR, Khan MQ (2021) Electrospun nanofiber-based viroblock/ZnO/PAN hybrid antiviral nanocomposite for personal protective applications. Nanomaterials 11:2208

Saleemi MA, Yong PVC, Wong EH (2020) Investigation of antimicrobial activity and cytotoxicity of synthesized surfactant-modified carbon nanotubes/polyurethane electrospun nanofibers. Nano-Struct Nano-Objects 24:100612

Santajit S, Indrawattana N (2016) Mechanisms of antimicrobial resistance in ESKAPE pathogens. Biomed Res Int 2016:2475067

Sarhan WA, Salem HG, Khalil MAF, El-Sherbiny IM, Azzazy HME (2021) Fabrication of gelatin/silk fibroin/phage nanofiber scaffold effective against multidrug resistant *Pseudomonas aeruginosa*. Drug Dev Ind Pharm 47:947–953. https://doi.org/10.1080/03639045.2021.1957915

Sekar AD, Kumar V, Muthukumar H, Gopinath P, Matheswaran M (2019) Electrospinning of Fe-doped ZnO nanoparticles incorporated polyvinyl alcohol nanofibers for its antibacterial treatment and cytotoxic studies. Eur Polym J 118:27–35

Seo KH, Lee KE, Yanilmaz M, Kim J (2022) Exploring the diverse morphology of porous poly(lactic acid) fibers for developing long-term controlled antibiotic delivery systems. Pharmaceutics 14:1272

Sethuram L, Thomas J (2023) Therapeutic applications of electrospun nanofibers impregnated with various biological macromolecules for effective wound healing strategy—a review. Biomed Pharmacother 157:113996

Shetty K, Bhandari A, Yadav KS (2022) Nanoparticles incorporated in nanofibers using electrospinning: a novel nano-in-nano delivery system. J Control Release 350:421–434

Shi Y, Cao X, Zhu Z, Ren J, Wang H, Kong B (2022) Fabrication of cellulose acetate/gelatin-eugenol core–shell structured nanofiber films for active packaging materials. Colloids Surf B Biointerfaces 218:112743

Shin JU, Gwon J, Lee SY, Yoo HS (2018) Silver-incorporated nanocellulose fibers for antibacterial hydrogels. ACS Omega 3:16150–16157. https://doi.org/10.1021/acsomega.8b02180

Singh S, Ashfaq M, Singh RK, Joshi HC, Srivastava A, Sharma A, Verma N (2013) Preparation of surfactant-mediated silver and copper nanoparticles dispersed in hierarchical carbon micro-nanofibers for antibacterial applications. N Biotechnol 30:656–665

Slavin YN, Asnis J, Hńfeli UO, Bach H, Häfeli UO, Bach H (2017) Metal nanoparticles: understanding the mechanisms behind antibacterial activity. J Nanobiotechnol 15:1–20. https://doi.org/10.1186/s12951-017-0308-z

Snetkov P, Rogacheva E, Kremleva A, Morozkina S, Uspenskaya M, Kraeva L (2022) In-vitro antibacterial activity of curcumin-loaded nanofibers based on hyaluronic acid against multidrug-resistant ESKAPE pathogens. Pharmaceutics 14:1186

Song K, Wu Q, Qi Y, Kärki T (2017) Electrospun nanofibers with antimicrobial properties. Electrospun Nanofibers:551–569

Soo JZ, Chai LC, Ang BC, Ong BH (2020) Enhancing the antibacterial performance of titanium dioxide nanofibers by coating with silver nanoparticles. ACS Appl Nano Mater 3:5743–5751. https://doi.org/10.1021/acsanm.0c00925

Spagnol C, Fragal EH, Pereira AGB, Nakamura CV, Muniz EC, Follmann HDM, Silva R, Rubira AF (2018) Cellulose nanowhiskers decorated with silver nanoparticles as an additive to antibacterial polymers membranes fabricated by electrospinning. J Colloid Interface Sci 531:705–715

Tabakoglu S, Kołbuk D, Sajkiewicz P (2022) Multifluid electrospinning for multi-drug delivery systems: pros and cons, challenges, and future directions. Biomater Sci 11:37–61

Teno J, Pardo-Figuerez M, Figueroa-Lopez KJ, Prieto C, Lagaron JM (2022) Development of multilayer ciprofloxacin hydrochloride electrospun patches for buccal drug delivery. J Funct Biomater 13:170

Toniatto TV, Rodrigues BVM, Marsi TCO, Ricci R, Marciano FR, Webster TJ, Lobo AO (2017) Nanostructured poly (lactic acid) electrospun fiber with high loadings of TiO2 nanoparticles: insights into bactericidal activity and cell viability. Mater Sci Eng C 71:381–385

Topcu B, Gultekinoglu M, Timur SS, Eroglu I, Ulubayram K, Eroglu H (2021) Current approaches and future prospects of nanofibers: a special focus on antimicrobial drug delivery. J Drug Target 29:563–575. https://doi.org/10.1080/1061186X.2020.1867991

Truskewycz A, Truong VK, Ball AS, Houshyar S, Nassar N, Yin H, Murdoch BJ, Cole I (2021) Fluorescent magnesium hydroxide nanosheet bandages with tailored properties for biocompatible antimicrobial wound dressings and pH monitoring. ACS Appl Mater Interfaces 13:27904–27919

Unnithan AR, Sasikala ARK, Sathishkumar Y, Lee YS, Park CH, Kim CS (2014) Nanoceria doped electrospun antibacterial composite mats for potential biomedical applications. Ceram Int 40:12003–12012

Ustundag CR, Piskin MB (2023) Investigation of electrospun poly (ε -caprolactone) fiber mats loaded with *Calophyllum inophyllum* essential oil for wound dressing applications: morphology, drug release and in vitro evaluation. Mater Technol 38

Varshosaz J, Arabloo K, Sarrami N, Ghassami E, Yazdani Kachouei E, Kouhi M, Jahanian-Najafabadi A (2020) RGD peptide grafted polybutylene adipate-co-terephthalate/gelatin electrospun nanofibers loaded with a matrix metalloproteinase inhibitor drug for alleviating of wounds: an in vitro/in vivo study. Drug Dev Ind Pharm 46:484–497

Vázquez-López R, Solano-Gálvez SG, Juárez Vignon-Whaley JJ, Abello Vaamonde JA, Padró Alonzo LA, Rivera Reséndiz A, Muleiro Álvarez M, Vega López EN, Franyuti-Kelly G, Álvarez-Hernández DA (2020) Acinetobacter baumannii resistance: a real challenge for clinicians. Antibiotics 9:205

Verma P, Chaturvedi T, Gupta V, Srivastava R, Srivastava A (2012) Evaluation of metronidazole nanofibers in patients with chronic periodontitis: a clinical study. Int J Pharm Investig 2:213

Wang R, Shi M, Xu F, Qiu Y, Zhang P, Shen K, Zhao Q, Yu J, Zhang Y (2020) Graphdiyne-modified TiO2 nanofibers with osteoinductive and enhanced photocatalytic antibacterial activities to prevent implant infection. Nat Commun 11:1–12

Weishaupt R, Zünd JN, Heuberger L, Zuber F, Faccio G, Robotti F, Ferrari A, Fortunato G, Ren Q, Maniura-Weber K, Guex AG (2020) Antibacterial, cytocompatible, sustainably sourced: cellulose membranes with bifunctional peptides for advanced wound dressings. Adv Healthc Mater 9:1901850

Wood SJ, Kuzel TM, Shafikhani SH (2023) *Pseudomonas aeruginosa*: infections, animal modeling, and therapeutics. Cells 12:199

World Health Organization (2017) Prioritization of pathogens to guide discovery, research and development of new antibiotics for drug-resistant bacterial infections, including tuberculosis. World Health Organization, Geneva, p 87. https://www.who.int/publications/i/item/WHO-EMP-IAU-2017.12

Wu J-Y, Wang C-Y, Chen K-H, Lai Y-R, Chiu C-Y, Lee H-C, Chang Y-K (2022) Electrospinning of Quaternized chitosan-poly(vinyl alcohol) composite nanofiber membrane: processing optimization and antibacterial efficacy. Membranes (Basel) 12:332

Yang G, Li X, He Y, Ma J, Ni G, Zhou S (2018) From nano to micro to macro: electrospun hierarchically structured polymeric fibers for biomedical applications. Prog Polym Sci 81:80–113

Yasin AS, Mohamed IMA, Mousa HM, Park CH, Kim CS (2018) Facile synthesis of TiO2/ZrO2 nanofibers/nitrogen co-doped activated carbon to enhance the desalination and bacterial inactivation via capacitive deionization. Sci Rep 8:1–14. https://www.nature.com/articles/s41598-017-19027-w

Yu L, Dou S, Ma J, Gong Q, Zhang M, Zhang X, Li M, Zhang W (2021) An antimicrobial peptide-loaded chitosan/polyethylene oxide nanofibrous membrane fabricated by electrospinning technology. Front Mater 8

Yue Y, Gong X, Jiao W, Li Y, Yin X, Si Y, Yu J, Ding B (2021) In-situ electrospinning of thymol-loaded polyurethane fibrous membranes for waterproof, breathable, and antibacterial wound dressing application. J Colloid Interface Sci 592:310–318

Zamani K, Allah-Bakhshi N, Akhavan F, Yousefi M, Golmoradi R, Ramezani M, Bach H, Razavi S, Irajian G-R, Gerami M (2021) Antibacterial effect of cerium oxide nanoparticle against *Pseudomonas aeruginosa*. BMC Biotechnol 21:1–11

Zargari A, Gill P, Rahimzadeh G, Ghadami F, Rafati A, Ebrahimnejad P (2022) Aloe vera nanofibers contained pseudomonas bacteriophages fabrication, characterization, and biofunction. BioNanoScience 12:1125–1135

Zhou K, Wang M, Zhou Y, Sun M, Xie Y, Yu DG (2022) Comparisons of antibacterial performances between electrospun polymer@drug nanohybrids with drug-polymer nanocomposites. Adv Compos Hybrid Mater 5:907–919. https://doi.org/10.1007/s42114-021-00389-9

The manufacturer's authorised representative in the EU is Springer Nature Customer Service Centre GmbH, Europaplatz 3, 69115 Heidelberg, Germany. If you have any concerns regarding our products, please contact ProductSafety@springernature.com

Printed and bound by CPI Group (UK) Ltd, Croydon, CR0 4YY

01/10/2026

02225864-0018